Organometallic Compounds of

Boron

Organometallic Compounds of

Boron

Edited by

K. Smith

University College of Swansea

LONDON NEW YORK

CHAPMAN AND HALL

First published 1985 by Chapman and Hall Ltd
11 New Fetter Lane, London EC4P 4EE
733 Third Avenue, New York NY 10017

Phototypeset in the United States of America by
Mack Printing Company, Easton, Pennsylvania 18042
Printed in Great Britain by J.W. Arrowsmith Ltd, Bristol

ISBN 0 412 26790 X

Library of Congress Cataloging in Publication Data

Main entry under title:

Organometallic compounds of boron.
(Chapman and Hall chemistry sourcebooks)
Includes index.
1. Organoboron compounds — Handbooks, manuals, etc.
I. Smith, K. II. Series.
QD412.B1'74 1985 547'.05671 84-29347

ISBN 0-412-26790-X

British Library Cataloguing in Publication Data

Organometallic compounds of boron. — (Chapman and Hall chemistry sourcebooks)
1. Organoboron compounds
I. Smith, K.
547'.05671 QD412.B1

ISBN 0-412-26790-X

Contents

Preface

This is one of the first volumes to be published in the series of *Chapman and Hall Chemistry Sourcebooks* which provides carefully tailored information to workers in specialized areas of chemistry. The information contained in this book is derived from the *Dictionary of Organometallic Compounds*, published in November 1984.

Organoboron compounds constitute an important class of organometallics: not only do they merit study in their own right but they are also frequently used as laboratory agents in organic synthesis, and have often been instrumental in the general development of structural theory. It is therefore anticipated that this particular compendium will reach a wide readership.

The databank on the properties of organometallic compounds, which is represented in its current form by the *Dictionary of Organometallic Compounds* and its subset publications such as this volume, will be kept continuously up-to-date. Supplements to the main *Dictionary* will appear annually and revised editions of this *Sourcebook* will be published from time to time as demands permits.

K. Smith

Introduction

1. Using the Sourcebook

The entries in this *Sourcebook* are arranged in order of molecular formula according to the Hill convention (i.e. C, then H, then other elements in alphabetical sequence of element symbol; where no carbon is present, the elements including H are ordered strictly alphabetically).

Every entry is numbered to assist ready location and the entry number consists of the symbol B followed by a five-digit number.

Indexes

There are three printed indexes: a name index which lists every compound name or synonym in alphabetical order; a molecular formula index which lists all molecular formulae, including those of derivatives, in Hill convention order; and a CAS registry number index listing all CAS numbers included in the *Sourcebook* in serial order. All indexes refer to the entry number. In the name index an entry number which follows immediately upon an index term means that the term itself is used as the entry name but an entry number which is preceded by the word 'see' means that the term is a synonym to an entry name. In all three indexes an entry number which is preceded by the word 'in' refers the reader to a specified stereoisomer or derivative which is to be found embedded within the particular entry.

In addition to the three printed indexes, a graphical structure index precedes the section containing the entries, allowing the rapid visual location of compounds of interest. The structure index reproduces all structure diagrams in reduced size and printed in entry number order.

The following paragraphs summarize important considerations in compiling the information in this *Sourcebook*. For more detailed information, see the Introduction to the *Dictionary of Organometallic Compounds* from which this *Sourcebook* derives.

2. Compound Selection

In compiling this *Sourcebook* the aim has been to include from the primary literature up to mid 1983:

(1) Compounds representative of all important structural types (typically, the parent member of each series, where known, together with a selection of its homologues).

(2) Any compound with an established use, such as in catalysis, as a synthetic reagent or starting material.

(3) Other compounds of particular chemical, structural, biological or historical interest, especially those thought to exhibit unusual bonding characteristics.

Some compounds which are not considered sufficiently important to justify separate entries of their own have been included as derivatives in the entries of other compounds. These may include for example:

(1) Organic derivatives in the classical sense.

(2) Donor-acceptor complexes.

(3) The various salts of an anion or cation. In nearly every case, the entry for an ionic substance refers to the naked anion or cation, and the molecular formula, molecular weight and CAS registry number given for the main entry are those of the ion, in agreement with current CAS practice. Salts of the ion with various counterions are then treated as derivatives and the molecular formulae of all of these are given.

(4) Oligomeric compounds. Where a compound is known in several states of molecular aggregation these are all included in the one entry, which usually refers to the monomer. Compounds which are known only in dimeric form are entered as such, but the hypothetical monomers are included as derivatives to ensure that the names and molecular formulae of the monomeric forms occur in the indexes.

All names and molecular formulae recorded for derivatives occur in the Name and Molecular Formula Indexes respectively.

3. Chemical Names and Synonyms

The naming of organometallic compounds is frequently problematic and so in selecting the range of alternative names to present for each compound or derivative, editorial policy has been to report names which are found in the literature, including *Chemical Abstracts*, and not to attempt to impose a system of nomenclature. The editorial generation of new names has therefore been kept to a minimum required by consistency. Most names given in the *Sourcebook* are those given in the original paper(s) and in *Chemical Abstracts*.

Names corresponding to those used by CAS during the 8th, 9th, and 10th collective index periods

(1967-71, 1972-6 and 1977-81 respectively) are labelled with the suffixes 8CI, 9CI and 10CI respectively.

4. Toxicity and Hazard Information

Toxicity and hazard information is highlighted by the sign ▷ which also appears in the indexes.

All organometallic compounds should be treated as if they have dangerous properties.

The information contained in the *Sourcebook* has been compiled from sources believed to be reliable. No warranty, guarantee or representation is made by the Publisher as to the correctness or sufficiency of any information herein, and the Publisher assumes no responsibility in connection therewith.

The specific information in this publication on the hazardous and toxic properties of certain compounds is intended to alert the reader to possible dangers associated with the use of those compounds. The absence of such information should not, however, be taken as an indication of safety in use or misuse.

5. Bibliographic References

The selection of references is made with the aim of facilitating entry into the literature for the user who wishes to locate more detailed information about a particular compound. Reference contents are frequently indicated using mnemonic suffixes. In general recent references are preferred to older ones, and the number of references quoted does not necessarily indicate the relative importance of a compound.

Journal abbreviations generally follow the practice of *Chemical Abstracts Service Source Index* (CASSI). In patent references, no distinction is made between patent applications and granted patents.

6. Sources of Further Information

The following books and review series provide more information about various aspects of organometallic chemistry. Lists of reviews specific to organic compounds of boron may be found in the next section.

General

Comprehensive Organometallic Chemistry, Wilkinson, G. *et al.* Eds, Pergamon, Oxford, 1982. This book represents the most complete and up to date review of the whole subject. In addition to sections for each element there are chapters on the use of organometallics in organic synthesis and catalysis.

Comprehensive Inorganic Chemistry, Trotman-Dickenson, A.F. *et al.* Eds, Pergamon, Oxford, 1973. Contains information about organometallics as well as discussions of oxidation states, coordination chemistry and analysis of the metals.

Gmelins Handbuch der Anorganischen Chemie, 8th Edn, Springer-Verlag, Berlin. Some volumes of Gmelin covering organometallic compounds have been updated relatively recently and can therefore be consulted for comprehensive data on some types of organometallics. Some Gmelin element sections, however, are many years out of date.

Houben-Weyls Methoden der Organischen Chemie, 4th Edn, Band XIII, *Metallorganische Verbindungen*, Thieme-Verlag, Stuttgart.

The Chemistry of the Carbon-Metal Bond, Hartley, F.R. and Patai, S. Eds, Wiley, New York, 1982-. Contains sections on the synthesis, analysis and thermochemistry of various classes of organometallic compounds.

Transition-Metal Complexes of Phosphorus, Arsenic and Antimony Ligands, McAuliffe, C.A. Ed., Macmillan, London, 1973.

Methods of Elemento-Organic Chemistry, Kocheshkov, K.A. Ed., North Holland, Amsterdam, 1967.

MTP International Review of Science: Inorganic Chemistry, Series 2, Emeléus, H.J. Ed., Butterworths, London; University Park Press, Baltimore, 1974-5.

Advances in Organometallic Chemistry, Academic Press, 1964-.

Annual Surveys of Organometallic Chemistry, Elsevier, 1964-7.

Organometallic Chemistry Reviews, Elsevier, 1966-7.

Organometallic Chemistry Reviews, Section A: Subject Reviews 1968-72.

Organometallic Chemistry Reviews, Section B: Annual Surveys 1968-74.

Journal of Organometallic Chemistry: This incorporates reviews and surveys after the discontinuation of the two series of *Organometallic Chemistry Reviews*.

Organometallic Chemistry, 1972-, (Specialist Periodical Reports), RSC.

Coordination Chemistry Reviews, Elsevier, 1966-.

Progress in Inorganic Chemistry, Interscience, 1959-.

Advances in Inorganic Chemistry and Radiochemistry, Academic Press, 1959-.

Analysis

Scott's Standard Methods of Chemical Analysis, Furman, N.H. Ed., 6th Edn, Van Nostrand, New York, 1962.

Crompton, T.R., *Chemical Analysis of Organometallic Compounds*, Academic Press, London, 1973.

Spectroscopy

Nuclear Magnetic Resonance Spectroscopy of Nuclei Other than the Proton, Axenrod, T. and Webb, G.A. Eds, Wiley, London, 1974.

NMR and the Periodic Table, Harris, R.K. and Mann, B.E. Eds, Academic Press, London, 1978.

^{13}C NMR Data for Organometallic Compounds, Mann, B.E. and Taylor, B.F. Eds, Academic Press, London, 1981.

Spectroscopic Properties of Inorganic and Organometallic Compounds, 1968-, (Specialist Periodical Reports), RSC.

Handling

Shriver, D.F., *The Manipulation of Air-Sensitive Compounds*, McGraw-Hill, 1969.

Organometallic Syntheses, Academic Press, New York, 1965, Vol. 1.

B Boron

K. Smith and J. B. Leach (boranes)

Bore (Fr.), Bor (Ger.), Boro (Sp., Ital.), Бор (Bor) (Russ.), ホウ素 (Japan.)

Atomic Number. 5

Atomic Weight. 10.811

Electronic Configuration. [He] $2s^2\,2p^1$

Valency. 3 for simple compounds, but so-called 'electron-deficient' polyhedral cage boron compounds (boranes, carboranes) show other formal valencies. The concept is of limited value for the latter compounds.

Coordination Number. Typically three or four in simple boron compounds, but again a greater variety is experienced in the polyhedral boranes and carboranes where typical values are five and six.

Colour. Colourless unless the organic attachments themselves impart colour. Boron-centered anion radicals are deeply coloured, however.

Availability. The basic raw material for synthesis of most boron compounds is borax, but this is converted into a wide range of secondary sources which are commercially available. Some of the commonest starting materials are BF_3—OEt_2, BCl_3, BH_3—SMe_2, $B(OMe)_3$ and $NaBH_4$, but many other compounds may be used as alternatives. $LiBH_4$, $NaBH_4$, $Me_4NB_3H_8$, B_2H_6, B_5H_9, $B_{10}H_{14}$, and 1,2-$C_2B_{10}H_{12}$ are common starting materials for polyhedral boranes and carboranes.

Handling. As a general rule, compounds containing B—H, B—halogen, B—OR, B—NR_2 or B—SR bonds are readily hydrolysed by moisture and must be handled in dry conditions. Compounds containing B—alkyl bonds are not particularly sensitive to moisture but are very sensitive to oxygen, volatile examples being pyrophoric. Thus, such compounds are handled in an inert atmosphere or under vacuum. Compounds in which boron is four-coordinate tend to be kinetically more stable than those in which it is three-coordinate, and four-coordinate compounds can often be handled in air for short periods without excessive decomposition. Such is true, for example, of $NaBH_4$, pyridine–borane, and related compounds. All organoboron compounds are best routinely handled in dry, aprotic solvents under an atmosphere of nitrogen or argon. The general methods of handling organoboron compounds involve use of syringes and double-ended needles to transfer liquids or solutions of solids into regular glass apparatus having side arms fitted with septa, and a nitrogen atmosphere is maintained by use of a small static pressure, venting through a mercury or paraffin oil bubbler. For small scale work magnetic stirrers are preferred, but mechanical stirrers may be used if appropriate seals are employed. Solids are handled using dry-boxes or atmos-bags. Very occasionally vacuum-line techniques may be required, but for many purposes this is unnecessary. References to techniques for handling air-sensitive compounds are given in the introduction to the *Sourcebook*.

Toxicity. Polyhedral boron compounds (boranes and carboranes) cause physiological problems through attack on the central nervous system and should be handled with caution. Simple trivalent boron compounds are often so readily hydrolysed or oxidized that their main toxicity results from the products of such hydrolysis/oxidation. Typically, the boron component of these product mixtures is boric acid, a moderately toxic compound. On occasions, the other components of the hydrolysis/oxidation reaction mixtures may contribute more to the toxicity (eg. production of HCl by hydrolysis of B—Cl compounds), and in any event it is wise to treat all boron compounds as toxic unless specific information is available to the contrary.

Isotopic Abundance. ^{10}B, 19.78%; ^{11}B, 80.22%.

Spectroscopy. Nmr is widely used to monitor boron compounds. Both ^{10}B ($I = 3$) and ^{11}B ($I = \frac{3}{2}$) nmr are possible, but ^{11}B nmr is far more common and easier to measure. Most compounds are diamagnetic. Multiple-resonance experiments, in which ^1H or ^{13}C nmr measurements are also made, are common. References are given at the end of the section.

Mass-spectra show characteristic boron-isotope patterns (see Gould, R. F., *Borax to Boranes*, *Adv. Chem. Ser.*, 1961, no. 32). Typically, boron-containing fragments contain a substantial proportion of the charge and show up well in the fragmentation pattern.

Infra-red spectra of boron compounds are significant, in particular for B—H bond frequencies which are different for bridging and terminal B—H bonds.

Analysis. This is conveniently achieved by total hydrolysis and/or oxidation to give boric acid. Mannitol is added to complex the boric acid which can then be titrated to a phenolphthalein end point with standard alkali. The method requires careful standardization of procedure and may give results which are inaccurate to 1–2% but is simple and reliable. An easy qualitative test for the presence of boron in a compound is obtained by burning a sample. A bright green flame indicates the presence of boron.

Nomenclature. The nomenclature of boron compounds is confused. *Chemical Abstracts* adopts names which are often out of line with IUPAC recommendations (eg. the continued use of "boronic acid" type nomenclature). Recently, *Chemical Abstracts* personnel have made proposals for changes in the nomenclature of polyboron hydrides which may result in changed usage in future (see Casey, J. B. *et al.*, *Inorg. Chem.*, 1983, **22**, 2228, 2236).

The convention for designation of ^{11}B nmr chemical shifts was changed in the late 1970's and they have been brought into line with designations for other nuclei (downfield shifts positive). Older reports of ^{11}B nmr chemical shifts have the opposite designations (downfield shifts negative).

References. In addition to the review sources common to many elements, listed in the introduction to the *Sourcebook*, a selection of important references is given below:

General

Brown, H. C., *Boranes in Organic Chemistry*, Cornell, 1972.

Brown, H. C., *Organic Synthesis via Boranes*, Wiley, 1975.

Cragg, G. M. L., *Organoboranes in Organic Synthesis*, Dekker, 1973.

Onak, T. P., *Organoboron Chemistry*, Academic Press, 1975.

Brown, H. C. and Smith, K., *Kirk-Othmer Encycl. Chem. Technol.*, 3rd Edn, 1980, **12,** 793.

Grimes, R. N., *Carboranes*, Academic Press, 1970.

Štibr, B. *et al.*, *Pure Appl. Chem.*, 1977, **49,** 803.

Boron Hydride Chemistry, Muetterties, E. L. Ed., Academic Press, 1975.

Beall, H., *Chem. Rev.*, 1973, **73,** 465.

Nmr

Eaton, G. R. and Lipscomb, W. N., *Nmr Studies of Boron Hydrides and Related Compounds*, Benjamin, 1969.

Noth, H. *et al.*, *Nmr Basic Principles and Progress*, Diehl, P. *et al.* Eds, Springer-Verlag, 1978, Chapt. 14.

Todd, L. J., *Prog. Nmr Spectrosc.* 1979, **3,** 87.

$Cl_3BH^{\ominus}$

B-00001

H_2BF

B-00002

F_2BH

B-00003

I_2BH

B-00004

BH_2I

B-00005

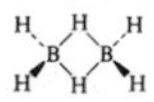
dimer

B-00006

H_2BOH

B-00007

$BH_4^{\ominus}$

B-00008

H_2BNH_2

B-00009

$HOBH_3^{\ominus}$

B-00010

$HB(NH_2)_2$

B-00011

$H_3B—PH_3$

B-00012

$H_2NNH_2—BH_3$

B-00013

$H_3B—NH_2—NH_2—BH_3$

B-00014

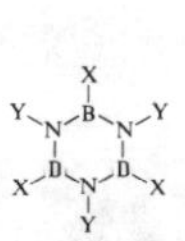
X = Br, Y = H

B-00015

As B-00015 with
X = Cl, Y = H

B-00016

As B-00015 with
X = F, Y = H

B-00017

As B-00019 with
R = F

B-00018

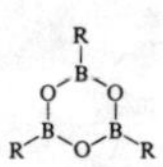
R = H

B-00019

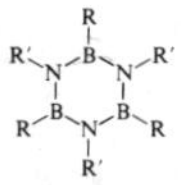
R = R′ = H

B-00020

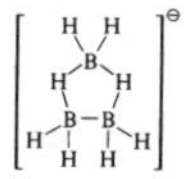

B-00021

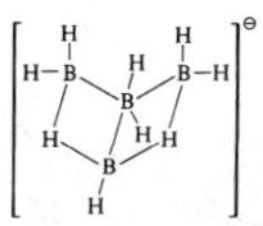

B-00022

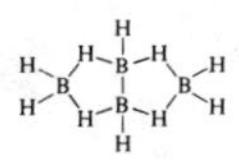

B-00023

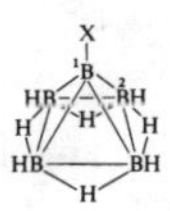
X = Br

B-00024

As B-00024 with
X = Cl

B-00025

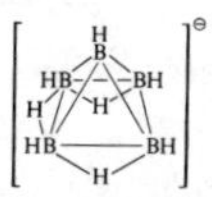

B-00027

As B-00024 with
X = H

B-00028

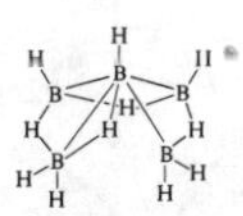

B-00029

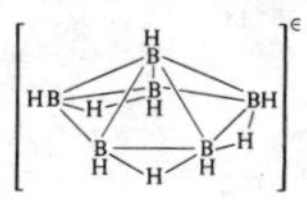

B-00030

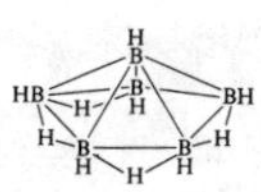

B-00031

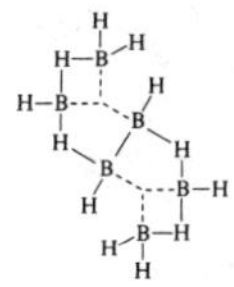

B-00032

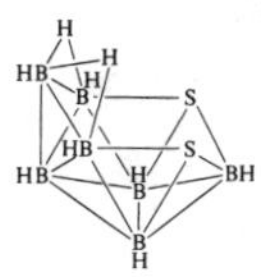

B-00033

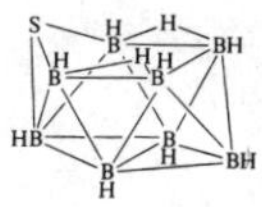

B-00034

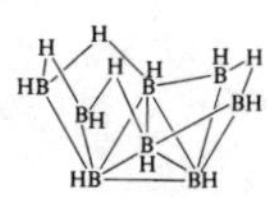

B-00035

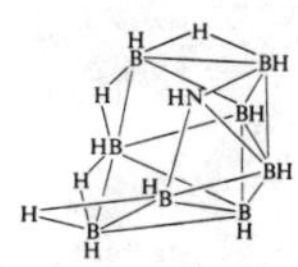

B-00036

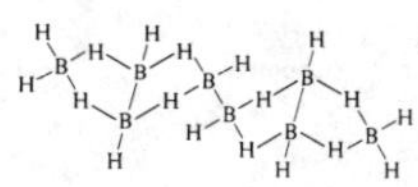

B-00037

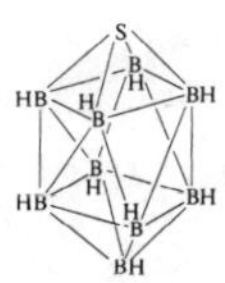

B-00038

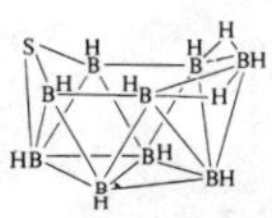

B-00039

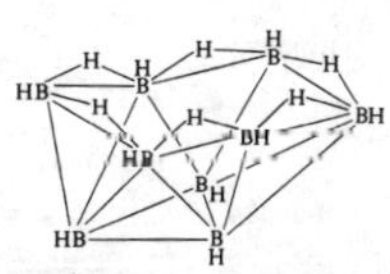

B-00040

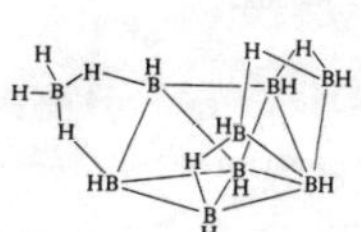

B-00041

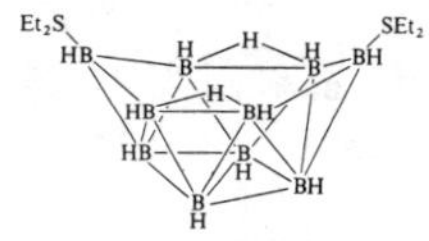
Bis (Et_2S) adduct

B-00042

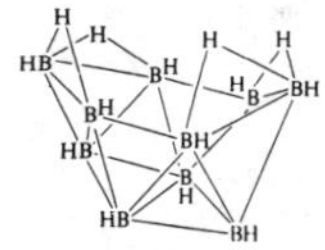

B-00043

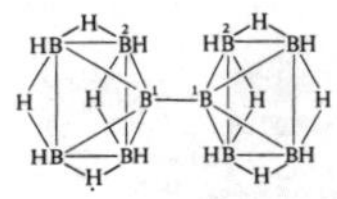

B-00044

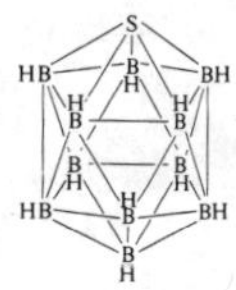

B-00047

$F_3B—CO$

B-00048

$F_3CBF_3^{\ominus}$

B-00049

$[BH_2CN]_n$

B-00050

MeB:

B-00051

$MeBBr_2$

B-00052

$MeBCl_2$

B-00053

$Cl_2CHB(OH)_2$

B-00054

F_2BMe

B-00055

F_2BOMe

B-00056

$MeBF_3^{\ominus}$

B-00057

$MeOBF_3^{\ominus}$

B-00058

$MeBI_2$

B-00059

$[H_3BCN]^{\ominus}$

B-00060

$H_3B—CO$

B-00061

H_2BCOOH

B-00062

$MeOH—BF_3$

B-00063

$MeBH_2$

B-00064

$H_3N—BH_2CN$

B-00065

$MeB(OH)_2$

B-00066

$(HO)_2BOMe$

B-00067

$MeBH_3^{\ominus}$

B-00068

$F_2BH—NH_2Me$

B-00069

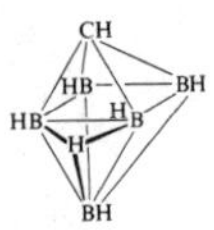

B-00070

$MeNH_2—BH_3$

B-00071

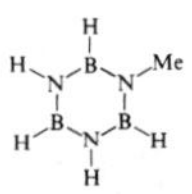

B-00072

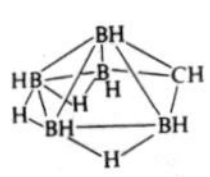

B-00073

$B_5H_8CH_2SiCl_3$

B-00074

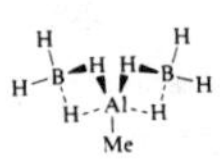

B-00075

$H_3B—H_2NNHMe—BH_3$

B-00076

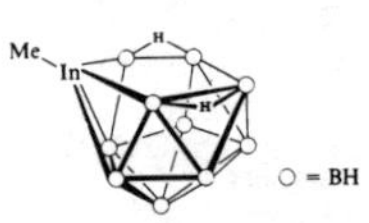

B-00077

$(F_3C)_2BF_2^{\ominus}$

B-00078

$F_2BC≡CH$

B-00079

$H_2BC≡CH$

B-00080

$H_2C=CHBF_2$

B-00081

$F_3CCOOBH_3^{\ominus}$

B-00082

$Cl_2BCH_2CH_2BCl_2$

B-00083

$H_2C=CHBH_2$

B-00084

$EtBBr_2$

B-00085

$EtBCl_2$

B-00086

$EtOBCl_2$

B-00087

$EtBF_2$

B-00088

$EtBI_2$

B-00089

$H_2C=CHB(OH)_2$

B-00090

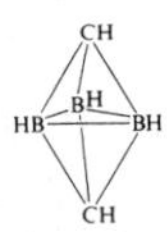

B-00091

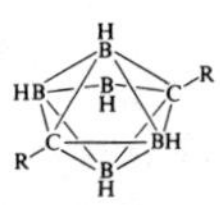

R = Li

B-00092

Me_2BBr

B-00093

Br_2BNMe_2

B-00094

$Me_2S—BBr_3$

B-00095

Me_2BCl

B-00096

$(MeO)_2BCl$

B-00097

$Me_2S—BCl_3$

B-00098

$FBMe_2$

B-00099

$(MeO)_2BF$

B-00100

$Me_2O—BF_3$

B-00101

$EtOH—BF_3$

B-00102

Me_2TlBF_4

B-00103

Me_2BI

B-00104

$MeNC—BH_3$

B-00105

$AcOBH_3^{\ominus}$

B-00106

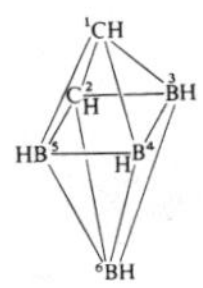

B-00107

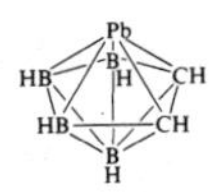

B-00109

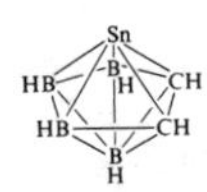

B-00110

Me_2BH

B-00111

$EtBH_2$

B-00112

$HBBr_2—SMe_2$

B-00113

$Cl_2BH—SMe_2$

B-00114

$Me_2NH—BCl_3$

B-00115

$EtNH_2—BF_3$

B-00116

$MeNH_2—BH_2CN$

B-00117

Me_2BOH

B-00118

$HB(OMe)_2$

B-00119

$EtB(OH)_2$

B-00120

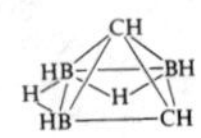

B-00121

$1,5-C_2B_3H_4SiH_3$

B-00122

As B-00092 with R = H

B-00123

$Me_2S—BH_2Br$

B-00124

$Me_2S—BH_2Cl$

B-00125

$F_2BH—NHMe_2$

B-00126

$IBH_2—SMe_2$

B-00127

Me_2BNH_2

B-00128

Me_2NBH_2

B-00129

Me_2PBH_2

B-00130

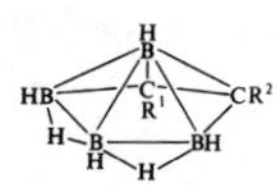

$R^1=R^2=H$

B-00131

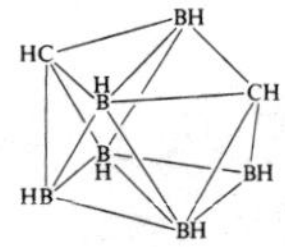

B-00132

$Me_2NH—BH_2Cl$

B-00133

$Me_2N—BH_3^{\ominus}$

B-00134

$Me_2S—BH_3$

B-00135

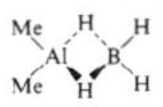

B-00136

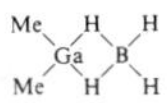

B-00137

$Me_2NH—BH_3$

B-00138

$Me_2PH—BH_3$

B-00139

$C_2B_4H_7(SiH_3)$

B-00140

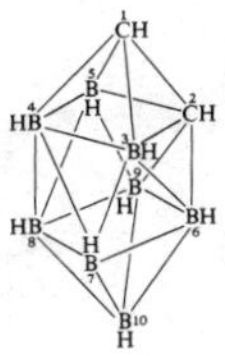

B-00141

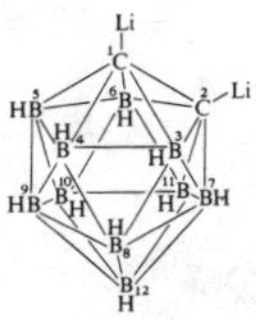

B-00144

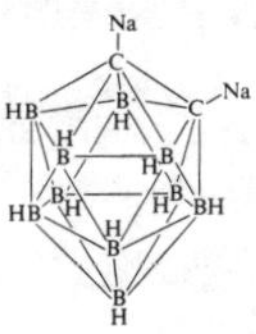

B-00147

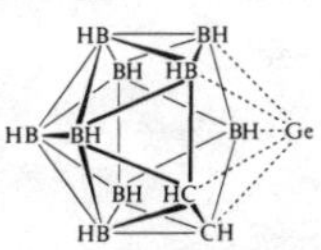

B-00148

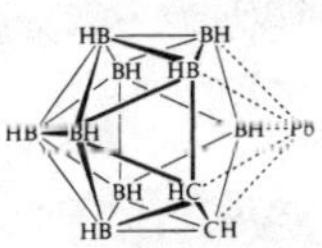

B-00149

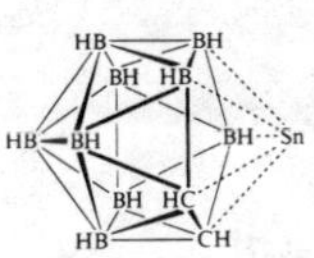

B-00150

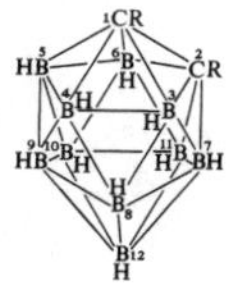

R = H

B-00151

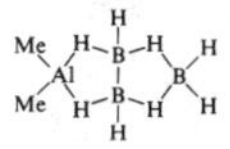

B-00154

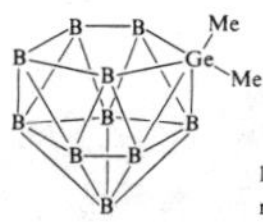

12 H atoms not shown

B-00155

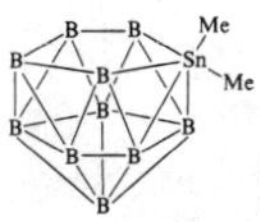

12 H atoms not shown

B-00156

$B_{10}H_{12}TlMe_2^{\ominus}$

B-00157

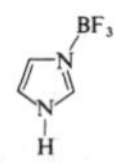

B-00158

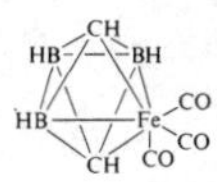

B-00159

$H_2C{=}CHCH_2BH_2$

B-00160

$Cl_2BCH(CH_3)_2$

B-00161

$H_3CCH_2CH_2BCl_2$

B-00162

$H_3CCH_2CH_2OBF_2$

B-00163

$(MeO)_2BCH_2HgCl$

B-00164

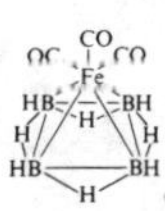

B-00165

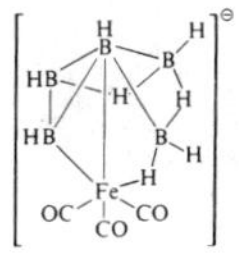

B-00166

BMe_3

B-00167

$MeBBrNMe_2$

B-00168

$Me_3N—BBr_3$

B-00169

$Me_3P—BBr_3$

B-00170

$Me_3N—BCl_3$

B-00171

$Me_3P—BCl_3$

B-00172

$Me_3N—BF_3$

B-00173

$Me_3P—BF_3$

B-00174

$Me_3P—BI_3$

B-00175

$Me_2NH—BH_2CN$

B-00176

Me_2BOMe

B-00177

$H_3CCH_2CH_2B(OH)_2$

B-00178

$MeB(OMe)_2$

B-00179

$(MeO)_3B$

B-00180

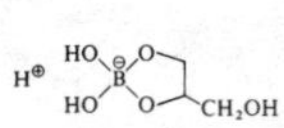

B-00181

Me_2BSMe

B-00182

$MeB(SMe)_2$

B-00183

$MeB(SeMe)_2$

B-00184

As B-00015 with
X = Br, Y = Me

B-00185

As B-00015 with
X = Cl, Y = Me

B-00186

As B-00015 with
X = F, Y = Me

B-00187

As B-00019 with
R = Me

B-00188

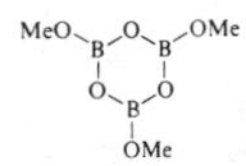

B-00189

1,5—$C_2B_3H_4(SiH_2Me)$

B-00190

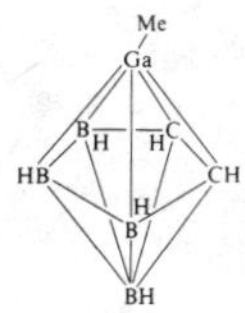

B-00191

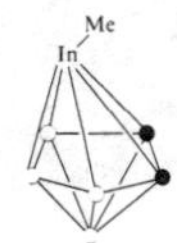

● = CH
○ = BH

B-00192

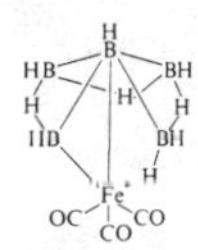

B-00193

$Me_3N—BHF_2$

B-00194

Me_2BNHMe

B-00195

$(MeO)_3BH^{\ominus}$

B-00196

$Me_3P—BH_2Br$

B-00197

$Me_3P—BH_2Cl$

B-00198

$Me_3N—BH_2F$

B-00199

$Me_3N—BH_2I$

B-00200

$IBH_2—PMe_3$

B-00201

$MeB(NHMe)_2$

B-00202

$Me_3N—BH_3$

B-00203

$Me_3B—NH_3$

B-00204

$(MeO)_3B—NH_3$

B-00205

$B(NHMe)_3$

B-00206

$(MeO)_3P—BH_3$

B-00207

$Me_3P—BH_3$

B-00208

As B-00015 with
X = H, Y = Me

B-00209

As B-00020 with
R = Me, R′ = H

B-00210

As B-00015 with
X = OMe, Y = H

B-00211

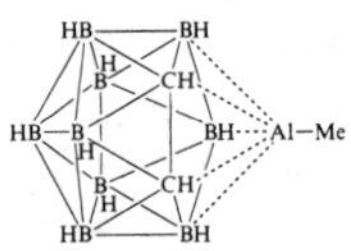

B-00212

$(B_5H_8)CH_2SiMe_2Cl$

B-00213

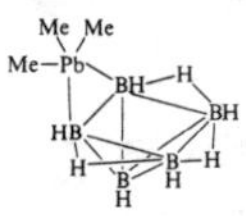

B-00214

$(F_3COOC)_2BH$

B-00215

$(ClCH_2COO)_2BF_2^{\ominus}$

B-00216

B-00217

B-00218

$(H_2C{=}CH)_2BCl$

B-00220

$(AcO)_2BF_2^{\ominus}$

B-00221

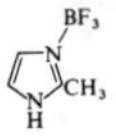

B-00222

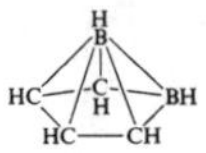

B-00223

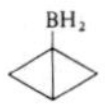

B-00224

$HC{\equiv}CB(OMe)_2$

B-00225

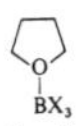

X = F

B-00226

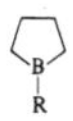

R = H

B-00227

$H_3CCH_2CH_2CH_2BBr_2$

B-00228

$Cl_2BC(CH_3)_3$

B-00229

$H_3CCH_2CH_2CH_2BCl_2$

B-00230

$F_2BCH_2CH_2CH_2CH_3$

B-00231

$F_2BOCH_2CH_2CH_2CH_3$

B-00232

$H_3C(CH_2)_3BF_3^{\ominus}$

B-00233

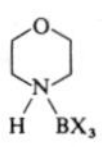

X = F

B-00234

Et_2BBr

B-00235

Br_2BNEt_2

B-00236

Et_2BCl

B-00237

B-00238

$(EtO)_2BCl$

B-00239

Cl_2BNEt_2

B-00240

$Et_2O—BCl_3$

B-00241

Et_2NBF_2

B-00242

$Et_2O—BF_3$

B-00243

$ICH_2CH_2CH_2CH_2B(OH)_2$

B-00244

Et_2BH

B-00245

$EtBMe_2$

B-00246

$Et_2O—BHBr_2$

B-00247

$Et_2O—BHCl_2$

B-00248

$(MeO)_3P—BH_2CN$

B-00249

$Me_3P—BH_2CN$

B-00250

$Me_3N—BH_2CN$

B-00251

As B-00226 with
X = H

B-00252

Et_2BOH

B-00253

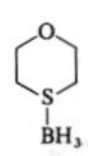

B-00254

$H_3C(CH_2)_3B(OH)_2$

B-00255

$EtB(OMe)_2$

B-00256

$(H_3C)_3CBH_3^{\ominus}$

B-00258

$Me_4B^{\ominus}$

B-00259

$BrB(NMe_2)_2$

B-00260

$Et_2O—BH_2Br$

B-00261

$Me_3N—BMeBr_2$

B-00262

$Me_3P—BBr_2Me$

B-00263

$ClB(NMe_2)_2$

B-00264

$Et_2O—BH_2Cl$

B-00265

$FB(NMe_2)_2$

B-00266

Me_2BNMe_2

B-00267

As B-00234 with
X = H

B-00268

$Me_2NBMeOMe$

B-00269

$(MeO)_2BNMe_2$

B-00270

$(MeS)_2BNMe_2$

B-00271

$(MeO)_3BMe^{\ominus}$

B-00272

$(MeO)_4B^{\ominus}$

B-00273

Me_2BOBMe_2

B-00274

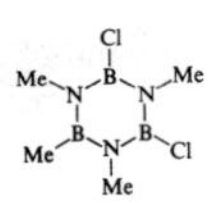

B-00275

As B-00131 with
$R^1 = R^2 = CH_3$

B-00276

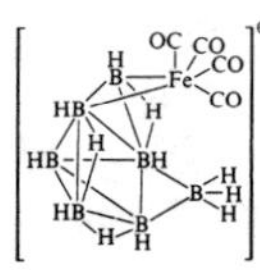

B-00277

$Me_2Al(C_2B_4H_7)$

B-00279

$HB(NMe_2)_2$

B-00280

$Me_2BNHBMe_2$

B-00281

$(H_3C)_3CNH_2—BH_3$

B-00282

$Et_2NH—BH_3$

B-00283

$Me_3PCH_2—BH_3$

B-00284

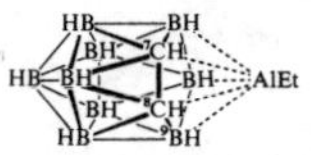

B-00285

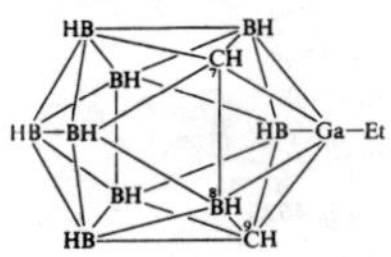

B-00288

As B-00151 with
R = $—CH_2OH$

B-00289

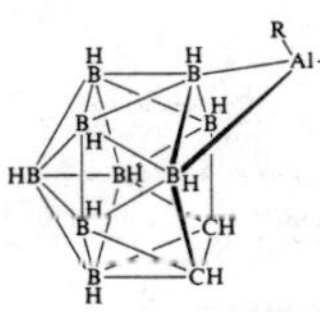

1 H atom not shown

B-00291

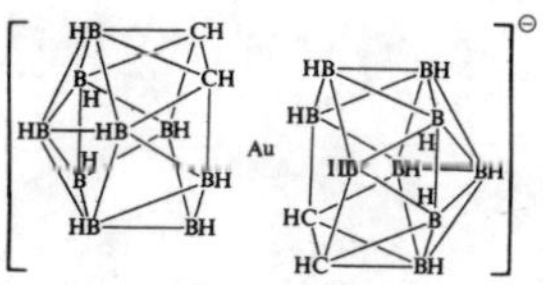

B-00292

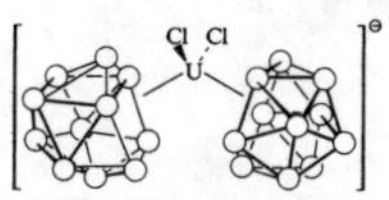

○ = BH or CH (position of CH's not rigorously established)

B-00294

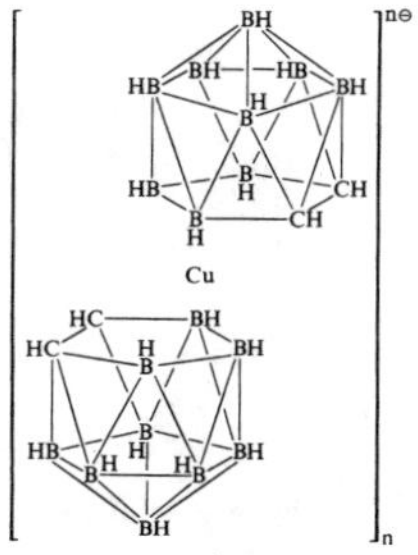

n = 1

B-00295

As B-00295 with
n = 2

B-00296

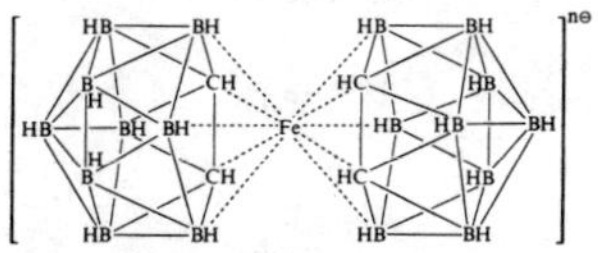

n = 1

B-00297

As B-00297 with
n = 2

B-00298

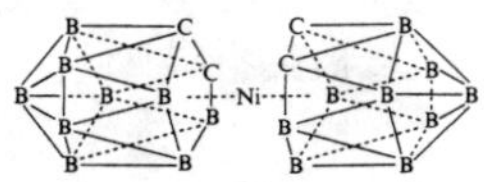

B-00299

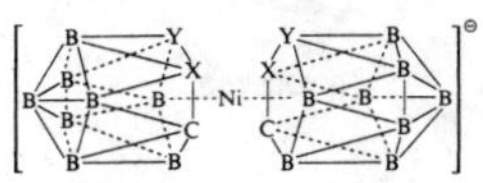

X = C,Y = B

B-00300

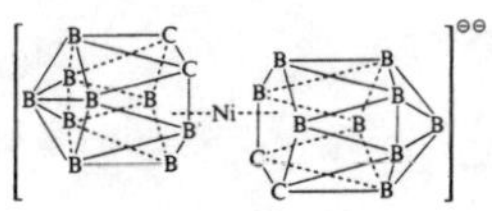

B-00301

As B-00300 with
X = B, Y = C

B-00302

$[Me_2In]^{\oplus}[B_{10}H_{12}InMe_2]^{\ominus}$

B-00303

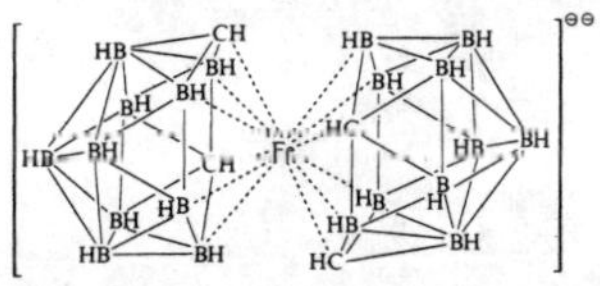

B-00304

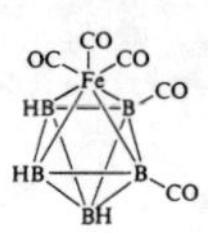

B-00305

B-00306

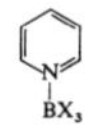

X = Br

B-00307

As B-00307 with
X = Cl

B-00308

As B-00307 with
X = F

B-00309

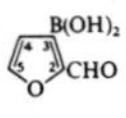

B-00310

B-00314

B-00316

$Me_2B—N{=}C(CF_3)_2$

B-00317

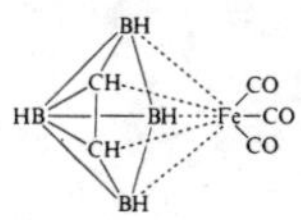

B-00318

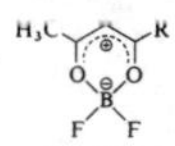

R = CH_3

B-00320

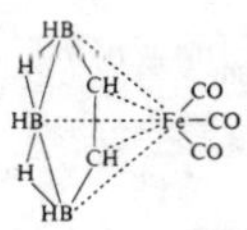

B-00321

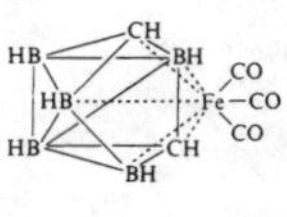

B-00322

As B-00307 with
X = H

B-00323

$(H_2C{=}CH)_2BMe$

B-00324

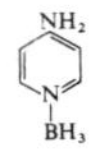

B-00325

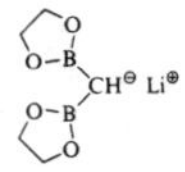

B-00326

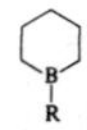

R = H

B-00327

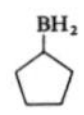

B-00328

As B-00227 with
R = Me

B-00329

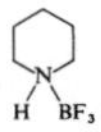

B-00330

As B-00227 with
R = OMe

B-00331

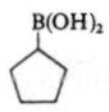

B-00332

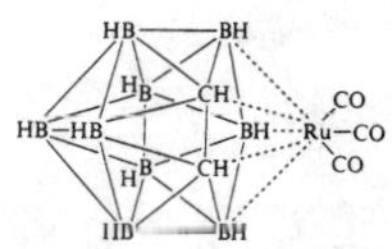

B-00333

Et_2BMe

B-00334

$(H_3C)_2CHCH(CH_3)BH_2$

B-00335

$(HOCH_2CH_2)_2NH—BH_2CN$

B-00336

Et_2BOMe

B-00337

$H_3CCH_2CH_2B(OMe)_2$

B-00338

$H_3C(CH_2)_4B(OH)_2$

B-00339

$H_3CCH_2CH_2B(SMe)_2$

B-00340

$H_3CCH(CH_3)CH(CH_3)BH_3^{\ominus}$

B-00341

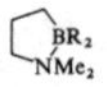

R = H

B-00342

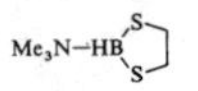

B-00343

$Me_2BBr—PMe_3$

B-00344

$MeB(NMe_2)_2$

B-00345

$MeOB(NMe_2)_2$

B-00346

$Me_2BNMeBMe_2$

B-00347

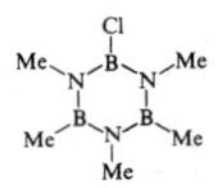

B-00348

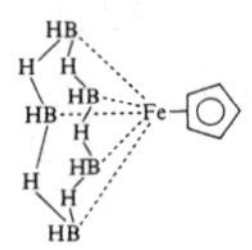

B-00349

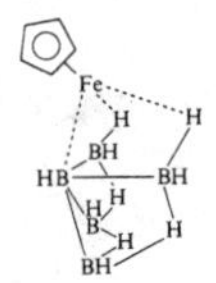

B-00350

$MeOCH_2CH_2NMe_2—BH_3$

B-00351

$Me_2BNHSiMe_3$

B-00352

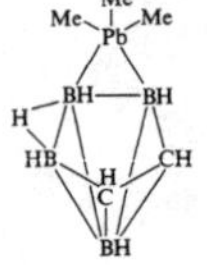

B-00353

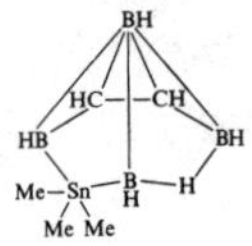

B-00354

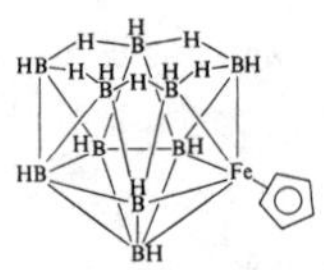

B-00355

$C_6F_5BCl_2$

B-00356

$B(OOCCF_3)_3$

B-00357

$(OC)_5CrCNBH_3^{\ominus}$ (O_h)

B-00358

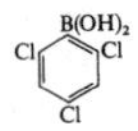

B-00359

Br_2BPh

B-00360

$PhBCl_2$

B-00361

B-00362

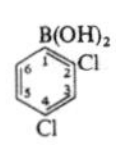

B-00363

$PhBCl_3^{\ominus}$

B-00366

$PhBF_2$

B-00367

$PhOBF_2$

B-00368

$PhBI_2$

B-00369

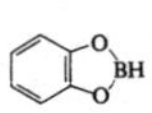

B-00370

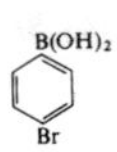

B-00371

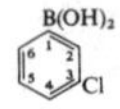

B-00372

$PhOH—BF_3$

B-00374

$Me_2NBFe(CO)_4$

B-00375

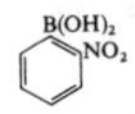

B-00376

$PhBH_2$

B-00378

B-00379

$PhB(OH)_2$

B-00380

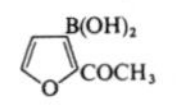

B-00381

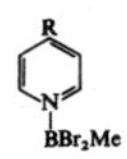

R = H

B-00382

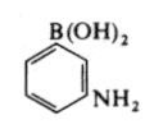

B-00383

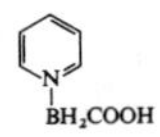

B-00384

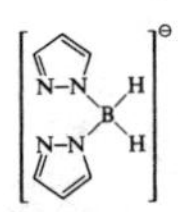

B-00385

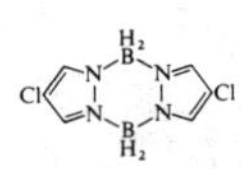

B-00386

$(H_2C{=}CH)_3B$

B-00387

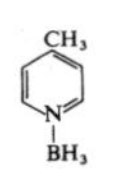

B-00388

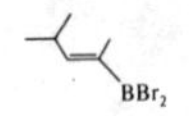

B-00389

$(H_3C)_3CCH{=}CHBBr_2$

B-00390

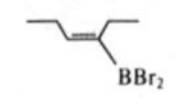

B-00391

$H_3C(CH_2)_3CH{=}CHBBr_2$

B-00392

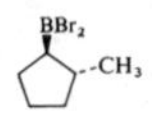

B-00393

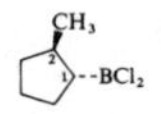

B-00394

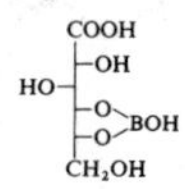

B-00395

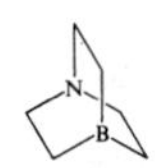

B-00396

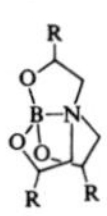

R = H

B-00397

$NCCH_2CH_2NMe_2—BH_2CN$

B-00398

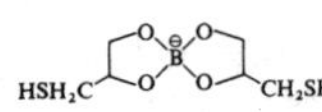

B-00399

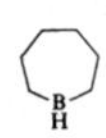

B-00400

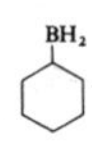

B-00401

As B-00227 with R = Et

B-00402

As B-00327 with R = Me

B-00403

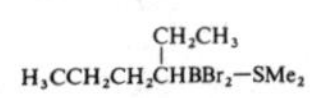

B-00404

$Br_2B(CH_2)_5CH_3$

B-00405

$H_3CCH_2CH_2CH(CH_3)CH_2BBr_2$

B-00406

$H_3C(CH_2)_5BF_2$

B-00407

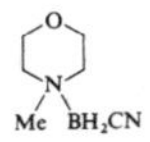

B-00408

$H_3C(CH_2)_3CH{=}CHB(OH)_2$

B-00409

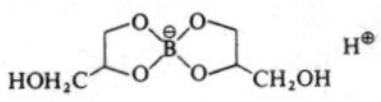

B-00410

$Br_2BN(CH_2CH_2CH_3)_2$

B-00411

$(H_3CCH_2CH_2)_2S{-}BBr_3$

B-00412

$(H_3CCH_2CH_2)_2Se{-}BBr_3$

B-00413

$(H_3CCH_2CH_2)_2BCl$

B-00414

$(H_3C)_2CHC(CH_3)_2BHCl$

B-00415

$[(H_3C)_2CH]_2NBCl_2$

B-00416

$H_3C(CH_2)_4CH_2OH{-}BF_3$

B-00417

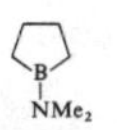

B-00418

BEt_3

B-00419

$(H_3C)_2CHC(CH_3)_2BH_2$

B-00420

$Et_3N{-}BBr_3$

B-00421

$Et_3N{-}BCl_3$

B-00422

$H_3CCH_2CH_2CH_2BFNMe_2$

B-00423

$Et_3O^+ BF_4^-$

B-00424

Et_2BOEt

B-00425

$H_3C(CH_2)_3B(OMe)_2$

B-00426

$EtB(OEt)_2$

B-00427

$H_3C(CH_2)_5B(OH)_2$

B-00428

$(EtO)_3B$

B-00429

$(MeS)_2BC(CH_3)_3$

B-00430

As B-00015 with
X = Cl, Y = Et

B-00431

As B-00019 with
R = Et

B-00432

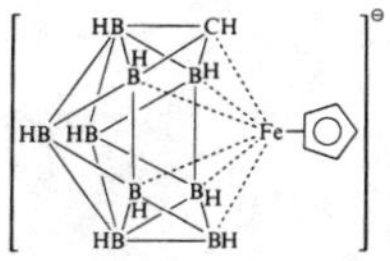

B-00433

$Et_3BH^{\ominus}$

B-00434

$(H_3CCH_2CH_2)_2BNH_2$

B-00435

Me_2BNEt_2

B-00436

Et_2BNMe_2

B-00437

$Me_3N{-}BH_2COOEt$

B-00438

$F_3B{-}Me_2NCH_2CH_2NMe_2{-}BF_3$

B-00439

$Et_3N{-}BH_2Cl$

B-00440

$Me_3N{-}BH_2CONHEt$

B-00441

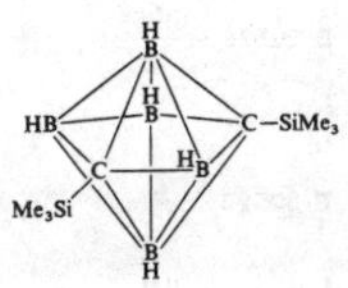

B-00442

$Et_3N{-}BH_3$

B-00443

$Me_3B{-}NMe_3$

B-00444

$EtOCH_2CH_2NMe_2{-}BH_3$

B-00445

$(HO)_3B{-}N(CH_2CH_2OH)_3$

B-00446

$B(NMe_2)_3$

B-00447

$(Me_2B)_3N$

B-00448

As B-00020 with
R = R′ = Me

B-00449

As B-00015 with
X = H, Y = Et

B-00450

As B-00020 with
R = Et, R′ = H

B-00451

As B-00019 with
$R = NMe_2$

B-00452

As B-00015 with
X = SMe, Y = Me

B-00453

$Me_2NCH_2CH_2NMe_2{-}BH_3$

B-00454

$Me_2NCH_2{-}BH_2{-}NMe_3$

B-00455

$(Me_3N)_2BH_2^{\oplus}$

B-00456

$(Me_3P)_2BH_2^{\oplus}$

B-00457

$(Me_2N)_3P{-}BH_3$

B-00458

As B-00291 with
R = Et

B-00459

$H_3B{-}Me_2NCH_2CH_2NMe_2{-}BH$

B-00460

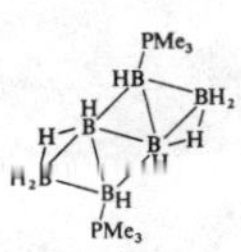

B-00461

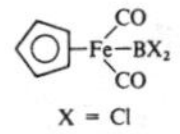

B-00462

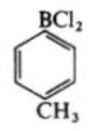

B-00463

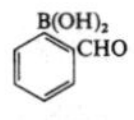

B-00464

$Cl_2BNMePh$

B-00465

$PhOMe{-}BF_3$

B-00466

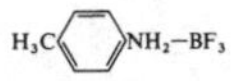

B-00467

$PhNH_2{-}BH_2CN$

B-00468

B-00469

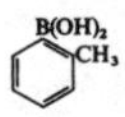

B-00472

$PhSCH_2B(OH)_2$

B-00475

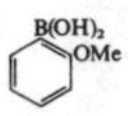

B-00476

As B-00382 with
$R = CH_3$

B-00479

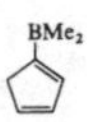

B-00481

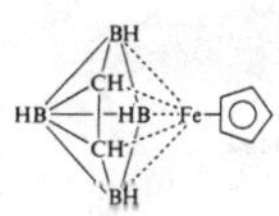

B-00482

$(Me_2N)_2BN{=}C(CF_3)_2$

B-00483

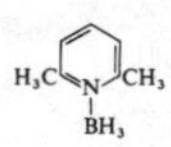

B-00484

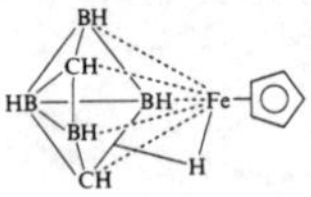

B-00485

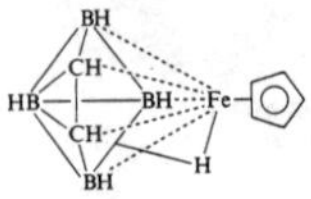

B-00486

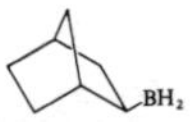

exo-form

B-00487

$(MeO)_2BCH{=}C{=}C(CH_3)_2$

B-00488

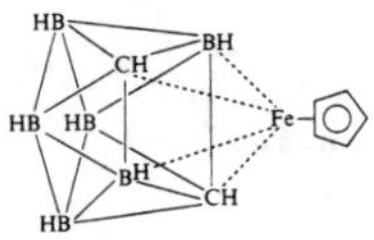

B-00489

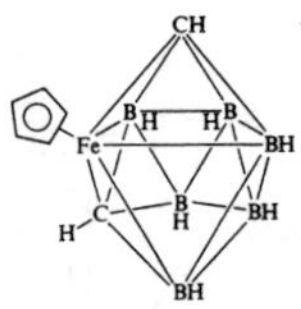

B-00490

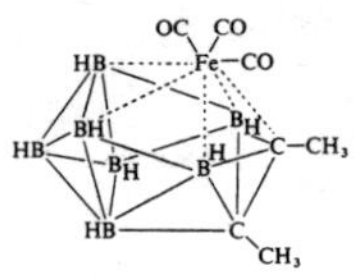

B-00491

$ClCH_2CH_2CH_2CH{=}CHBBr_2{-}SMe_2$

B-00492

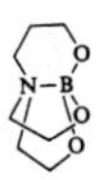

B-00493

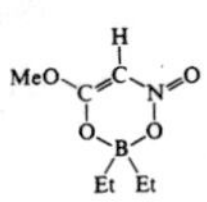

B-00494

$H_2C{=}CHCH_2BEt_2$

B-00496

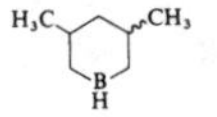

B-00497

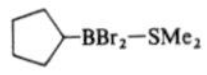

B-00498

$Et_3N{-}BCl_2CN$

B-00499

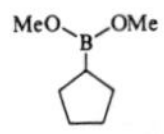

B-00500

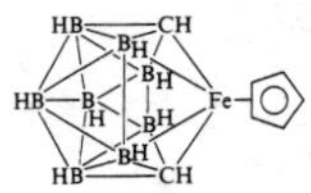

Proposed structure

B-00501

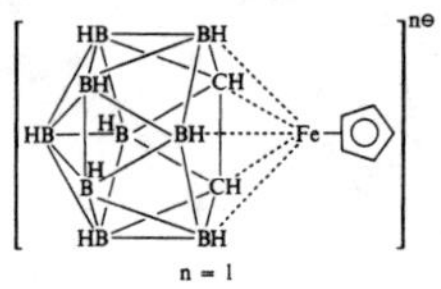

B-00502

$Et_3N{-}BH_2CN$

B-00504

$(H_3CCH_2CH_2)_2BOMe$

B-00505

$H_3C(CH_2)_6B(OH)_2$

B-00506

$(H_3CCH_2CH_2)_2BSMe$

B-00507

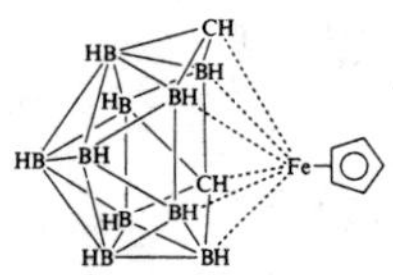

B-00508

$H_3C(CH_2)_3BMe_3^{\ominus}$

B-00509

$Et_3BMe^{\ominus}$

B-00510

$(H_3C)_2CHB(NMe_2)_2$

B-00511

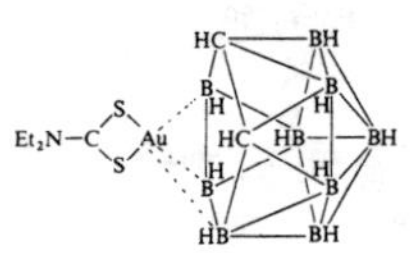

B-00512

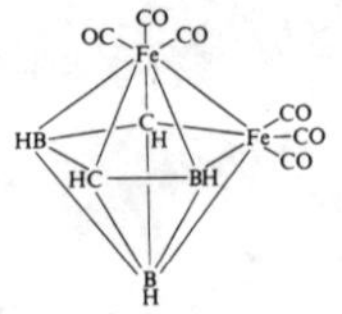

B-00513

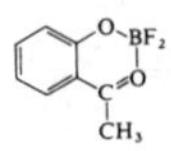

B-00514

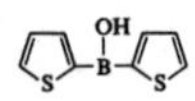

B-00515

$F_3B{-}O{=}CPhOMe$

B-00516

$F_3B{-}O{=}C(OPh)CH_3$

B-00517

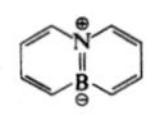

B-00518

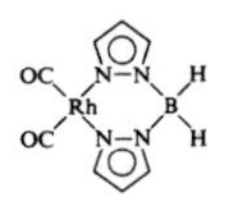

B-00519

$PhCH{=}CHB(OH)_2$

B-00520

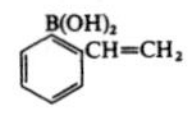

B-00521

$PhBBrNMe_2$

B-00524

$PhBClNMe_2$

B-00525

$PhN{=}C(CH_3)OBH_3^{\ominus}$

B-00526

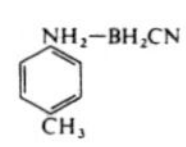

B-00527

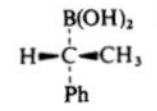

(R)-form

B-00528

$PhCH_2CH_2B(OH)_2$

B-00529

$PhB(OMe)_2$

B-00530

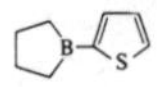

B-00531

$PhB(SMe)_2$

B-00532

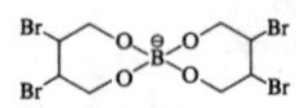

B-00533

$Et_2NB(C{\equiv}CH)_2$

B-00534

$(H_3CC{\equiv}C)_2BNMe_2$

B-00535

Me_2BNHPh

B-00536

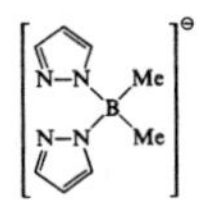

B-00537

$PhB(NHMe)_2$

B-00538

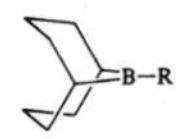

R = Br

B-00539

As B-00539 with
R = Cl

B-00540

$PhNMe_2{-}BH_3$

B-00541

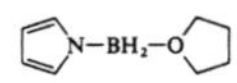

B-00542

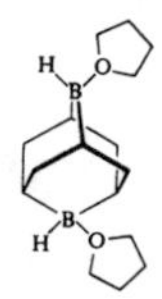

THF complex

B-00543

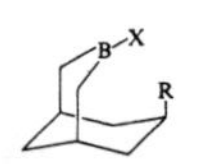

X = R = H

B-00544

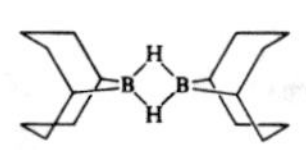

Dimer

B-00545

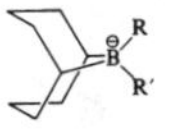
R = R′ = H

B-00546

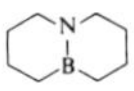

B-00547

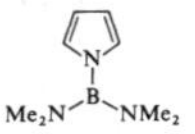

B-00548

$Et_2BCH_2CH{=}CHCH_3$

B-00549

$H_2C{=}C(CH_3)CH_2BEt_2$

B-00550

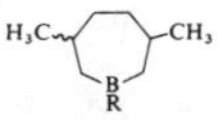
R = H

B-00551

$H_3C(CH_2)_7BCl_2$

B-00552

$H_3CCH_2CH{=}CHOBClNEt_2$

B-00553

$H_3C(CH_2)_7BCl_2$

B-00554

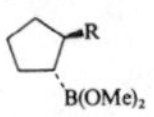
R = CH_3

B-00555

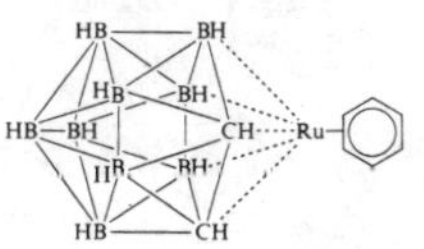

B-00556

$[H_3CCH_2CH(CH_3)]_2BBr$

B-00557

$(H_3CCH_2CH_2CH_2)_2BBr$

B-00558

$Br_2BN(CH_2CH_2CH_2CH_3)_2$

B-00559

$(H_3CCH_2CH_2CH_2)_2S_2\ \ BBr_3$

B-00560

$[(H_3C)_2CHCH_2]_2BCl$

B-00561

$(H_3CCH_2CH_2CH_2)_2BCl$

B-00562

$(H_3CCH_2CH_2CH_2)_2NBCl_2$

B-00563

$(H_3CCH_2CH_2CH_2)_2BF$

B-00564

$(H_3CCH_2CH_2CH_2)_2O{-}BF_3$

B-00565

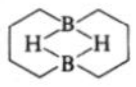

B-00566

$(H_3CCH_2CH_2CH_2)_2BH$

B-00567

$H_2C{=}CHCH(CH_3)B(NMe_2)_2$

B-00568

$H_3CCH{=}CHCH_2B(NMe_2)_2$

B-00569

$Et_2BO(CH_2)_3CH_3$

B-00570

$(H_3CCH_2CH_2CH_2)_2BOH$

B-00571

$(H_3CCH_2CH_2)_2BOEt$

B-00572

$H_3C(CH_2)_7B(OH)_2$

B-00573

$H_3C(CH_2)_5B(OMe)_2$

B-00574

$Et_4B^{\ominus}$

B-00575

$BrB(NEt_2)_2$

B-00576

$ClB(NEt_2)_2$

B-00577

$(H_3CCH_2CH_2CH_2)_2BNH_2$

B-00578

Et_2BNEt_2

B-00579

$(H_3CCH_2CH_2)_2BNMe_2$

B-00580

$NCBH_2{-}Me_2NCH_2CH_2NMe_2{-}H_2BCN$

B-00581

Et_2BOBEt_2

B-00582

$HOOCBH_2{-}NMe_2CH_2CH_2NMe_2{-}BH_2$
COOH

B-00583

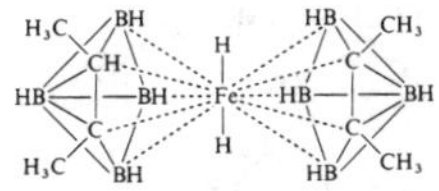

B-00584

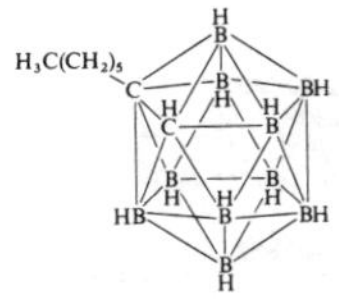

B-00585

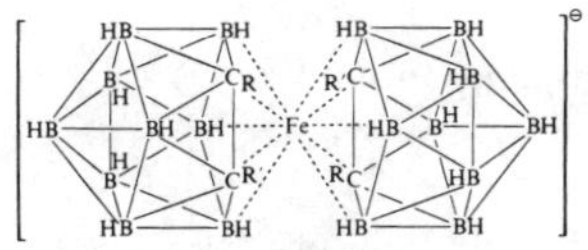
R = R′ = CH_3

B-00586

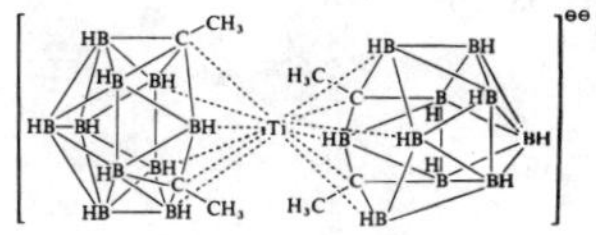

B-00587

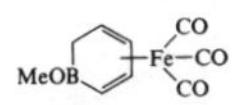

B-00588

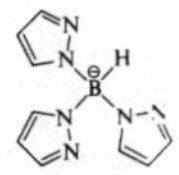

B-00589

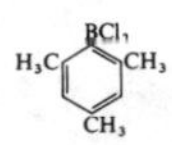

B-00590

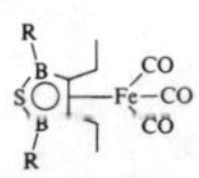
R = H

B-00591

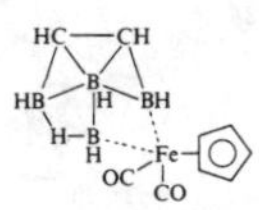

B-00592

$PhCH_2CH_2CH_2B(OH)_2$

B-00593

$PhBMe_3^{\ominus}$

B-00594

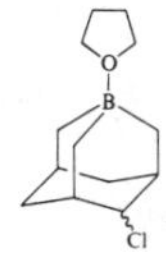
THF complex

B-00595

As B-00539 with
R = $-OSO_2CF_3$

B-00596

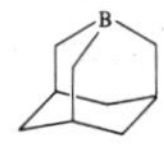

B-00597

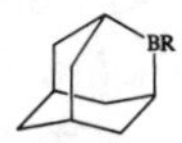
R = H

B-00598

$(H_2C{=}CHCH_2)_3B$

B-00599

As B-00546 with
R = CN, R′ = H

B-00600

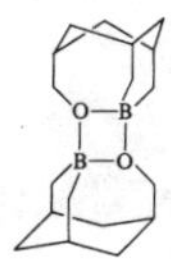
Dimer

B-00601

$(H_2C{=}CHCH_2O)_3B$

B-00602

As B-00544 with
X = Br, R = CH_3

B-00603

$H_3CCH_2CH_2B(CH_2CH{=}CH_2)_2$

B-00604

As B-00539 with
R = Me

B-00605

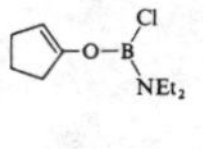

B-00606

As B-00544 with
X = OMe, R = H

B-00607

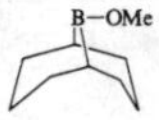

B-00608

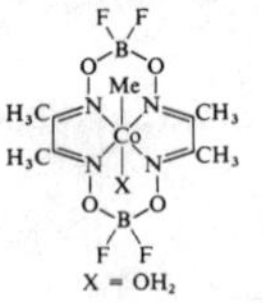

$X = OH_2$

B-00609

As B-00546 with
R = Me, R′ = H

B-00610

As B-00397 with
$R = CH_3$

B-00611

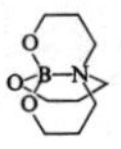

B-00612

B-00613

B-00614

$(H_3C)_2C{=}CHCH_2BEt_2$

B-00615

$H_2C{=}CHCH_2B(CH_2CH_2CH_3)_2$

B-00616

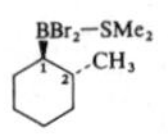

B-00617

As B-00555 with
$R = CH_2CH_3$

B-00618

$H_3C(CH_2)_3CH{=}CHBMe_3^{\ominus}$

B-00619

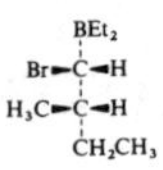

B-00620

$ICH_2B(OCH_2CH_2CH_2CH_3)_2$

B-00621

$B[CH(CH_3)_2]_3$

B-00622

$B(CH_2CH_2CH_3)_3$

B-00623

$(H_3CCH_2CH_2CH_2)_2BOMe$

B-00624

$[H_3CCH_2CH(CH_3)]_2BOMe$

B-00625

$[(H_3C)_2CHCH_2]_2BOMe$

B-00626

$(H_3CCH_2CH_2)_2BOCH_2CH_2CH_3$

B-00627

$B[OCH(CH_3)_2]_3$

B-00628

$(H_3CCH_2CH_2O)_3B$

B-00629

$(H_3CCH_2CH_2CH_2)_2BSMe$

B-00630

$Et_2BS(CH_2)_4CH_3$

B-00631

$[H_3CCH(CH_3)]_3BH^{\ominus}$

B-00632

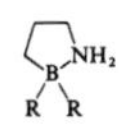

$R = CH_2CH_2CH_3$

B-00633

$[(H_3C)_2CHO]_3BH^{\ominus}$

B-00634

$(MeOCH_2CH_2O)_3BH^{\ominus}$

B-00635

As B-00015 with
X = Et, Y = Me

B-00636

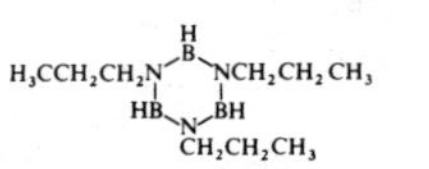

B-00637

$(Me_3SnO)_3B$

B-00638

As B-00015 with
$X = NMe_2$, Y = Me

B-00639

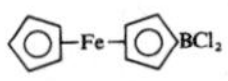

B-00640

As B-00320 with
R = Ph

B-00641

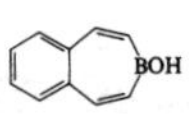

B-00642

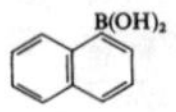

B-00643

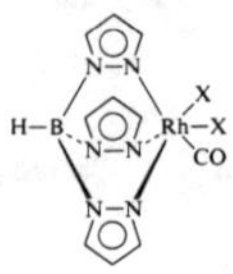

X = Cl

B-00644

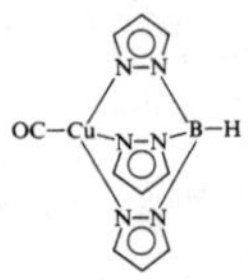

B-00645

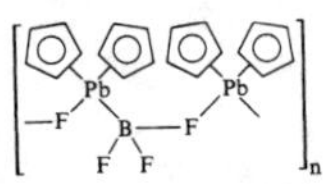

B-00646

$(C_5H_5)_2SnBF_3$

B-00647

As B-00644 with
X = I

B-00648

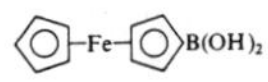

B-00649

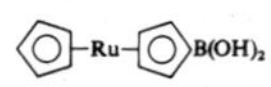

B-00650

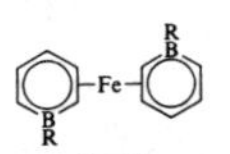

R = H

B-00651

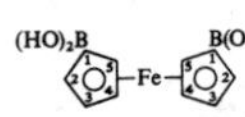

B-00652

As B-00227 with
R = Ph

B-00653

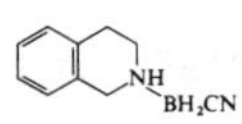

B-00654

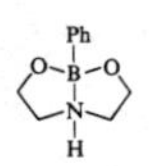

B-00655

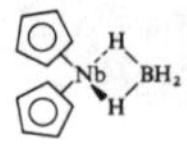

B-00656

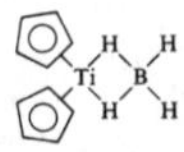

B-00657

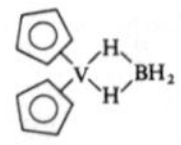

B-00658

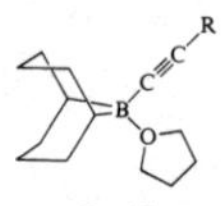

THF complex

R = H

B-00659

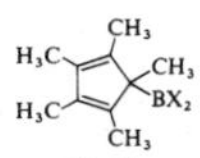

X = Br

B-00660

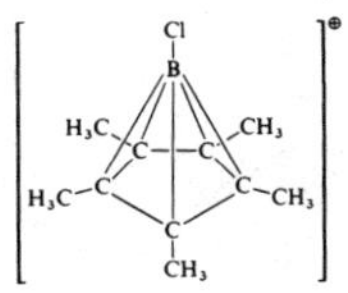

B-00661

$PhBClNEt_2$

B-00662

As B-00660 with
X = Cl

B-00663

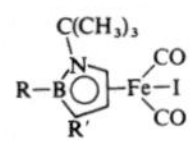

R = Me, R′ = H

B-00664

As B-00660 with
X = I

B-00665

$PhB(SEt)_2$

B-00666

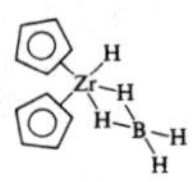

B-00667

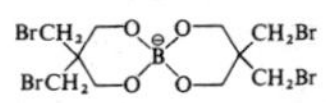

B-00668

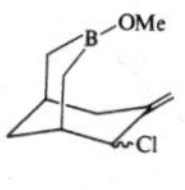

B-00669

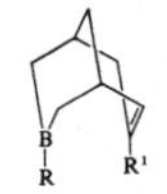

R = OMe, R[1] = CH_2Cl

B-00670

$Et_2NB(C{\equiv}CCH_3)_2$

B-00671

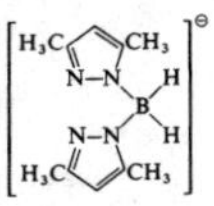

B-00672

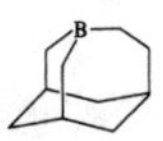

B-00673

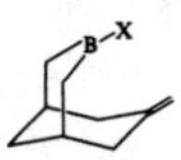

X = Me

B-00674

$PhB(NMe_2)_2$

B-00675

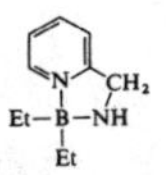

B-00676

As B-00598 with
R = OMe

B-00677

As B-00674 with
X = OMe

B-00678

As B-00539 with
R = —OAc

B-00679

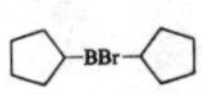

B-00680

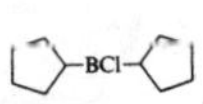

B-00681

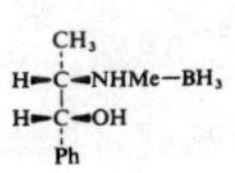

B-00682

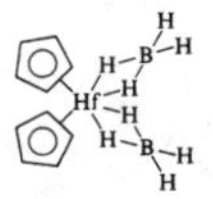

B-00683

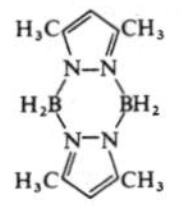

B-00684

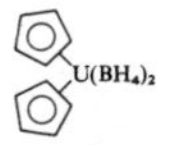

B-00685

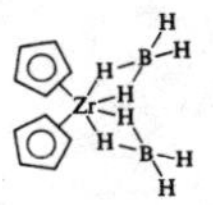

B-00686

$H_3CCH_2CH_2CH_2B(CH_2CH{=}CH_2)_2$

B-00687

As B-00544 with
X = R = CH_3

B-00688

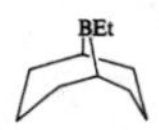

B-00689

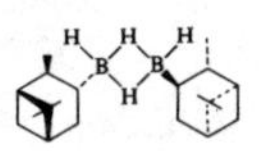

Dimer

B-00690

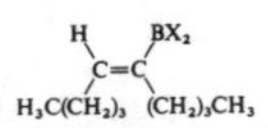

X = Cl

B-00691

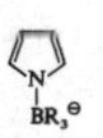

R = Et

B-00692

As B-00544 with
X = OMe, R = CH_3

B-00693

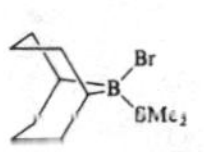

B-00694

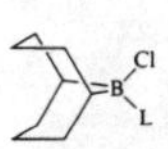

L = Me_2S

B-00695

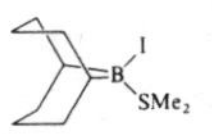

B-00696

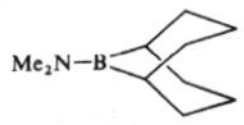

B-00697

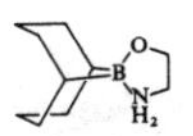

B-00698

As B-00544 with
X = OMe, R = $-CH_2BH_2$

B-00699

$(H_3CCH_2CH_2)_2BCH_2CH{=}CHCH_3$

B-00700

$H_3CCH{=}CHCH(BEt_2)CH_2CH_3$

B-00701

$H_2C{=}C(CH_3)CH_2B(CH_2CH_2CH_3)_2$

B-00702

$Et_2BOC(CH_3)_2CH_2CH{=}CH_2$

B-00703

$Et_2BOCH(CH_2CH_3)CH_2CH{=}CH_2$

B-00704

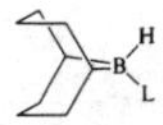

L = SMe_2

B-00705

$[H_3C(CH_2)_4]_2BBr$

B-00706

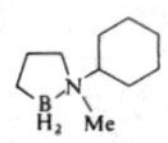

B-00707

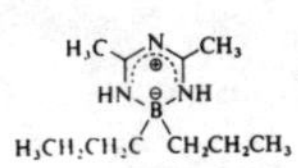

B-00708

$(H_3C)_2CH(CH_3)CHBHCH(CH_3)CH(CH_3)_2$

B-00709

$H_3C(CH_2)_7BI_2{-}SMe_2$

B-00710

$(H_3CCH_2CH_2)_2BO(CH_2)_3CH_3$

B-00711

$(H_3CCH_2CH_2CH_2)_2BOEt$

B-00712

$Et_2BO(CH_2)_5CH_3$

B-00713

$EtB(OCH_2CH_2CH_2CH_3)_2$

B-00714

$EtB[OC(CH_3)_3]_2$

B-00715

$(H_3CCH_2CH_2)_2BS(CH_2)_3CH_3$

B-00716

$Et_2BS(CH_2)_5CH_3$

B-00717

$MeB(CH_2CH_2CH_3)_3^{\ominus}$

B-00718

$(H_3CCH_2CH_2CH_2)_2BNMe_2$

B-00719

$[(H_3C)_3C]_2BNMe_2$

B-00720

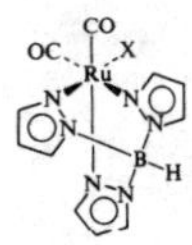

X = Br

B-00721

As B-00721 with
X = Cl

B-00722

As B-00721 with
X = I

B-00723

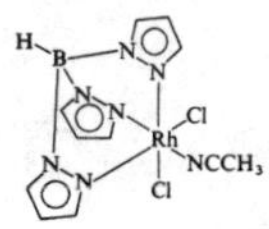

B-00724

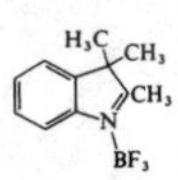

B-00725

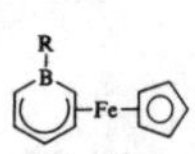

R = Me

B-00726

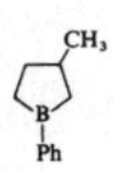

B-00727

As B-00327 with
R = Ph

B-00728

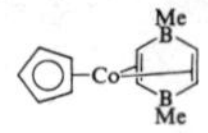

B-00729

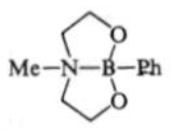

B-00730

As B-00591 with
R = Me

B-00731

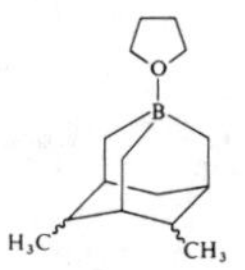

THF complex

B-00732

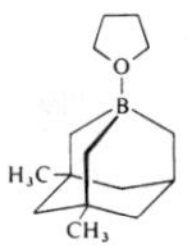

THF complex

B-00733

As B-00539 with
R = $—CH_2CH═CH_2$

B-00735

As B-00670 with
R = OMe, R′ = CH_2OMe

B-00736

As B-00539 with
R = $—CH(CH_3)_2$

B-00737

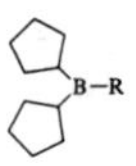

R = OMe

B-00738

As B-00544 with
X = Me, R = CH_2OMe

B-00739

As B-00544 with
X = OMe, R = $-CH_2OMe$

B-00740

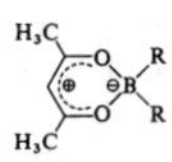

R = $CH_2CH_2CH_3$

B-00741

As B-00738 with
R = MeS—

B-00742

$(H_3CCH_2CH_2CH_2)_2BCH_2CH═CH_2$

B-00743

$(H_3C)_2C═CHCH_2B(CH_2CH_2CH_3)_2$

B-00744

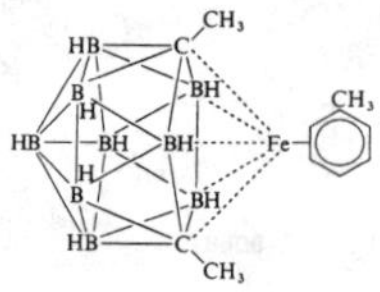

B-00745

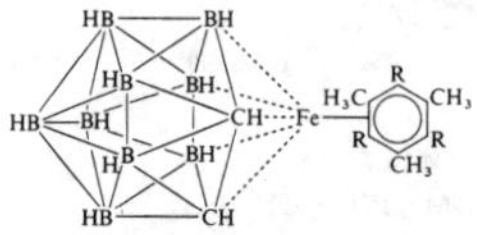

R = H

B-00746

As B-00705 with
L = $H_2NCH(CH_3)_2$

B-00747

As B-00705 with
L = $H_2NCH_2CH_2CH_3$

B-00748

As B-00705 with
L = Me_3N

B-00749

$(H_3C)_3CCH═NBEt_3^{\ominus}$

B-00750

$(H_3CCH_2CH_2CH_2)_2BOCH_2CH_2CH_3$

B-00751

As B-00633 with
R = $(CH_2)_3CH_3$

B-00752

As B-00342 with
R = $CH_2CH_2CH_3$

B-00753

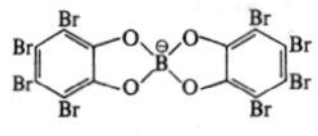

B-00754

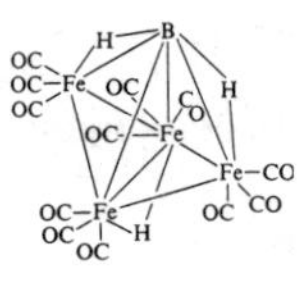

B-00755

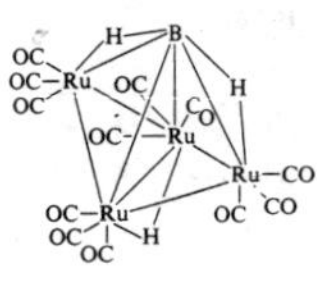

B-00756

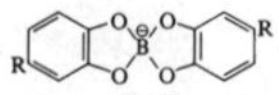

R = H

B-00757

Ph_2BBr

B-00758

Ph_2BCl

B-00759

Cl_2BNPh_2

B-00760

Ph_2BOH

B-00761

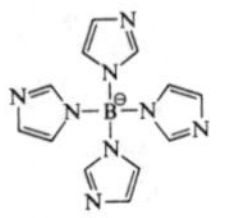

B-00762

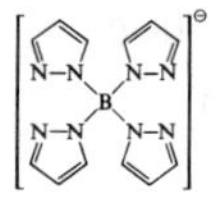

B-00763

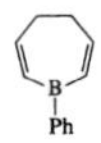

B-00764

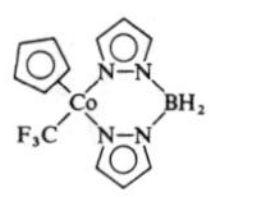

B-00765

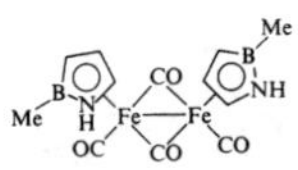

B-00766

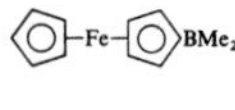

B-00767

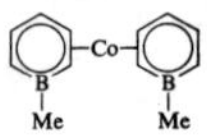

B-00768

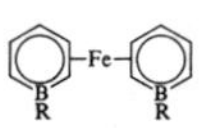

R = Me

B-00769

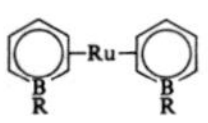

R = Me

B-00770

As B-00670 with
R = $—CH_2CH═CH_2$, R′ = CH_2Cl

B-00771

$PhBClN[CH(CH_3)_2]_2$

B-00772

$PhBClN(CH_2CH_2CH_3)_2$

B-00773

$[H_3C(CH_2)_3CBr═CH]_2BBr$

B-00774

$Et_3N\text{-}BPhCl_2$

B-00775

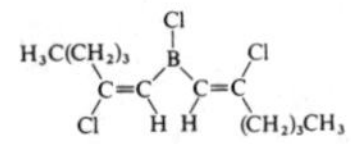

B-00776

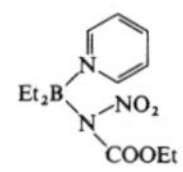

B-00777

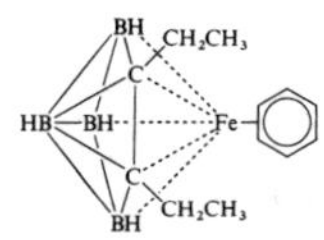

B-00778

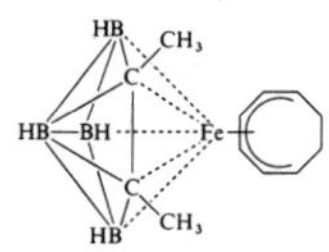

B-00779

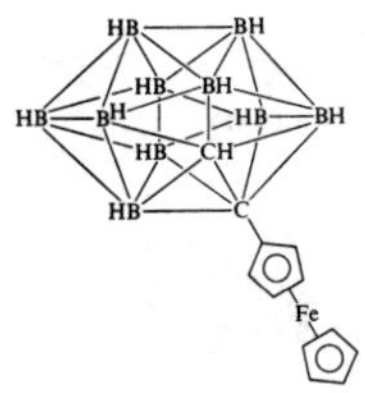

B-00780

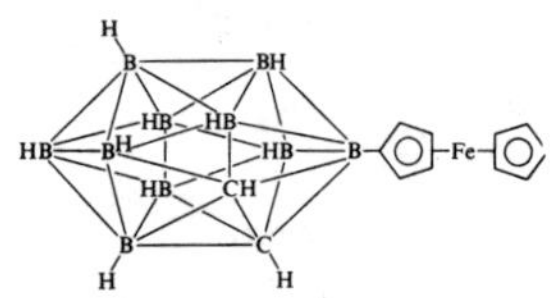

B-00781

As B-00539 with
R = $—CH_2CH═CHCH_3$

B-00782

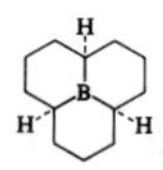

(3aα,6aα,9aα)-*form*

B-00783

$(H_3CCH{=}CHCH_2)_3B$

B-00784

$[H_2C{=}C(CH_3)CH_2]_3B$

B-00785

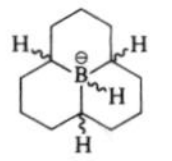

B-00786

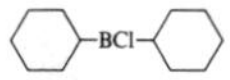

B-00787

$H_3C(CH_2)_3CH{=}CHBClCH{=}CH(CH_2)_3CH_3$

B-00788

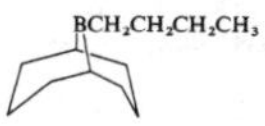

B-00789

As B-00539 with
$R = -C(CH_3)_3$

B-00790

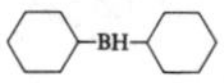

B-00791

$(H_2C{=}CHCH_2)_2B(CH_2)_5CH_3$

B-00792

As B-00539 with
$R = -CH(CH_3)CH_2CH_3$

B-00793

As B-00539 with
$R = -CH_2CH(CH_3)_2$

B-00794

$PhB(OSiMe_3)_2$

B-00795

As B-00546 with
$R = H_3C(CH_2)_3-$, $R' = H$

B-00796

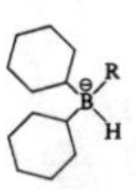

R = H

B-00797

$H_3CCH_2CH{=}C(CH_2CH_3)BClC(CH_3)_2CH(CH_3)_2$

B-00798

$H_3C(CH_2)_3CH{=}CHBClC(CH_3)_2CH(CH_3)_2$

B-00799

As B-00695 with
$L = Et_2O$

B-00800

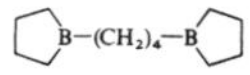

B-00801

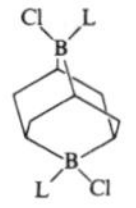

L = Me_2S

B-00802

$(H_3CCH_2CH_2CH_2)_2BCH_2CH{=}CHCH_3$

B-00803

As B-00551 with
$R = H_3C(CH_2)_3-$

B-00804

As B-00691 with
X = Me

B-00805

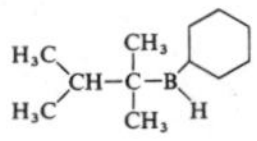

B-00806

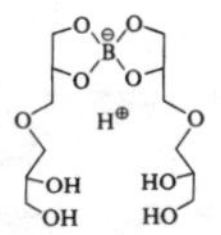

B-00807

$[H_3C(CH_2)_5]_2BBr$

B-00808

As B-00705 with
$L = H_2NC(CH_3)_3$

B-00809

As B-00705 with
$L = Et_2NH$

B-00810

$[(H_3C)_2CHC(CH_3)_2]_2BH$

B-00811

$B[C(CH_3)_3]_3$

B-00812

$B(CH_2CH_2CH_2CH_3)_3$

B-00813

$[H_3CCH_2CH(CH_3)]_3B$

B-00814

$[(H_3C)_2CHCH_2]_3B$

B-00815

$(H_3CCH_2CH_2CH_2)_3N{-}BBr_3$

B-00816

$(H_3CCH_2CH_2CH_2)_3N{-}BCl_3$

B-00817

$[H_3C(CH_2)_4]_2BOEt$

B-00818

$(H_3CCH_2CH_2)_2BO(CH_2)_5CH_3$

B-00819

$(H_3CCH_2CH_2CH_2O)_3B$

B-00820

$B[OC(CH_3)_3]_3$

B-00821

$[H_3CCH_2CH(CH_3)O]_3B$

B-00822

$[(H_3C)_2CHCH_2O]_3B$

B-00823

$(H_3CCH_2CH_2CH_2)_2BSCH_2CH_2CH_2CH_3$

B-00824

As B-00015 with
X = Cl, $Y = -(CH_2)_3CH_3$

B-00825

As B-00019 with
$R = -(CH_2)_3CH_3$

B-00826

As B-00019 with
$R = -O(CH_2)_3CH_3$

B-00827

$[H_3CCH_2CH(CH_3)]_3BH^{\ominus}$

B-00828

$[(H_3C)_2CHCH_2]_3BH^{\ominus}$

B-00829

$[(H_3C)_3C]_3BH^{\ominus}$

B-00830

$(H_3CCH_2CH_2CH_2)_3BH^{\ominus}$

B-00831

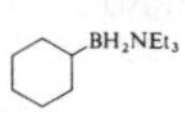

B-00832

$(H_3CCH_2CH_2CH_2)_2BNEt_2$

B-00833

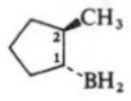

B-00834

$(H_3CCH_2CH_2)_2BOB(CH_2CH_2CH_3)_2$

B-00835

$[(H_3C)_3C]_3B{-}NH_3$

B-00836

$(H_3CCH_2CH_2)_3B{-}NMe_3$

B-00837

$B(NEt_2)_3$

B-00838

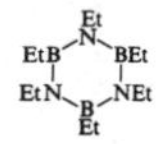

B-00839

$EtNHCOBH_2{-}Me_2NCH_2CH_2NMe_2{-}BH_2CONHEt$

B-00840

$[Pt(B_8H_{12})(PEt_3)_2]$

B-00841

$[Pt(B_{10}H_{12})(PEt_3)_2]$

B-00842

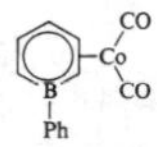

B-00843

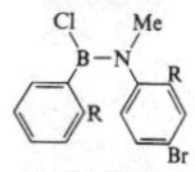

R = H

B-00844

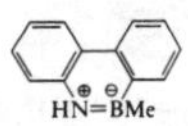

B-00845

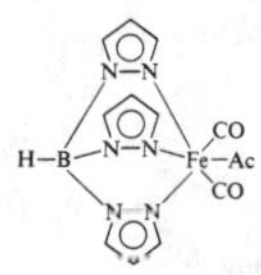

B-00846

Ph_2BOMe

B-00847

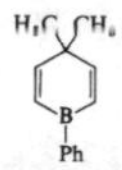

B-00848

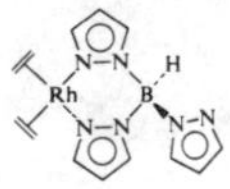

B-00850

As B-00695 with
L = Py

B-00851

As B-00539 with
$R = -C{\equiv}CCH_2CH_2CH_2Cl$

B-00852

As B-00705 with
L = pyridine

B-00853

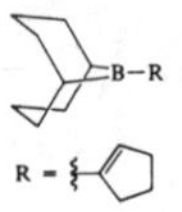

B-00854

$Rh(HCONMe_2)Py_2Cl_2(BH_4)$

B-00855

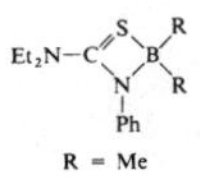

B-00856

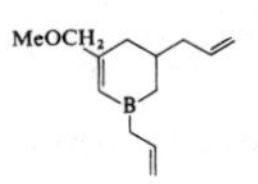

B-00857

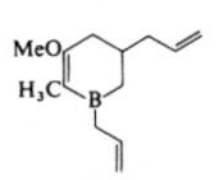

B-00858

As B-00670 with
R = $—CH_2CH{=}CH_2$, R′ = $—CH_2OMe$

B-00859

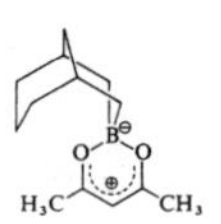

B-00860

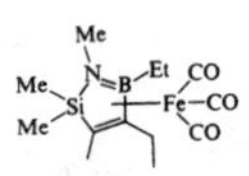

B-00861

As B-00854 with
R = cyclopentyl

B-00862

As B-00854 with
R = $—CH{=}CHCH_2CH_2CH_3$

B-00863

As B-00546 with
R = cyclopentyl, R′ = H

B-00864

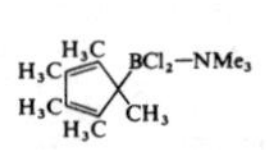

B-00865

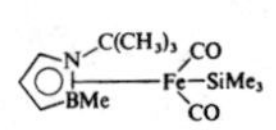

B-00866

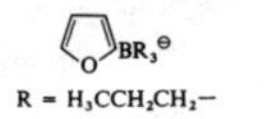

B-00867

As B-00544 with
X = $-C(CH_3)_3$, R = CH_3

B-00868

As B-00854 with
R = $—CH(CH_3)CH(CH_3)_2$

B-00869

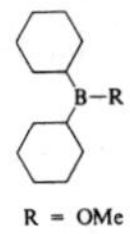

B-00870

As B-00870 with
R = SMe

B-00871

As B-00797 with
R = Me

B-00872

$(H_3CCH_2CH_2)_3BC{\equiv}COEt^{\ominus}$

B-00873

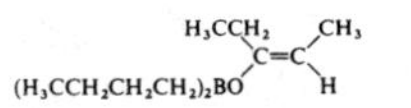

B-00874

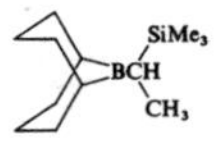

B-00875

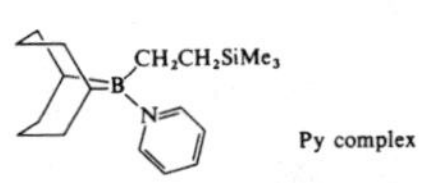

B-00876

$H_3C(CH_2)_4B(OCH_2CH_2CH_2CH_3)_2$

B-00877

$(H_3CCH_2CH_2CH_2)_2BS(CH_2)_4CH_3$

B-00878

$[(H_3C)_2CHCH_2]_3BMe^{\ominus}$

B-00879

$(H_3CCH_2CH_2CH_2)_3BMe^{\ominus}$

B-00880

$Me_2NCH_2CH_2CH_2B(CH_2CH_2CH_2CH_3)_2$

B-00881

$(H_3CCH_2CH_2CH_2)_2BCH_2CH_2CH_2NHEt$

B-00882

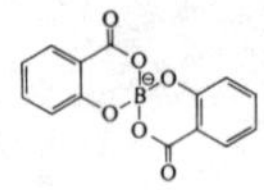

B-00883

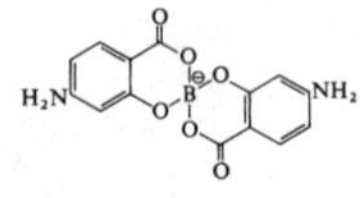

B-00884

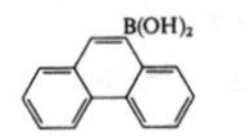

B-00885

Ph_2BEt

B-00886

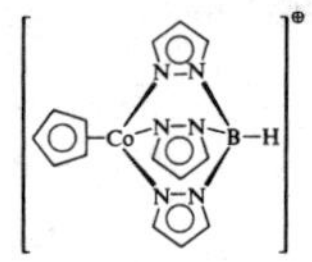

B-00887

Ph_2BOEt

B-00888

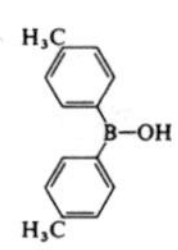

B-00889

$Me_2BPh_2^{\ominus}$

B-00890

Ph_2BNMe_2

B-00891

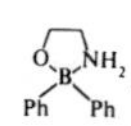

B-00892

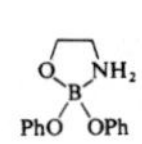

B-00893

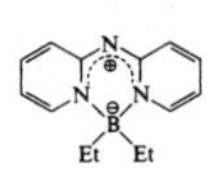

B-00894

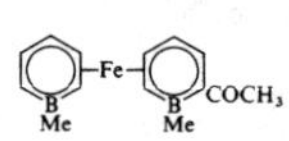

B-00895

As B-00854 with
R = Ph

B-00896

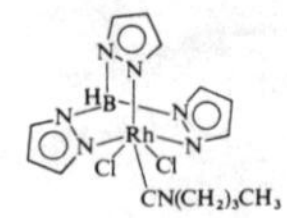

B-00897

As B-00546 with
R = Ph, R′ = H

B-00898

$PhCH{=}CHBClC(CH_3)_2CH(CH_3)_2$

B-00899

As B-00609 with
X = Py

B-00900

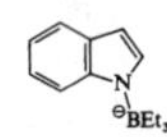

B-00901

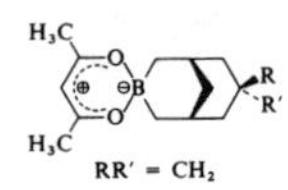

B-00902

As B-00705 with
L = 2-methylpyridine

B-00903

As B-00705 with
L = 4-methylpyridine

B-00904

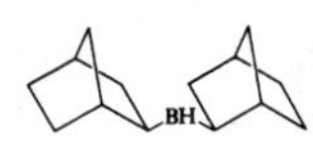

B-00905

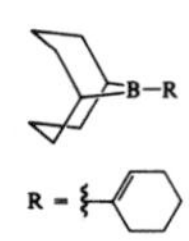

B-00906

As B-00906 with
R = $—C{\equiv}CC(CH_3)_3$

B-00907

As B-00906 with
R = $—C{\equiv}C(CH_2)_3CH_3$

B-00908

$PhBClN(CH_2CH_2CH_2CH_3)_2$

B-00909

As B-00902 with
R = H, R′ = CH_3

B-00910

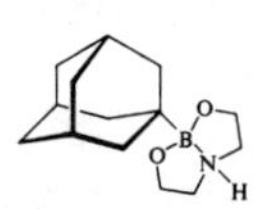

B-00911

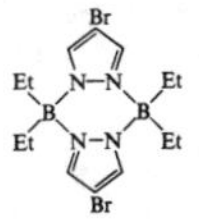

B-00912

As B-00906 with
R = cyclohexyl

B-00913

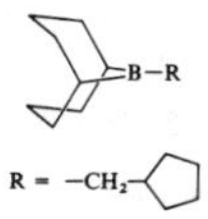

B-00914

As B-00914 with
R = $—CH{=}CHC(CH_3)_3$

B-00915

As B-00914 with
R = $—C(CH_2CH_3){=}CHCH_2CH_3$

B-00916

As B-00914 with
R = $—CH{=}CH(CH_2)_3CH_3$

B-00917

As B-00914 with
R = $—(CH_2)_3CH{=}CHCH_3$

B-00918

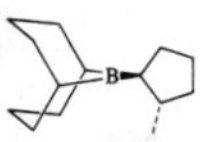

B-00919

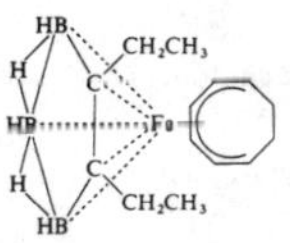

B-00920

As B-00914 with
R = $—CH(CH_2CH_3)CH_2CH_2CH_3$

B-00921

As B-00914 with
R = $—(CH_2)_5CH_3$

B-00922

As B-00914 with
R = $—C(CH_3)_2CH(CH_3)_2$

B-00923

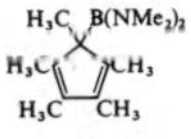

B-00924

As B-00705 with
L =

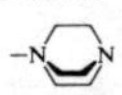

B-00925

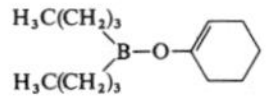

B-00926

$(H_3CCH_2CH_2CH_2)_3BC{\equiv}CH^{\ominus}$

B-00927

As B-00546 with
R = $-C(CH_3)_2CH(CH_3)_2$, R′ = H

B-00928

As B-00746 with
R = CH_3

B-00929

$(H_3C)_2CHC(CH_3)_2BCl(CH_2)_7CH_3$

B-00930

As B-00705 with
L = Et_3N

B-00931

$(H_3C)_2CHC(CH_3)_2B(CH_2CH_2CH_2CH_3)_2$

B-00933

As B-00705 with
L = $Me_2NCH_2CH_2NMe_2$

B-00934

$[H_3C(CH_2)_4]_2BOCH_2CH_2CH_2CH_3$

B-00935

$(H_3CCH_2CH_2CH_2)_2BO(CH_2)_5CH_3$

B-00936

$H_3C(CH_2)_5B(OCH_2CH_2CH_2CH_3)_2$

B-00937

$(H_3CCH_2CH_2CH_2)_2BS(CH_2)_5CH_3$

B-00938

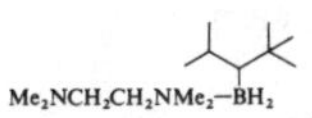

B-00939

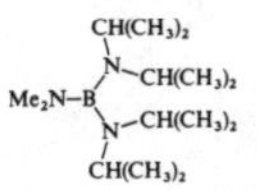

B-00940

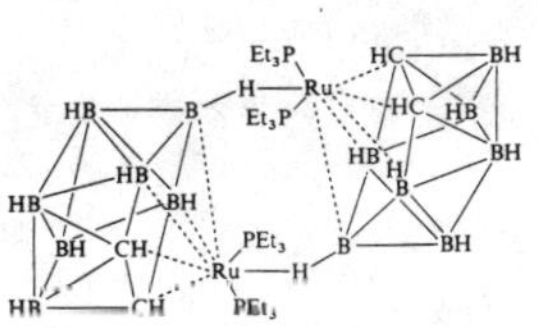

B-00941

$Ph_2B—N{=}C(CF_3)_2$

B-00942

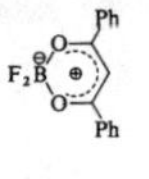

B-00943

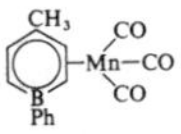

B-00944

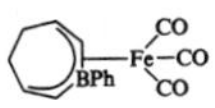

B-00945

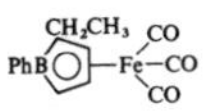

B-00946

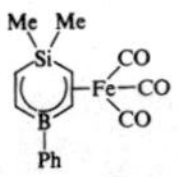

B-00947

As B-00844 with
R = CH_3

B-00948

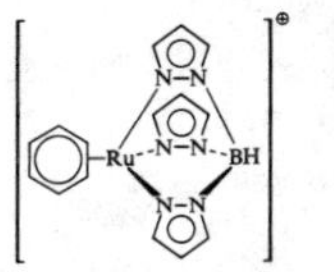

B-00949

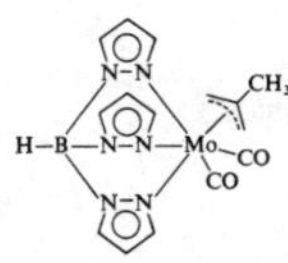

B-00950

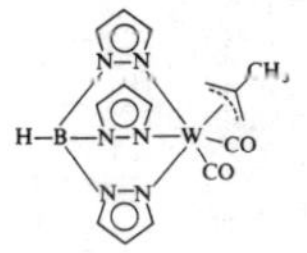

B-00951

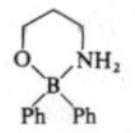

B-00952

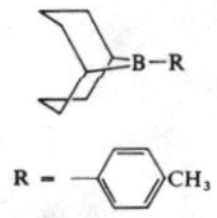

B-00953

As B-00544 with
X = OMe, R = Ph

B-00954

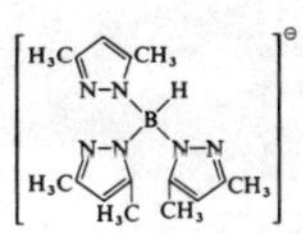

B-00955

As B-00705 with
L = 2,6-dimethylpyridine

B-00956

As B-00705 with
L = 2-Ethylpyridine

B-00957

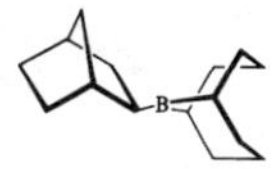

B-00958

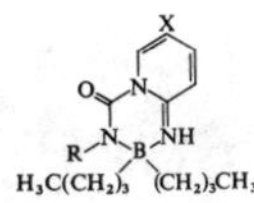

R = Me, X = Cl

B-00959

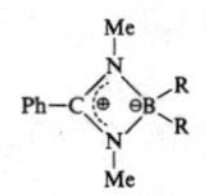

R = $—CH(CH_3)_2$

B-00960

As B-00960 with
R = $CH_2CH_2CH_3$

B-00961

As B-00856 with
R = Et

B-00962

As B-00902 with
R, R′ = H, $-CH_2OMe$

B-00963

As B-00959 with
R = Me, X = H

B-00964

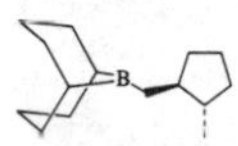

B-00965

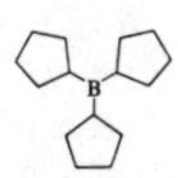

B-00966

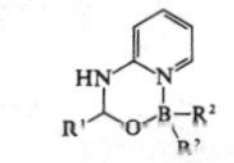

$R^1 = R^2 = CH_2CH_2CH_3$

B-00967

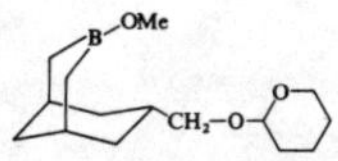

B-00968

As B-00546 with
$R = H_3C(CH_2)_3—$, $R' = H_2C{=}CHCH_2—$

B-00969

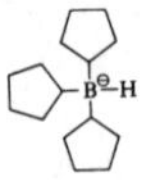

B-00970

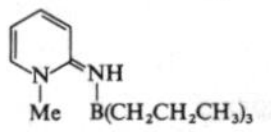

B-00971

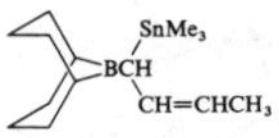

B-00972

$[(H_3C)_2CHCH(CH_3)]_2BCH{=}CHCH_2CH_2CH_2Cl$

B-00973

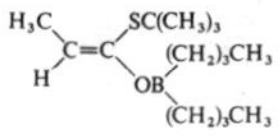

B-00975

$[H_3C(CH_2)_4]_3B$

B-00976

$[(H_3C)_2CHCH(CH_3)]_3B$

B-00977

$H_3CCH_2CH_2B[O(CH_2)_5CH_3]_2$

B-00978

$[H_3C(CH_2)_4O]_3B$

B-00979

$[(H_3C)_2CHCH(CH_3)]_3BH^{\ominus}$

B-00980

$Me_3N—B(CH_2CH_2CH_2CH_3)_3$

B-00981

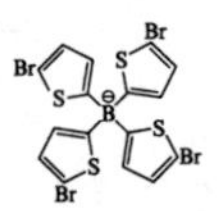

B-00982

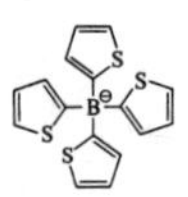

B-00983

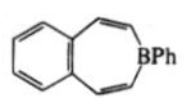

B-00984

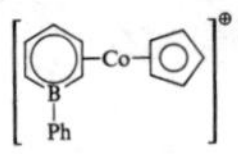

B-00985

As B-00726 with
R = Ph

B-00986

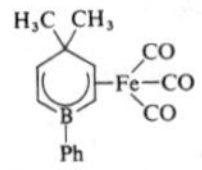

B-00987

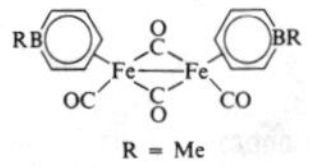

B-00988

As B-00659 with
R = Ph

B-00989

As B-00664 with
R = Ph, $R' = CH_3$

B-00990

$Ph_2BO(CH_2)_3CH_3$

B-00991

$Ph_2BOCH_2CH(CH_3)_2$

B-00992

$Ph_2BS(CH_2)_3CH_3$

B-00993

As B-00953 with
$R = —CH{=}CHPh$

B-00994

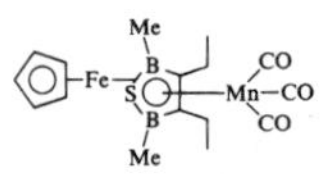

B-00995

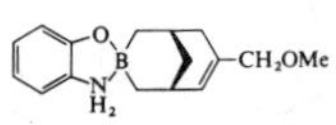

B-00996

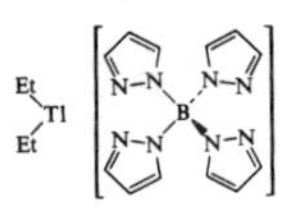

B-00997

As B-00953 with
$R = —CH_2CH_2Ph$

B-00998

As B-00953 with
$R = —C{\equiv}C—$cyclohexyl

B-00999

As B-00705 with
L = 2-Isopropylpyridine

B-01001

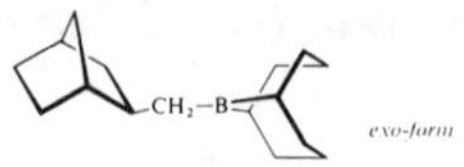

B-01002

As B-00659 with
$R = -(CH_2)_5CH_3$

B-01003

As B-00670 with
$R = —CH_2CH{=}CH_2$, $R' = -(CH_2)_4CH_3$

B-01004

As B-00953 with
R = cyclooctyl

B-01005

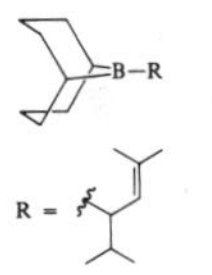

B-01006

As B-00953 with
$R = —CH{=}CH(CH_2)_5CH_3$

B-01007

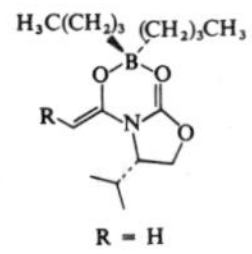

B-01008

As B-00867 with
$R = —CH(CH_3)CH_2CH_3$

B-01009

As B-00867 with
$R = -CH_2CH(CH_3)_2$

B-01010

As B-00586 with
R = Ph, R′ = H

B-01011

As B-00738 with
$R = -C(CH_3)_2CH(CH_3)_2$

B-01012

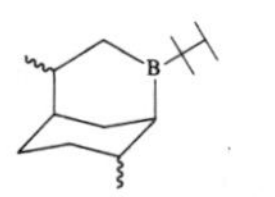

B-01013

As B-01006 with
$R = —(CH_2)_7CH_3$

B-01014

As B-00546 with
$R = R' = H_3C(CH_2)_3—$

B-01015

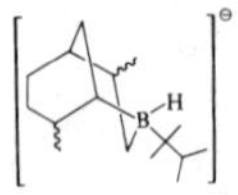

B-01016

$[(H_3C)_2CHCH(CH_3)]_2BCBr{=}CH(CH_2)_3CH_3$

B-01017

$[H_3C(CH_2)_3]_3BC{\equiv}COEt^{\ominus}$

B-01018

$[(H_3C)_2CHCH(CH_3)]_2BCH{=}CH(CH_2)_3CH_3$

B-01019

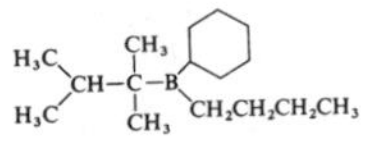

B-01020

As B-00551 with
$R = H_3C(CH_2)_7—$

B-01021

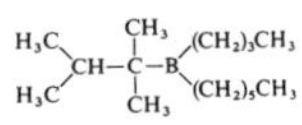

B-01022

$(H_3CCH_2CH_2CH_2)_4B^{\ominus}$

B-01023

$ClB[N(CH_2CH_2CH_2CH_3)_2]_2$

B-01024

$(H_3CCH_2CH_2CH_2)_2BOB(CH_2CH_2CH_2CH_3)_2$

B-01025

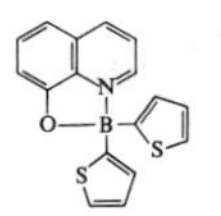

B-01026

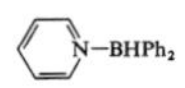

B-01027

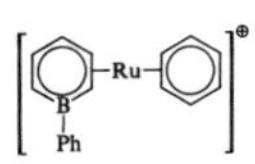

B-01028

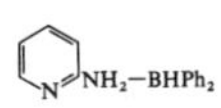

B-01029

As B-00741 with
R = Ph

B-01030

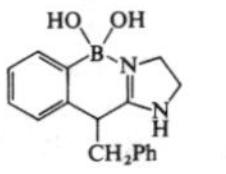

B-01031

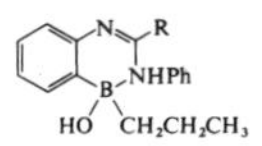

R = CH_3

B-01032

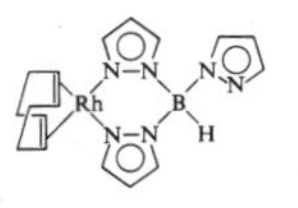

B-01033

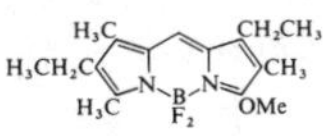

B-01034

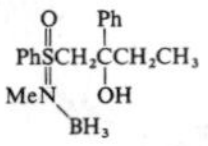

B-01035

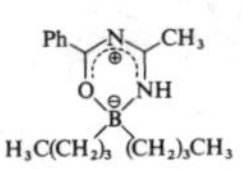

B-01036

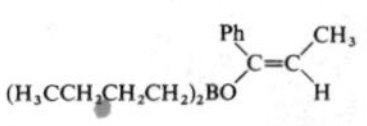

B-01037

$H_3CCH_2CPh{=}NB(CH_2CH_2CH_2CH_3)_2$

B-01038

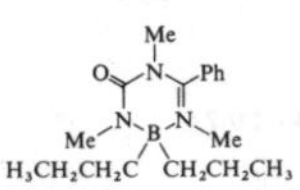

B-01039

As B-00870 with
R = $H_2C{=}C(CH_3)CH{=}CH-$

B-01040

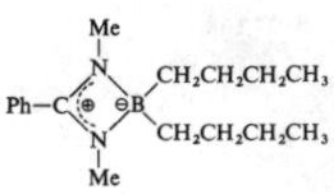

B-01041

As B-00960 with
R = $-CH_2CH(CH_3)_2$

B-01042

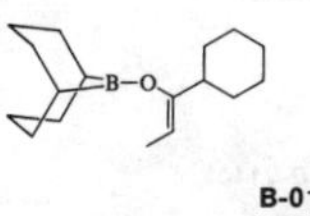

B-01043

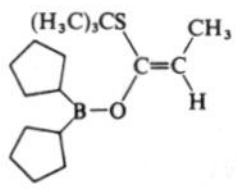

B-01044

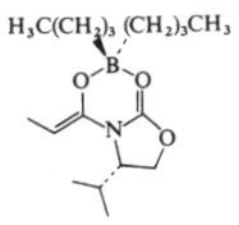

B-01045

As B-01008 with
R = MeS

B-01046

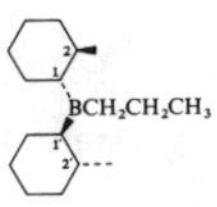

B-01047

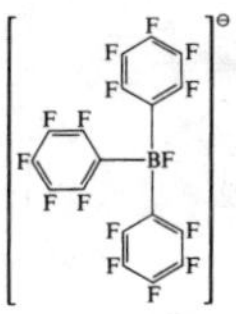

B-01048

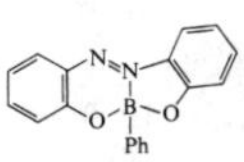

B-01049

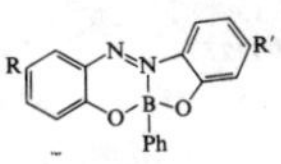

R = H, R′ = OH

B-01050

BPh_3

B-01051

$Ph_3PO{-}BBr_3$

B-01052

$Ph_3P{-}BBr_3$

B-01053

$Ph_3PO{-}BCl_3$

B-01054

$Ph_3P{-}BCl_3$

B-01055

$Ph_3PO{-}BF_3$

B-01056

$Ph_3P{-}BF_3$

B-01057

$Ph_3P{-}BI_3$

B-01058

$PhB(OPh)_2$

B-01059

$(PhO)_3B$

B-01060

As B-00015 with
X = Cl, Y = Ph

B-01061

As B-00019 with
R = Ph

B-01062

As B-00019 with
R = OPh

B-01063

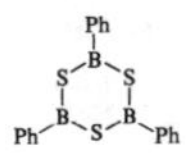

B-01064

$Ph_3BH^{\ominus}$

B-01065

$IBH_2{-}PPh_3$

B-01066

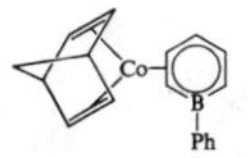

B-01067

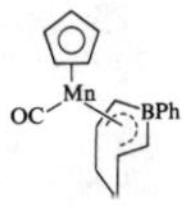

B-01068

$Ph_3N{-}BH_3$

B-01069

$Ph_3B{-}NH_2OH$

B-01070

$Ph_3P{-}BH_3$

B-01071

As B-00020 with
R = Ph, R′ = H

B-01072

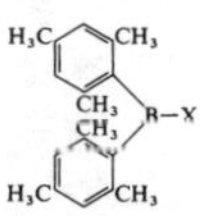

X = F

B-01073

As B-00802 with
L = Py

B-01074

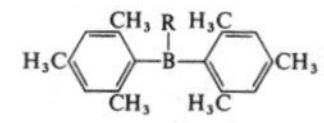

R = OH

B-01075

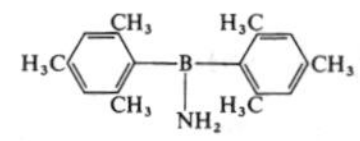

B-01076

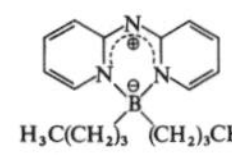

B-01077

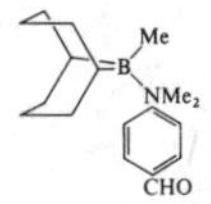

B-01078

As B-00651 with
R = $(H_3C)_3C-$

B-01079

$[(H_3C)_2CHCH(CH_3)]_2BCH{=}CHPh$

B-01080

$H_3CCH_2CH_2CPh{=}NB(CH_2CH_2CH_2CH_3)_2$

B-01081

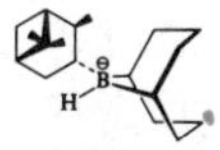

B-01082

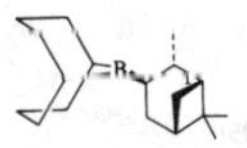

(1′R)-form

B-01083

$(H_3CCH_2CH_2CH_2)_3BPh^{\ominus}$

B-01084

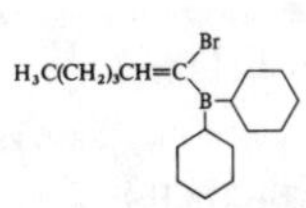

B-01085

$(H_3CCH_2CH_2CH_2)_3N{-}BPhCl_2$

B-01086

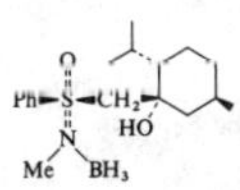

B-01087

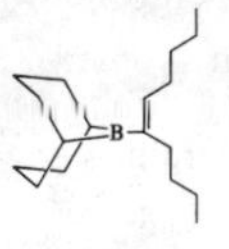

B-01088

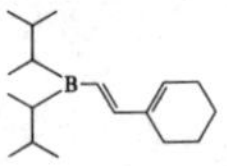

B-01089

As B-00870 with
X = $-CH{=}CH(CH_2)_3CH_3$

B-01090

B-01091

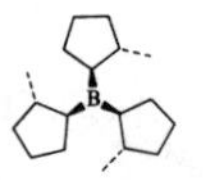

B-01092

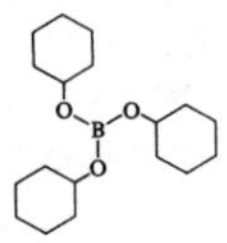

B-01093

As B-00670 with
R = $H_3C(CH_2)_4S$—, R′ = —$(CH_2)_4CH_3$

B-01094

As B-00019 with
R = cyclohexyl

B-01095

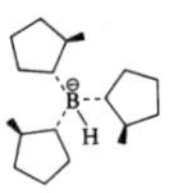

B-01096

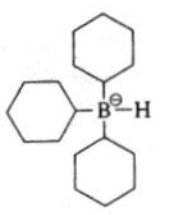

B-01097

As B-00870 with
R = $H_3CCH_2CH_2CH(CH_2CH_3)-$

B-01098

As B-00870 with
R = $-(CH_2)_5CH_3$

B-01099

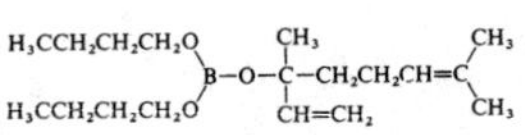

B-01100

$(H_3CCH_2CH_2CH_2)_3BC{\equiv}C(CH_2)_3CH_3^{\ominus}$

B-01101

$[H_3CCH_2CH_2CH(CH_3)CH_2]_2BCBr{=}CH(CH_2)_3CH_3$

B-01102

$[H_3C(CH_2)_5]_2BCBr{=}CH(CH_2)_3CH_3$

B-01103

$[(H_3C)_2CHCH(CH_3)]_2BCH{=}CH(CH_2)_5CH_3$

B-01104

$[(H_3C)_2CHCH(CH_3)]_2B(CH_2)_7CH_3$

B-01105

$(H_3C)_2CHC(CH_3)_2B[(CH_2)_5CH_3]_2$

B-01106

$[H_3C(CH_2)_5]_3B$

B-01107

$B[CH(CH_2CH_3)CH_2CH_2CH_3]_3$

B-01108

$[H_3CCH_2CH_2CH(CH_3)CH_2]_3B$

B-01109

$[H_3C(CH_2)_5O]_3B$

B-01110

B-01111

B-01112

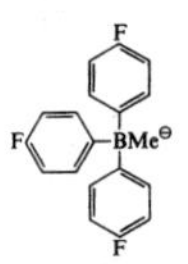

B-01113

As B-00462 with
X = Ph

B-01114

$[Ph_3BCN]^{\ominus}$

B-01115

As B-01050 with
R = CH_3, R′ = H

B-01116

$Ph_3P{-}BH_2CN$

B-01117

$MeBPh_3^{\ominus}$

B-01118

As B-00967 with
$R^1 = CH_3$, R^2 = Ph

B-01119

$HN{=}C(NH_2)NH_2{-}BPh_3$

B-01120

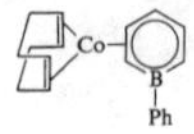

B-01121

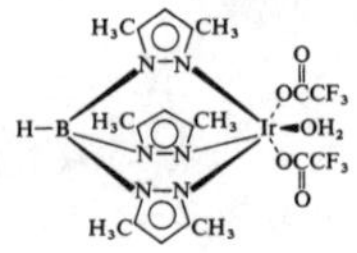

B-01122

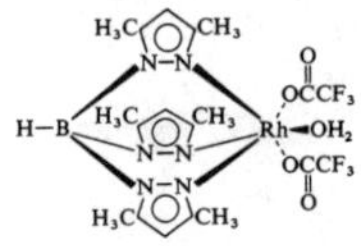

B-01123

As B-01073 with
X = Me

B-01124

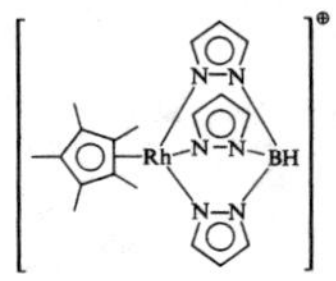

B-01125

As B-01075 with
R = OMe

B-01126

As B-01075 with
R = −SMe

B-01127

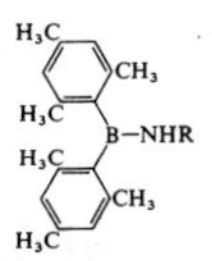

R = Me

B-01128

$(H_3CCH_2CH_2CH_2)_2BCH(SPh)CH_2CH_2CH_2CH_3$

B-01129

As B-00670 with
R = $-O(CH_2)_5CH_3$, R′ = —$(CH_2)_4CH_3$

B-01130

As B-00670 with
R = $-S(CH_2)_5CH_3$, R′ = —$(CH_2)_4CH_3$

B-01131

$(H_3CCH_2CH_2CH_2)_3BC{\equiv}C(CH_2)_4CH_3^{\ominus}$

B-01132

$[H_3C(CH_2)_4]_3BC{\equiv}COEt^{\ominus}$

B-01133

$[H_3C(CH_2)_5]_3BMe^{\ominus}$

B-01134

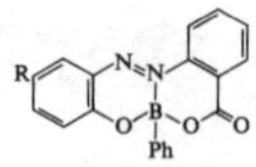

R = CH_3

B-01135

As B-01135 with
R = OMe

B-01136

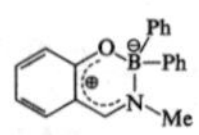

B-01137

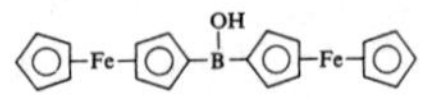

B-01138

As B-00757 with
R = $(H_3C)_3C-$

B-01139

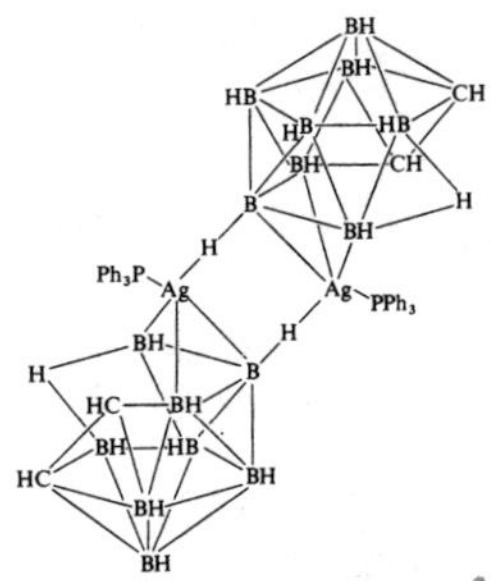

B-01140

$(Ph_3P)Cu(C_2B_8H_{11})$

B-01141

As B-01075 with
R = Et

B-01142

As B-00959 with
R = Ph, X = Cl

B-01143

As B-01075 with
R = −SEt

B-01144

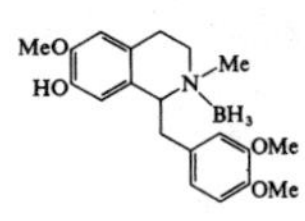

B-01145

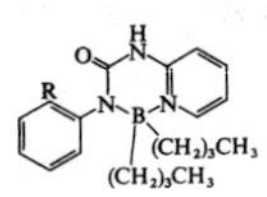

R = H

B-01146

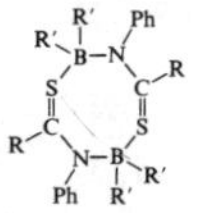

R = R′ = CH_3

B-01147

As B-00870 with
R = —CH=CHPh

B-01148

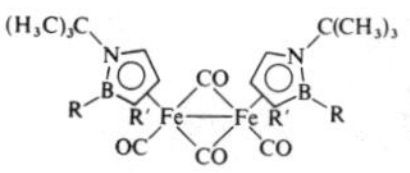

cis-form

R = Me, R′ = H

B-01149

$(H_3CCH_2CH_2CH_2)_3BC{\equiv}CPh^{\ominus}$

B-01150

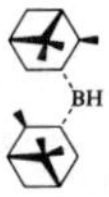

(+)-form

B-01151

As B-00539 with
R = $OCH{=}CH(CH_2)_8COOMe$

B-01152

As B-00870 with
R = $—CH{=}CH(CH_2)_5CH_3$

B-01153

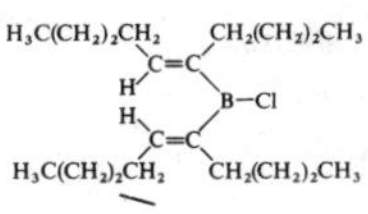

B-01154

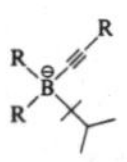

R = $—(CH_2)_3CH_3$

B-01155

$[H_3C(CH_2)_5]_3BC{\equiv}CH^{\ominus}$

B-01156

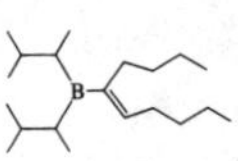

B-01157

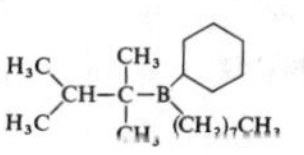

B-01158

$H_3C(CH_2)_3OCH_2CH_2OB$
$[(OCH_2CH_2)_3OMe]_2$

B-01159

$[(H_3C)_2CHCH_2CH_2]_4B^{\ominus}$

B-01160

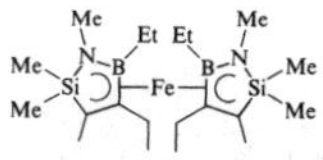

B-01161

As B-00015 with
X = C_6F_5, Y = Me

B-01162

$(PhCH_2)_3B$

B-01163

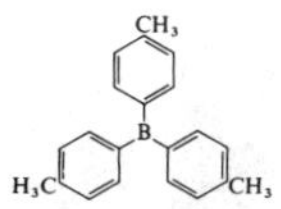

B-01164

$(PhCH_2O)_3B$

B-01165

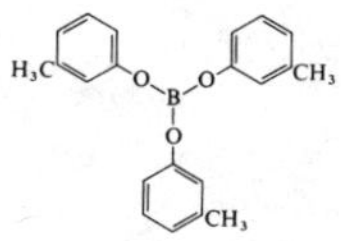

B-01166

$Ph_3P—BH_2COOEt$

B-01167

$Ph_3P—BH_2CONHEt$

B-01168

$Me_2NCONH_2—BPh_3$

B-01169

$Ph_3N—BMe_3$

B-01170

$Ph_3B—NMe_3$

B-01171

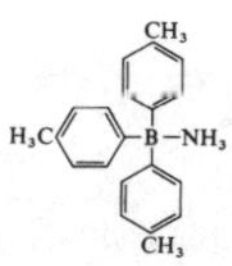

B-01172

As B-00015 with
X = Ph, Y = Me

B-01173

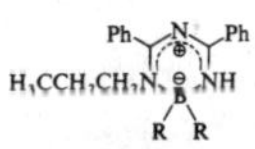

R = $H_3CCH_2CH_2—$

B-01174

As B-01073 with
X = $H_3CCH_2CH_2—$

B-01175

As B-01075 with
R = $—OCH(CH_3)_2$

B-01176

As B-01128 with
R = $CH(CH_3)_2$

B-01177

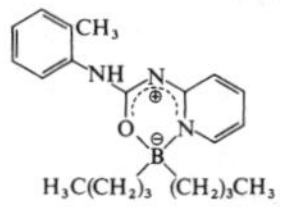

B-01178

As B-01146 with
R = CH_3

B-01179

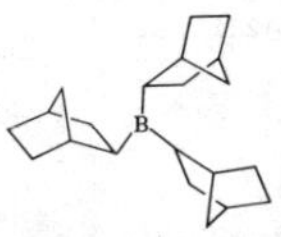

B-01180

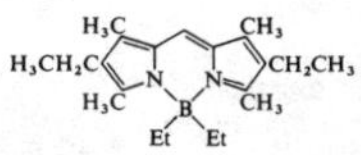

B-01181

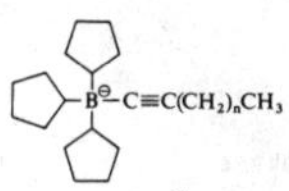

n = 3

B-01182

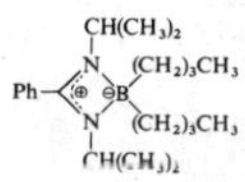

B-01183

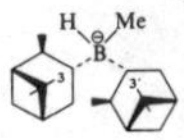

B-01184

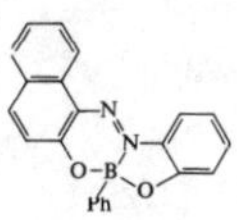

B-01185

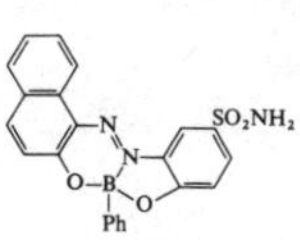

B-01186

As B-00692 with
R = Ph

B-01187

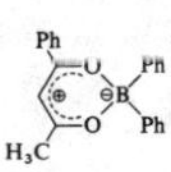

B-01188

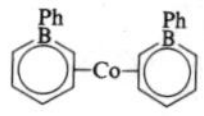

B-01189

As B-00769 with
R = Ph

B-01190

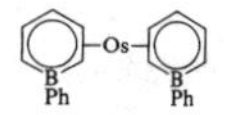

B-01191

As B-00770 with
R = Ph

B-01192

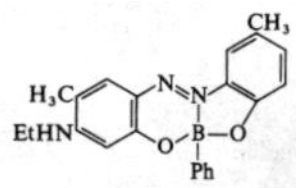

B-01193

As B-01032 with
R = Ph

B-01194

$H_3C(CH_2)_3BPh_3^{\ominus}$

B-01195

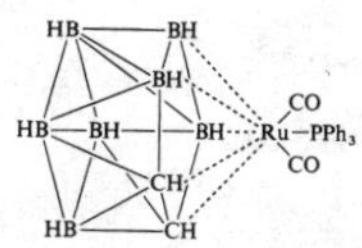

B-01196

$Ph_3B—NH_2(CH_2)_3CH_3$

B-01197

$(H_3C)_2CHCH_2NH_2—BPh_3$

B-01198

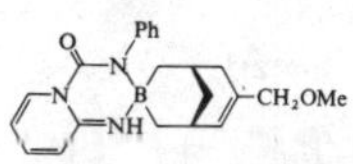

B-01199

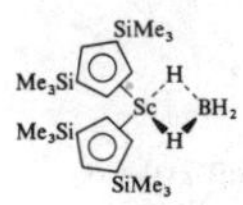

B-01200

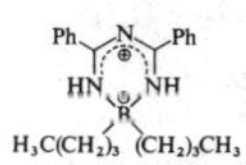

B-01201

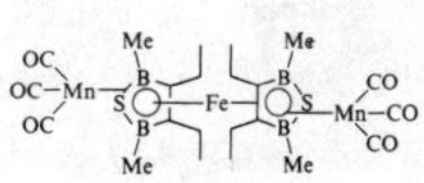

B-01202

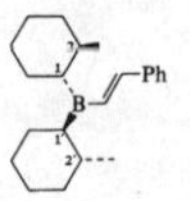

B-01203

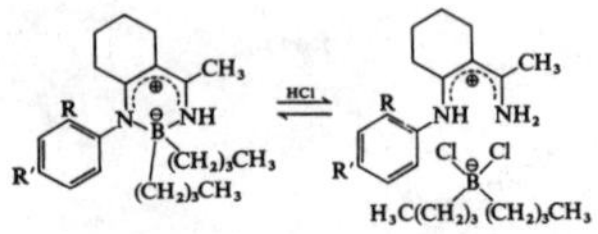

R = Cl, R′ = H

B-01204

As B-01204 with
R = H, R′ = Cl

B-01205

As B-01204 with
R = R′ = H

B-01206

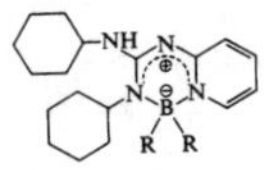

R = Et

B-01207

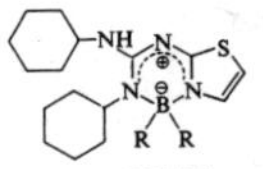

R = $-CH(CH_3)_2$

B-01208

As B-00691 with
X = Cyclohexyl

B-01209

As B-00867 with
R = $-(CH_2)_5CH_3$

B-01210

$(H_3C)_2CHC(CH_3)_2B[(CH{=}CH(CH_2)_5CH_3]_2$

B-01211

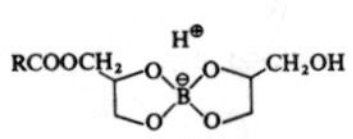

R = $H_3C(CH_2)_{14}-$

B-01212

$[H_3C(CH_2)_7]_2BCBr{=}CH(CH_2)_3CH_3$

B-01213

$(H_3C)_2CHC(CH_3)_2B[(CH_2)_7CH_3]_2$

B-01214

$[(H_3C)_2CHCH(CH_3)]_2BH{-}H_2NCH_2CH_2NH_2{-}BH[CH(CH_3)CH(CH_3)_2]_2$

B-01215

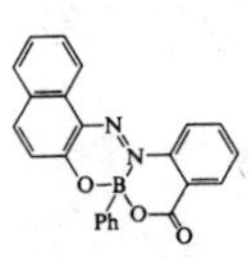

B-01216

As B-00307 with
X = Ph

B-01217

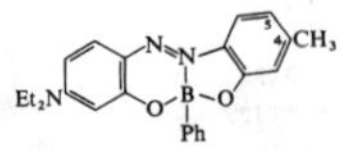

B-01218

As B-00856 with
R = Ph

B-01220

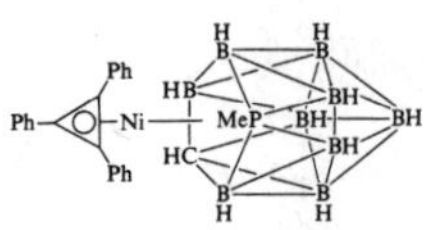

B-01221

As B-01204 with
R = H, R′ = CH_3

B-01222

As B-01182 with
n = 5

B-01223

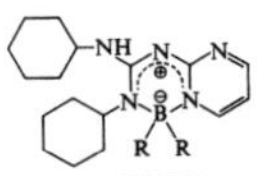

R = $-CH(CH_3)_2$

B-01224

$H_3C(CH_2)_5C{\equiv}CB[(CH_2)_4CH_3]_3^{\ominus}$

B-01225

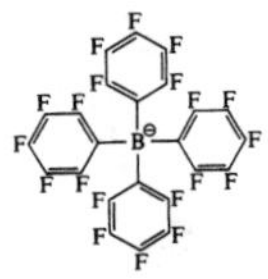

B-01226

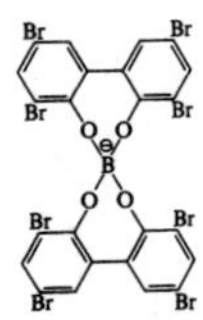

B-01227

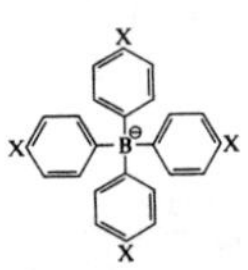

X = Cl

B-01228

As B-01228 with
X = F

B-01229

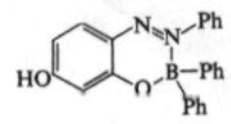

B-01230

$Ag^{\oplus}Ph_4B^{\ominus}$

B-01231

$Cs^{\oplus}Ph_4B^{\ominus}$

B-01232

$K^{\oplus}Ph_4B^{\ominus}$

B-01233

$[LiPh_4B]_n$

B-01234

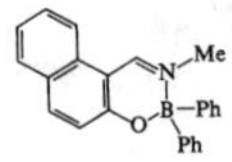

B-01235

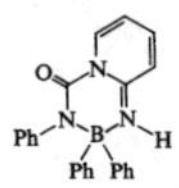

B-01236

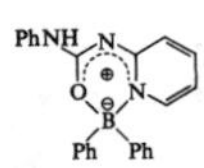

B-01237

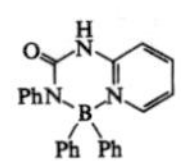

B-01238

$Rb^{\oplus}Ph_4B^{\ominus}$

B-01239

$Tl^{\oplus}Ph_4B^{\ominus}$

B-01240

Ph_2BOBPh_2

B-01241

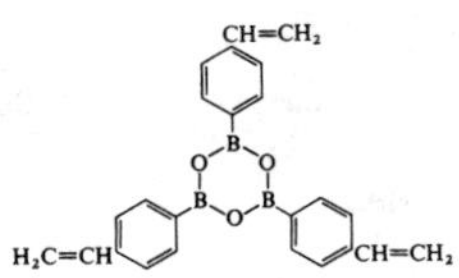

B-01242

$H_3C(CH_2)_3C{\equiv}CBPh_3^{\ominus}$

B-01243

$NH_4^{\oplus}Ph_4B^{\ominus}$

B-01244

As B-01075 with
R = 4-Chlorophenoxy

B-01245

As B-01075 with
R = —OPh

B-01246

As B-01075 with
R = —SPh

B-01247

As B-01128 with
R = Ph

B-01248

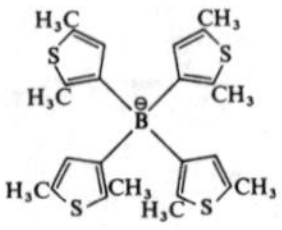

B-01249

$Et_3N{-}BPh_3$

B-01250

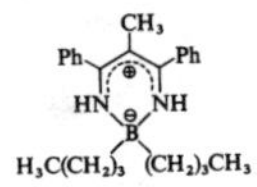

B-01251

As B-01207 with
R = $-CH_2CH_2CH_3$

B-01252

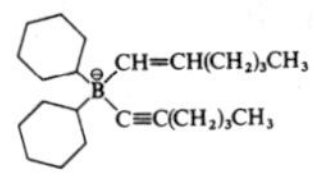

B-01253

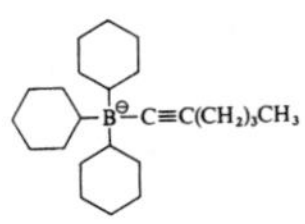

B-01254

As B-01212 with
R = $H_3C(CH_2)_7CH{=}CH(CH_2)_7-$

B-01255

As B-01212 with
R = $H_3C(CH_2)_{16}-$

B-01256

$[H_3C(CH_2)_5]_3BC{\equiv}C(CH_2)_3CH_3^{\ominus}$

B-01257

$[H_3C(CH_2)_7]_3B$

B-01258

$[H_3C(CH_2)_7O]_3B$

B-01259

$B[OCH_2CH(CH_2CH_3)CH_2CH_2CH_2CH_3]_3$

B-01260

$H_3C(CH_2)_3O(CH_2CH_2O)_3B[(OCH_2CH_2)_3OMe]_2$

B-01261

$H_3C(CH_2)_3OCH_2CH_2OB[(OCH_2CH_2)_4OMe]_2$

B-01262

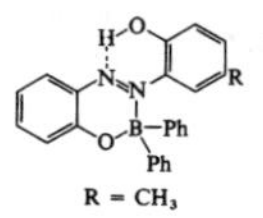

R = CH$_3$

B-01263

As B-01263 with
R = OMe

B-01264

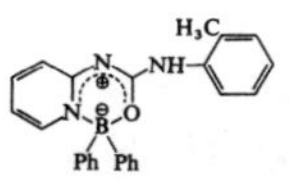

B-01265

$(Ph_2B)_2NMe$

B-01266

$H_3C(CH_2)_4C{\equiv}CBPh_3^{\ominus}$

B-01267

$(H_2N)_2C{=}NH_2^{\oplus}BPh_4^{\ominus}$

B-01268

As B-01075 with
R = 4-Methoxyphenoxy

B-01269

As B-01075 with
R = $-SCH_2Ph$

B-01270

As B-01075 with
R = 4-methylphenylthio

B-01271

As B-01128 with
R = $-CH_2Ph$

B-01272

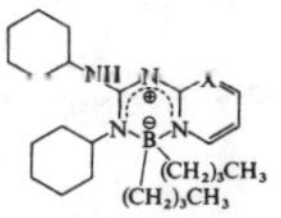

X = N

B-01273

$[H_3C(CH_2)_5]_3BC{\equiv}C(CH_2)_4CH_3^{\ominus}$

B-01274

$Ir(BH_4)(CO)[P(CH_2SiMe_3)_3]_2$

B-01275

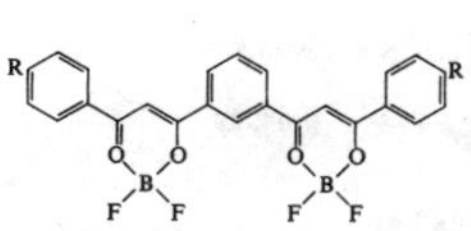

R = OMe

B-01276

As B-00988 with
R = Ph

B-01277

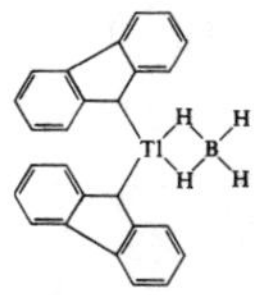

B-01278

Me_2TlBPh_4

B-01279

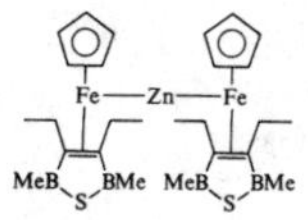

B-01280

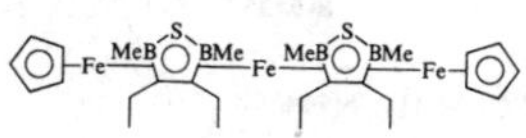

B-01281

$[H_3C(CH_2)_5]_3BC{\equiv}CPh^{\ominus}$

B-01282

As B-01273 with
X = CH

B-01283

As B-01155 with
R = $-(CH_2)_5CH_3$

B-01284

$[H_3C(CH_2)_5]_3BC{\equiv}C(CH_2)_5CH_3^{\ominus}$

B-01285

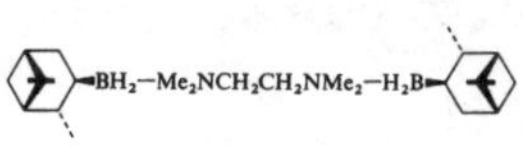

B-01286

$H_3C(CH_2)_3O(CH_2CH_2O)_2B[(OCH_2CH_2)_4OMe]_2$

B-01287

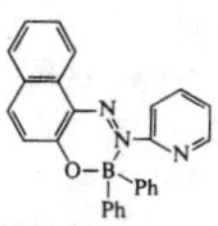

B-01288

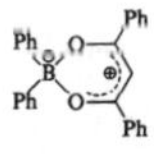

B-01289

As B-01174 with
R = Ph

B-01290

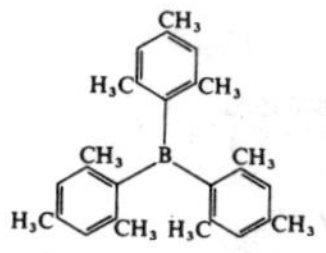

B-01291

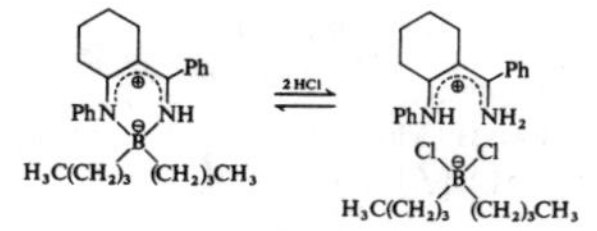

B-01292

$[H_3C(CH_2)_8]_3B$

B-01293

$[H_3C(CH_2)_8O]_3B$

B-01294

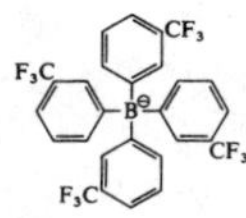

B-01295

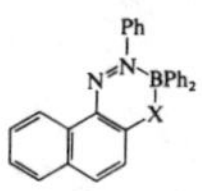

X = O

B-01297

As B-01297 with
X = NH

B-01298

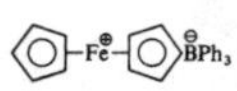

B-01299

As B-01276 with
R = CH_2CH_3

B-01300

$(PhCH_2O)_4B^{\ominus}$

B-01301

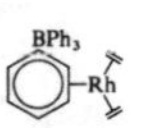

B-01302

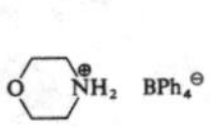

B-01303

$Me_4N^{\oplus}Ph_4B^{\ominus}$

B-01304

As B-01306 with
R^1 = Ph; R^2 = $-C(CH_3)_3$

B-01305

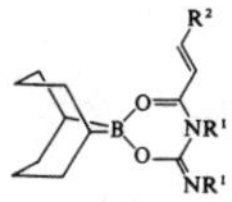

R^1 = Ph, R^2 = $(CH_2)_3CH_3$

B-01306

As B-01208 with
R = Ph

B-01307

As B-01075 with
R = 4-*tert*-butylphenoxy

B-01308

As B-01306 with
R^1 = cyclohexyl, R^2 = $-(CH_2)_3CH_3$

B-01309

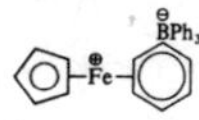

B-01310

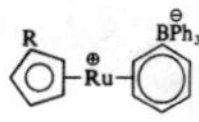

R = H

B-01311

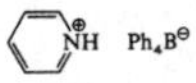

B-01312

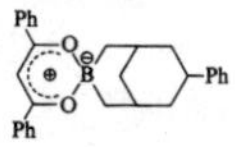

B-01313

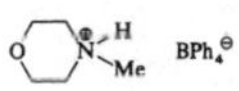

B-01314

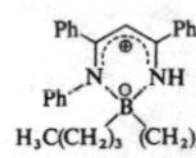

B-01315

As B-01224 with
R = Ph

B-01316

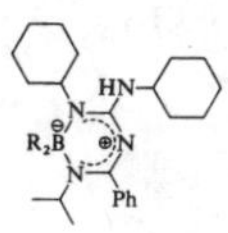

R = $H_3CCH_2CH_2-$

B-01317

B-01318

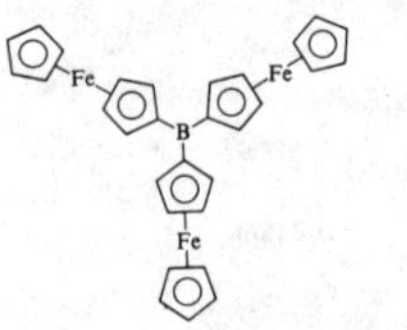

B-01319

As B-01311 with
R = CH_3

B-01320

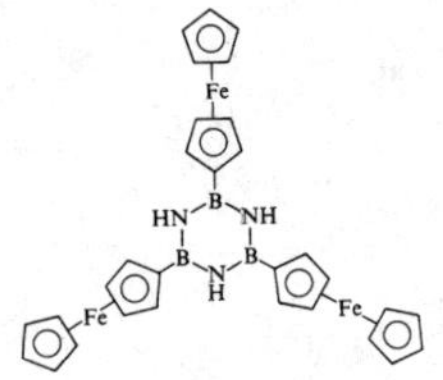

B-01321

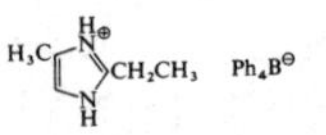

B-01322

As B-01147 with
R = Ph, R′ = Me

B-01323

$Et_3NH^{\oplus}Ph_4B^{\ominus}$

B-01324

As B-01207 with
R = Ph

B-01325

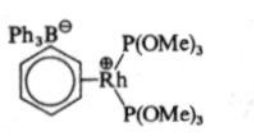

B-01326

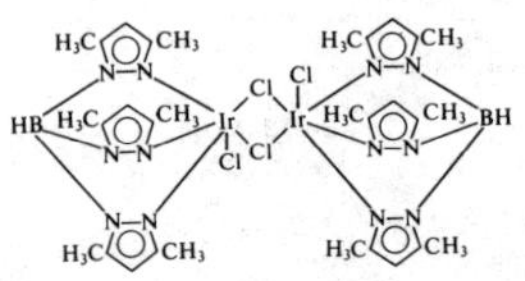

B-01327

$[H_3C(CH_2)_7]_3BC{\equiv}C(CH_2)_3CH_3^{\ominus}$

B-01328

$[H_3C(CH_2)_9]_3B$

B-01329

$[H_3C(CH_2)_9O]_3B$

B-01330

$PhN{=}C(NHPh)NH_2{-}BPh_3$

B-01331

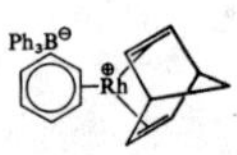

B-01332

$MePhTlBPh_4$

B-01333

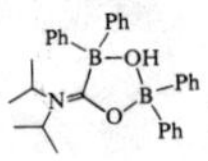

B-01334

As B-01317 with
R = $(H_3C)_2CHCH_2-$

B-01335

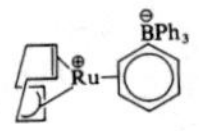

B-01336

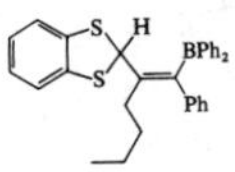

(E)-form

B-01337

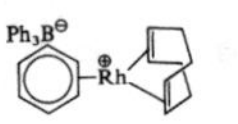

B-01338

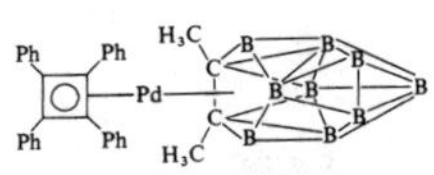

B-01339

As B-01149 with
R = Ph, R′ = CH_3

B-01340

$PhC{\equiv}CTlMeBPh_4$

B-01341

$PhNMe_3^{\oplus}Ph_4B^{\ominus}$

B-01342

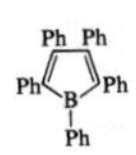

B-01343

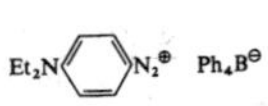

B-01344

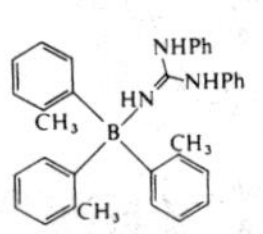

B-01345

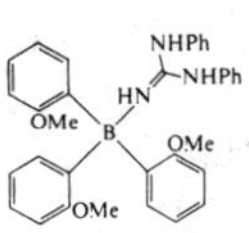

B-01348

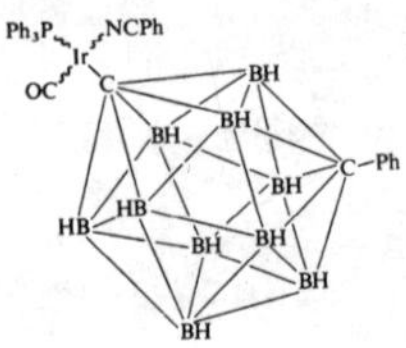

B-01350

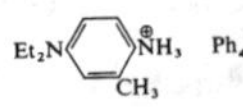

B-01351

$(Ph_3P)_2Cu^{\oplus}\ BH_4^{\ominus}$

B-01352

$(Ph_3P)_2CuBH_4$

B-01353

As B-01306 with
R^1 = 1-naphthyl, $R^2 = -(CH_2)_3CH_3$

B-01354

$(H_3CCH_2CH_2)_4N^{\oplus}Ph_4B^{\ominus}$

B-01355

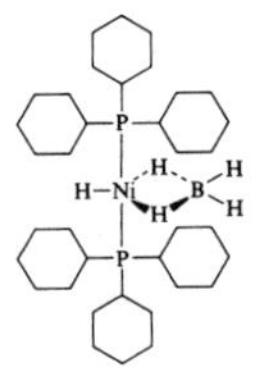

B-01356

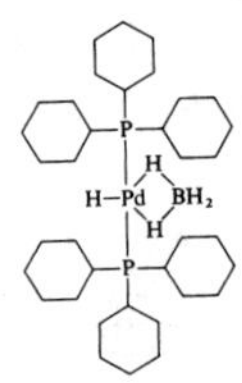

B-01357

$HOB[O(CH_2)_{17}CH_3]_2$

B-01358

$[H_3C(CH_2)_{11}O]_3B$

B-01359

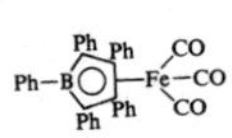

B-01360

$(Ph_3P)_2Rh(CO)(BH_4)$

B-01361

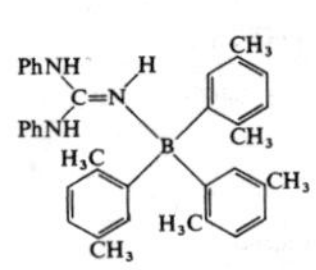

B-01362

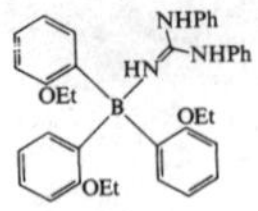

B-01363

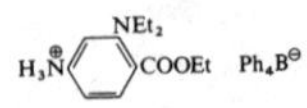

B-01365

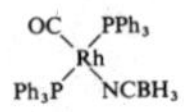

B-01366

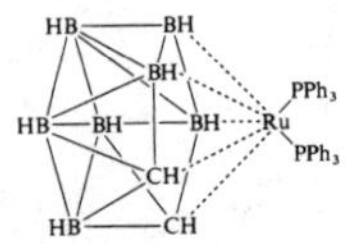

B-01367

$(Ph_3P)_2Ag(C_2B_8H_{11})$

B-01368

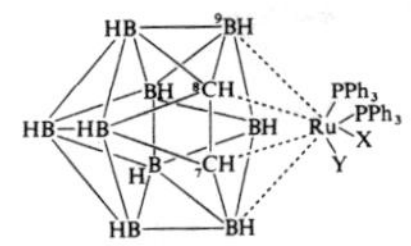

X = Cl, Y = H

B-01369

As B-01369 with
X = Y = H

B-01371

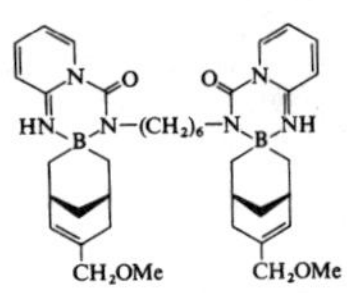

B-01373

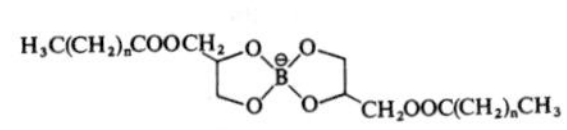

n = 14

B-01374

As B-01369 with
XY = CO

B-01375

$H_3B{-}H{-}Ag(PMePh$...

B-01377

$H_3B{-}H{-}Cu(PMePh_2)_3$

B-01378

$H_3C(CH_2)_{11}NMe_3^{\oplus}Ph_4B^{\ominus}$

B-01379

As B-01147 with
R = CH_3, R′ = Ph

B-01380

$(H_3CCH_2CH_2CH_2)_4N^{\oplus}Ph_4B^{\ominus}$

B-01381

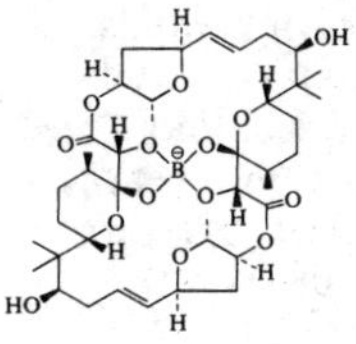

B-01382

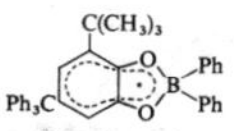

B-01383

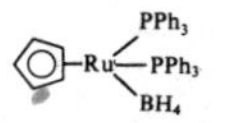

B-01384

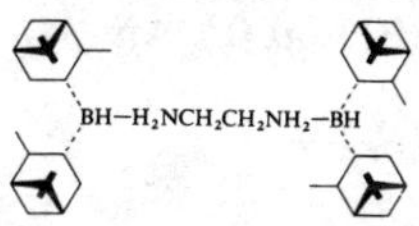

B-01385

As B-01374 with
n = 16

B-01386

$Ph_3PMe^{\oplus}Ph_4B^{\ominus}$

B-01387

$[(H_3C)_2CH_2CH_2CH_2]_3N(CH_2)_3CH_3^{\oplus}$-
$Ph_4B^{\ominus}$

B-01388

$[(H_3C)_2CHCH_2CH_2]_4N^{\oplus}Ph_4B^{\ominus}$

B-01389

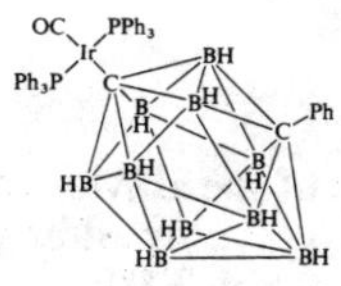

B-01390

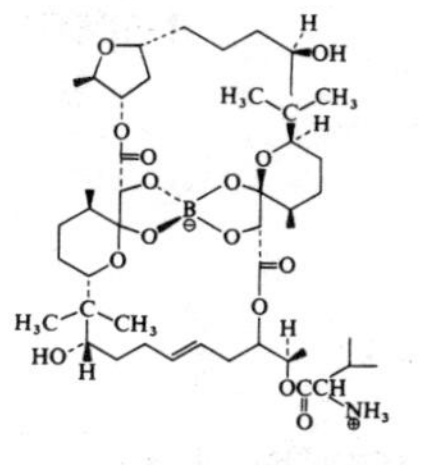

B-01391

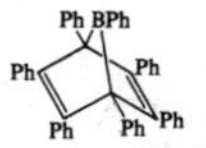

B-01392

$Ph_4As^{\oplus}$ $Ph_4B^{\ominus}$

B-01393

$Ph_4P^{\oplus}Ph_4B^{\ominus}$

B-01394

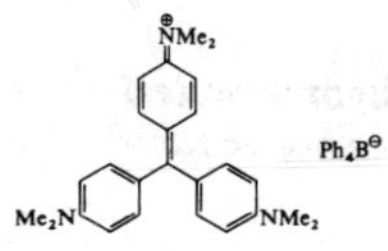

B-01395

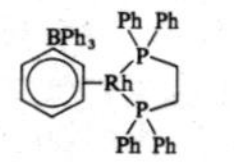

B-01396

$[H_3C(CH_2)_{17}O]_3B$

B-01397

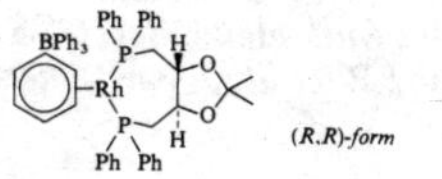

B-01398

$BCl_3H^{\ominus}$ B-00001

Trichlorohydroborate(1−), 10CI
Trichloroborohydride

$Cl_3BH^{\ominus}$

M 118.177 (ion)

Tetrabutylammonium salt: [69969-98-8].
$C_{16}H_{37}BCl_3N$ M 360.644
Selective reducing agent. Moisture-sensitive cryst. solid. Sol. THF, diglyme, arom. hydrocarbons; insol. alkanes.
▷Dec. slowly at 110°, rapidly at 200°

Titov, L.V. *et al*, *Izv. Akad. Nauk SSSR, Ser. Khim.*, 1978, 1722; *Bull. Acad. Sci. USSR, Div. Chem. Sci.*, **27**, 1504 (*synth, ir, props*)
Kryshtal, G.V. *et al*, *Izv. Akad. Nauk SSSR, Ser. Khim.*, 1981, 1146; *Bull. Acad. Sci. USSR, Div. Chem. Sci.*, **30**, 907 (*use*)
Leach, J.B. *et al*, *J. Am. Chem. Soc.*, 1981, **103**, 988 (*synth, nmr*)

BFH_2 B-00002

Fluoroborane
Fluoroborine. Fluorodihydroboron
[27944-03-2]

H_2BF

M 31.824
Not prepd. in pure form, but known in a variety of complexes.

Me_3P complex: [67226-44-2]. *Fluorodihydro(trimethylphosphine)boron, 10CI.*
$C_3H_{11}BFP$ M 107.902
Less stable than trimethylamine complex.

Ph_3P complex: [67194-38-1]. *Fluorodihydro(triphenylphosphine)boron, 10CI.*
$C_{18}H_{17}BFP$ M 294.114
Much less stable than the trimethylamine complex.

$MeNH_2$ complex: [67194-41-6]. *Fluorodihydro(methylamine)boron.*
CH_7BFN M 62.881
Substantially more stable than trimethylamine complex.

Me_2NH complex: [67194-40-5]. *Fluorodihydro(dimethylamine)boron, 10CI.*
C_2H_9BFN M 76.908
More stable than trimethylamine complex.

Me_3N complex: [35449-24-2]. *Fluorodihydro(trimethylamine)boron.*
$C_3H_{11}BFN$ M 90.935

Walsh, A.D., *J. Chem. Soc.*, 1953, 2260 (*struct*)
Van Paasschen, J.M. *et al*, *Inorg. Nucl. Chem. Lett.*, 1972, **8**, 879.
Dill, J.D. *et al*, *J. Am. Chem. Soc.*, 1975, **97**, 3402.
Van Paasschen, J.M. *et al*, *J. Inorg. Nucl. Chem.*, 1976, **38**, 2321.
Bernardi, F. *et al*, *J. Am. Chem. Soc.*, 1978, **100**, 7205.
Cassoux, P. *et al*, *J. Mol. Struct.*, 1978, **48**, 25 (*microwave*)
Van Paasschen, J.M. *et al*, *Inorg. Chem.*, 1978, **17**, 3302.
Hinde, A.L. *et al*, *J. Comput. Chem.*, 1980, **1**, 118.

BF_2H B-00003

Difluoroborane, 10CI, 9CI
Difluoroborine
[13709-83-6]

F_2BH

M 49.815
Prototype of the alkyldifluoroborane series of compds. Prepd. from solid boroxine and BF_3 (gas). Air- and moisture-sensitive gas. ΔH_{f298K} −175.5 Kcalmol.$^{-1}$.

Mikhailov, B.M. *et al*, *Zh. Obshch. Khim.*, 1959, **29**, 3443 (*use*)
Coyle, T.D. *et al*, *Proc. Chem. Soc., London*, 1964, 25 (*synth*)
Porter, R.F. *et al*, *J. Phys. Chem.*, 1965, **69**, 2208 (*synth*)
Tuchagues, J.P. *et al*, *J. Organomet. Chem.*, 1973, **54**, 69 (*ir, nmr*)
Destofano, A.J. *et al*, *Thermochim. Acta*, 1976, **16**, 236 (*use*)
Fussstetter, H. *et al*, *Chem. Ber.*, 1977, **110**, 3172 (*nmr*)

BHI_2 B-00004

Diiodoborane, 10CI, 9CI
[43338-13-2]

I_2BH

M 265.627
Prototype of the alkyldiiodoborane series of compds. Prepd. *in situ* by adding I_2/THF to BH_3 at 0°. Air-sensitive. Sol. THF.

Et_3N complex: [56598-29-9]. *Diiodoborane triethylamine.*
$C_6H_{16}BI_2$ M 352.811
Prepd. by the reaction of I_2 with triethylamine borane. Extremely sensitive to moisture, dec. on standing. Sol. toluene.

Me_3N complex: [25741-84-8]. *Diiodoborane trimethylamine.*
$C_3H_{10}BI_2$ M 310.731
Prepd. by the reaction of I_2 with trimethylamine borane. Extremely sensitive to moisture, dec. on standing. Sol. toluene.

Dimethyl sulfide complex: [28920-40-3]. *Hydrodiiodo[thiobis[methane]]boron, 10CI.*
$C_2H_7BI_2S$ M 327.756
Useful means of storing diiodoborane; hydroborates alkenes to give alkyldiiodoboranes. Moisture-sensitive liq.

Phosphine complex: [41593-72-0].
BH_4I_2P M 299.624
Sol. MeI, CH_2Cl_2. pmr δ −6.30, J_{PH} 4.18Hz; ^{31}P nmr δ + 95, J_{PH} 424Hz.

Brown, M.P. *et al*, *J. Chem. Soc. (A)*, 1970, 410 (*ir*)
Lowe, J.R. *et al*, *Inorg. Chem.*, 1970, **9**, 1423 (*props*)
Rapp, B. *et al*, *Inorg. Chem.*, 1973, **12**, 2868 (*nmr*)
Kinberger, K. *et al*, *Z. Naturforsch., B*, 1975, **30**, 55 (*synth*)
Myers, W.H. *et al*, *Inorg. Chem.*, 1975, **14**, 2874 (*nmr*)
Ashby, E.L. *et al*, *Tetrahedron Lett.*, 1976, 3865 (*synth, use*)
Brown, H.C. *et al*, *J. Org. Chem.*, 1980, **45**, 384 (*use*)

BH_2I B-00005

Iodoborane
Dihydroiodoboron

BH_2I

M 139.730

4-Methylpyridine complex: [59499-23-9]. *Dihydroiodo(4-methylpyridine)boron, 10CCI. Iodoborane γ-picoline.*
C_6H_9BIN M 232.858
Identified by nmr only.

Denniston, M.L. *et al*, *J. Inorg. Nucl. Chem.*, 1976, **38**, 379 (*synth, nmr*)
Vidal, J.L. *et al*, *Inorg. Chem.*, 1977, **16**, 1673.

BH_3 B-00006

Borane

Borine. Boron hydride

[13283-31-3]

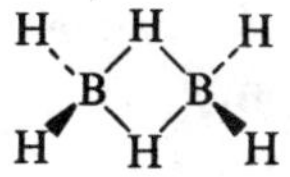

dimer

M 13.834

Exists as the dimer (diborane) unless complexed with a nucleophilic sp. The THF soln., dimethyl sulfide complex, and other complexes are important hydroborating and reducing agents.

Dimer: [19287-45-7]. *Diborane(6)*, 10CI. *Boroethane.*

B_2H_6 M 27.667

Hydroborating and reducing reagent as soln. in THF or ethers. Deuterated diborane useful for incorporation of deuterium into org. molecules. Thermolysis is a major route to higher boranes. Colourless air- and moisture-sensitive gas. Sol. THF as BH_3-THF complex. Mp −165.5°. Bp −92.5°.

▷HQ9275000.

Schlesinger, H. *et al*, *J. Am. Chem. Soc.*, 1953, **75**, 199 (*synth*)
Bartell, L. *et al*, *J. Chem. Phys.*, 1965, **42**, 1135 (*ed*)
Naiman, K. *et al*, *Inorg. Nucl. Chem. Lett.*, 1970, **6**, 74 (*pmr*)
Leach, J.B. *et al*, *J. Magn. Reson.*, 1972, **6**, 74 (*nmr*)
Solomon, J.L. *et al*, *J. Am. Chem. Soc.*, 1972, **94**, 1443 (*ms*)
Odom, J.D. *et al*, *Inorg. Chem.*, 1973, **12**, 95 (*nmr*)
Murano, Y. *et al*, *Radiochem. Radioanal. Lett.*, 1980, **44**, 315 (*deriv, synth*)
Ducan, J.L. *et al*, *J. Mol. Spectrosc.*, 1981, **85**, 16 (*ir*)
Leach, J.B. *et al*, *J. Am. Chem. Soc.*, 1981, **103**, 988 (*synth*)

BH_3O B-00007

Hydroxyborane

Borinic acid, 10CI, 9CI

[35825-58-2]

H_2BOH

M 29.833

Prototype member of the dialkyl(hydroxy)boranes. Unstable to moisture and to self-decomposition. Dipole moment μ_{total} = 1.506D.

Gropen, O. *et al*, *J. Mol. Struct.*, 1975, **25**, 161 (*props*)
Hinde, A.L. *et al*, *J. Comput. Chem.*, 1980, **1**, 118 (*props*)
Boggs, J.E. *et al*, *Theochem.*, 1981, **1**, 329; *CA*, **95**, 103720a (*struct*)
Kawashima, Y. *et al*, *J. Chem. Phys.*, 1981, **74**, 5430 (*microwave*)
Sliznev, V.V. *et al*, *CA*, 1982, **97**, 118997m.

$BH_4^{\ominus}$ B-00008

Tetrahydroborate(1−)

Borohydride(1−)

$BH_4^{\ominus}$

M 14.842 (ion)

Li salt: [16949-15-8]. *Lithium borohydride. Lithium tetrahydroborate.*

BH_4Li M 21.783

Reducing agent, reagent for synth. of other boranes. White cryst., stable in dry air, unstable >280°. Sol. ethers. Mp 268°.

▷OJ6427000.

Na salt. [16940-66-2]. *Sodium borohydride. Sodium tetrahydroborate.*

BH_4Na M 37.831

White cryst., stable *in vacuo* to 400°. Sol. liq. NH_3, amines, ethers. Mp 505°.

▷ED3325000.

Onak, T. *et al*, *J. Phys. Chem.*, 1959, **63**, 1533 (*nmr*)
James, B. *et al*, *Prog. Inorg. Chem.*, 1970, **11**, 124 (*synth*)

BH_4N B-00009

Aminoborane, 8CI

Boranamine, 10CI, 9CI. *Aminoborine*

[14720-35-5]

H_2BNH_2

M 28.848

Parent system for a range of organic aminoborane derivs. Air- and moisture-sensitive gas (solid aminoborane polymerizes between −196° and −155° to yield inert white material). Dipole moment = 1.844D.

Schaeffer, G.W. *et al*, *J. Am. Chem. Soc.*, 1955, **77**, 331 (*synth*)
Kuznesof, P.M. *et al*, *J. Am. Chem. Soc.*, 1968, **90**, 2557 (*ms*)
McGee, H.A. *et al*, *Inorg. Chem.*, 1970, **9**, 2458 (*ms, ir, synth*)
Sugie, M. *et al*, *Chem. Phys. Lett.*, 1979, **64**, 573.
Dewar, M.J.S. *et al*, *J. Mol. Struct.*, 1980, **68**, 105 (*ir*)
Haddon, R.C., *Pure Appl. Chem.*, 1982, **54**, 1129.

$BH_4O^{\ominus}$ B-00010

Trihydrohydroxyborate(1−), 10CI, 9CI, 8CI

Trihydrohydroxyborate(1−), 9CI

[24492-48-6]

$HOBH_3^{\ominus}$

M 30.841 (ion)

Intermed. in hydrol. of tetrahydroborate anion.

Li salt: [63717-91-9].

BH_4LiO M 37.782

Reducing agent.

Na salt: [16903-32-5].

BH_4NaO M 53.831

K salt: [63706-85-4].

BH_4KO M 69.939

Reducing agent.

Wang, F.T. *et al*, *Inorg. Chem.*, 1972, **11**, 1933 (*props*)
Reed, J.W. *et al*, *J. Am. Chem. Soc.*, 1974, **96**, 1248 (*synth*)
Reed, J.W. *et al*, *J. Org. Chem.*, 1977, **42**, 3963 (*use*)
Strouf, O. *et al*, *Acta Chem. Scand., Ser. A*, 1977, **31**, 391 (*deriv, props*)
Reed, J., *CA*, 1978, **89**, 49491n (*synth, props*)

BH_5N_2 B-00011

Diaminoborane, 8CI

Boranediamine, 10CI

[18447-53-5]

$HB(NH_2)_2$

M 43.863

Parent compd. for the series of alkyldiaminoboranes. Vapour can be kept for several days at r.t. but liq. decs. to form glassy solid with evolution of NH_3.

Baird, N.C. *et al*, *Can. J. Chem.*, 1967, **45**, 2059 (*props*)
Briggs, T.S. *et al*, *J. Am. Chem. Soc.*, 1978, **100**, 7762 (*synth, nmr, ir*)
Fjeldberg, T. *et al*, *Acta Chem. Scand., Ser. A*, 1980, **34**, 547 (*props, struct*)
Thorne, L.R., *CA*, 1980, **92**, 138026u (*ir, struct*)
Briggs, T.S., *CA*, 1981, **94**, 163335t (*props*)

BH_6P B-00012

Borane phosphine

Trihydro(phosphine)boron, 10CI. *Phosphine compd. with borane(1:1)*, 9CI, 8CI. *Borane compd. with PH_3*

[19121-56-3]

$H_3B—PH_3$

M 47.831

Rudolph, R.W. *et al*, *Inorg. Chem.*, 1966, **5**, 723 (*nmr, ir, raman*)
Mayer, E., *Monatsh. Chem.*, 1971, **102**, 940 (*synth*)
Rapp, B. *et al*, *Inorg. Chem.*, 1973, **12**, 2868 (*nmr*)
Umeyama, H. *et al*, *Chem. Pharm. Bull.*, 1981, **29**, 287 (*struct*)
Cowley, A.H. *et al*, *Inorg. Chem.*, 1982, **21**, 85 (*pe*)
Sebestyen, A. *et al*, *Theochem.*, 1982, **6**, 259 (*struct*)

BH_7N_2 B-00013

Borane hydrazine

(Hydrazine-N)trihydroboron, 10CI, 9CI. *Hydrazine compd. with borane*, 9CI, 8CI. *Borane compd. with hydrazine*, 9CI

[14931-40-9]

$H_2NNH_2—BH_3$

M 45.879

Previously considered as rocket fuel. Large translucent cryst. Mp 61° (difficult to reproduce). Dec. >50°.

▷Shock sensitive, nonflammable shielding necessary in synth.

Gunderloy, F.C., *Inorg. Chem.*, 1963, **2**, 221 (*synth*)
Hermanek, S. *et al*, *Collect. Czech. Chem. Commun.*, 1966, **31**, 1281 (*chromatog*)
Inorg. Synth., **9**, 13 (*synth*)
Andrianov, L.O. *et al*, *Zh. Strukt. Khim.*, 1967, **8**, 303 (*struct*)
Zakharov, V.V. *et al*, *Zh. Fiz. Khim.*, 1974, **48**, 492.
Lagodzinskaya, G.V. *et al*, *Zh. Fiz. Khim.*, 1975, **49**, 2238; 2624 (*nmr*)

$B_2H_{10}N_2$ B-00014

Hydrazine bis(borane)

Borane compd. with hydrazine(2:1), 9CI, 8CI. *Hydrazine compd. with borane(1:2)*, 9CI. *[μ-(Hydrazine-N:N′)]-hexahydrodiboron*

[13730-91-1]

$H_3B—NH_2—NH_2—BH_3$

M 59.712

Previously considered as rocket fuel. Fine white cryst. Does not melt below 100°.

▷May explode violently, especially on heating. Shielding and remote-control manipulation are recommended

Hermanek, S. *et al*, *Collect. Czech. Chem. Commun.*, 1966, **31**, 1281 (*chromatog*)
Hanousek, F. *et al*, *Collect. Czech. Chem. Commun.*, 1973, **38**, 2692 (*ir, raman*)
Zakharov, V.V. *et al*, *Zh. Fiz. Khim.*, 1974, **48**, 499.
Inorg. Synth., **9**, 13 (*synth*)

The symbol ▷ in Entries highlights hazard or toxicity information

$B_3Br_3H_3N_3$ B-00015

2,4,6-Tribromoborazine, 10CI, 9CI, 8CI

B,B,B-*Tribromoborazole*

[13703-88-3]

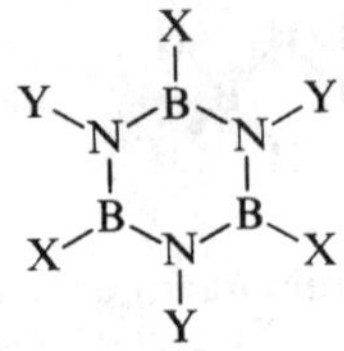

X = Br, Y = H

M 317.186

Moisture-sensitive solid. Mp 128-9°.

Emeléus, H.J. *et al*, *J. Chem. Soc.*, 1959, 1306 (*synth*)
Scaife, C.W.J., *CA*, 1967, **66**, 121650h (*props*)
Hensen, K. *et al*, *Theor. Chim. Acta*, 1967, **9**, 17 (*nmr, uv*)
Graffeuil, M. *et al*, *J. Chim. Phys. Phys. Chim. Biol.*, 1971, **68**, 1379; *CA*, **76**, 19778a (*struct*)
Wrackmeyer, B. *et al*, *Chem. Ber.*, 1976, **109**, 3480 (*nmr*)

$B_3Cl_3H_3N_3$ B-00016

2,4,6-Trichloroborazine, 10CI, 9CI, 8CI

B,B,B-*Trichloroborazole*

[933-18-6]

As 2,4,6-Tribromoborazine, B-00015 with

X = Cl, Y = H

M 183.833

Curing agent for epoxy resins. Colourless orthorhombic cryst. slowly dec. under N_2 at 0°; sublimes readily; moisture sensitive. Sol. C_6H_6, CCl_4, Et_2O, CS_2, PhCl, cyclohexane, $CHCl_3$, insol. Py, $PhNO_2$. d_4^{25} 1.58. Mp 84°. Vigorous reaction with H_2O and alcohols. Above 100° slow irreversible dec. occurs with formation of HCl.

Brown, C.A. *et al*, *J. Am. Chem. Soc.*, 1955, **77**, 3699 (*synth, props*)
Brennan, G.L. *et al*, *J. Am. Chem. Soc.*, 1960, **82**, 6248 (*props*)
Inorg. Synth., 1972, **13**, 41 (*synth, props, ir, pmr*)
Kroner, J. *et al*, *Tetrahedron*, 1972, **28**, 1585 (*pe*)
Melcher, L.A. *et al*, *Inorg. Chem.*, 1973, **12**, 601 (*ms*)
Wrackmeyer, B. *et al*, *Chem. Ber.*, 1976, **109**, 3480 (*nmr*)
Molt, K. *et al*, *Z. Anorg. Allg. Chem.*, 1981, **474**, 182 (*ir, raman*)

$B_3F_3H_3N_3$ B-00017

2,4,6-Trifluoroborazine, 10CI, 9CI, 8CI

B-*Trifluoroborazine*

[13779-24-3]

As 2,4,6-Tribromoborazine, B-00015 with

X = F, Y = H

M 134.469

White cryst. Sol. Et_2O, C_6H_6, CCl_4, CS_2, toluene, spar. sol. cyclohexane, pet. ether. Mp 118° (sealed tube). Bp 122° subl. Subl. *in vacuo*. Reacts with moist air.

Niedenzu, K., *Inorg. Chem.*, 1962, **1**, 943 (*synth*)
Laubengayer, A.W. *et al*, *Inorg. Chem.*, 1963, **2**, 519 (*ir, uv, ms*)
Kroner, J. *et al*, *Tetrahedron*, 1972, **28**, 1585 (*pe*)
Melcher, L.A. *et al*, *Inorg. Chem.*, 1973, **12**, 601 (*ms*)
Wrackmeyer, B. *et al*, *Chem. Ber.*, 1976, **109**, 3480 (*nmr*)
Molt, K. *et al*, *Z. Anorg. Allg. Chem.*, 1981, **474**, 182 (*ir*)

$B_3F_3O_3$ B-00018

Trifluoroboroxin, 10CI, 9CI, 8CI

[13703-95-2]

As Boroxin, B-00019 with

R = F

M 137.423

Used in boronizing process for surface treatment of steels. Disproportionates above −135°.

Fisher, H.D. *et al*, *J. Phys. Chem.*, 1961, **65**, 1166 (*synth*, *ir*, *struct*)
Fisher, H.D., *CA*, 1963, **59**, 2376b.
Laubengayer, A.W. *et al*, *Inorg. Chem.*, 1963, **2**, 519 (*synth*, *ir*)
Brookes, C.W. *et al*, *Org. Mass. Spectrom.*, 1970, **3**, 231 (*glc*, *ms*)
Dewar, M.J. *et al*, *J. Am. Chem. Soc.*, 1978, **100**, 58.
Goeuriot, P. *et al*, *Thin Solid Films*, 1981, **78**, 67.

$B_3H_3O_3$ B-00019

Boroxin, 10CI, 9CI, 8CI

[289-56-5]

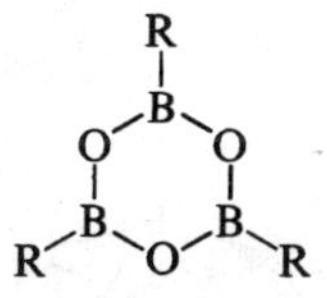

R = H

M 83.452

Prototype of an important class of organoboron compds. Hydrolyses in water, unstable at r.t.

Lee, G.H. *et al*, *J. Phys. Chem.*, 1963, **67**, 1742 (*synth*, *ir*)
Barton, L. *et al*, *Inorg. Chem.*, 1966, **5**, 2076 (*synth*)
Kaldor, A. *et al*, *Inorg. Chem.*, 1969, **8**, 1418 (*props*)
Kaldor, A. *et al*, *J. Am. Chem. Soc.*, 1971, **93**, 2140 (*ir*)
Serafini, A. *et al*, *J. Mol. Struct.*, 1975, **26**, 129 (*struct*)

$B_3H_6N_3$ B-00020

Borazine, 10CI, 9CI

Triaminotriborane. Borazole. Hexahydro-s-triazatriborine

[6569-51-3]

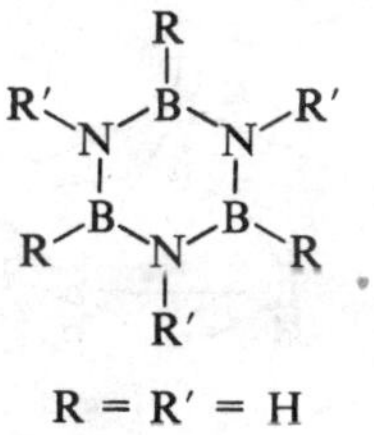

R = R′ = H

M 80.498

Inorganic benzene analogue. Liq. d^{55} 0.808. Mp −58°. Bp 55°. n_D^{20} 1.3821. Dipole moment 0.50D. Dec. slowly at r.t. to give hydrogen and a nonvolatile residue. Hydrol. slowly in water to $B(OH)_3$.

2,4,6-Tri-Me: see 2,4,6-Trimethylborazine, B-00210
Hexa-Me: see Hexamethylborazine, B-00449
2,4,6-Tri-Et. see 2,4,6-Triethylborazine, B-00451
2,4,6-Tri-Ph: see 2,4,6-Triphenylborazine, B-01072

Stock, A. *et al*, *Ber.*, 1926, **59**, 2215 (*synth*)
Schaeffer, R. *et al*, *J. Am. Chem. Soc.*, 1954, **76**, 3303 (*synth*)
Niedenzu, W.S. *et al*, *Inorg. Chem.*, 1967, **6**, 1453 (*ir*, *raman*)
Harshbarger, W. *et al*, *Inorg. Chem.*, 1969, **8**, 1683 (*cryst struct*)
Porter, R.F. *et al*, *J. Am. Chem. Soc.*, 1971, **93**, 56 (*nmr*)
Wrackmeyer, B. *et al*, *Chem. Ber.*, 1976, **109**, 3480 (*nmr*)

$B_3H_8^{\ominus}$ B-00021

Octahydrotriborate(1−), 10CI

Triborohydride(1−). Di-μ-hydrohexahydrotriborate(1−)

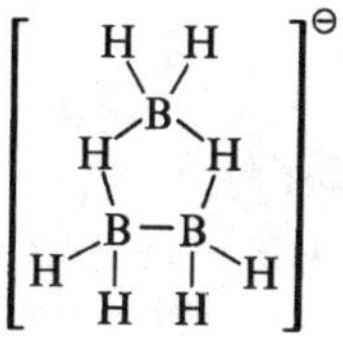

M 40.493 (ion)

arachno-anion.

Tetrabutylammonium salt: [53451-55-1].
$C_{16}H_{44}B_3N$ M 282.960
Used in synth. of higher boranes, borane anions, carboranes and transition metal complexes. White cryst., stable to air and water (CH_2Cl_2/Et_2O). Sol. CH_2Cl_2, MeCN, spar. sol. C_6H_6, Et_2O, hexane, H_2O. Mp 212° dec.

Inorg. Synth., 1974, **15**, 113 (*synth*, *nmr*)

$B_4H_9^{\ominus}$ B-00022

Nonahydrotetraborate(1−), 10CI

[12429-85-5]

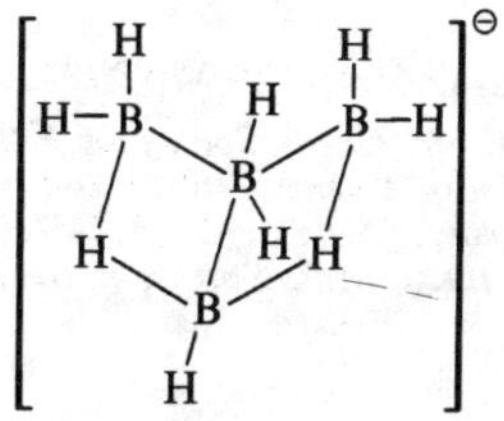

M 52.311 (ion)

arachno-anion. Intermed. for synth. of tetraborane(10) derivs. and of pentaborane(11).

K salt:
B_4H_9K M 91.409
Colourless cryst. stable at −78°. Sol. ethers.

Tetrabutylammonium salt:
$C_{16}H_{40}B_4N$ M 289.739
Colourless cryst.

Johnson, H. *et al*, *J. Am. Chem. Soc.*, 1970, **92**, 7586 (*deriv*)
Reitz, R. *et al*, *Inorg. Chem.*, 1972, **11**, 1242 (*nmr*)
Remmel, R. *et al*, *J. Am. Chem. Soc.*, 1975, **97**, 5395 (*synth*, *nmr*)

B_4H_{10} B-00023

Tetraborane(10), 10CI

[18283-93-7]

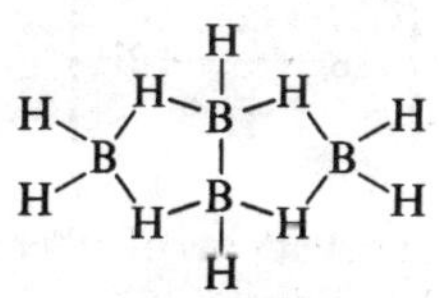

M 53.319

arachno-. Colourless gas or liq. Mp −120°. Bp 18°.

Leach, J. *et al*, *Inorg. Chem.*, 1970, **9**, 2170 (*nmr*)
Taft, M. *et al*, *Inorg. Chem.*, 1982, **21**, 1952 (*synth*)

B_5BrH_8 B-00024

1-Bromopentaborane(9), 10CI

[23753-67-5]

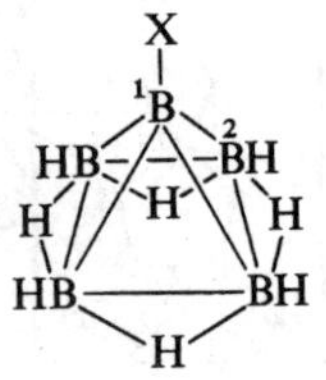

X = Br

M 142.017

nido-. Reactive compd. used for synthesis of other boranes. White solid. Mp 34°. Bp_{34} 82°.

▷Spontaneously inflammable. Explosive with air

Tucker, P. *et al*, *Inorg. Chem.*, 1970, **9**, 1430 (*nmr*)
Inorg. Synth., 1979, **19**, 247 (*synth, haz*)

B_5ClH_8 B-00025

1-Chloropentaborane(9), 10CI

[19469-13-7]

As 1-Bromopentaborane(9), B-00024 with

X = Cl

M 97.566

nido-. Colourless liq. or solid. Mp 27°.

Gaines, D. *et al*, *J. Am. Chem. Soc.*, 1966, **88**, 4528 (*ir*)
Gaines, D. *et al*, *Inorg. Chem.*, 1968, **7**, 704 (*synth*)
Tucker, P. *et al*, *Inorg. Chem.*, 1970, **9**, 1430 (*nmr*)
Murphy, C. *et al*, *Int. J. Mass Spectrom. Ion Phys.*, 1971, **7**, 327 (*ms*)

B_5ClH_8 B-00026

2-Chloropentaborane(9), 10CI

[19469-14-8]

M 97.566

nido-. Colourless liq. Mp −41° (−40°).

Burg, A., *J. Am. Chem. Soc.*, 1968, **90**, 1407 (*ir*)
Gaines, D. *et al*, *Inorg. Chem.*, 1968, **7**, 704 (*synth*)
Tucker, P. *et al*, *Inorg. Chem.*, 1970, **9**, 1430 (*nmr*)
Murphy, C. *et al*, *Int. J. Mass. Spectrom. Ion Phys.*, 1971, **7**, 327 (*ms*)

$B_5H_8^{\ominus}$ B-00027

Octahydropentaborate(1−), 10CI

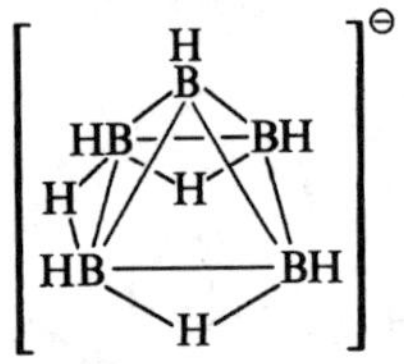

M 62.113 (ion)

nido anion. Used for synthesis of other borane species.

K salt: [56009-95-1].

B_5H_8K M 101.212

Colourless cryst. Sol. Me_2O, Et_2O, glyme, THF.

Johnson, H. *et al*, *Inorg. Chem.*, 1970, **4**, 908 (*synth, nmr, ir, cryst struct*)

B_5H_9 B-00028

Pentaborane(9), 10CI

[19624-22-7]

As 1-Bromopentaborane(9), B-00024 with

X = H

M 63.121

nido-. Stable intermed. used for prepn. of higher boranes. Many derivs. formed. Liq. Mp −46.6°. Bp 50.3°.

▷Spontaneously inflammable (explosive) with air. Solvs. with reactive carbonyl groups and halogenated solvs. may form shock-sensitive solns. RY8925000.

Dibeler, V. *et al*, *J. Res. Natl. Bur. Std.*, 1949, **43**, 97 (*ms*)
McCarty, L. *et al*, *J. Am. Chem. Soc.*, 1951, **73**, 3138 (*synth*)
Hrostowski, H. *et al*, *J. Am. Chem. Soc.*, 1954, **76**, 998 (*ir*)
Onak, T., *J. Chem. Soc., Chem. Commun.*, 1972, 351 (*nmr*)
Inorg. Synth., 1974, **15**, 118 (*synth*)
Muetterties, E.L., *Boron Hydrides*, Academic Press, New York, 1975, 101 (*nmr*)

B_5H_{11} B-00029

Pentaborane(11), 10CI

[18433-84-6]

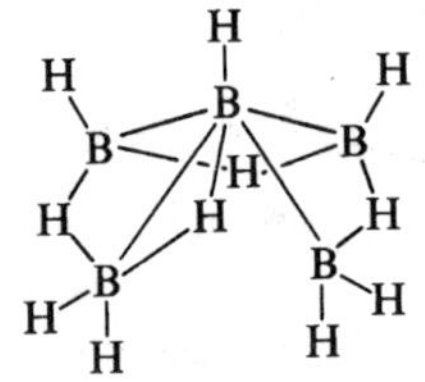

M 65.137

arachno-. Colourless liq. Mp −123°. Bp 63°.

Klein, M. *et al*, *J. Am. Chem. Soc.*, 1958, **80**, 4149 (*synth*)
Lutz, C. *et al*, *Can. J. Chem.*, 1963, **41**, 1344 (*ms*)
Spielman, J. *et al*, *Inorg. Chem.*, 1963, **2**, 1139 (*ir*)
Fehlner, T. *et al*, *J. Am. Chem. Soc.*, 1964, **86**, 581 (*ms*)
Leach, J. *et al*, *Inorg. Chem.*, 1970, **9**, 2170 (*nmr*)
Toft, M. *et al*, *Inorg. Chem.*, 1982, **21**, 1952 (*synth*)

$B_6H_9^{\ominus}$ B-00030

Nonahydrohexaborate(1−), 10CI

[31426-85-4]

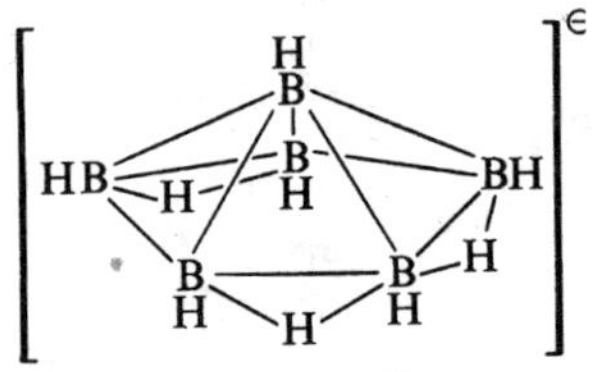

M 73.931 (ion)

nido-anion.

K salt:

B_6H_9K M 113.029

Colourless cryst., unstable at r.t. Sol. ethers.

Johnson, H. *et al*, *Inorg. Chem.*, 1970, **9**, 908 (*synth, cryst struct, nmr*)
Johnson, H. *et al*, *J. Am. Chem. Soc.*, 1972, **94**, 6711 (*pmr, ir*)

B_6H_{10} **B-00031**

Hexaborane(10)

[23777-80-2]

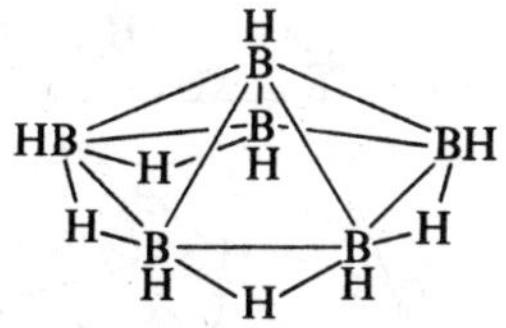

M 74.939

nido-. Liq. Mp −62.3°. Bp 108°.

Rossmann, M.G. *et al*, *Acta Crystallogr.*, 1959, **12**, 530 (*cryst struct*)
Shapiro, I. *et al*, *Adv. Chem. Ser.*, 1961, **32**, 127 (*ms*)
Burg, A.B. *et al*, *Inorg. Chem.*, 1962, **1**, 725 (*ir*)
Leach, J. *et al*, *Inorg. Chem.*, 1970, **9**, 2170 (*nmr*)
Brice, V. *et al*, *J. Am. Chem. Soc.*, 1973, **95**, 6629 (*nmr*)
Inorg. Synth., 1979, **19**, 247 (*synth*)

B_6H_{12} **B-00032**

Hexaborane(12), 10CI

[12008-19-4]

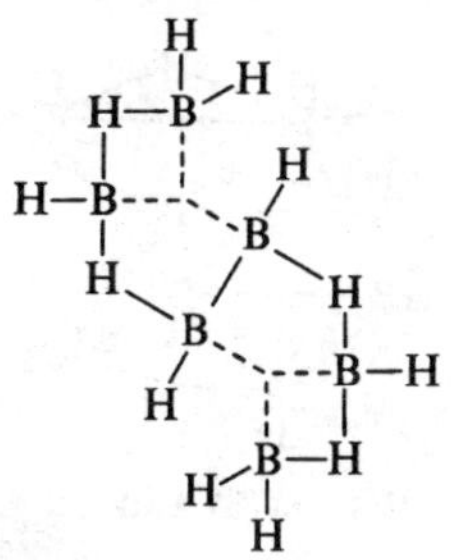

M 76.955

arachno-. Colourless liq., dec. slowly at 25°. Mp −82°. Bp 80-90°.

Gaines, D. *et al*, *Inorg. Chem.*, 1964, **3**, 438 (*synth, ir*)
Leach, J. *et al*, *Inorg. Chem.*, 1970, **9**, 2170 (*nmr*)
McLaughlan, E. *et al*, *J. Phys. Chem.*, 1971, **75**, 3106 (*ms*)
Remmel, R. *et al*, *J. Am. Chem. Soc.*, 1971, **93**, 3798 (*synth*)

$B_7H_9S_2$ **B-00033**

6,8-Dithianonaborane(9), 10CI

[63115-77-5]

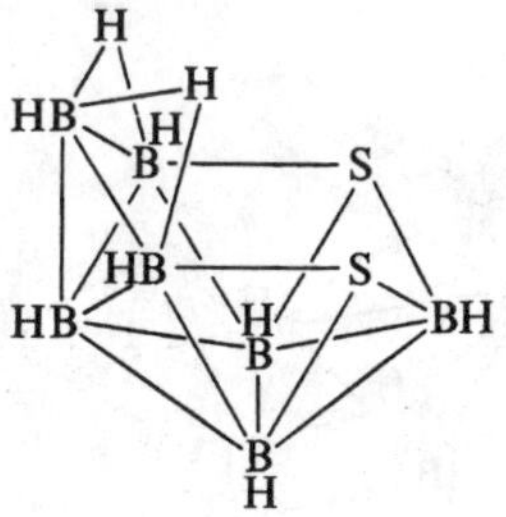

M 148.861

nido-. Pale-yellow cryst. Sol. hexane. Mp 103°. Subl. at 50°/0.01 mm.

Plešek, J. *et al*, *Collect. Czech. Chem. Commun.*, 1977, **42**, 785 (*synth, nmr, ms, uv*)

$B_8H_{10}S$ **B-00034**

7-Thianonaborane(10), 9CI

4-Thianonaborane(10)

[59351-07-4]

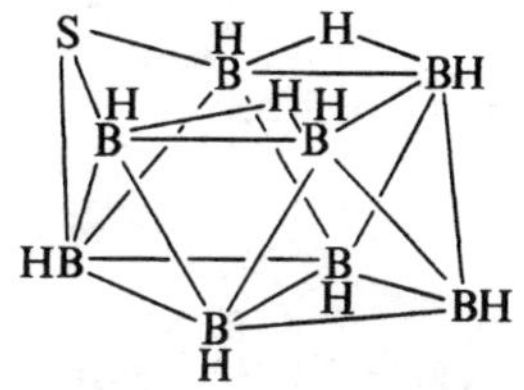

M 128.619

nido-. Colourless solid, air- and water-sensitive. Sol. C_6H_6, CH_2Cl_2, pet. ether. Subl. at 40° *in vacuo*. Stored at −78°.

Pretzer, W. *et al*, *J. Am. Chem. Soc.*, 1976, **98**, 1441 (*synth, nmr, ir, ms*)

B_8H_{12} **B-00035**

Octaborane(12), 10CI

[19499-16-0]

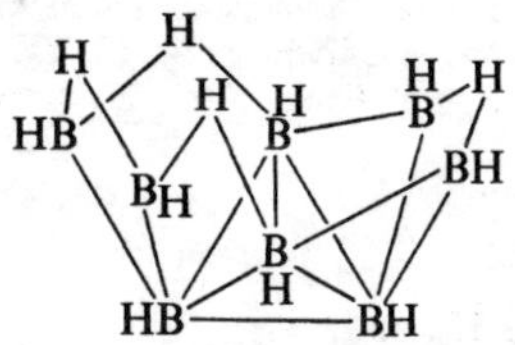

M 98.575

nido-. Sol. pentane. Mp −20°. Unstable above Mp.

Enrione, R. *et al*, *Inorg. Chem.*, 1964, **3**, 1659 (*cryst struct*)
Dobson, J. *et al*, *Inorg. Chem.*, 1968, **7**, 402 (*synth*)
Steck, S, *et al*, *Inorg. Chem.*, 1970, **9**, 2452 (*ms*)
Rietz, R. *et al*, *Inorg. Chem.*, 1972, **11**, 1242 (*nmr*)

$B_8H_{13}N$ **B-00036**

4-Azanonaborane(13), 10CI

[58920-21-1]

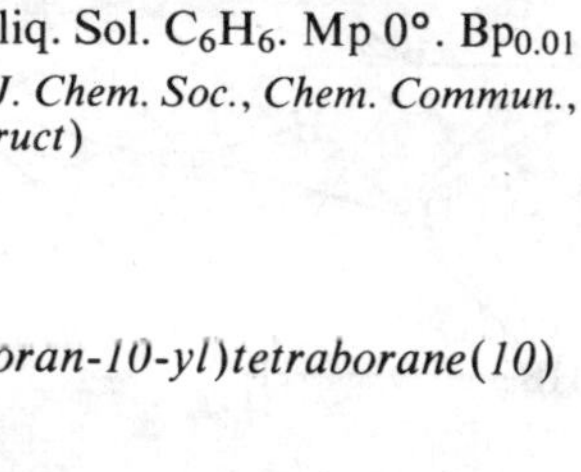

M 113.589

Air-sensitive liq. Sol. C_6H_6. Mp 0°. $Bp_{0.01}$ 60°.

Base, K. *et al*, *J. Chem. Soc., Chem. Commun.*, 1975, 934 (*synth, ir, struct*)

B_8H_{18} **B-00037**

Octaborane(18)

2-(2′-Tetraboran-10-yl)tetraborane(10)

[12421-66-8]

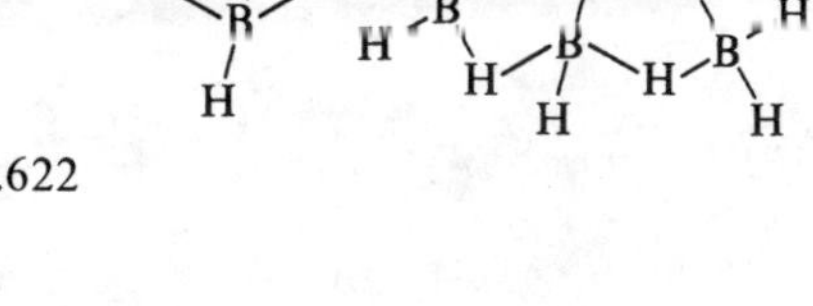

M 104.622

2,2-Linked *arachno-*. Thermally unstable.

Steck, S. *et al*, *Inorg. Chem.*, 1969, **8**, 830 (*ms*)
Rietz, R. *et al*, *Inorg. Chem.*, 1972, **11**, 1242 (*nmr*)
Glore, J. *et al*, *Inorg. Chem.*, 1973, **12**, 2175 (*synth*)

B_9H_9S B-00038

1-Thiadecaborane(9)

[41646-56-4]

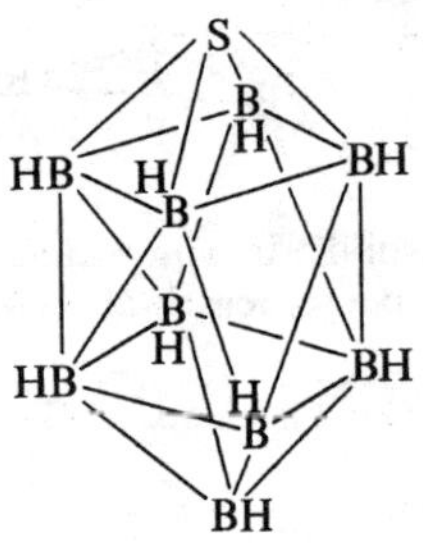

M 138.421

closo-. White cryst. by subl. *in vacuo*. Mp 216.5° (sealed tube).

Pretzer, W. *et al*, *J. Am. Chem. Soc.*, 1973, **95**, 931 (*ms*)
Pretzer, W. *et al*, *Inorg. Chem.*, 1974, **13**, 2779 (*nmr*)
Pretzer, W. *et al*, *J. Am. Chem. Soc.*, 1976, **98**, 1441 (*synth, pmr, uv*)

$B_9H_{11}S$ B-00039

6-Thiadecaborane(11), 10CI

[12447-77-7]

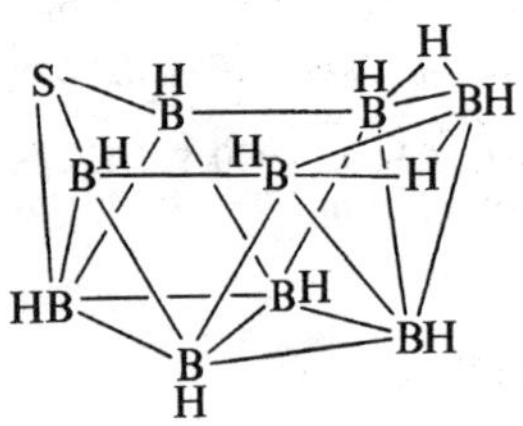

M 140.437

nido-. Transparent cryst. Sol. C_6H_6. Mp 76.5°. Subl. at 70-90°/0.1 mm.

Pretzer, W. *et al*, *J. Am. Chem. Soc.*, 1976, **98**, 1441 (*synth, struct, nmr*)

B_9H_{15} B-00040

Isononaborane(15), 10CI

[19465-30-6]

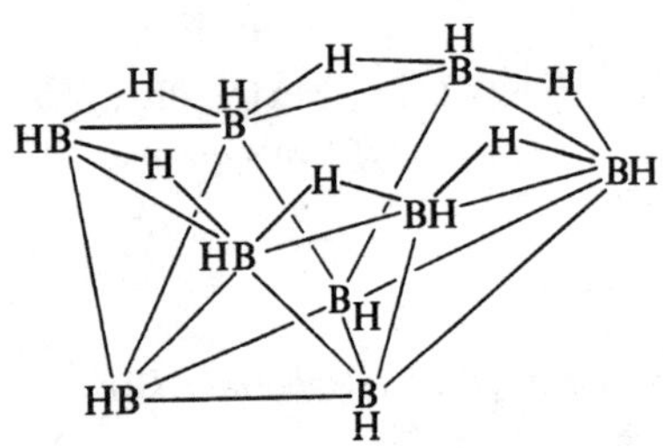

M 112.409

arachno-. Solid, dec. above −30°. Sol. pentane.

Dobson, J. *et al*, *Inorg. Chem.*, 1968, **7**, 399 (*synth*)
Moody, D. *et al*, *Inorg. Chem.*, 1976, **15**, 233 (*nmr*)

B_9H_{15} B-00041

Nonaborane(15), 10CI

[19465-30-6]

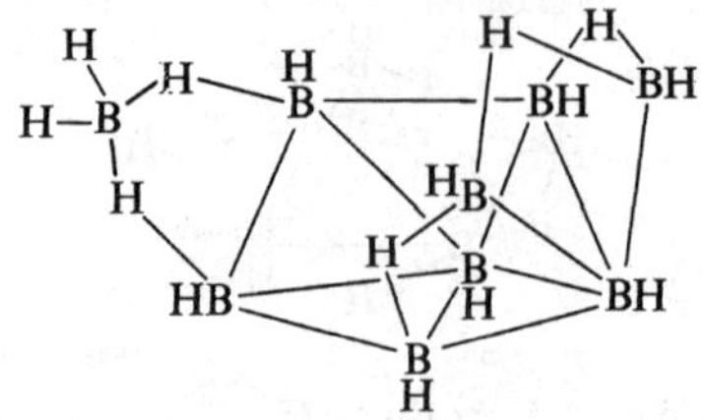

M 112.409

arachno-. Cryst. solid (pentane at −30°), liq. at r.t. Sol. pentane. Mp 2.5°.

Simpson, P. *et al*, *J. Chem. Phys.*, 1961, **35**, 1340 (*cryst struct*)
Burg, A. *et al*, *Inorg. Chem.*, 1962, **1**, 725 (*ir*)
Ditter, J. *et al*, *Inorg. Chem.*, 1966, **5**, 118 (*ms*)
Schaeffer, R. *et al*, *Inorg. Chem.*, 1972, **11**, 3102 (*synth, nmr*)

$B_{10}H_{12}$ B-00042

Decaborane(12)

[32124-79-1]

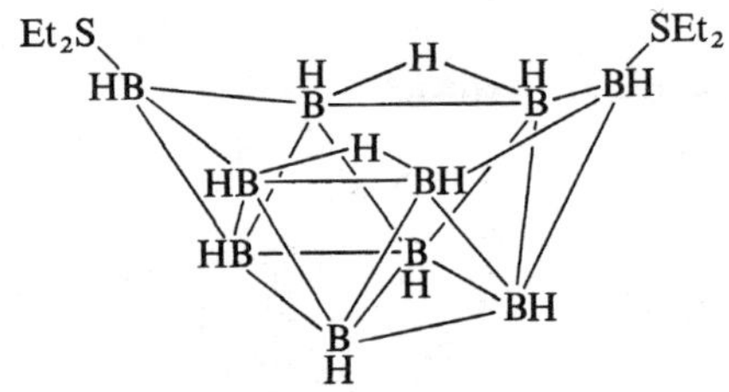

Bis (Et_2S) adduct

M 120.195

Isol. as adduct with Et_2S.

Bis (Et_2S) adduct: [32174-79-1]. *Decaborane(12) compd. with 1,1′-thiobis[ethane](1:2)*, 10CI.

$C_8H_{32}B_{10}S_2$ M 300.561

Important precursor to 1,2-dicarbadodecaborane(12) derivs. White solid. Mp 91°.

Heyling, T.L. *et al*, *Inorg. Chem.*, 1963, **2**, 1089 (*synth*)
Hyatt, D.E. *et al*, *Inorg. Chem.*, 1967, **6**, 630 (*nmr*)
Inorg. Synth., 1967, **10**, 95 (*synth*)
Cragg, R.H. *et al*, *J. Chem. Soc.* (*A*), 1970, 1817 (*uv*)

$B_{10}H_{14}$ B-00043

Decaborane(14), 10CI

[17702-41-9]

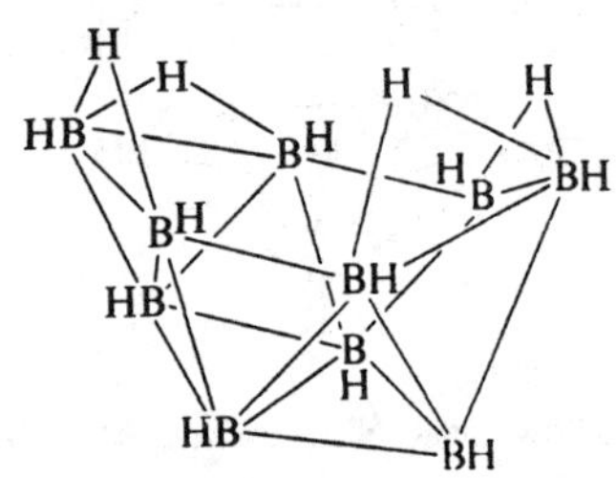

M 122.211

nido-. Precursor to production of carboranes. Cryst. solid by subl. Sol. CS_2. Mp 99.6°. Bp 213° (221°).

▷Toxic. HD1400000.

Stock, A. *et al*, *Ber.*, 1913, **46**, 3353 (*synth*)
Kasper, J. *et al*, *Acta Crystallogr.*, 1950, **3**, 436 (*cryst struct*)
Keller, W. *et al*, *J. Chem. Phys.*, 1952, **20**, 1749 (*uv, ir*)
Quayle, A., *J. Appl. Chem.* (*London*), 1959, **9**, 395 (*ms*)

Williams, R., *Inorg. Chem.*, 1965, **4**, 1504 (*nmr*)
Holzmann, R. *et al*, *Production of the Boranes and Related Research*, Academic Press, New York, 1967, 102 (*synth*)
Bodner, G. *et al*, *Inorg. Chem.*, 1970, **9**, 1421 (*nmr*)
Toft, M. *et al*, *Inorg. Chem.*, 1982, **21**, 1952 (*synth*)

$B_{10}H_{16}$ B-00044

1,1′-Bipentaborane(9), 10CI
1-(1-Pentaboran(9)yl)pentaborane(9). Decaborane(16)
[71595-75-0]

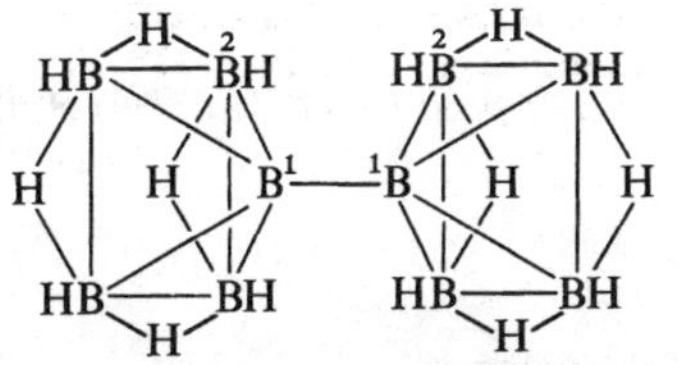

M 124.226
Linked *nido*-. One of three isomeric decaborane(16)'s. See also 1,2′-Bipentaborane(9), B-00045 and 2,2′-Bipentaborane(9), B-00046 . Solid, stable at 25°. Mp 81°. Sublimes.

Grimes, R. *et al*, *Proc. Natl. Acad. Sci. U.S.A.*, 1962, **48**, 496 (*synth*, *ir*)
Ardini, L. *et al*, *Inorg. Chem.*, 1973, **12**, 798 (*synth*)

$B_{10}H_{16}$ B-00045

1,2′-Bipentaborane(9), 10CI
1-(2-Pentaboran(9)yl)pentaborane(9). Decaborane(16)
[31831-99-9]
M 124.226
Linked *nido*-. See also 1,1′-Bipentaborane(9), B-00044 and 2,2′-Bipentaborane(9), B-00046 . Solid or liq. Mp 18.4°.

Gaines, D. *et al*, *Inorg. Chem.*, 1971, **10**, 1096 (*nmr*, *ir*, *ms*)
Gaines, D. *et al*, *J. Chem. Soc.*, *Chem. Commun.*, 1979, 380 (*synth*)

$B_{10}H_{16}$ B-00046

2,2′-Bipentaborane(9), 10CI
2-(2-Pentaboran(9)yl)pentaborane(9). Decaborane(16)
[28775-81-7]
M 124.226
Linked *nido*-. See also 1,1′-Bipentaborane(9), B-00044 and 1,2′-Bipentaborane(9), B-00045 . Liq. Mp −20.5° (−22.8°).

Gaines, D. *et al*, *Inorg. Chem.*, 1971, **10**, 1096 (*nmr*, *ir*, *ms*)
Rietz, R. *et al*, *Inorg. Chem.*, 1972, **11**, 1242 (*nmr*)
Gaines, D. *et al*, *J. Chem. Soc.*, *Chem. Commun.*, 1979, 380 (*synth*)

$B_{11}H_{11}S$ B-00047

1-Thiadodecaborane(11)
[56464-75-6]

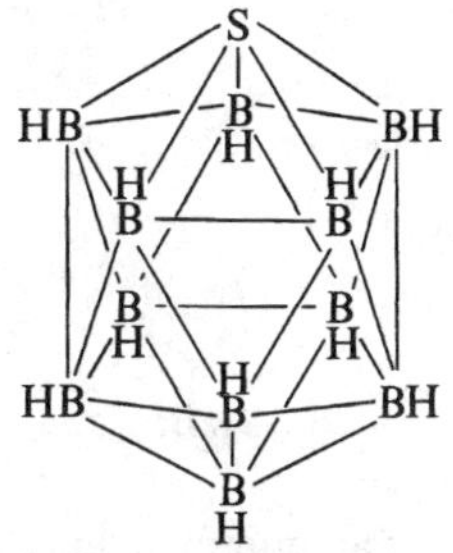

M 162.057
closo-. White cryst. Sol. CS_2, CH_2Cl_2. Mp 325° (sealed tube). Subl. >240° (>50° *in vacuo*).

Plešek, J. *et al*, *J. Chem. Soc.*, *Chem. Commun.*, 1975, 127 (*synth*, *ms*, *ir*, *nmr*)
Pretzer, W. *et al*, *J. Am. Chem. Soc.*, 1976, **98**, 1441 (*synth*, *ms*, *uv*, *nmr*)

CBF_3O B-00048

Trifluoroborane carbonyl
Carbonyltrifluoroboron, 10CI. *Trifluoroborane, compd. with carbon monoxide(1:1)*, 9CI. *Boron trifluoride carbonyl*
[51006-66-7]

$F_3B{-}CO$

M 95.816

Archibald, R.M. *et al*, *J. Chem. Soc.*, *Faraday Trans. 2*, 1973, **69**, 1793 (*struct*)
Novick, S.E. *et al*, *J. Am. Chem. Soc.*, 1973, **95**, 8547.
Janda, K.C. *et al*, *J. Am. Chem. Soc.*, 1978, **100**, 8074 (*struct*)
Armstrong, D.R., *Inorg. Chim. Acta*, 1979, **33**, 177 (*struct*)
Clark, D.T. *et al*, *Chem. Phys.*, 1980, **46**, 43.

$CBF_6^{\ominus}$ B-00049

Trifluoro(trifluoromethyl)borate(1−), 10CI, 9CI
(Trifluoromethyl)trifluoroborate

$F_3CBF_3^{\ominus}$

M 136.811 (ion)
K salt: [42298-15-7].
CBF_6K M 175.910
Monoclinic cryst. (EtOH).

Jackovitz, J.F. *et al*, *Appl. Spectrosc.*, 1973, **27**, 209 (*ir*, *raman*)
Brauer, D.J. *et al*, *Inorg. Chem.*, 1977, **16**, 2305 (*ir*, *raman*, *synth*, *cryst struct*)
Pawelke, G. *et al*, *J. Organomet. Chem.*, 1979, **178**, 1 (*synth*)

CH_2BN B-00050

Cyanoborane
Boranecarbonitrile, 10CI, 9CI
[60633-76-3]

$[BH_2CN]_n$

M 38.844
Forms homopolymer. Prepd. from $NaBH_3CN$ in Et_2O by addition of HCl gas. Air- and moisture-sensitive white semi-solid. Sol. HMPA, CH_2Cl_2. ir 2460, 2411, 2429, 2295 cm^{-1} (neat).

Brändström, A. *et al*, *Tetrahedron Lett.*, 1972, 3173 (*synth*)
Hutchins, R.O. *et al*, *J. Org. Chem.*, 1977, **42**, 82 (*synth*, *ir*, *use*)

Vidal, J.C. *et al*, *Inorg. Chem.*, 1977, **16**, 1898 (*synth*, *props*)
Martin, D.R. *et al*, *J. Inorg. Nucl. Chem.*, 1978, **40**, 9 (*synth*, *use*)
Pelter, A. *et al*, *Compr. Org. Chem.*, (*Barton, D.H.R. et al*, Ed.), 1979, **3**, 695 (*rev*)

CH_3B B-00051

Methylborylidene, 10CI
Methylborylene
[62785-41-5]

MeB:

M 25.845
Said to be formed as an intermed. in reaction of $MeBCl_2$ with potassium, but evidence unconvincing.

Van der Kerk, S.M. *et al*, *Tetrahedron Lett.*, 1979, 4765 (*synth*)
Van der Kerk, S.M. *et al*, *J. Organomet. Chem.*, 1980, **190**, C8 (*synth*, *use*)

CH_3BBr_2 B-00052

Dibromomethylborane, 10CI, 9CI
Methylboron dibromide. Methyldibromoborane
[17933-16-3]

$MeBBr_2$

M 185.653
Liq., sensitive to air and moisture. Bp 60-2°.

Nöth, H. *et al*, *J. Org. Chem.*, 1968, **12**, 23 (*pmr*)
Berger, H. *et al*, *Chem. Ber.*, 1976, **109**, 2266 (*nmr*)
Haubold, W. *et al*, *Z. Anorg. Allg. Chem.*, 1976, **420**, 251 (*synth*)
Meler, A. *et al*, *Chem. Ber.*, 1981, **114**, 2519 (*use*)
Schulze, J. *et al*, *Chem. Ber.*, 1981, **114**, 495 (*use*)

CH_3BCl_2 B-00053

Dichloromethylborane, 10CI, 9CI, 8CI
Methyldichloroborane. Dichloromethylborine
[7318-78-7]

$MeBCl_2$

M 96.751
Easily condensed gas, sensitive to air and moisture. Bp 11°.

Lengyel, B. *et al*, *Z. Anorg. Allg. Chem.*, 1963, **322**, 103 (*synth*)
Timms, P. *et al*, *J. Chem. Soc., Chem. Commun.*, 1968, 1525 (*use*)
Haubold, W. *et al*, *Z. Anorg. Allg. Chem.*, 1976, **420**, 251 (*synth*)
Wrackmeyer, B. *et al*, *Chem. Ber.*, 1976, **109**, 2266 (*nmr*)
Barlos, K. *et al*, *Chem. Ber.*, 1977, **110**, 2774 (*nmr*)
Becher, H. *et al*, *Spectrochim. Acta, Part A*, 1979, **35**, 59 (*ir*)

$CH_3BCl_2O_2$ B-00054

(Dichloromethyl)dihydroxyborane
(Dichloromethyl)boronic acid, 10CI. *Dichloromethaneboronic acid*
[62260-98-4]

$Cl_2CHB(OH)_2$

M 128.750
White air-sensitive solid (Me_2CO aq.). Sol. EtOH, Me_2CO, THF, H_2O. Insol. CH_2Cl_2, C_6H_6, $CHCl_3$, MeCN. Mp 75-100° (impure).

Rathke, M.W. *et al*, *J. Organomet. Chem.*, 1976, **122**, 145 (*synth*, *nmr*)

CH_3BF_2 B-00055

Difluoromethylborane, 10CI, 9CI
Methyldifluoroborane
[373-64-8]

F_2BMe

M 63.842
Air- and moisture-sensitive gas. V.p. 69.5 mm at −97.4°; 286.9 mm at −78.5°.

Burg, A.B. *et al*, *J. Am. Chem. Soc.*, 1961, **83**, 2667 (*synth*)
Steele, W.C. *et al*, *J. Am. Chem. Soc.*, 1962, **84**, 1154 (*ms*)
Barlos, K. *et al*, *Chem. Ber.*, 1977, **110**, 2774 (*nmr*)
Fussstetter, H. *et al*, *Chem. Ber.*, 1977, **110**, 3172 (*nmr*)
McKean, D.C. *et al*, *Spectrochim. Acta, Part A*, 1977, **33**, 951 (*ir*, *raman*)

CH_3BF_2O B-00056

Difluoromethoxyborane, 9CI
Difluoromethoxyborine
[381-59-9]

F_2BOMe

M 79.841
Air- and moisture-sensitive solid. Sol. THF. Mp 41-2°.

Goubeau, J. *et al*, *Z. Phys. Chem.*, 1959, **20**, 15 (*raman*)
Sternach, B. *et al*, *J. Am. Chem. Soc.*, 1961, **83**, 3384 (*synth*)
deMoor, J.E. *et al*, *Ber. Bunsenges. Phys. Chem.*, 1963, 429 (*ir*)
deMoor, J.E. *et al*, *J. Organomet. Chem.*, 1966, **6**, 235 (*nmr*)
Anosov, V.I. *et al*, *CA*, 1980, **93**, P221317f (*use*)

$CH_3BF_3^{\ominus}$ B-00057

Trifluoromethylborate(1−), 10CI
Methyltrifluoroborate

$MeBF_3^{\ominus}$

M 82.840 (ion)
Cu salt: [65139-98-2].
CH_3BCuF_3 M 146.386
Proposed intermed. in BF_3 - catalysed reacns. of CuMe.

Maruyama, K. *et al*, *J. Am. Chem. Soc.*, 1977, **99**, 8068 (*props*)
Yamamoto, Y. *et al*, *J. Am. Chem. Soc.*, 1978, **100**, 3240; 1980, **102**, 2318 (*props*)

$CH_3BF_3O^{\ominus}$ B-00058

Trifluoromethoxyborate(1−), 10CI, 9CI
Methoxytrifluoroborate
[51259-50-8]

$MeOBF_3^{\ominus}$

M 98.839 (ion)
Na salt: [21655-10-7].
CH_3BF_3NaO M 121.829
K salt: [15696-17-0].
CH_3BF_3KO M 137.938

Kolditz, L. *et al*, *Z. Anorg. Allg. Chem.*, 1968, **360**, 25 (*props*, *synth*)
Ryss, I.G. *et al*, *Zh. Neorg. Khim.*, 1968, **13**, 1535 (*synth*, *props*)
Plakhotnik, V.N. *et al*, *Dokl. Akad. Nauk SSSR, Ser. Sci. Khim.*, 1976, **228**, 1112 (*nmr*)
Plakhotnik, V.N. *et al*, *Koord. Khim.*, 1976, **2**, 855 (*props*)
Chernyshov, B.N. *et al*, *Koord. Khim.*, 1980, **6**, 1520 (*nmr*)
Vasilyuk, N.S. *et al*, *Koord. Khim.*, 1981, **7**, 78 (*props*)

CH_3BI_2 B-00059

Diiodomethylborane

Methyldiiodoborane

[17933-08-3]

MeBI₂

M 279.654

Prepd. from tetramethylstannane and BI_3. Air-, moisture- and light-sensitive red liq.; stored under N_2 at −10°. Sol. CH_2Cl_2. $Bp_{0.01}$ 50°.

Nöth, H. *et al*, *J. Organomet. Chem.*, 1968, **11**, 399; 1968, **12**, 23 (*synth, nmr*)
Siebert, W. *et al*, *Chem. Ber.*, 1973, **106**, 1012 (*use*)
Renk, T. *et al*, *J. Organomet. Chem.*, 1976, **120**, 1 (*use*)
Haubold, W. *et al*, *Z. Anorg. Allg. Chem.*, 1978, **446**, 45.
Eisch, J.J. *et al*, *J. Organomet. Chem.*, 1979, **171**, 141.

$CH_3BN^{\ominus}$ B-00060

(Cyano-*C*)trihydroborate(1−), 10CI, 9CI

[33195-00-5]

$[H_3BCN]^{\ominus}$

M 39.851 (ion)

Na salt: [25895-60-7]. *Sodium cyanoborohydride.*
CH_3BNNa M 62.841
Reducing agent for org. functional groups, particularly for reductive amination of ketones. Cryst. ($MeNO_2$). Mp >242° dec. Commercially available.

▷Severe poison, flammable solid

Berschied, J.R. *et al*, *Inorg. Chem.*, 1970, **9**, 624 (*ir, nmr, raman*)
Wade, R.C. *et al*, *Inorg. Chem.*, 1970, **9**, 2146 (*synth, ir, nmr*)
Berschied, J.R. *et al*, *Inorg. Chem.*, 1972, **11**, 930 (*ir, raman, struct*)
Hall, L.W. *et al*, *Inorg. Chem.*, 1975, **14**, 580 (*nmr*)
Hutchins, R.O. *et al*, *Org. Prep. Proced. Int.*, 1979, **11**, 201 (*rev, use*)
Yamamoto, Y. *et al*, *CA*, 1980, **92**, 139884j (*synth, props, rev*)

CH_3BO B-00061

Borane carbonyl

Carbonyltrihydroboron, 10CI. *Carbon monoxide compd. with borane(1:1)*, 9CI, 8CI. *Borane compd. with carbon monoxide(1:1). Borine carbonyl*

[13205-44-2]

$H_3B—CO$

M 41.844

Wason, S. *et al*, *Inorg. Chem.*, 1966, **5**, 161 (*synth*)
Mayer, E., *Monatsh. Chem.*, 1971, **102**, 940 (*synth*)
Finn, P. *et al*, *J. Am. Chem. Soc.*, 1972, **94**, 1540 (*nmr, props*)
Lambert, L. *et al*, *J. Mol. Spectrosc.*, 1972, **44**, 578 (*ir*)
Lloyd, D.R. *et al*, *J. Chem. Soc., Faraday Trans. 2*, 1972, **68**, 947 (*pe*)
Carter, J.C. *et al*, *Appl. Spectrosc.*, 1974, **28**, 427 (*nmr*)
Pepin, C. *et al*, *J. Mol. Spectrosc.*, 1976, **59**, 43 (*ir*)

CH_3BO_2 B-00062

Boranecarboxylic acid, 10CI, 9CI

H_2BCOOH

M 57.844

NH_4 salt: [74861-59-9].
CH_6BNO_2 M 74.874
Prepd. by reacn. of Me_3NBH_2COOH with NH_3. Shows antihypolipidaemic activity. Air-sensitive cryst. solid. Sol. $CHCl_3$. Mp 116°.

NMe_3 complex: [60788-33-2].
$C_4H_{12}BNO_2$ M 116.954
Shows antiinflammatory and antihypolipidaemic activity. Air-sensitive white cryst. (H_2O). Sol. $CHCl_3$. Mp 147°.

Py complex: See Pyridine carboxyborane, B-00384

*Tetramethylethylenediamine complex: See N,N,N′,N′-*Tetramethylethylenediamine bis(carboxyborane), B-00583

Spielvogel, B.F. *et al*, *J. Am. Chem. Soc.*, 1976, **98**, 5702; 1980, **102**, 6343 (*synth, nmr, ir, cryst struct, use*)
Hall, I.H. *et al*, *J. Pharm. Sci.*, 1979, **68**, 685 (*synth, pharmacol*)
Pelter, A. *et al*, *Compr. Org. Chem.*, (Barton, D.H.R. *et al*, Ed.), 1979, **3**, 695 (*rev*)
Hall, I.H. *et al*, *J. Pharm. Sci.*, 1980, **69**, 1025; 1981, **70**, 339 (*pharmacol*)

CH_4BF_3O B-00063

Trifluoroborane methanol

Trifluoro(methanol)boron, 10CI. *Methanol compd. with trifluoroborane(1:1)*, 9CI. *Methanol compd. with boron fluoride(BF_3)(1:1)*, 8CI

[373-57-9]

$MeOH—BF_3$

M 99.847

Catalyst for a variety of reactions. Moisture-sensitive.

Fratiello, A. *et al*, *J. Org. Chem.*, 1972, **37**, 2237 (*nmr*)
Jao, L.K.Y., *Diss. Abstr. Int. B.*, 1973, **34**, 2524; *CA*, **80**, 94809x (*nmr*)
Burya, G.F. *et al*, *Zh. Fiz. Khim.*, 1975, **49**, 1227 (*ir*)
Derouault, J. *et al*, *J. Mol. Struct.*, 1978, **47**, 59 (*ir, raman*)
Derouault, J. *et al*, *Spectrochim. Acta, Part A*, 1979, **35**, 773 (*ir, raman*)

CH_5B B-00064

Methylborane, 10CI, 9CI, 8CI

Methylborine

[12538-96-4]

$MeBH_2$

M 27.861

Exists as B-H bridged dimer in the uncomplexed state. Air- and moisture-sensitive; readily redistributes to other compds. of the type Me_nBH_{3-n}. Sol. non-hydroxylic solvs.

Me_3N complex: [52920-77-1]. N,N-*Dimethylmethanamine methylborane.*
$C_4H_{14}BN$ M 86.971
Prepd. from $MeBH_3Li$ and NMe_3·HCl. Sol. pentane. Mp 0°.

Me_2NH complex: [52920-78-2]. *Methylborane* N-*methylmethanamine.*
$C_3H_{12}BN$ M 72.945
Prepd. by amine exchange of $MeBH_2$·NMe_3 with excess Me_2NH. Colourless involatile liq. Sol. $CHCl_3$.

Schlesinger, H.I. *et al*, *J. Am. Chem. Soc.*, 1939, **61**, 1078 (*deriv*)
Steck, S.J. *et al*, *J. Phys. Chem.*, 1969, **73**, 1000 (*ms*)
Nöth, H. *et al*, *Chem. Ber.*, 1974, **107**, 3039 (*nmr*)
Beachley, O.T. *et al*, *Inorg.Chem.*, 1975, **14**, 120 (*deriv*)
Dill, J.D. *et al*, *J. Am. Chem. Soc.*, 1976, **98**, 1663 (*struct*)
Pross, A. *et al*, *J. Comput. Chem.*, 1980, **1**, 295 (*struct*)

CH_5BN_2 B-00065

Cyanoborane ammonia

Ammine(cyano-C)dihydroboron, 10CI, 9CI. *Boranecarbonitrile monoammoniate*, 9CI

[59301-48-3]

$H_3N—BH_2CN$

M 55.874

White solid. Mp 89°.

Bratt, P.J. *et al*, *J. Chem. Soc., Dalton Trans.*, 1976, 353 (*synth, ir*)
McPhail, A.T. *et al*, *J. Chem. Res. (S)*, 1978, 205 (*cryst struct*)
Hall, I.H. *et al*, *J. Pharm. Sci.*, 1980, **69**, 1025; 1981, **70**, 339 (*pharmacol*)

CH_5BO_2 B-00066

Dihydroxy(methyl)borane

Methylboronic acid, 10CI, 9CI. *Methylboric acid. Methaneboric acid. Methyldihydroxyborane*

[13061-96-6]

$MeB(OH)_2$

M 59.859

Solid. Mp 95-100° dec. (freshly prepared). Readily dehydrates to trimethylboroxin.

Burg, A.B., *J. Am. Chem. Soc.*, 1940, **62**, 2228 (*synth*)
de Moor, J.E. *et al*, *J. Organomet. Chem.*, 1967, **9**, 31 (*ir, raman*)
Nöth, H. *et al*, *J. Organomet. Chem.*, 1968, **12**, 23 (*pmr*)
McFarlane, W. *et al*, *Chem. Ber.*, 1975, **108**, 2112 (*nmr*)
Rensch, H. *et al*, *Chem. Ber.*, 1977, **110**, 2189 (*synth*)
Kossa, W.C., *Chem. Deriv. Anal. Chem.*, (*Frei, R.W. et al*, Ed.), 1981, Plenum, N.Y., 99 (*rev*)
Fieser, M. *et al*, *Reagents for Organic Synthesis*, Wiley, 1967-83, **1**, 1231.

CH_5BO_3 B-00067

Dihydroxymethoxyborane

Boric acid (H_3BO_3) *monomethyl ester*, 10CI. *Methoxydihydroxyborane*

[68498-84-0]

$(HO)_2BOMe$

M 75.859

Moisture-sensitive; readily dec.

Guthrie, J.P., *Can. J. Chem.*, 1978, **56**, 2342 (*props*)
Kamars, A. *et al*, *CA*, 1979, **91**, 19380z (*pmr*)
Steinfeldt, K.P. *et al*, *Z. Naturforsch., B*, 1981, **36**, 691 (*raman*)

$CH_6B^{\ominus}$ B-00068

Trihydromethylborate(1−)

Methylborohydride

[71844-38-7]

$MeBH_3^{\ominus}$

M 28.868 (ion)

Li salt: [52950-75-1].
CH_6BLi M 35.809
Air- and moisture-sensitive.

Mg salt: [68193-75-9]. *Bis[trihydromethylborato(1−)-H,H′]magnesium.*
$C_2H_{12}B_2Mg$ M 82.042
Air- and moisture-sensitive.

Goel, A.B., *Indian J. Chem., Sect. A*, 1978, **16**, 491.
Umeyama, H. *et al*, *Chem. Pharm. Bull.*, 1979, **27**, 1626 (*props*)
Pross, A. *et al*, *Tetrahedron*, 1980, **36**, 673 (*props*)
Biffar, W. *et al*, *Chem. Ber.*, 1981, **114**, 49 (*nmr*)

CH_6BF_2N B-00069

Difluoroborane methylamine

Difluorohydro(methanamine)boron, 10CI. *Difluoroborane methanamine. Methylamine difluoroborane*

[42525-40-6]

$F_2BH—NH_2Me$

M 80.872

White solid below −50°, phys. state not recorded at r.t. Sol. CCl_4, CH_2Cl_2.

Tuchagues, J.P. *et al*, *J. Organomet. Chem.*, 1973, **54**, 69 (*synth, ir, nmr*)
Van Paasschen, J.M. *et al*, *Inorg. Chem.*, 1978, **17**, 3302.

CH_7B_5 B-00070

1-Carbahexaborane(7), 10CI

1-Carbahexaborane(7)-2,3,6-μ_3-H

[55188-36-8]

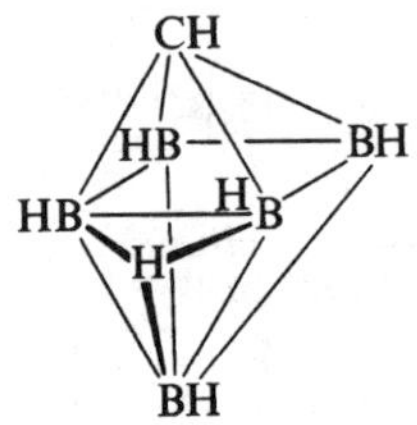

M 73.116

Exceptional *closo*-. Colourless liq. by vacuum fractionation.

Groszek, E. *et al*, *Inorg. Chem.*, 1971, **10**, 2770 (*synth, nmr*)
Onak, T. *et al*, *J. Chem. Soc., Chem. Commun.*, 1971, 76 (*nmr*)
Onak, T. *et al*, *J. Chem. Soc., Dalton Trans.*, 1974, 665.
McNeill, E.A. *et al*, *Inorg. Chem.*, 1975, **13**, 1081 (*ed*)
McKown, G. *et al*, *J. Am. Chem. Soc.*, 1976, **98**, 6909 (*microwave*)

CH_8BN B-00071

Borane methylamine

Trihydro(methanamine)boron, 10CI. *Methanamine compd. with borane(1:1)*, 9CI. *Methylamine compd. with borane*, 8CI. *Methylamine borane*

[1722-33-4]

$MeNH_2—BH_3$

M 44.891

Solid. Mp 56°. Dipole moment 5.19D.

Nöth, H. *et al*, *Chem. Ber.*, 1960, **93**, 928 (*synth*)
Heitsch, C.W., *Inorg. Chem.*, 1965, **4**, 1019 (*nmr*)
Lloyd, D.R. *et al*, *J. Chem. Soc., Faraday Trans. 2*, 1972, **68**, 947 (*pe*)
Nöth, H. *et al*, *Chem. Ber.*, 1974, **107**, 3070 (*nmr*)
Umeyama, H. *et al*, *J. Am. Chem. Soc.*, 1976, **98**, 7208.
Van Paasschen, J.M. *et al*, *Inorg. Chem.*, 1978, **17**, 3302.

$CH_8B_3N_3$ B-00072

1-Methylborazine, 10CI, 9CI, 8CI

1-Methylborazole

[21127-94-6]

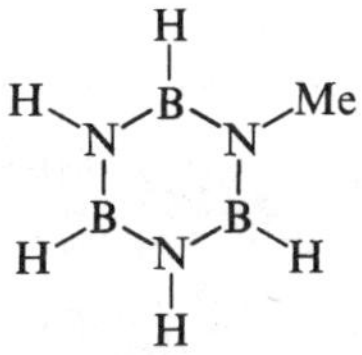

M 94.524

Stable compound. V.p. 27mm at r.t., 23.5mm at 0°.

Hoffmann, R., *J. Chem. Phys.*, 1964, **40**, 2474 (*struct*)
Powell, P. *et al*, *J. Chem. Soc.*, 1964, 280 (*synth*)
Beachley, O.T., *Inorg. Chem.*, 1969, **8**, 981 (*nmr, ms, ir*)
Beaumelou, G. *et al*, *Org. Magn. Reson.*, 1973, **5**, 585 (*nmr*)
Turbini, L.J. *et al*, *Org. Magn. Reson.*, 1974, **6**, 456 (*pmr*)
Kline, G.A. *et al*, *Inorg. Chem.*, 1977, **16**, 11 (*props, nmr*)

CH_9B_5 B-00073

2-Carbahexaborane(9), 10CI

[12385-35-2]

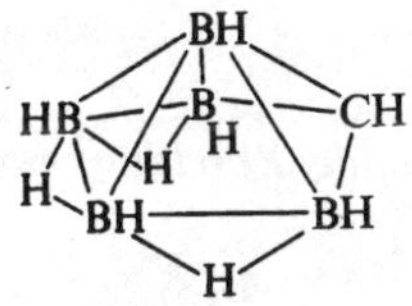

M 75.132

nido-. Colourless liq. by vacuum fractionation.

Dunks, G. *et al*, *Inorg. Chem.*, 1969, **8**, 2667 (*synth, nmr, ir*)
Cheung, C. *et al*, *Inorg. Chem.*, 1971, **10**, 1144 (*microwave struct*)
Groszek, E. *et al*, *Inorg. Chem.*, 1971, **10**, 2770 (*synth*)
Leach, J.B. *et al*, *J. Chem. Soc., Dalton Trans.*, 1977, 819 (*synth*)

$CH_{10}B_5Cl_3Si$ B-00074

1-[(Trichlorosiiyl)methyl]pentaborane(9), 10CI

[63788-21-6]

$B_5H_8CH_2SiCl_3$

M 210.585

Leach, J.B. *et al*, *J. Chem. Soc., Dalton Trans.*, 1977, 819 (*nmr, ms, synth*)

$CH_{11}AlB_2$ B-00075

Methylbis(tetrahydroborato)aluminum

Methylaluminum bis(tetrahydroborate)

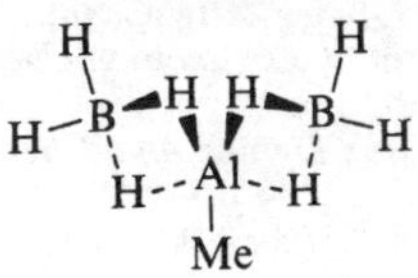

M 71.699

Monomeric. Mp −76°. Extremely sensitive to traces of oxygen or moisture. Forms adducts with NMe_3, PMe_3, $AsMe_3$, Me_2O, Et_2O, Me_2S, Me_2NH.

Oddy, P.R. *et al*, *J. Chem. Soc., Dalton Trans.*, 1976, 869 (*synth, ir*)
Barlow, M.T. *et al*, *J. Chem. Soc., Dalton Trans.*, 1980, 1374 (*synth, struct*)

$CH_{12}B_2N_2$ B-00076

Methylhydrazine bis(borane)

Hexahydro[μ-(methylhydrazine-N:N′)]diboron, 10CI.
Methylhydrazine compd. with borane(1:2), 9CI

[58978-79-3]

$H_3B—H_2NNHMe—BH_3$

M 73.739

Used in photographic image development.

Hamilton, J.F., *J. Vac. Sci. Technol.*, 1976, **13**, 319.
Gysling, H.J. *et al*, *Res. Discl.*, 1976, **142**, 34.
U.S.P., 4 042 392, (*1977*); *CA*, **88**, P30399w
Lelental, M. *et al*, *J. Photogr. Sci.*, 1978, **26**, 135.
Ger. Pat., 2 819 885, (*1978*); *CA* **90**, P178152x

$CH_{15}B_{10}In$ B-00077

[η^4-Dodecahydrodecaborato(2−)]methylindium, 9CI

[55988-46-0]

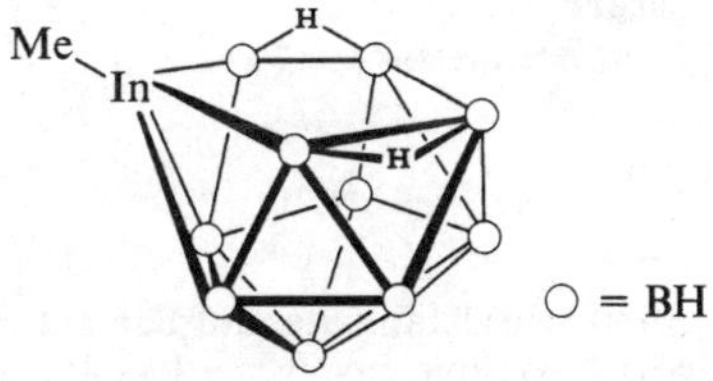

M 250.050

Pale-yellow solid. Sol. Me_2CO.

Greenwood, N.N. *et al*, *J. Chem. Soc., Dalton Trans.*, 1975, 299 (*synth, ir, nmr*)

$C_2BF_8^{\ominus}$ B-00078

Difluorobis(trifluoromethyl)borate(1−), 10CI

$(F_3C)_2BF_2^{\ominus}$

M 186.819 (ion)

H deriv.: [72035-26-8].
C_2HBF_8 M 187.827
No phys. props. reported.

K salt: [72035-25-7].
C_2BF_8K M 225.918
No phys. props. reported.

Cs salt: [72035-27-9].
C_2BCsF_8 M 319.725
No phys. props. reported.

Pawelke, G. *et al*, *J. Organomet. Chem.*, 1979, **178**, 1 (*synth*)
Brauer, D.J. *et al*, *J. Organomet. Chem.*, 1980, **192**, 305 (*struct, raman, ir, nmr*)

C_2HBF_2 B-00079

Ethynyldifluoroborane, 10CI, 9CI

Difluoroethynylborane

[23728-64-5]

$F_2BC≡CH$

M 73.837

Prepd. from ethynyltrimethyltin and BF_3 at −80° to −63°. Air- and moisture-sensitive. Dipole moment 1.87D.

Lafferty, W.J. *et al*, *J. Chem. Soc., Dalton Trans.*, 1969, 909 (*spectra*)
Ritter, J.J. *et al*, *J. Chem. Soc., Dalton Trans.*, 1969, 908 (*synth, ms*)
Lafferty, W.J. *et al*, *J. Mol. Spectrosc.*, 1971, **38**, 181 (*spectra*)
Durig, J.M. *et al*, *Spectrochim. Acta, Part A*, 1974, **30**, 993 (*raman*)
Grindley, T.B. *et al*, *J. Chem. Soc., Perkin Trans. 2*, 1974, 276 (*props*)

Tupitsyn, I.F. *et al*, *Zh. Strukt. Khim.*, 1979, **20**, 987.

C_2H_3B B-00080

Ethynylborane, 10CI

[72507-59-6]

$H_2BC{\equiv}CH$

M 37.856

Never prepd. as a free compd. V. unstable, presumably both air- and moisture-sensitive.

Dill, J.D. *et al*, *J. Am. Chem. Soc.*, 1979, **101**, 6814.
Pelter, A. *et al*, *Comprehensive Organic Chemistry*, (*Barton, D.H.R. et al*, Ed.), Pergamon, Oxford, 1979, **3**, 695 (*rev*)
Radom, L. *et al*, *Isr. J. Chem.*, 1980, **19**, 305.

$C_2H_3BF_2$ B-00081

Difluorovinylborane, 8CI

Ethenyldifluoroborane, 10CI, 9CI

[358-95-2]

$H_2C{=}CHBF_2$

M 75.853

Prepd. from tetravinylstannane and B_2F_4 at −80°. Air- and moisture-sensitive gas. Mp −133.4°. Bp −38.8°. Trouton const. 23.0 e.u., ΔH_{vap} 5.4 Kcal/mol.

Brinckman, F.E. *et al*, *J. Am. Chem. Soc.*, 1960, **82**, 6218 (*synth*)
Good, C.D. *et al*, *J. Chem. Eng. Data*, 1962, **7**, 416 (*chromatog, props*)
Steele, W.C. *et al*, *J. Am. Chem. Soc.*, 1962, **84**, 1154 (*ms*)
Holliday, A.K. *et al*, *J. Chem. Soc.*, 1964, 2731 (*synth*)
Hall, L.W. *et al*, *J. Am. Chem. Soc.*, 1975, **97**, 4527 (*nmr*)
Barlos, K. *et al*, *Chem. Ber.*, 1977, **110**, 2774 (*nmr*)

$C_2H_3BF_3O_2^{\ominus}$ B-00082

Trihydro(trifluoroacetato-*O*)borate(1−), 10CI

Trifluoroacetoxytrihydroborate. Trifluoroacetoxyborohydride

$F_3CCOOBH_3^{\ominus}$

M 126.850 (ion)

Na salt: [61666-40-8].

$C_2H_3BF_3NaO_2$ M 149.839

Used for redn. of nitriles and of oxime ethers to amines.

Umino, N. *et al*, *Tetrahedron Lett.*, 1976, 2875.
Umino, N. *et al*, *Chem. Pharm. Bull.*, 1978, **26**, 2897 (*synth*)
Okushi, T., *CA*, 1981, **95**, 23491m (*rev, props*)

$C_2H_4B_2Cl_4$ B-00083

1,2-Bis(dichloroboryl)ethane

1,2-Ethanediylbis[dichloroborane], 10CI, 9CI. *Ethylenebis[dichloroborane]*, 8CI

[20816-71-1]

$Cl_2BCH_2CH_2BCl_2$

M 191.486

Prepd. from B_2Cl_4 and ethylene. Stable liq., sensitive to air and moisture. Sol. THF. Mp −28.5°. Bp 142°. ΔH_{evap} = 10,140 calmol.$^{-1}$ Trouton const. 24.4 e.u.

Grant, U. *et al*, *J. Am. Chem. Soc.*, 1954, **76**, 5299 (*synth*)
Wadsley, A.D. *et al*, *Acta Crystallogr.*, 1957, **10**, 261 (*cryst struct*)
Massey, A.G. *et al*, *J. Chem. Soc.*, 1960, 5264 (*use*)
Haubold, W. *et al*, *Z. Allorg. Allg. Chem.*, 1974, **406**, 171 (*ir, nmr, raman*)
Reason, M.S. *et al*, *J. Organomet. Chem.*, 1974, **77**, C9 (*use*)
Haubold, W. *et al*, *Chem. Ber.*, 1980, **113**, 3352 (*use*)

C_2H_5B B-00084

Vinylborane, 8CI

Ethenylborane, 10CI, 9CI. *Vinylborine*

[5856-70-2]

$H_2C{=}CHBH_2$

M 39.872

Probably exists as H-bridged dimer. Prepd. from BCl_3 and $(H_2C{=}CH)_4Sn$ at r.t. Air- and moisture-sensitive liq. Sol. THF. Mp −111.1°. Bp 45.1°. Trouton const. 22.9 e.u., ΔH_{vap} 6.63 Kcalmol.$^{-1}$.

Brinckman, F.E. *et al*, *J. Am. Chem. Soc.*, 1960, **82**, 6218 (*synth*)
Zweifel, G. *et al*, *J. Am. Chem. Soc.*, 1967, **89**, 5086 (*use*)
Fisher, R.P., *Diss. Abstr. Int. B*, 1975, **35**, 4833 (*use*)
Hall, L.W., *Diss. Abstr. Int. B*, 1975, **36**, 2794 (*use*)
Apeloig, Y. *et al*, *J. Am. Chem. Soc.*, 1977, **99**, 1291 (*props*)
Bonacic, K.V. *et al*, *J. Am. Chem. Soc.*, 1977, **99**, 842 (*use*)

$C_2H_5BBr_2$ B-00085

Dibromoethylborane, 9CI

Ethylboron dibromide. Ethyldibromoborane

[17933-10-7]

$EtBBr_2$

M 199.680

Liq., air- and moisture-sensitive. d^{20} 1.771. Bp 95°.

Vahrenkamp, H. *et al*, *J. Organomet. Chem.*, 1968, **11**, 399; 1968, **12**, 23 (*synth, nmr, pmr*)
Köster, R. *et al*, *Justus Liebigs Ann. Chem.*, 1969, **719**, 169 (*synth*)
Storch, W. *et al*, *Synth. Inorg. Met.-Org. Chem.*, 1971, **1**, 197 (*synth*)
Haubold, G. *et al*, *Z. Anorg. Allg. Chem.*, 1976, **420**, 251 (*ir*)

$C_2H_5BCl_2$ B-00086

Dichloroethylborane, 10CI, 9CI

Ethylboron dichloride. Ethyldichloroborane

[1739-53-3]

$EtBCl_2$

M 110.778

Used to aid copolymerisations. Liq., sensitive to air and moisture. Bp 51.5°, Bp_{100} 25°.

▷Highly toxic, powerful irritant

Lengyel, B. *et al*, *Z. Anorg. Allg. Chem.*, 1963, **322**, 103 (*synth*)
Vahrenkamp, H. *et al*, *J. Organomet. Chem.*, 1968, **11**, 399; 1968, **12**, 23 (*synth, pmr*)
Köster, R. *et al*, *Justus Liebigs Ann. Chem.*, 1969, **719**, 169 (*synth*)
Storch, W. *et al*, *Synth. Inorg. Metal-Org. Chem.*, 1971, **1**, 197 (*synth*)
Haubold, G. *et al*, *Z. Anorg. Allg. Chem.*, 1976, **420**, 251 (*ir*)
Barlos, K. *et al*, *Chem. Ber.*, 1977, **110**, 2774 (*nmr*)
Leleu, J., *Cah. Notes Doc.*, 1977, **88**, 361 (*haz*)
Sax, N.I., *Dangerous Properties of Industrial Materials*, 5th Ed., Van Nostrand-Reinhold, 1979, 561, 623.

$C_2H_5BCl_2O$ B-00087

Dichloroethoxyborane, 10CI, 9CI, 8CI

Dichloroethoxyborine

[16339-28-9]

$EtOBCl_2$

M 126.777

Liq., sensitive to moisture. Bp 79-80°. ΔH_f −158.3 kcal mol^{-1}. $\Delta H_{Hydrolysis}$ −42.5 kcal mol^{-1}.

Gerrard, W. *et al*, *J. Chem. Soc.*, 1952, 1486 (*synth*)
Skinner, H.A. *et al*, *J. Chem. Soc.*, 1954, 3930 (*props*)
Abel, E.W. *et al*, *J. Chem. Soc.*, 1957, 501 (*synth, props*)
Onak, T.P. *et al*, *J. Phys. Chem.*, 1959, **63**, 1533 (*nmr*)
Gerrard, W. *et al*, *J. Inorg. Nucl. Chem.*, 1967, **29**, 943 (*synth*)
Lappert, M.F., *J. Chem. Soc. (A)*, 1970, 2320 (*props*)

$C_2H_5BF_2$ B-00088

Ethyldifluoroborane, 10CI, 9CI
Ethyldifluoroborine. Difluoroethylborane
[430-41-1]

$EtBF_2$

M 77.868

Prepd. from difluoroborane and ethylene or from Et_4Sn and BF_3. Air- and moisture-sensitive gas. Mp −101°. Bp −25°.

Py complex: [22405-14-7]. *Difluoroethylborane pyridine. Pyridine ethyldifluoroborane.*
$C_7H_{10}BF_2N$ M 156.970

Me_3N complex: [381-57-7].
$C_5H_{14}BF_2N$ M 136.979
Prepd. by addn. of Me_3N to $EtBF_2$. Solid. Sol. CH_2Cl_2. Mp 25.3°. pmr δ 2.88 (CH_3) ppm.

Brinckman, F.E. *et al*, *J. Am. Chem. Soc.*, 1960, **82**, 6218 (*synth*)
Coyle, T.D. *et al*, *J. Am. Chem. Soc.*, 1961, **83**, 4138 (*deriv*)
Steele, W.C. *et al*, *J. Am. Chem. Soc.*, 1962, **84**, 1154 (*ms*)
Coyle, T.D. *et al*, *Proc. Chem. Soc., London*, 1964, 25 (*synth*)
Köster, R. *et al*, *Justus Liebigs Ann. Chem.*, 1968, **717**, 1 (*deriv*)
Nöth, H. *et al*, *J. Organomet. Chem.*, 1968, **12**, 23 (*nmr*)
Barlos, K. *et al*, *Chem. Ber.*, 1977, **110**, 2774 (*nmr*)

$C_2H_5BI_2$ B-00089

Ethyldiiodoborane, 10CI, 9CI
Diiodoethylborane
[17933-07-2]

$EtBI_2$

M 293.681

Prepd. from tetraethylstannane and BI_3. Air-, moisture- and light-sensitive. Sol. CCl_4. pmr δ −1.59 (CH_2), −1.15 (CH_3) ppm.

Nöth, H. *et al*, *J. Organomet. Chem.*, 1968, **11**, 399 (*synth*)
Nöth, H. *et al*, *J. Organomet. Chem.*, 1968, **12**, 23 (*nmr*)
Haubold, W. *et al*, *Z. Anorg. Allg. Chem.*, 1978, **446**, 45 (*raman, ir*)

$C_2H_5BO_2$ B-00090

Dihydroxyvinylborane
Ethenylboronic acid, 10CI, 9CI. *Etheneboronic acid,* 8CI. *Vinylboronic acid. Ethenyldihydroxyborane*
[4363-34-2]

$H_2C{=}CHB(OH)_2$

M 71.870

Synthetic intermed. Cryst. (C_6H_6). Polymerises almost immediately in air. Sol. THF. Mp 80-4° (74°).

Polymer: Obt. spontaneously from monomer in air. Grey gum. Sol. v. dilute aq. alkali, THF, insol. other org. solvs.

Normant, H. *et al*, *C.R. Hebd. Seances Acad. Sci.*, 1959, **248**, 828 (*synth*)
Fr. Pat., 1 224 181, (*1960*); *CA*, **56**, 12944h (*synth*)
Seiji, K. *et al*, *Bull. Chem. Soc. Jpn.*, 1966, **39**, 2018 (*synth, polymers*)
Braun, J. *et al*, *Bull. Soc. Chim. Fr.*, 1966, 2557 (*synth*)
Solkan, V.N. *et al*, *Izv. Akad. Nauk SSSR, Ser. Khim.*, 1974, 2623; *Bull. Acad. Sci. USSR, Div. Chem. Sci.*, **23**, 2532 (*nmr*)
Kabalka, G.W. *et al*, *CA*, 1981, **85**, 132208v (*use*)

$C_2H_5B_3$ B-00091

1,5-Dicarbapentaborane(5), 10CI
[20693-66-7]

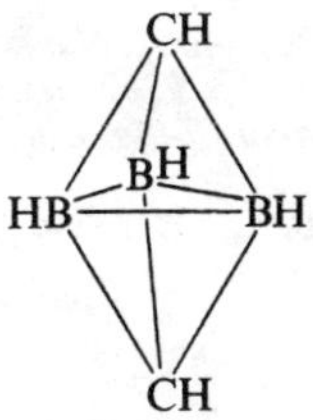

M 61.492

closo-. Colourless gas. Mp −126.4°. Bp −3.7°.

Shapiro, I. *et al*, *J. Am. Chem. Soc.*, 1962, **84**, 3837 (*ir, ms, nmr*)
Ditter, J. *et al*, *Inorg. Chem.*, 1970, **9**, 889 (*synth*)

$C_2H_5B_5Li_2$ B-00092

μ-2,4-Dicarbaheptaborane(7)-2,4-diyldilithium, 9CI
2,4-Dilithio-2,4-dicarbaheptaborane(7)
[38386-34-4]

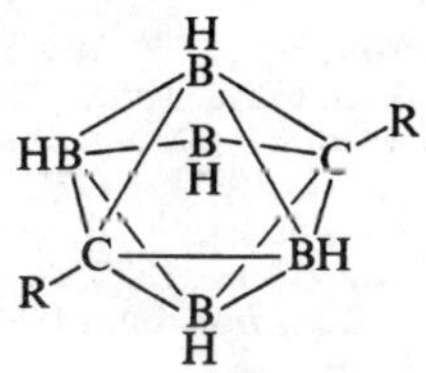

R = Li

M 96.994

closo-. Precursor for the preparation of σ-substituted derivatives eg. for polymer formation. Sol. Et_2O, C_6H_6.

Kesting, R. *et al*, *J. Appl. Polymer Sci.*, 1970, **14**, 2525 (*synth*)
Olsen, R. *et al*, *J. Am. Chem. Soc.*, 1970, **92**, 5072 (*synth*)

C_2H_6BBr B-00093

Bromodimethylborane, 10CI, 9CI, 8CI
Bromodimethylborine. Dimethylbromoborane
[5158-50-9]

Me_2BBr

M 120.783

Air- and moisture-sensitive liq. Bp_{720} 30°.

Schabacher, W. *et al*, *Z. Anorg. Allg. Chem.*, 1958, **294**, 183 (*synth*)
Nöth, H. *et al*, *J. Organomet. Chem.*, 1968, **11**, 399; **12**, 23 (*synth, pmr*)
Nöth, H. *et al*, *Chem. Ber.*, 1974, **107**, 3089 (*pmr*)
Wrackmeyer, B. *et al*, *Chem. Ber.*, 1976, **109**, 1075 (*pmr*)
Haubold, W. *et al*, *Z. Anorg. Allg. Chem.*, 1976, **420**, 251 (*synth*)
Gragg, B.R. *et al*, *J. Organomet. Chem.*, 1977, **132**, 29 (*cmr*)

$C_2H_6BBr_2N$ B-00094

Dibromo(dimethylamino)borane, 8CI

1,1-Dibromo-N,N-dimethylboranamine, 10CI, 9CI. *Dibromo(dimethylamino)borine*

[7360-64-7]

Br_2BNMe_2

M 214.694

Moisture-sensitive liq. Bp_1 49-51°.

Dimer: [7392-26-9].

$C_4H_{12}B_2Br_4N_2$ M 429.388

Characterized spectroscopically.

Banister, A.J. *et al, J. Chem. Soc.*, 1964, 995 (*synth, ir*)
Nöth, H. *et al, Chem. Ber.*, 1967, **100**, 3353 (*dimer, nmr*)
Beck, W. *et al, Chem. Ber.*, 1972, **105**, 2883 (*nmr*)
King, G.H. *et al, Faraday Discuss. Chem. Soc.*, 1972, **54**, 70 (*pe*)
Bryukhova, E.V. *et al, Izv. Akad. Nauk SSSR, Ser. Khim.*, 1976, 1614 (*Engl. trans. p. 1529*) (*nmr*)
Haubold, W. *et al, Z. Anorg. Allg. Chem.*, 1976, **421**, 105 (*synth, nmr*)
Nöth, H. *et al, Chem. Ber.*, 1981, **114**, 1150 (*nmr*)

$C_2H_6BBr_3S$ B-00095

Tribromoborane dimethyl sulfide

Tribromo[thiobis[methane]]boron, 10CI. *Thiobis[methane] compd. with tribromoborane(1:1),* 9CI. *Methyl sulfide compd. with boron bromide(BBr_3)(1:1),* 8CI

[29957-59-3]

$Me_2S—BBr_3$

M 312.651

Convenient means of handling the otherwise highly moisture-sensitive compd., BBr_3. White amorph. powder, stable to brief exp. to air or long-term storage under N_2. Mp 108°.

Schmidt. M. *et al, Chem. Ber.*, 1970, **103**, 3705 (*pmr*)
Peach, M.E., *Inorg. Nucl. Chem. Lett.*, 1971, **7**, 75.
Bula, M.J. *et al, J. Chem. Soc., Dalton Trans.*, 1973, 1047 (*nmr, synth*)
Kinberger, K. *et al, Z. Naturforsch., B*, 1975, **30**, 55 (*synth*)
Brown, H.C. *et al, Inorg. Chem.*, 1977, **16**, 2938 (*synth, pmr*)
Williard, P.G. *et al, Tetrahedron Lett.*, 1980, **21**, 3731.

C_2H_6BCl B-00096

Chlorodimethylborane, 10CI, 9CI

Chlorodimethylborine. Dimethylchloroborane

[1803-36-7]

Me_2BCl

M 76.332

Not easily separable from dichloromethylborane, generally formed in admixture with it. Easily liquified gas, sensitive to air and moisture. Bp 5°.

▷Prob. pyrophoric in air

Becher, H.J., *Z. Anorg. Allg. Chem.*, 1953, **271**, 243 (*synth, ir*)
Barnes, R.L. *et al, J. Am. Chem. Soc.*, 1963, **85**, 360 (*synth*)
de Moor, J.D. *et al, J. Organomet. Chem.*, 1966, **6**, 235 (*nmr*)
Haubold, W. *et al, Z. Anorg. Allg. Chem.*, 1976, **420**, 251 (*synth, ir*)
Wrackmeyer, B. *et al, Chem. Ber.*, 1976, **109**, 1075 (*nmr*)
Barlos, K. *et al, Chem. Ber.*, 1977, **110**, 2774 (*nmr*)

$C_2H_6BClO_2$ B-00097

Chlorodimethoxyborane, 10CI, 9CI

Chlorodimethoxyborine

[868-81-5]

$(MeO)_2BCl$

M 108.331

V. easily hydrolysed by moisture. Not stable for prolonged periods.

Lehmann, W.J. *et al, J. Chem. Phys.*, 1959, **30**, 1219 (*synth, ir*)
de Moor, J.D. *et al, J. Organomet. Chem.*, 1966, **6**, 235 (*nmr*)
Nöth, H. *et al, Chem. Ber.*, 1966, **99**, 1049 (*nmr*)
de Moor, J.D. *et al, J. Organomet. Chem.*, 1967, **9**, 31 (*ir*)
Katz, J.J. *et al, J. Org. Chem.*, 1974, **39**, 2817.
Barlos, K. *et al, Chem. Ber.*, 1977, **110**, 2774 (*nmr*)

$C_2H_6BCl_3S$ B-00098

Trichloroborane dimethyl sulfide

Trichloro[thiobis[methane]]boron, 10CI. *Thiobis[methane] compd. with trichloroborane(1:1),* 9CI. *Methyl sulfide compd. with boron chloride(BCl_3)(1:1),* 8CI

[5523-19-3]

$Me_2S—BCl_3$

M 179.298

Convenient means of handling trichloroborane, which is otherwise much more volatile and moisture-sensitive. White solid, stable on brief exp. to air or long-term storage under N_2. Mp 86-7° (85-90°). Bp 181.9°.

Morris, H.L. *et al, Inorg. Chem.*, 1966, **5**, 124.
Schmidt, M. *et al, Chem. Ber.*, 1970, **103**, 3705 (*pmr*)
Bula, M.J. *et al, J. Chem. Soc., Dalton Trans.*, 1973, 1047 (*nmr, synth*)
Brown, H.C. *et al, Inorg. Chem.*, 1977, **16**, 2938 (*synth, pmr*)
Williard, P.G. *et al, Tetrahedron Lett.*, 1980, **21**, 3731.
Glavincevski, B. *et al, Can. J. Chem.*, 1981, **59**, 3012 (*nmr*)

C_2H_6BF B-00099

Fluorodimethylborane, 10CI, 9CI, 8CI

Dimethylfluoroborane. Fluorodimethylborine

[353-46-8]

$FBMe_2$

M 59.878

Prepd. from Me_2BCl and SbF_3 or from BF_3 and Me_2BNH_2. Air- and moisture-sensitive gas. Sol. $CHCl_3$. Dipole moment 1.34D. ^{11}Bnmr δ 42 ppm.

Becher, H.J., *Z. Anorg. Allg. Chem.*, 1957, **291**, 151 (*raman*)
Nöth, H. *et al, Chem. Ber.*, 1974, **107**, 3089 (*nmr*)
Fuss, W. *et al, CA*, 1975, **82**, 138775f (*uv*)
Berger, H.O. *et al, Chem. Ber.*, 1976, **109**, 2266 (*pe*)
Haubold, W. *et al, Z. Anorg. Allg. Chem.*, 1976, **420**, 251 (*synth*)
Barlos, K. *et al, Chem. Ber.*, 1977, **110**, 2774 (*nmr*)

$C_2H_6BFO_2$ B-00100

Fluorodimethoxyborane, 10CI, 9CI, 8CI

Dimethoxyfluoroborane. Fluorodimethoxyborine

[367-46-4]

$(MeO)_2BF$

M 91.877

Prepd. from tetramethoxysilane and BF_3 at −80°. Stable as a gas, but on condensation undergoes partial redistribution at −30°, giving F_2BOMe and $B(OMe)_3$. Sol. THF. Bp 52.7°. Dipole moment 1.91D.

Wiberg, E. *et al, Z. Naturforsch., B*, 1953, **8**, 608 (*synth*)
de Moor, J.E. *et al, J. Organomet. Chem.*, 1966, **6**, 235 (*nmr*)

Burya, G.F. *et al*, *Zh. Fiz. Khim.*, 1970, **44**, 1596 (*ir*)
Lappert, M.F. *et al*, *Colloq. Int. Cent. Nat. Rech. Sci.*, 1970, **191**, 241; *CA*, **73**, 103772g (*props*)
Rauchschwalbe, G. *et al*, *Helv. Chim. Acta*, 1975, **58**, 1094 (*use*)
Hoffmann, R.W. *et al*, *J. Org. Chem.*, 1981, **46**, 1309 (*use*)

$C_2H_6BF_3O$ B-00101

Trifluoroborane dimethyl ether

Trifluoro[oxybis[methane]]boron, 10CI. *Oxybis[methane] compd. with trifluoroborane(1:1)*, 9CI. *Methyl ether compd. with boron fluoride(BF_3)(1:1)*, 8CI. *Trifluoroborane compd. with oxybis[methane](1:1)*. *Boron trifluoride dimethyl etherate*

[353-42-4]

$Me_2O—BF_3$

M 113.874

Catalyst for epoxide reactions. Comly. available. Moisture-sensitive liq. d 1.239. Mp −14°. Bp 126-7°. n_D^{20} 1.3079. Dipole moment 7.6D.

▷Corrosive, flammable. ED8400000.

Derouault, J. *et al*, *Spectrochim. Acta, Part A*, 1972, **28**, 359 (*raman, ir*)
Fratiello, A. *et al*, *J. Org. Chem.*, 1972, **37**, 2237 (*nmr*)
Fratiello, A. *et al*, *J. Chem. Soc., Perkin Trans. 2*, 1975, 1415 (*cmr*)
Borovikov, Y.Y. *et al*, *Zh. Obshch. Khim.*, 1980, **50**, 2531.
Umeyama, H., *Chem. Pharm. Bull.*, 1980, **28**, 1633 (*struct*)
Hirota, F. *et al*, *J. Mol. Struct.*, 1981, **70**, 305 (*struct*)

$C_2H_6BF_3O$ B-00102

Trifluoroborane ethanol

(Ethanol)trifluoroboron, 10CI. *Ethanol compd. with trifluoroborane(1:1)*, 9CI. *Ethyl alcohol compd. with boron fluoride(1:1)*, 8CI

[353-41-3]

$EtOH—BF_3$

M 113.874

Diehl, P. *et al*, *Nature (London)*, 1957, **180**, 1114 (*synth, nmr*)
Babushkin, A.A., *Izv. Akad. Nauk SSSR, Ser. Khim.*, 1958, 1131 (*ir, raman*)
Babushkin, A.A. *et al*, *Optika i Spectroskopiya*, 1958, **5**, 256; *CA*, **53**, 1919g (*struct*)
Diehl, P., *Helv. Phys. Acta*, 1958, **31**, 685 (*nmr*)
Burya, G.F. *et al*, *Zh. Fiz. Khim.*, 1975, **49**, 1227 (*synth, ir*)

$C_2H_6BF_4Tl$ B-00103

Dimethyl(tetrafluoroborato)thallium

Dimethylthallium tetrafluoroborate

Me_2TlBF_4

M 321.256

Solid (THF).

▷Toxic

Shies, G.D. *et al*, *J. Organomet. Chem.*, 1966, **5**, 330 (*synth, pmr, ir*)
Burke, P.J. *et al*, *J. Chem. Res. (S)*, 1981, 124 (*pmr, cmr, nmr*)

C_2H_6BI B-00104

Iododimethylborane, 10CI, 9CI, 8CI

Dimethyliodoborane

[17933-09-4]

Me_2BI

M 167.784

Prepd. from BI_3 and $SnMe_4$. Air- and moisture-sensitive liq. Sol. THF. Bp_{720} 65°. ^{11}Bnmr δ 79.1 ppm.

Nöth, H. *et al*, *J. Organomet. Chem.*, 1968, **11**, 399 (*synth*)
Nöth, H. *et al*, *J. Organomet. Chem.*, 1968, **12**, 23 (*nmr*)
Riegel, F. *et al*, *Z. Naturforsch., B*, 1974, **29**, 719 (*use*)
Siebert, W. *et al*, *Z. Anorg. Allg. Chem.*, 1975, **418**, 273 (*use*)
Haubold, W. *et al*, *Z. Anorg. Allg. Chem.*, 1978, **446**, 45 (*ir*)
Haubold, W. *et al*, *Z. Naturforsch., B*, 1978, **33**, 140 (*synth*)

C_2H_6BN B-00105

Borane isocyanomethane

Trihydro(isocyanomethane)boron, 10CI. *Methyl isocyanide borane*

[62630-46-0]

$MeNC—BH_3$

M 54.886

Dipole moment 6.73D.

Stevans, J.F. *et al*, *J. Am. Chem. Soc.*, 1977, **99**, 1442 (*struct, microwave, synth*)
Umeyama, H. *et al*, *Chem. Pharm. Bull.*, 1980, **28**, 2279.
Bevan, J.W. *et al*, *Spectrochim. Acta, Part A*, 1981, **37**, 601 (*pe*)
Watari, F., *Inorg. Chem.*, 1982, **21**, 1442 (*ir, raman*)

$C_2H_6BO_2^{\ominus}$ B-00106

(Acetato-*O*)trihydroborate(1−), 10CI

Acetoxytrihydroborate. Acetoxyborohydride

$AcOBH_3^{\ominus}$

M 72.878 (ion)

Na salt: [63947-71-7].
$C_2H_6BO_2Na$ M 95.868
Reducing agent. Usually prepd. *in situ*.

Japan. Pat., 77 36 606, (*1977*); *CA*, **87**, 134467n (*use*)
Umino, N. *et al*, *Chem. Pharm. Bull.*, 1978, **26**, 2897 (*synth, props*)
Japan. Pat., 79 70 262, (*1979*); *CA*, **91**, 211258g

$C_2H_6B_4$ B-00107

1,2-Dicarbahexaborane(6)

[20693-68-9]

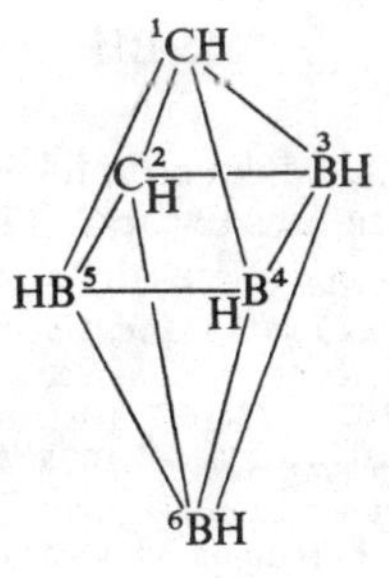

M 73.309

closo-. Colourless liq. Isom. at 250° to 1,6-Dicarbahexaborane(6), B-00108 .

Shapiro, I. *et al*, *J. Am. Chem. Soc.*, 1963, **85**, 3167 (*synth, nmr, ms*)
Onak, T. *et al*, *Inorg. Chem.*, 1964, **3**, 1686 (*nmr*)
Beaudet, R. *et al*, *J. Chem. Phys.*, 1970, **53**, 1899 (*microwave*)

$C_2H_6B_4$ B-00108

1,6-Dicarbahexaborane(6), 10CI

[20693-67-8]

M 73.309
closo-. Colourless liq. Mp −29.5° (−32.6°). Bp 22.7°.

Shapiro, I. *et al*, *J. Am. Chem. Soc.*, 1963, **85**, 3167 (*synth, ir, nmr*)
Ditter, J. *et al*, *Inorg. Chem.*, 1970, **9**, 889 (*synth*)

$C_2H_6B_4Pb$ B-00109

2,3-Dicarba-1-plumbaheptaborane(6), 10CI
[63181-05-5]

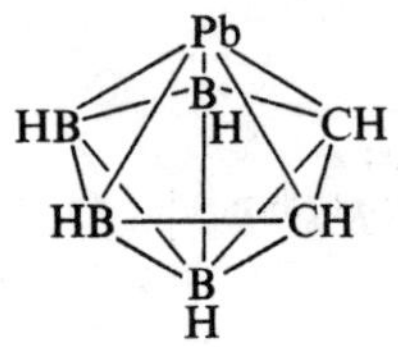

M 280.509
closo-. Light-yellow solid. Sol. CH_2Cl_2. Dec. at r.t. but stable at 0°. Subl. at 20°/0.0001 mm.

Wong, K. *et al*, *Inorg. Chem.*, 1977, **16**, 2053 (*synth, ir, nmr, ms*)

$C_2H_6B_4Sn$ B-00110

2,3-Dicarba-1-stannaheptaborane(6), 10CI
[63181-06-6]

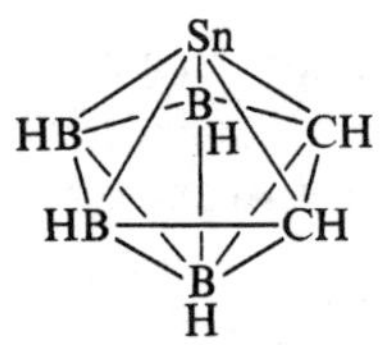

M 191.999
closo-. Off-white solid. Sol. CH_2Cl_2. Subl. at 70-120°/0.0001 mm onto cold finger.

Wong, K. *et al*, *Inorg. Chem.*, 1977, **16**, 2053 (*synth, ir, nmr, ms*)

C_2H_7B B-00111

Dimethylborane, 10CI, 9CI
Dimethylborine
[7216-97-9]

Me_2BH

M 41.887
Exists as the H-bridged dimer in the uncomplexed state. Air- and moisture-sensitive. Sol. THF.

Nöth, H. *et al*, *Chem. Ber.*, 1974, **107**, 3089 (*nmr*)
Köster, R. *et al*, *Justus Liebigs Ann. Chem.*, 1975, 1339 (*props*)
Spielvogel, B.F. *et al*, *J. Am. Chem. Soc.*, 1975, **97**, 1609 (*nmr*)
U.S.P., 4 012 230, (*1977*); *CA*, **87**, 10124d (*synth*)
Dewar, M.J.S. *et al*, *Inorg. Chem.*, 1978, **17**, 1075 (*use*)
Pelter, A. *et al*, *Comprehensive Organic Chemistry*, (*Barton, D.H.R. et al*, Ed.), Pergamon, Oxford, 1979, **3**, 695 (*rev*)

C_2H_7B B-00112

Ethylborane, 10CI, 9CI
Ethylborine
[25070-50-2]

$EtBH_2$

M 41.887
Exists as H-bridged dimer in the uncomplexed state. Air- and moisture-sensitive. Sol. THF. Easily disproportionates to borane, diethylborane etc.

Ph_3P complex: Ethylborane triphenylphosphine.
$C_{20}H_{22}BP$ M 304.178
Formed in redn. of triphenylphosphine oxide by ethylborane. Sol. THF. Mp 115°.

Et_2PH complex: [22864-44-4]. *Diethylphosphine ethylborane. Ethylborane diethylphosphine.*
$C_6H_{18}BP$ M 131.992
Prepd. from Et_2PH and R_3NBEtH_2. Sol. THF.

Dipropylphosphine complex: [22864-45-5]. *Ethylborane dipropylphosphine.*
$C_8H_{22}BP$ M 160.046
Sol. THF.

B.P., 823 327, (*1959*); *CA*, **54**, 12995 (*synth*)
Falbe, J. *et al*, *Chem. Ber.*, 1965, **98**, 2312 (*deriv*)
Jugie, G. *et al*, *C.R. Hebd. Seances Acad. Sci.*, 1969, **268**, 1377 (*derivs*)
Steck, S.J. *et al*, *J. Phys. Chem.*, 1969, **73**, 1000 (*synth*)
Gallais, F. *et al*, *J. Chim. Phys. Phys. Chim. Biol.*, 1970, **67**, 934 (*deriv*)
Köster, R. *et al*, *Am. Chem. Soc. Symp. Ser.*, 1976, **39**, 1 (*rev*)

$C_2H_7BBr_2S$ B-00113

Dibromoborane dimethyl sulfide
Dibromohydro[thiobis[methane]]boron, 10CI
[55671-55-1]

$HBBr_2—SMe_2$

M 233.755
Hydroborating agent for alkenes and alkynes with a high regiospecifity of hydroboration; provides a route to alkyldibromoboranes and alkyldihydroxyboranes. Air- and moisture-sensitive solid. Mp 30-3°.

Kinberger, K. *et al*, *Z. Naturforsch.*, *B*, 1975, **30**, 55.
Brown, H.C. *et al*, *Inorg. Chem.*, 1977, **16**, 2938 (*synth, nmr*)
Brown, H.C. *et al*, *J. Am. Chem. Soc.*, 1977, **99**, 7097.
Brown, H.C. *et al*, *J. Org. Chem.*, 1980, **45**, 384, 389.

$C_2H_7BCl_2S$ B-00114

Dichloroborane dimethyl sulfide
Dichlorohydro[thiobis[methane]]boron, 10CI. Dichloroborane methyl sulfide
[63462-42-0]

$Cl_2BH—SMe_2$

M 144.853
Stable hydroborating agent with high reaction regiospecificity. Clear viscous liq., remarkably stable though slowly dec. by moisture.

Brown, H.C. *et al*, *Inorg. Chem.*, 1977, **16**, 2938 (*synth, nmr*)
Brown, H.C. *et al*, *J. Org. Chem.*, 1977, **42**, 2533; 1980, **45**, 384 (*use*)
Dodds, A.R. *et al*, *Inorg. Chem.*, 1979, **18**, 1465 (*synth, nmr*)

$C_2H_7BCl_3N$ B-00115

Trichloroborane dimethylamine
Trichloro(N-methylmethanamine)boron, 10CI. N-Methylmethanamine compd. with trichloroborane(1:1), 9CI. Dimethylamine boron trichloride. Dimethylamine trichloroborane
[13072-44-1]

$Me_2NH—BCl_3$

M 162.253
Fuming white solid or cryst. ($CHCl_3$), relatively stable to short air exp.

Kiesel, R. *et al*, *Inorg. Chem.*, 1973, **12**, 1090 (*ir*)

Myers, W.H. *et al*, *Inorg. Chem.*, 1975, **14**, 2874 (*pmr*)
Ryschkewitsch, G.E. *et al*, *Synth. React. Inorg. Metal-Org. Chem.*, 1975, **5**, 123 (*synth*)
Myers, W.H. *et al*, *Inorg. Chem.*, 1978, **17**, 1157 (*nmr*)
Haubold, W. *et al*, *Chem. Ber.*, 1980, **113**, 2333 (*synth*)

$C_2H_7BF_3N$ B-00116

Trifluoroborane ethylamine

(Ethanamine)trifluoroboron, 10CI. *Ethanamine compd. with trifluoroborane*, 9CI. *Ethylamine compd. with boron fluoride* (BF_3)*(1:1)*, 8CI. *Boron trifluoride ethylamine. Ethylamminoboron trifluoride*

[75-23-0]

$EtNH_2—BF_3$

M 112.889

Widely used as a catalyst in industrial processes. Cryst. Et_2O, dec. rapidly when heated in air above Mp. Sol. C_6H_6, sl. sol. Et_2O. Mp 75-6°, 89°. Bp_1 147-8°.

▷KQ8060000.

Kraus, C.A. *et al*, *J. Am. Chem. Soc.*, 1929, **51**, 2690 (*synth*)
Gallais, F. *et al*, *CA*, 1960, **54**, 16064d (*struct*, *mord*)
Spitsyn, V.I. *et al*, *Dokl. Akad. Nauk SSSR*, *Ser. Sci. Khim.*, 1965, **160**, 1101 (*synth*)
Meek, D.W. *et al*, *Inorg. Chem.*, 1966, **5**, 445 (*nmr*)
Elter, G. *et al*, *J. Organomet. Chem.*, 1972, **36**, 257.
Barber, M. *et al*, *J. Chem. Soc.*, *Faraday Trans. 2*, 1973, **69**, 551 (*pe*)

$C_2H_7BN_2$ B-00117

Cyanoborane methylamine

(Cyano-C)dihydro(methanamine)boron, 10CI. *Boranecarbonitrile, compd. with methanamine(1:1)*

[66632-42-6]

$MeNH_2—BH_2CN$

M 69.901

Sol. Et_2O. All attempts to crystallise or sublime this compd. were unsuccessful.

Wisian-Neilson, P. *et al*, *Inorg. Chem.*, 1978, **17**, 2327 (*synth*, *pmr*, *ir*)

C_2H_7BO B-00118

Hydroxydimethylborane

Dimethylborinic acid, 10CI, 9CI. *Dimethylboric acid. Dimethylhydroxyborane*

[13061-97-7]

Me_2BOH

M 57.887

Liq. Bp_{36} 0°.

Burg, A.B. *et al*, *J. Am. Chem. Soc.*, 1940, **62**, 2228; 1953, **75**, 3872 (*synth*)
Goubeau, J. *et al*, *Z. Phys. Chem. (Wiesbaden)*, 1959, **20**, 15 (*ir*, *raman*)
de Moor, J.E. *et al*, *J. Organomet. Chem.*, 1967, **9**, 3 (*ir*)
Vahrenkamp, H. *et al*, *J. Organomet. Chem.*, 1968, **12**, 23 (*pmr*)
Sax, N.I., *Dangerous Properties of Industrial Materials*, 5th Ed., Van Nostrand-Reinhold, 1979, 600.

$C_2H_7BO_2$ B-00119

Dimethoxyborane, 10CI, 9CI

Dimethoxyborine. Boronic acid dimethyl ester

[4542-61-4]

$HB(OMe)_2$

M 73.886

Prepd. from diborane and MeOH. Colourless liq., rapidly hydrol. in air, rapidly dec. at high temp. Sol. THF. Bp 25.9°. V.p. 275 mm at 0°. $\Delta H_{f(liq.)}$ −145.3 $Kcalmol.^{-1}$.

Burg, A. *et al*, *J. Am. Chem. Soc.*, 1933, **55**, 4020 (*synth*)
Law, R.W. *et al*, *J. Chem. Phys.*, 1956, **25**, 1086 (*ms*)
Onak, T.P. *et al*, *J. Phys. Chem.*, 1959, **63**, 1533 (*pmr*)
Uchida, H.S. *et al*, *J. Phys. Chem.*, 1959, **63**, 1414 (*props*)
Nöth, H. *et al*, *Chem. Ber.*, 1974, **107**, 3089 (*nmr*)
Jeffers, P.M. *et al*, *Inorg. Chem.*, 1981, **20**, 1698 (*synth*)

$C_2H_7BO_2$ B-00120

Ethyldihydroxyborane

Ethylboronic acid, 10CI, 9CI. *Ethylboric acid. Ethaneboronic acid*

[4433-63-0]

$EtB(OH)_2$

M 73.886

Fungicide and fuel additive. Acetylcholinesterase inhibitor. Useful for protection of diols. Cryst. (Et_2O). Subl. at 40°.

Khotinsky, E. *et al*, *Ber.*, 1909, **42**, 3095 (*synth*)
Ger. Pat., 1 111 181, (*1957*); *CA*, **56**, 8742 (*synth*)
Washburn, R.M. *et al*, *Adv. Sci. Ser.*, 1961, **32**, 208 (*synth*)
Nöth, H. *et al*, *J. Organomet. Chem.*, 1968, **12**, 23 (*pmr*)
Jarv, J. *et al*, *Bioorg. Khim.*, 1978, **4**, 1364 (*pharmacol*)

$C_2H_7B_3$ B-00121

1,2-Dicarbapentaborane(7), 10CI

[26249-71-8]

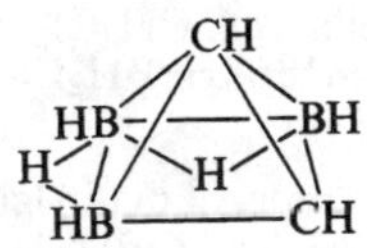

M 63.507

nido-. Colourless liq. by vacuum fractionation. Polymerises at 50°.

Franz, D. *et al*, *J. Am. Chem. Soc.*, 1970, **92**, 1438 (*synth*, *ir*, *nmr*)

$C_2H_7B_3Si$ B-00122

1-Silyl 1,5-dicarbapentaborane(5), 9CI

[55906-09-7]

$1,5-C_2B_3H_4SiH_3$

M 91.593

Thompson, M.L. *et al*, *Inorg. Chem.*, 1972, **11**, 1925 (*ir*)
Leach, J.B. *et al*, *J. Chem. Soc.*, *Dalton Trans.*, 1975, 1018 (*synth*, *nmr*, *ms*)

$C_2H_7B_5$ B-00123

2,4-Dicarbaheptaborane(7), 10CI

[20693-69-0]

As μ-2,4-Dicarbaheptaborane(7)-2,4-diyldilithium, B-00092 with

R = H

M 85.127

closo-. Parent compd. incorporated into high temp. polymers as silicon derivs. Colourless liq. by vacuum fractionation.

Beaudet, R. *et al*, *J. Chem. Phys.*, 1965, **43**, 2166 (*struct, microwave*)
Onak, T. *et al*, *J. Am. Chem. Soc.*, 1966, **88**, 4622 (*nmr, ir*)
Ditter, J. *et al*, *Inorg. Chem.*, 1968, **7**, 1748; 1970, **9**, 899 (*synth*)
Jotham, R. *et al*, *J. Chem. Soc. (A)*, 1971, 3181 (*ir, raman*)
Onak, T. *et al*, *J. Chem. Soc., Dalton Trans.*, 1973, 76 (*ms*)
Ditter, J. *et al*, *Inorg. Chem.*, 1976, **15**, 1063 (*nmr*)

C_2H_8BBrS B-00124

Bromoborane dimethyl sulfide

Bromodihydro[thiobis[methane]]boron, 10CI. *Bromoborane compd. with thiobis[methane](1:1)*, 9CI

[55652-52-3]

$Me_2S—BH_2Br$

M 154.859

V. useful, stable reagent for selective hydroborations and prepn. of dialkylbromoboranes. Clear liq., sensitive to moisture. Sol. many aprotic solvs., particularly ethers and chlorinated hydrocarbons.

Kinberger, K. *et al*, *Z. Naturforsch., B*, 1975, **30**, 55 (*synth*)
Brown, H.C. *et al*, *Synthesis*, 1977, 695 (*synth*)
Brown, H.C. *et al*, *J. Org. Chem.*, 1979, **44**, 2417, 2422.
Wisian-Neilson, P. *et al*, *J. Inorg. Nucl. Chem.*, 1979, **41**, 1545 (*synth*)
Foret, C.J. *et al*, *J. Inorg. Nucl. Chem.*, 1979, **41**, 1661 (*synth*)

C_2H_8BClS B-00125

Chloroborane dimethyl sulfide

Chlorodihydro[thiobis(methane)]boron, 10CI. *Chloroborane compd. with thiobis(methane)(1:1)*

[63348-81-2]

$Me_2S—BH_2Cl$

M 110.408

A useful source for prepn. of dialkylchloroboranes and dialkyl(hydroxy)boranes. Colourless viscous liq., indefinite stability at r.t. in the absence of moisture; hydrol. by H_2O. As usually prepared, contains a few per cent of BH_3SMe_2 and $BHCl_2SMe_2$.

Brown, H.C. *et al*, *Inorg. Chem.*, 1977, **16**, 2938 (*synth, pmr, nmr*)
Brown, H.C. *et al*, *J. Org. Chem.*, 1977, **42**, 2533 (*synth*)
Kulkarni, S.V. *et al*, *J. Org. Chem.*, 1979, **44**, 1747.
Brown, H.C. *et al*, *J. Org. Chem.*, 1979, **44**, 2417, 2422; 1980, **45**, 4540.
Paget, W.E. *et al*, *J. Chem. Soc., Chem. Commun.*, 1980, 1169 (*synth*)

$C_2H_8BF_2N$ B-00126

Difluoroborane dimethylamine

*Difluorohydro(*N*-methylmethanamine)boron*, 10CI. *Difluoroborane* N*-methylmethanamine. Dimethylamine difluoroborane*

[42525-42-8]

$F_2BH—NHMe_2$

M 94.899

White solid below −50°; phys. state not recorded at r.t. Sol. CCl_4, CH_2Cl_2.

Tuchagues, J.P. *et al*, *J. Organomet. Chem.*, 1973, **54**, 69 (*synth, ir, nmr*)
Van Paasschen, J.M. *et al*, *Inorg. Chem.*, 1978, **17**, 3302.

C_2H_8BIS B-00127

Iodoborane dimethyl sulfide

Dihydroiodo[thiobis[methane]]boron, 10CI. *Iodoborane complex with thiobis[methane]*, 9CI. *Iodoborane methyl sulfide. Dimethyl sulfide iodoborane. MIBS*

[55652-50-1]

$IBH_2—SMe_2$

M 201.860

Synth. from borane dimethylsulfide, iodine and triodoborane. Useful hydroborating agent for the synth. of dialkyliodoboranes. Stable cryst., slowly hydrol. in moist air.

Kinberger, K. *et al*, *Z. Naturforsch., B*, 1975, **30**, 55 (*synth, nmr*)
Brown, H.C. *et al*, *J. Org. Chem.*, 1979, **44**, 2417, 2422 (*use*)

C_2H_8BN B-00128

Aminodimethylborane, 8CI

1,1-Dimethylboranamine, 10CI, 9CI. *Aminodimethylborine*

[15973-91-8]

Me_2BNH_2

M 56.902

Air- and moisture-sensitive.

Dimer: [30442-90-1].
$C_4H_{16}B_2N_2$ M 113.804
Characterised spectroscopically.

Hoffmann, R., *J. Chem. Phys.*, 1964, **40**, 2474.
Chakrabarty, M.R. *et al*, *Inorg. Chem.*, 1967, **6**, 518.
Nöth, H. *et al*, *Chem. Ber.*, 1967, **100**, 3353 (*dimer*)
Nöth, H. *et al*, *J. Organomet. Chem.*, 1968, **12**, 23 (*pmr*)
Nöth, H. *et al*, *Chem. Ber.*, 1981, **114**, 1150 (*nmr*)

C_2H_8BN B-00129

(Dimethylamino)borane, 8CI

N,N-*Dimethylboranamine*, 10CI, 9CI. *(Dimethylamino)borine*

[1838-13-7]

Me_2NBH_2

M 56.902

Dimeric at r.t., dissociates at higher temps. Reducing agent. Used in photographic development and nickel plating.

Dimer: [26085-03-0].
$C_4H_{16}B_2N_2$ M 113.804
Solid by subl. Mp 74.3-74.5°. V.p. 9.1mm at 23°.

Hoffmann, R., *J. Chem. Phys.*, 1964, **40**, 2474.
Nöth, H. *et al*, *Chem. Ber.*, 1967, **100**, 3353 (*nmr*)
Spielman, J.R., *J. Chem. Educ.*, 1970, **47**, 225 (*synth*)
Bock, H. *et al*, *Chem. Ber.*, 1971, **104**, 1687 (*pe*)
Burg, A.B., *Inorg. Chem.*, 1973, **12**, 1448.
Nöth, H. *et al*, *Inorg. Chem.*, 1982, **21**, 706 (*struct*)

C_2H_8BP B-00130

(Dimethylphosphino)borane, 10CI, 9CI, 8CI

Boryldimethylphosphine, 9CI. *(Dimethylphosphino)borine*

[43625-73-6]

Me_2PBH_2

M 73.869

Prepd. by thermolysis of $Me_2PH\text{-}BH_3$. Unstable with respect to trimer, tetramer and not isol. in monomeric form.

Trimer:
$C_6H_{24}B_3P_3$ M 221.607
Sol. THF. Mp 86°. Bp 235° (est.).
Tetramer:
$C_8H_{32}B_4P_4$ M 295.476
Sol. THF. Mp 161°. Bp 310° (est.).
Homopolymer: [69773-26-8]. Used in electrophotographic plate insulating layers.

Burg, A.B. *et al*, *J. Am. Chem. Soc.*, 1953, **75**, 3872 (*synth*)
Florin, R.E. *et al*, *J. Am. Chem. Soc.*, 1954, **76**, 3344 (*ms*)
Hamilton, W.C. *et al*, *Acta Crystallogr.*, 1955, **8**, 199 (*struct*)
Burg, A.B. *et al*, *J. Inorg. Nucl. Chem.*, 1959, **11**, 258.
Wall, L.A. *et al*, *J. Res. Nat. Bur. Stand.*, 1959, **63**, 63 (*ms*)
Nöth, H. *et al*, *Z. Naturforsch., B*, 1960, **15**, 327 (*synth, use*)

$C_2H_8B_4$ B-00131

2,3-Dicarbahexaborane(8)
[18972-20-8]

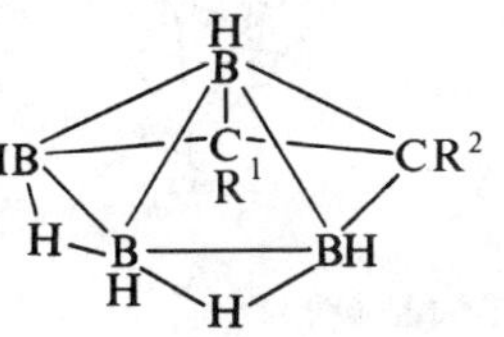

$R^1=R^2=H$

M 75.325
closo-. Colourless liq. Bp 56.9°.

Onak, T. *et al*, *J. Am. Chem. Soc.*, 1962, **84**, 2830 (*nmr*)
Streib, W. *et al*, *J. Am. Chem. Soc.*, 1963, **85**, 2331 (*struct*)
Onak, T. *et al*, *Inorg. Chem.*, 1966, **5**, 439 (*synth, ir, nmr*)

$C_2H_8B_6$ B-00132

1,7-Dicarbaoctaborane(8), 10CI
[23732-90-3]

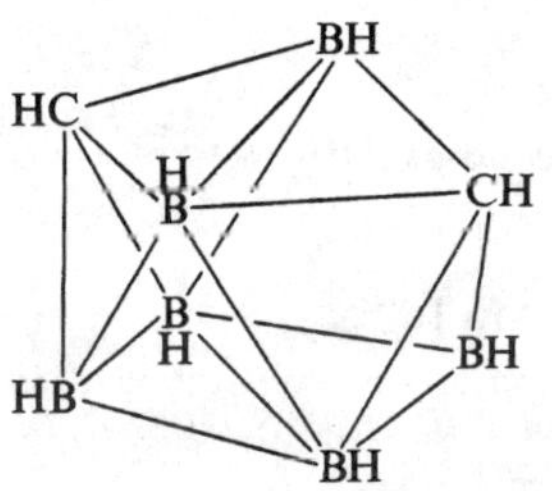

M 96.945
closo-. Sol. pentane. Mp 30°. Purif. by vacuum dist.

Garrett, P. *et al*, *Inorg. Chem.*, 1969, **8**, 1907 (*ir, nmr*)
Muetterties, E. *et al*, *Inorg. Chem.*, 1975, **14**, 950 (*nmr*)
Rogers, H. *et al*, *Inorg. Chem.*, 1976, **15**, 1775.

C_2H_9BClN B-00133

Chloroborane dimethylamine
Chlorodihydro(N-*methylmethanamine*)*boron*, 10CI. *Chloroborane compd. with* N-*methylmethanamine*(*1:1*), 9CI. N-*Methylmethanamine compd. with chloroborane*(*1: 1*)
[52920-74-8]

$Me_2NH—BH_2Cl$

M 93.363
Moisture-sensitive solid. Mp 18°.

Nöth, H. *et al*, *Chem. Ber.*, 1960, **93**, 225 (*synth*)
Beachley, O.T. *et al*, *Inorg. Chem.*, 1975, **14**, 120 (*pmr*)
Myers, W.H. *et al*, *Inorg. Chem.*, 1975, **14**, 2874 (*synth, pmr*)
Myers, W.H. *et al*, *Inorg. Chem.*, 1978, **17**, 1157 (*nmr*)
Balulescu, C.R. *et al*, *Inorg. Chem.*, 1978, **17**, 3707.

$C_2H_9BKN^{\ominus}$ B-00134

Dimethylaminotrihydroborate(1−)
Trihydro(N-*methylmethanaminato*)*borate*(*1*−), 10CI, 9CI. *Dimethylaminoborohydride*(*1*−)

$Me_2N—BH_3^{\ominus}$

M 97.008 (ion)
Li salt: [53042-33-4].
C_2H_9BLiN M 64.851
Na salt: [20596-52-5].
C_2H_9BNNa M 80.900
Has anticholesteremic and hypolipidemic props.
K salt: [43210-87-3].
C_2H_9BKN M 97.008

Keller, P.C., *Synth. React. Inorg. Metal-Org. Chem.*, 1973, **3**, 307 (*props*)
Keller, P.C., *Inorg. Chem.*, 1975, **14**, 438 (*props*)
Keller, P.C., *Inorg. Chem.*, 1975, **14**, 440 (*synth*)
Strouf, O. *et al*, *Acta Chem. Scand., Ser. A*, 1977, **31**, 391 (*props*)
Balulescu, C.R. *et al*, *Inorg. Chem.*, 1978, **17**, 3707.
Hall, I.H. *et al*, *J. Pharm. Sci.*, 1981, **70**, 339 (*pharmacol*)

C_2H_9BS B-00135

Borane dimethyl sulfide
Trihydro[*thiobis*[*methane*]]*boron*, 10CI. *Borane compd. with thiobis*[*methane*](*1:1*), 9CI. *Methyl sulfide compd. with borane*, 8CI. *Borine compd. with Me_2S*
[13292-87-0]

$Me_2S—BH_3$

M 75.963
Reducing agent. Hydroborating agent which may be used in a variety of solvs. Produced comly. Moisture-sensitive colourless liq. at r.t. Fp −27°, −38°.
▷Flammable

Heal, H.G., *J. Inorg. Nucl. Chem.*, 1960, **12**, 255 (*synth, ir*)
Young, D.E. *et al*, *J. Am. Chem. Soc.*, 1966, **88**, 4390 (*nmr*)
Lane, C.F., *Chem. Rev.*, 1976, **76**, 773 (*rev, use*)
Yamamoto, Y. *et al*, *Yuki Gosei Kagaku Kyokaishi*, 1979, **37**, 1008; *CA*, **92**, 139884j (*synth*)
Krishnamurthy, S. *et al*, *J. Org. Chem.*, 1980, **45**, 849.
Brown, H.C. *et al*, *Synthesis*, 1981, 996.
Mancilla, T. *et al*, *Tetrahedron Lett.*, 1982, **23**, 1561.

$C_2H_{10}AlB$ B-00136

Dimethyl[tetrahydroborato(1−)*H,H′*]aluminum, 10CI
Dimethylaluminum borohydride. Dimethyl[*tetrahydroborato*(*1*−)H,H′]*alane*
[17713-04-1]

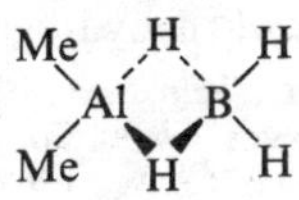

M 71.893
Mp 14-5°.
Me_3N complex: [60236-48-8]. (N,N-*Dimethylmethanamine*)*dimethyl*[*tetrahydroborato*(*1*−)-H,H′]*aluminum*, 9CI.
$C_5H_{19}AlBN$ M 131.003
Not fully descr.
Me_3P complex: [60236-49-9]. *Dimethyl*[*tetrahydroborato*(*1*−)-H,H′](*trimethylphosphine*)*aluminum*, 9CI.
$C_5H_{19}AlBP$ M 147.970
Not fully descr.

Oddy, P.R. *et al*, *J. Chem. Soc., Dalton Trans.*, 1976, 869 (*synth, ir, pmr, nmr, complexes*)
Barlow, M.T. *et al*, *J. Chem. Soc., Dalton Trans.*, 1979, 1793 (*synth, struct*)

$C_2H_{10}BGa$ B-00137

Dimethyl[tetrahydroborato(1−)-*H,H'*]gallium, 10CI

[68264-40-4]

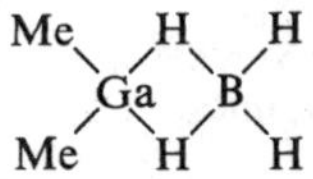

M 114.631

Monomeric. Liq. Mp 1°.

Downs, A.J. *et al*, *J. Chem. Soc., Dalton Trans.*, 1978, 809 (*synth, ms, ir, raman, nmr*)
Barlow, M.T. *et al*, *J. Chem. Soc., Dalton Trans.*, 1979, 1793.

$C_2H_{10}BN$ B-00138

Borane dimethylamine

Trihydro(N-*methylmethanamine*)*boron*, *10CI*. N-*Methylmethanamine compd. with borane*(*1:1*), *9CI*. *Dimethylamine compd. with borane*(*1:1*), *8CI*

[74-94-2]

$Me_2NH—BH_3$

M 58.918

Used in electroless plating of nickel and copper and in photographic developers. Reducing agent. Solid. Mp 37°. $Bp_{0.01}$ 49°. Dec. at 150°.

▷IP9450000.

Nöth, H. *et al*, *Chem. Ber.*, 1960, **93**, 928 (*props*)
Brown, M.P. *et al*, *J. Chem. Soc. (A)*, 1970, 410 (*ir*)
Lloyd, D.R. *et al*, *J. Chem. Soc., Faraday Trans. 2*, 1972, **68**, 947 (*pe*)
Nöth, H. *et al*, *Chem. Ber.*, 1974, **107**, 3070 (*nmr*)
Beachley, O.T. *et al*, *Inorg. Chem.*, 1975, **14**, 120 (*pmr*)
Kovar, R.A. *et al*, *Inorg. Chem.*, 1975, **14**, 2239 (*synth, ir*)
Myers, W.H. *et al*, *Inorg. Chem.*, 1975, **14**, 2874 (*pmr*)

$C_2H_{10}BP$ B-00139

Borane dimethylphosphine

(*Dimethylphosphine*)*trihydroboron*, *10CI*. *Dimethylphosphine, compd. with borane*(*1:1*), *9CI*, *8CI*. *Dimethylphosphine borine*

[4268-35-3]

$Me_2PH—BH_3$

M 75.885

Liq. Mp −22.6°. Slowly dehydrogenates on htg. at 150°.

Burg, A.B. *et al*, *J. Am. Chem. Soc.*, 1953, **75**, 3872 (*synth*)
Phillips, W.D. *et al*, *J. Am. Chem. Soc.*, 1959, **81**, 4496 (*nmr*)
Verkade, J.G. *et al*, *Inorg. Chem.*, 1964, **3**, 884 (*pmr*)
Cowley, A.H. *et al*, *J. Am. Chem. Soc.*, 1971, **93**, 6815 (*nmr*)
Drake, J.E. *et al*, *J. Chem. Soc. (A)*, 1971, 2074 (*pmr, ir, raman*)
Rapp, B. *et al*, *Inorg. Chem.*, 1973, **12**, 2868 (*nmr*)

$C_2H_{10}B_4Si$ B-00140

μ-Silyl-2,3-dicarbahexaborane(8), 9CI

[38999-41-6]

$C_2B_4H_7(SiH_3)$

M 105.427

Colourless liq.

Thompson, M.L. *et al*, *Inorg. Chem.*, 1972, **11**, 1925 (*ir, nmr, ms, struct, synth*)

$C_2H_{10}B_8$ B-00141

1,2-Dicarbadecaborane(10), 10CI

[41655-27-0]

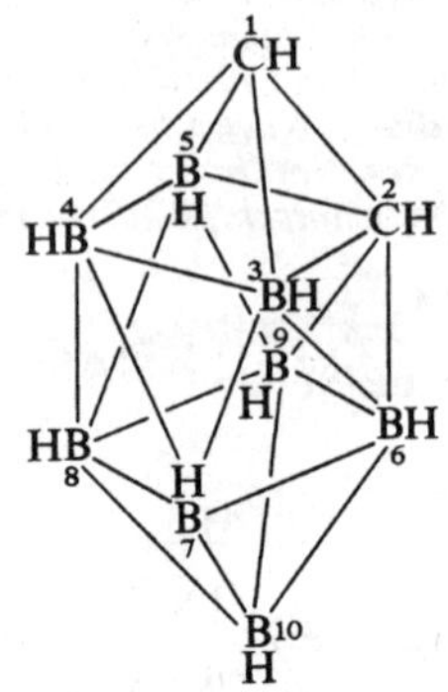

M 120.581

Cryst. Mp 182° (174°).

Rietz, R.R. *et al*, *J. Organomet. Chem.*, 1973, **63**, 1 (*nmr, synth*)
Plešek, J. *et al*, *Collect. Czech. Chem. Commun.*, 1974, **39**, 821 (*ms, synth*)

$C_2H_{10}B_8$ B-00142

1,6-Dicarbadecaborane(10), 10CI

[23704-81-6]

M 120.581

closo-. Cryst. Mp 146°. Dipole moment 2.22D.

Garrett, P.M. *et al*, *Inorg. Chem.*, 1969, **8**, 1907 (*synth, ir, nmr*)

$C_2H_{10}B_8$ B-00143

1,10-Dicarbadecaborane(10), 10CI

[23653-23-8]

M 120.581

closo-. Cryst. Mp 162°.

Garrett, P.M. *et al*, *Inorg. Chem.*, 1969, **8**, 1907 (*synth, ir, nmr*)
Leites, L.A. *et al*, *Bull. Acad. Sci. USSR, Div. Chem. Sci.*, 1975, 1866 (*ir, raman*)
Atavin, E.G. *et al*, *J. Mol. Struct.*, 1980, **65**, 259 (*ed*)

$C_2H_{10}B_{10}Li_2$ B-00144

μ-1,2-Dicarbadodecaborane(12)-1,2-diyldilithium, 10CI

Dilithium-1,2-dicarbadodecaborate(*2*−). *1,2-Dilithio-1,2-dicarbadodecaborane*(*12*)

[22220-85-5]

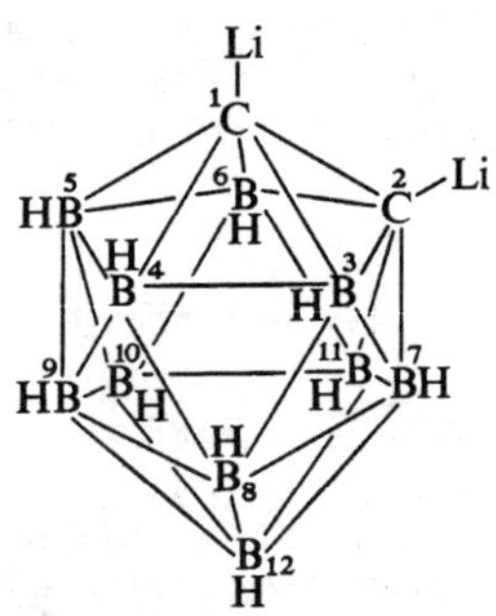

M 156.083

closo-. Precursor to many σ-substituted derivs. of 1,2-$C_2B_{10}H_{12}$.

Grafstein, D. *et al*, *Inorg. Chem.*, 1963, **2**, 1120 (*synth*)
Heying, T. *et al*, *Inorg. Chem.*, 1963, **2**, 1097 (*synth*)

$C_2H_{10}B_{10}Li_2$ B-00145

μ-1,7-Dicarbadodecaborane(12)-1,7-diyldilithium, 10CI

Dilithium-1,7-dicarbadodecaborate(2−). 1,7-Dilithio-1,7-dicarbadodecaborane(12)

[17217-89-9]

M 156.083

closo-. Precursor to many σ-substituted derivs. of 1,7-$C_2B_{10}H_{12}$. Sol. Et_2O, C_6H_6, THF.

Grafstein, D. *et al*, *Inorg. Chem.*, 1963, **2**, 1128.
Stanko, V. *et al*, *J. Gen. Chem., USSR* (*Engl. Transl.*), 1971, **41**, 2053

$C_2H_{10}B_{10}Li_2$ B-00146

μ-1,12-Dicarbadodecaborane(12)-1,12-diyldilithium, 10CI

Dilithium-1,12-dicarbadodecaborate(2−). 1,12-Dilithio-1,12-dicarbadodecaborane(12)

[23087-97-0]

M 156.083

closo-. Precursor to many σ-substituted derivs. of 1,12-$C_2B_{10}H_{12}$. Sol. liq. NH_3.

Zakharkin, L. *et al*, *Bull. Acad. Sci. USSR, Div. Chem. Sci.*, 1968, 2532 (*synth*)
Stanko, V. *et al*, *J. Gen. Chem. USSR* (*Engl. Transl.*), 1971, **41**, 2053 (*synth*)

$C_2H_{10}B_{10}Na_2$ B-00147

μ-1,2-Dicarbadodecaborane(12)-1,2-diyldisodium, 10CI

Disodium-1,2-dicarbadodecaborate(2−). 1,2-Disodio-1,2-dicarbadodecaborane(12)

[56310-81-7]

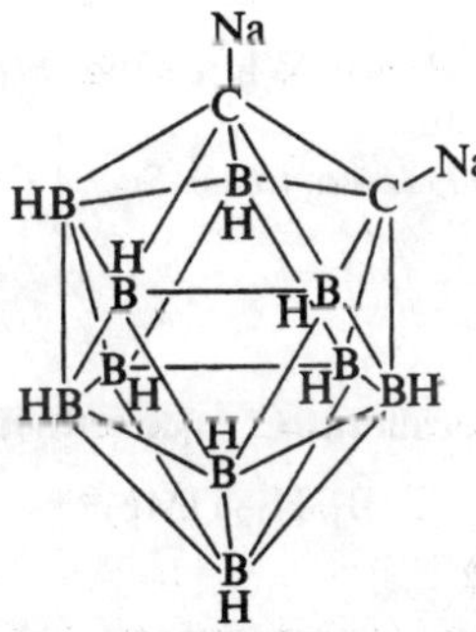

M 188.181

closo-. Precursor to many σ-substituted 1,2-$C_2B_{10}H_{12}$ derivs. Sol. Et_2O.

Zakharkin, L. *et al*, *Bull. Akad. Sci. USSR, Div. Chem. Sci.*, 1964, 546 (*synth*)
Zakharkin, L., *Tetrahedron Lett.*, 1964, 2255 (*synth*)

$C_2H_{11}B_9Ge$ B-00148

1,2-Dicarba-3-germadodecaborane, 8CI

*1-Germa-2,3-dicarba-*closo*-dodecacarborane*

[27071-52-9]

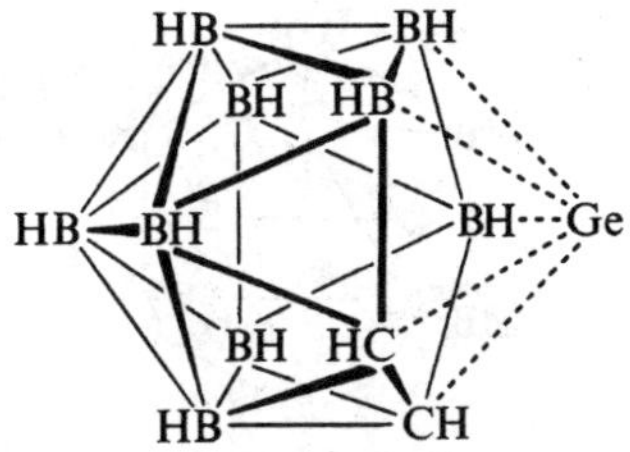

M 204.989

White solid (C_6H_6). Sol. C_6H_6. Bp 80° subl. (*in vacuo*). Dec. at 352°.

Rudolph, R.W. *et al*, *J. Am. Chem. Soc.*, 1970, **92**, 3351.

$C_2H_{11}B_9Pb$ B-00149

1,2-Dicarba-3-plumbadodecacarborane(11), 8CI

*1-Plumba-2,3-dicarba-*closo*-dodecacarborane*

[27071-51-8]

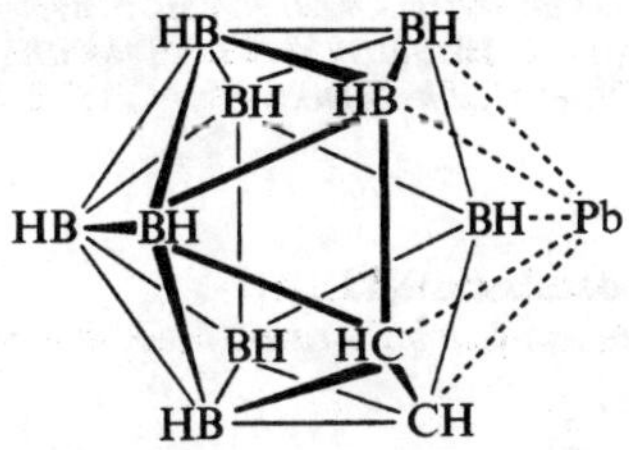

M 339.599

Yellow solid (C_6H_6). Bp 250° subl. (*in vacuo*). Dec. at 250°.

Rudolph, R.W. *et al*, *J. Am. Chem. Soc.*, 1970, **92**, 3351.

$C_2H_{11}B_9Sn$ B-00150

1,2-Dicarba-3-stannadodecaborane

1-Stanna-2,3-dicarbadodecacarborane

[23151-46-4]

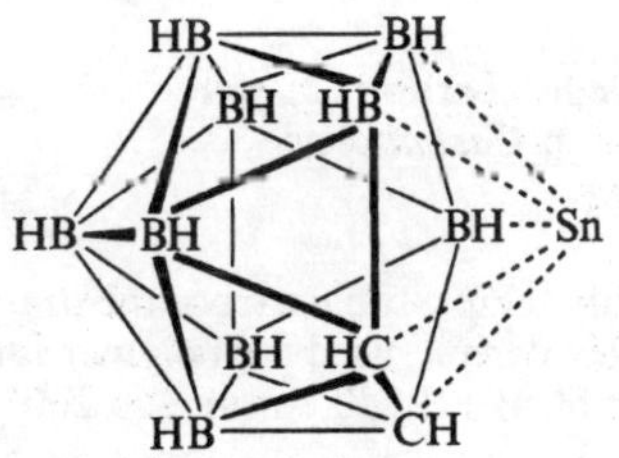

M 251.089

White solid (C_6H_6). Sol. C_6H_6. Bp 140° subl. (*in vacuo*). Dec. at 210°.

Rudolph, R.W. *et al*, *J. Am. Chem. Soc.*, 1970, **92**, 3351.

$C_2H_{12}B_{10}$ B-00151

1,2-Dicarbadodecaborane(12), 10CI

o-Carborane. o-Carbaborane

[16872-09-6]

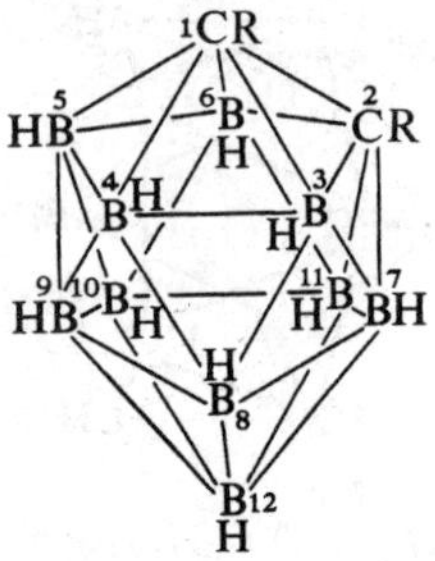

R = H

M 144.217

closo-. Shows high temp. stability; forms many C- and B-substituted derivs. White solid (pentane). Sol. aromatic solvs. Mp 320° (evacuated sealed tube).

Schroeder, H. *et al*, *Inorg. Chem.*, 1963, **2**, 1092 (*ir*)
Potenza, J. *et al*, *Inorg. Chem.*, 1964, **3**, 1673 (*cryst struct*)
Schroeder, H. *et al*, *Inorg. Chem.*, 1966, **5**, 693 (*nmr*)
Inorg. Synth., 1967, **10**, 93; 1968, **11**, 19 (*synth*)
Tupciauskas, V. *et al*, *Zh. Strukt. Khim.*, 1972, **13**, 823 (*pmr*)

$C_2H_{12}B_{10}$ B-00152

1,7-Dicarbadodecaborane(12), 10CI

m-Carborane. m-Carbaborane. Neocarborane

[16986-24-6]

M 144.217

closo-. Highly temp. stable carborane unit which forms many stable derivs., used for polymer formation. White cryst. Mp 273°.

▷HS9290000.

Grafstein, D. *et al*, *Inorg. Chem.*, 1963, **2**, 1128 (*synth, ms*)
Schroeder, H. *et al*, *Inorg. Chem.*, 1966, **5**, 693 (*nmr*)
Beall, H. *et al*, *Inorg. Chem.*, 1967, **6**, 874 (*cryst struct*)
Ditter, J. *et al*, *Inorg. Chem.*, 1970, **9**, 889 (*synth*)
Tupciauskas, V. *et al*, *Zh. Struct. Khim.*, 1972, **13**, 823 (*pmr*)

$C_2H_{12}B_{10}$ B-00153

1,12-Dicarbadodecaborane(12), 10CI

p-Carborane. p-Carbaborane

[20644-12-6]

M 144.217

closo-. Highly temp. stable carborane unit which forms many stable derivs., used for polymer formation. White solid by subl. at 110°/0.1 mm. Mp 260°.

Papetti, S. *et al*, *J. Am. Chem. Soc.*, 1964, **86**, 2295 (*synth, nmr*)
Schröder, H. *et al*, *Inorg. Chem.*, 1966, **5**, 693 (*nmr*)
Tupciauskas, V. *et al*, *Zh. Strukt. Khim.*, 1972, **13**, 823 (*pmr*)

$C_2H_{14}AlB_3$ B-00154

Dimethyl[octahydrotriborato(1–)]aluminum

Dimethylaluminum octahydrotriborate(1–)

[12654-95-4]

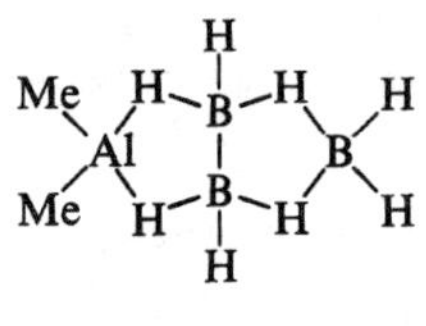

M 97.544

Extremely sensitive to air and moisture. Thermally unstable, $t_{1/2}$ ~10 min. at 20°/10-20 mm.

Borlin, J.-J. *et al*, *J. Am. Chem. Soc.*, 1972, **94**, 1367 (*synth, pmr, nmr*)
Dain, C.J. *et al*, *J. Chem. Soc., Dalton Trans.*, 1981, 2465 (*synth, struct*)

$C_2H_{18}B_{10}Ge$ B-00155

7,7-Dimethyl-7-germaundecaborane(14), 9CI

[34514-72-2]

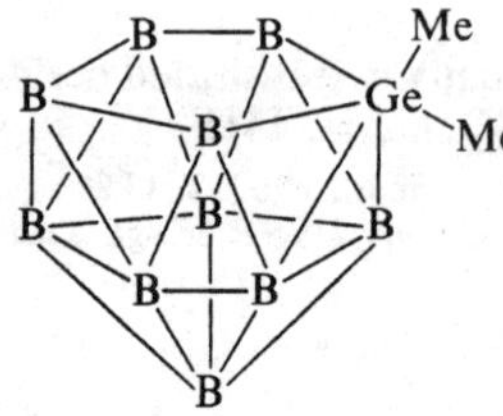

12 H atoms not shown

M 222.854

Proposed struct. shown. White solid. Mp 82-3°. Reasonably stable in air.

Loffredo, R.E. *et al*, *J. Am. Chem. Soc.*, 1971, **93**, 5587 (*synth, pmr, nmr, ms*)

$C_2H_{18}B_{10}Sn$ B-00156

7,7-Dimethyl-7-stannaundecaborane(14), 9CI

[34514-71-1]

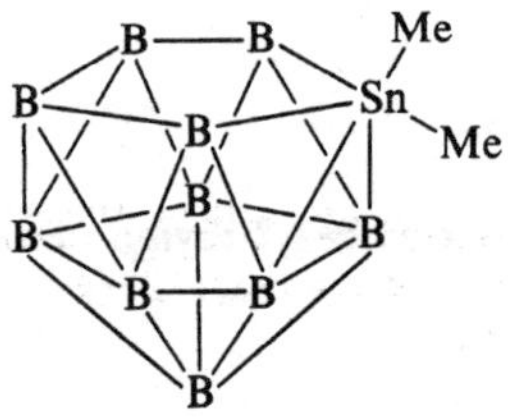

12 H atoms not shown

M 268.954

Proposed struct. shown. White solid. Mp 123-4°. Reasonably stable in air.

Loffredo, R.E. *et al*, *J. Am. Chem. Soc.*, 1971, **93**, 5587 (*synth, pmr, nmr, ms*)

$C_2H_{18}B_{10}Tl^{\ominus}$ B-00157

[η^4-Dodecahydrodecaborato(2–)]dimethylthallate(1–), 9CI

$B_{10}H_{12}TlMe_2^{\ominus}$

M 354.647 (ion)

Methyltriphenylphosphonium salt: [58776-50-4]. [*η^4-Dodecahydrodecaborato(2–)]dimethylmethyltriphenylphosphoniumthallate(1–)*, 9CI.
$C_{21}H_{36}B_{10}PTl$ M 631.972
Yellow solid, cryst. (CH_2Cl_2, Et_2O).

Greenwood, N.N. *et al*, *J. Chem. Soc., Dalton Trans.*, 1975, 299.

Having problems with locating a compound? Have you checked the indexes?

$C_3H_4BF_3N_2$ B-00158

Trifluoroborane imidazole

Trifluoro(1H-imidazole-N³)boron, 10CI. *BF_3-Imidazole. 1-H-Imidazole compd. with trifluoroborane*

[69080-38-2]

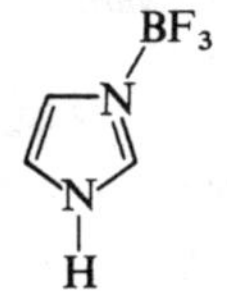

M 135.883

Crosslinking catalyst for epoxy resins.

Japan. Pat., 78 114 900, (*1978*); *CA*, **90**, 72868v (*use*)
Japan. Pat., 81 04 648, (*1981*); *CA*, **95**, 8267z

$C_3H_5B_3FeO_3$ B-00159

Tricarbonyl[η^4-pentahydrodicarbapentaborato(2−)]iron, 10CI

(1,2,4-$C_2B_3H_5$)tricarbonyliron

[53363-10-3]

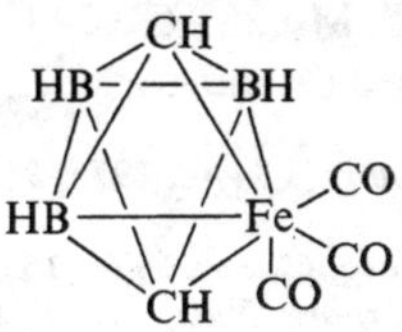

M 177.348

Orange liq.

Miller, V.R. *et al*, *J. Am. Chem. Soc.*, 1974, **96**, 3090 (*synth, ms, ir, nmr, pmr*)
Ulman, J.A. *et al*, *J. Am. Chem. Soc.*, 1978, **100**, 456 (*pe*)

C_3H_7B B-00160

2-Propenylborane, 10CI, 9CI

Allylborane

[51531-21-6]

$H_2C{=}CHCH_2BH_2$

M 53.898

Presumably would form B-H bridged dimer in the uncomplexed state, but not isol. Synthetic intermed. generated *in situ*. Air- and moisture-sensitive; cannot be kept since it undergoes self-hydroboration.

Mikhailov, B.M., *Intra-Sci. Chem. Rep.*, 1973, **7**, 191 (*rev*)
Mikhailov, B.M., *Usp. Khim.*, 1976, **45**, 1102 (*rev*)
Pelter, A. *et al*, *Compr. Org. Chem.*, (*Barton, D.H.R. et al*, Ed.), 1979, **3**, 695 (*rev*)
Mikhailov, B.M., *Sov. Sci. Rev., Sect. B*, 1980, **2**, 283 (*rev*)
Henriksen, U. *et al*, *J. Org. Chem.*, 1981, **46**, 3767.

$C_3H_7BCl_2$ B-00161

Dichloroisopropylborane

Dichloro(1-methylethyl)borane, 10CI. *Dichloroisopropylborine. Dichloro(2-propyl)borane*

[7680-99-1]

$Cl_2BCH(CH_3)_2$

M 124.804

Fuming liq., rapidly hydrol. by H_2O. Bp_{749} 72°.

▷Ignites in air

McCusker, P. *et al*, *J. Am. Chem. Soc.*, 1957, **79**, 5185 (*synth, props*)
Mikhailov, B.M. *et al*, *Izv. Akad. Nauk SSSR, Ser. Khim.*, 1957, 1080 (*synth*)
Lynds, L. *et al*, *J. Am. Chem. Soc.*, 1959, **81**, 5006 (*synth*)
Gorbunov, A. *et al*, *CA*, 1966, **64**, 16626g (*chromatog*)
Barlos, K. *et al*, *Chem. Ber.*, 1977, **110**, 2774 (*nmr*)
Bogatskii, A. *et al*, *CA*, 1981, **94**, 121624f (*props*)

$C_3H_7BCl_2$ B-00162

Dichloropropylborane, 10CI, 9CI, 8CI

Dichloropropylborine. Propyldichloroborane

[1739-52-2]

$H_3CCH_2CH_2BCl_2$

M 124.804

Strongly fuming liq., rapidly hydrol. in moist air. Sol. Et_2O. Bp_{739} 78.2°.

▷Ignites in air

McCusker, P.A. *et al*, *J. Am. Chem. Soc.*, 1957, **79**, 5182 (*synth*)
Mikhailov, B.M. *et al*, *Izv. Akad. Nauk SSSR, Ser. Khim.*, 1957, 1080 (*synth*)
Gerrard, W. *et al*, *J. Chem. Soc.*, 1964, 740 (*synth*)
Gorbunov, A.I. *et al*, *Gaz. Kromatogr. Moscow*, 1964, 35 (*chromatog*)
Köster, R. *et al*, *Justus Liebigs Ann. Chem.*, 1969, 34 (*use*)
Costes, J.P. *et al*, *J. Organomet. Chem.*, 1979, **175**, 257 (*use*)

$C_3H_7BF_2O$ B-00163

Difluoropropoxyborane, 10CI, 9CI

[20957-79-3]

$H_3CCH_2CH_2OBF_2$

M 107.895

Catalyst for polymerisation reactions. Moisture-sensitive.

B.P., 2 030 993, (*1980*); *CA*, **93**, P221317f (*use*)
Fr. Pat.,2 438 665, (*1980*); *CA*, **93**, P240977n (*use*)

$C_3H_8BClHgO_2$ B-00164

(Chloromercurimethyl)dimethoxyboron

(Chloromercuri)(dimethoxyboryl)methane. [(Chloromercuri)methyl]borinic acid dimethyl ester

$(MeO)_2BCH_2HgCl$

M 322.918

Cryst. (MeOH).

Matteson, D.S. *et al*, *J. Organomet. Chem.*, 1973, **54**, 35 (*synth, pmr*)

$C_3H_8B_4FeO_3$ B-00165

Tricarbonyl[(1,2,3,4-η)-tetraborane(8)]iron, 10CI

[54748-47-9]

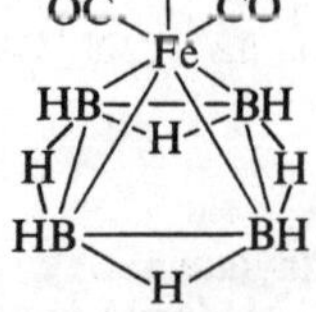

M 191.181

Orange liq. Mp ca. 5°.

Greenwood, N.N. *et al*, *J. Chem. Soc., Chem. Commun.*, 1974, 718 (*synth, nmr, pmr, ms, ir*)

Salahub, D.R., *J. Chem. Soc., Chem. Commun.*, 1978, 385 (*pe*)

$C_3H_8B_5FeO_3^{\ominus}$ B-00166

Tricarbonyl[(1,2,5-η)-octahydropentaborato(3−)-H^2]ferrate(1−), 10CI

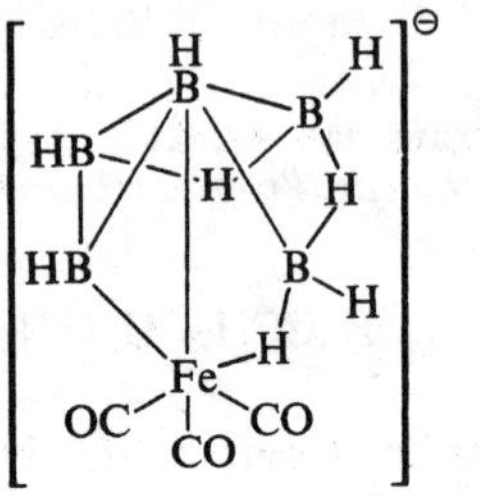

M 201.991 (ion)

Tetrabutylammonium salt: [61491-75-6].
$C_{19}H_{44}B_5FeNO_3$ M 444.459
No phys. props. reported.

Fehlner, T.P. *et al*, *J. Am. Chem. Soc.*, 1976, **98**, 7085.

C_3H_9B B-00167

Trimethylborane, 10CI, 9CI

Boron trimethyl. Trimethylborine

[593-90-8]

BMe_3

M 55.914

Gas. Mp −161.5°. Bp −20.2°. Forms addn. compd. with ammonia, Me_3BNH_3 (Mp 56°, Bp 110°).

▷Highly toxic. Spontaneously flammable

Katal'nikov, S.G. *et al*, *CA*, 1962, **57**, 7678 (*synth*)
U.S.P., 3 329 485, (*1967*); *CA*, **67**, 108741 (*synth*)
Nöth, H. *et al*, *J. Organomet. Chem.*, 1968, **12**, 23 (*pmr*)
Casanova, J. *et al*, *Org. Prep. Proced.*, 1969, **1**, 57 (*synth*)
Nöth, H. *et al*, *Chem. Ber.*, 1974, **107**, 3089 (*nmr*)
Becher, H.J. *et al*, *Spectrochim. Acta, Part A*, 1977, **33**, 951 (*ir, raman*)
Sax, N.I., *Dangerous Properties of Industrial Materials*, 5th Ed., Van Nostrand-Reinhold, 1979, 429.

C_3H_9BBrN B-00168

Bromo(dimethylamino)methylborane, 8CI

*1-Bromo-*N,N,*1-trimethylboranamine, 10CI, 9CI*

[18494-96-7]

$MeBBrNMe_2$

M 149.825

Air- and moisture-sensitive liq. d_{20} 1.21. Bp_{22} 42°. n_D^{20} 1.4602.

Nöth, H. *et al*, *Z. Anorg. Allg. Chem.*, 1963, **322**, 297; **324**, 270 (*synth*)
Nöth, H. *et al*, *J. Organomet. Chem.*, 1968, **12**, 23 (*nmr, pmr*)
Haubold, W. *et al*, *Z. Anorg. Allg. Chem.*, 1976, **421**, 105.
Paetzold, P. *et al*, *Chem. Ber.*, 1977, **110**, 3678; 1979, **112**, 663.

$C_3H_9BBr_3N$ B-00169

Tribromoborane trimethylamine

*Tribromo(*N,N*-dimethylmethanamine)boron, 10CI.* N,N*-Dimethylmethanamine compd. with tribromoborane(1:1), 9CI. Trimethylamine compd. with boron bromide(BBr_3)(1:1), 8CI. Boron tribromide trimethylamine*

[1516-54-7]

$Me_3N—BBr_3$

M 309.633

White cryst. (EtOH or by subl.).

Benton-Jones, B. *et al*, *J. Chem. Soc., Dalton Trans.*, 1972, 2603 (*nmr*)
Van Paasschen, J.M. *et al*, *Synth. React. Inorg. Metal-Org. Chem.*, 1974, **4**, 11 (*synth*)
Myers, W.H. *et al*, *Inorg. Chem.*, 1975, **14**, 2874 (*pmr*)
Miller, J.M. *et al*, *Inorg. Chem.*, 1976, **15**, 284 (*nmr*)
Okuda, T. *et al*, *Bull. Chem. Soc. Jpn.*, 1979, **52**, 307 (*nqr, nmr, synth*)
Iijima, K. *et al*, *Bull. Chem. Soc. Jpn.*, 1980, **53**, 1908 (*struct*)

$C_3H_9BBr_3P$ B-00170

Tribromoborane trimethylphosphine

Tribromo(trimethylphosphine)boron, 10CI. Trimethylphosphine compd. with tribromoborane(1:1), 9CI

[13292-85-8]

$Me_3P—BBr_3$

M 326.600

Solid, dec. at 180°. Mp 255-6° (248°).

Black, D.L. *et al*, *Acta Crystallogr., Sect. B*, 1975, **31**, 1116 (*struct*)
Mente, D.C. *et al*, *Inorg. Chem.*, 1975, **14**, 1862 (*nmr, ir, raman, ms*)
Drake, J.E. *et al*, *Inorg. Chem.*, 1977, **16**, 2289 (*ir, raman*)
Sisler, H.H. *et al*, *J. Inorg. Nucl. Chem.*, 1977, **39**, 1745 (*pmr, synth*)
Apel, J. *et al*, *Z. Anorg. Allg. Chem.*, 1979, **453**, 28.
Iijima, K. *et al*, *Bull. Chem. Soc. Jpn.*, 1981, **54**, 2255 (*struct*)

$C_3H_9BCl_3N$ B-00171

Trichloroborane trimethylamine

*Trichloro(*N,N*-dimethylmethanamine)boron, 10CI.* N,N*-Dimethylmethanamine compd. with trichloroborane(1:1), 9CI. Trimethylamine compd. with boron chloride(BCl_3)(1:1), 8CI. Boron trichloride trimethylamine*

[1516-55-8]

$Me_3N—BCl_3$

M 176.280

Cryst. (EtOH or by subl.).

Miller, J.M. *et al*, *J. Chem. Soc. (A)*, 1971, 346 (*ms*)
Benton-Jones, B. *et al*, *J. Chem. Soc., Dalton Trans.*, 1972, 2603 (*nmr*)
Myers, W.H. *et al*, *Inorg. Chem.*, 1975, **14**, 2874 (*pmr*)
Miller, J.M. *et al*, *Inorg. Chem.*, 1976, **15**, 284 (*nmr*)
Okuda, T. *et al*, *Bull. Chem. Soc. Jpn.*, 1979, **52**, 307 (*nqr, nmr, synth*)
Iijima, K. *et al*, *Bull. Chem. Soc. Jpn.*, 1980, **53**, 1908 (*struct*)

$C_3H_9BCl_3P$ B-00172

Trichloroborane trimethylphosphine

Trichloro(trimethylphosphine)boron, 10CI. Trimethylphosphine compd. with trichloroborane(1:1), 9CI

[13292-83-6]

$Me_3P—BCl_3$

M 193.247

Cryst. (C_6H_6), stable to heat and water. Mp 241-2°. Bp 275° subl.

Phillips, G.M. *et al*, *J. Chem. Soc.*, 1945, 146 (*synth*)

Black, D.L. *et al*, *Acta Crystallogr., Sect. B*, 1975, **31**, 1116 (*struct*)
Mente, D.C. *et al*, *Inorg. Chem.*, 1975, **14**, 1862 (*nmr, ms, ir, raman*)
Drake, J.E. *et al*, *Inorg. Chem.*, 1977, **16**, 2289 (*ir, raman*)
Sisler, H.H. *et al*, *J. Inorg. Nucl. Chem.*, 1977, **39**, 1745 (*pmr, synth*)
Symons, M.C.R. *et al*, *J. Chem. Res. (S)*, 1978, 122
Apel, J. *et al*, *Z. Anorg. Allg. Chem.*, 1979, **453**, 28 (*nmr*)

$C_3H_9BF_3N$ B-00173

Trifluoroborane trimethylamine

(N,N-*Dimethylmethanamine*)*trifluoroboron*, 10CI. *Trimethylamine compd. with boron fluoride*(BF_3)(*1:1*), 8CI. N,N-*Dimethylmethanamine, compd. with trifluoroborane*(*1:1*), 9CI. *Trimethylamine boron trifluoride*
[420-20-2]

$Me_3N—BF_3$

M 126.916
White cryst. solid. Mp 140-6°. Bp 238°.

Brown, F., *J. Am. Chem. Soc.*, 1952, **74**, 1219 (*synth*)
Clippard, P.H. *et al*, *J. Cryst. Mol. Struct.*, 1971, **1**, M363 (*cryst struct*)
Miller, J.A. *et al*, *J. Chem. Soc. (A)*, 1971, 346 (*ms*)
Van Paasschen, J.M. *et al*, *J. Am. Chem. Soc.*, 1972, **94**, 2680 (*synth*)
Geanangel, R.A., *Inorg. Chem.*, 1975, **14**, 696 (*cmr, nmr*)
Myers, W.H. *et al*, *Inorg. Chem.*, 1975, **14**, 2874 (*synth, pmr*)
Miller, J.M. *et al*, *Inorg. Chem.*, 1976, **15**, 284 (*nmr*)

$C_3H_9BF_3P$ B-00174

Trifluoroborane trimethylphosphine

Trifluoro(*trimethylphosphine*)*boron*, 10CI. *Trimethylphosphine compd. with trifluoroborane*(*1:1*), 9CI. *Trimethylphosphine compd. with* BF_3(*1:1*)
[420-21-3]

$Me_3P—BF_3$

M 143.883
Solid. Mp 139-42°.

Heitsch, C.W., *Inorg. Chem.*, 1965, **4**, 1019 (*nmr*)
Tuchagues, J.P. *et al*, *Bull. Soc. Chim. Fr.*, 1971, 4246.
Markham, R.T. *et al*, *J. Inorg. Nucl. Chem.*, 1974, **36**, 503 (*synth*)
Bula, M.J. *et al*, *Can. J. Chem.*, 1975, **53**, 326 (*nmr*)
Van Paasschen, J.M. *et al*, *Inorg. Chem.*, 1978, **17**, 3302.
Apel, J. *et al*, *Z. Anorg. Allg. Chem.*, 1979, **453**, 28 (*nmr*)

$C_3H_9BI_3P$ B-00175

Triiodoborane trimethylphosphine

Triiodo(*trimethylphosphine*)*boron*, 10CI. *Trimethylphosphine compd. with triiodoborane*, 9CI. *Trimethylphosphine triiodoborane*
[54215-81-5]

$Me_3P—BI_3$

M 467.601
Solid, turns light yellow at 125°. Mp 299-302°.

Denniston, M.L. *et al*, *J. Inorg. Nucl. Chem.*, 1974, **36**, 1461 (*nmr*)
Black, D.L. *et al*, *Acta Crystallogr., Sect. B*, 1975, **31**, 1116 (*struct*)
Drake, J.E. *et al*, *Inorg. Chem.*, 1977, **16**, 2289 (*ir, raman*)
Sisler, H.H. *et al*, *J. Inorg. Nucl. Chem.*, 1977, **39**, 1745 (*synth, pmr, ir*)
Iijima, K. *et al*, *Bull. Chem. Soc. Jpn.*, 1982, **55**, 255 (*struct*)

$C_3H_9BN_2$ B-00176

Cyanoborane dimethylamine

(*Cyano*-C)*dihydro*(N-*methylmethanamine*)*boron*, 10CI. *Boranecarbonitrile, compd. with* N-*methylmethanamine*(*1:1*), 9CI
[51329-61-4]

$Me_2NH—BH_2CN$

M 83.928
Shows a range of useful pharmacological props. White hygroscopic cryst. solid. Sol. Et_2O. Mp 46-8°, 54-5°.

Weidig, C. *et al*, *Inorg. Chem.*, 1974, **13**, 1763 (*synth, ir*)
Beachley, O.T. *et al*, *Inorg. Chem.*, 1975, **14**, 120 (*pmr*)
Wisian-Nielson, P. *et al*, *Inorg. Chem.*, 1978, **17**, 2327 (*synth, pmr, ir*)
Hall, I.J. *et al*, *J. Pharm. Sci.*, 1979, **68**, 685; 1980, **69**, 1025; 1981, **70**, 339 (*pharmacol*)

C_3H_9BO B-00177

Methoxydimethylborane

Dimethylborinic acid methyl ester, 10CI, 9CI, 8CI. *Dimethyl*(*methoxy*)*boron*
[4443-43-0]

Me_2BOMe

M 71.914
Air- and moisture-sensitive liq. Bp 21°.
▷Pyrophoric

Burg, A.B. *et al*, *J. Am. Chem. Soc.*, 1953, **75**, 3872 (*synth, props*)
Petry, R.C. *et al*, *J. Am. Chem. Soc.*, 1956, **78**, 6416 (*synth*)
Nöth, H. *et al*, *J. Organomet. Chem.*, 1968, **12**, 23 (*nmr*)
Kroner, J. *et al*, *Z. Naturforsch., B*, 1973, **28**, 416 (*pe*)
Berger, H.O. *et al*, *Chem. Ber.*, 1976, **109**, 2266 (*props*)
Biffar, W. *et al*, *Chem. Ber.*, 1980, **113**, 333 (*nmr*)

$C_3H_9BO_2$ B-00178

Dihydroxypropylborane

Propylboronic acid, 10CI, 9CI. *1-Propaneboronic acid*, 8CI. *Propyldihydroxyborane*
[17745-45-8]

$H_3CCH_2CH_2B(OH)_2$

M 87.913
Inhibits cholinesterase, lipase and proteinase. Cryst. solid with characteristic smell. V. sol. hot H_2O; less sol. cold H_2O. Mp 107° (105-6°). Sublimes.

Long, L.H. *et al*, *J. Chem. Soc.*, 1953, 3902 (*synth*)
Laurent, J.P., *C.R. Hebd. Seances Acad. Sci.*, 1961, **253**, 1812 (*synth*)
Stibr, B. *et al*, *Collect. Czech. Chem. Commun.*, 1968, **33**, 976 (*props*)
Jarv, J. *et al*, *Bioorg. Khim.*, 1978, **4**, 1364 (*use*)
Mikhailov, B.M. *et al*, *Izv. Akad. Nauk SSSR, Ser. Khim.*, 1978, 1586; *Bull. Acad. Sci. USSR, Div. Chem. Sci.*, 1385 (*synth*)
Mikhailov, B.M. *et al*, *J. Organomet. Chem.*, 1978, **154**, 113 (*synth*)

$C_3H_9BO_2$ B-00179

Methyldimethoxyborane

Methylboronic acid dimethyl ester, 10CI, 9CI. *Methaneboronic acid dimethyl ester*, 8CI. *Dimethoxymethylborane*
[7318-81-2]

$MeB(OMe)_2$

M 87.913
Air- and moisture-sensitive liq. Sol. non-hydroxylic solvs. Bp 53.5°.

Petry, R.C. *et al*, *J. Am. Chem. Soc.*, 1956, **78**, 6416 (*synth*)
Goubeau, J. *et al*, *Z. Phys. Chem.*, 1959, **20**, 15 (*ir, raman*)
Kroner, J. *et al*, *Z. Naturforsch., B*, 1973, **28**, 416 (*pe*)
McFarlane, W. *et al*, *Chem. Ber.*, 1975, **108**, 3831 (*nmr*)
Goetze, R. *et al*, *Chem. Ber.*, 1981, **114**, 1884 (*nmr*)
Gundersen, G. *et al*, *Acta Chem. Scand., Ser. A*, 1981, **35**, 325 (*struct*)

$C_3H_9BO_3$ B-00180

Trimethoxyborane

Boric acid (H_3BO_3) trimethyl ester, 10CI, 9CI, 8CI. *Methyl borate. Trimethyl borate. Trimethyl orthoborate*

[121-43-7]

$(MeO)_3B$

M 103.912
Vulcanizing agent, insect sterilant, catalyst, gelling agent heat stabilizer, fire extinguisher for hydrogen fires; used in prepn. of reactive silica and in cotton smoulder-resistant treatment. Liq., fumes in air, moisture-sensitive. Sol. polar non-hydroxylic solvs. d_4^{20} 0.920. Mp −29.3°. Bp 68-9°. n_D^{25} 1.3543, 1.3548. Forms azeotrope with MeOH, Bp 55°.

▷Mod. toxic. Flammable. ED5600000.

Webster, S.H. *et al*, *J. Am. Chem. Soc.*, 1933, **55**, 3233 (*synth*)
Schlesinger, H.I. *et al*, *J. Am. Chem. Soc.*, 1953, **75**, 213 (*synth, props*)
Fletton, R.A. *et al*, *Analyst (London)*, 1971, **96**, 370 (*pmr*)
Cragg, R.H. *et al*, *J. Chem. Soc., Dalton Trans.*, 1972, 1373 (*ms*)
Kroner, J. *et al*, *Z. Naturforsch., B*, 1973, **28**, 416 (*pe*)
Nöth, H. *et al*, *Chem. Ber.*, 1974, **107**, 3089 (*nmr*)
Gundersen, G., *J. Mol. Struct.*, 1976, **33**, 79 (*struct*)
Rogstad, A. *et al*, *J. Mol. Struct.*, 1976, **35**, 121 (*ir, raman*)
Sax, N.I., *Dangerous Properties of Industrial Materials*, 5th Ed., Van Nostrand-Reinhold, 1979, 810.

$C_3H_9BO_5$ B-00181

Hydrogen dihydroxy[1,2,3-propanetriolato(2−)-O^1,O^2]borate(1−), 10CI

[64466-24-6]

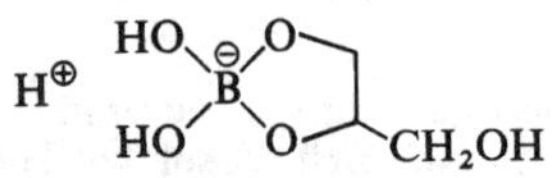

M 135.911

U.S.S.R. Pat., 566 824-5, (*1977*); *CA*, **87**, 151712f, 151713g (*props*)

C_3H_9BS B-00182

Dimethyl(methylthio)borane

Dimethylthioborinic acid methyl ester, 10CI, 9CI, 8CI

[19163-05-4]

Me_2BSMe

M 87.974
Air- and moisture-sensitive liq., dec. over several months at 20°. Mp −84°. Bp 71°.

Burg, A.B. *et al*, *J. Am. Chem. Soc.*, 1954, **76**, 3307 (*synth, props*)
Nöth, H. *et al*, *J. Organomet. Chem.*, 1968, **12**, 23 (*nmr*)
Vahrenkamp, H., *J. Organomet. Chem.*, 1971, **28**, 181 (*raman, ir*)
Kroner, J. *et al*, *Z. Naturforsch., B*, 1973, **28**, 416 (*pe*)
McFarlane, W. *et al*, *Chem. Ber.*, 1975, **108**, 3831 (*nmr*)
Berger, H.O. *et al*, *Chem. Ber.*, 1976, **109**, 2266 (*props*)

$C_3H_9BS_2$ B-00183

Methylbis(methylthio)borane

Methyldithioboronic acid dimethyl ester, 10CI, 9CI. *Dithiomethaneboronic acid dimethyl ester*, 8CI. *Bis(methanethio)methylborine*

[19163-08-7]

$MeB(SMe)_2$

M 120.034
Air- and moisture-sensitive liq. Sol. non-hydroxylic solvs. Mp −59°. Bp 155° (est.).

Burg, A.B. *et al*, *J. Am. Chem. Soc.*, 1956, **78**, 1523 (*synth*)
Vahrenkamp, H., *J. Organomet. Chem.*, 1971, **28**, 181 (*ir, raman*)
Kroner, J. *et al*, *Z. Naturforsch., B*, 1973, **28**, 416 (*pe*)
McFarlane, W. *et al*, *Chem. Ber.*, 1975, **108**, 3831 (*nmr*)
Lindoey, S. *et al*, *Acta Chem. Scand., Ser. A*, 1976, **30**, 54 (*struct*)
Goetze, R. *et al*, *Chem. Ber.*, 1981, **114**, 1884 (*nmr*)

$C_3H_9BSe_2$ B-00184

Methylbis(methylseleno)borane

Methyldiselenoboronic acid dimethyl ester, 10CI, 9CI

[35450-18-1]

$MeB(SeMe)_2$

M 213.834
Air- and moisture-sensitive liq. Sol. non-hydroxylic solvs. Bp_{10} 63°.

Siebert, W. *et al*, *Chem. Ber.*, 1972, **105**, 454 (*synth*)
McFarlane, W. *et al*, *Chem. Ber.*, 1975, **108**, 3831 (*nmr*)
Fussstetter, H. *et al*, *Chem. Ber.*, 1977, **110**, 3172 (*nmr*)

$C_3H_9B_3Br_3N_3$ B-00185

2,4,6-Tribromo-1,3,5-trimethylborazine, 10CI, 9CI, 8CI

B-*Tribromo*-N-*trimethylborazole*

[703-85-5]

As 2,4,6-Tribromoborazine, B-00015 with

X = Br, Y = Me

M 359.266
Solid, sensitive to moisture. Mp 143-6°.

Emeléus, H.J. *et al*, *J. Chem. Soc.*, 1959, 1306 (*synth*)
Riley, R.F. *et al*, *Inorg. Chem.*, 1964, **3**, 1651 (*props*)
Anderson, G.A. *et al*, *Spectrochim. Acta, Part A*, 1970, **26**, 2013 (*ir*)
Melcher, L.A. *et al*, *Inorg. Chem.*, 1973, **12**, 601 (*ms*)
Wrackmeyer, B. *et al*, *Chem. Ber.*, 1976, **109**, 3480.
Nöth, H. *et al*, *Chem. Ber.*, 1981, **114**, 1150 (*nmr*)

$C_3H_9B_3Cl_3N_3$ B-00186

2,4,6-Trichloro-1,3,5-trimethylborazine, 10CI, 9CI, 8CI

B,B,B-*Trichloro*-N,N,N-*trimethylborazole*

[703-86-6]

As 2,4,6-Tribromoborazine, B-00015 with

X = Cl, Y = Me

M 225.913
Curing agent for epoxy resins. Long white needles (pet. ether/C_6H_6). Sol. C_6H_6, Et_2O. Mp 153-6°, 162-3°. $Bp_{0.1-1}$ 120-30° subl. Moisture-sensitive.

Hohnstedt, L.F. *et al*, *J. Am. Chem. Soc.*, 1960, **82**, 89 (*synth, props*)
Butcher, I.M. *et al*, *Spectrochim. Acta*, 1962, **18**, 1487 (*ir, struct*)
Butcher, I.M. *et al*, *J. Chem. Soc.*, 1964, 4528 (*pmr*)
Inorg. Synth., 1972, **13**, 43 (*synth, props, ir, nmr*)
Kroner, J. *et al*, *Tetrahedron*, 1972, **28**, 1585 (*pe*)
Melcher, L.A. *et al*, *Inorg. Chem.*, 1972, **11**, 1247; 1973, **12**, 601 (*ir, ms*)
Wrackmeyer, B. *et al*, *Chem. Ber.*, 1976, **109**, 3480 (*nmr*)
Nöth, H. *et al*, *Chem. Ber.*, 1981, **114**, 1150 (*nmr*)

$C_3H_9B_3F_3N_3$ B-00187

2,4,6-Trifluoro-1,3,5-trimethylborazine, 10CI, 9CI, 8CI
B,B,B-*Trifluoro*-N,N,N-*trimethylborazole*
[13722-15-1]
As 2,4,6-Tribromoborazine, B-00015 with

X = F, Y = Me

M 176.549
Moisture-sensitive solid. Mp 90.5°. Bp_9 70° subl.

Wiberg, E. *et al*, *Z. Naturforsch., B*, 1951, **6**, 338 (*synth*)
Niedenzu, K., *Inorg. Chem.*, 1962, **1**, 943 (*synth*)
Anderson, G.A. *et al*, *Spectrochim. Acta, Part A*, 1970, **26**, 2013 (*synth, ir*)
Kroner, J. *et al*, *Tetrahedron*, 1972, **28**, 1585 (*pe*)
Miller, J.M. *et al*, *Inorg. Chem.*, 1974, **13**, 498 (*ms*)
Wrackmeyer, B. *et al*, *Chem. Ber.*, 1976, **109**, 3480 (*nmr*)
Nöth, H. *et al*, *Chem. Ber.*, 1981, **114**, 1150 (*nmr, cmr*)

$C_3H_9B_3O_3$ B-00188

Trimethylboroxin, 10CI, 9CI, 8CI
Trimethyltriborotrioxane
[823-96-1]
As Boroxin, B-00019 with

R = Me

M 125.532
Used in identification and separation of isomeric *cis-trans* diols. Moisture-sensitive liq., disproportionates at 200°. Mp −38°. Bp_{745} 79-80°. n_D^{20} 1.3638.

Lehmann, W.J. *et al*, *J. Inorg. Nucl. Chem.*, 1961, **21**, 25 (*ms*)
Rathke, M.W. *et al*, *J. Am. Chem. Soc.*, 1966, **88**, 2606 (*synth, nmr*)
Wester, D.W., *Org. Prep. Proced. Int.*, 1971, **3**, 191 (*synth*)
Barton, L. *et al*, *J. Inorg. Nucl. Chem.*, 1977, **39**, 1845 (*pmr, ms*)
McKean, D.C. *et al*, *Spectrochim. Acta, Part A*, 1977, **33**, 951 (*ir*)
Biffar, W. *et al*, *Chem. Ber.*, 1980, **113**, 333 (*nmr*)

$C_3H_9B_3O_6$ B-00189

Trimethoxyboroxin, 9CI
Trimethoxyboroxole
[102-24-9]

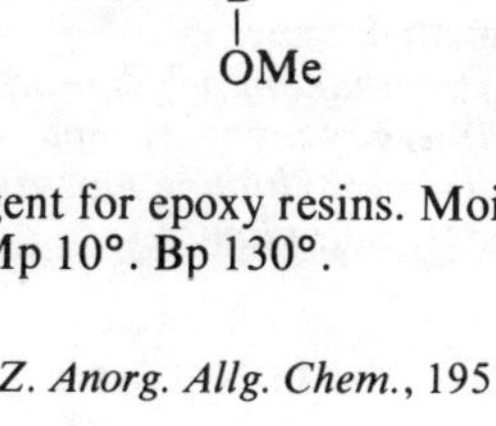

M 173.531
Cross-linking agent for epoxy resins. Moisture-sensitive liq. d 1.195. Mp 10°. Bp 130°.
▷ED8720000.

Goubeau, J. *et al*, *Z. Anorg. Allg. Chem.*, 1951, **267**, 1 (*synth*)

Nöth, H. *et al*, *NMR Spectroscopy of Boron Compounds* (*NMR Basic Principles and Progress*, Part 14), 1978, Springer (*nmr*)
Lin, S.C. *et al*, *J. Polym. Sci., Polym. Chem. Ed.*, 1979, **17**, 3095 (*use*)
Biffar, W. *et al*, *Chem. Ber.*, 1980, **113**, 333 (*nmr*)

$C_3H_9B_3Si$ B-00190

1-(Methylsilyl)-1,5-dicarbapentaborane(5), 9CI
[55906-10-0]

1,5—$C_2B_3H_4(SiH_2Me)$

M 105.620

Leach, J.B. *et al*, *J. Chem. Soc., Dalton Trans.*, 1975, **11**, 1018 (*ir, nmr, ms, synth*)

$C_3H_9B_4Ga$ B-00191

[(2,3,4,5,6-η)-Hexahydro-2,3-dicarbahexaborato(2−)]methylgallium, 9CI
[36550-13-7]

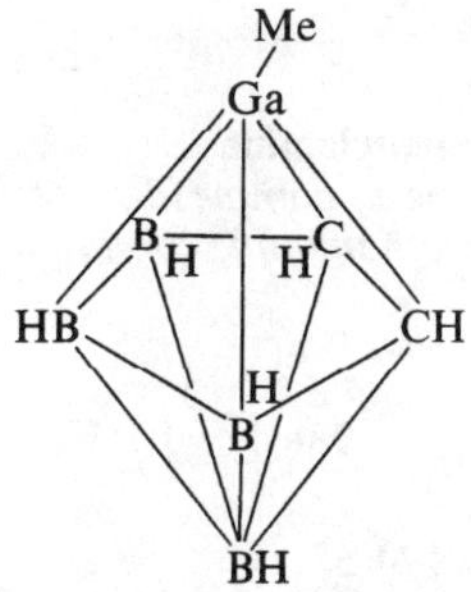

M 158.064
Colourless, sl. volatile cryst. slowly degraded by air. Stable to 100°.

Grimes, R.N. *et al*, *J. Am. Chem. Soc.*, 1972, **94**, 1865 (*synth, cryst struct, ir, ms, nmr*)

$C_3H_9B_4In$ B-00192

[(2,3,4,5,6-η)-Hexahydro-2,3-dicarbahexaborato(2−)]methylindium, 9CI
[36550-14-8]

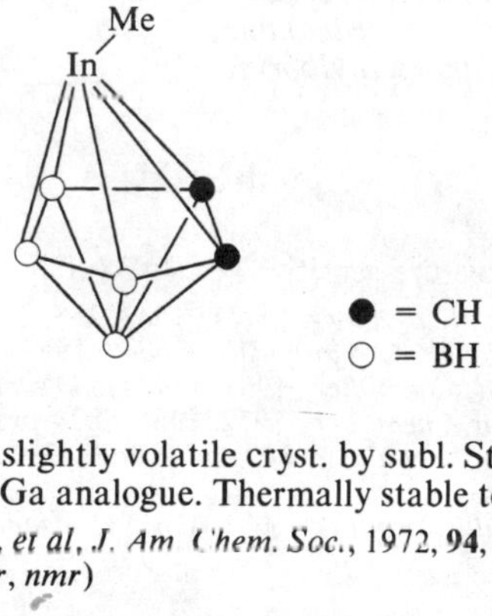

M 203.164
Colourless, slightly volatile cryst. by subl. Struct. by analogy with Ga analogue. Thermally stable to ~100°.

Grimes, R.N. *et al*, *J. Am. Chem. Soc.*, 1972, **94**, 1865 (*synth, ir, ms, pmr, nmr*)

$C_3H_9B_5FeO_3$ B-00193

Tricarbonyl[nonahydropentaborate(2−)]iron, 10CI

[61403-41-6]

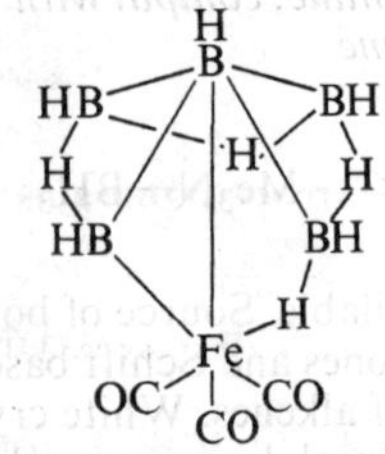

M 202.999

Fluxional. Deep-red mobile liq.

▷Extremely pyrophoric

Fehlner, T.P. *et al*, *J. Am. Chem. Soc.*, 1976, **98**, 7085 (*synth, nmr, ms, pmr, cmr*)
Ulman, J.A. *et al*, *J. Am. Chem. Soc.*, 1978, **100**, 456 (*pe*)
Shore, S.G. *et al*, *Inorg. Chem.*, 1979, **18**, 670 (*synth, ms, pmr, cmr, nmr*)

$C_3H_{10}BF_2N$ B-00194

Difluoroborane trimethylamine

(N,N-*Dimethylmethanamine*)*difluorohydroboron*, 10CI. N,N-*Dimethylmethanamine compd. with difluoroborane*(*1:1*), 9CI

[35237-72-0]

$Me_3N—BHF_2$

M 108.926

White solid. Mp 124-5°.

Mogeot, H. *et al*, *J. Inorg. Nucl. Chem.*, 1972, **34**, 825 (*ir, raman*)
Van Paasschen, J.M. *et al*, *J. Am. Chem. Soc.*, 1972, **94**, 2680 (*synth, ir, nmr, pmr*)
Tuchagues, J.-P. *et al*, *J. Organomet. Chem.*, 1973, **54**, 69 (*ir, nmr*)
Geanangel, R.A., *Inorg. Chem.*, 1975, **14**, 696 (*nmr*)
Van Paasschen, J.M. *et al*, *J. Inorg. Nucl. Chem.*, 1976, **38**, 2321.
Van Paasschen, J.M. *et al*, *Inorg. Chem.*, 1978, **17**, 3302.

$C_3H_{10}BN$ B-00195

Dimethyl(methylamino)borane, 8CI

N,*1,1-Trimethylboranamine*, 10CI, 9CI. *Dimethylaminomethylborine*

[4023-40-9]

Me_2BNHMe

M 70.929

Air- and moisture-sensitive liq. Bp 38.3°.

Wiberg, E. *et al*, *Z. Anorg. Chem.*, 1947, **255**, 141 (*synth*)
Becher, H.J. *et al*, *Z. Anorg. Allg. Chem.*, 1952, **268**, 133 (*ir*)
Nöth, H. *et al*, *Chem. Ber.*, 1967, **100**, 3353 (*synth, nmr*)
Beck, W. *et al*, *Chem. Ber.*, 1972, **105**, 2883 (*nmr*)
Almenningen, A. *et al*, *Acta Chem. Scand., Ser. A*, 1981, **35**, 341 (*struct*)
Nöth, H. *et al*, *Chem. Ber.*, 1981, **114**, 1150 (*nmr*)

$C_3H_{10}BO_3^{\ominus}$ B-00196

Hydrotrimethoxyborate(1−), 10CI, 9CI, 8CI

Trimethoxyhydroborate. Trimethoxyborohydride

[54787-85-8]

$(MeO)_3BH^{\ominus}$

M 104.920 (ion)

Li salt: [39659-89-7].

$C_3H_{10}BLiO_3$ M 111.861

No phys. props. reported.

Na salt: [16940-17-3].

$C_3H_{10}BNaO_3$ M 127.910

Intermed. in methanolysis of $NaBH_4$. Selective reducing agent. Catalyst for manuf. of foamed polyamides. Used as sterilant for control of screwworm flies. Solid, stable in dry air. Sol. dioxan, isopropylamine, NH_3. Mp ca. 230° dec. Reacts with H_2O, alcohols.

K salt: [63637-24-1].

$C_3H_{10}BKO_3$ M 144.019

No phys. props. reported.

Brown, H.C. *et al*, *J. Am. Chem. Soc.*, 1953, **75**, 192 (*synth*)
Japan. Pat., 62 10 567, (*1962*); *CA*, **58**, 11545a (*use*)
Mikheeva, V.I. *et al*, *Zh. Neorg. Khim.*, 1967, **12**, 2019; *CA*, **67**, 121875f (*synth*)
Japan. Pat., 72 39 297, (*1972*); *CA*, **78**, 137304g (*use*)
Kadlec, V. *et al*, *Collect. Czech. Chem. Commun.*, 1974, **39**, 3200 (*synth*)
Mindl, J. *et al*, *Collect. Czech. Chem. Commun.*, 1976, **41**, 3314 (*use, props*)
Niehues, K.J., *CA*, 1980, **93**, 196748n (*rev*)

$C_3H_{11}BBrP$ B-00197

Bromoborane trimethylphosphine

Bromodihydro(*trimethylphosphine*)*boron*, 10CI. *Trimethylphosphine compd. with bromoborane*, 9CI

[60228-69-5]

$Me_3P—BH_2Br$

M 168.808

Mp 61.0-62.5°. Erroneous coupling constants have been reported for this compd.

Dennison, M.L. *et al*, *J. Inorg. Nucl. Chem.*, 1976, **38**, 979 (*nmr, synth*)
Jugie, G. *et al*, *J. Inorg. Nucl. Chem.*, 1977, **39**, 900 (*nmr*)
Sisler, H.H. *et al*, *J. Inorg. Nucl. Chem.*, 1977, **39**, 1745 (*ir, pmr*)
Schmidbaur, H. *et al*, *Angew. Chem., Int. Ed. Engl.*, 1981, **20**, 283.

$C_3H_{11}BClP$ B-00198

Chloroborane trimethylphosphine

Chlorodihydro(*trimethylphosphine*)*boron*, 10CI. *Trimethylphosphine, compd. with chloroborane*(*1:1*)

[64160-46-9]

$Me_3P—BH_2Cl$

M 124.357

Cryst. (Et_2O/pentane). Mp 73°.

Jugie, G. *et al*, *J. Inorg. Nucl. Chem.*, 1977, **39**, 900.
Sisler, H.H. *et al*, *J. Inorg. Nucl. Chem.*, 1977, **39**, 1745 (*synth, ir, pmr*)
Kodama, G. *et al*, *Inorg. Chem.*, 1979, **18**, 3302 (*synth*)
Kameda, M. *et al*, *Inorg. Chem.*, 1980, **19**, 2288 (*synth, nmr*)
Schmidbaur, H. *et al*, *Angew. Chem., Int. Ed. Engl.*, 1981, **20**, 283 (*synth, ir, nmr*)

$C_3H_{11}BFN$ B-00199

Fluoroborane trimethylamine

(N,N-*Dimethylmethanamine*)*fluorodihydroboron*, 10CI. N,N-*Dimethylmethanamine, compd. with fluoroborane*(*1:1*), 9CI. *Trimethylamine monofluoroborane. Monofluoroborane trimethylamine*

[35449-24-2]

$Me_3N—BH_2F$

M 90.935
Solid by subl. Mp 117-8°.

Van Paasschen, J.M. *et al*, *J. Am. Chem. Soc.*, 1972, **94**, 2680 (*synth, pmr, nmr, ir*)
Geanangel, R.A., *Inorg. Chem.*, 1975, **14**, 696 (*cmr*)
Denniston, M.L. *et al*, *J. Inorg. Nucl. Chem.*, 1976, **38**, 979 (*nmr*)
Weidig, C. *et al*, *Inorg. Chem.*, 1976, **15**, 1785.
Cassoux, P. *et al*, *J. Mol. Struct.*, 1978, **48**, 25 (*struct, ir*)
Van Paasschen, J.M. *et al*, *Inorg. Chem.*, 1978, **17**, 3302.

$C_3H_{11}BIN$ B-00200

Iodoborane trimethylamine

(N,N-*Dimethylmethanamine*)*dihydroiodoboron*, 10CI. N,N-*Dimethylmethanamine compd. with iodoborane*(*1:1*), 9CI. *Iodoborane compd. with trimethylamine*(*1:1*), 8CI
[25741-81-5]

$Me_3N—BH_2I$

M 198.841
White cryst. solid (CCl_4/cyclohexane), moisture- and light-sensitive. Sol. CCl_4, C_6H_6, insol. alkanes. Mp 73°. Subl. near Mp, moisture- and light-sensitive.
▷YH2720000.

Brown, M.P. *et al*, *J. Chem. Soc.* (*A*), 1970, 410 (*ir*)
Lowe, J.R. *et al*, *Inorg. Chem.*, 1970, **9**, 1423.
Inorg. Synth., 1970, **12**, 116 (*synth*)
Bratt, P.J. *et al*, *Inorg. Chim. Acta*, 1974, **10**, 191 (*ir*)
Myers, W.H. *et al*, *Inorg. Chem.*, 1975, **14**, 2874 (*pmr*)
Denniston, M.L. *et al*, *J. Inorg. Nucl. Chem.*, 1976, **38**, 379 (*pmr, nmr*)

$C_3H_{11}BIP$ B-00201

Iodoborane trimethylphosphine

Dihydroiodo(*trimethylphosphine*)*boron*, 10CI. *Iodoborane complex with trimethylphospine*, 9CI. *Trimethylphosphine iodoborane*
[60228-73-1]

$IBH_2—PMe_3$

M 215.808
Moisture-sensitive cryst. (CH_2Cl_2/pet. ether). Mp 69.5-71°.

Denniston, M.L. *et al*, *J. Inorg. Nucl. Chem.*, 1976, **38**, 979 (*synth*)
Sisler, H.H. *et al*, *J. Inorg. Nucl. Chem.*, 1977, **39**, 1745 (*synth, nmr, ir*)
Pennington, B.T. *et al*, *J. Inorg. Nucl. Chem.*, 1978, **40**, 389 (*synth, pmr*)

$C_3H_{11}BN_2$ B-00202

Methylbis(methylamino)borane, 8CI

N,N′,*1-Trimethylboranediamine*, 10CI, 9CI
[18029-12-4]

$MeB(NHMe)_2$

M 85.943
Liq., sensitive to air and moisture. Sol. Et_2O. Bp_{20} 52°.

Nöth, H. *et al*, *Chem. Ber.*, 1968, **101**, 969 (*synth*)
Nöth, H. *et al*, *J. Organomet. Chem.*, 1968, **12**, 23 (*pmr*)
Beck, W. *et al*, *Chem. Ber.*, 1972, **105**, 2883 (*nmr*)
Arne, A. *et al*, *Acta Chem. Scand., Ser. A*, 1981, **35**, 341 (*struct*)
Goetze, R. *et al*, *Chem. Ber.*, 1981, **114**, 1884.

$C_3H_{12}BN$ B-00203

Borane trimethylamine

(N,N-*Dimethylmethanamine*)*trihydroboron*, 10CI. N,N-*Dimethylmethanamine, compd. with borane*(*1:1*). *Trimethylamine borane*
[75-22-9]

$Me_3N—BH_3$

M 72.945
Commercially available. Source of borane for the reduction of ketones and Schiff bases and for hydroboration of alkenes. White cryst. solid. Stable, easy to handle. Insol. hexane, v. sol. most org. solvs., sl. sol. H_2O, cyclohexane. Mp 92-4°. Bp 172°.
▷PA0525000.

Burg, A.B. *et al*, *J. Am. Chem. Soc.*, 1937, **59**, 780 (*synth*)
Billman, J.H. *et al*, *J. Org. Chem.*, 1962, **27**, 2640 (*use*)
Larsen, L.A. *et al*, *Inorg. Chem.*, 1974, **13**, 2284 (*ms*)
Beachley, O.T. *et al*, *Inorg. Chem.*, 1975, **14**, 120 (*ir, pmr*)
Cassoux, P. *et al*, *Inorg. Chem.*, 1975, **14**, 126 (*far ir*)
Geanangel, R.A., *Inorg. Chem.*, 1975, **14**, 696 (*nmr, cmr*)

$C_3H_{12}BN$ B-00204

Trimethylborane ammonia

Amminetrimethylboron, 10CI. *Trimethylborane monoammoniate*, 9CI, 8CI. *Trimethylborane ammoniate. Ammonia trimethylborane complex*
[1830-95-1]

$Me_3B—NH_3$

M 72.945
Cryst. Mp 73-4°. Heat of dissociation = 13.8 kcal. $mol.^{-1}$.

Brown, H.C. *et al*, *J. Am. Chem. Soc.*, 1944, **66**, 435 (*synth, props*)
Resing, H.A., *J. Chem. Phys.*, 1965, **63**, 669 (*nmr*)
Sawodny, W. *et al*, *Z. Physik. Chem.*, 1965, **44**, 227 (*spectra*)
Gol'dshtein, I.P. *et al*, *Zh. Obshch. Khim.*, 1968, **38**, 1984.
Purser, J.M. *et al*, *Inorg. Chem.*, 1968, **7**, 2156 (*nmr*)
Nöth, H. *et al*, *Chem. Ber.*, 1974, **107**, 3070 (*nmr*)
Paetzold, P. *et al*, *Chem. Ber.*, 1979, **112**, 654.

$C_3H_{12}BNO_3$ B-00205

Trimethoxyborane ammonia

Amminetrimethoxyboron, 10CI. *Boric acid trimethyl ester, monoammoniate. Methyl borate ammonia complex*
[66794-43-2]

$(MeO)_3B—NH_3$

M 120.943
Solid. Mp 45° subl., 72°. Stable to at least 375°, does not react. with NaOH.

Goubeau, J. *et al*, *Z. Anorg. Allg. Chem.*, 1951, **266**, 161; **267**, 27 (*synth*)
U.S.P., 2 629 732, (*1953*); *CA*, **47**, 7171i
U.S.P., 4 196 177, (*1980*); *CA*, **93**, P49645y

$C_3H_{12}BN_3$ B-00206

Tris(methylamino)borane, 8CI

N,N′,N″-*Trimethylboranetriamine*, 10CI, 9CI
[7397-44-6]

$B(NHMe)_3$

M 100.958
Liq., sensitive to moisture. Sol. nonpolar solvs. d_4^{20} 0.887. Bp_{48} 70-80°. n_D^{25} 1.4432.

Aubrey, D.W. *et al*, *J. Chem. Soc.*, 1959, 2927 (*synth, props*)
English, W.D. *et al*, *J. Am. Chem. Soc.*, 1961, **83**, 2122 (*synth*)
Davis, F.A. *et al*, *J. Org. Chem.*, 1971, **36**, 1300 (*nmr*)
Beck, W. *et al*, *Chem. Ber.*, 1972, **105**, 2883 (*nmr*)
Niedenzu, K. *et al*, *Z. Anorg. Allg. Chem.*, 1972, **387**, 107 (*nmr*)
Beachley, O.T. *et al*, *Inorg. Chem.*, 1977, **16**, 2935 (*pmr, props*)
Almenningen, A. *et al*, *Acta Chem. Scand., Ser. A*, 1981, **35**, 341 (*struct*)

$C_3H_{12}BO_3P$ B-00207

Borane trimethyl phosphite

Trihydro(trimethyl phosphite-P)boron, 10CI. *Phosphorous acid trimethyl ester compd. with borane(1:1)*, 9CI, 8CI. *Methyl phosphite compd. with borane*

[6867-39-6]

$(MeO)_3P—BH_3$

M 137.910

Liq. Bp_{23} 86°. n_D^{25} 1.4162.

Reetz, T., *J. Am. Chem. Soc.*, 1960, **82**, 5039 (*synth*)
McFarlane, H.C.E. *et al*, *J. Chem. Soc., Faraday Trans. 2*, 1972, **68**, 1300 (*nmr*)
Fussstetter, H. *et al*, *Chem. Ber.*, 1977, **110**, 3172 (*nmr*)
Griend, V. *et al*, *J. Am. Chem. Soc.*, 1977, **99**, 2459 (*ir, nmr*)
Hassler, K., *Ber. Bunsenges. Phys. Chem.*, 1980, **84**, 285 (*ir, raman*)
Cowley, A.H. *et al*, *Inorg. Chem.*, 1982, **21**, 85 (*pe*)

$C_3H_{12}BP$ B-00208

Borane trimethylphosphine

Trihydro(trimethylphosphine)boron, 10CI. *Trimethylphosphine compd. with borane(1:1)*, 8CI. *Trimethylphosphine borine*

[1898-77-7]

$Me_3P—BH_3$

M 89.912

Solid. Antiknock agent in petrol. Mp 103-103.5°.

Burg, A.B. *et al*, *J. Am. Chem. Soc.*, 1953, **75**, 3872.
Sisler, H.H. *et al*, *J. Inorg. Nucl. Chem.*, 1977, **39**, 1745 (*ir*)
Van Paasschen, J.M. *et al*, *Inorg. Chem.*, 1978, **17**, 3302.
Apel, J. *et al*, *Z. Anorg. Allg. Chem.*, 1979, **453**, 28 (*nmr*)
Cowley, A.H. *et al*, *Inorg. Chem.*, 1982, **21**, 85 (*pe*)
Kameda, M. *et al*, *Inorg. Chem.*, 1982, **21**, 1267 (*nmr*)

$C_3H_{12}B_3N_3$ B-00209

1,3,5-Trimethylborazine, 10CI, 9CI, 8CI

N,N,N-*Trimethylborazole. Hexahydro-1,3,5-trimethyl-1,3,5-triazatriborine*

[1004-35-9]

As 2,4,6-Tribromoborazine, B-00015 with

X = H, Y = Me

M 122.578

Colourless liq. Mp −8.2° to −7.5°. Bp 131-3°, Bp_{15} 28°. n_D^{20} 1.4404, 1.4375.

Schaeffer, G.W. *et al*, *J. Am. Chem. Soc.*, 1949, **71**, 2143 (*synth*)
Hough, W.V. *et al*, *J. Am. Chem. Soc.*, 1955, **77**, 864 (*synth*)
Steinberg, H. *et al*, *Organoboron Chemistry, Vol. 2*, Wiley, 1966, 175 (*synth, props*)
Inorg. Synth., 1967, **9**, 8 (*synth*)
Kroner, J. *et al*, *Tetrahedron*, 1972, **28**, 1585 (*pe*)
Beaumelou, G. *et al*, *Org. Magn. Reson.*, 1973, **5**, 585 (*pmr*)
Melcher, L.A. *et al*, *Inorg. Chem.*, 1973, **12**, 601 (*ms*)
Blick, K.E. *et al*, *Z. Anorg. Allg. Chem.*, 1975, **417**, 19 (*ir, raman*)
Destefano, A. *et al*, *Inorg. Chem.*, 1975, **14**, 2882 (*deriv, props, ms*)
Wrackmeyer, B. *et al*, *Chem. Ber.*, 1976, **109**, 3480 (*nmr*)
Nöth, H. *et al*, *Chem. Ber.*, 1981, **114**, 1150 (*nmr, cmr*)

$C_3H_{12}B_3N_3$ B-00210

2,4,6-Trimethylborazine, 10CI, 9CI, 8CI

B,B,B-*Trimethylborazole*

[5314-85-2]

As Borazine, B-00020 with

R = Me, R′ = H

M 122.578

Solid with unpleasant odour. Sol. most org. solvs. d_{35} 0.84. Mp 29-31°. Bp 129°, Bp_{20} 35°. V.p. 16.3 mm at 29.8°. Trouton constant 23.3 e.u. High thermal stability; v. slowly hydrol. by water.

Schlesinger, H.I. *et al*, *J. Am. Chem. Soc.*, 1936, **58**, 409 (*synth*)
Haworth, D.T. *et al*, *J. Am. Chem. Soc.*, 1960, **82**, 3860 (*ir, synth*)
Wagner, R.I. *et al*, *Inorg. Chem.*, 1962, **1**, 93 (*ir, synth*)
Anzenhofer, K., *Mol. Phys.*, 1966, **11**, 495 (*cryst struct*)
Nöth, H. *et al*, *J. Organomet. Chem.*, 1968, **12**, 23 (*pmr*)
Wrackmeyer, B. *et al*, *Chem. Ber.*, 1976, **109**, 3480 (*nmr*)
Blick, K.E. *et al*, *Z. Anorg. Allg. Chem.*, 1978, **442**, 183 (*ir*)
Nöth, H. *et al*, *Chem. Ber.*, 1981, **114**, 1150 (*nmr*)

$C_3H_{12}B_3N_3O_3$ B-00211

2,4,6-Trimethoxyborazine, 10CI, 8CI

Tri-B-*methoxyborazole*

[15185-53-2]

As 2,4,6-Tribromoborazine, B-00015 with

X = OMe, Y = H

M 170.576

Moisture-sensitive solid. Mp 110-2° (102-5°). Bp 60° subl. *in vacuo*.

Haworth, D.T. *et al*, *J. Am. Chem. Soc.*, 1959, **81**, 842 (*synth, props*)
Mikhailov, B.M. *et al*, *Izv. Akad. Nauk SSSR, Ser. Khim.*, 1961, 371.
Meller, A. *et al*, *Monatsh. Chem.*, 1967, 390 (*ir*)
Wrackmeyer, B. *et al*, *Chem. Ber.*, 1976, **109**, 3480 (*nmr*)
Biffar, W. *et al*, *Chem. Ber.*, 1980, **113**, 333 (*nmr*)

$C_3H_{14}AlB_9^{\ominus\ominus}$ B-00212

Methyl(7,8,9,10,11-η)undecahydro-7,8-dicarbaundecaborato(2−)aluminum

*3-Methyl-3-alumino-1,2-dicarba*closo*dodecaborane(12)*

[34540-16-4]

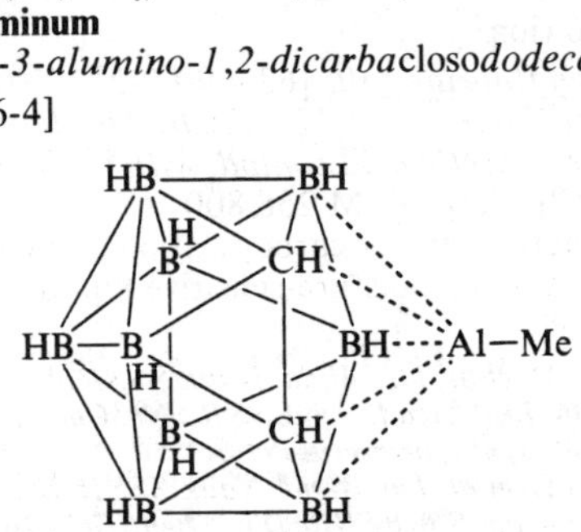

M 174.415 (ion)

White cryst. Mp 97-100°. $Bp_{0.01}$ 90° subl.

▷Explodes with water

Young, D.A.T. *et al*, *J. Am. Chem. Soc.*, 1971, **93**, 5687 (*synth, ir, pmr, nmr*)

$C_3H_{16}B_5ClSi$ B-00213

2-[(Chlorodimethylsilyl)methyl]pentaborane(9), 10CI

[63838-59-5]

$(B_5H_8)CH_2SiMe_2Cl$

M 169.748

Leach, J.B. *et al*, *J. Chem. Soc., Dalton Trans.*, 1977, 819 (*nmr, ms, synth*)

$C_3H_{17}B_5Pb$ B-00214

2,3-μ-(Trimethylplumbyl)pentaborane(9), 8CI

[22142-47-8]

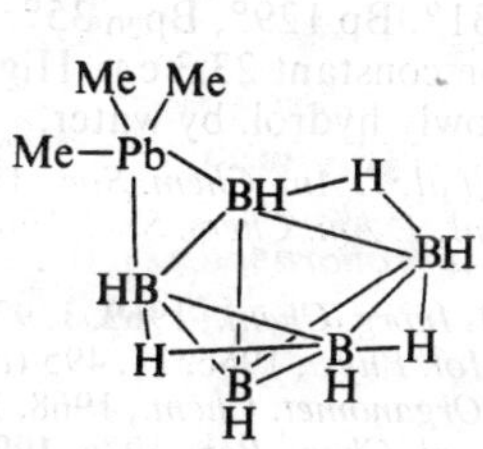

M 314.417

Presumed struct. Unstable liq. Mp −5° dec.

Gaines, D.F. *et al*, *J. Am. Chem. Soc.*, 1968, **90**, 6617.

$C_4HBF_6O_4$ B-00215

Bis(trifluoroacetoxy)borane

Trifluoroacetic acid dianhydride with boronic acid, 10CI

[67177-55-3]

$(F_3COOC)_2BH$

M 237.850

Formed as THF complex (1:1) and normally used *in situ*. Acid-stable reducing agent.

Maryanoff, B.E. *et al*, *J. Org. Chem.*, 1978, **43**, 2733 (*synth, use*)
U.S.P., 4 210 590, (*1980*); *CA*, **93**, 204454f (*use*)
Fieser, M. *et al*, *Reagents for Organic Synthesis*, Wiley, 1967-83, **8**, 45 (*use*)

$C_4H_4BCl_2F_2O_4^\ominus$ B-00216

Bis(chloroacetato-*O*)difluoroborate(1−), 10CI

$(ClCH_2COO)_2BF_2^\ominus$

M 235.786 (ion)

Dihydrogen fluoride salt: [63399-62-2]. *Bis(chloroacetato-O)difluoroborate(1−)hydrogen fluoride(1:2:1)*, 10CI. *Chloroacetic acid compd. with trifluoroborane.*

$C_4H_6BCl_2F_3O_4$ M 256.800

Highly active cat. for carbonylation of alcohols and alkenes. Stable, moisture sensitive liq. d_4^{20} 1.585. Bp 53-4°.

Pirozhkov, S.D. *et al*, *Izv. Akad. Nauk SSSR, Ser. Khim.*, 1976, 1314, 1534; *Bull. Acad. Sci. USSR, Div. Chem. Sci.*, **25**, 1264, 1459 (*synth, props, use*)
Pirozhkov, S.D. *et al*, *Izv. Akad. Nauk SSSR, Ser. Khim.*, 1977, 817; *Bull. Acad. Sci. USSR, Div. Chem. Sci.*, **26**, 744.
Pirozhkov, S.D. *et al*, *Izv. Akad. Nauk SSSR, Ser. Khim.*, 1978, 110; *Bull. Acad. Sci. USSR, Div. Chem. Sci.*, **27**, 96.

C_4H_5B B-00217

Borole, 10CI, 9CI

1-H-Borole

[287-87-6]

M 63.894

Parent member of the borole series of organoboron compds., but not known in the free state. See Pentaphenylborole, B-01343 .

Eisch, J.J., *CA*, 1973, **78**, 96987m.
Allinger, N.L. *et al*, *J. Am. Chem. Soc.*, 1975, **97**, 752.
Van der Kerk, S.M., *J. Organomet. Chem.*, 1981, **215**, 315.

$C_4H_5BO_3$ B-00218

2-Furanyldihydroxyborane

(2-Furanyl)boronic acid, 10CI, 9CI. *2-Furanboronic acid. 2-Furyldihydroxyborane*

[13331-23-2]

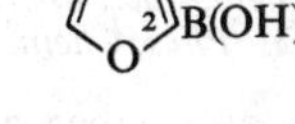

M 111.892

Sludge inhibitor for fuels. Cryst. (H_2O). Mp 110° dec.

B.P., 1 031 533, (*1966*); *CA*, **65**, 10404a
Breuer, S.W. *et al*, *Tetrahedron Lett.*, 1974, 3719.
Roques, B.P. *et al*, *J. Heterocycl. Chem.*, 1975, **12** 195.
Florentin, D. *et al*, *Bull. Soc. Chim. Fr.*, 1976, 1999 (*synth*)
Florentin, D. *et al*, *J. Heterocycl. Chem.*, 1976, **13**, 1265.

$C_4H_5BO_3$ B-00219

3-Furanyldihydroxyborane

(3-Furanyl)boronic acid, 10CI, 9CI. *3-Furyldihydroxyborane. 3-Furanboronic acid*

[55552-70-0]

M 111.892

Cryst. (H_2O). Mp 126-8°.

Roques, B.P. *et al*, *J. Heterocycl. Chem.*, 1975, **12**, 195.
Florentin, D. *et al*, *Bull. Soc. Chim. Fr.*, 1976, 1999 (*synth*)
Florentin, D. *et al*, *J. Heterocycl. Chem.*, 1976, **13**, 1265.

C_4H_6BCl B-00220

Chlorodivinylborane

Chlorodiethenylborane, 10CI, 9CI. *Chlorodivinylborine*

[10147-89-4]

$(H_2C{=}CH)_2BCl$

M 100.354

Liq.; under N_2 undergoes slow dec. at 20°; sensitive to air and moisture. Mp −109.6°. Bp 71.1°. Trouton constant 22.9eu ΔH_{vap} 7.88 Kcal mol^{-1}.

Coyle, T.D. *et al*, *J. Chem. Soc.*, 1961, 3108 (*ir*)
Brinckman, F.E. *et al*, *J. Am. Chem. Soc.*, 1962, **82**, 6218 (*synth, props*)
Good, C.D. *et al*, *J. Am. Chem. Soc.*, 1962, **84**, 1162 (*synth*)
Hall, L.W. *et al*, *J. Am. Chem. Soc.*, 1975, **97**, 4527 (*nmr*)
Wrackmeyer, B. *et al*, *Chem. Ber.*, 1976, **109**, 1075 (*nmr*)
Odom, J.D. *et al*, *J. Organomet. Chem.*, 1980, **201**, 351 (*props*)

$C_4H_6BF_2O_4^{\ominus}$ B-00221

Bis(acetato-*O*)difluoroborate(1−), 10CI

Diacetoxydifluoroborate

$(AcO)_2BF_2^{\ominus}$

M 166.896 (ion)

Dihydrogen fluoride salt: [373-61-5]. *Acetic acid compd. with trifluoroborane(2:1)*, 9CI.

$C_4H_8BF_3O_4$ M 187.910

Highly active cat. for carbonylation of alcohols and alkenes. Moisture-sensitive liq. d_4^{20} 1.366.

Muetterties, E.L., *J. Am. Chem. Soc.*, 1958, **80**, 4526 (*synth*)
Taillandier, M. *et al*, *J. Mol. Struct.*, 1971, **10**, 471 (*struct, ir*)
Pirozhkov, S.D. *et al*, *Izv. Akad. Nauk SSSR, Ser. Khim.*, 1976, 1314, 1534; *Bull. Acad. Sci. USSR, Div. Chem. Sci.*, **25**, 1264, 1459 (*synth, props, use*)
Pirozhkov, S.D. *et al*, *Izv. Akad. Nauk SSSR, Ser. Khim.*, 1977, 817; *Bull. Acad. Sci. USSR, Div. Chem. Sci.*, **26**, 744.
Pirozhkov, S.D. *et al*, *Izv. Akad. Nauk SSSR, Ser. Khim.*, 1978, 110; *Bull. Acad. Sci. USSR, Div. Chem. Sci.*, **27**, 96.

$C_4H_6BF_3N_2$ B-00222

Trifluoroborane 2-methylimidazole

*Trifluoro(2-methyl-1*H*-imidazole-*N[3]*)boron*, 10CI. *2-Methyl-1*H*-imidazole compd. with trifluoroborane. Boron trifluoride 2-methylimidazole*

[69080-35-9]

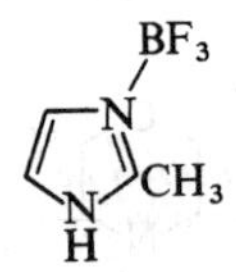

M 149.910

Crosslinking catalyst for epoxy resins.

Japan. Pat., 78 114 900, (*1978*); *CA*, **90**, 72868v (*use*)
Japan. Pat., 80 25 474, (*1980*); *CA*, **93**, 48074z
Japan. Pat., 80 142 023, (*1980*); *CA*, **94**, 140669r

$C_4H_6B_2$ B-00223

2,3,4,5-Tetracarbahexaborane(6), 10CI

[28323-17-3]

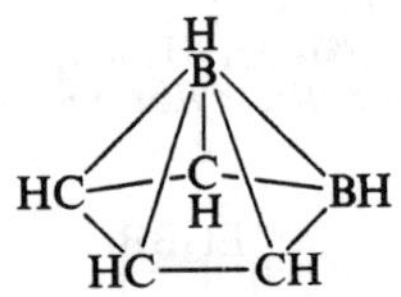

M 75.711

nido-. Isoelectronic with B_6H_{10} without any bridge H atoms. Liq. (by dist. *in vacuo*).

Groszek, E. *et al*, *Inorg. Chem.*, 1971, **10**, 2770 (*synth, ir, ms, nmr*)
Miller, V.R. *et al*, *Inorg. Chem.*, 1972, **11**, 862 (*synth*)
Pasinski, J.P. *et al*, *J. Chem. Soc., Chem. Commun.*, 1973, 928 (*microwave*)

C_4H_7B B-00224

Bicyclo[1.1.0]but-1-ylborane, 10CI

[69804-77-9]

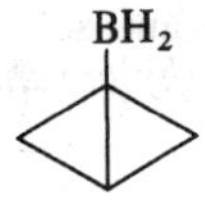

M 65.909

Uncomplexed material would exist as a bridged dimer. Air- and moisture-sensitive.

Greenberg, A., *Tetrahedron Lett.*, 1978, **38**, 3509 (*synth*)
Dill, J.D. *et al*, *J. Am. Chem. Soc.*, 1979, **101**, 6814.

$C_4H_7BO_2$ B-00225

Ethynyldimethoxyborane

Ethynylboronic acid dimethyl ester, 10CI. *Ethyneboronic acid dimethyl ester*, 8CI. *Dimethyl ethynylboronate*

[13722-62-8]

$HC{\equiv}CB(OMe)_2$

M 97.908

Mobile liq., air- and moisture-sensitive, v. volatile. Bp_{262} 59-60°.

B.P., 884 547, (*1961*); *CA*, **57**, 7308f (*synth*)
Soulié, J. *et al*, *Bull. Soc. Chim. Fr.*, 1966, 3846 (*synth, ir, uv*)
Wrackmeyer, B. *et al*, *Chem. Ber.*, 1977, **110**, 1086 (*nmr*)

$C_4H_8BF_3O$ B-00226

Trifluoroborane tetrahydrofuran

Trifluoro(tetrahydrofuran)boron, 10CI. *Tetrahydrofuran compd. with trifluoroborane(1:1)*, 9CI. *Tetrahydrofuran compd. with boron fluoride (BF_3)(1:1)*, 8CI

[462-34-0]

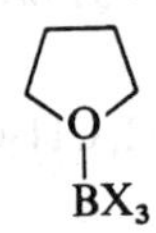

X = F

M 139.912

Catalyst. Moisture-sensitive white solid or colourless liq. Mp 12°.

McLaughlin, D.E. *et al*, *J. Am. Chem. Soc.*, 1960, **82**, 5621 (*pmr*)
Begun, G.M. *et al*, *J. Chem. Phys.*, 1963, **38**, 2112 (*raman, ir*)
Fratiello, A. *et al*, *Inorg. Chem.*, 1969, **8**, 480 (*pmr*)
Krysin, A.P. *et al*, *Izv. Sib. Otd. Akad. Nauk SSSR, Ser. Khim. Nauk*, 1977, 151.
Grinevich, T.V. *et al*, *Vysokomol. Soedin., Ser. A*, 1980, **22**, 1576 (*use*)
Hui, B.C., *Inorg. Chem.*, 1980, **19**, 3185 (*use*)

C_4H_9B B-00227

Borolane, 10CI, 9CI, 8CI

Tetrahydroborole. Boracyclopentane

[1692-31-5]

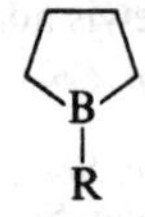

R = H

M 67.925

Unknown. The compd. previously thought to be the bridged dimer of borolane is 1,6-Diboracyclodecane, B-00566 .

Köster, R. *et al*, *Angew. Chem., Int. Ed. Engl.*, 1963, **2**, 219.
Breuer, E. *et al*, *J. Am. Chem. Soc.*, 1969, **91**, 4164 (*struct*)
Brown, H.C. *et al*, *J. Am. Chem. Soc.*, 1971, **93**, 6682.
Inorg. Synth., 1979, **19**, 239 (*synth, struct, ir, nmr*)

$C_4H_9BBr_2$ B-00228

Dibromobutylborane, 9CI

Butylboron dibromide. Butyldibromoborane

[1001-49-6]

$H_3CCH_2CH_2CH_2BBr_2$

M 227.733

Unstable liq., air- and moisture-sensitive. Bp_{23} 65°, $Bp_{0.11}$ 49-50°.

Gerrard, W. *et al*, *J. Chem. Soc.*, 1955, 1470; 3084 (*synth*)
Schmidt, M. *et al*, *Z. Naturforsch., B*, 1970, **25**, 1344 (*synth*)

$C_4H_9BCl_2$ B-00229

tert-Butyldichloroborane

Dichloro(1,1-dimethylethyl)borane, 10CI

[76873-78-4]

$Cl_2BC(CH_3)_3$

M 138.831

Air- and moisture-sensitive.

Bogatskii, A.V. *et al*, *CA*, 1981, **94**, 121624f (*props*)

$C_4H_9BCl_2$ B-00230

Butyldichloroborane, 10CI, 9CI

Butylboron dichloride. Dichloro(butyl)borane

[14090-22-3]

$H_3CCH_2CH_2CH_2BCl_2$

M 138.831

Liq., sensitive to air and moisture. Bp 106-8°.

▷Pyrophoric

McCusker, P.A. *et al*, *J. Am. Chem. Soc.*, 1957, **79**, 5192 (*synth*)
Inorg. Synth., 1967, **10**, 126 (*synth*)
Brown, H.C. *et al*, *J. Organomet. Chem.*, 1972, **44**, 233 (*synth*)
Barlos, K. *et al*, *Chem. Ber.*, 1977, **110**, 2774 (*nmr*)
Leleu, J., *Cah. Notes Doc.*, 1977, **88**, 361 (*hazs, rev*)

$C_4H_9BF_2$ B-00231

Butyldifluoroborane, 10CI, 9CI

Butyldifluoroborine

[372-53-2]

$F_2BCH_2CH_2CH_2CH_3$

M 105.922

Catalyst for polymerisation of vinyl compds. Fuming liq., sensitive to air and moisture. d_{20} 0.851. Bp_{742} 36.3-36.5°. n_D^{19} 1.3300. Lewis acid.

McCusker, P.A. *et al*, *J. Am. Chem. Soc.*, 1955, **77**, 4253 (*synth*)
Brindley, P.B. *et al*, *J. Chem. Soc.*, 1956, 824 (*synth*)
Buls, V.W. *et al*, *J. Am. Chem. Soc.*, 1957, **79**, 337 (*synth*)
Coyle, T.D. *et al*, *J. Chem. Soc.*, 1961, 3103 (*nmr, ir, uv*)
Barlos, K. *et al*, *Chem. Ber.*, 1977, **110**, 2774 (*nmr*)
Costes, J.P. *et al*, *J. Organomet. Chem.*, 1979, **175**, 257 (*props*)

$C_4H_9BF_2O$ B-00232

Butoxydifluoroborane, 10CI, 9CI, 8CI

Butoxydifluoroborine

[372-84-9]

$F_2BOCH_2CH_2CH_2CH_3$

M 121.921

Exists as dimer and trimer. Fuming liq., sensitive to moisture. d_{20} 1.122. Bp_{11} 43°. n_D^{22} 1.3858. Lewis acid.

Lappert, M.F., *J. Chem. Soc.*, 1955, 784 (*synth*)
Brindley, P.B. *et al*, *J. Chem. Soc.*, 1956, 824 (*synth, props*)
Gerrard, W. *et al*, *J. Chem. Soc.*, 1957, 381 (*synth*)
Lappert, M.F. *et al*, *J. Chem. Soc.*, 1958, 3256.
Nöth, H. *et al*, *Chem. Ber.*, 1966, **99**, 1049 (*nmr*)
Gerrard, W. *et al*, *J. Inorg. Nucl. Chem.*, 1967, **29**, 943 (*synth, ir, nmr, ms, props*)
B.P., 2 030 993, (*1980*); *CA*, **93**, 221317 (*props*)

$C_4H_9BF_3^{\ominus}$ B-00233

Butyltrifluoroborate(1–), 10CI

$H_3C(CH_2)_3BF_3^{\ominus}$

M 124.920 (ion)

Cu(I) salt: [65139-99-3].

$C_4H_9BCuF_3$ M 188.466

Representative of intermeds. produced in BF_3-mediated reacns. of Cu(I) alkyls.

Maruyama, K. *et al*, *J. Am. Chem. Soc.*, 1977, **99**, 8068 (*props*)
Yamamoto, Y. *et al*, *J. Am. Chem. Soc.*, 1978, **100**, 3240; 1980, **102**, 2318 (*props*)

$C_4H_9BF_3NO$ B-00234

Trifluoroborane morpholine

Trifluoro(morpholine-N⁴)boron, 10CI. *Morpholine, compd. with trifluoroborane(1:1)*, 9CI

[660-92-4]

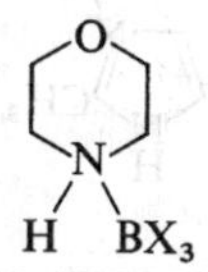

X = F

M 154.926

Pisarev, M.A. *et al*, *CA*, 1973, **79**, 89955s.
Kugai, N. *et al*, *Adv. Exp. Med. Biol.*, 1977, **103**, 537.
Syutkin, E.A. *et al*, *Probl. Endokrinol.*, 1980, **26**, 53.

$C_4H_{10}BBr$ B-00235

Bromodiethylborane, 10CI, 9CI, 8CI

Bromodiethylborine. Diethylbromoborane

[19162-10-8]

Et_2BBr

M 148.837

Air- and moisture-sensitive. Sol. THF.

Wiberg, E. *et al*, *Z. Anorg. Allg. Chem.*, 1933, **213**, 77 (*synth*)
Nöth, H. *et al*, *J. Organomet. Chem.*, 1968, **12**, 23 (*pmr*)
Köster, R. *et al*, *Justus Liebigs Ann. Chem.*, 1969, **719**, 169 (*synth*)
Nöth, H. *et al*, *Synth. React. Inorg. Metal-Org. Chem.*, 1971, 1197 (*synth*)
Haubold, W. *et al*, *Z. Anorg. Allg. Chem.*, 1976, **420**, 251 (*synth, ir*)
Paetzold, P. *et al*, *Chem. Ber.*, 1979, **112**, 663.

$C_4H_{10}BBr_2N$ B-00236

Dibromo(diethylamino)borane, 8CI

1,1-Dibromo-N,N-diethylboranamine, 10CI, 9CI

[868-29-1]

Br_2BNEt_2

M 242.748

Moisture-sensitive liq. d_4^{12} 1.608. Bp_5 90°. n_4^{18} 1.4946.

Wolfgang, B. *et al*, *Chem. Ber.*, 1972, **105**, 2883 (*nmr*)
Spitsyn, V.I. *et al*, *Izv. Akad. Nauk SSSR, Ser. Khim.*, 1974, 2626; *Bull. Acad. Sci. USSR, Div. Chem. Sci.*, 2535 (*synth*)
Bryukhova, E.V. *et al*, *Izv. Akad. Nauk SSSR, Ser. Khim.*, 1976, 1614; *Bull. Acad. Sci. USSR, Div. Chem. Sci.*, 1529 (*nmr*)
Balabanova, E.A. *et al*, *Izv. Akad. Nauk SSSR, Ser. Khim.*, 1978, 2638; *Bull. Acad. Sci. USSR, Div. Chem. Sci.*, 2360 (*nmr*)
Nöth, H. *et al*, *Chem. Ber.*, 1981, **114**, 1150 (*nmr*)
Nöth, H. *et al*, *Inorg. Chem.*, 1982, **21**, 706.

$C_4H_{10}BCl$ B-00237

Chlorodiethylborane, 10CI, 9CI

Chlorodiethylborine. Diethylchloroborane

[5314-83-0]

Et_2BCl

M 104.386

Catalyst for polymerisation of alkenes and butadiene oligomerisation. Liq., sensitive to air and moisture. Bp_{100} 25°°.

Buls, V.W. *et al*, *J. Am. Chem. Soc.*, 1957, **79**, 337 (*synth*)
Gipstein, E. *et al*, *J. Org. Chem.*, 1966, **31**, 1087 (*synth*)
Nöth, H. *et al*, *Chem. Ber.*, 1966, **99**, 1049 (*nmr*)
Nöth, H. *et al*, *J. Organomet. Chem.*, 1968, **12**, 23 (*nmr*)
Köster, R. *et al*, *Justus Liebigs Ann. Chem.*, 1969, **719**, 187 (*props*)
Nöth, H. *et al*, *Synth. React. Inorg. Metal-Org. Chem.*, 1971, **1**, 197 (*synth*)
Barlos, K. *et al*, *Chem. Ber.*, 1977, **110**, 2774 (*nmr*)
Sax, N.I., *Dangerous Properties of Industrial Materials*, 5th Ed., Van Nostrand-Reinhold, 1979, 491.

$C_4H_{10}BClO$ B-00238

Chloroborane tetrahydrofuran

Chlorodihydro(tetrahydrofuran)boron, 10CI. Chloroborane compd. with tetrahydrofuran(1:1), 9CI

[55606-72-9]

M 120.385

Not likely to be a v. stable material and certainly moisture sensitive.

Kodama, G. *et al*, *Inorg. Chem.*, 1975, **14**, 2243.
Brown, H.C. *et al*, *J. Am. Chem. Soc.*, 1976, **98**, 1785 (*synth, nmr*)
Drummond, A. *et al*, *Inorg. Chim. Acta*, 1977, **24**, 191.
Dodds, A.R. *et al*, *Inorg. Chem.*, 1979, **18**, 1465.

$C_4H_{10}BClO_2$ B-00239

Chlorodiethoxyborane, 10CI, 9CI

Chlorodiethoxyborine

[20905-32-2]

$(EtO)_2BCl$

M 136.385

Catalyst for polymerisation of alkenes. Liq. Easily hydrolysed by moisture. d_4^{20} 0.972. Bp_{60} 45.5-46°. n_D^{20} 1.3878.

Kinney, C.R. *et al*, *J. Am. Chem. Soc.*, 1935, **57**, 2396 (*props*)
Skinner, H.A. *et al*, *J. Chem. Soc.*, 1954, 3930 (*synth, props*)
Abel, E.W. *et al*, *J. Chem. Soc.*, 1957, 501 (*synth*)
Onak, T.P. *et al*, *J. Phys. Chem.*, 1959, **63**, 1533 (*nmr*)
Köster, R. *et al*, *Justus Liebigs Ann. Chem.*, 1969, **719**, 169 (*synth*)
Ger. Pat., 2 951 673, (*1980*); *CA*, **93**, 240269h (*use*)

$C_4H_{10}BCl_2N$ B-00240

Dichloro(diethylamino)borane, 8CI

*1,1-Dichloro-*N,N*-diethylboranamine, 10CI, 9CI. Dichloro(diethylamino)borine. Diethylaminodichloroborane*

[868-30-4]

Cl_2BNEt_2

M 153.846

Moisture-sensitive liq. Bp 140-6°, Bp_{20} 51-2°.

Py complex (1:3): [15590-32-6].

$C_{19}H_{25}BCl_2N_4$ M 391.149

White cryst. solid. Mp 82-6°. Hydrolyses readily in atm.

Lappert, M.F. *et al*, *J. Chem. Soc. (A)*, 1967, 602 (*synth*)
Greenwood, N.N. *et al*, *J. Chem. Soc. (A)*, 1967, 959 (*ir, synth*)
Beck, W. *et al*, *Chem. Ber.*, 1972, **105**, 2883 (*nmr*)
Lochmueller, C.H. *et al*, *J. Chromatogr.*, 1973, **85**, 132 (*chromatog*)
Baudler, M. *et al*, *Z. Anorg. Allg. Chem.*, 1981, **474**, 18 (*synth*)
Nöth, H. *et al*, *Chem. Ber.*, 1981, **114**, 1150 (*cmr, nmr*)

$C_4H_{10}BCl_3O$ B-00241

Trichloroborane diethyl ether

Trichloro[1,1′-oxybis[ethane]]boron, 10CI. Ethyl ether compd. with boron chloride(BCl_3)(1:1), 8CI. Boron trichloride etherate

[2102-03-6]

$Et_2O—BCl_3$

M 191.291

Used in catalysts for manufacture of butadiene copolymer. Moisture-sensitive.

Schuster, R.E. *et al*, *J. Chem. Soc., Chem. Commun.*, 1967, 1038 (*pmr*)
Fratiello, A. *et al*, *J. Am. Chem. Soc.*, 1968, **90**, 1194 (*nmr*)
Ger. Pat., 2 023 405, (*1970*); *CA*, **74**, 54901r (*use*)
Smith, J.A. *et al*, *J. Chem. Soc. (A)*, 1971, 178 (*nqr*)
U.S.S.R. Pat., 621 802, (*1978*); *CA*, **90**, 42899p
Hu, M.G. *et al*, *Inorg. Chem.*, 1979, **18**, 3297.

$C_4H_{10}BF_2N$ B-00242

(Diethylamino)difluoroborane, 8CI

N,N*-Diethyl-1,1-difluoroboranamine, 10CI, 9CI. (Diethylamino)difluoroborine. Diethylaminoboron difluoride*

[392-73-4]

Et_2NBF_2

M 120.937

Cryst., dec. quickly in air. Mp 65.5°. Bp 92-92.5°. Subl. at 30-5° *in vacuo*.

Dimer: [7416-85-5].

$C_8H_{20}B_2F_4N_2$ M 241.873

Characterized spectroscopically.

Gerrard, W. *et al*, *J. Chem. Soc.*, 1957, 381 (*synth*)
Spitsyn, V.I. *et al*, *Vestn. Mosk. Univ. Ser. 2: Khim*, 1966, **21**, 46; *CA*, **66**, 11184b (*polymers*)
Greenwood, N.N. *et al*, *J. Chem. Soc. (A)*, 1967, 959 (*synth, ir, pmr, nmr*)
Spitsyn, V.I. *et al*, *Zh. Neorg. Khim.*, 1967, **12**, 1158 (*ir*)
Barlos, K. *et al*, *Chem. Ber.*, 1977, **110**, 2774 (*nmr*)
Nöth, H. *et al*, *Chem. Ber.*, 1981, **114**, 1150 (*nmr*)

$C_4H_{10}BF_3O$ B-00243

Trifluoroborane diethyl ether

Trifluoro[1,1'-oxybis[ethane]]boron, 10CI. 1,1'-Oxybis[ethane] compd. with trifluoroborane(1:1), 9CI. Ethyl ether compd. with boron fluoride(BF_3)(1:1), 8CI. Trifluoroborane etherate. Boron trifluoride etherate

[109-63-7]

$Et_2O—BF_3$

M 141.928

V. important catalyst for many types of reactions. Lewis acid. Comly. available. Moisture-sensitive liq. d^{25} 1.125. Bp 126°. n_D^{25} 1.348. Dipole moment 7.7D.

▷Corrosive. KX7375000.

Fratiello, A. *et al*, *J. Am. Chem. Soc.*, 1968, **90**, 1194 (*nmr, pmr*)
Fratiello, A. *et al*, *Inorg. Chem.*, 1969, **8**, 480 (*nmr*)
Topchii, V.A. *et al*, *Zh. Obshch. Khim.*, 1969, **39**, 2689.
Voloshchuk, A.M. *et al*, *Kernenergie*, 1974, **17**, 121 (*ir*)
Sugawara, T. *et al*, *Bull. Chem. Soc. Jpn.*, 1979, **52**, 3391 (*nmr*)
Fieser, M. *et al*, *Reagents for Organic Synthesis*, Wiley, 1967-83, **1**, 70 (*use*)

$C_4H_{10}BIO_2$ B-00244

Dihydroxy(4-iodobutyl)borane

(4-Iodobutyl)boronic acid, 10CI. 4-Iodobutaneboronic acid

[71081-49-7]

$ICH_2CH_2CH_2CH_2B(OH)_2$

M 227.836

Used in trace analysis of physiological fluids and for analysis of bifunctional compds. Cryst. (MeOH aq.). Mp 96-9°.

Poole, C.F. *et al*, *Analyst.* (*London*), 1979, **104**, 82 (*use, ms, ir, synth*)
Poole, C.F. *et al*, *J. Chromatogr.*, 1979, **186**, 307 (*use*)

$C_4H_{11}B$ B-00245

Diethylborane, 10CI, 9CI

Diethylborine

[5518-25-2]

Et_2BH

M 69.941

Bridged dimer in uncomplexed state. Prepd. *in situ* from 2-(hydroxyphenoxy)diethylborane and aluminum hydride in THF. Air- and moisture-sensitive. Sol. THF. n_D^{20} 1.5065. Readily disproportionates to $EtBH_2$ and Et_3B on standing.

▷Probably pyrophoric

Köster, R. *et al*, *Angew. Chem.*, 1964, **76**, 649 (*use*)
Köster, R. *et al*, *Justus Liebigs Ann. Chem.*, 1975, 1339 (*use*)
Wolfgang, F. *et al*, *Justus Liebigs Ann. Chem.*, 1975, 2201 (*use*)
Gupta, S.K., *J. Organomet. Chem.*, 1978, **156**, 95 (*synth*)
Midland, M.M. *et al*, *J. Am. Chem. Soc.*, 1979, **101**, 248 (*use*)
Dahlhoff, W.V. *et al*, *Angew. Chem.*, 1980, **92**, 552 (*use*)

$C_4H_{11}B$ B-00246

Ethyldimethylborane, 10CI, 9CI

Dimethylethylborane. Ethyldimethylborine

[1113-22-0]

$EtBMe_2$

M 69.941

Prepd. from ethylene and $H_2B_2Me_4$. Originally thought to be highly unstable towards disproportionation into Me_3B, Et_3B etc. but shown not to be so. V. volatile liq. highly sensitive to air. Sol. THF. Mp −160°. Bp 25.5°. ΔH_{vap} 5.86 Kcal/mol.$^{-1}$ Trouton const. 19.6 e.u.

▷Pyrophoric

Birchall, J.M. *et al*, *Chem. Ind.* (*London*), 1961, 1080 (*synth*)
Henneburg, D. *et al*, *Justus Liebigs Ann. Chem.*, 1961, **640**, 52 (*ms*)
Köster, R. *et al*, *Justus Liebigs Ann. Chem.*, 1965, **682**, 21 (*synth*)
Haag, A. *et al*, *Justus Liebigs Ann. Chem.*, 1971, **751**, 95 (*use*)
Spielvogel, B.F. *et al*, *J. Am. Chem. Soc.*, 1975, **97**, 1609 (*nmr*)
Stampf, E.J. *et al*, *J. Organomet. Chem.*, 1977, **131**, 171 (*synth*)

$C_4H_{11}BBr_2O$ B-00247

Dibromoborane diethyl ether

Dibromohydro[1,1'-oxybis[ethane]]boron, 10CI. Dibromoborane etherate

[71648-08-3]

$Et_2O—BHBr_2$

M 245.748

Evidence of this compd. provided only by ^{11}B nmr. Moisture-sensitive.

Hu, M.G. *et al*, *Inorg. Chem.*, 1979, **18**, 3297 (*synth, nmr*)

$C_4H_{11}BCl_2O$ B-00248

Dichloroborane diethyl ether

Dichlorohydro[1,1'-oxybis[ethane]]boron, 10CI. Dichloroborane[1,1'-oxybis(ethane)], 9CI. Dichloroborane ethyl etherate, 8CI. Dichloroborine ethyl etherate

[41156-66-5]

$Et_2O—BHCl_2$

M 156.846

Reagent for general synth. of alkyl- and alkenyldichloroboranes *via* hydroboration. Air- and moisture-sensitive liq. Mp −27° to −23°.

Onak, T. *et al*, *J. Phys. Chem.*, 1958, **62**, 1605 (*synth, nmr, ir*)
Brown, H.C. *et al*, *J. Inorg. Nucl. Chem.*, 1959, **9**, 51 (*synth*)
Brown, H.C. *et al*, *J. Am. Chem. Soc.*, 1973, **95**, 2396; 1976, **98**, 1798, 4917 (*use*)
Hu, M.G. *et al*, *Inorg. Chem.*, 1979, **18**, 3297.

$C_4H_{11}BNO_3P$ B-00249

Cyanoborane trimethyl phosphite

(Cyano-C)dihydro(trimethyl phosphite-P)boron, 10CI, 9CI. Phosphorous acid trimethyl ester, compd. with boranecarbonitrile(1:1)

[67374-61-2]

$(MeO)_3P—BH_2CN$

M 162.920

Liq.

Martin, D.R. *et al*, *J. Inorg. Nucl. Chem.*, 1978, **40**, 9 (*synth, nmr, ir*)

$C_4H_{11}BNP$ B-00250

Cyanoborane trimethylphosphine

(Cyano-C)dihydro(trimethylphosphine)boron, 10CI. *Boranecarbonitrile, compd. with trimethylphosphine(1:1)*

[67374-59-8]

$Me_3P—BH_2CN$

M 114.921

Martin, D.R. *et al*, *J. Inorg. Nucl. Chem.*, 1978, **40**, 9 (*synth, nmr, ir*)

$C_4H_{11}BN_2$ B-00251

Cyanoborane trimethylamine

(Cyano-C)(N,N-dimethylmethanamine)dihydroboron, 10CI. *Boranecarbonitrile, compd. with N,N-dimethylmethanamine(1:1)*, 9CI. *Boranecarbonitrile, compd. with trimethylamine*, 8CI

[30353-61-8]

$Me_3N—BH_2CN$

M 97.954

Boron analogue of an α-aminonitrile. Shows a range of pharmacological effects. White cryst. (THF/pet. ether). Mp 63°.

▷ED1100000.

Kelly, H.C. *et al*, *J. Chem. Soc., Chem. Commun.*, 1970, 1619 (*synth, nmr, ir*)
Lavrent'eva, E.A. *et al*, *Zh. Obshch. Khim.*, 1973, **43**, 295.
Weidig, C. *et al*, *Inorg. Chem.*, 1974, **13**, 1763 (*synth, nmr, ir*)
Bratt, P.J. *et al*, *J. Chem. Soc., Dalton Trans.*, 1974, 2161 (*synth, ir*)
Beachley, O.T. *et al*, *Inorg. Chem.*, 1975, **14**, 120 (*pmr*)
Wisian-Neilson, P. *et al*, *Inorg. Chem.*, 1978, **17**, 2327 (*synth, pmr, ir*)
Hall, I.H. *et al*, *J. Pharm. Sci.*, 1979, **68**, 685; 1980, **69**, 1025; 1981, **70**, 339 (*pharmacol*)

$C_4H_{11}BO$ B-00252

Borane tetrahydrofuran

Trihydro(tetrahydrofuran)boron, 10CI. *Tetrahydrofuran compd. with borane(1:1)*, 9CI, 8CI

[14044-65-6]

As Trifluoroborane tetrahydrofuran, B-00226 with

$X = H$

M 85.940

Hydroborating and reducing agent. Produced comly. Used only in soln., unstable in pure state, moisture-sensitive.

▷Flammable

Coyle, T.D. *et al*, *J. Am. Chem. Soc.*, 1959, **81**, 2989 (*synth*)
Fratiello, T.P.O. *et al*, *J. Am. Chem. Soc.*, 1968, **90**, 1194 (*nmr*)
Dodds, A.R. *et al*, *Inorg. Chem.*, 1977, **16**, 2900.
Pelter, A. *et al*, *Comprehensive Org. Chem.*, Barton, D.H.R. *et al*, Ed., 1979, Pergamon, Oxford, **3**, 695 (*rev*)
Kabalka, G.W. *et al*, *J. Org. Chem.*, 1981, **46**, 4582.
Nöth, H. *et al*, *Inorg. Chem.*, 1981, **20**, 1062.
Stanfield, C.F. *et al*, *J. Org. Chem.*, 1981, **46**, 4799.
Marsella, J.A. *et al*, *J. Am. Chem. Soc.*, 1982, **104**, 2361.

$C_4H_{11}BO$ B-00253

Diethylhydroxyborane

Diethylborinic acid, 10CI, 9CI. *Diethylboric acid. Hydroxydiethylborane*

[4426-31-7]

Et_2BOH

M 85.940

Cryst. Mp 95-7°. Dehydrates on heating → $(Et_2B)_2O$.

Köster, R. *et al*, *Justus Liebigs Ann. Chem.*, 1969, **724**, 34; 1974, 54 (*synth*)

$C_4H_{11}BOS$ B-00254

Borane 1,4-oxathiane

Trihydro(1,4-oxathiane-O[1])boron, 10CI. *Borane compd. with 1,4-oxathiane(1:1)*

[72525-30-5]

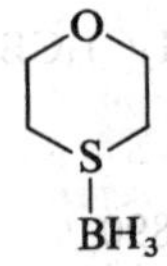

M 118.000

Hydroborates alkenes to trialkylboranes efficiently; a less-noxious alternative to borane-dimethyl sulfide as a conc. source of borane; comly. available. Stable liq. which cryst. on cooling, moisture-sensitive. Mp 11-5°.

Brown, H.C. *et al*, *J. Org. Chem.*, 1980, **45**, 916.
Brown, H.C. *et al*, *Synthesis*, 1980, 153 (*synth, nmr*)
U.S.P., 4 298 750, (*1981*); *CA*, **96**, 52322h

$C_4H_{11}BO_2$ B-00255

Butyldihydroxyborane

Butylboronic acid, 10CI, 9CI. *Butylboric acid. Butaneboronic acid*

[4426-47-5]

$H_3C(CH_2)_3B(OH)_2$

M 101.940

Used for protection of diols and in GC/MS of bifunctional compds. Cryst. (toluene). Mp 92-4°.

Letsinger, R.L. *et al*, *J. Am. Chem. Soc.*, 1954, **76**, 864 (*synth*)
Mirviss, S.B., *J. Org. Chem.*, 1967, **32**, 1713 (*synth*)
Brown, H.C. *et al*, *J. Am. Chem. Soc.*, 1976, **98**, 1785, 1798 (*synth*)
Ger. Pat., 2 627 675, (*1977*); *CA*, **86**, 155254k (*use*)
Mazurek, M. *et al*, *Org. Magn. Reson.*, 1977, **9**, 193 (*nmr*)
Kossa, W.C., *Chem. Deriv. Anal. Chem.* (*Frei, R.W. et al*, Ed.), 1981, Plenum, N.Y., 99 (*rev*)

$C_4H_{11}BO_2$ B-00256

Ethyldimethoxyborane

Ethylboronic acid dimethyl ester, 10CI, 9CI. *Ethaneboronic acid dimethyl ester*, 8CI. *Dimethyl ethaneboronate*

[7318-82-3]

$EtB(OMe)_2$

M 101.940

Fungicide; fuel additive. Liq., air- and moisture-sensitive. Bp 75-7°.

Ger. Pat., 1 111 181, (*1958*); *CA*, **56**, 8742i (*synth*)
U.S.P., 3 027 396, (*1962*); *CA*, **57**, 11232q (*synth, use*)
Nöth, H. *et al*, *Chem. Ber.*, 1966, **99**, 1049 (*pmr*)
Nöth, H. *et al*, *J. Organomet. Chem.*, 1968, **12**, 23 (*pmr*)
Dahlhoff, W.V. *et al*, *Justus Liebigs Ann. Chem.*, 1976, 387 (*props*)
Dahlhoff, W.V. *et al*, *Angew. Chem., Int. Ed. Engl.*, 1980, **19**, 548 (*props*)

$C_4H_{11}B_7FeO_4$ B-00257

Tetracarbonyl(η^2-heptaborane(11))iron, 9CI

[54190-57-7]

M 254.646

Structure unknown but see derived anion, Tetracarbonyl[η^2-dodecahydroheptaborato(1−)]ferrate(1−), B-00277 . Mp ~20°.

Hollander, O. *et al*, *J. Chem. Soc.*, *Chem. Commun.*, 1974, 604 (*synth, nmr, ms*)

$C_4H_{12}B^{\ominus}$ B-00258

***tert*-Butyltrihydroborate**

(1,1-Dimethylethyl)trihydroborate(1−), 10CI. *tert-Butylborohydride*

$$(H_3C)_3CBH_3^{\ominus}$$

M 70.949 (ion)

Li salt: [76826-51-2].

$C_4H_{12}BLi$ M 77.890

Air- and moisture-sensitive.

Biffar, W. *et al*, *Chem. Ber.*, 1981, **114**, 49 (*pmr, nmr*)

$C_4H_{12}B^{\ominus}$ B-00259

Tetramethylborate(1−), 10CI, 9CI, 8CI

$$Me_4B^{\ominus}$$

M 70.949 (ion)

Li salt: [2169-38-2].

$C_4H_{12}BLi$ M 77.890

Used in battery electrolytes. White needles (C_6H_6), fairly stable to dry air. Sol. Et_2O, H_2O, C_6H_6, toluene. Mp 189°. Subl. at 140° *in vacuo*.

▷May ignite spontaneously in moist air

Hurd, D.T., *J. Org. Chem.*, 1948, **13**, 711 (*synth*)
Williams, K.C. *et al*, *J. Am. Chem. Soc.*, 1966, **88**, 4134 (*pmr, nmr*)
Oliver, J.P. *et al*, *J. Am. Chem. Soc.*, 1967, **89**, 163 (*nmr*)
Stucky, G.D. *et al*, *J. Am. Chem. Soc.*, 1971, **93**, 1553 (*struct, pmr*)
Rhine, W.E. *et al*, *J. Am. Chem. Soc.*, 1975, **97**, 6401 (*struct*)
Horowitz, H.H. *et al*, *Proc.-Electrochem. Soc.*, 1981, **81**, 131 (*use*)

$C_4H_{12}BBrN_2$ B-00260

Bromobis(dimethylamino)borane, 8CI

1-Bromo-N,N,N′,N′-tetramethylboranediamine, 10CI, 9CI

[6990-27-8]

$$BrB(NMe_2)_2$$

M 178.866

Liq., sensitive to moisture. Bp_{10} 70-80°, $Bp_{0.5}$ 20-8°.

Brotherton, R.J. *et al*, *J. Am. Chem. Soc.*, 1960, **82**, 6242 (*synth*)
Nöth, H., *Z. Naturforsch.*, *B*, 1961, **16**, 618 (*synth, props*)
Nöth, H. *et al*, *Chem. Ber.*, 1966, **99**, 2197.
Beck, W. *et al*, *Chem. Ber.*, 1972, **105**, 2883 (*nmr*)
King, G.H. *et al*, *Faraday Discuss. Chem. Soc.*, 1972, **54**, 70 (*pe*)
Fuss, W., *Z. Naturforsch.*, *B*, 1974, **29**, 514 (*uv*)
Davidson, G. *et al*, *J. Chem. Soc.*, *Dalton Trans.*, 1981, 306 (*ir, raman*)

$C_4H_{12}BBrO$ B-00261

Bromoborane diethyl ether

Bromodihydro[1,1′-oxybis[ethane]]boron, 10CI. *Bromoborane compd. with 1,1′-oxybis[ethane](1:1)*

[69765-80-6]

$$Et_2O—BH_2Br$$

M 166.852

Moisture sensitive.

Dodds, A.R. *et al*, *Inorg. Chem.*, 1979, **18**, 1465 (*nmr*)
Hu, M.G. *et al*, *Inorg. Chem.*, 1979, **18**, 3297.

$C_4H_{12}BBr_2N$ B-00262

Dibromomethylborane trimethylamine

Dibromo(N,N-dimethylmethanamine)methylboron, 10CI. *Trimethylamine dibromomethylborane*

[69521-22-8]

$$Me_3N—BMeBr_2$$

M 244.764

Solid, air- and moisture-sensitive. Mp 256-8°.

Martin, D.R. *et al*, *J. Inorg. Nucl. Chem.*, 1978, **40**, 1289 (*synth, pmr*)

$C_4H_{12}BBr_2P$ B-00263

Dibromomethylborane trimethylphosphine

Dibromomethyl(trimethylphosphine)boron, 10CI. *Methyldibromoborane trimethylphosphine*

[69513-15-1]

$$Me_3P—BBr_2Me$$

M 261.731

Solid. Mp 236-7°.

Martin, D.R. *et al*, *J. Inorg. Nucl. Chem.*, 1978, **40**, 1289 (*synth, pmr*)

$C_4H_{12}BClN_2$ B-00264

Chlorobis(dimethylamino)borane, 8CI

1-Chloro-N,N,N′,N′-tetramethylboranediamine, 10CI, 9CI. *Bisdimethylaminochloroborine. Bisdimethylaminoboron chloride*

[6562-41-0]

$$ClB(NMe_2)_2$$

M 134.415

Liq., sensitive to moisture. d_4^{18} 0.832. Bp 145-8°, Bp_{15} 44°. n_D^{24} 1.4491.

Skinner, H.A. *et al*, *J. Chem. Soc.*, 1954, 2324 (*synth, props*)
Brotherton, R.J. *et al*, *J. Am. Chem. Soc.*, 1960, **82**, 6242 (*synth*)
Cros, G. *et al*, *Bull. Soc. Chim. Fr.*, 1969, 2601 (*ir, raman*)
King, G.H. *et al*, *Faraday Discuss. Chem. Soc.*, 1972, 70 (*pe*)
Blick, K.E. *et al*, *Z. Anorg. Allg. Chem.*, 1980, **467**, 177 (*ir*)
Hoffmann, R.W. *et al*, *J. Org. Chem.*, 1981, **46**, 1309 (*synth, props, pmr*)

$C_4H_{12}BClO$ B-00265

Chloroborane diethyl ether

Chlorodihydro[1,1′-oxybis[ethane]]boron, 10CI. *Chloroborane compd. with 1,1′-oxybis[ethane](1:1)*, 9CI. *Monochloroborane diethyl etherate*

[36594-41-9]

$$Et_2O—BH_2Cl$$

M 122.401

Used in prepn. of dialkylchloroboranes by hydroboration of alkenes. Stable at 0° for several weeks (under N_2); moisture sensitive.

Brown, H.C. *et al*, *J. Am. Chem. Soc.*, 1972, **94**, 2112 (*synth*)
Brown, H.C. *et al*, *J. Org. Chem.*, 1973, **38**, 182; 1976, **98**, 1785; 1979, **44**, 2422.
Carlson, B.A. *et al*, *J. Am. Chem. Soc.*, 1973, **95**, 6876.
Hu, M.G. *et al*, *Inorg. Chem.*, 1979, **18**, 3297 (*nmr*)
Ger. Pat., 2 740 017, (*1979*); *CA*, **91**, P76290u

$C_4H_{12}BFN_2$ B-00266

Bis(dimethylamino)fluoroborane, 8CI

1-Fluoro-N,N,N′,N′-*tetramethylboranediamine*, 10CI, 9CI. *Bisdimethylaminoboron fluoride*

[383-90-4]

$FB(NMe_2)_2$

M 117.961

Liq., sensitive to moisture; readily dec. Mp −44.3°. Bp 106°, Bp_{18} 27°.

Burg, A.B. *et al*, *J. Am. Chem. Soc.*, 1954, **76**, 3903 (*synth, props*)
Nöth, H. *et al*, *Chem. Ber.*, 1966, **99**, 1049 (*nmr*)
Nöth, H. *et al*, *Chem. Ber.*, 1966, **99**, 2757.
Bock, H. *et al*, *Chem. Ber.*, 1971, **104**, 1687 (*pe*)
Fuss, W., *Z. Naturforsch, B*, 1974, **29**, 514 (*uv*)
Fussstetter, H. *et al*, *Chem. Ber.*, 1977, **110**, 3172 (*nmr*)

$C_4H_{12}BN$ B-00267

(Dimethylamino)dimethylborane, 8CI

N,N,*1,1-Tetramethylboranamine*, 9CI

[1113-30-0]

Me_2BNMe_2

M 84.956

Air- and moisture-sensitive liq. d 0.729. Bp 63-5°. μ = 1.40D.

Erickson, C.E. *et al*, *J. Org. Chem.*, 1959, **24**, 1161 (*synth*)
Bullen, G.J. *et al*, *J. Chem. Soc.* (*A*), 1970, 992 (*cryst struct*)
Haubold, W. *et al*, *Z. Anorg. Allg. Chem.*, 1976, **421**, 105 (*synth, nmr*)
Nöth, H. *et al*, *Chem. Ber.*, 1976, **109**, 1075; 1981, **114**, 49, 1150 (*nmr*)

$C_4H_{12}BNO$ B-00268

Borane morpholine

Trihydro(*morpholine*-N[4])*boron*, 10CI. *Morpholine compd. with borane*(*1:1*), 9CI, 8CI. *Morpholine borane*

[4856-95-5]

As Trifluoroborane morpholine, B-00234 with

X = H

M 100.955

Used in photographic colour processing. Preservative for metal-working fluid. Shows anticholesteremic, hypolipidemic and . antiflammatory activity. Made comly. Cryst. (H_2O). Mp 82-5°, 97-9°.

▷Toxic. QD8530000.

U.S.P., 2 927 133, (*1960*); *CA*, **55**, 3433i
Hill, E.A. *et al*, *Tetrahedron Lett.*, 1976, 3013.
Bennett, E.O. *et al*, *Mater. Perform.*, 1979, **18**, 40 (*use*)
Hall, I.H. *et al*, *J. Pharm. Sci.*, 1981, **70**, 339 (*use*)
Wilson, I. *et al*, *Inorg. Chem.*, 1982, **21**, 1622.
U.S.P., 4 312 989, (*1982*); *CA*, **97**, 85322g

$C_4H_{12}BNO$ B-00269

Dimethylaminomethoxymethylborane

1-Methoxy-N,N,*1-trimethylboranamine*, 10CI, 9CI. *Methyl*(*dimethylamino*)*methoxyborane*, 8CI

[7318-91-4]

$Me_2NBMeOMe$

M 100.955

Air- and moisture-sensitive liq. Bp_{725} 87-9°. n_D^{20} 1.3949.

Nöth, H. *et al*, *Chem. Ber.*, 1966, **99**, 1049; 1973, **106**, 1145 (*nmr*)
Nöth, H. *et al*, *J. Organomet. Chem.*, 1968, **12**, 23 (*synth, pmr*)
Nöth, H. *et al*, *Chem. Ber.*, 1974, **107**, 3089.
McFarlane, W. *et al*, *Chem. Ber.*, 1975, **108**, 3831 (*nmr*)

$C_4H_{12}BNO_2$ B-00270

(Dimethylamino)dimethoxyborane

1,1-Dimethoxy-N,N-*dimethylboranamine*, 10CI, 9CI

[7318-74-3]

$(MeO)_2BNMe_2$

M 116.954

Moisture-sensitive.

Nöth, H. *et al*, *Chem. Ber.*, 1966, **99**, 1049 (*nmr*)
Ebraheem, K.A.K. *et al*, *Org. Magn. Reson.*, 1977, **10**, 258.
Biffar, W. *et al*, *Chem. Ber.*, 1980, **113**, 333 (*nmr*)
Nöth, H. *et al*, *Chem. Ber.*, 1981, **114**, 1150 (*nmr*)

$C_4H_{12}BNS_2$ B-00271

Dimethylaminobis(methylthio)borane

N,N-*Dimethyl-1,1-bis*(*methylthio*)*boranamine*, 10CI, 9CI

[55580-29-5]

$(MeS)_2BNMe_2$

M 149.076

Moisture-sensitive.

Costes, J.P. *et al*, *C.R. Hebd. Seances Acad. Sci.*, 1975, **280**, 665 (*pmr*)
Niedenzu, K. *et al*, *Synth. React. Inorg. Metal-Org. Chem.*, 1980, **10**, 373 (*synth, nmr*)
Nöth, H. *et al*, *Chem. Ber.*, 1981, **114**, 1150 (*nmr*)

$C_4H_{12}BO_3^{\ominus}$ B-00272

Trimethoxymethylborate(1−), 10CI

$(MeO)_3BMe^{\ominus}$

M 118.947 (ion)

Cu(I) salt: [65139-97-1].

$C_4H_{12}BCuO_3$ M 182.493

Air- and moisture-sensitive.

Maruyama, K. *et al*, *J. Am. Chem. Soc.*, 1977, **99**, 8068 (*props*)
Yamamoto, Y. *et al*, *J. Am. Chem. Soc.*, 1980, **102**, 2318 (*props*)

$C_4H_{12}BO_4^{\ominus}$ B-00273

Tetramethoxyborate(1−), 10CI, 9CI, 8CI

$(MeO)_4B^{\ominus}$

M 134.946 (ion)

Li salt: [6867-35-2].

$C_4H_{12}BLiO_4$ M 141.887

Product of total methanolysis of $LiBH_4$. Cryst. solid. Sol. MeOH, Me_2CO, insol. pet. ether, $CHCl_3$. Mp >310° dec.

Schlesinger, H.I. *et al*, *J. Am. Chem. Soc.*, 1940, **62**, 3429.
Onak, T.P. *et al*, *J. Phys. Chem.*, 1959, **63**, 1533 (*nmr*)
Lehmann, H.A. *et al*, *Z. Anorg. Allg. Chem.*, 1960, **304**, 88 (*props*)
Steinberg, H., *Organoboron Chemistry*, Wiley-Interscience, 1964, **1**, 609 (*synth, props*)

Funck, E., *Ber. Bunsenges. Phys. Chem.*, 1967, **71**, 170 (*ir, struct*)
Heller, G. *et al*, *Z. Anorg. Allg. Chem.*, 1980, **460**, 228 (*synth*)

$C_4H_{12}B_2O$ B-00274

Oxybis[dimethylborane], 8CI

Dimethylborinic acid anhydride, 10CI, 9CI. *Dimethylboric anhydride*

[19163-04-3]

Me_2BOBMe_2

M 97.758

Air-sensitive liq. Mp −21.2°. Bp 34.8°.

▷Pyrophoric

Goubeau, J. *et al*, *Chem. Ber.*, 1957, **90**, 171 (*synth*)
Nöth, H. *et al*, *J. Organomet. Chem.*, 1968, **12**, 23 (*pmr*)
Lanthier, G.F. *et al*, *Can. J. Chem.*, 1969, **47**, 569 (*synth, ms, ir, nmr, props*)
Nöth, H. *et al*, *Chem. Ber.*, 1974, **107**, 3089 (*nmr*)
Gundersen, G. *et al*, *J. Mol. Struct.*, 1976, **33**, 97 (*struct*)
Durig, J.R. *et al*, *J. Mol. Struct.*, 1977, **42**, 13 (*ir, raman*)

$C_4H_{12}B_3Cl_2N_3$ B-00275

2,4-Dichloro-1,3,5,6-tetramethylborazine, 10CI, 9CI, 8CI

*Tetramethyldi-*B*-chloroborazine*

[7387-21-5]

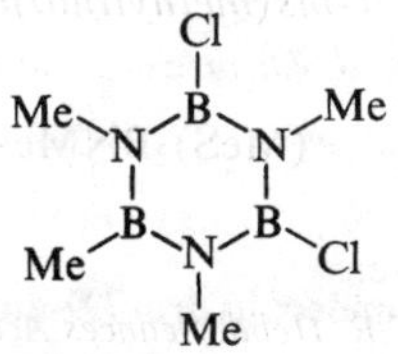

M 205.495

Forms polymers. Solid by subl. Mp 145-6°.

Homopolymer: [30946-46-4]. Component of flame-resistant compositions, plasticizers, fuel additives, neutron absorbers, laminating resins, high temp. insulation, heat transfer media. Light-brown glassy polymer. Mp 180-200°.

Wagner, R.I. *et al*, *Inorg. Chem.*, 1962, **1**, 93 (*synth*)
U.S.P., 3 341 474, (*1967*); *CA*, **67**, 109171r (*use*)
Melcher, L.A. *et al*, *Inorg. Chem.*, 1972, **11**, 1247 (*synth, ir*)
Adcock, J.L. *et al*, *Inorg. Chem.*, 1973, **12**, 788 (*pmr*)
Melcher, L.A. *et al*, *Inorg. Chem.*, 1973, **12**, 601 (*ms*)
Nöth, H. *et al*, *Chem. Ber.*, 1981, **114**, 1150 (*props*)

$C_4H_{12}B_4$ B-00276

2,3-Dimethyl-2,3-dicarbahexaborane(8), 10CI

[20741-68-8]

As 2,3-Dicarbahexaborane(8), B-00131 with

$R^1 = R^2 = CH_3$

M 103.379

Colourless liq. by vacuum fractionation. Mp −64°.

Onak, T. *et al*, *J. Am. Chem. Soc.*, 1963, **85**, 3378 (*synth, nmr*)
Boer, F. *et al*, *Inorg. Chem.*, 1964, **3**, 1666 (*struct*)
Jotham, R. *et al*, *J. Chem. Soc., Dalton Trans.*, 1972, 473 (*ir, raman*)
Grimes, R. *et al*, *Inorg. Chem.*, 1979, **18**, 3294 (*synth*)

$C_4H_{12}B_7FeO_4^{\ominus}$ B-00277

Tetracarbonyl[η^2-dodecahydroheptaborato(1−)]ferrate(1−), 9CI

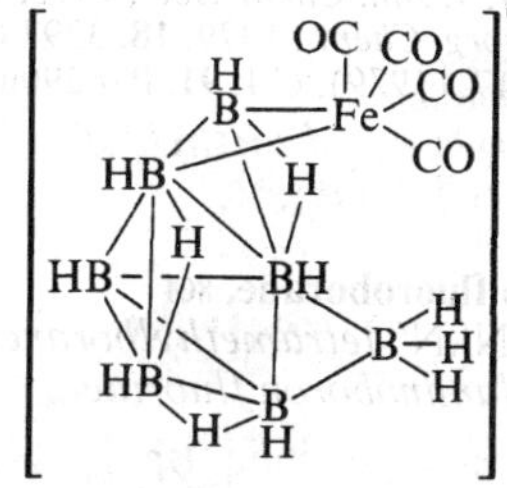

M 255.653 (ion)

Tetrabutylammonium salt: [54083-17-9]. $C_{20}H_{48}B_7FeNO_4$ M 498.121
Deep-amber cryst. (CH_2Cl_2/pentane).

Hollander, O. *et al*, *J. Chem. Soc., Chem. Commun.*, 1974, 604 (*synth, struct*)
Mangion, M. *et al*, *Inorg. Chem.*, 1977, **16**, 2110 (*struct*)

$C_4H_{12}B_{10}O_4$ B-00278

1,7-Dicarbadodecaborane(12)-1,7-dicarboxylic acid, 10CI

1,7-Dicarboxy-1,7-dicarbadodecaborane(12). m-*Carboranedicarboxylic acid*

[50571-15-8]

M 232.236

closo-. Precursor to carborane polymers. White solid (CH_2Cl_2). Mp 207°. pK_a 6.8.

Grafstein, D. *et al*, *Inorg. Chem.*, 1963, **2**, 1129 (*synth*)
Rabilloud, G. *et al*, *J. Organomet. Chem.*, 1979, **182**, 275 (*synth, ir, ms*)

$C_4H_{13}AlB_4$ B-00279

[(2,6-η)Heptahydro-4,5-dicarbahexaborato(1−)]dimethylaluminum

μ-Dimethylaluminum-2,3-η-dicarbahexaborane

[55663-16-6]

$Me_2Al(C_2B_4H_7)$

M 131.368

Extremely unstable in soln. or liq. phase, characterised by ms. Sl. volatile liq.

Magee, C.P. *et al*, *J. Organomet. Chem.*, 1975, **86**, 159 (*synth*)

$C_4H_{13}BN_2$ B-00280

Bis(dimethylamino)borane, 8CI

N,N,N′,N′-*Tetramethylboranediamine*, 10CI, 9CI. *Bis(dimethylamino)borine*

[2386-98-3]

$HB(NMe_2)_2$

M 99.970

Liq., sensitive to moisture; quite stable on storage under N_2. d_4^{20} 0.767. Mp −56.8° −45°. Bp_{720} 106-8°. n_D^{20} 1.4232; Trouton constant 20.7 cal. deg.$^{-1}$ mole^{-1}.

▷Toxic

Burg, A.B. *et al*, *J. Am. Chem. Soc.*, 1951, **73**, 953 (*synth, props*)
Nöth, H. *et al*, *Z. Anorg. Allg. Chem.*, 1962, **318**, 293 (*synth, props*)
Goubeau, J. *et al*, *Z. Anorg. Allg. Chem.*, 1967, **352**, 285 (*ir, raman*)
Bock, H. *et al*, *Chem. Ber.*, 1971, **104**, 1687 (*pe*)

Fussstetter, H. *et al*, *Chem. Ber.*, 1977, **110**, 3172 (*nmr*)
Inorg. Synth., 1977, **17**, 30 (*synth, props, ir, nmr*)

$C_4H_{13}B_2N$ B-00281

Bis(dimethylboryl)amine

N-(*Dimethylboryl*)-*1,1-dimethylboranamine*, 10CI, 9CI. *Iminobis*[*dimethylborane*], 8CI

[19163-14-5]

$Me_2BNHBMe_2$

M 96.773

Air- and moisture-sensitive liq. Bp_{720} 71-2°.

N-*Me: See N,N-Bis*(*dimethylboryl*)*methylamine, B-00347*

Nöth, H. *et al*, *J. Organomet. Chem.*, 1968, **12**, 23; 1969, **16**, 357 (*synth, pmr*)
Nöth, H. *et al*, *Chem. Ber.*, 1974, **107**, 3089; 1977, **110**, 2783; 1978, **111**, 3596 (*nmr*)

$C_4H_{14}BN$ B-00282

Borane *tert*-butylamine

Trihydro(*2-methyl-2-propanamine*)*boron*, 10CI. *2-Methyl-2-propanamine compd. with borane*(*1:1*), 9CI. tert-*Butylamine compd. with borane*, 8CI

[7337-45-3]

$(H_3C)_3CNH_2—BH_3$

M 86.971

Used in photographic colour development. Water-soluble borane reducing agent for organic and inorganic compds. Comly. produced. Powder. Mp 98-100° dec., 96°. Dipole moment 4.64D.

▷Severe poison. EO3900000.

Nöth, H. *et al*, *Chem. Ber.*, 1960, **93**, 928; 1964, **97**, 110 (*synth*)
Purser, J.M. *et al*, *Inorg. Chem.*, 1968, **7**, 2156 (*nmr*)
Lane, C.F., *Aldrichimica Acta*, 1973, **6**, 51 (*rev*)
Pelter, A. *et al*, *Comprehensive Organic Chemistry*, Barton, D.H.R. *et al*, Ed., 1979, Pergamon, Oxford, **3**, 695 (*rev*)

$C_4H_{14}BN$ B-00283

Borane diethylamine

(N-*Ethylethanamine*)*trihydroboron*, 10CI. N-*Ethylethanamine, compd. with borane*(*1:1*), 9CI. *Diethylamine compd. with borane*(*1:1*), 8CI

[2670-68-0]

$Et_2NH—BH_3$

M 86.971

Reducing agent in electrodeless coating procedures. Liq.

Laurent, J.P. *et al*, *C.R. Hebd. Seances Acad. Sci.*, 1966, **263**, 965 (*mord*)
Kovar, R.A. *et al*, *Inorg. Chem.*, 1971, **10**, 900.
Frick, W., *Obeflaechentechnik*, 1973, **50**, 450; *CA*, **80**, 43394a.
Kovar, R.A. *et al*, *Inorg. Chem.*, 1975, **14**, 2239 (*synth*)
Myers, W.H. *et al*, *Inorg. Chem.*, 1975, **14**, 2874 (*synth, ir, nmr*)
Ryschkewitsch, G.E. *et al*, *Synth. React. Inorg. Met.-Org. Chem.*, 1975, **5**, 123 (*synth*)

$C_4H_{14}BP$ B-00284

Borane trimethylphosphonium methylide

Trihydro(*trimethylphosphonium-η-methylide*)*boron*, 10CI

[73946-45-9]

$Me_3PCH_2—BH_3$

M 103.938

Cryst. (CH_2Cl_2/pentane or THF). Mp 133°.

Schmidbaur, H. *et al*, *Chem. Ber.*, 1980, **113**, 1480 (*synth, ms, nmr*)
Schmidbaur, H. *et al*, *Chem. Ber.*, 1980, **113**, 2575 (*struct*)

$C_4H_{16}AlB_9$ B-00285

Ethyl(7,8,9,10,11-η)undecahydro-7,8-dicarbaundecaborato(2−)aluminum

3-Ethyl-3-alumino-1,2-dicarba-closo-*dodecaborane*(*12*)

[34540-15-3]

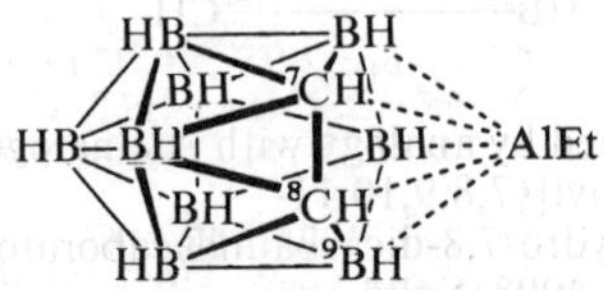

M 188.442

Colourless cryst. (C_6H_6/hexane). Mp 97-9°.

▷Explodes with water

Bis-THF adduct: [34540-18-6].
$C_{12}H_{32}AlB_9O_2$ M 332.655
Mp 119-20°.

Young, D.A.T. *et al*, *J. Am. Chem. Soc.*, 1971, **93**, 5687 (*synth, ir, pmr, nmr*)
Churchill, M.R. *et al*, *J. Chem. Soc., Dalton Trans.*, 1972, 1317 (*struct*)

$C_4H_{16}AlB_9$ B-00286

Ethyl(7,8,9,10,11-η)undecahydro-7,9-dicarbaundecaborato(2−)aluminum

3-Ethyl-3-alumino-1,7-dicarba-closo-*dodecaborane*(*12*)

[34606-90-1]

M 188.442

Mp 100-2°.

▷Explodes with water

Young, D.A.T. *et al*, *J. Am. Chem. Soc.*, 1971, **93**, 5687 (*synth, ir, pmr, nmr*)

$C_4H_{16}B_9Ga$ B-00287

Ethyl[(7,8,9,10,11-η)undecahydro-7,8-dicarbaundecaborato(2−)]gallium, 9CI

3-Ethyl-3-gallo-1,2-dicarbadodecaborane(*12*)

[34540-17-5]

M 231.180

Struct. by analogy with Al analogue. Extremely air-sensitive solid. Mp 114-5° (sealed tube). Partly isom. on heating at 400° to Ethyl[(7,8,9,10,11-η)undecahydro-7,9-dicarbaundecaborato(2−)]gallium, B-00288 .

Young, D.A.T. *et al*, *J. Am. Chem. Soc.*, 1971, **93**, 5687 (*synth, nmr, ir, pmr*)

$C_4H_{16}B_9Ga$ B-00288

Ethyl[(7,8,9,10,11-η)undecahydro-7,9-dicarbaundecaborato(2−)]gallium, 9CI

3-Ethyl-3-gallo-1,7-dicarbadodecaborane(12)

[34540-19-7]

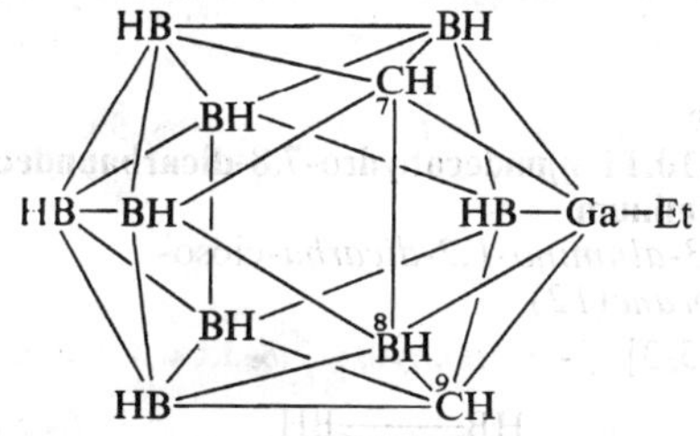

M 231.180

Struct. assumed by analogy with Al analogue. Formed on heating Ethyl[(7,8,9,10,11-η)undecahydro-7,8-dicarbaundecaborato(2−)]gallium, B-00287 to 400°. Solid.

Young, D.A.T. *et al*, *J. Am. Chem. Soc.*, 1971, **93**, 5687 (*synth, nmr, ir, pmr*)

$C_4H_{16}B_{10}O_2$ B-00289

1,2-Dicarbadodecaborane(12)-1,2-dimethanol, 10CI

1,2-Dimethylol-1,2-dicarbadodecaborane(12)

[19610-37-8]

As 1,2-Dicarbadodecaborane(12), B-00151 with

R = $-CH_2OH$

M 204.269

closo-. Used in polyester carborane and polyurethane carborane syntheses. White solid (toluene). Mp 304°.

Fein, M.M. *et al*, *Inorg. Chem.*, 1963, **2**, 1115 (*synth*)
Heying, T.L. *et al*, *Inorg. Chem.*, 1963, **2**, 1097 (*synth*)
Green, J. *et al*, *J. Polym. Sci., Part A*, 1964, **2**, 3113 (*synth*)
Korshak, V.V. *et al*, *Vysokomol. Soedin, Ser. A*, 1970, **13**, 2137 (*synth*)
Bartet, B. *et al*, *C.R. Hebd. Seances Acad. Sci.*, 1974, **279**, 283, 593 (*ir, nmr*)

$C_4H_{16}B_{10}O_2$ B-00290

1,7-Dicarbadodecaborane(12)-1,7-dimethanol, 10CI

1,7-Dimethylol-1,7-dicarbadodecaborane

[23924-78-9]

M 204.269

closo-. Used in polyester carborane and polyurethane carborane syntheses. White solid (toluene). Mp 194-6°.

Grafstein, D. *et al*, *Inorg. Chem.*, 1963, **2**, 1129 (*synth*)
Bartet, B. *et al*, *C.R. Hebd. Seances Acad. Sci.*, 1974, **279**, 283 (*ir*)

Handle all chemicals with care

$C_4H_{18}AlB_9$ B-00291

Dodecahydro-7,8-dicarbaundecaborato(1−)-H^9,H^{10} dimethylaluminum

7,8-μ-Dimethylalumino-1,2-dicarba-nido-undecaborane(13)

[11098-56-9]

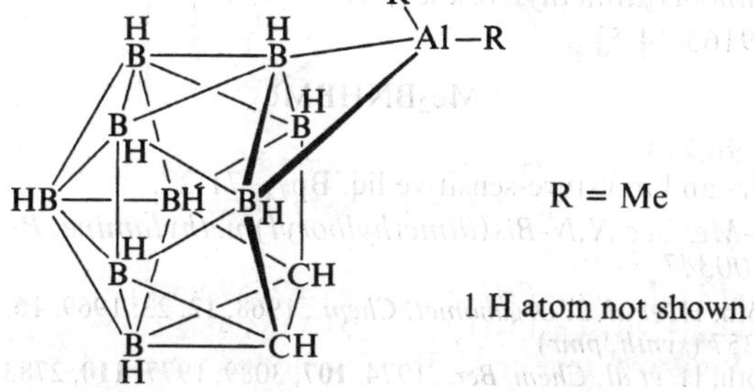

M 190.458

Colourless cryst. (C_6H_6). Mp 120-2°. Subl. *in vacuo*.

▷Explodes with water

Young, D.A.T. *et al*, *J. Am. Chem. Soc.*, 1971, **93**, 5687 (*synth, pmr, nmr*)
Churchill, M.R. *et al*, *J. Chem. Soc., Dalton Trans.*, 1972, 1314 (*struct*)

$C_4H_{22}AuB_{18}^{\ominus}$ B-00292

Bis[undecahydro-1,2-dicarbaundecaborato(2−)]aurate(1−), 8CI

Bis(dicarbollide)aurate(1−)

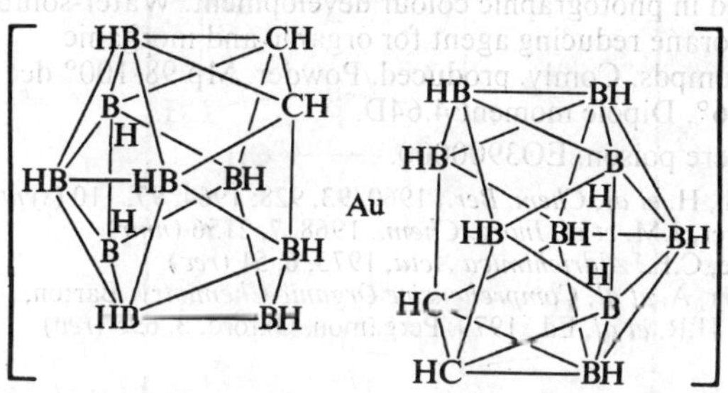

M 461.764 (ion)

"Slipped" sandwich struct. Large red cryst. (CH_2Cl_2/hexane).

Tetraethylammonium salt:
$C_{12}H_{42}AuB_{18}N$ M 592.017
Glittering red plates (CH_2Cl_2/hexane). Sol. Me_2CO, CH_2Cl_2, MeCN.

Methyltriphenylphosphonium salt:
$C_{23}H_{40}AuB_{18}P$ M 739.089
Large red cryst. (CH_2Cl_2/hexane).

Warren, L.F. *et al*, *J. Am. Chem. Soc.*, 1968, **90**, 4823 (*synth, nmr, pmr, ir*)
Wing, R.M., *J. Am. Chem. Soc.*, 1968, **90**, 4828 (*cryst struct*)

$C_4H_{22}AuB_{18}^{\ominus\ominus}$ B-00293

Bis[undecahydro-1,2-dicarbaundecaborato(2−)]aurate(2−), 8CI

Bis(dicarbollide)aurate(2−)

M 461.764 (ion)

"Slipped" sandwich struct.

Bis(tetraethylammonium) salt:
$C_{20}H_{62}AuB_{18}N_2$ M 722.270
Blue, paramagnetic microcryst. Sol. THF, MeCN. μ_{eff} = 1.79 BM.

Warren, L.F. *et al*, *J. Am. Chem. Soc.*, 1968, **90**, 4823 (*synth*)

$C_4H_{22}B_{18}Cl_2U^{\ominus\ominus}$ B-00294

Dichlorobis[(7,8,9,10,11-η)undecahydro-7,8-dicarbaundeca-borato]uranate(2−)

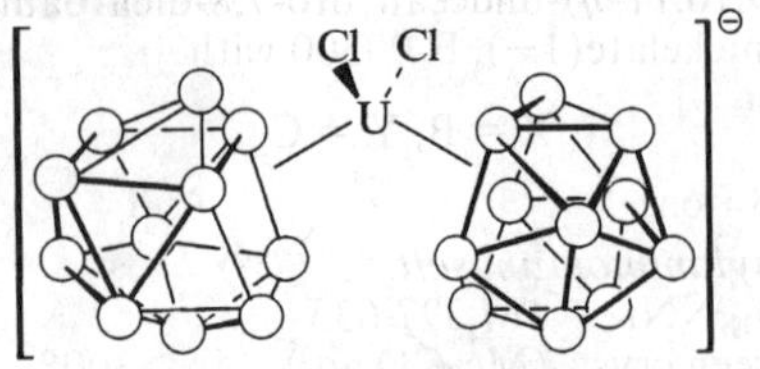

○ = BH or CH (position of CH's not rigorously established)

M 573.733 (ion)

Di-Li salt, octa-THF complex: [62559-95-9].
$C_{36}H_{86}B_{18}Cl_2Li_2O_8U$ M 1164.468
Yellow-orange cryst. Dec. by air and water.

Fronczek, F.R. *et al*, *J. Am. Chem. Soc.*, 1977, **99**, 1769 (*synth, uv, cryst struct*)

$C_4H_{22}B_{18}Cu^{\ominus}$ B-00295

Bis[(7,8,9,10,11-η)-undecahydro-7,8-dicarbaundecabora-to(2−)]cuprate(1−), 9CI

Bis(dicarbollide)cuprate(1−)

[51805-68-6]

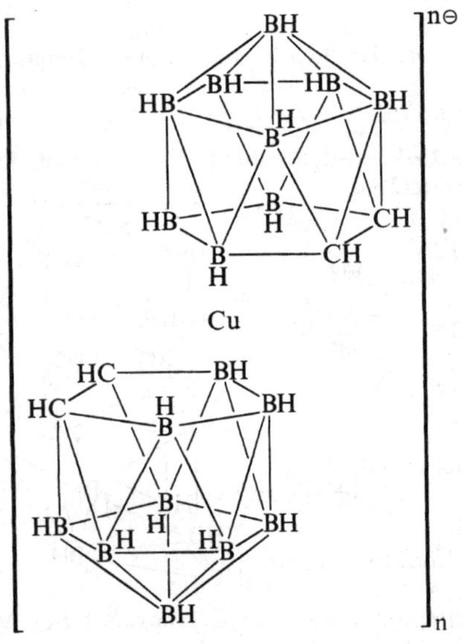

n = 1

M 328.344 (ion)
An example of a "slipped" sandwich complex.

Methyltriphenylphosphonium salt:
$C_{23}H_{40}B_{18}CuP$ M 605.669
Dark blood-red cryst. (CH_2Cl_2/hexane). Sol. CH_2Cl_2, Me_2CO.

Warren, L.F. *et al*, *J. Am. Chem. Soc.*, 1968, **90**, 4823 (*synth*)
Wing, R.M. *et al*, *J. Am. Chem. Soc.*, 1968, **90**, 4828 (*cryst struct*)
Mingos, D.M.P. *et al*, *J. Organomet. Chem.*, 1978, **146**, C37.

$C_4H_{22}B_{18}Cu^{\ominus\ominus}$ B-00296

Bis[(7,8,9,10,11-η)-undecahydro-7,8-dicarbaundecabora-to(2−)]cuprate(2−), 9CI

Bis(dicarbollide)cuprate(2−)

As Bis[(7,8,9,10,11-η)-undecahydro-7,8-dicarbaundecaborato(2−)]cuprat-e(1−), B-00295 with

n = 2

M 328.344 (ion)
An example of a "slipped" sandwich complex.

Di-K salt: [42992-84-7].
$C_4H_{22}B_{18}CuK_2$ M 406.540
Blue cryst. (Me_2CO/hexane). Sol. CH_2Cl_2, Me_2CO.

Wing, R.M., *J. Am. Chem. Soc.*, 1967, **89**, 5599 (*cryst struct*)
Hawthorne, M.F. *et al*, *J. Am. Chem. Soc.*, 1968, **90**, 879 (*synth*)
Warren, L.F. *et al*, *J. Am. Chem. Soc.*, 1968, **90**, 4823 (*synth*)
Mingos, D.M.P. *et al*, *J. Organomet. Chem.*, 1978, **146**, C37.

$C_4H_{22}B_{18}Fe^{\ominus}$ B-00297

Bis[(7,8,9,10,11-η)-undecahydro-7,8-dicarbaundecabora-to(2−)]ferrate(1−)

Bis-π-(3)-1,2-Dicarballyliron(1−)

[65465-56-7]

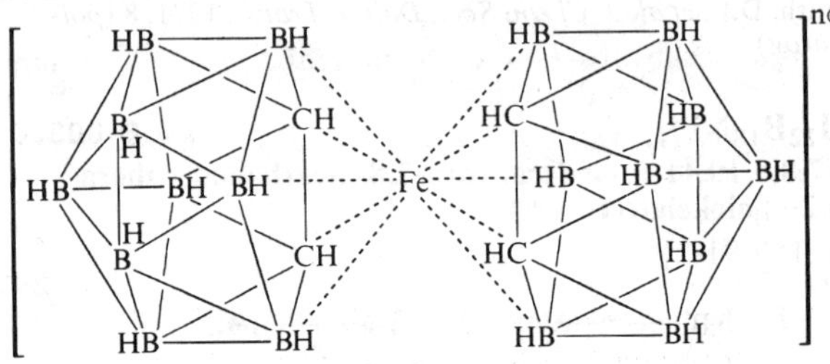

n = 1

M 320.645 (ion)
Paramagnetic.

Cs salt: [12304-97-1].
$C_4H_{22}B_{18}CsFe$ M 453.550
Black cryst. (H_2O).

Tetramethylammonium salt: [12304-95-9].
$C_8H_{34}B_{18}FeN$ M 394.790
Red plates (Me_2CO aq.). Mp >300°.

Inorg. Synth., 1967, **10**, 91 (*synth*)
Hawthorne, M.F. *et al*, *J. Am. Chem. Soc.*, 1968, **90**, 879 (*synth, ir, uv, nmr*)
Inorg. Synth., 1970, **10**, 111 (*synth*)
Birchall, T. *et al*, *Inorg. Chem.*, 1971, **10**, 399 (*mössbauer*)
Wiersema, R.J. *et al*, *J. Am. Chem. Soc.*, 1974, **96**, 761 (*cmr, nmr*)
Ammeter, J.H., *J. Magn. Reson.*, 1978, **30**, 299 (*esr*)

$C_4H_{22}B_{18}Fe^{\ominus\ominus}$ B-00298

Bis[(7,8,9,10,11-η)-undecahydro-7,8-dicarbaundecabora-to(2−)]ferrate(2−)

Bis-π-(3)-1,2-dicarballyliron(2−)

[51868-94-1]

As Bis[(7,8,9,10,11-η)-undecahydro-7,8-dicarbaundecaborato(2−)]ferrate-(1−), B-00297 with

n = 2

M 320.645 (ion)
Diamagnetic.

Bis(tetramethylammonium) salt:
$C_{12}H_{46}B_{18}FeN_2$ M 468.936
Deep-pink cryst. (Me_2CO/toluene).

Hawthorne, M.F. *et al*, *J. Am. Chem. Soc.*, 1968, **90**, 879 (*synth, ir, pmr, nmr, uv*)

$C_4H_{22}B_{18}Ni$ **B-00299**

Bis[(7,8,9,10,11-η)-undecahydro-7,8-dicarbaundecabora-to(2−)]nickel, 9CI

[36548-55-7]

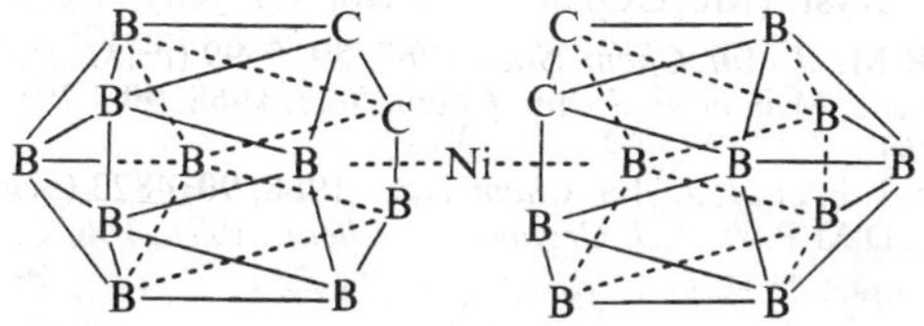

M 323.488

Air-stable orange cryst. Sol. most org. solvs. Subl. unchanged *in vacuo* at 150°, dec. ~265°.

Hawthorne, M.F. *et al*, *J. Am. Chem. Soc.*, 1968, **90**, 879 (*synth*, *ir*, *pmr*)
St Clair, D. *et al*, *J. Am. Chem. Soc.*, 1970, **92**, 1173 (*struct*)
Smith, D.E. *et al*, *J. Chem. Soc., Dalton Trans.*, 1971, 8 (*polarog*)

$C_4H_{22}B_{18}Ni^{\ominus}$ **B-00300**

Bis[(7,8,9,10,11-η)-undecahydro-7,8-dicarbaundecabora-to(2−)]nickelate(1−), 9CI

[51159-91-2]

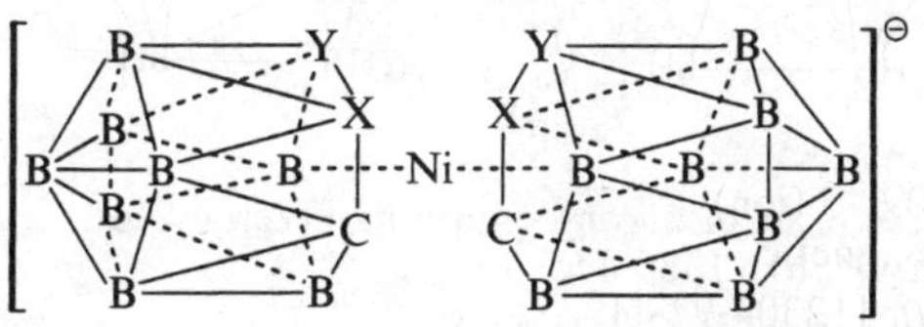

X = C,Y = B

M 323.488 (ion)

Rb salt:
$C_4H_{22}B_{18}NiRb$ M 408.956
Green-black lustrous cryst.

Tetramethylammonium salt: [51159-92-3].
$C_8H_{34}B_{18}NNi$ M 397.633
Yellow-brown solid. Mp >300°.

Hawthorne, M.F. *et al*, *J. Am. Chem. Soc.*, 1968, **90**, 879 (*synth*, *ir*, *pmr*)
Hansen, F.V. *et al*, *Acta Chem. Scand.*, 1973, **27**, 1210 (*struct*)

$C_4H_{22}B_{18}Ni^{\ominus\ominus}$ **B-00301**

Bis[(7,8,9,10,11-η)-undecahydro-7,8-dicarbaundecabora-to(2−)]nickelate(2−), 9CI

[36733-09-2]

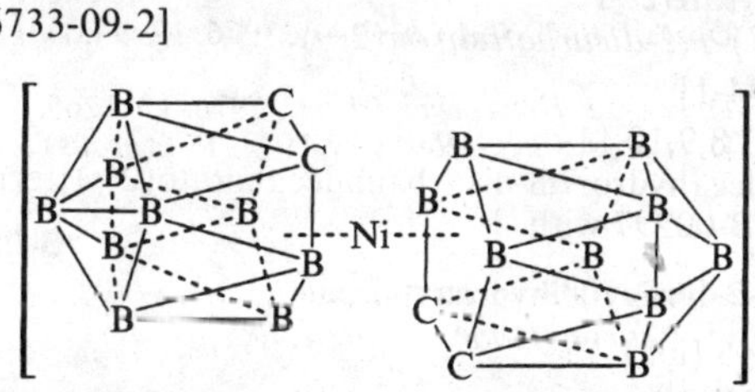

M 323.488 (ion)

Bis(tetramethylammonium) salt:
$C_{12}H_{46}B_{18}N_2Ni$ M 471.779
Pale-brown needles. Extremely air-sensitive.

Hawthorne, M.F. *et al*, *J. Am. Chem. Soc.*, 1968, **90**, 879 (*synth*, *ir*, *pmr*)
Wing, R.M., *J. Am. Chem. Soc.*, 1968, **90**, 4828 (*struct*)
Pavlik, I. *et al*, *CA*, 1976, **85**, 93221 (*pe*)

$C_4H_{22}B_{18}Ni^{\ominus}$ **B-00302**

Bis[(7,8,9,10,11-η)-undecahydro-7,9-dicarbaundecabora-to(2−)]nickelate(1−), 9CI

As Bis[(7,8,9,10,11-η)-undecahydro-7,8-dicarbaundecaborato(2−)]nickelate(1−), B-00300 with

X = B, Y = C

M 323.488 (ion)

Tetramethylammonium salt:
$C_8H_{34}B_{18}NNi$ M 397.633
Olive-green cryst. (Me_2CO aq.). Mp >300°.

Rb salt: [42443-78-7].
$C_4H_{22}B_{18}NiRb$ M 408.956
No phys. details recorded.

Hawthorne, M.F. *et al*, *J. Am. Chem. Soc.*, 1968, **90**, 862, 879 (*synth*, *ir*, *pmr*)
Pont, L.O. *et al*, *Inorg. Chem.*, 1974, **13**, 483 (*pe*)

$C_4H_{24}B_{10}In_2$ **B-00303**

Dimethylindium[η^4-dodecahydrodecaborato(2−)dimethyl-indate(1−)], 9CI

[55988-45-9]

$[Me_2In]^{\oplus}[B_{10}H_{12}InMe_2]^{\ominus}$

M 409.974

Solid, air-sensitive. Sol. Me_2CO.

Greenwood, N.N. *et al*, *J. Chem. Soc., Dalton Trans.*, 1975, **4**, 299 (*synth*, *ir*, *nmr*)

$C_4H_{24}B_{20}Fe^{\ominus\ominus}$ **B-00304**

Bis[(7,8,9,10,11,12-η)-dodecahydro-7,9-dicarbadodecabora-to(2−)]ferrate(2−), 9CI

Bis(π-$B_{10}C_2H_{12}$)ferrate(2−)

[55073-27-3]

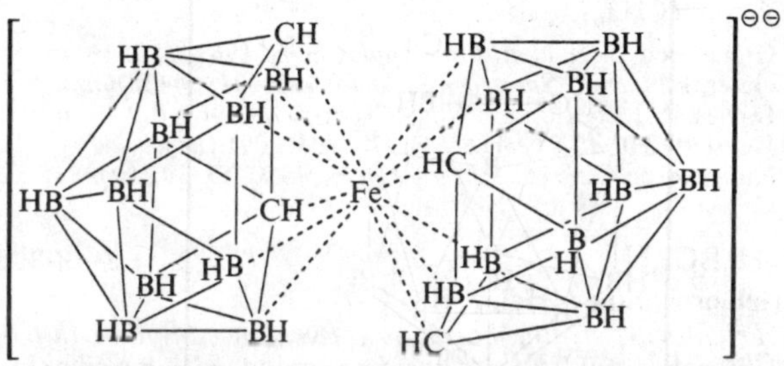

M 344.281 (ion)

Bis(tetraethylammonium) salt: [40904-18-5].
$C_{20}H_{64}B_{20}FeN_2$ M 604.786
Purple cryst. (Me_2CO/EtOH). Mp >240° dec.

Dustin, D.F. *et al*, *J. Am. Chem. Soc.*, 1973, **95**, 1109 (*synth*, *polarog*, *pmr*, *nmr*, *ir*, *uv*)

$C_5H_3B_5FeO_5$ **B-00305**

Tricarbonyl[(2,3,4,5-η)-dicarbonyltrihydropentabora-to(2−)]iron, 10CI

[61525-93-7]

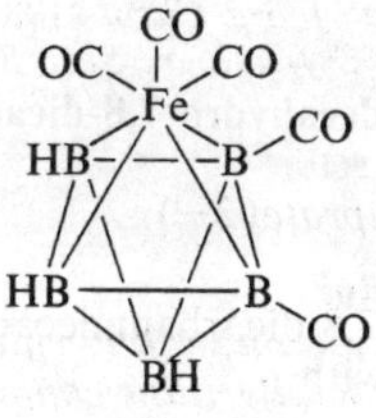

M 252.973

Orange solid or red liq. Mp ~5°.

Ulman, J.A. *et al*, *J. Chem. Soc., Chem. Commun.*, 1976, 632 (*synth, ms, ir, nmr, pmr*)
Ulman, J.A. *et al*, *J. Am. Chem. Soc.*, 1978, **100**, 456 (*pe*)

C_5H_5B B-00306

Borabenzene

Borin, 10CI, 9CI

[31029-61-5]

M 75.905

Parent member of a series of organoboron compds. known only in the form of anionic derivatives and metal complexes thereof. Never prepd. as the free parent compd. See also Boratabenzene, B-00316 .

Allen, C.W. *et al*, *J. Chem. Educ.*, 1978, **55**, 497 (*rev*)

$C_5H_5BBr_3N$ B-00307

Tribromoborane pyridine

Tribromo(pyridine)boron, 10CI. *Pyridine compd. with tribromoborane(1:1)*, 9CI. *Pyridine compd. with boron bromide(BBr_3)(1:1)*, 8CI

[3022-54-6]

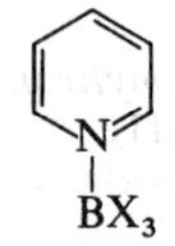

X = Br

M 329.623

Cryst. (C_6H_6).

Greenwood, N.N. *et al*, *Pure Appl. Chem.*, 1961, **2**, 55.
Gates, P.N. *et al*, *Spectrochim. Acta*, 1965, **21**, 1445 (*nmr*)
Davies, P.G. *et al*, *Inorg. Nucl. Chem. Lett.*, 1967, **3**, 249 (*ir*)
Hensen, K. *et al*, *Chem. Ber.*, 1970, **103**, 2091 (*uv*)
Rubenacker, G.V. *et al*, *Inorg. Chem.*, 1980, **19**, 392 (*nqr*)

$C_5H_5BCl_3N$ B-00308

Trichloroborane pyridine

Trichloro(pyridine)boron, 10CI. *Pyridine compd. with trichloroborane(1:1)*, 9CI. *Pyridine compd. with boron chloride(BCl_3)(1:1)*, 8CI

[2903-67-5]

As Tribromoborane pyridine, B-00307 with

X = Cl

M 196.270

Crosslinking catalyst for epoxy resins. Cryst. (C_6H_6). Mp 110-3°.

Nöth, H. *et al*, *Chem. Ber.*, 1966, **99**, 1089 (*synth*)
Davies, P.G. *et al*, *Inorg. Nucl. Chem. Lett.*, 1967, **3**, 249 (*ir*)
Fratiello, A. *et al*, *Inorg. Chem.*, 1968, **7**, 1581 (*nmr*)
Hensen, K. *et al*, *Chem. Ber.*, 1970, **103**, 2091 (*uv*)
Rubenacker, G.V. *et al*, *Inorg. Chem.*, 1980, **19**, 392 (*nqr*)
Toepal, K. *et al*, *Acta Crystallogr., Sect. B*, 1981, **37**, 969 (*struct*)

$C_5H_5BF_3N$ B-00309

Trifluoroborane pyridine

Trifluoro(pyridine)boron, 10CI. *Pyridine compd. with trifluoroborane(1:1)*, 9CI. *Pyridine compd. with boron fluoride(BF_3)(1:1)*, 8CI

[407-76-1]

As Tribromoborane pyridine, B-00307 with

X = F

M 146.906

Crosslinking agent for epoxy resins. White cryst. with very faint pyridine odour. Sol. C_6H_6, Py, hot EtOH, insol. Et_2O. Mp 45°. Bp 300°.

Van der Meulen *et al*, *J. Am. Chem. Soc.*, 1932, **54**, 4404 (*synth*)
Perkampus, H.H. *et al*, *Bunsenges Phys. Chem.*, 1967, **71**, 439 (*nmr*)
Hensen, K. *et al*, *Theor. Chim. Acta*, 1968, **12**, 206 (*uv*)
Taillandier, M. *et al*, *Spectrochim. Acta, Part A*, 1969, **25**, 1807 (*ir*)
Hensen, K. *et al*, *Chem. Ber.*, 1970, **103**, 2091 (*uv*)
Barber, M. *et al*, *J. Chem. Soc., Faraday Trans. 2*, 1973, **69**, 551 (*pe*)

$C_5H_5BO_4$ B-00310

(2-Formyl-3-furanyl)dihydroxyborane

(2-Formyl-3-furanyl)boronic acid, 10CI, 9CI. *2-Formyl-3-furanboronic acid*, 8CI

[27339-37-3]

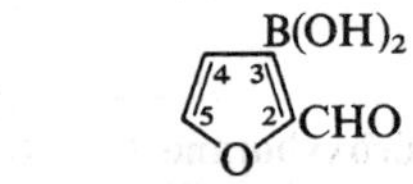

M 139.902

Used in prep. of boron-containing heterocycles. Cryst. (H_2O). Mp 138° dec.

Florentin, D. *et al*, *C.R. Hebd. Seances Acad. Sci.*, 1970, **270**, 1608.
Ger. Pat., 2 264 363, (*1973*); *CA*, **79**, 92369j (*use*)
Florentin, D. *et al*, *Bull. Soc. Chim. Fr.*, 1974, **11**, 2620 (*props*)
Florentin, D. *et al*, *Bull. Soc. Chim. Fr.*, 1976, 1999 (*synth, nmr*)
Florentin, D. *et al*, *J. Heterocycl. Chem.*, 1976, **13**, 1265.
Florentin, D. *et al*, *J. Chem. Res. (S)*, 1977, 158.

$C_5H_5BO_4$ B-00311

(3-Formyl-2-furanyl)dihydroxyborane

(3-Formyl-2-furanyl)boronic acid, 10CI, 9CI. *3-Formyl-2-furanboronic acid*, 8CI

[27339-38-4]

M 139.902

Cryst. (H_2O). Mp 155° dec.

Florentin, D. *et al*, *C.R. Hebd. Seances Acad. Sci.*, 1970, **270**, 1608.
Florentin, D. *et al*, *Bull. Soc. Chim. Fr.*, 1976, 1999 (*synth, nmr*)
Florentin, D. *et al*, *J. Heterocycl. Chem.*, 1976, **13**, 1265.
Florentin, D. *et al*, *J. Chem. Res. (S)*, 1977, 158 (*props*)

$C_5H_5BO_4$ B-00312

(4-Formyl-2-furanyl)dihydroxyborane

(4-Formyl-2-furanyl)boronic acid, 10CI

[62306-78-9]

M 139.902

Cryst. (H_2O). Mp 150° dec.

Florentin, D. *et al*, *Bull. Soc. Chim. Fr.*, 1976, 1999 (*synth, nmr*)
Florentin, D. *et al*, *J. Heterocycl. Chem.*, 1976, **13**, 1265 (*props*)

$C_5H_5BO_4$ B-00313

(4-Formyl-3-furanyl)dihydroxyborane

(4-Formyl-3-furanyl)boronic acid, 10CI. *4-Formyl-3-furanboronic acid*, 8CI

[27339-36-2]
M 139.902
Cryst. (H_2O). Mp 136-8° dec.

Florentin, D. *et al*, *C.R. Hebd. Seances Acad. Sci.*, 1970, **270**, 1608.
Florentin, D. *et al*, *Bull. Soc. Chim. Fr.*, 1976, 1999 (*synth*, *nmr*)
Florentin, D. *et al*, *J. Heterocycl. Chem.*, 1976, **13**, 1265 (*props*)
Florentin, D. *et al*, *J. Chem. Res.* (*S*), 1977, 158 (*props*)

$C_5H_5BO_4$ B-00314

(5-Formyl-2-furanyl)dihydroxyborane
(5-Formyl-2-furanyl)boronic acid, 10CI
[27329-70-0]

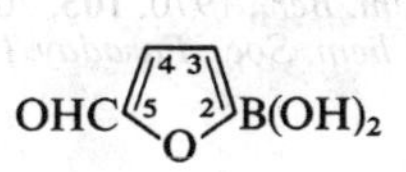

M 139.902
Cryst. (H_2O). Mp 145° dec.

Florentin, D. *et al*, *Bull. Soc. Chim. Fr.*, 1976, 1999 (*synth*)
Florentin, D. *et al*, *J. Heterocycl. Chem.*, 1976, **13**, 1265.

$C_5H_5BO_4$ B-00315

(5-Formyl-3-furanyl)dihydroxyborane
(5-Formyl-3-furanyl)boronic acid, 10CI
[62306-80-3]
M 139.902
Cryst. (H_2O). Dec. at 116°.

Florentin, D. *et al*, *Bull. Soc. Chim. Fr.*, 1976, 1999 (*synth*)
Florentin, D. *et al*, *J. Heterocycl. Chem.*, 1976, **13**, 1265.

$C_5H_6B^{\ominus}$ B-00316

Boratabenzene, 10CI, 9CI, 8CI
Borin, boratabenzene deriv., 10CI, 9CI
[55926-39-1]

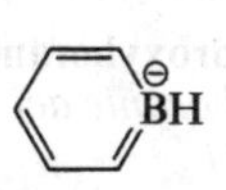

M 76.912 (ion)
The simple metal salts (e.g. Na+) of boratabenzene anions have only been prepd. for compds. containing an organic substituent on boron. However, transition metal complexes of this, the parent system, are known.
Fe complex: see Bis(η^6-boratabenzene)iron, B-00651

Allen, C.W. *et al*, *J. Chem. Educ.*, 1978, **55**, 497 (*rev*)
Ashe, A.J. *et al*, *J. Am. Chem. Soc.*, 1979, **101**, 7066 (*synth*, *cryst struct*, *nmr*, *ms*, *deriv*)
Clack, D.W. *et al*, *Inorg. Chem.*, 1979, **18**, 513.
Clack, D.W. *et al*, *J. Organomet. Chem.*, 1981, **208**, 183.

$C_5H_6BF_6N$ B-00317

1,1-Dimethyl-*N*-[2,2,2-trifluoro-1-(trifluoromethyl)ethylidene]boranamine, 10CI
[61707-35-5]

$$Me_2B{-}N{=}C(CF_3)_2$$

M 204.910
Yellow liq., sensitive to air. Bp_{99} 70-3°.

Gragg, B.R. *et al*, *Synth. React. Inorg. Metal-Org. Chem.*, 1976, **6**, 275 (*synth*, *props*)

Emerick, D.P. *et al*, *J. Organomet. Chem.*, 1978, **153**, 9 (*synth*, *ms*, *nmr*, *ir*)
Emerick, D.P. *et al*, *Z. Anorg. Allg. Chem.*, 1980, **468**, 44.

$C_5H_6B_4FeO_3$ B-00318

Tricarbonyl[(2,3,4,5,6-η)-hexahydro-2,3-dicarbahexaborato(2−)]iron, 9CI
(π-2,3-$C_2B_4H_6$)tricarbonyliron. 1,2,3-Tricarbonyliron-$C_2B_4H_6$
[51177-52-7]

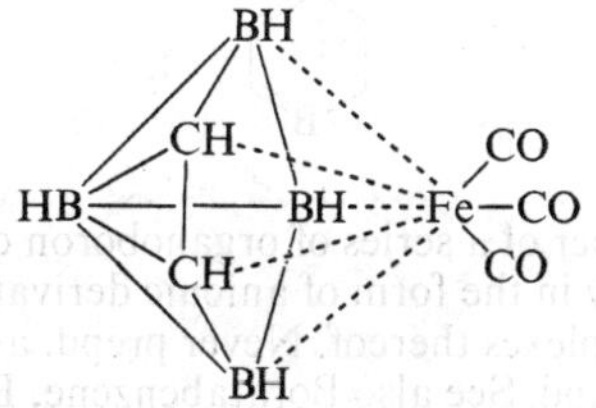

M 213.188
Deep yellow-orange liq.

Grimes, R.N., *J. Am. Chem. Soc.*, 1971, **93**, 261 (*synth*, *ms*, *nmr*, *pmr*, *ir*)
Sneddon, L.G. *et al*, *J. Am. Chem. Soc.*, 1973, **95**, 6623 (*synth*, *ms*, *pmr*, *nmr*, *ir*)

$C_5H_6B_4FeO_3$ B-00319

Tricarbonyl[(2,3,4,5,6-η)-hexahydro-2,4-dicarbahexaborato(2−)]iron
(π-2,4-$C_2B_4H_6$)tricarbonyliron. 1,2,4-(Tricarbonyliron)-$C_2B_4H_6$
M 213.188
Orange.

Sneddon, L.G. *et al*, *J. Am. Chem. Soc.*, 1973, **95**, 6623 (*synth*, *nmr*, *pmr*, *ir*)

$C_5H_7BF_2O_2$ B-00320

Difluoro(2,4-pentanedionato-*O*,*O′*)boron, 10CI
Difluoroboryl acetylacetonate. Boron acetylacetone difluoride
[15390-25-7]

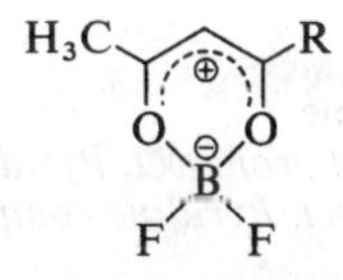

R = CH_3

M 147.916
Catalyst for curing epoxy resins. Cryst. ($CHCl_3$/pet. ether); air- and moisture-sensitive. Sol. C_6H_6; insol. pet. ether. Mp 43°. Dec. by warm EtOH.
▷ED7600000.

Morgan, G.T. *et al*, *J. Chem. Soc.*, 1924, **125**, 1963 (*synth*)
Toporcer, L.H. *et al*, *Inorg. Chem.*, 1965, **4**, 1649 (*synth*, *nmr*, *uv*, *ir*)
Angelos, N., *Justus Liebigs Ann. Chem.*, 1966, **700**, 29 (*synth*)
Trestianu, A. *et al*, *Tetrahedron*, 1968, **24**, 2499 (*pmr*)
Brown, N.M.D. *et al*, *J. Chem. Soc.* (*A*), 1969, 526 (*ir*, *uv*, *nmr*)

Hammel, J.C. *et al*, *J. Chem. Soc.* (*A*), 1969, 1461 (*nmr*, *ir*, *uv*)

$C_5H_7B_3FeO_3$ B-00321

Tricarbonyl[(1,2,3,4,5-η)-heptahydro-1,2-dicarbapentaborato(2−)]iron, 10CI

Dicarbacyclopentaboranyliron tricarbonyl. (2,3-$C_2B_3H_7$)tricarbonyliron

[33362-89-9]

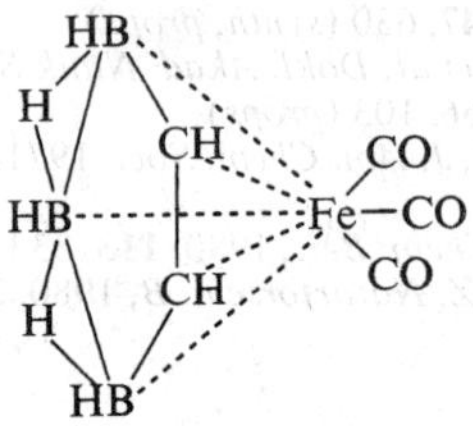

M 203.386

Pale-yellow liq. Mp 0-15°.

Grimes, R.N., *J. Am. Chem. Soc.*, 1971, **93**, 261 (*synth*, *pmr*, *ms*, *ir*)
Franz, D.A. *et al*, *J. Am. Chem. Soc.*, 1972, **94**, 412 (*synth*)
Brennan, J.P. *et al*, *Inorg. Chem.*, 1973, **12**, 2266 (*struct*)
Sneddon, L.G. *et al*, *J. Am. Chem. Soc.*, 1973, **95**, 6623 (*synth*, *nmr*, *ir*)
Ulman, J.A. *et al*, *J. Am. Chem. Soc.*, 1978, **100**, 456 (*pe*)

$C_5H_7B_5FeO_3$ B-00322

Tricarbonyl[(3,4,5,6,7-η)-heptahydro-3,5-dicarbaheptaborato(2−)]iron, 9CI

(4)-1,8-(Tricarbonyliron)-$C_2B_5H_7$

[51203-46-4]

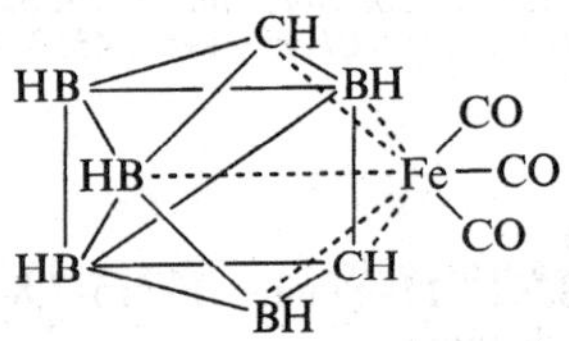

M 225.006

Orange.

Sneddon, L.G. *et al*, *J. Am. Chem. Soc.*, 1973, **95**, 6623 (*synth*, *nmr*, *pmr*, *ir*)

C_5H_8BN B-00323

Borane pyridine

Trihydro(pyridine)boron, 10CI. *Pyridine compd. with borane(1:1),* 9CI, 8CI

[110-51-0]

As Tribromoborane pyridine, B-00307 with

X = H

M 92.935

Germicide, fungicide, and protozoicide, used in photographic colour development of reversal films. Reducing agent. Produced comly. Liq. Mp 10-1°. n_D^{20} 1.5315.

▷Toxic, flammable. US3675000.

Brey, W.S. *et al*, *Adv. Chem. Ser.*, 1964, **42**, 100 (*nmr*)
Hoffmann, R., *J. Chem. Phys.*, 1964, **40**, 2474 (*struct*)
Hermanek, S. *et al*, *Collect. Czech. Chem. Commun.*, 1966, **31**, 1281 (*chromotog*)
Andrews, G.C., *Tetrahedron Lett.*, 1980, **21**, 697.
Pelter, A. *et al*, *J. Chem. Soc., Chem. Commun.*, 1981, 1014.
Schaefer, T. *et al*, *Can. J. Chem.*, 1981, **59**, 3026 (*struct*)

C_5H_9B B-00324

Methyldivinylborane, 8CI

Diethenylmethylborane, 10CI, 9CI. *Methyldivinylborine*

[5856-68-8]

$(H_2C{=}CH)_2BMe$

M 79.936

Volatile liq., easily oxid. in air. Sol. nonpolar solvs. V.p. 126 mm at 0°.

▷Probably pyrophoric

Parsons, T.D. *et al*, *J. Am. Chem. Soc.*, 1954, **76**, 1710; 1957, **79**, 5091 (*synth*)
Good, C.D. *et al*, *J. Am. Chem. Soc.*, 1962, **84**, 1162 (*synth*)
Good, C.D. *et al*, *J. Chem. Eng. Data*, 1962, **7**, 416 (*props*, *chromatog*)
Wrackmeyer, B. *et al*, *Chem. Ber.*, 1976, **109**, 1075 (*nmr*)
Durig, J.R. *et al*, *J. Mol. Struct.*, 1981, **72**, 85 (*raman*, *ir*, *nmr*)

$C_5H_9BN_2$ B-00325

Borane 4-aminopyridine

Trihydro(4-pyridinamine-N[1])boron, 10CI. *4-Pyridinamine compd. with borane(1:1),* 9CI

[59900-43-5]

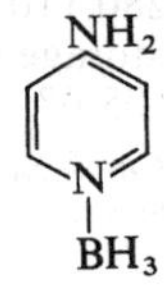

M 107.950

Used in photographic developer solns.

May, C.E. *et al*, *Synth. React. Inorg. Metal-Org. Chem.*, 1977, **7**, 509.
U.S.P., 4 042 392, (*1977*); *CA*, **88**, 30399w
Lelental, M. *et al*, *J. Photogr. Sci.*, 1978, **26**, 135 (*use*)

$C_5H_9B_2LiO_4$ B-00326

[Bis(1,3,2-dioxaborolan-2-yl)methyl]lithium, 9CI

Lithium bis(ethylenedioxyboryl)methide

[57404-82-7]

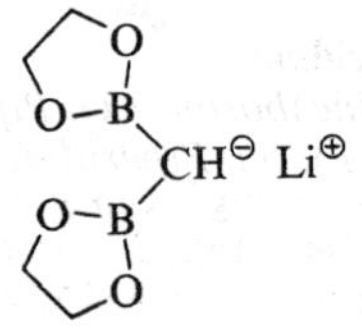

M 161.685

Reagent for conversion of RCOR′ to RR′CHCHO.

Matteson, D.S. *et al*, *J. Am. Chem. Soc.*, 1975, **97**, 5608 (*synth*)
Fieser, M. *et al*, *Reagents for Organic Synthesis*, Wiley, 1967-83, **6**, 328.

$C_5H_{11}B$ B-00327

Borinane, 10CI, 9CI, 8CI

Boracyclohexane

[289-53-2]

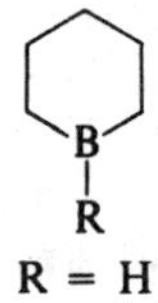

R = H

M 81.952

Known as H-bridged dimer or as complexes. Hydroborating agent. Air- and moisture-sensitive.

Tetraethylethylenediamine complex: [79069-87-7]. No phys. props. reported.

Brown, H.C. *et al*, *J. Am. Chem. Soc.*, 1971, **93**, 3777 (*props*)
Zweifel, G. *et al*, *Synthesis*, 1972, 557 (*props, use*)
Negishi, E. *et al*, *J. Am. Chem. Soc.*, 1973, **95**, 6757 (*deriv*)
Brown, H.C. *et al*, *J. Org. Chem.*, 1981, **46**, 4713 (*props*)
Brown, H.C. *et al*, *Heterocycles*, 1982, **17**, 77 (*props*)
Brown, H.C. *et al*, *J. Organomet. Chem.*, 1982, **225**, 63 (*use*)

$C_5H_{11}B$ B-00328
Cyclopentylborane

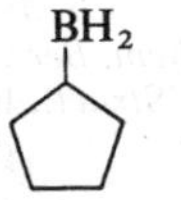

M 81.952

Tetramethylethylenediamine complex (2:1): [69319-86-4]. *Dicyclopentyltetrahydro[μ-(N,N,N′,N′-tetramethyl-1,2-ethanediamine-N:N′)]diboron*, 10CI. *Bis-(cyclopentylborane)tetramethylethylenediamine.*
$C_{16}H_{38}B_2N_2$ M 280.110
Useful for long-term storage of cyclopentylborane. Air-stable solid. Mp 105-6°.

Brown, H.C. *et al*, *J. Org. Chem.*, 1979, **44**, 465 (*synth*)

$C_5H_{11}B$ B-00329
1-Methylborolane, 10CI

[67963-90-0]

As Borolane, B-00227 with

R = Me

M 81.952

Air-sensitive. Sol. nonhydroxylic solvs.

▷Prob. pyrophoric

Schmidt, A. *et al*, *Z. Naturforsch., B*, 1978, **33**, 855 (*props*)

$C_5H_{11}BF_3N$ B-00330
Trifluoroborane piperidine

Trifluoro(piperidine)boron, 10CI. *Piperidine compd. with trifluoroborane(1:1)*, 9CI. *Piperidine compd. with boron fluoride(BF₃)(1:1)*, 8CI

[592-39-2]

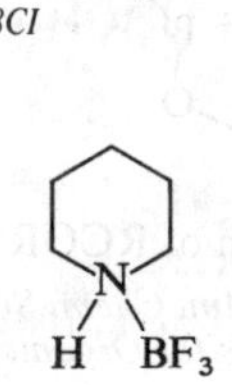

M 152.954

Crosslinking agent for epoxy resins, catalyst. Cryst. Mp 71-3°.

Bowlus, H. *et al*, *J. Am. Chem. Soc.*, 1931, **53**, 3835.
Lawesson, S.O., *Ark. Kemi*, 1956, **10**, 171 (*synth*)
Onak, T.P. *et al*, *J. Phys. Chem.*, 1959, **63**, 1533 (*nmr*)
Meek, D.W. *et al*, *Inorg. Chem.*, 1966, **5**, 445.
Japan. Pat., 73 39 154, (*1973*); *CA*, **81**, 30228v
Japan. Pat., 78 114 900, (*1978*); *CA*, **90**, 72868v (*use*)

$C_5H_{11}BO$ B-00331
1-Methoxyborolane, 10CI, 9CI, 8CI

B-*Methoxyboracyclopentane*

[6166-10-5]

As Borolane, B-00227 with

R = OMe

M 97.951

Air- and moisture-sensitive liq. Sol. nonhydroxylic solvs. d_{20} 0.842. Bp_{83} 40.5-41°. n_D^{20} 1.4172.

▷Prob. pyrophoric

Köster, R., *Angew. Chem.*, 1959, **71**, 520 (*synth*)
Mikhailov, B.M. *et al*, *Dokl. Akad. Nauk SSSR, Ser. Sci. Khim.*, 1962, **147**, 630 (*synth, props*)
Gal'chenko, G.L. *et al*, *Dokl. Akad. Nauk SSSR, Ser. Sci. Khim.*, 1966, **166**, 103 (*props*)
Brown, H.C. *et al*, *J. Am. Chem. Soc.*, 1971, **93**, 3400, 6682 (*synth*)
Biffar, W. *et al*, *Chem. Ber.*, 1980, **113**, 333 (*nmr*)
Wrackmeyer, B., *Z. Naturforsch., B*, 1980, **35**, 439 (*nmr*)

$C_5H_{11}BO_2$ B-00332
Cyclopentyldihydroxyborane

Cyclopentylboronic acid, 10CI. *Cyclopentaneboronic acid*

[63076-51-7]

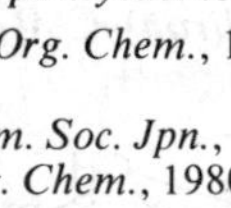

M 113.951

Cryst. (H_2O); somewhat sensitive to air. Mp 90-2°.

O,O-*Di-Me: see Cyclopentyldimethoxyborane, B-00500*

Thaisrivongs, S. *et al*, *J. Org. Chem.*, 1977, **42**, 3243 (*synth, ir, pmr*)
Taniguchi, H., *Bull. Chem. Soc. Jpn.*, 1979, **52**, 2946 (*ester*)
Brown, H.C. *et al*, *J. Org. Chem.*, 1980, **45**, 384 (*ester, props*)

$C_5H_{11}B_9O_3Ru$ B-00333
Tricarbonyl[(7,8,9,10,11-η)undecahydro-7,8-dicarbaundecaborato(2−)]ruthenium, 9CI

[56465-06-6]

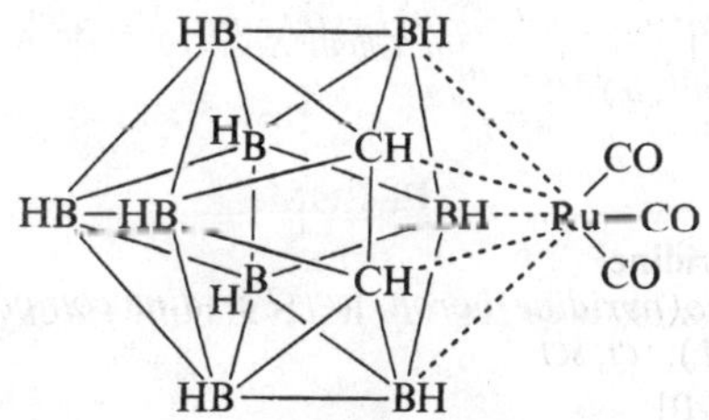

M 317.500

Yellow cryst. + $\frac{1}{2}C_6H_6$ (C_6H_6/cyclohexane). Mp >350°.

Siedle, A.R., *J. Organomet. Chem.*, 1975, **90**, 249 (*synth, ir, uv, nmr, pmr, ms*)

$C_5H_{13}B$ B-00334
Diethylmethylborane, 10CI, 9CI

Methyldiethylborane. Diethylmethylborine

[1115-07-7]

Et_2BMe

M 83.968

Liq., air-sensitive. Sol. THF. Mp −135.5°. Bp 61°.

▷Pyrophoric

Birchall, J.M. *et al*, *Chem. Ind.* (*London*), 1961, 1080 (*synth*)
Henneberg, D. *et al*, *Justus Liebigs Ann. Chem.*, 1961, **640**, 52 (*ms*)
Nöth, H., *Chem. Ber.*, 1973, **106**, 1508 (*use*)
Smith, M.B., *J. Organomet. Chem.*, 1974, **76**, 171 (*props*)
Spielvogel, B.F. *et al*, *J. Am. Chem. Soc.*, 1975, **97**, 1609 (*nmr*)
Stampf, E.J. *et al*, *J. Organomet. Chem.*, 1977, **131**, 171 (*synth, props*)

$C_5H_{13}B$ B-00335

3-Methyl-2-butylborane

(1,2-Dimethylpropyl)dihydroboron. 1,2-Dimethylpropylborane. Monosiamylborane. Siamylborane

$(H_3C)_2CHCH(CH_3)BH_2$

M 83.968

Et_3N complex: [56118-68-4]. (N,N-*Diethylethanamine*)-(*1,2-dimethylpropyl*)*dihydroboron*, 10CI. *1,2-Dimethylpropylborane* N,N-*diethylethanamine*, 9CI. *Siamylborane triethylamine.*
$C_{11}H_{28}BN$ M 185.159
Useful means of storage of monosiamylborane. Stable to disproportion at r.t., unlike the free borane.

Brown, H.C. *et al*, *J. Am. Chem. Soc.*, 1975, **97**, 2791.
Brown, H.C. *et al*, *J. Organomet. Chem.*, 1977, **135**, C10 (*synth*)

$C_5H_{13}BN_2O_2$ B-00336

Cyanoborane diethanolamine

(*Cyano*-C)*dihydro*[*2,2′*-(*methylimino*)*bis*[*ethanol*]-N]*boron*, 10CI. *Boranecarbonitrile, compd. with 2,2′-*(*methylimino*)*bis*[*ethanol*](*1:1*)

[75519-12-9]

$(HOCH_2CH_2)_2NH—BH_2CN$

M 143.980

Hall, I.J. *et al*, *J. Pharm. Sci.*, 1980, **69**, 1025 (*pharmacol*)

$C_5H_{13}BO$ B-00337

Diethylmethoxyborane

Diethylborinic acid methyl ester, 10CI, 9CI, 8CI. *Methoxydiethylborane*

[7397-46-8]

Et_2BOMe

M 99.967

Air- and moisture-sensitive liq. Sol. nonhydroxylic solvs. Bp 90°.

▷Pyrophoric

Nöth, H. *et al*, *Chem. Ber.*, 1966, **99**, 1049 (*nmr*)
Nöth, H. *et al*, *J. Organomet. Chem.*, 1968, **12**, 23 (*nmr*)
Köster, R. *et al*, *Justus Liebigs Ann. Chem.*, 1975, 352 (*synth, pmr, ms*)
Mikhailov, B.M. *et al*, *Izv. Akad. Nauk SSSR, Ser. Khim.*, 1978, 1586; *Bull. Acad. Sci. USSR, Div. Chem. Sci.*, **27**, 1385 (*synth*)
Paetzold, P. *et al*, *Z. Naturforsch.*, *B*, 1980, **35**, 568.

$C_5H_{13}BO_2$ B-00338

Dimethoxypropylborane

Propylboronic acid dimethyl ester, 10CI. *1-Propaneboronic acid dimethyl ester. Propyldimethoxyborane. Dimethyl propaneboronate*

[2938-87-6]

$H_3CCH_2CH_2B(OMe)_2$

M 115.967

Air- and moisture-sensitive liq. Sol. non-hydroxylic solvs. d^{20} 0.833. Bp 102-3°, Bp_{95} 45-56°. n_D^{20} 1.3847.

Ger. Pat., 1 061 781, (*1959*); *CA*, **55**, 20960f (*synth*)
Mikhailov, B.M. *et al*, *Izv. Akad. Nauk SSSR, Ser. Khim.*, 1962, 827 (*synth, props*)
Mikhailov, B.M. *et al*, *Zh. Obshch. Khim.*, 1965, **35**, 1073.
Costes, J.P. *et al*, *J. Organomet. Chem.*, 1979, **175**, 257 (*props, pmr, nmr*)

$C_5H_{13}BO_2$ B-00339

Pentyldihydroxyborane

Pentylboronic acid, 10CI, 9CI. *1-Pentaneboronic acid*, 8CI. *Amyldihydroxyborane. Dihydroxypentylborane*

[4737-50-2]

$H_3C(CH_2)_4B(OH)_2$

M 115.967

Inhibits cholinesterase, proteinase and lipase. Mod. air-sensitive cryst. (pet. ether). Mp 93-4°.

Laurent, J.P., *C.R. Hebd. Seances Acad. Sci.*, 1961, **253**, 1812 (*synth, props*)
Niedenzu, K. *et al*, *J. Org. Chem.*, 1961, **26**, 3037 (*synth*)
Antonov, V.K. *et al*, *Mol. Biol.*, 1970, **4**, 558.
Jarv, J. *et al*, *Bioorg. Khim.*, 1978, **4**, 1364; *CA*, **90**, 2320m (*use*)

$C_5H_{13}BS_2$ B-00340

Bis(methylthio)propylborane

Propyldithioboronic acid dimethyl ester, 10CI. *Propylbis*(*methylthio*)*borane*

[72334-76-0]

$H_3CCH_2CH_2B(SMe)_2$

M 148.088

No phys. props. reported, but presumed to be air- and moisture-sensitive liq. Sol. nonhydroxylic solvs.

Costes, J.P. *et al*, *J. Organomet. Chem.*, 1979, **175**, 257 (*props, nmr, pmr*)

$C_5H_{14}B^{\ominus}$ B-00341

(1,2-Dimethylpropyl)trihydroborate(1−), 10CI

(*3-Methyl-2-butyl*)*borohydride. Siamylborohydride*

$H_3CCH(CH_3)CH(CH_3)BH_3^{\ominus}$

M 84.976 (ion)

Li salt: [76430-53-0].
$C_5H_{14}BLi$ M 91.917
No phys. props. reported.

Brown, H.C. *et al*, *J. Org. Chem.*, 1981, **46**, 2712, 4541 (*synth, nmr*)

$C_5H_{14}BN$ B-00342

(3-Dimethylaminopropyl)borane

[3-(Dimethylamino)propyl-C,N]dihydroboron, 10CI. *3-Boryl-N,N-dimethylpropylamine. 1,1-Dimethyl-1,2-azaborolidine*

[17141-43-4]

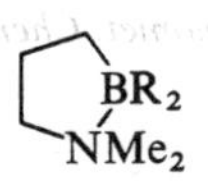

R = H

M 98.982

Exists in internally coordinated form. Colourless liq. d_{25} 0.816. Bp_{745} 185°, Bp_{13} 72-4°.

Adams, R.M. *et al*, *Inorg. Chem.*, 1963, **2**, 640 (*ms, ir, synth*)
Polivka, Z. *et al*, *Collect. Czech. Chem. Commun.*, 1969, **34**, 3009 (*synth, ir, nmr*)
Polivka, Z. *et al*, *Collect. Czech. Chem. Commun.*, 1970, **35**, 1131.
Baboulene, M. *et al*, *Bull. Soc. Chim. Fr.*, 1980, 565 (*synth, ir, raman*)

$C_5H_{14}BNS_2$ B-00343

1,3,2-Dithiaborolane trimethylamine

(N,N-*Dimethylmethanamine*)[*1,2-ethanedithiolato(2−)-S,S′*]*hydroboron*, 10CI. N,N-*Dimethylmethanamine compd. with 1,3,2-dithiaborolane(1:1)*

[13291-21-9]

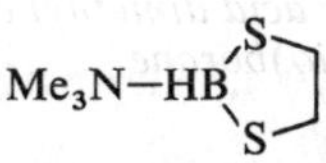

M 163.102

White cryst., stable *in vacuo* at r.t. Reported solubility data contradictory. Mp 100-1°. Subl. undec.

Egan, B.Z. *et al*, *Inorg. Chem.*, 1964, **3**, 1024 (*synth, ir*)
Young, D.E. *et al*, *J. Am. Chem. Soc.*, 1966, **88**, 4390 (*pmr*)
Thaisrivongs, S. *et al*, *J. Org. Chem.*, 1977, **42**, 3243 (*synth, ir, pmr*)

$C_5H_{15}BBrP$ B-00344

Bromodimethylborane trimethylphosphine

Bromodimethyl(trimethylphosphine)boron, 10CI. *Trimethylphosphine compd. with bromodimethylborane(1:1). Dimethylbromoborane trimethylphosphine complex*

[69513-18-4]

$Me_2BBr—PMe_3$

M 196.861

White solid, stable at Mp. Mp 211-4°.

Martin, D.R. *et al*, *J. Inorg. Nucl. Chem.*, 1978, **40**, 1289 (*synth, pmr*)
Apel, J. *et al*, *Z. Anorg. Allg. Chem.*, 1979, **453**, 28 (*nmr*)

$C_5H_{15}BN_2$ B-00345

Bis(dimethylamino)methylborane, 8CI

Pentamethylboranediamine, 10CI, 9CI

[6914-63-2]

$MeB(NMe_2)_2$

M 113.997

Liq., sl. air- and moisture-sensitive. Sol. THF. d^{20} 0.799. $n_D{}^{20}$ 1.4360.

Nöth, H. *et al*, *Z. Anorg. Allg. Chem.*, 1963, **322**, 297 (*synth*)
Goubeau, J. *et al*, *Z. Anorg. Allg. Chem.*, 1967, **352**, 285 (*raman, ir*)
Bock, H. *et al*, *Chem. Ber.*, 1971, **104**, 1687 (*pe*)
Fuss, W., *Z. Naturforsch, B*, 1974, **29**, 514 (*uv*)
Wrackmeyer, B. *et al*, *Chem. Ber.*, 1976, **109**, 1075 (*nmr*)
Fussstetter, H. *et al*, *Chem. Ber.*, 1977, **110**, 3172 (*nmr*)

$C_5H_{15}BN_2O$ B-00346

Bis(dimethylamino)methoxyborane

*1-Methoxy-*N,N,N′,N′*-tetramethylboranediamine*, 10CI

[7318-73-2]

$MeOB(NMe_2)_2$

M 129.996

Prob. liq., sensitive to moisture.

Nöth, H. *et al*, *Chem. Ber.*, 1966, **99**, 1049 (*nmr*)
Biffar, W. *et al*, *Chem. Ber.*, 1980, **113**, 333 (*nmr*)

$C_5H_{15}B_2N$ B-00347

***N,N*-Bis(dimethylboryl)methylamine**

N-*(Dimethylboryl)-*N*,1,1-trimethylboranamine*, 10CI, 9CI. *(Methylimino)bis[dimethylborane]*, 8CI

[19163-15-6]

$Me_2BNMeBMe_2$

M 110.800

Air- and moisture-sensitive liq. Bp_{720} 103°.

Nöth, H. *et al*, *J. Organomet. Chem.*, 1968, **12**, 23; 1969, **16**, 357 (*nmr*)
Nöth, H. *et al*, *Chem. Ber.*, 1974, **107**, 3089; 1977, **110**, 2790; 1980, **113**, 728 (*synth, nmr*)

$C_5H_{15}B_3ClN_3$ B-00348

2-Chloro-1,3,4,5,6-pentamethylborazine, 10CI, 9CI

*Pentamethyl-*B*-chloroborazine*

[13058-05-4]

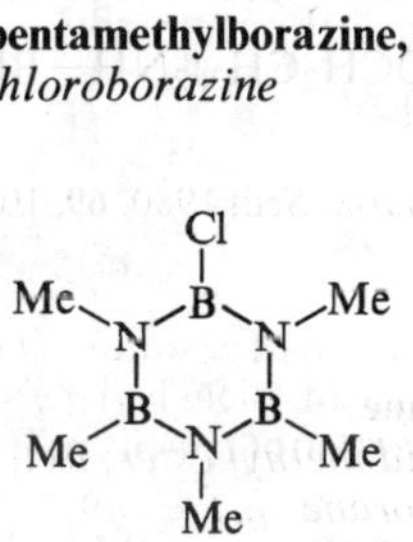

M 185.077

Colourless cryst. by subl. Mp 127-128.5°.

Wagner, R.I. *et al*, *Inorg. Chem.*, 1962, **1**, 93 (*synth*)
Melcher, L.A. *et al*, *Inorg. Chem.*, 1972, **11**, 1247 (*synth, ir*)
Adcock, J.L. *et al*, *Inorg. Chem.*, 1973, **12**, 788 (*pmr*)
Melcher, L.A. *et al*, *Inorg. Chem.*, 1973, **12**, 601 (*ms*)
Enterling, D. *et al*, *Z. Naturforsch.*, *B*, 1978, **33**, 527 (*props*)
Meller, A. *et al*, *Chem. Ber.*, 1979, **112**, 1252 (*props*)

$C_5H_{15}B_5Fe$ B-00349

(η^5-2,4-Cyclopentadien-1-yl)[η^5-decahydropentaborato(1−)]-iron, 10CI

1-(η^5-Cyclopentadienyl)-FeB_5H_{10}. *Pentaboraferrocene*

[65188-99-0]

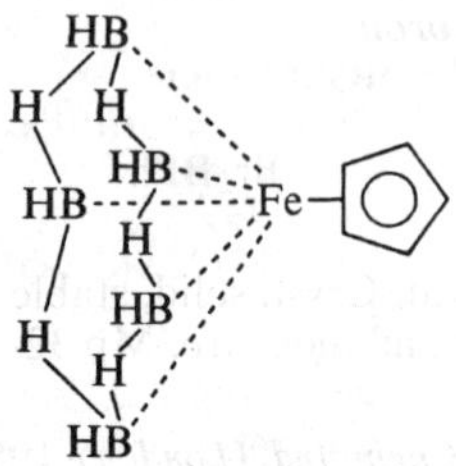

M 185.071

Dark-violet air-sensitive solid.

Weiss, R. *et al*, *J. Am. Chem. Soc.*, 1977, **99**, 8087; *Inorg. Chem.*, 1979, **18**, 3291 (*synth, pmr, nmr, ms*)

$C_5H_{15}B_5Fe$ B-00350

(η^5-2,4-Cyclopentadien-1-yl)[η^5-decahydropentaborato(1−)]-iron, 10CI

2-(η^5-Cyclopentadienyl)-FeB_5H_{10}

[71661-61-5]

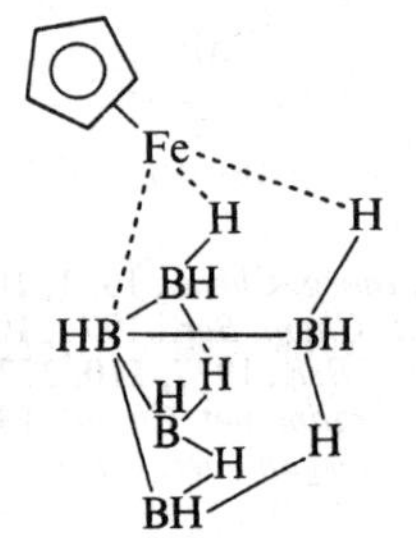

M 185.071

Air-sensitive violet cryst (hexane). Thermally isomerises to (η^5-2,4-Cyclopentadien-1-yl)[η^5-decahydropentaborato(1−)]iron, B-00349 .

Weiss, R. *et al*, *J. Am. Chem. Soc.*, 1977, **99**, 8087; *Inorg. Chem.*, 1979, **18**, 3291 (*synth, pmr, nmr, ms*)

$C_5H_{16}BNO$ B-00351

Borane *N,N*-dimethyl(2-methoxyethyl)amine

*Trihydro(2-methoxy-*N,N*-dimethylethanamine-*N*)boron, 10CI. Borane 2-methoxy-*N,N*-dimethylethylamine*

[67036-07-1]

$MeOCH_2CH_2NMe_2—BH_3$

M 116.998

Used in electrolysis bath for Au and Pd coating.

Ger. Pat., 2 750 932, (*1978*); *CA*, **89**, 202734r (*use*)
U.S.P., 4 080 381, (*1978*); *CA*, **89**, 23753z
U.S.P., 4 142 902, (*1979*); *CA*, **91**, 25752p (*use*)
Ger. Pat., 3 000 526, (*1980*); *CA*, **94**, 19420z

$C_5H_{16}BNSi$ B-00352

Dimethyl(trimethylsilylamino)borane

*1,1-Dimethyl-*N*-(trimethylsilyl)boranamine, 10CI, 9CI*

[51860-59-4]

$Me_2BNHSiMe_3$

M 129.084

Air- and moisture-sensitive.

Nöth, H. *et al*, *Chem. Ber.*, 1974, **107**, 518 (*nmr*)
Barlos, K. *et al*, *J. Magn. Reson.*, 1978, **31**, 363.
Fussstetter, H. *et al*, *Chem. Ber.*, 1978, **111**, 3596; 1980, **113**, 728 (*props*)
Nöth, H. *et al*, *Chem. Ber.*, 1981, **114**, 1150 (*nmr*)

$C_5H_{16}B_4Pb$ B-00353

2,3-μ-(Trimethylplumbyl)-4,5-dicarbahexaborane(8), 9CI

[38117-57-6]

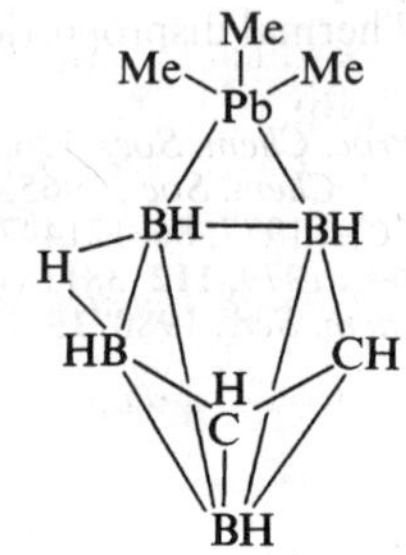

M 326.621

Proposed struct. shown. Colourless liq. of low volatility. Unstable at r.t.

Tabereaux, A. *et al*, *Inorg. Chem.*, 1973, **12**, 792 (*synth, nmr, pmr, ir*)

$C_5H_{16}B_4Sn$ B-00354

2,3-μ-(Trimethylstannyl)-4,5-dicarbahexaborane(8), 9CI

[38117-56-5]

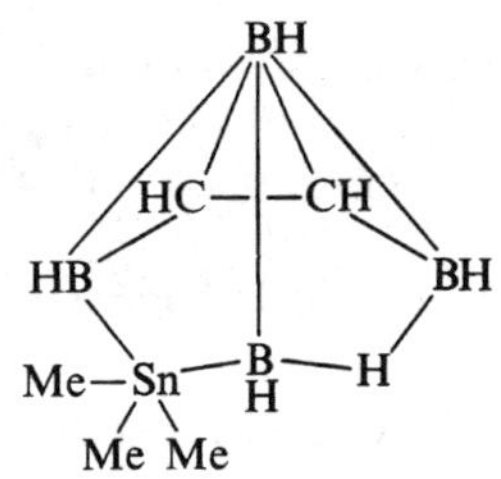

M 238.111

Colourless liq. Stable indefinitely at 120°, but dec. rapidly at 220°.

Tabereaux, A. *et al*, *Inorg. Chem.*, 1973, **12**, 792 (*synth, ir, pmr, nmr, ms*)

$C_5H_{20}B_{10}Fe$ B-00355

(η^5-2,4-Cyclopentadien-1-yl)[η^5-pentadecahydrodecaborato(2−)]iron, 10CI

2-(η^5-Cyclopentadienyl)-$FeB_{10}H_{15}$

[71661-62-6]

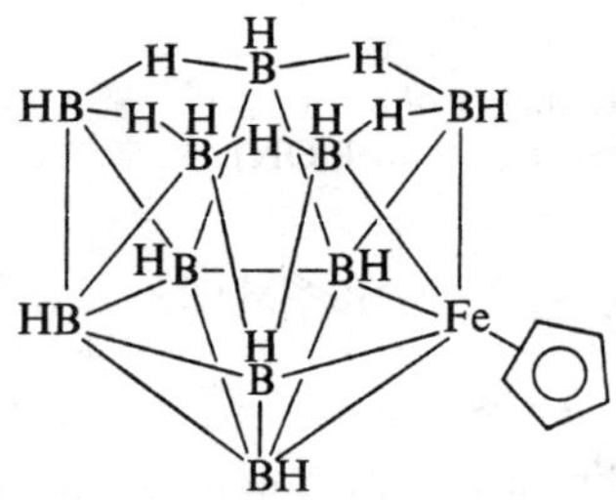

M 244.160

Yellow.

Weiss, R. *et al*, *Inorg. Chem.*, 1979, **18**, 3291 (*synth, pmr, nmr, ms*)

$C_6BCl_2F_5$ B-00356

Dichloro(pentafluorophenyl)borane, 10CI, 9CI, 8CI
Pentafluorophenyldichloroborane
[830-48-8]

$C_6F_5BCl_2$

M 248.774
Volatile liq., readily hydrol. even at −80°. Sol. dibutyl ether. Bp 155°. Thermal disproportionation occurs at 220°.

Brown, A.J.R. *et al*, *Proc. Chem. Soc.*, 1963, 200 (*nmr*)
Chambers, R.D. *et al*, *J. Chem. Soc.*, 1965, 3933 (*nmr, synth*)
Glanville, W.K. *et al*, *CA*, 1977, **86**, 121447k (*use*)
Paetzold, P., *Chem. Ber.*, 1979, **112**, 3811 (*use*)
Gronowitz, S. *et al*, *Chem. Scr.*, 1980, **15**, 135 (*use*)

$C_6BF_9O_6$ B-00357

Tris(trifluoroacetoxy)borane
Trifluoroacetic acid trianhydride with boric acid (H_3BO_3), 10CI, 9CI. Boron tris(trifluoroacetate)
[350-70-9]

$B(OOCCF_3)_3$

M 349.858
Deprotecting agent in amino acid and peptide synth. Moisture-sensitive. Mp 88° dec.

Gerrard, W. *et al*, *J. Chem. Soc.*, 1958, 3648 (*synth*)
Pless, J. *et al*, *Angew. Chem., Int. Ed. Engl.*, 1973, **12**, 147 (*synth*)
Olah, G.A. *et al*, *J. Am. Chem. Soc.*, 1975, **97**, 5481 (*nmr*)

$C_6H_3BCrNO_5^{\ominus}$ B-00358

Pentacarbonyl[(cyano-*C*)trihydroborato(1−)-*N*]chromate(1−), 9CI

$(OC)_5CrCNBH_3^{\ominus}$ (O_h)

M 231.899 (ion)
Mp >170° dec.
Tetramethylammonium salt: [52138-78-0]. $C_{10}H_{15}BCrN_2O_5$ M 306.045
Yellow cryst.

King, R.B. *et al*, *J. Organomet. Chem.*, 1974, **65**, 71 (*synth, ir, nmr*)

$C_6H_4BCl_3O_2$ B-00359

Dihydroxy(2,4,6-trichlorophenyl)borane
(2,4,6-Trichlorophenyl)boronic acid, 10CI. 2,4,6-Trichlorobenzeneboronic acid
[73852-18-3]

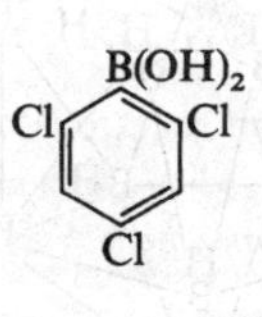

M 225.265
Derivs. used for trace analysis of physiological fluids. Electron-capturing props. useful for selective analysis of bifunctional compds. Cryst. (hexane). Mp 54-5°.

Poole, C.F. *et al*, *J. Chromatogr.*, 1979, **186**, 307 (*synth, use*)

$C_6H_5BBr_2$ B-00360

Dibromophenylborane, 10CI, 9CI, 8CI
Dibromophenylborine. Phenyldibromoborane. Phenyldibromoboron
[4151-77-3]

Br_2BPh

M 247.724
Synthetic intermed. Cryst. solid, stable under N_2 but sensitive to air and moisture. Mp 32-4° (25-8°). Bp_{14} 89-91°.

Bujwid, Z.J. *et al*, *Chem. Ind. (London)*, 1959, 1091
Frenkin, E.I. *et al*, *Izv. Akad. Nauk SSSR, Ser. Khim.*, 1960, 1507 (*synth*)
Muetterties, E.L., *J. Am. Chem. Soc.*, 1960, **82**, 4163 (*synth*)
Nöth, H. *et al*, *J. Organomet. Chem.*, 1968, **11**, 399 (*synth, nmr*)
Nahm, F.C., *J. Organomet. Chem.*, 1972, **35**, 9 (*ir*)
Odom, J.D. *et al*, *J. Organomet. Chem.*, 1979, **173**, 15 (*nmr*)

$C_6H_5BCl_2$ B-00361

Dichlorophenylborane, 10CI, 9CI
Phenylboron dichloride. Phenyldichloroborane
[873-51-8]

$PhBCl_2$

M 158.822
Fuming liq., sensitive to moisture. d 1.194. Mp 7°. Bp 175°.

Treichel, P.M. *et al*, *Inorg. Chem.*, 1971, **10**, 32 (*synth*)
Wrackmeyer, B. *et al*, *Chem. Ber.*, 1976, **109**, 1075 (*nmr*)
Barlos, K. *et al*, *Chem. Ber.*, 1977, **110**, 2774 (*nmr*)
Gragg, B.R. *et al*, *J. Organomet. Chem.*, 1977, **132**, 29 (*nmr*)
Ramsey, B.G. *et al*, *J. Organomet. Chem.*, 1977, **141**, 257 (*uv, pmr, struct*)
Odom, J.D. *et al*, *J. Organomet. Chem.*, 1979, **173**, 15 (*nmr*)
Ramsey, B.G. *et al*, *J. Org. Chem.*, 1980, **45**, 1332 (*nmr, struct*)
Sax, N.I., *Dangerous Properties of Industrial Materials*, 5th Ed., Van Nostrand-Reinhold, 1979, 900.

$C_6H_5BCl_2N_2$ B-00362

Dichlorocyanoborane pyridine
Dichloro(cyano-C)(pyridine)boron, 10CI. Dichloroboranecarbonitrile pyridine. Pyridine dichlorocyanoborane
[67374-63-4]

M 186.835
White solid. Mp 106-8°.

Martin, D.R. *et al*, *J. Inorg. Nucl. Chem.*, 1978, **40**, 9 (*synth, ir, pmr*)

$C_6H_5BCl_2O_2$ B-00363

(2,4-Dichlorophenyl)dihydroxyborane
(2,4-Dichlorophenyl)boronic acid, 10CI. *2,4-Dichlorobenzeneboronic acid*
[68716-47-2]

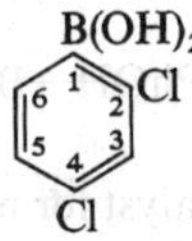

M 190.820
Used as derivatisation reagent in gas chromatog. of bifunctional org. compds. Cryst. (toluene). Mp 242-5°. Commercially available.

Poole, C.F. *et al*, *J. Chromatogr.*, 1978, **158**, 33 (*use, synth, props*)
Poole, C.F. *et al*, *Analyst*, 1979, **104**, 82 (*props*)
Poole, C.F. *et al*, *J. Chromatogr.*, 1979, **186**, 307 (*props*)
Singhawangcha, S. *et al*, *J. High Resolut. Chromatogr. Chromatogr. Commun.*, 1979, **2**, 77 (*props, use*)

$C_6H_5BCl_2O_2$ B-00364

(2,6-Dichlorophenyl)dihydroxyborane
(2,6-Dichlorophenyl)boronic acid, 10CI. *2,6-Dichlorobenzeneboronic acid*
[73852-17-2]
M 190.820
Used as derivatisation reagent for gas chromatog. of bifunctional compds. Cryst. (MeOH). Mp 149-51°.

Poole, C.F. *et al*, *J. Chromatogr.*, 1979, **186**, 307 (*synth, props, use*)

$C_6H_5BCl_2O_2$ B-00365

(3,5-Dichlorophenyl)dihydroxyborane
(3,5-Dichlorophenyl)boronic acid, 10CI. *3,5-Dichlorobenzeneboronic acid*
[67492-50-6]
M 190.820
Used as derivatising agent in gas chromatog. of bifunctional org. compds. Cryst. (toluene). Mp 310-5°.

Poole, C.F. *et al*, *Chromatographia*, 1978, **11**, 347 (*use*)
Poole, C.F. *et al*, *J. Chromatogr.*, 1978, **158**, 33 (*synth, props, use*)

$C_6H_5BCl_3^{\ominus}$ B-00366

Trichlorophenylborate(1–), 10CI
Phenyltrichloroborate

$PhBCl_3^{\ominus}$

M 194.275 (ion)
Tetraethylammonium salt: [23273-29-2]. N,N,N-*Triethylethanaminium trichlorophenylborate(1–)*, 10CI.
$C_{14}H_{25}BCl_3N$ M 324.527
Electrolyte. White solid. Mp 99-103° dec.
Tetrabutylammonium salt: [61132-55-6]. N,N,N-*Tributyl-1-butanaminium trichlorophenylborate(1–)*, 10CI.
$C_{22}H_{41}BCl_3N$ M 436.742
Used in formn. of B-N polymers.

Ahmed, I.Y., *Inorg. Chem.*, 1969, **8**, 1411 (*synth, props, use*)
Sheikh, S.U., *Bull. Chem. Soc. Jpn.*, 1976, **49**, 2879 (*props*)
Sheikh, S.U., *Therm. Anal.* [*Proc. Int. Conf.*], *5th*, 1977, 509; *CA*, **90**, 71519v.
Sheikh, S.U., *J. Therm. Anal.*, 1980, **18**, 299 (*synth*)
Sheikh, S.U., *Pak. J. Sci. Ind. Res.*, 1980, **23**, 21; *CA*, **94**, 113678m.

$C_6H_5BF_2$ B-00367

Difluorophenylborane, 10CI, 9CI, 8CI
Difluorophenylborine. Phenyldifluoroborane
[368-98-9]

$PhBF_2$

M 125.912
Air- and moisture-sensitive liq. d^{25} 1.087. Mp −36.2°. Bp_{747} 97.8°. n_D^{25} 1.4441.

Torssell, K., *Acta Chem. Scand.*, 1954, **8**, 1779 (*props*)
McCusker, P.A. *et al*, *J. Am. Chem. Soc.*, 1957, **79**, 5185 (*synth*)
Holliday, A.K. *et al*, *J. Chem. Soc.*, 1964, 2731 (*synth*)
Odom, J.D. *et al*, *J. Organomet. Chem.*, 1979, **173**, 15 (*nmr*)
Paetzold, P. *et al*, *Chem. Ber.*, 1980, **113**, 3724 (*use*)

$C_6H_5BF_2O$ B-00368

Difluorophenoxyborane, 10CI, 9CI
[14117-84-1]

$PhOBF_2$

M 141.912
Catalyst for polymerisation reactions. Sensitive to moisture.

Ger. Pat., 1 217 066, (*1966*); *CA*, **65**, PC13885d (*use*)
B.P., 2 030 993, (*1980*); *CA*, **93**, P221317f (*use*)
Fr. Pat., 2 438 665, (*1980*); *CA*, **93**, P240977n (*use*)

$C_6H_5BI_2$ B-00369

Diiodophenylborane, 10CI, 9CI
Phenyldiiodoborane. Diiodophenylborine
[17381-60-1]

$PhBI_2$

M 341.725
Prepd. from BI_3 and iodobenzene or tetraphenylstannane. Air- and moisture-sensitive liq. Sol. CH_2Cl_2. $Bp_{0.5}$ 70-2°. ^{11}B nmr δ −47.8 ppm.

Nöth, H. *et al*, *J. Organomet. Chem.*, 1969, **20**, 35 (*synth*)
Siebert, W. *et al*, *J. Organomet. Chem.*, 1970, **25**, 305 (*synth*)
Nahm, F.C. *et al*, *J. Organomet. Chem.*, 1972, **35**, 9 (*ir*)
Renk, T. *et al*, *J. Organomet. Chem.*, 1976, **120**, 1 (*use*)
Haubold, W. *et al*, *Z. Anorg. Allg. Chem.*, 1978, **446**, 45 (*raman*)
Odom, J.D. *et al*, *J. Organomet. Chem.*, 1979, **173**, 15 (*nmr*)

$C_6H_5BO_2$ B-00370

1,3,2-Benzodioxaborole, 10CI, 9CI
Catecholborane
[274-07-7]

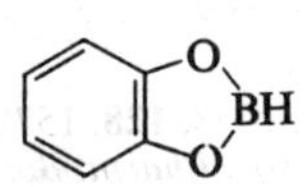

M 119.914
Used in hydroboration reactions; reducing agent. d_4^{20} 1.27. Mp 12°. Bp_{156} 88°, Bp_{80} 66°. Moisture sensitive.

Brown, H.C. *et al*, *J. Am. Chem. Soc.*, 1975, **97**, 5249 (*synth, ir, ms*)
Brown, H.C., *Organic Syntheses via Boranes*, 1975, Wiley, N.Y., 63.
Kabalka, G.W., *Org. Prep. Proced. Int.*, 1977, **9**, 131 (*rev*)
Alves, A.C.P. *et al*, *J. Mol. Spectrosc.*, 1979, **77**, 124 (*uv, struct*)

Org. Synth., 1980, **59**, 42 (*use*)
Goetze, R. *et al*, *Chem. Ber.*, 1981, **114**, 1884 (*nmr*)
Fieser, M. *et al*, *Reagents for Organic Synthesis*, Wiley, 1967-83, **8**, 79.

$C_6H_6BBrO_2$ B-00371

(4-Bromophenyl)dihydroxyborane

4-Bromophenylboronic acid, 10CI, 9CI. p-*Bromophenylboric acid. 4-Bromobenzeneboronic acid*

[5467-74-3]

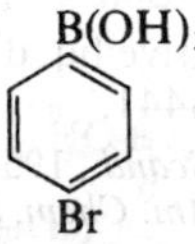

M 200.826

Useful as a derivatizing agent for bifunctional mols., partic. for glc. Needles (EtOH). Sol. hot H_2O. Mp 191°, 286-9°, 312-5°. pK_a 9.14 (25°, 25% EtOH aq.).

▷CY8650000.

Bean, F. *et al*, *J. Am. Chem. Soc.*, 1932, **54**, 4415 (*synth*)
Zvonkova, Z.V., *Kristallografiya*, 1957, **2**, 408; *CA*, **52**, 3431 (*struct*)
Nöth, H. *et al*, *NMR Spectroscopy of Boron Compounds* (*NMR Basic Principles and Progress*, Part 14), 1978, Springer (*nmr*)
Poole, C.F. *et al*, *J. Chromatogr.*, 1978, **158**, 33 (*use*)
Poole, C.F. *et al*, *J. Chromatogr.*, 1979, **186**, 307 (*use*)

$C_6H_6BClO_2$ B-00372

(3-Chlorophenyl)dihydroxyborane

(*3-Chlorophenyl*)*boronic acid*, 10CI. m-*Chlorobenzeneboronic acid.* m-*Chlorophenylboric acid*

[63503-60-6]

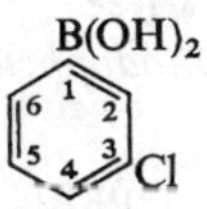

M 156.375

Cryst. (C_6H_6 or H_2O). Mp 183°. pK_a 8.9.

Bettman, B. *et al*, *J. Am. Chem. Soc.*, 1934, **56**, 1865.
Branch, G.E.K. *et al*, *J. Am. Chem. Soc.*, 1934, **56**, 937.
Hanai, K. *J. Biochem.* (*Tokyo*), 1977, **81**, 1273
Kajimoto, O. *et al*, *J. Phys. Chem.*, 1977, **81**, 1712 (*pmr*)
Nakatani, H. *et al*, *Biochim. Biophys. Acta*, 1978, **525**, 423.

$C_6H_6BClO_2$ B-00373

(4-Chlorophenyl)dihydroxyborane

(*4-Chlorophenyl*)*boronic acid*, 10CI, 9CI. p-*Chlorobenzeneboronic acid*, 8CI

[1679-18-1]

M 156.375

Used as sterilant for insect control. Has antispasmodic props. Sludge inhibitor for fuels. Used for penetration of brain tumours. Solid.

▷Toxic. CY8950000.

Soloway, A.H., *Science*, 1958, **128**, 1572 (*use*)
Ploquin, J., *Bull. Trav. Soc. Pharm. Bordeaux*, 1962, **101**, 229 (*struct*)
Beachell, H.C. *et al*, *Inorg. Chem.*, 1964, **3**, 1028 (*synth*, *nmr*)
Gout, R. *et al*, *Bull. Trav. Soc. Pharm. Lyon*, 1965, **9**, 236 (*tox*)
Breuer, S.W. *et al*, *Tetrahedron Lett.*, 1974, 3719 (*synth*)
Thorpe, F.G. *et al*, *J. Organomet. Chem.*, 1978, **145**, C26.

$C_6H_6BF_3O$ B-00374

Trifluoroborane phenol

Trifluoro(*phenol*)*boron*, 10CI. *Phenol compd. with trifluoroborane*(*1:1*), 9CI. *Phenol compd. with boron fluoride*(BF_3)(*1:1*), 8CI

[372-44-1]

PhOH—BF_3

M 161.918

Moisture sensitive. Catalyst for many types of reactions.

Muetterties, E.L., *J. Am. Chem. Soc.*, 1958, **80**, 4526.
Palko, A.A. *et al*, *J. Chem. Phys.*, 1967, **47**, 2561.
Kohichi, Y. *et al*, *J. Polym. Sci.*, *Polym. Lett. Ed.*, 1976, **14**, 517.
Ger. Pat., 2 834 126, (*1979*); *CA*, **90**, 170446a (*use*)
Chmiel-Pela, J., *CA*, 1981, **95**, 189369 (*rev*)

$C_6H_6BFeNO_4$ B-00375

Tetracarbonyl[*N*,*N*-dimethylboranaminato(2−)-*B*]iron, 9CI

Tetracarbonyl[(*dimethylamino*)*borylene*]*iron*, 8CI. *Dimethylaminoborinetetracarbonyliron*

[31388-53-1]

$Me_2NBFe(CO)_4$

M 222.775

Volatile, colourless cryst., dec. at r.t. Mp −25°. Forms brown oligomer.

Schmid, G. *et al*, *Inorg. Chim. Acta*, 1970, **4**, 423 (*synth*, *ir*)
Schmid, G., *Angew. Chem.*, *Int. Ed. Engl.*, 1972, **11**, 1034.

$C_6H_6BNO_4$ B-00376

Dihydroxy(2-nitrophenyl)borane

(*2-Nitrophenyl*)*boronic acid*, 10CI, 9CI. o-*Nitrobenzeneboronic acid*, 8CI

[5570-19-4]

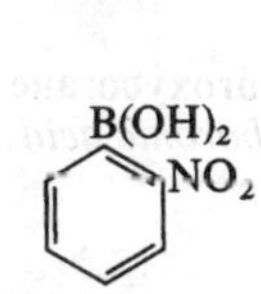

M 166.928

Chemosterilant for insects; enzyme inhibitor. Yellow needles. Sol. EtOH, Et_2O, mod. sol. H_2O. Mp 147.5° (140-2°).

Seaman, W. *et al*, *J. Am. Chem. Soc.*, 1931, **53**, 714 (*synth*)
Kuivila, H.G. *et al*, *J. Am. Chem. Soc.*, 1957, **79**, 5659 (*props*, *use*)
Verbit, L. *et al*, *Tetrahedron Lett.*, 1966, 1053 (*synth*)
Settepani, J.A. *et al*, *J. Med. Chem.*, 1970, **13**, 128 (*use*)
Carna, J., *F.E.B.S. Lett.*, 1980, **119**, 343 (*use*)

$C_6H_6BNO_4$ B-00377

Dihydroxy(3-nitrophenyl)borane

(*3-Nitrophenyl*)*boronic acid*, 10CI. m-*Nitrobenzeneboronic acid*

[13331-27-6]

M 166.928

Used in detection of bifunctional compds. by glc; enzyme inhibitor and chemosterilant for insects. Yellow cryst. (H_2O). Mod. sol. hot H_2O, sol. EtOH, Et_2O. Mp 275°, 285° 319°.

▷CY8980000.

Torsell, K., *CA*, 1958, **52**, 1455f (*synth*)
Skowronska-Serafinowa, B. *et al*, *Rocz. Chem.*, 1961, **35**, 937 (*ir*)
Ramsey, B.G., *J. Phys. Chem.*, 1970, **74**, 2464 (*props*)
Barker, S.A. *et al*, *Carbohydr. Res.*, 1973, **26**, 33 (*use*, *props*)

Brown, B.J. *et al*, *Aust. J. Chem.*, 1976, **29**, 941 (*props*)
Poole, C.F. *et al*, *J. Chromatogr.*, 1978, **158**, 33 (*use*)

C_6H_7B B-00378

Phenylborane, 10CI, 9CI, 8CI

[45481-30-9]

$PhBH_2$

M 89.931

Exists as B-H bridged dimer in uncomplexed state. Synthetic intermed., generated *in situ*. Air- and moisture-sensitive. Sol. THF.

Me_3N complex: [1194-24-7]. N,N-*Dimethylmethanamine phenylborane.*
$C_9H_{16}BN$ M 149.042
Cryst. (Et_2O). Sol. C_6H_6. Mp 68-9°.

Et_3N complex: [15068-65-2]. N,N-*Diethylethanamine phenylborane.*
$C_{12}H_{22}BN$ M 191.123
Sol. C_6H_6.

Wilhelm, K. *et al*, *Z. Anorg. Allg. Chem.*, 1963, **325**, 225 (*derivs*)
Faulks, J.N.G. *et al*, *J. Inorg. Nucl. Chem.*, 1967, **29**, 329 (*nmr, deriv*)
Miller, J.J. *et al*, *Inorg. Chem.*, 1970, **4**, 69 (*deriv, synth*)
Hawthorne, M.F. *et al*, *J. Am. Chem. Soc.*, 1971, **93**, 350 (*deriv, props*)
McCormick, C.L. *et al*, *J. Org. Chem.*, 1976, **41**, 2803 (*deriv*)
Greenberg, A. *et al*, *Tetrahedron*, 1979, **35**, 2623 (*struct*)

$C_6H_7BN_2$ B-00379

Cyanoborane pyridine

(*Cyano*-C)*dihydro*(*pyridine*)*boron*, 10CI. *Boranecarbonitrile, compd. with pyridine*(*1:1*)

[66632-43-7]

M 117.945

Shows anticholesteraemic and antihypopolipidaemic acitivity. White cryst. Mp 34-5°.

Martin, D.R. *et al*, *J. Inorg. Nucl. Chem.*, 1978, **40**, 9 (*synth, nmr, ir*)
Wisian-Nielson, P. *et al*, *Inorg. Chem.*, 1978, **17**, 2327 (*synth, pmr, ir*)
Spielvogel, B.F. *et al*, *J. Inorg. Nucl. Chem.*, 1979, **41**, 1223 (*props*)
Hall, I.H. *et al*, *J. Pharm. Sci.*, 1980, **69**, 1025; 1981, **70**, 339 (*pharmacol*)

$C_6H_7BO_2$ B-00380

Phenyldihydroxyborane

Phenylboronic acid, 10CI, 9CI. *Phenylboric acid. Benzeneboronic acid. Dihydroxy*(*phenyl*)*borane*

[98-80-6]

$PhB(OH)_2$

M 121.930

Used as a derivatizing agent for bifunctional molecules, esp. for glc. Needles (H_2O). Sol. MeOH, EtOH, spar. sol. H_2O, Et_2O. Mp 216°. pK_a 8.86 (25°, H_2O). Forms anhydride below Mp.

▷CY8575000.

Morlyan, N.M. *et al*, *CA*, 1972, **76**, 14619 (*synth*)
Gragg, B.R. *et al*, *J. Organomet. Chem.*, 1977, **132**, 29 (*nmr*)
Rettig, S.J. *et al*, *Can. J. Chem.*, 1977, **55**, 3071 (*struct*)
Shapiro, B.L. *et al*, *J. Phys. Chem. Ref. Data*, 1977, **6**, 919 (*pmr*)
Nöth, H. *et al*, *NMR Spectroscopy of Boron Compounds* (*NMR Basic Principles and Progress*, Part 14), 1978, Springer (*nmr*)
Bullen, N.P. *et al*, *J. Organomet. Chem.*, 1980, **195**, 147 (*synth*)
Ramsey, B.G. *et al*, *J. Org. Chem.*, 1980, **45**, 1322 (*nmr, struct*)

$C_6H_7BO_4$ B-00381

(2-Acetyl-3-furanyl)dihydroxyborane

(*2-Acetyl-3-furanyl*)*boronic acid*, 10CI, 9CI

[49777-66-4]

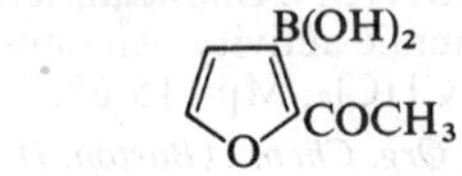

M 153.929

Yellow cryst. (EtOH aq.). Mp 150°.

Ger. Pat., 2 264 363, (*1973*); *CA*, **79**, 92369j
Ger. Pat., 2 248 690, (*1973*); *CA*, **79**, 92371d
Florentin, D. *et al*, *J. Heterocycl. Chem.*, 1976, **13**, 1265.
Florentin, D. *et al*, *Bull. Soc. Chim. Fr.*, 1976, 1999 (*synth*)
Florentin, D. *et al*, *J. Chem. Res.* (*S*), 1977, 158.

$C_6H_8BBr_2N$ B-00382

Dibromomethylborane pyridine

Dibromomethyl(*pyridine*)*boron*, 10CI. *Methyldibromoborane pyridine*

[69513-16-2]

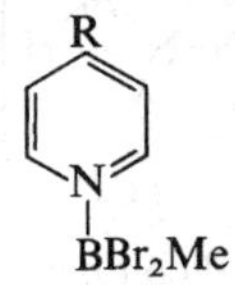

R = H

M 264.754

Solid. Mp 120-3° dec.

Martin, D.R. *et al*, *J. Inorg. Nucl. Chem.*, 1978, **40**, 1289 (*synth, nmr, ir*)

$C_6H_8BNO_2$ B-00383

(3-Aminophenyl)dihydroxyborane

3-Aminophenylboronic acid, 10CI, 9CI. m-*Aminophenylboric acid. 3-Aminobenzeneboronic acid*

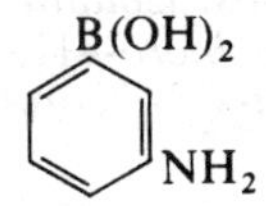

M 136.945

Used in affinity chromatogr. of biologically important mols. Hexagonal plates (H_2O). Sol. H_2O, EtOH, spar. sol. Et_2O. Mp 176°. pK_{a1} 4.47, pK_{a2} 8.81.

B,½H_2SO_4: Cryst. (MeOH).

Seaman, W. *et al*, *J. Am. Chem. Soc.*, 1931, **53**, 711.
Bean, F.R. *et al*, *J. Am. Chem. Soc.*, 1932, **54**, 4415.
Singhal, R.P. *et al*, *Anal. Biochem.*, 1980, **109**, 1 (*use*)
Ger. Pat., 2 933, 832, (*1981*); *CA*, **94**, 205033a (*use*)
Wielckens, K. *et al*, *Eur. J. Biochem.*, 1981, **117**, 69 (*use*)

$C_6H_8BNO_2$ B-00384

Pyridine carboxyborane

Boranecarboxylic acid compd. with pyridine(1:1), 10CI. *Boranecarboxylic acid pyridine*

[72695-27-3]

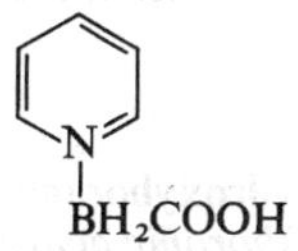

M 136.945

Prepd. from $C_6H_5N.BH_2CN$. Shows anticholesteraemic and antihypolipidaemic activity. Air-sensitive, white cyrst. (H_2O). Sol. $CHCl_3$. Mp 115-6°.

Pelter, A. *et al*, *Compr. Org. Chem.*, (*Barton, D.H.R. et al*, Ed.), 1979, **3**, 695 (*rev*)
Spielvogel, B.F. *et al*, *J. Inorg. Nucl. Chem.*, 1979, **41**, 1223 (*synth*)
Hall, I.H. *et al*, *J. Pharm. Sci.*, 1980, **69**, 1025; 1981, **70**, 339 (*pharmacol*)

$C_6H_8BN_4^{\ominus}$ B-00385

Dihydrobis(1*H*-pyrazolato-*N'*)borate(1–), 10CI, 9CI

Dihydrobis(1-pyrazolyl)borate

[46147-74-4]

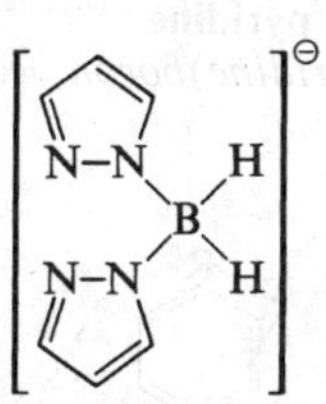

M 146.966 (ion)

Used in formn. of metal complexes.

K salt: [18583-59-0].
C_6H_8BKN M 144.044
Cryst. (anisole). Sol. H_2O. Mp 171-2°.

Free acid: [18583-65-8]. Mp 128-9°. Evolves H_2 at Mp.

Trofimenko, S., *J. Am. Chem. Soc.*, 1967, **89**, 3170 (*synth*)
Inorg. Synth., 1970, **12**, 100 (*synth*)
Trofimenko, S., *Chem. Rev.*, 1972, **72**, 497 (*rev*)
McCurdy, W.H., *Inorg. Chem.*, 1975, **14**, 1939 (*pmr*)

$C_6H_8B_2Cl_2N_4$ B-00386

Bis(4-chloro-1-pyrazolylborane)

*Bis[μ-(4-chloro-1*H*-pyrazolato-*N¹:N²*)]tetrahydrodiboron*, 10CI, 9CI. *2,6-Dichloro-4*H*,4*H*,8*H*,8*H*-7*a*,8*a*-diaza-3*a*,4*a*-diazonia-4,8-diborata-*s*-indacene*, 8CI

[18601-55-3]

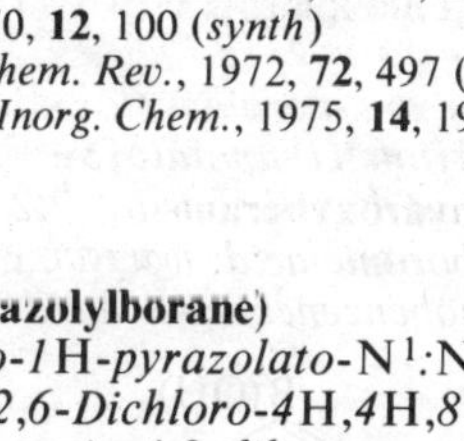

M 228.682

Cryst. (heptane). Mp 119-20°.

Trofimenko, S., *J. Am. Chem. Soc.*, 1967, **89**, 3165 (*synth, pmr*)
May, C.E. *et al*, *Z. Naturforsch., B*, 1976, **31**, 1662 (*ms*)
Hodgkins, T.G. *et al*, *Inorg. Chem.*, 1981, **20**, 2097.

C_6H_9B B-00387

Trivinylborane

Triethenylborane, 10CI, 9CI, 8CI

[7217-01-8]

$(H_2C{=}CH)_3B$

M 91.947

Prepd. from BCl_3 and tetravinylplumbane. Synthetic intermed. Air- and moisture-sensitive gas. Sol. non-hydroxylic solvs. V.p. 34mm at 0°, 80mm at 22°.

Holliday, A.K. *et al*, *J. Organomet. Chem.*, 1967, **10**, 295 (*synth*)
Odom, J.D. *et al*, *Inorg. Chem.*, 1974, **13**, 170 (*spectra*)
Hall, L.W. *et al*, *Diss. Abstr. Int. B*, 1975, **36**, 2794 (*props*)
Hall, L.W. *et al*, *J. Am. Chem. Soc.*, 1975, **97**, 4527 (*cmr*)
Wrackmeyer, B. *et al*, *Chem. Ber.*, 1976, **109**, 1075 (*nmr*)
Odom, J.D. *et al*, *J. Organomet. Chem.*, 1980, **201**, 351 (*use*)

$C_6H_{10}BN$ B-00388

Borane 4-methylpyridine

Trihydro(4-methylpyridine)boron, 10CI. *4-Methylpyridine compd. with borane(1:1)*, 9CI. *4-Picoline compd. with BH_3*

[3999-39-1]

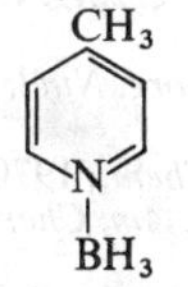

M 106.962

Used in photographic direct-positive emulsions. Shows antiflammatory activity.

Bigot, J.A. *et al*, *Recl. Trav. Chim. Pays-Bas*, 1957, **76**, 996 (*synth*)
Brey, W.S. *et al*, *Adv. Chem. Ser.*, 1964, **42**, 100 (*nmr*)
Hoffmann, R., *J. Chem. Phys.*, 1964, **40**, 2474 (*struct*)
Mathur, M.A. *et al*, *Inorg. Chem.*, 1980, **19**, 887.
U.S.P., 4 312 989, (*1982*); *CA*, **97**, 85322g (*use*)

$C_6H_{11}BBr_2$ B-00389

Dibromo(1,3-dimethyl-1-butenyl)borane

4-Methyl-2-penten-2-yldibromoborane

M 253.771

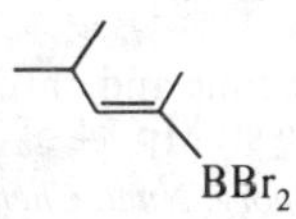

(*Z*)-*form* [72228-62-7]

Me_2S complex: Dibromo(1,3-dimethyl-1-butenyl)[thiobis[methane]]boron, 10CI.
$C_8H_{17}BBr_2S$ M 315.900
Air- and moisture-sensitive solid. Mp 57-8°. $Bp_{0.08}$ 56-8°.

Brown, H.C. *et al*, *J. Org. Chem.*, 1980, **45**, 389 (*synth, pmr, nmr*)

$C_6H_{11}BBr_2$ B-00390

Dibromo(3,3-dimethyl-1-butenyl)borane

3,3-Dimethyl-1-buten-1-yldibromoborane

$(H_3C)_3CCH{=}CHBBr_2$

M 253.771

(*E*)-*form*

Me_2S complex: Dibromo(3,3-dimethyl-1-butenyl)[thiobis[methane]]boron, 10CI.
$C_8H_{17}BBr_2S$ M 315.900
Air- and moisture-sensitive cryst. ($CHCl_3$/hexane). Mp 95-96.5°.

Brown, H.C. *et al*, *J. Org. Chem.*, 1980, **45**, 389 (*synth, nmr*)

$C_6H_{11}BBr_2$ B-00391

Dibromo(1-ethyl-1-butenyl)borane

3-Hexen-3-yldibromoborane. Dibromo(3-hexen-3-yl)-borane

M 253.771

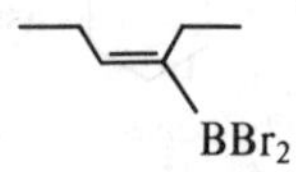

(Z)-form

Me_2S complex: Dibromo(1-ethyl-1-butenyl)[thiobis[methane]]boron, 10CI.

$C_8H_{17}BBr_2S$ M 315.900

Air- and moisture-sensitive liq. Mp 5-8°. $Bp_{0.5}$ 53-4°.

Brown, H.C. *et al*, *J. Org. Chem.*, 1980, **45**, 389 (*synth, nmr*)

$C_6H_{11}BBr_2$ B-00392

Dibromo-1-hexenylborane, 10CI

1-Hexenyldibromoborane

$H_3C(CH_2)_3CH{=}CHBBr_2$

M 253.771

(E)-form [72228-55-8]

Prepd. *in situ* as the Me_2S adduct by hydroboration of 1-hexyne with $Br_2BH{-}SMe_2$. Synthetic intermed.

Me_2S adduct: [72228-56-9].

$C_8H_{17}BBr_2S$ M 315.900

Clear oil, sensitive to air and moisture. Sol. $CHCl_3$. $Bp_{0.12}$ 59-60°. Slowly dec. under N_2 at 25°.

Brown, H.C. *et al*, *J. Org. Chem.*, 1980, **45**, 389 (*synth, use*)

$C_6H_{11}BBr_2$ B-00393

Dibromo(2-methylcyclopentyl)borane

2-Methylcyclopentyldibromoborane

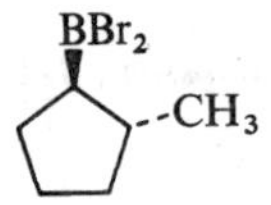

M 253.771

(1RS,2SR)-form

(±)-trans-*form*

Me_2S complex: [72205-99-3]. *Dibromo(2-methylcyclopentyl)[thiobis[methane]]boron, 10CI.*

$C_8H_{17}BBr_2S$ M 315.900

Synthetic intermediate. Air- and moisture-sensitive liq. $Bp_{0.5}$ 68-9°.

Brown, H.C. *et al*, *J. Org. Chem.*, 1980, **45**, 384 (*synth*)

$C_6H_{11}BCl_2$ B-00394

2-Methylcyclopentyldichloroborane

Dichloro(2-methylcyclopentyl)boron

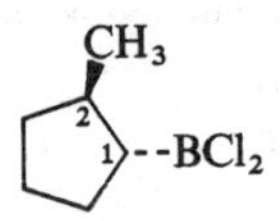

M 164.869

(1RS,2SR)-form

(±)-trans-*form*

Me_2S complex: [72205-95-9]. Prepd. by hydroboration of 1-methylcyclopentene with $BHCl_2$-SMe_2. Air- and moisture-sensitive liq.

Brown, H.C. *et al*, *J. Org. Chem.*, 1980, **45**, 384 (*synth*)

$C_6H_{11}BO_8$ B-00395

Borogluconic acid

D-Gluconic acid cyclic 4,5-ester with boric acid (H_3BO_3)-, 10CI, 9CI

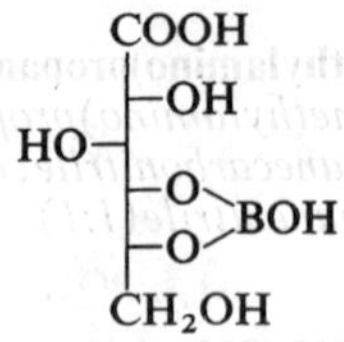

M 221.958

Ca salt: [5743-34-0]. Used to treat hypocalcemic states in cattle (bovine milk fever). Sol. H_2O. Dec. on heating.

Na salt: [62185-81-3]. Corrosion inhibitor.

Dryerre, H. *et al*, *Vet. Rec.*, 1935, **15**, 456; *CA* **29**, 7472 (*synth*)
Macpherson, H.T. *et al*, *Biochem. J.*, 1938, **32**, 76 (*struct*)
Merck Index, 9th Ed., No. 1644.
Mullen, P.A., *Vet. Rec.*, 1975, **97**, 87 (*use*)
Gyrd-Hansen, N. *et al*, *J. Vet. Pharmacol. Ther.*, 1981, **4**, 15 (*use*)

$C_6H_{12}BN$ B-00396

1-Aza-4-borabicyclo[2.2.2]octane

(Nitrilotri-2,1-ethanediyl)boron, 10CI

[61061-78-7]

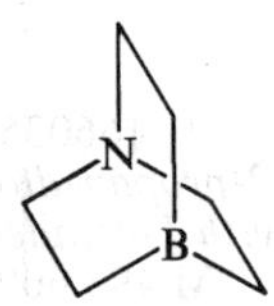

M 108.978

Paddon-Row, M.N. *et al*, *J. Chem. Soc., Chem. Commun.*, 1976, 427 (*struct*)
Gregory, A.R. *et al*, *Aust. J. Chem.*, 1977, **30**, 473 (*struct*)

$C_6H_{12}BNO_3$ B-00397

[[2,2′,2″-Nitrilotris[ethanolato]](3−)-*N,O,O′,O″*]boron, 10CI, 9CI

2,8,9-Trioxa-5-aza-1-borabicyclo[3.3.3]undecane, 10CI, 9CI. [2,2′,2″-Nitrilotriethanolato(3−)]boron, 8CI. 2,2′-,2″-Nitrilotriethanol cyclic borate. 2,2′,2″-Nitrilotriethylborate, 8CI. Triethanolamine borate

[15277-97-1]

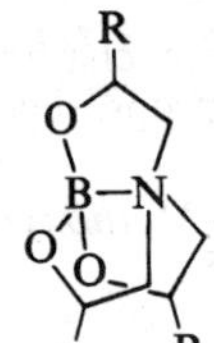

R = H

M 156.976

Used in colour reagent test strip for urine anal. Reagent for ensuring monoalkylation of ketones, curing agent for epoxy resins. Cryst. (Py). Sol. Py, MeCN, H_2O, alcohols, AcOH, $PhNO_2$, insol. THF, C_6H_6, CCl_4. Mp 236.5-237.5°.

Brown, H.C. *et al*, *J. Am. Chem. Soc.*, 1951, **73**, 2808 (*synth*)
Taira, Z. *et al*, *Inorg. Nucl. Chem. Lett.*, 1971, **7**, 509 (*struct*)
Aldrich Library of IR Spectra, 1975, 2nd Ed., 1326D (*ir*)
Mazheika, I. *et al*, *Zh. Obshch. Khim.*, 1978, **48**, 2728 (*ms*)
Rathke, M.W. *et al*, *Synth. Commun.*, 1978, **8**, 9 (*synth, use*)
Zeldin, M. *et al*, *J. Inorg. Nucl. Chem.*, 1978, **40**, 597 (*pmr*)

Shatz, V.D. *et al*, *J. Chromatogr.*, 1980, **200**, 105 (*chromatog*)
Fieser, M. *et al*, *Reagents for Organic Synthesis*, Wiley, 1967-83, **8**, 500.

$C_6H_{12}BN_3$ B-00398

Cyanoborane 3-(dimethylamino)propanenitrile

*(Cyano-*C*)[3-(dimethylamino)propanenitrile-*N^3*]dihydroboron*, IOCI. *Boranecarbonitrile, compd. with 3-(dimethylamino)propanenitrile(1:1)*

[72695-25-1]

$NCCH_2CH_2NMe_2—BH_2CN$

M 136.991

Exhibits anticholesteraemic and antihypolipidaemic activity. Oil obt. by evapn.; solidifies on standing. Sol. $CHCl_3$. Mp 42-3°.

Spielvogel, B.F. *et al*, *J. Inorg. Nucl. Chem.*, 1979, **41**, 1223 (*synth, pmr, ir*)
Hall, I.H. *et al*, *J. Pharm. Sci.*, 1980, **69**, 1025; 1981, **70**, 339 (*pharmacol*)

$C_6H_{12}BO_4S_2^{\ominus}$ B-00399

Bis[3-mercapto-1,2-propanediolato(2−)-*O,O′*]borate(1−), IOCI

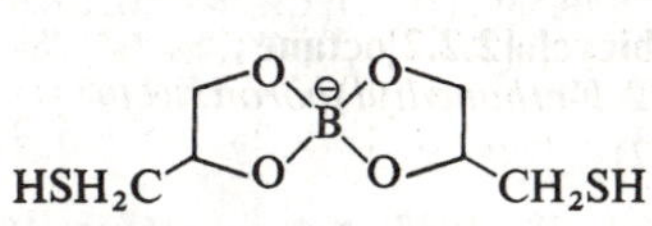

M 223.088 (ion)

Octadecylammonium salt: [66035-11-8]. *Hydrogen bis[3-mercapto-1,2-propanediolato(2−)-O,O′]borate(1−) compd. with 1-octadecanamine (1:1)*, IOCI.

$C_{24}H_{52}BNO_4S_2$ M 493.609

Lubricating oil antiwear additive.

Baldwin, B.A., *Wear*, 1977, **45**, 345; *CA*, **88**, 123498t (*use*)
U.S.P., 4 115 286, (*1978*); *CA*, **90**, 41188u (*use*)

$C_6H_{13}B$ B-00400

Borepane, IOCI, 9CI

Boracycloheptane

[4480-56-2]

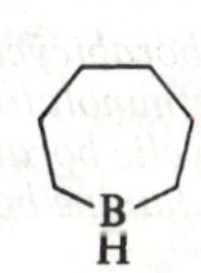

M 95.979

Dimeric; not isol. in pure state; characterised as *B*-methoxy deriv. No phys. props. reported. Mod. sol. THF.

B-*methoxy:* [5747-29-5]. *1-Methoxyborepane.*

$C_7H_{15}BO$ M 126.005

Sol. THF. Bp_{32} 74-8°.

Brown, H.C. *et al*, *J. Am. Chem. Soc.*, 1972, **94**, 3561 (*synth*)
Burke, P.L. *et al*, *J. Am. Chem. Soc.*, 1973, **95**, 3654 (*synth, props, ir, glc*)
Negishi, E. *et al*, *J. Am. Chem. Soc.*, 1973, **95**, 6757 (*deriv, ir, pmr, synth*)

$C_6H_{13}B$ B-00401

Cyclohexylborane

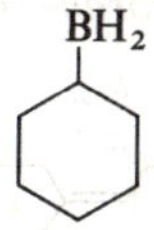

M 95.979

Tetramethylethylenediamine complex (2:1): [69319-84-2]. *Dicyclohexyltetrahydro[μ-(N,N,N′,N′-tetramethyl-1,2-ethanediamine-N:N′)]diboron*, IOCI.

$C_{18}H_{42}B_2N_2$ M 308.163

Used for long-term storage of cyclohexylborane. Air-stable solid. Mp 100-1°.

Brown, H.C. *et al*, *J. Org. Chem.*, 1979, **44**, 465 (*synth*)

$C_6H_{13}B$ B-00402

1-Ethylborolane, IOCI

[75126-77-1]

As Borolane, B-00227 with

R = Et

M 95.979

Air-sensitive liq. Sol. nonhydroxylic solvs.

▷Prob. spont. inflammable

Wrackmeyer, B. *et al*, *Z. Naturforsch.*, *B*, 1980, **35**, 439 (*nmr*)

$C_6H_{13}B$ B-00403

1-Methylborinane, IOCI

1-Methyl-1-boracyclohexane

[75126-76-0]

As Borinane, B-00327 with

R = Me

M 95.979

Characterised spectroscopically. Sensitive to air. Sol. THF and $CHCl_3$.

Wrackmeyer, B., *Z. Naturforsch.*, *B*, 1980, **35**, 439 (*nmr*)

$C_6H_{13}BBr_2$ B-00404

Dibromo(1-ethylbutyl)borane

Dibromo(3-hexyl)borane. 3-Hexyldibromoborane

$H_3CCH_2CH_2CH(CH_2CH_3)BBr_2—SMe_2$

M 255.787

Me_2S complex: [64770-06-5]. *Dibromo(1-ethylbutyl)-[thiobis[methane]]boron*, IOCI.

$C_8H_{19}BBr_2S$ M 317.916

Prepared *in situ* from 3-hexene and $HBBr_2—SMe_2$. Synthetic intermediate. Air- and moisture-sensitive liq. $Bp_{2.2}$ 73-5°.

Brown, H.C. *et al*, *J. Am. Chem. Soc.*, 1977, **99**, 7097 (*synth*)
Brown, H.C. *et al*, *J. Org. Chem.*, 1980, **45**, 384.

$C_6H_{13}BBr_2$ B-00405

Dibromohexylborane, IOCI

Hexyldibromoborane

[64770-03-2]

$Br_2B(CH_2)_5CH_3$

M 255.787

Generally prepd. *in situ* by hydroboration of 1-hexene with dibromoborane. Can be isol. as its dimethylsulfide complex. Useful source of alkyldihydroxyboranes or alkyldialkoxyboranes *via* hydrolysis or alcoholysis. Air- and moisture-sensitive liq. $Bp_{0.9}$ 56-8°.

Me_2S adduct: [64770-04-3].
$C_8H_{19}BBr_2S$ M 317.916
Oil, air- and moisture-sensitive. Bp_1 97-100°. Slowly dec. under N_2 at 25°.

Brown, H.C. *et al*, *J. Am. Chem. Soc.*, 1977, **99**, 7097 (*synth*)
Brown, H.C. *et al*, *J. Org. Chem.*, 1980, **45**, 384 (*synth, use*)

$C_6H_{13}BBr_2$ B-00406

Dibromo(2-methylpentyl)borane
2-Methyl-1-pentyldibromoborane

$$H_3CCH_2CH_2CH(CH_3)CH_2BBr_2$$

M 255.787
Me_2S complex: [72205-97-1]. *Dibromo(2-methylpentyl)-[thiobis[methane]]boron*, 10CI.
$C_8H_{19}BBr_2S$ M 317.916
Air- and moisture-sensitive liq. $Bp_{1.6}$ 92-5°.

Brown, H.C. *et al*, *J. Org. Chem.*, 1980, **45**, 384 (*synth*)

$C_6H_{13}BF_2$ B-00407

Difluorohexylborane, 10CI, 9CI, 8CI
Hexyldifluoroborane. Difluorohexylborine
[592-71-2]

$$H_3C(CH_2)_5BF_2$$

M 133.976
Prepd. from hexyldihydroxyborane and BF_3. Air- and moisture-sensitive oil. d^{25} 0.859. Bp_{744} 91.9-92°.

McCusker, P.A. *et al*, *J. Am. Chem. Soc.*, 1955, **77**, 4253 (*synth*)
Brother, C.C. *et al*, *J. Am. Chem. Soc.*, 1957, **79**, 5188 (*props*)
Davies, A.G. *et al*, *Chem. Ind. (London)*, 1957, 1622 (*use*)
McCusker, P.A., *CA*, 1958, **52**, P11120b (*synth*)
Mikhailov, B.M. *et al*, *Zh. Obshch. Khim.*, 1959, **29**, 3443 (*synth*)
Anosov, V.I. *et al*, *CA*, 1980, **93**, P221317f (*use*)

$C_6H_{13}BN_2O$ B-00408

Cyanoborane 4-methylmorpholine
(Cyano-C)dihydro(4-methylmorpholine-N^4)boron, 10CI. *Boranecarbonitrile, compd. with 4-methylmorpholine(1:1)*, 9CI
[51329-60-3]

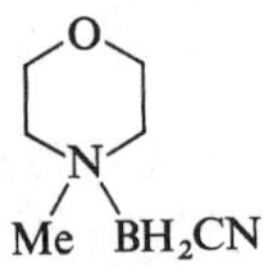

M 139.992
Shows inhibition of tumour growth in mice, but is v. toxic. White cryst. solid (H_2O). Mp 34-5°.

▷Toxic, LD_{50} 23 mg/kg

Weidig, C. *et al*, *Inorg. Chem.*, 1974, **13**, 1763 (*synth, nmr, ir*)
Spielvogel, B.F. *et al*, *J. Inorg. Nucl. Chem.*, 1979, **41**, 1223 (*synth, nmr, ir*)
Hall, I.H. *et al*, *J. Pharm. Sci.*, 1981, **70**, 339 (*pharmacol*)

$C_6H_{13}BO_2$ B-00409

1-Hexenyldihydroxyborane
1-Hexenylboronic acid, 10CI, 9CI. *1-Hexene-1-boronic acid*
[78461-60-6]

$$H_3C(CH_2)_3CH{=}CHB(OH)_2$$

M 127.978

(E)-form [42599-18-8]
Cryst.

(Z)-form [54354-55-1]
Characterised spectroscopically.

Dieck, H.A. *et al*, *J. Org. Chem.*, 1975, **40**, 1083.
Raucher, S. *et al*, *J. Org. Chem.*, 1978, **43**, 4885.
Brown, H.C. *et al*, *J. Org. Chem.*, 1980, **45**, 389 (*synth*)
Kabalka, G.W. *et al*, *Synth. Commun.*, 1981, **11**, 247.
Kabalka, G.W. *et al*, *Heterocycles*, 1982, **18**, 157.
Kabalka, G.W. *et al*, *J. Labelled Compd. Radiopharm.*, 1982, **19**, 795.

$C_6H_{13}BO_6$ B-00410

Hydrogen bis[1,2,3-propanetriolato(2−)-O^1,O^2]borate(1−), 10CI, 9CI
Bisglycerol borate. Di(glycerin)borate. Diglyceryl borate
[49625-59-4]

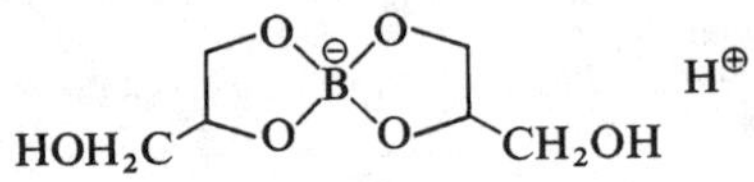

M 191.975
Paper adhesive, anticorrosion additive for cements and brake fluids, heat stabiliser for polycarbonates.

Japan. Pat., 75 79 536, (*1975*); *CA*, **83**, 207052c (*use*)
Japan. Pat., 75 121 320, (*1975*); *CA*, **84**, 94628v (*use*)
Japan. Pat., 78 90 361, (*1978*); *CA*, **89**, 216341x (*use*)
Hamanaka, H., *Yukagaku*, 1979, **28**, 285; *CA*, **91**, 59208f (*props*)
Ger. Pat., 2 831 538, (*1980*); *CA*, **93**, 134858q (*use*)
U.S.P., 4 192 759, (*1980*); *CA*, **94**, 86976y (*synth, use*)

$C_6H_{14}BBr_2N$ B-00411

Dibromo(dipropylamino)borane
1,1-Dibromo-N,N-dipropylboranamine, 10CI, 9CI
[54336-82-2]

$$Br_2BN(CH_2CH_2CH_3)_2$$

M 270.801
Colourless mobile liq., fumes in air due to moisture-sensitivity. d_4^{18} 1.474. Bp_1 48°. n_D^{18} 1.4918.

Spitsyn, V.I. *et al*, *Izv. Akad. Nauk SSSR, Ser. Khim.*, 1974, 2626; *Bull. Acad. Sci. USSR, Div. Chem. Sci.*, 2535 (*synth*)
Bryukhova, E.V. *et al*, *Izv. Akad. Nauk SSSR, Ser. Khim.*, 1976, 1614; *Bull. Acad. Sci. USSR, Div. Chem. Sci.*, 1529 (*nmr*)
Balabanova, E.A. *et al*, *Izv. Akad. Nauk SSSR, Ser. Khim.*, 1978, 2638; *Bull. Acad. Sci. USSR, Div. Chem. Sci.*, 2360 (*nmr*)
Balabanova, E.A. *et al*, *Zh. Neorg. Khim.*, 1980, **25**, 1793.

$C_6H_{14}BBr_3S$ B-00412

Tribromoborane dipropyl sulfide
Tribromo[1,1'-thiobis[propane]]boron, 10CI. *Tribromoborane compd. with 1,1'-thiobis[propane](1:1). Boron tribromide dipropyl sulfide*
[61552-15-6]

$$(H_3CCH_2CH_2)_2S{-}BBr_3$$

M 368.759

Romm, I.P. *et al*, *Zh. Obshch. Khim.*, 1976, **46**, 2445.
Gol'dshtein, I.P. *et al*, *Dokl. Akad. Nauk SSSR, Ser. Sci. Khim.*, 1977, **237**, 1392.
Gol'dshtein, I.P. *et al*, *Zh. Obshch. Khim.*, 1978, **48**, 2568.
Gol'dshtein, I.P. *et al*, *Dokl. Akad. Nauk SSSR, Ser. Sci. Khim.*, 1980, **252**, 909.

$C_6H_{14}BBr_3Se$ B-00413

Tribromoborane dipropyl selenide

Tribromo[1,1′-selenobis[propane]]boron, 10CI. *Tribromoborane compd. with 1,1′-selenobis[propane](1:1)*. *Boron tribromide dipropyl selenide*

[65907-94-0]

$(H_3CCH_2CH_2)_2Se—BBr_3$

M 415.659

Gol'dshtein, I.P. *et al*, *Dokl. Akad. Nauk SSSR, Ser. Sci. Khim.*, 1977, **237**, 1392; 1980, **252**, 909.
Gol'dshtein, I.P. *et al*, *Zh. Obshch. Khim.*, 1978, **48**, 2568.

$C_6H_{14}BCl$ B-00414

Chlorodipropylborane, 10CI, 9CI, 8CI

Chlorodipropylborine. Dipropylchloroborane

[22086-53-9]

$(H_3CCH_2CH_2)_2BCl$

M 132.440

Synthetic intermed. Colourless liq., sensitive to air and moisture. d_4^{20} 0.848. Mp <−125°. Bp 127°.

▷Ignites in air

Dollimore, D. *et al*, *J. Chem. Soc.*, 1953, 3906 (*synth*)
U.S.P., 2 930 743, (*1960*); *CA*, **54**, 16386e (*synth*)
B.P., 853 063, (*1960*); *CA*, **55**, 10317h (*synth*)
Long, L.H. *et al*, *J. Chem. Soc.*, 1963, 2181 (*synth*)
Köster, R. *et al*, *Justus Liebigs Ann. Chem.*, 1969, **719**, 169 (*synth*)
Köster, R. *et al*, *Justus Liebigs Ann. Chem.*, 1969, **724**, 34.
Qureshi, A.R. *et al*, *Pak. J. Sci.*, 1977, **29**, 129 (*synth, props*)

$C_6H_{14}BCl$ B-00415

Chloro(1,1,2-trimethylpropyl)borane, 10CI

Thexylchloroborane. Chlorothexylborane

[75030-54-5]

$(H_3C)_2CHC(CH_3)_2BHCl$

M 132.440

Used in ketone synth. Air- and moisture-sensitive. Known only in ether soln. or as Me_2S adduct.

Me_2S adduct: [75067-06-0].

$C_8H_{20}BClS$ M 194.569

Reagent for ketone synth. Formed as CH_2Cl_2 or ether soln. and used as such.

Brown, H.C. *et al*, *J. Org. Chem.*, 1980, **45**, 4540 (*synth, use*)
Kulkarni, S.V. *et al*, *J. Org. Chem.*, 1980, **45**, 4542 (*use*)
Zweifel, G. *et al*, *J. Am. Chem. Soc.*, 1980, **102**, 5919 (*synth, use*)

$C_6H_{14}BCl_2N$ B-00416

Dichloro(diisopropylamino)borane

1,1-Dichloro-N,N-bis(1-methylethyl)boranamine, 10CI, 9CI

[44873-49-6]

$[(H_3C)_2CH]_2NBCl_2$

M 181.899

Moisture-sensitive.

Py complex: [15590-33-7].

$C_{11}H_{19}BCl_2N_2$ M 261.001

White cryst. solid. Bp_{10} 60° subl. Hydrolyses readily in atm.

4-Methylpyridine complex: [15590-35-9].

$C_{12}H_{21}BCl_2N_2$ M 275.027

White cryst. solid. Mp 74°. Hydrolyses readily in atm.

Lappert, M.F. *et al*, *J. Chem. Soc. (A)*, 1967, 602 (*deriv*)
Baudler, M. *et al*, *Z. Naturforsch., B*, 1978, **33**, 355.
Baudler, M. *et al*, *Z. Anorg. Allg. Chem.*, 1981, **474**, 18 (*synth*)

$C_6H_{14}BF_3O$ B-00417

Trifluoroborane 1-hexanol

Trifluoro(1-hexanol)boron, 10CI. *1-Hexanol compd. with trifluoroborane(1:1)*, 9CI. *Hexyl alcohol compd. with boron fluoride(BF_3)(1:1)*, 8CI

[1536-90-9]

$H_3C(CH_2)_4CH_2OH—BF_3$

M 169.981

Catalyst for dimerisation of alkenes. Moisture-sensitive.

Ogg, R.A. *et al*, *Chem. Co-ord. Compds., Symp., Rome*, 1957, 468; *CA*, **58**, 2041d (*nmr*)
Romadne, T., *Latvijas PSR Zinatnu Akad. Vestis, Kim. Ser.*, 1963, 343; *CA*, **60**, 2531a.
Yoneda, N. *et al*, *Kogyo Kagaku Zasshi*, 1971, **74**, 908; *CA*, **75**, 48075d (*use*)
Can. Pat., 1 034 967, (*1978*); *CA*, **90**, 137235h (*synth, use*)

$C_6H_{14}BN$ B-00418

1-(Dimethylamino)borolane, 8CI

N,N-*Dimethylborolanamine*, 10CI

[1192-61-6]

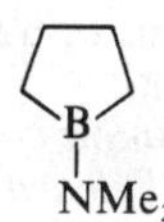

M 110.993

Air- and moisture-sensitive liq. Sol. nonhydroxylic solvs. d_{20} 0.817. Bp 48-50°, Bp_{24} 38-9°. n_D^{20} 1.4460.

Ger. Pat., 1 118 200, (*1961*); *CA*, **56**, 14327h (*synth, ir*)
Mikhailov, B.M. *et al*, *Dokl. Akad. Nauk SSSR, Ser. Sci. Khim.*, 1965, **160**, 615 (*synth, props*)
Nöth, H. *et al*, *Chem. Ber.*, 1981, **114**, 1150 (*nmr*)

$C_6H_{15}B$ B-00419

Triethylborane, 10CI, 9CI

Boron triethyl. Triethylborine

[97-94-9]

BEt_3

M 97.995

Fuming liq. d_4^{23} 0.70. Mp −92.5°. Bp 95°.

▷Spontaneously flammable; toxic. ED2100000.

Ital. Pat., 758 247, (*1967*); *CA*, **68**, 49728 (*synth*)
Nöth, H. *et al*, *J. Organomet. Chem.*, 1968, **12**, 23 (*pmr*)
Nöth, H. *et al*, *Synth. Inorg. Met.-Org. Chem.*, 1971, **1**, 197 (*synth*)
Japan Pat., 60 226, (*1974*); *CA*, **80**, 108666 (*synth*)
Harbold, W. *et al*, *Z. Anorg. Allg. Chem.*, 1976, **420**, 251 (*ir, raman*)
Wrackenmeyer, B. *et al*, *Chem. Ber.*, 1976, **109**, 1075 (*nmr*)
Bretherick, L., *Handbook of Reactive Chemical Hazards*, 2nd Ed., Butterworths, London and Boston, 1979, 608.
Sax, N.I., *Dangerous Properties of Industrial Materials*, 5th Ed., Van Nostrand-Reinhold, 1979, 428.
Hazards in the Chemical Laboratory, (Bretherick, L., Ed.), 3rd Ed., Royal Society of Chemistry, London, 1981, 518.

$C_6H_{15}B$ B-00420

(1,1,2-Trimethylpropyl)borane, 10CI, 9CI, 8CI

2,3-Dimethyl-2-butylborane. Thexylborane

[3688-24-2]

$(H_3C)_2CHC(CH_3)_2BH_2$

M 97.995

Exists as dimer in THF soln. or in neat form. Prepd. from tetramethylethylene and borane in THF at 0°. Reagent in organic synthesis; hydroborating agent generally prepd. *in situ*. Colourless liq. Stable for 10 days in refrigerator but isomerizes over prolonged periods at 25°. Air- and moisture-sensitive. Sol. THF, Et_2O. Mp −34.7° to −32.3°. ir 2540, 1565 cm^{-1} (BHR, B—H—B). V.p. <0.5mm at 0°.

Me_2S complex: [68297-75-6]. *Dihydro[thiobis[methane]](1,1,2-trimethylpropyl)boron*, 10CI.
$C_8H_{21}BS$ M 160.124
Hydroboration agent comparable to thexylborane in THF. Air- and moisture-sensitive liq. Sol. THF.

NEt_3 complex: [56118-59-3].
$C_{12}H_{30}BN$ M 199.186
Useful store for thexylborane.

Tetramethylethylenediamine complex(1:1): [67826-89-5]. *Dihydro(N,N,N',N'-tetramethyl-1,2-ethanediamine-N)(1,1,2-trimethylpropyl)boron*, 10CI.
$C_{12}H_{31}BN_2$ M 214.200
Useful store of thexylborane. Air-stable viscous liq., can be stored for several weeks at r.t. Bp_{12} 38-40°. n_D^{20} 1.4240. Treatment with BF_3. Et_2O precipitates. TMEDA. $2BF_3$ leaving thexylborane.

Tetramethylethylenediamine complex(2:1): Bis(thexylborane)tetramethylethylenediamine.
$C_{18}H_{46}B_2N_2$ M 312.195
Stable thexylborane complex useful for its storage. Cryst. (pentane). Mp 43-5°.

$PhNEt_2$ complex: [70868-92-7].
$C_{16}H_{30}BN$ M 247.230
Active reducing and hydroborating agent with similar props. to free thexylborane. Liq., stable for long periods at 0°. Misc. hexane, C_6H_6, CH_2Cl_2, ethers.

Brown, H.C. *et al*, *Inorg. Chem.*, 1962, **1**, 204 (*synth*)
Zweifel, G. *et al*, *J. Am. Chem. Soc.*, 1963, **85**, 2066 (*synth, use*)
Pelter, A. *et al*, *J. Chem. Soc., Chem. Commun.*, 1971, 1048 (*use*)
Brown, H.C. *et al*, *J. Am. Chem. Soc.*, 1972, **94**, 3567; 1975, **97**, 2791; 1977, **99**, 5514 (*synth, use*)
Brown, H.C. *et al*, *J. Org. Chem.*, 1972, **37**, 2942 (*use*)
Negishi, E. *et al*, *Synthesis*, 1974, 77 (*rev, nmr, ir*)
Brown, H.C. *et al*, *J. Am. Chem. Soc.*, 1975, **97**, 2791, 2799 (*use*)
Brown, H.C. *et al*, *J. Org. Chem.*, 1977, **42**, 1392 (*deriv*)
Pelter, A. *et al*, *Tetrahedron Lett.*, 1978, 4715.
Brown, H.C. *et al*, *Inorg. Chem.*, 1979, **18**, 51 (*synth, nmr*)
Yamamoto, Y. *et al*, *CA*, 1980, **92**, 139884 (*rev*)

$C_6H_{15}BBr_3N$ B-00421

Tribromoborane triethylamine

Tribromo(N,N-diethylethanamine)boron, 10CI. *N,N-Diethylethanamine compd. with tribromoborane(1:1). Triethylamine compd. with boron bromide(BBr_3)(1:1)*, 8CI. *Boron tribromide triethylamine*

[2903-69-7]

$Et_3N—BBr_3$

M 351.713

Gates, P.N. *et al*, *Spectrochim. Acta*, 1965, **21**, 1445 (*nmr*)
Massey, A.G. *et al*, *J. Organomet. Chem.*, 1966, **5**, 218 (*nmr*)
Davies, P.G. *et al*, *Inorg. Nucl. Chem. Lett.*, 1967, **3**, 249 (*ir*)
Romm, I.P. *et al*, *Zh. Obshch. Khim.*, 1976, **46**, 2445.
Sheikh, S.U., *J. Thermal. Anal.*, 1980, **18**, 299 (*synth*)

$C_6H_{15}BCl_3N$ B-00422

Trichloroborane triethylamine

Trichloro(N,N-diethylethanamine)boron, 10CI. *N,N-Diethylethanamine compd. with trichloroborane(1:1)*, 8CI. *Triethylamine compd. with boron chloride(BCl_3)(1:1)*, 8CI. *Boron trichloride triethylamine*

[2890-88-2]

$Et_3N—BCl_3$

M 218.360

Davies, P.G. *et al*, *Inorg. Nucl. Chem. Lett.*, 1967, **3**, 249 (*ir*)
Kaplansky, M. *et al*, *Can. J. Chem.*, 1970, **48**, 697 (*nqr*)
Ryschkewitsch, G.E. *et al*, *J. Magn. Reson.*, 1969, **1**, 584 (*nmr*)
Sheikh, S.U., *Bull. Chem. Soc. Jpn.*, 1976, **49**, 2879 (*nqr*)
Sheikh, S.U., *J. Therm. Anal.*, 1980, **18**, 299 (*synth*)

$C_6H_{15}BFN$ B-00423

Butyl(dimethylamino)fluoroborane

1-Butyl-1-fluoro-N,N-dimethylboranamine, 10CI

[72334-78-2]

$H_3CCH_2CH_2CH_2BFNMe_2$

M 131.000

Air- and moisture-sensitive.

Costes, J.P. *et al*, *J. Organomet. Chem.*, 1979, **175**, 257 (*nmr*)

$C_6H_{15}BF_4O$ B-00424

Triethyloxonium tetrafluoroborate, 10CI, 9CI

Triethyloxonium fluoroborate

[368-39-8]

$Et_3O^{\oplus}\ BF_4^{\ominus}$

M 189.988

Powerful alkylating agent. Hygroscopic cryst. (Et_2O). Sol. CH_2Cl_2, sl. sol. Et_2O. Mp 91-2° dec.

▷RR4584700.

Org. Synth., 1966, **46**, 113 (*synth*)
Fieser, M. *et al*, *Reagents for Organic Synthesis*, Wiley, 1967-83, **8**, 500.
Jones, F.R. *et al*, *J. Chem. Soc., Dalton Trans.*, 1979, 927 (*synth*)

$C_6H_{15}BO$ B-00425

Ethoxydiethylborane

Diethylborinic acid ethyl ester, 10CI, 9CI, 8CI. *Diethyl(ethoxy)borane*

[15706-69-1]

Et_2BOEt

M 113.994

Air- and moisture-sensitive liq. Sol. nonhydroxylic solvs. Bp 105-7°.

Ger. Pat., 1 125 923, (*1962*); *CA*, **57**, 9880f (*synth*)
Ital. Pat., 704 906, (*1966*); *CA*, **68**, 22370b (*use*)
Shitov, O.P. *et al*, *Zh. Obshch. Khim.*, 1973, **43**, 1266.

Köster, R. *et al*, *Justus Liebigs Ann. Chem.*, 1975, 352 (*synth, props, pmr, ms*)
Paetzold, P. *et al*, *Z. Naturforsch.*, *B*, 1980, **35**, 568.

$C_6H_{15}BO_2$ B-00426

Butyldimethoxyborane

Butylboronic acid dimethyl ester, 10CI, 9CI. *1-Butaneboronic acid dimethyl ester. Dimethyl butaneboronate*

[2117-94-4]

$$H_3C(CH_2)_3B(OMe)_2$$

M 129.993

Liq., sensitive to air and moisture. d_{20} 0.828. Bp_{20} 39°. n_D^{20} 1.3935.

Laurent, J.P. *et al*, *C.R. Hebd. Seances Acad. Sci.*, 1961, **253**, 1812 (*synth*)
Vasil'ev, L.S. *et al*, *Izv. Akad. Nauk SSSR, Ser. Khim.*, 1977, 1126; *Bull. Acad. Sci. USSR, Div. Chem. Sci.*, 1031.
Mikhailov, B.M. *et al*, *Izv. Akad. Nauk SSSR, Ser. Khim.*, 1978, 1586; *Bull. Acad. Sci. USSR, Div. Chem. Sci.*, 1385.
Mikhailov, B.M. *et al*, *J. Organomet. Chem.*, 1978, **154**, 113.
Mendoza, A. *et al*, *J. Org. Chem.*, 1979, **44**, 1352.

$C_6H_{15}BO_2$ B-00427

Diethoxyethylborane

Ethylboronic acid diethyl ester, 10CI. *Ethyldiethoxyborane. Diethyl ethaneboronate. Ethaneboronic acid diethyl ester*

[53907-92-9]

$$EtB(OEt)_2$$

M 129.993

Catalyst for polymn. of ethylene. Probably liq., air- and moisture-sensitive. Sol. non-hydroxylic solvs.

Ger. Pat., 2 523 167, (*1975*); *CA*, **84**, 60251b (*use*)
Japan. Pat., 79 30 425, (*1979*); *CA*, **92**, 77161k (*use*)
Paetzold, P. *et al*, *Z. Naturforsch.*, *B*, 1980, **35**, 568.

$C_6H_{15}BO_2$ B-00428

Hexyldihydroxyborane

Hexylboronic acid, 10CI, 9CI. *1-Hexaneboronic acid*, 8CI

[16343-08-1]

$$H_3C(CH_2)_5B(OH)_2$$

M 129.993

Used to inhibit cholinesterase, proteinase, lipase and chymotrypsin. White cryst. (H_2O or hexane/Me_2CO). Air-sensitive. Spar. sol. H_2O. Mp 88-90°. V. weak acid, almost neutral to litmus.

Snyder, H.R. *et al*, *J. Am. Chem. Soc.*, 1938, **60**, 105 (*synth, props*)
McCusker, P.A. *et al*, *J. Am. Chem. Soc.*, 1957, **79**, 5182 (*props*)
Laurent, J.P., *C.R. Hebd. Seances Acad. Sci.*, 1962, **254**, 866 (*synth*)
Zweifel, G., *J. Organomet. Chem.*, 1967, **9**, 215 (*synth*)
Antonov, V.K. *et al*, *Mol. Biol.*, 1970, **4**, 558 (*props*)
Jarv, J. *et al*, *CA*, 1978, **89**, 125125p (*use*)

$C_6H_{15}BO_3$ B-00429

Triethoxyborane

Boric acid (H_3BO_3) triethyl ester, 10CI, 9CI, 8CI. *Triethyl borate. Triethyl orthoborate. Ethyl borate*

[150-46-9]

$$(EtO)_3B$$

M 145.993

Used in manuf. of oxalate esters by carbonylation; catalyst for polymn. of alkenes and dienes and in polyester manuf.; stabilizing agent, insect sterilant. Liq. readily hydrolysed. Sol. non-hydroxylic solvs. d_4^{26} 0.864. Mp −84.8°. Bp 120°, 118.6°. n_D^{25} 1.3721. Forms azeotrope (1:7) with EtOH.

▷ED5075000.

Webster, S.H. *et al*, *J. Am. Chem. Soc.*, 1933, **55**, 3233 (*synth, props*)
Inorg. Synth., 1957, **5**, 29 (*synth, props*)
Fallon, P.J. *et al*, *Int. J. Mass Spectrom. Ion Phys.*, 1968, **1**, 133 (*ms*)
Cradock, S. *et al*, *J. Chem. Soc., Dalton Trans.*, 1975, 25 (*pe*)
Gmelins Handbuch der Anorg. Chem., 8th Ed., 1975, **13**, 149 (*bibl*)
Furlei, I.I. *et al*, *Izv. Akad. Nauk SSSR, Ser. Khim.*, 1976, 2127; *Bull. Acad. Sci. USSR, Div. Chem. Sci.*, **25**, 2127 (*ms*)
Pola, M. *et al*, *Collect. Czech. Chem. Commun.*, 1979, **44**, 3688 (*ir, props*)

$C_6H_{15}BS_2$ B-00430

***tert*-Butylbis(methylthio)borane**

(1,1-Dimethylethyl)dithioboronic acid dimethyl ester, 10CI

[65568-73-2]

$$(MeS)_2BC(CH_3)_3$$

M 162.115

Liq., sensitive to air and moisture. Bp_{15} 74-7°.

Nöth, H. *et al*, *J. Organomet. Chem.*, 1977, **142**, 281 (*synth, nmr*)

$C_6H_{15}B_3Cl_3N_3$ B-00431

2,4,6-Trichloro-1,3,5-triethylborazine, 10CI, 9CI, 8CI

B,B,B-*Trichloro*-N,N,N-*triethylborazole*

[2608-99-3]

As 2,4,6-Tribromoborazine, B-00015 with

$$X = Cl, Y = Et$$

M 267.994

Solid. Mp 55-9°. $Bp_{0.1}$ 66°.

Turner, H.S. *et al*, *Chem. Ind. (London)*, 1958, 526 (*synth*)
Butcher, I.M. *et al*, *Spectrochim. Acta*, 1962, **18**, 1487 (*ir, struct*)
Butcher, I.M. *et al*, *J. Inorg. Nucl. Chem.*, 1965, **27**, 817 (*props*)
Turner, H.S. *et al*, *J. Chem. Soc.*, 1965, 6421 (*synth, props*)
Cros, G. *et al*, *Bull. Soc. Chim. Fr.*, 1969, 2601 (*nmr*)
Wrackmeyer, B. *et al*, *Chem. Ber.*, 1976, **109**, 3480.

$C_6H_{15}B_3O_3$ B-00432

Triethylboroxin, 10CI, 9CI, 8CI

[3043-60-5]

As Boroxin, B-00019 with

$$R = Et$$

M 167.613

Used in polymerisation of cyclic ethers. Moisture-sensitive liq., disproportionates above 200°. d^{25} 0.896. Bp_{736} 153°. n_D^{25} 1.3958.

McCusker, P.A. *et al*, *J. Am. Chem. Soc.*, 1957, **79**, 5179 (*synth*)
Ashby, E. *et al*, *J. Am. Chem. Soc.*, 1959, **81**, 4791 (*synth*)
Laurent, P.J. *et al*, *Bull. Soc. Chim. Fr.*, 1970, **3**, 836 (*nmr*)
Köster, R. *et al*, *Angew. Chem., Int. Ed. Engl.*, 1980, **92**, 553 (*use*)

$C_6H_{15}B_9Fe^{\ominus}$ B-00433

(η^5-2,4-Cyclopentadien-1-yl)[η^6-decahydrocarbadecaborato(2−)]ferrate(1−), 10CI
closo-[*1-η^5-Cyclopentadienylferrate-2-CB_9H_{10}*]

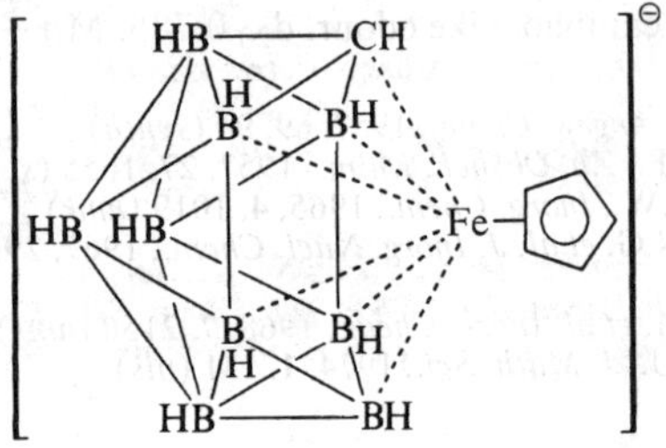

M 240.322 (ion)
Paramagnetic.
Tetramethylammonium salt: [62580-70-5].
$C_{10}H_{27}B_9FeN$ M 314.467
Violet needles (MeOH).

Dolansky, J. *et al*, *Chem. Ind. (London)*, 1976, 853 (*synth, nmr*)
Subrtova, V. *et al*, *Acta Crystallogr., Sect. B*, 1978, **34**, 2720 (*struct*)

$C_6H_{16}B^{\ominus}$ B-00434

Triethylhydroborate(1−), 10CI, 9CI, 8CI
Triethylborohydride

$Et_3BH^{\ominus}$

M 99.002 (ion)
Li salt: [22560-16-3]. *Lithium triethylhydroborate*, 10CI, 9CI, 8CI. *Lithium triethylborohydride. Superhydride.*
$C_6H_{16}BLi$ M 105.943
Powerful nucleophile for S_N2 displacements; reducing agent. Monodeuterio deriv. is used for deuterium incorporation. Air- and moisture-sensitive cryst. (C_6H_6). Mp 66-7°. Commercially available as THF soln.
▷Flammable
Na salt: [17979-81-6]. *Sodium triethylhydroborate*, 10CI, 9CI, 8CI. *Sodium triethylborohydride.*
$C_6H_{16}BNa$ M 121.992
Air- and moisture-sensitive colourless cryst. when pure (often obt. as dark oil). Mp 30°.
K salt: [22560-21-0]. *Potassium triethylhydroborate*, 10CI, 9CI, 8CI. *Potassium triethylborohydride.*
$C_6H_{16}BK$ M 138.101
Air- and moisture-sensitive.

Brown, H.C. *et al*, *J. Am. Chem. Soc.*, 1973, **95**, 1669 (*use*)
Mielczarek, I., *Wiad. Chem.*, 1973, **27**, 814; *CA*, **80**, 46964x (*rev*)
Inorg. Synth., 1974, **15**, 136 (*synth*)
Brown, C.A., *J. Organomet. Chem.*, 1978, **156**, C17 (*props*)
Brown, H.C. *et al*, *J. Org. Chem.*, 1979, **44**, 467 (*props*)
Brown, H.C. *et al*, *J. Organomet. Chem.*, 1980, **188**, 1 (*synth, nmr, ir*)
Dervan, P.B. *et al*, *J. Am. Chem. Soc.*, 1980, **102**, 3863 (*props*)
Krishnamurthy, S. *et al*, *J. Org. Chem.*, 1980, **45**, 849 (*use*)
Yamamoto, Y. *et al*, *CA*, 1980, **92**, 139884j (*rev*)
Selover, J.C. *et al*, *J. Organomet. Chem.*, 1981, **206**, 317 (*props*)

$C_6H_{16}BN$ B-00435

Aminodipropylborane, 8CI
1,1-Dipropylboranamine, 10CI, 9CI
[28872-72-2]

$(H_3CCH_2CH_2)_2BNH_2$

M 113.009
Air- and moisture-sensitive liq. d_{20} 0.748. Bp 117°, Bp_{96} 64-5°. n_D^{20} 1.4170.

Mikhailov, B.M. *et al*, *Zh. Obshch. Khim.*, 1961, **31**, 160 (*synth*)
Laurent, J.P., *Bull. Soc. Chim. Fr.*, 1963, 558 (*synth*)
Hohnstedt, L.F. *et al*, *J. Chem. Soc. (A)*, 1970, 2455 (*ir, synth*)

$C_6H_{16}BN$ B-00436

(Diethylamino)dimethylborane
N,N-*Diethyl-1,1-dimethylboranamine*, 10CI, 9CI. *(Diethylamino)dimethylborine*
[7318-87-8]

Me_2BNEt_2

M 113.009
Air- and moisture-sensitive liq. Bp_{712} 90-2°.

Goubeau, J. *et al*, *Chem. Ber.*, 1960, **93**, 1379.
Nöth, H., *Z. Naturforsch., B*, 1961, **16**, 470 (*synth*)
Nöth, H. *et al*, *Chem. Ber.*, 1966, **99**, 1049; 1981, **114**, 1150 (*nmr*)
Beck, W. *et al*, *Chem. Ber.*, 1972, **105**, 2883 (*nmr*)

$C_6H_{16}BN$ B-00437

(Dimethylamino)diethylborane, 8CI
*1,1-Diethyl-*N,N*-dimethylboranamine*, 10CI, 9CI. *(Dimethylamino)diethylborine*
[7397-47-9]

Et_2BNMe_2

M 113.009
Air- and moisture-sensitive.

Erickson, C.E. *et al*, *J. Org. Chem.*, 1959, **24**, 1161.
Nöth, H. *et al*, *J. Organomet. Chem.*, 1964, **1**, 401.
Beck, W. *et al*, *Chem. Ber.*, 1972, **105**, 2883 (*nmr*)
Emerick, D.P. *et al*, *J. Organomet. Chem.*, 1978, **154**, 147.
Paetzold, P. *et al*, *Chem. Ber.*, 1979, **112**, 654.
Nöth, H. *et al*, *Chem. Ber.*, 1981, **114**, 1150 (*nmr*)

$C_6H_{16}BNO_2$ B-00438

(Ethoxycarbonyl)borane trimethylamine
*(Ethoxycarbonyl)(*N,N*-dimethylmethanamine)dihydroboron*, 10CI. *Boranecarboxylic acid ethyl ester compd. with* N,N*-dimethylmethanamine(1:1)*
[75519-14-1]

$Me_3N—BH_2COOEt$

M 145.008

Hall, I.H. *et al*, *J. Pharm. Sci.*, 1980, **69**, 1025; 1981, **70**, 339 (*synth, pharmacol*)

$C_6H_{16}B_2F_6N_2$ B-00439

Bis(trifluoroborane) tetramethylethylenediamine
*Hexafluoro[μ-(*N,N,N′,N′*-tetramethyl-1,2-ethanediamine-*N,N′*)]diboron*, 10CI. N,N,N′,N′*-Tetramethyl-1,2-ethanediamine compd. with trifluoroborane(1:2)*

[67813-45-0]

$F_3B—Me_2NCH_2CH_2NMe_2—BF_3$

M 251.816

Large cryst. (MeCN). Form. of this complex allows removal of TMEDA from alkylboranes. Insol. nonpolar org. solvs. Mp 210-2°.

Singaram, B. *et al*, *J. Organomet. Chem.*, 1978, **156**, C1 (*synth*)
Brown, H.C. *et al*, *Inorg. Chem.*, 1979, **18**, 51 (*synth*)
Brown, H.C. *et al*, *Inorg. Chem.*, 1980, **19**, 455 (*nmr*)

$C_6H_{17}BClN$ B-00440

Chloroborane triethylamine

Chloro(N,N-*diethylethanamine*)*dihydroboron*, *10CI*. *Chloroborane compd. with* N,N-*diethylethanamine*(*1:1*), *9CI*. *Chloroborane compd. with triethylamine*(*1:1*), *8CI*. *Triethylamine compd. with chloroborane*(*1:1*)

[13240-39-6]

$Et_3N—BH_2Cl$

M 149.470

White cryst. solid. Mp 43°.

Nöth, H. *et al*, *Chem. Ber.*, 1960, **93**, 2251 (*synth*)
Faulks, J.N.G. *et al*, *J. Inorg. Nucl. Chem.*, 1967, **29**, 329 (*ir*, *nmr*)
Baker, C.S.L., *J. Organomet. Chem.*, 1969, **19**, 287 (*synth*)
Zhigach, A.F. *et al*, *Zh. Obshch. Khim.*, 1973, **43**, 1966; *CA*, **79**, 146579y.
Myers, W.H. *et al*, *Inorg. Chem.*, 1975, **14**, 2874 (*pmr*)
Drummond, A. *et al*, *Inorg. Chim. Acta*, 1977, **24**, 191.
Schmidbaur, H. *et al*, *Chem. Ber.*, 1979, **112**, 1448.

$C_6H_{17}BN_2O$ B-00441

***N*-Ethylcarbamoylborane trimethylamine**

(N,N-*Dimethylmethanamine*)[(*ethylamino*)*carbonyl*]-*dihydroboron*, *10CI*. N-*Ethylboranecarboxamide, compd. with* N,N-*dimethylmethanamine*(*1:1*), *9CI*

[60788-35-4]

$Me_3N—BH_2CONHEt$

M 144.023

Shows a broad spectrum of physiological activity. Clear liq. $Bp_{0.001}$ 80°.

Spielvogel, B.F. *et al*, *J. Am. Chem. Soc.*, 1976, **98**, 5702 (*synth*, *nmr*, *ms*, *ir*)
Hall, I.H. *et al*, *J. Pharm. Sci.*, 1979, **68**, 685; 1980, **69**, 1025; 1981, **70**, 339 (*synth*, *pharmacol*)

$C_6H_{17}B_5Cl_2Si_2$ B-00442

2,4-Bis(chlorodimethylsilyl)-2,4-dicarbaheptaborane(7), 10CI

[28699-83-4]

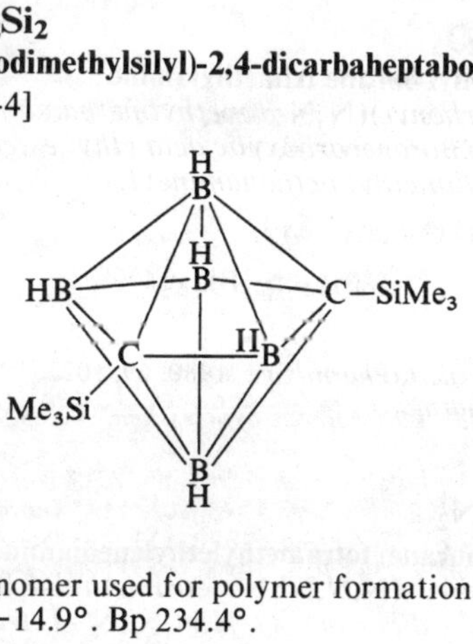

M 270.327

closo-. Monomer used for polymer formation. Colourless liq. Mp −14.9°. Bp 234.4°.

Kesting, R. *et al*, *J. Appl. Polymer Sci.*, 1970, **14**, 2525.
Zhigach, A. *et al*, *J. Gen. Chem. USSR* (*Engl. Transl.*), 1978, **48**, 536 (*ir*, *ms*)
Golosova, R. *et al*, *Russ. J. Phys. Chem.*, 1979, **53**, 435.

$C_6H_{18}BN$ B-00443

Borane triethylamine

(N,N-*Diethylethanamine*)*trihydroboron*, *10CI*. *Triethylamine borane*

[1722-26-5]

$Et_3N—BH_3$

M 115.025

Liq. with camphor-like odour. d_{20} 0.785. Mp −2°. Bp_{12} 100-1°.

Köster, R., *Angew. Chem.*, 1957, **69**, 94 (*synth*)
Zigach, A.F., *Zh. Obshch. Khim.*, 1957, **27**, 1655 (*synth*)
Heitsch, C.W., *Inorg. Chem.*, 1965, **4**, 1019 (*nmr*)
Faulks, J.N.G. *et al*, *J. Inorg. Nucl. Chem.*, 1967, **29**, 329 (*ir*, *nmr*)
Purser, J.M. *et al*, *Inorg. Chem.*, 1968, **7**, 2156 (*nmr*)
Sheikh, S.U., *J. Math. Sci.*, 1974, **1**, 131 (*ms*)

$C_6H_{18}BN$ B-00444

Trimethylborane trimethylamine

(N,N-*Dimethylmethanamine*)*trimethylboron*, *10CI*. N,N-*Dimethylmethanamine, compd. with trimethylborane*(*1:1*), *9CI*

[1704-27-4]

$Me_3B—NMe_3$

M 115.025

Cryst. Mp 120°.

Schlesinger, H.I. *et al*, *J. Am. Chem. Soc.*, 1939, **61**, 1078 (*synth*)
Purser, J.M. *et al*, *Inorg. Chem.*, 1968, **7**, 2156 (*nmr*, *cmr*)
Ang, T.T. *et al*, *J. Chem. Soc., Faraday Trans. 2*, 1972, **68**, 1331 (*pmr*)
Nöth, H. *et al*, *Chem. Ber.*, 1974, **107**, 3070 (*nmr*)
McFarlane, W. *et al*, *Chem. Ber.*, 1975, **108**, 3831 (*nmr*)
Fussstetter, H. *et al*, *Chem. Ber.*, 1977, **110**, 3172 (*cmr*)
Kuznesof, P.M. *et al*, *Inorg. Chem.*, 1978, **17**, 2308 (*struct*, *ir*)

$C_6H_{18}BNO$ B-00445

Borane 2-ethoxy-*N*,*N*-dimethylethylamine

(*2-Ethoxy*-N,N-*dimethylethanamine*-N)*trihydroboron*, *10CI*. *2-Ethoxy*-N,N-*dimethylethanamine compd. with borane*(*1:1*)

[67036-04-8]

$EtOCH_2CH_2NMe_2—BH_3$

M 131.024

U.S.P., 4 080 381, (*1978*); *CA*, **89**, P23753z
U.S.P., 4 142 902, (*1979*); *CA*, **91**, P25752p

$C_6H_{18}BNO_6$ B-00446

Trihydroxyborane tris(2-hydroxyethyl)amine

Boric acid (H_3BO_3) *compd. with 2,2′,2″-nitrilotris*[*ethanol*], *10CI*, *9CI*. *Triethanolamine boric acid*

[10049-36-2]

$(HO)_3B—N(CH_2CH_2OH)_3$

M 211.021

Corrosion inhibitor for water-based cutting fluids. Stabilizer for enzymes in aq. detergent compositions. Anticorrosion agent for tin.

Fr. Pat., 1 317 963, (*1963*); *CA*, **58**, 13532e (*use*)
Ger. Pat., 1 953 714, (*1970*); *CA*, **73**, 69323g (*use*)
U.S.P., 3 645 901, (*1972*); *CA*, **76**, 129698g (*use*)
Ger. Pat., 2 748 212, (*1978*); *CA*, **89**, 45386j (*use*)

$C_6H_{18}BN_3$ B-00447

Tris(dimethylamino)borane, 8CI

Hexamethylboranetriamine, 10CI, 9CI. Tris(dimethylamino)borine. Tris(dimethylamino)boron

[4375-83-1]

$B(NMe_2)_3$

M 143.038

Sterilant for control of *Cochliomyia hominivorax* (screwworm fly). Reagent for synth. of dimethylamides from carboxylic acids. Liq., sensitive to moisture. Sol. Et_2O. Mp −16.5° to −16.1°. Bp 147°, Bp_9 38°, $Bp_{5.5}$ 33°. n_D^{24} 1.4435.

Burg, A.B. *et al*, *J. Am. Chem. Soc.*, 1951, **73**, 953 (*synth*)
Ruff, J.K., *J. Organomet. Chem.*, 1962, **27**, 1020 (*synth, ir, nmr*)
Dewar, M.J.S. *et al*, *J. Am. Chem. Soc.*, 1965, **87**, 5510 (*ms*)
Clark, A.H. *et al*, *J. Chem. Soc., Chem. Commun.*, 1969, 1082 (*struct*)
King, G.H. *et al*, *Faraday Discuss. Chem. Soc.*, 1972, **54**, 70 (*pe*)
Nöth, H. *et al*, *Chem. Ber.*, 1974, **107**, 3089 (*nmr*)
McFarlane, W. *et al*, *Chem. Ber.*, 1975, **108**, 3831 (*nmr*)
Davidson, G. *et al*, *J. Chem. Soc., Dalton Trans.*, 1981, 306 (*ir, raman*)
Nöth, H. *et al*, *Chem. Ber.*, 1981, **114**, 1150 (*nmr*)

$C_6H_{18}B_3N$ B-00448

Tris(dimethylboryl)amine

N,N-*Bis(dimethylboryl)-1,1-dimethylboranamine, 10CI. Nitrilotris[dimethylborane], 8CI*

[19163-16-7]

$(Me_2B)_3N$

M 136.645

Liq., unstable and fumes in air. Bp_{730} 132-3°.

Nöth, H. *et al*, *J. Organomet. Chem.*, 1968, **12**, 23 (*pmr, nmr*)
Storch, W. *et al*, *Chem. Ber.*, 1977, **110**, 1636 (*nmr, pmr, synth*)
Fussstetter, H. *et al*, *Chem. Ber.*, 1977, **110**, 3172 (*nmr*)
Nöth, H. *et al*, *Chem. Ber.*, 1981, **114**, 3024 (*pmr, nmr, ir*)

$C_6H_{18}B_3N_3$ B-00449

Hexamethylborazine, 10CI, 9CI

Hexahydrohexamethyl-s-triazatriborine. Hexamethyltriborinetriamine. Hexamethylborazole

[877-07-6]

As Borazine, B-00020 with

R = R′ = Me

M 164.658

Air-stable cryst. (MeCN). Sol. hexane, $CHCl_3$. Mp 98-9°. Bp 221°. High thermal stability; mod. hydrolytic stability in neutral soln.

Wiberg, E. *et al*, *Z. Anorg. Allg. Chem.*, 1947, **255**, 141 (*synth*)
Haworth, D.T. *et al*, *J. Am. Chem. Soc.*, 1960, **82**, 3860 (*synth*)
Butcher, I.M. *et al*, *J. Chem. Soc.*, 1964, 4528 (*synth, pmr*)
Melcher, L.A. *et al*, *Inorg. Chem.*, 1972, **11**, 1247 (*ir*)
Melcher, L.A. *et al*, *Inorg. Chem.*, 1973, **12**, 601 (*ms*)
Wrackmeyer, B. *et al*, *Chem. Ber.*, 1976, **109**, 3480 (*nmr*)
Nöth, H. *et al*, *Chem. Ber.*, 1981, **114**, 1150 (*nmr*)

$C_6H_{18}B_3N_3$ B-00450

1,3,5-Triethylborazine, 10CI, 9CI, 8CI

N,N,N-*Triethylborazole. 1,3,5-Triethylhexahydro-1,3,5-triazatriborine*

[7360-03-4]

As 2,4,6-Tribromoborazine, B-00015 with

X = H, Y = Et

M 164.658

Rocket propellant component. Synthetic intermed. and reducing agent. Clear colourless liq. with odour similar to ethylamine. Mp −49.6°. Bp 184°, Bp_{20} 66-8°, $Bp_{2.5}$ 21°. V.p. 6mm at 34°, n_D^{20} 1.4344.

Hough, W.V. *et al*, *J. Am. Chem. Soc.*, 1955, **77**, 864 (*synth, props*)
Hohnstadt, L.K. *et al*, *J. Am. Chem. Soc.*, 1960, **82**, 89 (*synth*)
Phillips, C.S.G. *et al*, *Z. Anal. Chem.*, 1963, **197**, 202 (*synth, chromatog*)
Meller, A., *J. Inorg. Nucl. Chem.*, 1966, **2**, 41 (*ir, props*)
Ashby, E.C. *et al*, *Inorg. Chem.*, 1971, **10**, 1524 (*ir, ms, nmr, synth*)
Powell, P. *et al*, *J. Chem. Soc. (A)*, 1971, 2951 (*ms*)
Wrackmeyer, B. *et al*, *Chem. Ber.*, 1976, **109**, 3480 (*nmr*)

$C_6H_{18}B_3N_3$ B-00451

2,4,6-Triethylborazine, 10CI, 9CI, 8CI

B,B,B-*Triethylborazole*

[7443-22-3]

As Borazine, B-00020 with

R = Et, R′ = H

M 164.658

Liq. with unpleasant odour. Sol. most org. solvs. d_4^{20}0.866. Fp −54° to −53°, −46.4°. Bp_7 66-7°. Viscosity 1.48 centistokes at 20°. Molar heat of vaporisation 9.88kcal/mol. Trouton const. 22.1 e.u. Dec. >100°; stable to water at r.t. V.p. 3.35mm at 26.6°.

Haworth, D.T. *et al*, *J. Am. Chem. Soc.*, 1960, **82**, 3860 (*ir, synth*)
Hawthorne, M.F. *et al*, *J. Am. Chem. Soc.*, 1961, **83**, 833 (*uv, ir, nmr, synth*)
Köster, R. *et al*, *Justus Liebigs Ann. Chem.*, 1961, **644**, 1 (*ir, synth*)
Powell, P. *et al*, *J. Chem. Soc.*, 1964, 280 (*synth, chomatog*)
Nöth, H. *et al*, *J. Organomet. Chem.*, 1968, **12**, 23 (*pmr*)
Wrackmeyer, B. *et al*, *Chem. Ber.*, 1976, **109**, 3480 (*nmr*)
Nöth, H. *et al*, *Chem. Ber.*, 1981, **114**, 1150 (*nmr*)

$C_6H_{18}B_3N_3O_3$ B-00452

Tris(dimethylamino)boroxin

Hexamethylboroxintriamine, 10CI, 9CI, 8CI. Dimethylaminoboroxole

[17197-55-6]

As Boroxin, B-00019 with

$R = NMe_2$

M 212.657

Volatile moisture-sensitive solid.

Massey, A.G. *et al*, *J. Inorg. Nucl. Chem.*, 1963, **25**, 175 (*synth, ir*)
Meller, A., *Monatsh. Chem.*, 1967, **98**, 2014 (*ir*)
Nöth, H. *et al*, *Chem. Ber.*, 1973, **106**, 1145 (*nmr*)
Biffar, W. *et al*, *Chem. Ber.*, 1980, **113**, 333 (*nmr*)
Nöth, H. *et al*, *Chem. Ber.*, 1981, **114**, 1150 (*cmr*)

$C_6H_{18}B_3N_3S_3$ B-00453

1,3,5-Trimethyl-2,4,6-tris(methylthio)borazine, 10CI

[61423-62-9]

As 2,4,6-Tribromoborazine, B-00015 with

X = SMe, Y = Me

M 260.838

No phys. data given, but presumably moisture-sensitive.

Wrackmeyer, B. *et al*, *Chem. Ber.*, 1976, **109**, 3480 (*nmr*)
Nöth, H. *et al*, *Chem. Ber.*, 1981, **114**, 1150 (*nmr, cmr*)

$C_6H_{19}BN_2$ B-00454

Borane tetramethylethylenediamine

Trihydro(N,N,N′,N′-*tetramethyl-1,2-ethanediamine-*N)*boron*, 10CI. N,N,N′,N′-*Tetramethylethylenediamine compd. with borane(1:1). TMED BH_3*

[5843-32-3]

$Me_2NCH_2CH_2NMe_2—BH_3$

M 130.040

White cryst. solid. Can be stored below m.p. at least 24 hrs. Above m.p. quickly dec. in absence of excess TMEDA.

Gatti, A.R. *et al*, *Inorg. Chem.*, 1966, **5**, 329 (*synth, nmr*)
Brown, H.C. *et al*, *Inorg. Chem.*, 1980, **19**, 455.

$C_6H_{19}BN_2$ B-00455

(Dimethylaminomethyl)borane trimethylamine

[(*Dimethylamino*)*methyl*](N,N-*dimethylmethanamine*)-*dihydroboron*, 10CI. *Boryltrimethylamine compd. with trimethylamine(1:1)*, 8CI

[14175-47-4]

$Me_2NCH_2—BH_2—NMe_3$

M 130.040

Colourless liq. of low volatility and strong basicity. Misc. H_2O. Fp −26° to −24°. Slowly dec. at r.t.

Miller, N.E., *J. Am. Chem. Soc.*, 1966, **88**, 4284.
Miller, N.E. *et al*, *Inorg. Chem.*, 1969, **8**, 275 (*synth, ir, pmr*)
Miller, N.E., *J. Organomet. Chem.*, 1977, **137**, 131 (*synth, pmr*)
Byrne, M.J. *et al*, *Inorg. Chem.*, 1981, **20**, 1328.

$C_6H_{20}BN_2^{\oplus}$ B-00456

Dihydrobis(trimethylamine)boron(1+)

Bis(N,N-*dimethylmethanamine*)*dihydroboron(1+)*, 10CI. *Bis(trimethylamine)boronium(1+)*

$(Me_3N)_2BH_2^{\oplus}$

M 131.047 (ion)

Chloride: [18517-67-4].
$C_6H_{20}BClN_2$ M 166.500
No phys. props. reported.

Fr. Pat., 1 394 178, (*1965*); *CA*, **63**, 15897a
Miller, N.E., *Inorg. Chem.*, 1977, **16**, 2664 (*synth*)
Miller, N.E., *J. Organomet. Chem.*, 1977, **137**, 131.

$C_6H_{20}BP_2^{\oplus}$ B-00457

Dihydrobis(trimethylphosphine)boron(1+), 10CI

Bis(trimethylphosphine)boronium(1+). Boratobis(trimethylphosphonium)(1+)

$(Me_3P)_2BH_2^{\oplus}$

M 164.982 (ion)

Bromide: [65293-66-5].
$C_6H_{20}BBrP_2$ M 244.886

Schmidbaur, H. *et al*, *Angew. Chem., Int. Ed. Engl.*, 1978, **17**, 126.
Müller, G. *et al*, *Chem. Ber.*, 1979, **112**, 3302.
Schmidbaur, H. *et al*, *Chem. Ber.*, 1979, **112**, 1448.
Schmidbaur, H. *et al*, *Monatsh. Chem.*, 1980, **111**, 1233.

$C_6H_{21}BN_3P$ B-00458

Borane tris(dimethylamino)phosphine

Hexamethylphosphorous triamide compd. with borane(1:1), 9CI, 8CI. *(Hexamethylphosphorous triamide)-trihydroboron*, 10CI. *Hexamethylphosphorous triamide borine*

[7319-05-3]

$(Me_2N)_3P—BH_3$

M 177.036

Cryst. with camphoraceous odour. Mp 32.5°. Bp 49-50°.

Reetz, T. *et al*, *J. Am. Chem. Soc.*, 1960, **82**, 5036 (*synth*)
Laurent, J.P. *et al*, *Bull. Soc. Chim. Fr.*, 1969, 26.
Laurant, J.P. *et al*, *J. Inorg. Nucl. Chem.*, 1969, **31**, 1353 (*nmr*)
Cowley, A.H. *et al*, *J. Am. Chem. Soc.*, 1971, **93**, 6815 (*pmr, nmr*)
Jouany, C. *et al*, *Bull. Soc. Chim. Fr.*, 1972, 880 (*pmr, nmr*)
Elegant, L. *et al*, *Can. J. Chem.*, 1978, **56**, 857 (*pmr, nmr*)

$C_6H_{22}AlB_9$ B-00459

Dodecahydro-7,8-dicarbaundecaborato(1−)-H^9,H^{10}-diethylaluminum

*7,8-μ-Diethylalumino-1,2-dicarba-*nido*-undecaborane-(13)*

[11098-58-1]

As Dodecahydro-7,8-dicarbaundecaborato(1−)-H^9,H^{10} dimethylaluminum, B-00291 with

R = Et

M 218.511

Colourless cryst. (hexane). Mp 34.5-35.5°.

▷Explodes with water

Young, D.A.T. *et al*, *J. Am. Chem. Soc.*, 1971, **93**, 5687 (*synth, ir, pmr, nmr*)

$C_6H_{22}B_2N_2$ B-00460

Tetramethylethylenediamine bis(borane)

Hexahydro[μ-(N,N,N′,N′-*tetramethyl-1,2-ethanediamine-N:N′*)]*diboron*, 10CI. N,N,N′,N′-*Tetramethyl-1,2-ethanediamine, compd. with borane(1:2)*, 9CI

[5843-33-4]

$H_3B—Me_2NCH_2CH_2NMe_2—BH_3$

M 143.873

White solid. Mp 181-3°.

Miller, N.E. *et al*, *Inorg. Chem.*, 1964, **3**, 866.
Miller, H.C. *et al*, *Inorg. Chem.*, 1964, **3**, 1456 (*synth*)
Gatti, A.R. *et al*, *Inorg. Chem.*, 1966, **5**, 329 (*synth, pmr, nmr*)
Nainan, K.C. *et al*, *Inorg. Chem.*, 1969, **8**, 2671 (*synth, pmr, nmr*)
Hu, M.G. *et al*, *Inorg. Chim. Acta*, 1974, **10**, 83.
Brown, H.C. *et al*, *Inorg. Chem.*, 1980, **19**, 455.

$C_6H_{28}B_6P_2$ B-00461

Hexaborane(10) bis(trimethylphosphine)

Trimethylphosphine compd. with hexaborane(10)(2:1)

[57034-29-4]

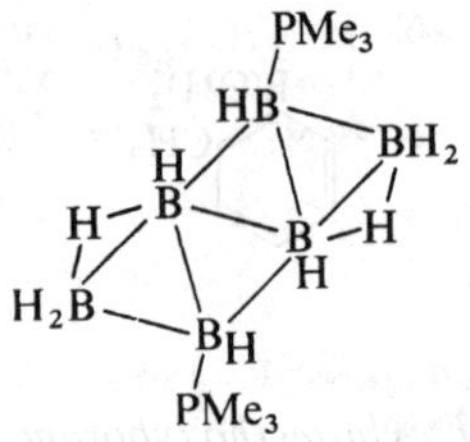

M 227.095

hypho-. Stable hypho-cage structure. Monoclinic cryst. Sol. CH_2Cl_2, THF. Mp 125° dec.

Mangion, M. *et al*, *J. Am. Chem. Soc.*, 1976, **98**, 449 (*synth, cryst struct, nmr*)

$C_7H_5BCl_2FeO_2$ B-00462

Dicarbonyl(η^5-2,4-cyclopentadien-1-yl)dichloroboraneiron

Dicarbonyl-π-cyclopentadienyl(dichloroboryl)iron, 8CI. *(Cyclopentadienyldicarbonyliron)boron dichloride. Dicarbonyl(η^5-cyclopentadienyl)irondichloroborane*

[12306-68-2]

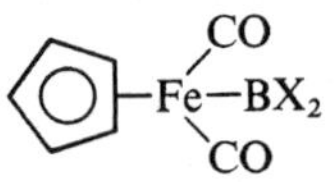

X = Cl

M 258.678

Brown.

Schmid, G. *et al*, *Chem. Ber.*, 1968, **101**, 1205 (*synth*)

$C_7H_7BCl_2$ B-00463

Dichloro(4-methylphenyl)borane, 10CI, 9CI

Dichloro-p-*tolylborane*, 8CI. p-*Tolyldichloroborane*

[4250-45-7]

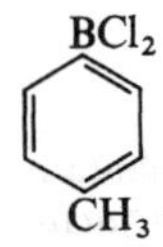

M 172.848

Moisture-sensitive liq. Sol. Et_2O. d_{30} 1.160. Bp_{747} 205°, $Bp_{0.2}$ 42°. n_D^{30} 1.5452.

McCusker, P. *et al*, *J. Am. Chem. Soc.*, 1957, **79**, 5185 (*synth, props*)
Butcher, F. *et al*, *Spectrochim. Acta*, 1963, **19**, 905 (*ir*)
Gerrard, W. *et al*, *J. Chem. Soc.*, 1963, 1582 (*synth*)
Smith, J. *et al*, *J. Chem. Soc. (A)*, 1971, 178 (*nmr*)
Hooz, J. *et al*, *Org. Prep. Proced. Int.*, 1972, **4**, 219 (*synth*)
Paetzold, P. *et al*, *Chem. Ber.*, 1980, **113**, 3724 (*use*)

$C_7H_7BO_3$ B-00464

(2-Formylphenyl)dihydroxyborane

(2-Formylphenyl)boronic acid, 10CI, 9CI

[40138-16-7]

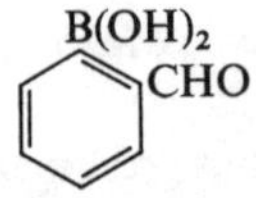

M 149.941

Fr. Pat., 95 127, (*1970*); *CA*, **78**, 72108z
Ger. Pat., 1 670 494, (*1976*); *CA*, **86**, 90009n
Ger. Pat., 2 533 918, (*1977*); *CA*, **87**, 23334x
Ger. Pat., 2 620 776, (*1977*); *CA*, **88**, 89835p
Ger. Pat., 2 620 777, (*1977*); *CA*, **88**, 89836q
U.S.P., 4 199 573, (*1980*); *CA*, **93**, 186544d

$C_7H_8BCl_2N$ B-00465

Dichloro(methylphenylamino)borane

1,1-Dichloro-N-*methyl*-N-*phenylboranamine*, 10CI, 9CI. *Dichloro*(N-*methylanilino*)*borane*, 8CI. *Dichloro*(N-*methylanilino*)*borine*

[1125-73-1]

$Cl_2BNMePh$

M 187.863

Moisture-sensitive liq. Bp_7 94-8°.

Niedenzu, K. *et al*, *J. Am. Chem. Soc.*, 1959, **81**, 5553 (*synth*)
Baechle, H.T. *et al*, *Spectrochim. Acta*, 1965, **21**, 579 (*ir, raman*)
Becher, H.J. *et al*, *Chem. Ber.*, 1965, **98**, 526 (*synth*)
Scott, K.N. *et al*, *Inorg. Chem.*, 1969, **8**, 1703 (*nmr*)
Lochmueller, C.H. *et al*, *J. Chromatogr.*, 1973, **85**, 132 (*chromatog*)
Baudler, M. *et al*, *Z. Anorg. Allg. Chem.*, 1981, **474**, 18 (*synth*)

$C_7H_8BF_3O$ B-00466

Trifluoroborane methoxybenzene

Trifluoro(methoxybenzene)boron, 10CI. *Trifluoroborane compd. with methoxybenzene(1:1)*, 9CI. *Anisole compd. with boron fluoride(BF_3)(1:1)*, 8CI. *Trifluoroborane anisole*

[456-31-5]

PhOMe—BF_3

M 175.945

Catalyst for oligomerisation of benzonitrile. Moisture-sensitive liq., dec. before dist. Mp −13°.

Meerwein, H. *et al*, *J. Prakt. Chem.*, 1932, **134**, 51.
Taillandier, E. *et al*, *C.R. Hebd. Seances Acad. Sci.*, 1966, **263**, 1265 (*ir*)
Romm, I.P. *et al*, *Zh. Fiz. Khim.*, 1968, **42**, 220.
Voloshchuk, A.M. *et al*, *Kernenergie*, 1974, **17**, 150 (*ir*)
Borovikov, Y.Y. *et al*, *Zh. Obshch. Khim.*, 1980, **50**, 2531.
Polevoi, A.S. *et al*, *Zh. Fiz. Khim.*, 1981, **55**, 1929.

$C_7H_9BF_3N$ B-00467

Trifluoroborane 4-methylaniline

Trifluoro(4-methylbenzenamine)boron, 10CI. *4-Methylbenzenamine compd. with trifluoroborane(1:1)*, 9CI. p-*Toluidine compd. with boron fluoride(BF_3)*, 8CI. *Boron trifluoride* p-*toluidine*

[1978-51-4]

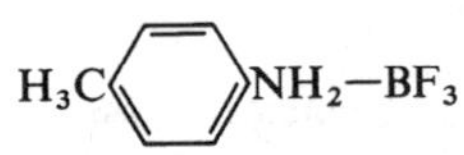

M 174.960

Crosslinking catalyst for epoxy resins. Yellow solid. Mp 250°.

Martin, D.R., *Chem. Rev.*, 1948, **42**, 581 (*rev*)
Smith, C.H., *Mod. Plast.*, 1969, **46**, 118.
Zubkova, Z.A. *et al*, *Plast. Massy*, 1969, 24; *CA*, **71**, 4054j (*use*)
Zubkova, Z.A. *et al*, *Vysokomol. Soedin., Ser. A*, 1971, **13**, 2559 (*use*)

$C_7H_9BN_2$ B-00468

Cyanoborane aniline

(*Benzenamine*)(*cyano*-C)*dihydroboron*, *10CI*

[66632-44-8]

$PhNH_2—BH_2CN$

M 131.972

Cryst. (EtOH aq.). Mp 152-4°.

McPhail, A.T. *et al*, *J. Chem. Res. (S)*, 1978, 205.
Wisian-Neilson, P. *et al*, *Inorg. Chem.*, 1978, **17**, 2327 (*synth, pmr, ir*)

$C_7H_9BN_2$ B-00469

Cyanoborane 2-methylpyridine

(*Cyano*-C)*dihydro*(*2-methylpyridine*)*boron*, *10CI*. *Boranecarbonitrile, compd. with 2-methylpyridine*(*1:1*)

[67374-57-6]

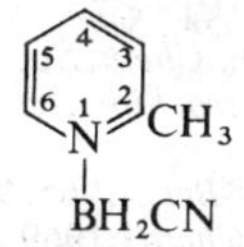

M 131.972

White solid. Mp 32-4°.

Martin, D.R. *et al*, *J. Inorg. Nucl. Chem.*, 1978, **40**, 9 (*synth, nmr, ir*)

$C_7H_9BN_2$ B-00470

Cyanoborane 3-methylpyridine

(*Cyano*-C)*dihydro*(*3-methylpyridine*)*boron*, *10CI*. *Boranecarbonitrile, compd. with 3-methylpyridine*(*1:1*)

[67374-56-5]

M 131.972

Liq.

Martin, D.R. *et al*, *J. Inorg. Nucl. Chem.*, 1978, **40**, 9 (*synth, ir, nmr*)

$C_7H_9BN_2$ B-00471

Cyanoborane 4-methylpyridine

(*Cyano*-C)*dihydro*(*4-methylpyridine*)*boron*, *10CI*. *Boranecarbonitrile, compd. with 4-methylpyridine*(*1:1*), *9CI*. *Boranecarbonitrile, compd. with 4-picoline*(*1:1*), *8CI*

[30353-62-9]

M 131.972

White cryst. by subl. Mp 60-1°.

Kelly, H.C. *et al*, *J. Chem. Soc., Chem. Commun.*, 1970, 1619 (*synth*)
Weidig, C. *et al*, *Inorg. Chem.*, 1974, **13**, 1763 (*synth, pmr, ir*)
Martin, D.R. *et al*, *J. Inorg. Nucl. Chem.*, 1978, **40**, 9 (*synth, nmr, ir*)

$C_7H_9BO_2$ B-00472

Dihydroxy(2-methylphenyl)borane

(*2-Methylphenyl*)*boronic acid*, *10CI*. o-*Tolueneboronic acid*, *8CI*

[16419-60-6]

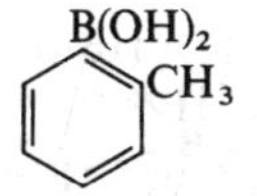

M 135.957

Sol. hexane, H_2O.

O,O-*Di-Me*: o-*Tolyldimethoxyborane*.
$C_9H_{13}BO_2$ M 164.011
Obt. by methanolysis. Moisture-sensitive cryst. Sol. C_6H_6. d_4^{25} 0.980. Bp_9 74-5°. n_D^{25} 1.4870.

Skowronska-Serafinowa, B. *et al*, *Rocz. Chem.*, 1961, **937**, 35 (*ir*)
Washburn, R.M. *et al*, *Adv. Chem. Ser.*, 1961, **32**, 208.
Beachell, H.C. *et al*, *Inorg. Chem.*, 1964, **3**, 1028 (*nmr, pmr*)
Dang, Q.Q., *C.R. Hebd. Seances Acad. Sci.*, 1969, **24**, 269 (*uv*)
Ramsey, B.G., *J. Phys. Chem.*, 1970, **74**, 2464 (*props*)
Bullen, N.P. *et al*, *J. Organomet. Chem.*, 1980, **195**, 147 (*synth*)

$C_7H_9BO_2$ B-00473

Dihydroxy(3-methylphenyl)borane

(*3-Methylphenyl*)*boronic acid*, *10CI*, *9CI*. m-*Tolueneboronic acid*, *8CI*

[17933-03-8]

M 135.957

Cryst. (Et_2O). Sol. Et_2O.

▷Toxic

O,O-*Di-Me*: [40881-44-5]. m-*Tolyldimethoxyborane*.
$C_9H_{13}BO_2$ M 164.011
Methanolysis product. Moisture-sensitive cryst. Sol. C_6H_6. d_{25} 0.985. Bp_{10} 85-6°. n_D^{25} 1.4926.

Kuivila, H.G. *et al*, *Can. J. Chem.*, 1963, **41**, 3081 (*props, use*)
Beachell, H.C. *et al*, *Inorg. Chem.*, 1964, **3**, 1028 (*nmr, pmr*)
Ramsey, B.G., *J. Phys. Chem.*, 1970, **74**, 2464 (*props*)
Kajimoto, O. *et al*, *J. Phys. Chem.*, 1977, **81**, 1712 (*props*)
Thorpe, F.G. *et al*, *J. Organomet. Chem.*, 1977, **128**, 305 (*synth*)
Bullen, N.P. *et al*, *J. Organomet. Chem.*, 1980, **195**, 147 (*synth*)

$C_7H_9BO_2$ B-00474

Dihydroxy(4-methylphenyl)borane

(*4-Methylphenyl*)*boronic acid*, *10CI*, *9CI*. p-*Tolueneboronic acid*, *8CI*

[5720-05-8]

M 135.957

Corrosion inhibitor; protecting group for synthesis. Cryst. (H_2O or Et_2O). Sol. Et_2O, EtOH, C_6H_6, $CDCl_3$, hexane. Mp 259-60°.

▷Toxic; affects the central nervous system. XS7400000.

Bean, F.R. *et al*, *J. Am. Chem. Soc.*, 1932, **54**, 4415 (*synth*)
Beachell, H.C. *et al*, *Inorg. Chem.*, 1964, **3**, 1028 (*nmr, pmr*)
Ramsey, B.G., *J. Phys. Chem.*, 1970, **74**, 2464.
Pham, H.C. *et al*, *Agressologie*, 1974, **15**, 131; *CA*, **81**, 86179x (*use*)
Kemp, D.S. *et al*, *Tetrahedron Lett.*, 1975, 4629 (*use*)
Bullen, N.P. *et al*, *J. Organomet. Chem.*, 1980, **195**, 147 (*synth*)

$C_7H_9BO_2S$ B-00475

Dihydroxy(phenylthiomethyl)borane

[(*Phenylthio*)*methyl*]*boronic acid*, *10CI*

[67975-88-6]

$PhSCH_2B(OH)_2$

M 168.017

Synthetic intermed. Cryst. (Et_2O/pet. ether). Mp 109-10°.

Matteson, D.S. *et al*, *J. Am. Chem. Soc.*, 1978, **100**, 1325 (*synth, use*)
Mendoza, A. *et al*, *J. Organomet. Chem.*, 1978, **156**, 149 (*use*)
Matteson, D.S. *et al*, *J. Organomet. Chem.*, 1979, **170**, 259 (*use, props*)
Ray, R., *CA*, 1981, **95**, 42531r.

$C_7H_9BO_3$ B-00476

Dihydroxy(2-methoxyphenyl)borane

(2-Methoxyphenyl)boronic acid, 10CI. o-*Methoxybenzeneboronic acid*. o-*Anisylboronic acid*

[5720-06-9]

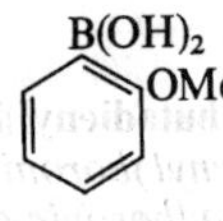

M 151.956

Solid.

▷Affects the central nervous system and causes vasodilation

Dandegaonker, S.H. *et al*, *J. Chem. Soc.*, 1959, 2076 (*synth*)
Kuivila, H.G. *et al*, *Can. J. Chem.*, 1963, **41**, 3081.
Beachell, H.C. *et al*, *Inorg. Chem.*, 1964, **3**, 1028 (*nmr, pmr*)
Eggers, C.A. *et al*, *Inorg. Chem.*, 1967, **6**, 160 (*synth*)
Caujolle, F. *et al*, *Agressologie*, 1969, **10**, 155 (*use*)
Bullen, N.P. *et al*, *J. Organomet. Chem.*, 1980, **195**, 147 (*synth*)

$C_7H_9BO_3$ B-00477

Dihydroxy(3-methoxyphenyl)borane

(3-Methoxyphenyl)boronic acid, 10CI. m-*Methoxybenzeneboronic acid*, 8CI. m-*Anisylboronic acid*

[10365-98-7]

M 151.956

Solid. Sol. hexane, H_2O.

Dandegaonker, S.H. *et al*, *J. Chem. Soc.*, 1959, 2076 (*synth*)
Kuivila, H.G. *et al*, *Can. J. Chem.*, 1963, **41**, 3081 (*props, use*)
Beachell, H.C. *et al*, *Inorg. Chem.*, 1964, **3**, 1028 (*nmr, pmr*)
Ramsey, B.G., *J. Phys. Chem.*, 1970, **74**, 2464 (*props*)
Kajimoto, O. *et al*, *J. Phys. Chem.*, 1977, **81**, 1712.

$C_7H_9BO_3$ B-00478

Dihydroxy(4-methoxyphenyl)borane

(4-Methoxyphenyl)boronic acid, 10CI. p-*Methoxybenzeneboronic acid*. p-*Anisylboronic acid*

[5720-07-0]

M 151.956

Used as a bactericide and chemosterilant for insects. Cryst. (H_2O). Sol. hexane, MeCN, H_2O. Mp 208.5-209.5°, 229-30°.

▷CY8975000.

Dandegaonker, S.H. *et al*, *J. Chem. Soc.*, 1959, 2076 (*synth*)
Beachell, H.C. *et al*, *Inorg. Chem.*, 1964, **3**, 1028 (*nmr, pmr*)
Ramsey, B.G. *et al*, *Theor. Chim. Acta*, 1970, **17**, 188 (*props*)
Settepani, J.A. *et al*, *J. Med. Chem.*, 1970, **3**, 128 (*use*)
Thorpe, F.G. *et al*, *J. Organomet. Chem.*, 1977, **128**, 305 (*synth*)
Floris, B. *et al*, *J. Organomet. Chem.*, 1978, **150**, 101.

$C_7H_{10}BBr_2N$ B-00479

Dibromomethylborane 4-methylpyridine

Dibromomethyl(4-methylpyridine)boron, 10CI. *Dibromo(methyl)borane γ-picoline*

[69513-17-3]

As Dibromomethylborane pyridine, B-00382 with

$R = CH_3$

M 278.781

Solid. Mp 124-6° dec.

Martin, D.R. *et al*, *J. Inorg. Nucl. Chem.*, 1978, **40**, 1289 (*synth, nmr, ir*)

$C_7H_{11}B$ B-00481

1,3-Cyclopentadien-1-yldimethylborane, 10CI

1-Dimethylboryl-1,3-cyclopentadiene

[68250-27-1]

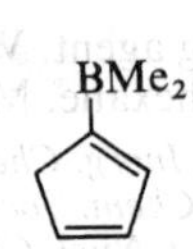

M 105.974

Fluxional behavior seen in nmr below −15°. Characterized spectroscopically. Air-sensitive.

Johnson, H.D. *et al*, *J. Chem. Soc., Chem. Commun.*, 1978, 242 (*synth, nmr*)

$C_7H_{11}B$ B-00480

1,4-Cyclopentadien-1-yldimethylborane, 9CI

2-Dimethylboryl-1,3-cyclopentadiene

[68250-29-3]

M 105.974

Fluxional behaviour seen in nmr below −15°. Characterised spectroscopically. Air-sensitive.

Johnson, H.D. *et al*, *J. Chem. Soc., Chem. Commun.*, 1978, 242 (*synth, nmr*)

$C_7H_{11}B_4Fe$ B-00482

(η^5-2,4-Cyclopentadien-1-yl)[(2,3,4,5,6-η)-hexahydro-2,3-dicarbahexaborato(2−)]iron, 9CI

[38930-35-7]

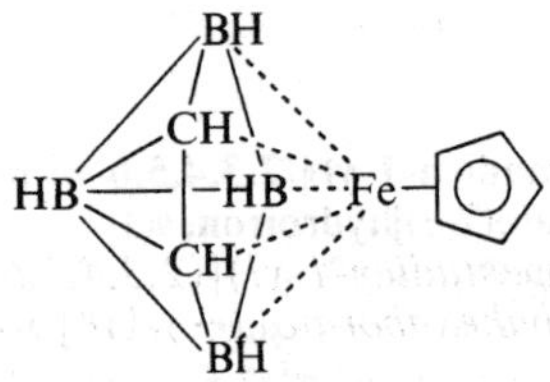

M 194.251

Brown cryst. Paramagnetic.

Sneddon, L.G. *et al*, *J. Am. Chem. Soc.*, 1972, **94**, 7161; 1973, **95**, 6623 (*synth, ms, pmr, nmr*)

$C_7H_{12}BF_6N_3$ B-00483

Hexafluoroisopropylideniminobis(dimethylamino)borane

N,N,N′,N′-*Tetramethyl*-N″-*[2,2,2-trifluoro-1-(trifluoromethyl)ethylidene]boranetriamine*, 10CI

[67395-53-3]

$(Me_2N)_2BN{=}C(CF_3)_2$

M 262.992

Yellow liq., sensitive to moisture. Sol. hexane. Bp_{99} 70-3°.

Emerick, D.P. *et al*, *J. Organomet. Chem.*, 1978, **153**, 9 (*synth, ms, ir, nmr, raman, cmr*)
Emerick, D.P. *et al*, *Z. Anorg. Allg. Chem.*, 1980, **468**, 44 (*nmr, ir*)

$C_7H_{12}BN$ B-00484

Borane 2,6-dimethylpyridine

(2,6-Dimethylpyridine)trihydroboron, 10CI. *2,6-Dimethylpyridine, compd. with borane(1:1)*, 9CI. *2,6-Lutidine, compd. with borane(1:1)*, 8CI

[3999-42-6]

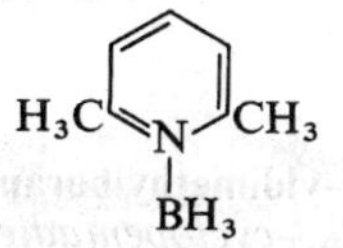

M 120.989

Stereoselective reducing agent. White cryst. solid. V. sol. THF, MeOH, insol. hexane. Mp 112-4° (106-7°).

Ryschkewitsch, G.E. *et al*, *Inorg. Chem.*, 1965, **4**, 575 (*props*)
Brown, H.C. *et al*, *J. Am. Chem. Soc.*, 1966, **88**, 2520.
Mooney, E.F. *et al*, *J. Inorg. Nucl. Chem.*, 1968, **30**, 1439 (*synth, nmr*)
Nöth, H. *et al*, *Chem. Ber.*, 1974, **107**, 3070 (*nmr*)
Andrews, G.C. *et al*, *Tetrahedron Lett.*, 1980, **21**, 693, 697 (*use*)

$C_7H_{12}B_4Fe$ B-00485

(η^5-2,4-Cyclopentadien-1-yl)[(2,3,4,5,6-η)-heptahydro-2,5-dicarbahexaborato(1−)-H^μ]iron, 9CI

π-Cyclopentadienyliron-2,4-$C_2B_4H_6$ hydride

[51177-77-6]

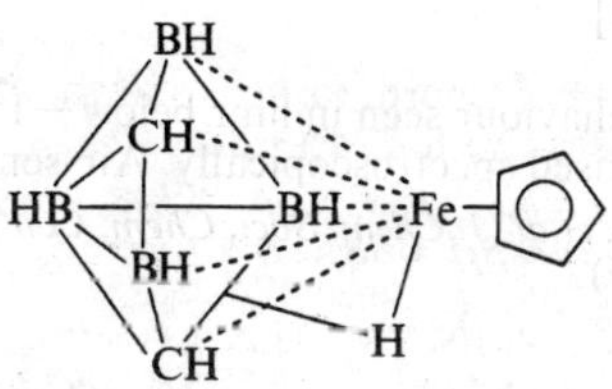

M 195.259

Red-orange.

Sneddon, L.G. *et al*, *J. Am. Chem. Soc.*, 1973, **95**, 6623 (*synth, pmr, ms, nmr, ir*)

$C_7H_{12}B_4Fe$ B-00486

(η^5-2,4-Cyclopentadien-1-yl)[(2,3,4,5,6-η)-hexahydro-2,3-dicarbahexaborato(2−)]hydroiron, 9CI

(η^5-2,4-Cyclopentadien-1-yl)[(2,3,4,5,6-η)-heptahydro-4,5-dicarbahexaborato(1−)-H^μ]iron, 9CI

[38980-34-6]

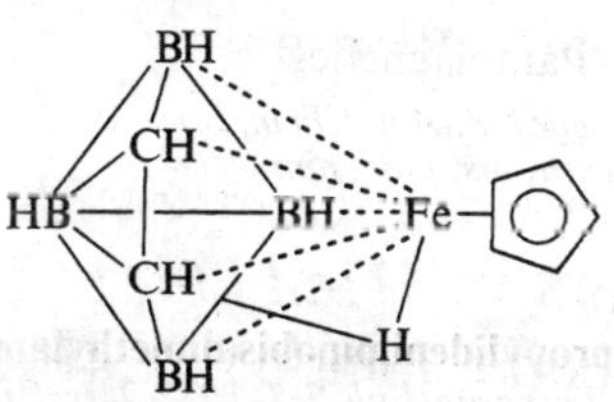

M 195.259

Diamagnetic. Orange solid. Sublimes.

Sneddon, L.G. *et al*, *J. Am. Chem. Soc.*, 1972, **94**, 7161; 1973, **95**, 6623 (*synth, ms, pmr, nmr*)

$C_7H_{13}B$ B-00487

Bicyclo[2.2.1]hept-2-ylborane, 10CI, 9CI

2-Norbornylborane

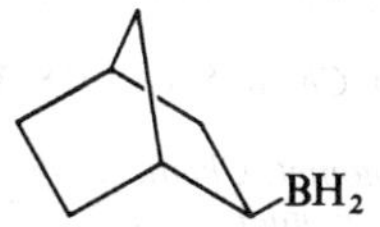

M 107.990

Uncomplexed material would exist as H-bridged dimer.

Exo-form [56118-55-9]

Intermed. in hydroboration of norbornene with BH_3. Air- and moisture-sensitive. Sol. THF.

Taniguchi, H., *Bull. Chem. Soc. Jpn.*, 1979, **52**, 2946.
Taniguchi, H., *Nippon Kagaku Kaishi*, 1979, 906; *CA*, **91**, 210504x.
Brown, H.C. *et al*, *J. Org. Chem.*, 1981, **46**, 4599.

$C_7H_{13}BO_2$ B-00488

Dimethoxy(3-methyl-1,2-butadienyl)borane

(3-Methyl-1,2-butadienyl)boronic acid dimethyl ester, 10CI. *3,3-Dimethylallenylboronic acid dimethyl ester*. *(3,3-Dimethylallenyl)dimethoxyborane*

[61357-30-0]

$$(MeO)_2BCH{=}C{=}C(CH_3)_2$$

M 139.989

Synthetic intermed. Air- and moisture-sensitive liq. Sol. non-hydroxylic solvs. Bp_{12} 51.5-52°. n_D^{20} 1.4502.

Mikhailov, B.M. *et al*, *Izv. Akad. Nauk SSSR, Ser. Khim.*, 1976, 2062; *Bull. Acad. Sci. USSR, Div. Chem. Sci.*, **25**, 1932 (*synth, ir, pmr*)
Mikhailov, B.M. *et al*, *Izv. Akad. Nauk SSSR, Ser. Khim.*, 1976, 2402; *Bull. Acad. Sci. USSR, Div. Chem. Sci.*, **25**, 2246 (*props*)

$C_7H_{13}B_6Fe$ B-00489

(η^5-2,4-Cyclopentadien-1-yl)[(1,3,6,8-η)-octahydro-1,8-dicarbaoctaborato(2−)]iron, 9CI

4-(η-Cyclopentadienyl)-4,1,8-$FeC_2B_6H_8$

[42808-85-5]

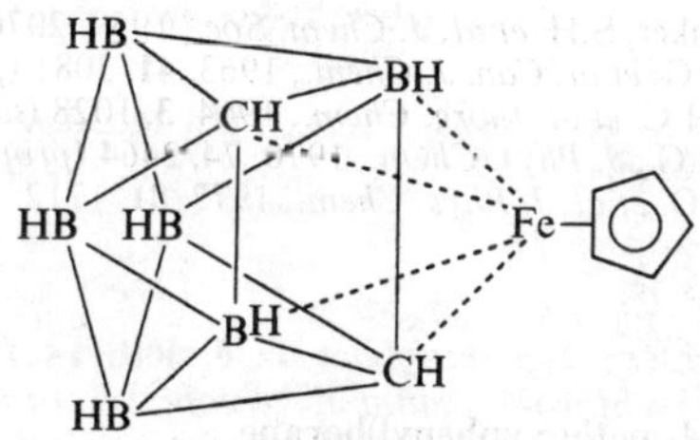

M 217.887

Red cryst. (hexane). Mp 132-3°. Paramagnetic. Reduced to a diamagnetic purple Fe(II) deriv.

Evans, W.J. *et al*, *J. Am. Chem. Soc.*, 1973, **95**, 4565 (*synth, ms, uv, ir, polarog, pmr, nmr*)

$C_7H_{13}B_6Fe$ B-00490

(η^5-2,4-Cyclopentadien-1-yl)[(3,5,6,7,8-η)-octahydro-5,6-dicarbaoctaborato(2–)]iron, 9CI

1-(η^5-Cyclopentadienyl)-1,4,6-$FeC_2B_6H_8$. 4,6-π-Cyclopentadienyliron-$C_2B_6H_8$

[42968-39-8]

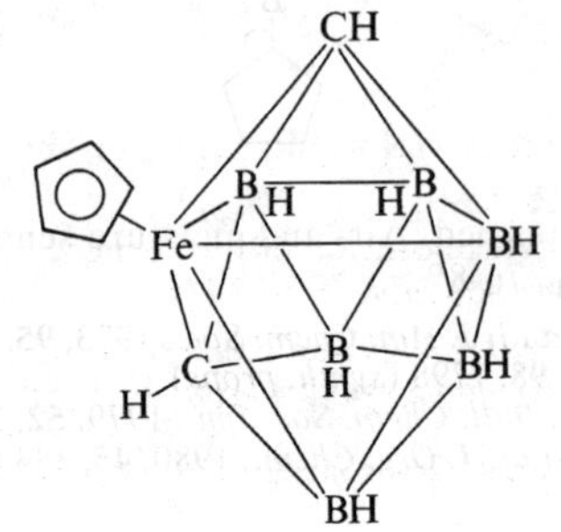

M 217.887

Brown cryst. (hexane). Paramagnetic. Reduced to a diamagnetic Fe(II) species.

Evans, W.J. *et al*, *J. Am. Chem. Soc.*, 1973, **95**, 4565 (*synth, ms, uv, ir, polarog, pmr, nmr*)

$C_7H_{13}B_7FeO_3$ B-00491

Tricarbonyl[η^5-heptahydro-*C,C'*-dimethyldicarbanonaborato(2–)]iron, 10CI

2,1,4-(Tricarbonyliron)-1,4-dimethyl-$C_2B_7H_7$

[64103-53-3]

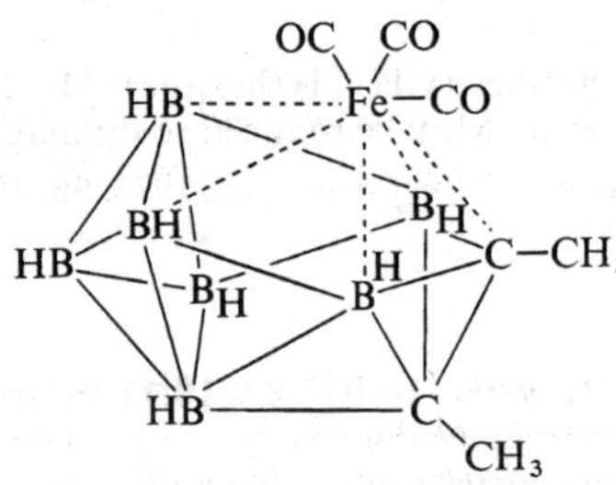

M 276.695

Air-stable, light-sensitive bright-yellow cryst. (hexane). $Bp_{0.00001}$ 25° subl.

Maxwell, W.M. *et al*, *Inorg. Chem.*, 1977, **16**, 3094 (*synth, ir, ms, nmr, pmr*)

$C_7H_{14}BBr_2ClS$ B-00492

Dibromo(5-chloro-1-pentenyl)borane dimethyl sulfide

Dibromo(5-chloro-1-pentenyl)[thiobis[methane]]boron, 10CI. *5-Chloro-1-penten-1-yldibromoborane dimethyl sulfide*

$ClCH_2CH_2CH_2CH{=}CHBBr_2{-}SMe_2$

M 336.319

(*E*)-*form* [72228-64-9]

Prepd. *in situ* by hydroboration of 1-pentane with $Br_2BH{-}SMe_2$. Synthetic intermediate. Air- and moisture-sensitive liq. $Bp_{0.08}$ 60-2° part. dec.

Brown, H.C. *et al*, *J. Org. Chem.*, 1980, **45**, 389 (*synth, nmr*)

$C_7H_{14}BNO_3$ B-00493

[3-[Bis(2-hydroxyethyl)amino]-1-propanolato(3–)-*N,O,O',O''*]boron, 10CI, 9CI

[3-[Bis(2-hydroxyethyl)amino]-1-propanolato(3–)]-boron, 8CI

[27952-41-6]

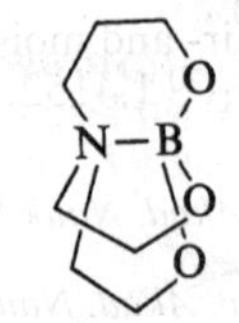

M 171.003

Ger. Pat., 1 908 844, (*1969*); *CA*, **72**, P4052c

Lukevics, E. *et al*, *CA*, 1975, **83**, 193247f.

U.S.S.R. Pat., 504 776, (*1976*); *CA*, **85**, P33020g

Mazeika, I. *et al*, *Zh. Obshch. Khim.*, 1978, **48**, 2728 (*ms*)

Shatz, V.D. *et al*, *J. Chromatogr.*, 1980, **200**, 105.

$C_7H_{14}BNO_4$ B-00494

Diethyl(methylnitroacetato)borane

*Diethyl(methyl nitroacetato-*O[1'],O[2]*)boron,* 10CI. *[Diethylboryl]-aci-nitroacetic acid methyl ester,* 9CI

[62945-91-9]

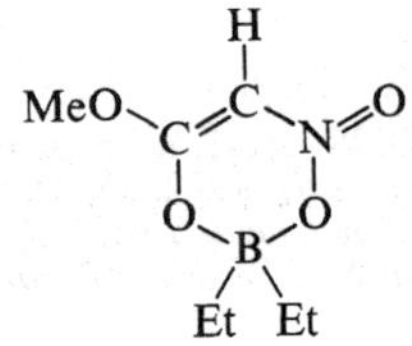

M 187.002

Exists in intramol. coord. form as shown. Air- and moisture-sensitive liq. Mp –30°.

Shitov, O.P. *et al*, *Izv. Akad. Nauk SSSR, Ser. Khim.*, 1974, 2782; *Bull. Acad. Sci. USSR, Div. Chem. Sci.*, 2684 (*synth, nmr*)

Ioffe, S.L. *et al*, *CA*, 1977, **87**, 23357g (*synth*)

Ioffe, S.L. *et al*, *Izv. Akad. Nauk SSSR, Ser. Khim.*, 1978, 1172; *Bull. Acad. Sci. USSR, Div. Chem. Sci.*, 1019 (*nmr*)

$C_7H_{15}B$ B-00496

Diethyl-2-propenylborane, 10CI, 9CI
Allyldiethylborane
[18759-99-4]

$H_2C{=}CHCH_2BEt_2$

M 110.006
Synthetic intermed. Air- and moisture-sensitive liq. Sol. THF. Bp_{38} 40-2°. n_D^{20} 1.4152. ^{11}B nmr δ −83.8 ppm.
▷Probably pyrophoric

Bubnov, Y.N. *et al*, *Izv. Akad. Nauk SSSR, Ser. Khim.*, 1976, 2842 (*use*)
Mikhailov, B.M. *et al*, *Izv. Akad. Nauk SSSR, Ser. Khim.*, 1978, 1586, 1594, 1892 (*synth, use*)
Mikahilov, B.M. *et al*, *J. Organomet. Chem.*, 1978, **154**, 113, 131 (*synth, use, props*)

$C_7H_{15}B$ B-00497

3,5-Dimethylborinane, 9CI, 8CI
[39546-62-8]

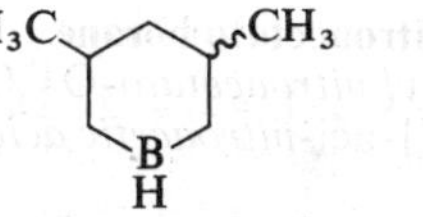

M 110.006
Used in hydroboration. Bp_1 79-82°. Dimeric, mixt. of *cis-* and *trans*-isomers.

Zweifel, G. *et al*, *Synthesis*, 1972, 557 (*use*)
Negishi, E. *et al*, *J. Am. Chem. Soc.*, 1973, **95**, 6757 (*synth*)
Brown, H.C., *Organic Syntheses via Boranes*, 1975, Wiley, N.Y., 34 (*synth*)

$C_7H_{15}BBr_2S$ B-00498

Dibromocyclopentylborane dimethyl sulfide
Dibromocyclopentyl[thiobis[methane]]boron, 10CI. *Cyclopentyldibromoborane dimethyl sulfide*
[64770-10-1]

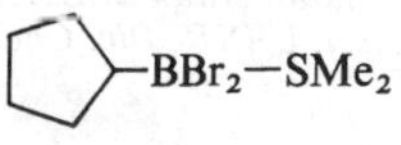

M 301.874
Solid at 25°, but may contain uncomplexed material; air- and moisture-sensitive. $Bp_{2.1}$ 140-4°.

Brown, H.C. *et al*, *J. Am. Chem. Soc.*, 1977, **99**, 7097 (*synth*)
Brown, H.C. *et al*, *J. Org. Chem.*, 1980, **45**, 384.

$C_7H_{15}BCl_2N_2$ B-00499

Dichlorocyanoborane triethylamine
Dichloro(cyano-C)(N,N-diethylethanamine)boron, 10CI. *Dichloroboranecarbonitrile N,N-diethylethanamine. Triethylamine dichlorocyanoborane*
[67374-62-3]

$Et_3N{-}BCl_2CN$

M 208.925
Air- and moisture-sensitive solid. Mp 59-63°.

Martin, D.R. *et al*, *J. Inorg. Nucl. Chem.*, 1978, **40**, 9 (*synth, ir, nmr*)

$C_7H_{15}BO_2$ B-00500

Cyclopentyldimethoxyborane
Cyclopentylboronic acid dimethyl ester, 10CI, 9CI. *Dimethyl cyclopentaneboronate. Dimethyl cyclopentylboronate*
[41156-60-9]

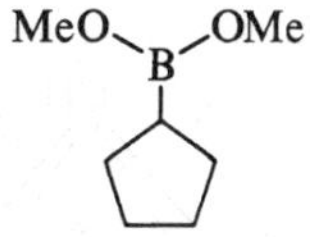

M 142.004
Synthetic intermed. Air- and moisture-sensitive liq. Sol. THF. Bp_{40} 76-8°.

Brown, H.C. *et al*, *J. Am. Chem. Soc.*, 1973, **95**, 2396; 1975, **97**, 2791; 1976, **98**, 1798 (*synth, props*)
Taniguchi, H., *Bull. Chem. Soc. Jpn.*, 1979, **52**, 2946 (*nmr*)
Brown, H.C. *et al*, *J. Org. Chem.*, 1980, **45**, 384 (*synth, props*)

$C_7H_{15}B_8Fe$ B-00501

(η^5-2,4-Cyclopentadien-1-yl)[(5,6,7,8,9,10-η)-decahydro-6,9-dicarbadecaborato(2−)]iron, 10CI
1,2,3-π-Cyclopentadienyliron-$C_2B_8H_{10}$
[42769-15-3]

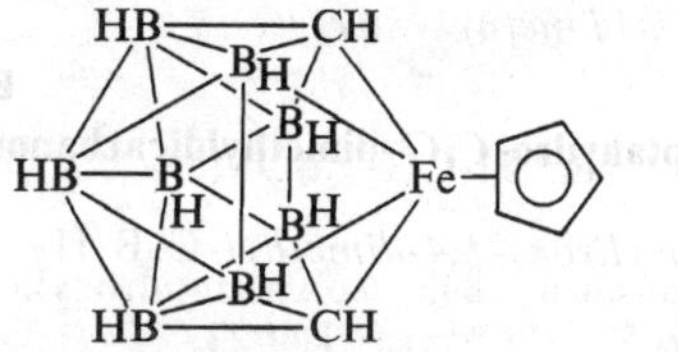

Proposed structure

M 241.523
Red-brown needles (CH_2Cl_2/hexane). Mp 275-7°. Paramagnetic. May be reduced to diamagnetic species.

Evans, W.J. *et al*, *J. Am. Chem. Soc.*, 1973, **95**, 4565 (*synth, uv, ir, ms, polarog*)

$C_7H_{16}B_9Fe^{\ominus}$ B-00502

(η^5-2,4-Cyclopentadien-1-yl)[(7,8,9,10,11-η)-undecahydro-7,8-dicarbaundecaborato(2−)]ferrate(1−), 9CI
[Undecahydrodicarbaundecaborato(2−)]cyclopentadienylferrate(2−)
[37185-23-2]

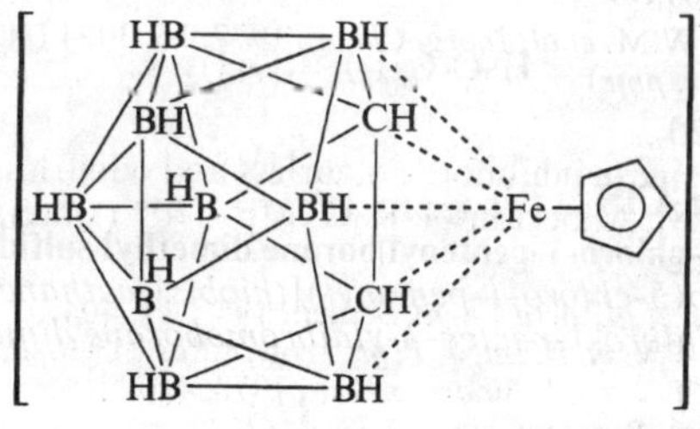

M 253.340 (ion)
Diamagnetic.

Tetramethylammonium salt:
$C_{11}H_{28}B_9FeN$ M 327.486
Orange plates (EtOH). Mp 158°.

Hawthorne, M.F. *et al*, *J. Am. Chem. Soc.*, 1968, **90**, 879 (*synth, pmr, uv, ir*)

$C_7H_{16}B_9Fe$ B-00503

(η^5-2,4-Cyclopentadien-1-yl)[(7,8,9,10,11-η)-undecahydro-7,9-dicarbaundecaborato(2−)]iron
π-Cyclopentadienyl-(π-7,9-$B_9C_2H_{11}$)iron

[40958-61-0]
M 253.340
Green. Paramagnetic. Can be reduced to diamagnetic Fe(II) species.

Dustin, D.F. *et al*, *J. Am. Chem. Soc.*, 1973, **95**, 1109 (*synth, pmr, ir, nmr*)
Wiersema, R.J. *et al*, *J. Am. Chem. Soc.*, 1974, **96**, 761 (*esr*)

$C_7H_{17}BN_2$ B-00504

Cyanoborane triethylamine
(*Cyano*-C)(N,N-*diethylethanamine*)*dihydroboron*, 10CI. *Boranecarbonitrile, compd. with* N,N-*diethylethanamine*(*1:1*)
[67374-58-7]

$Et_3N—BH_2CN$

M 140.035

Martin, D.R. *et al*, *J. Inorg. Nucl. Chem.*, 1978, **40**, 9 (*synth, nmr, ir*)

$C_7H_{17}BO$ B-00505

Methoxydipropylborane
Dipropylborinic acid methyl ester, 10CI, 8CI
[2938-89-8]

$(H_3CCH_2CH_2)_2BOMe$

M 128.021
Air- and moisture-sensitive liq. Sol. nonhydroxylic solvs. d_{20} 0.773. Bp_{44} 55.5-56°. n_D^{20} 1.4023.

Mikhailov, B.M. *et al*, *Izv. Akad. Nauk SSSR, Ser. Khim.*, 1959, 172 (*synth, props*)
Mikhailov, B.M. *et al*, *Zh. Obshch. Khim.*, 1959, **29**, 1648 (*synth, props*)
Mikhailov, B.M. *et al*, *Izv. Akad. Nauk SSSR, Ser. Khim.*, 1960, 1872; 1962, 628, 1756.
Mikhailov, B.M. *et al*, *J. Organomet. Chem.*, 1978, **154**, 113 (*synth, props*)

$C_7H_{17}BO_2$ B-00506

Heptyldihydroxyborane
Heptylboronic acid, 10CI, 9CI. *1-Heptaneboronic acid*, 8CI
[28741-07-3]

$H_3C(CH_2)_6B(OH)_2$

M 144.020
Alk. proteinase inhibitor. Colourless and odourless solid (toluene). Almost insol. H_2O. Mp ~70° (sinters), 81-3° dec.

Torssell, K. *et al*, *Acta Chem. Scand.*, 1957, **11**, 404.
Rotanova, T.V. *et al*, *Int. J. Pept. Protein Res.*, 1976, **8**, 225.
Rotanova, T.V. *et al*, *Bioorg. Khim.*, 1978, **4**, 694.
Jarv, J. *et al*, *Bioorg. Khim.*, 1978, **4**, 1364.
Gmoshinskii, I.V. *et al*, *Bioorg. Khim.*, 1982, **8**, 822.

$C_7H_{17}BS$ B-00507

Dipropyl(methylthio)borane
Dipropylthioborinic acid methyl ester, 10CI. (*Methylthio*)-*dipropylborane*
[3765-14-8]

$(H_3CCH_2CH_2)_2BSMe$

M 144.081
Air- and moisture-sensitive liq. Sol. nonhydroxylic solvs. Bp_7 44-9°. n_D^{20} 1.4659.

Mikhailov, B.M. *et al*, *Izv. Akad. Nauk SSSR, Ser. Khim.*, 1965, 1487.
Costes, J.P. *et al*, *J. Organomet. Chem.*, 1979, **175**, 257 (*props*)

$C_7H_{17}B_{10}Fe$ B-00508

(η^5-2,4-Cyclopentadien-1-yl)[η^6-dodecahydrodicarbadodecaborato(2−)]iron, 9CI
(*η^5-2,4-Cyclopentadien-1-yl*)[(*7,8,9,10,11,12-η*)-*dodecahydro-7,9-dicarbadodecaborato*(*2−*)]*iron*
[41077-34-3]

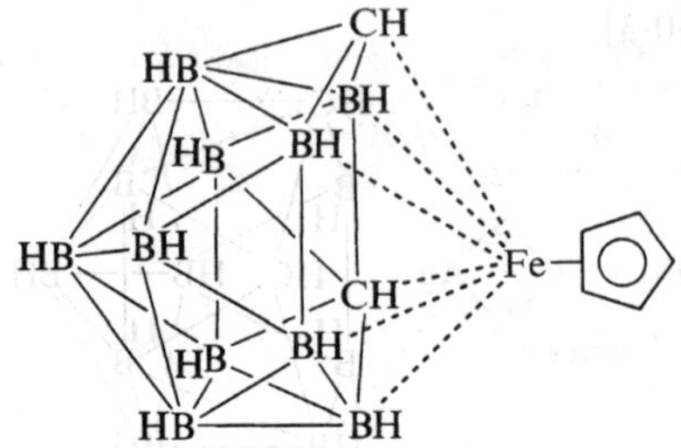

M 265.158
Paramagnetic. Violet. μ_{eff} = 2.12BM. Easily reduced to an orange diamagnetic anionic Fe(II) species.

Dustin, D.F. *et al*, *J. Am. Chem. Soc.*, 1973, **95**, 1109 (*synth, polarog, pmr, nmr, uv, ir*)

$C_7H_{18}B^{\ominus}$ B-00509

Butyltrimethylborate(1−), 10CI

$H_3C(CH_2)_3BMe_3^{\ominus}$

M 113.029 (ion)
Li salt: [65859-85-0].
$C_7H_{18}BLi$ M 119.970
Used in battery electrolytes.

U.S.P., 4 060 674, (*1977*); *CA*, **88**, 107897s (*use*)
Klemann, L.P. *et al*, *Proc. Electrochem. Soc.*, 1981, **81-4**, 179; *CA*, **95**, 153786x (*use*)

$C_7H_{18}B^{\ominus}$ B-00510

Triethylmethylborate(1−), 10CI

$Et_3BMe^{\ominus}$

M 113.029 (ion)
Li salt: [26282-82-6].
$C_7H_{18}BLi$ M 119.970
Prepd. *in situ* from Et_3B + MeLi.
Cu(I) salt: [65167-44-4].
$C_7H_{18}BCu$ M 176.575
Obt. by reacn. of Li salt with CuI. Synthetic intermed.

Suzuki, A. *et al*, *Tetrahedron Lett.*, 1969, 2955.
Maruyama, K. *et al*, *J. Am. Chem. Soc.*, 1977, **99**, 8068.
Yamamoto, Y. *et al*, *J. Am. Chem. Soc.*, 1980, **102**, 2318.
U.S.P., 4 216 192, (*1980*); *CA*, **93**, 188617d
Horowitz, H.H. *et al*, *Proc. Electrochem. Soc.*, 1981, **81**, 131; *CA*, **95**, 118331y.

$C_7H_{19}BN_2$ B-00511

Isopropylbis(dimethylamino)borane
N,N,N′,N′-*Tetramethyl-1-*(*1-methylethyl*)*boranediamine*, 10CI
[65478-40-2]

$(H_3C)_2CHB(NMe_2)_2$

M 142.051
Liq., mod. air- and moisture-sensitive. Bp 145.5°, Bp_{27} 62°.

Paetzold, P. *et al*, *Chem. Ber.*, 1977, **110**, 3678 (*synth, pmr, cmr, nmr, ms*)
Hoffmann, R.W. *et al*, *J. Org. Chem.*, 1981, **46**, 1309 (*synth, pmr*)

$C_7H_{21}AuB_9NS_2$ **B-00512**

(Diethylcarbamodithioato-*S,S'*)[(9,10,11-η)-undecahydro-7,8-dicarbaundecaborato(2−)]gold
[62572-50-3]

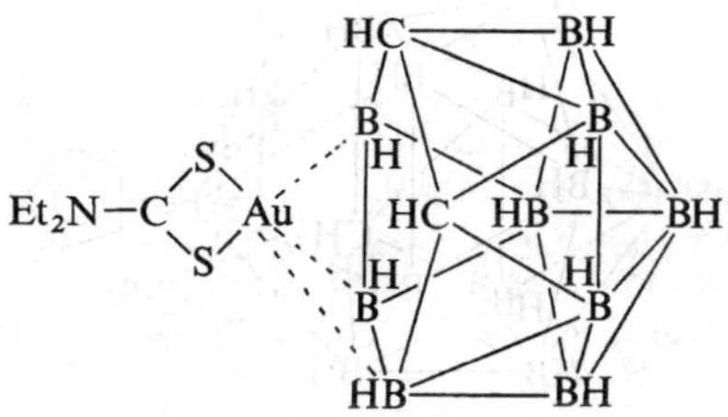

M 477.626
"Slipped" dicarbollide ligand in struct. Orange cryst. (CH_2Cl_2/pentane). Sol. CH_2Cl_2. Mp 115° dec.

Colquhoun, H.M. *et al*, *J. Chem. Soc., Dalton Trans.*, 1978, 303 (*synth, cryst struct, nmr*)

$C_8H_5B_3Fe_2O_6$ **B-00513**

Hexacarbonyl[μ-[(1,4,5-η:2,3,4,5-η)-pentahydro-2,4-dicarbapentaborato(2−)]]diiron, 9CI
1,2,3,5-(Hexacarbonyldiiron)-$C_2B_3H_5$
[59421-56-0]

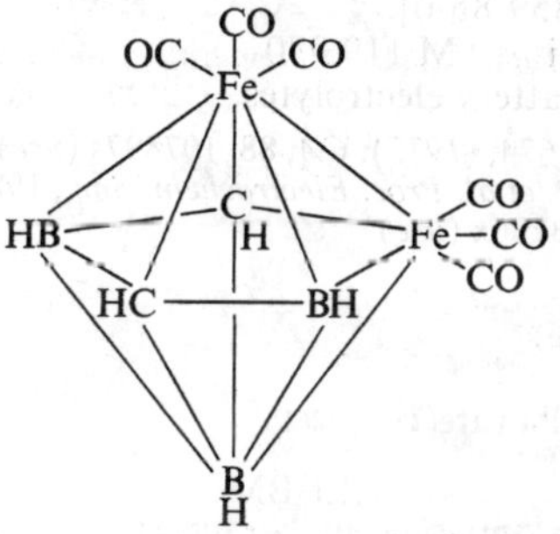

M 341.248
Orange liq.

Miller, V.R. *et al*, *J. Am. Chem. Soc.*, 1974, **96**, 3090 (*synth, ir, pmr, nmr*)

$C_8H_7BF_2O_2$ **B-00514**

Difluoro(2-acetylphenoxy)borane
Difluoro[1-(2-hydroxyphenyl)ethanone-O,O']boron, 10CI. *Difluoro(2'-hydroxyacetophenonato)boron*, 8CI
[14974-24-4]

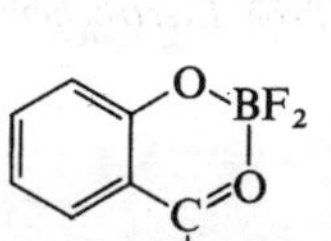

M 183.949
Exists in intramol. coord. form as shown. Used as a catalyst for curing epoxy resins. Yellow cryst. (PhCl/pet. ether), with bluish-grey fluorescence. Mp 142-5°.

Ried, W. *et al*, *Tetrahedron Lett.*, 1962, 247 (*synth, ir*)
Durden, J.A. *et al*, *J. Org. Chem.*, 1965, **30**, 1684 (*synth*)
Umland, F. *et al*, *Chem. Ber.*, 1973, **106**, 2427 (*synth, uv, ir*)
Hohaus, E., *Z. Naturforsch., B*, 1976, **31**, 324 (*ms*)
Shapet'ko, N.N. *et al*, *Dokl. Akad. Nauk SSSR, Ser. Sci. Khim.*, 1977, **234**, 876 (*cmr*)
Reynolds, G.A. *et al*, *J. Heterocycl. Chem.*, 1979, **16**, 369.

$C_8H_7BOS_2$ **B-00515**

Hydroxybis(2-thienyl)borane
Di-2-thienylborinic acid, 10CI, 9CI
[41816-82-4]

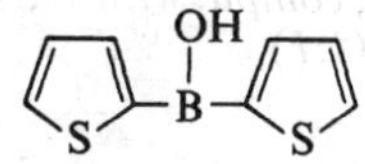

M 194.073
Bactericide.

Fr. Pat., 2 140 127, (*1973*); *CA*, **79**, 42677n (*synth, use*)
Bailey, P.J. *et al*, *Antimicrob. Agents Chemother.*, 1980, **17**, 549 (*props, use*)

$C_8H_8BF_3O_2$ **B-00516**

Trifluoroborane methyl benzoate
Trifluoro(methyl benzoate-O')boron, 10CI. *Benzoic acid methyl ester compd. with boron fluoride(BF_3)(1:1)*, 8CI. *Benzoic acid methyl ester compd. with BF_3(1:1)*, 7CI
[325-51-9]

F_3B—O=CPhOMe

M 203.955
Moisture-sensitive.

Csuros, Z. *et al*, *Periodica Polytechnica*, 1964, **8**, 1; *CA*, **61**, 12632h.
Taillandier, M. *et al*, *J. Mol Struct.*, 1968, **2**, 437 (*nmr, ir*)
Gal, J.F. *et al*, *Bull. Soc. Chim. Fr.*, 1979, 311.
Hartman, J.S. *et al*, *Inorg. Chim. Acta*, 1980, **44**, L39; 1981, **49**, 63 (*nmr*)

$C_8H_8BF_3O_2$ **B-00517**

Trifluoroborane phenyl acetate
Trifluoro(phenyl acetate-O')boron, 10CI. *Acetic acid phenyl ester compd. with boron fluoride(BF_3)(1:1)*, 8CI
[30844-81-6]

F_3B—O=C(OPh)CH_3

M 203.955
Moisture-sensitive solid. Mp 40-55°.

Topchii, V.A. *et al*, *Zh. Obshch. Khim.*, 1970, **40**, 1334.
Borovikov, Y.Y. *et al*, *Zh. Obshch. Khim.*, 1980, **50**, 2531.
Hartman, J.S. *et al*, *Inorg. Chim. Acta*, 1981, **49**, 63.

C_8H_8BN **B-00518**

10,9-Borazaronaphthalene
Azaborino[1,2-a][1,2]azaborine, 9CI
[1425-58-7]

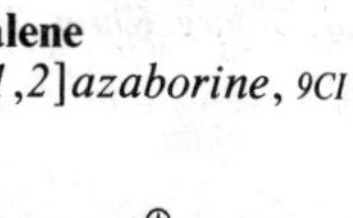

M 128.968
Flakes with naphthalenic odour. Mp 48-9°. Subl. λ_{max} 236 (log ε 4.509), 242 (4.657) and 248 nm (4.487) (EtOH).
Picrate: Yellow powder (EtOH). Mp 133.5-134.5°.

Davis, F.A. *et al*, *J. Am. Chem. Soc.*, 1968, **90**, 706.
Dewar, M.J.S. *et al*, *J. Am. Chem. Soc.*, 1968, **90**, 2137 (*synth*, *pmr*, *ms*)
Davis, F.A. *et al*, *J. Am. Chem. Soc.*, 1969, **91**, 2094.

$C_8H_8BN_4O_2Rh$ B-00519

Dicarbonyl[dihydrobis(1*H*-pyrazolato-*N'*)borato(1−)-*N*^{3},*N*$^{3'}$]rhodium, 10CI
Dicarbonyl[bis(pyrazolyl)borato]rhodium
[33058-09-2]

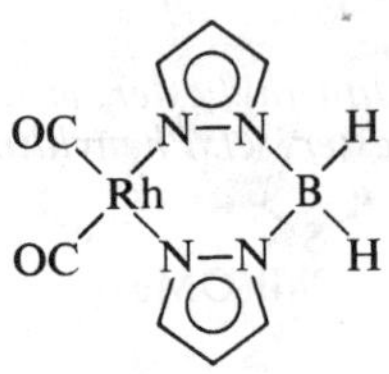

M 305.892
Catalyst for the hydroformylation of alkenes. Yellow cryst. (hexane). Mp 71° (66°). ir ν_{CO} 2088s, 2022s cm^{-1} (hexane).

King, R.B. *et al*, *J. Organomet. Chem.*, 1974, **73**, 115 (*synth*, *ir*, *pmr*)
Bonati, F. *et al*, *J. Organomet. Chem.*, 1975, **87**, 365 (*synth*, *pmr*, *ir*, *uv*, *ms*)
Cocivera, M. *et al*, *Organometallics*, 1982, **1**, 1125 (*ir*, *synth*)

$C_8H_9BO_2$ B-00520

Dihydroxy(2-phenylethenyl)borane
(2-Phenylethenyl)boronic acid, 10CI, 9CI, 8CI. *β-Styreneboronic acid. 2-Phenyletheneboronic acid. β-Styryldihydroxyborane*
[4363-35-3]

PhCH=CHB(OH)$_2$

M 147.968
Synth. intermed.
(***E***)***-form*** [6783-05-7]
Solid. Mp 168-9°.
(***Z***)***-form*** [60806-02-2]

Sazonova, V.A. *et al*, *Zh. Obshch. Khim.*, 1956, **26**, 1876 (*synth*)
Joy, F. *et al*, *J. Organomet. Chem.*, 1966, **5**, 506 (*synth*, *ir*)
Brown, H.C. *et al*, *J. Am. Chem. Soc.*, 1972, **94**, 4370; 1973, **95**, 5786; 1975, **97**, 5249 (*synth*, *use*)
Matteson, D.S. *et al*, *J. Organomet. Chem.*, 1976, **110**, 25 (*synth*)
Matteson, D.S. *et al*, *J. Organomet. Chem.*, 1976, **114**, 1 (*pmr*)
Raucher, S. *et al*, *J. Org. Chem.*, 1978, **43**, 4885 (*synth*)
Rao, V.V.R. *et al*, *J. Organomet. Chem.*, 1979, **179**, C7 (*use*)

$C_8H_9BO_2$ B-00521

(2-Ethenylphenyl)dihydroxyborane
(2-Ethenylphenyl)boronic acid, 10CI. o-*Vinylbenzeneboronic acid.* o-*Styrylboronic acid.* o-*Vinylphenylboronic acid*
[15016-42-9]

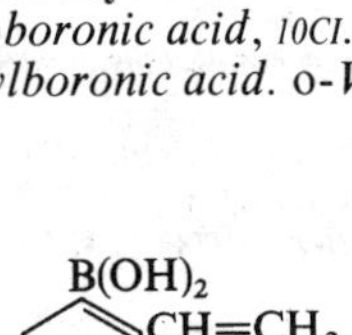

M 147.968
Useful as binder for paper, glass, wood and rubber. White plates (H_2O). Mp 111-2° (108-9°).

Letsinger, R.L. *et al*, *J. Org. Chem.*, 1961, **26**, 1271 (*synth*, *ir*)
Wesley, J.D. *et al*, *J. Org. Chem.*, 1962, **27**, 2598 (*ir*, *synth*, *props*)
U.S.P., 3 025 274, (*1962*); *CA*, **57**, 6143b (*use*)
Ger. Pat., 1 147 230, (*1963*); *CA*, **59**, 10117e (*use*, *synth*)
Armstrong, D.R. *et al*, *J. Chem. Soc.* (*A*), 1967, 123 (*struct*, *uv*, *props*)

$C_8H_9BO_2$ B-00522

(3-Ethenylphenyl)dihydroxyborane
(3-Ethenylphenyl)boronic acid, 10CI. m-*Vinylbenzeneboronic acid*, 8CI. m-*Styrylboronic acid.* m-*Vinylphenylboronic acid*
[15016-43-0]
M 147.968
Binder for paper, glass, wood and rubber. Used as adhesive and in production of optical devices. White plates (H_2O). Mp 145-60° (144-5°).

Wesley, J.D. *et al*, *J. Org. Chem.*, 1962, **27**, 2598 (*synth*, *ir*)
U.S.P., 3 025 274, (*1962*); *CA*, **57**, 6143b (*use*)
Ger. Pat., 1 147 230, (*1963*); *CA*, **59**, 10117e (*use*, *synth*)
Armstrong, D.R. *et al*, *J. Chem. Soc.* (*A*), 1967, 123 (*struct*, *uv*, *props*)
Rajwar, D.C. *et al*, *CA*, 1980, **93**, 174751f (*props*)

$C_8H_9BO_2$ B-00523

(4-Ethenylphenyl)dihydroxyborane
(4-Ethenylphenyl)boronic acid, 10CI, 9CI. p-*Vinylbenzeneboronic acid*, 8CI. p-*Styrylboronic acid.* p-*Vinylphenylboronic acid*
[2156-04-9]
M 147.968
Binder for paper, glass, wood and rubber. Used as adhesive and in production of optical devices; forms copolymers with styrene. Cryst. (H_2O). Mp 188-9°.

Hawkins, R.T. *et al*, *J. Am. Chem. Soc.*, 1960, **82**, 3053 (*pmr*)
Wesley, J.D. *et al*, *J. Org. Chem.*, 1962, **27**, 2598 (*synth*, *ir*)
U.S.P., 3 025 274, (*1962*); *CA*, **57**, 6143b (*use*)
Ger. Pat., 1 147 230, (*1963*); *CA*, **59**, 10117e (*use*, *synth*)
Armstrong, D.R. *et al*, *J. Chem. Soc.* (*A*), 1967, 123 (*struct*, *uv*, *props*)
Hartmann, M. *et al*, *Makromol. Chem.*, 1976, **177**, 131; *CA*, **84**, 74704p (*use*)

$C_8H_{11}BBrN$ B-00524

Bromo(dimethylamino)phenylborane, 8CI
*1-Bromo-*N,N*-dimethyl-1-phenylboranamine*, 10CI
[16548-71-3]

PhBBrNMe$_2$

M 211.896
Air- and moisture-sensitive liq. $Bp_{0.8}$ 63°. n_D^{20} 1.5554.

Lappert, M.F. *et al*, *J. Chem. Soc.* (*A*), 1967, 854; 1968, 554 (*synth*, *nmr*, *pmr*)
Barfield, P.A. *et al*, *Trans. Faraday Soc.*, 1968, **64**, 2571 (*nmr*)
Meller, A. *et al*, *Monatsh. Chem.*, 1977, **108**, 1279.
Paetzold, P. *et al*, *Chem. Ber.*, 1977, **110**, 3678 (*ms*, *nmr*)
Maringgele, W., *Z. Anorg. Allg. Chem.*, 1980, **468**, 99.

$C_8H_{11}BClN$ B-00525

Chloro(dimethylamino)phenylborane, 8CI
*1-Chloro-*N,N*-dimethyl-1-phenylboranamine*, 10CI, 9CI
[1196-44-7]

PhBClNMe$_2$

M 167.445

Air- and moisture-sensitive liq. Bp_1 40°. n_D^{20} 1.5324.

Brotherton, R.J. *et al*, *Inorg. Chem.*, 1962, **1**, 749 (*synth*)
Friebolin, H. *et al*, *Org. Magn. Reson.*, 1976, **8**, 287; 1977, **10**, 157 (*nmr*)
Odom, J.D. *et al*, *J. Organomet. Chem.*, 1979, **173**, 15 (*nmr*)
Cragg, R.H. *et al*, *J. Organomet. Chem.*, 1981, **217**, 1; 1982, **232**, 201 (*synth, nmr, pmr*)

$C_8H_{11}BNO^{\ominus}$ B-00526

Trihydro(*N*-phenylacetamidato-*O*)borate(1−), 10CI, 9CI

$PhN{=}C(CH_3)OBH_3^{\ominus}$

M 147.991 (ion)

Na salt: [60542-50-9]. *Sodium acetanilidoborohydride.* $C_8H_{11}BNNaO$ M 170.981
Reducing agent, reduces esters, aldehydes and ketones to alcohols. Cryst. solid.

Kikugawa, Y., *Chem. Pharm. Bull.*, 1976, **24**, 1059, 1939.
Kikugawa, Y., *CA*, 1979, **91**, 38304u; *CA*, 1981, **95**, 150080b.

$C_8H_{11}BN_2$ B-00527

Cyanoborane 4-methylaniline

(*Cyano-C*)*dihydro*(*4-methylbenzenamine*)*boron*, 10CI. *Boranecarbonitrile, compd. with 4-methylbenzenamine*(*1:1*). *Cyanoborane* p-*toluidine*
[66632-45-9]

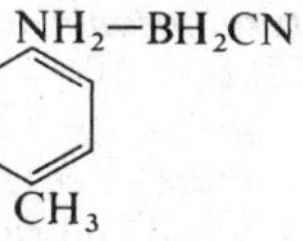

M 145.998
White cryst. solid (THF/pet. ether). Mp 134°.

Wisian-Neilson, P. *et al*, *Inorg. Chem.*, 1978, **17**, 2327 (*synth, pmr, ir*)

$C_8H_{11}BO_2$ B-00528

Dihydroxy(1-phenylethyl)borane

(*1-Phenylethyl*)*boronic acid*, 10CI. *1-Phenylethaneboronic acid*

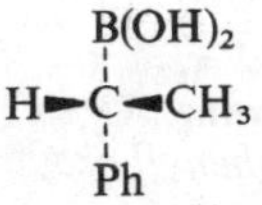

(*R*)-*form*

M 149.984
Powerful bifunctional reversible inhibitor of enzyme systems. Some papers referring to this compd. are indexed incorrectly under 2-phenylethylboronic acid in *CA*.
▷Toxic

(*R*)-*form*
$[\alpha]_D$ −14.8° (C_6H_6).

Matteson, D.S. *et al*, *J. Am. Chem. Soc.*, 1965, **87**, 2587 (*use, props*)
Davies, A.G. *et al*, *J. Chem. Soc., Chem. Commun.*, 1966, 298 (*use, props*)

$C_8H_{11}BO_2$ B-00529

Dihydroxy(2-phenylethyl)borane

(*2-Phenylethyl*)*boronic acid*, 10CI. (*2-Phenylethane*)*boronic acid*, 8CI
[34420-17-2]

$PhCH_2CH_2B(OH)_2$

M 149.984
Enzyme inhibitor. Sol. $CHCl_3$.
▷Toxic

Ger. Pat., 1 111 181, (*1958*); *CA*, **56**, 8742i (*synth*)
Koehler, K.A. *et al*, *Biochemistry*, 1974, **13**, 5345 (*use*)
Byers, L.D., *Bioorg. Chem.*, 1978, 7, 15 (*use*)
Rotonova, T.V., *Bioorg. Khim.*, 1978, **4**, 694 (*use*)

$C_8H_{11}BO_2$ B-00530

Dimethoxyphenylborane

Phenylboronic acid dimethyl ester, 10CI, 9CI. *Benzeneboronic acid dimethyl ester*, 8CI. *Phenyldimethoxyborane*
[13471-35-7]

$PhB(OMe)_2$

M 149.984
Liq., sensitive to moist air. Sol. non-hydroxylic solvs. Bp_9 71-2°. n_D^{20} 1.478.

Burch, J.E. *et al*, *Spectrochim. Acta*, 1962, **18**, 1403 (*ir, struct*)
Laurent, J.P., *C.R. Hebd. Seances Acad. Sci.*, 1962, **254**, 866 (*synth, props*)
Jefferson, R. *et al*, *J. Chem. Soc.* (*A*), 1966, 1584 (*props*)
Wieber, M. *et al*, *Z. Anorg. Allg. Chem.*, 1974, **403**, 107 (*synth*)
Odom, J.D. *et al*, *J. Organomet. Chem.*, 1979, **173**, 15 (*nmr*)
Goetze, R. *et al*, *Chem. Ber.*, 1981, **114**, 1884 (*nmr*)

$C_8H_{11}BS$ B-00531

1-(2-Thienyl)borolane, 10CI, 9CI
[59073-91-5]

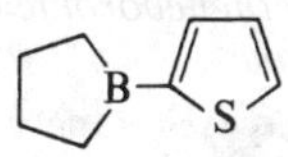

M 150.045
Air-sensitive.

Wrackmeyer, B. *et al*, *Chem. Ber.*, 1976, **109**, 1075 (*nmr*)
Schmidt, A. *et al*, *Z. Naturforsch., B*, 1978, **33**, 855.
Odom, J.D. *et al*, *J. Organomet. Chem.*, 1979, **173**, 15 (*nmr*)

$C_8H_{11}BS_2$ B-00532

Bis(methylthio)phenylborane

Phenyldithioboronic acid dimethyl ester, 10CI, 9CI. *Dithiobenzeneboronic acid dimethyl ester*, 8CI. *Phenylbis*(*methylthio*)*borane*
[17381-61-2]

$PhB(SMe)_2$

M 182.105
Liq., sensitive to moist air. Sol. non-hydroxylic solvs. $Bp_{4.5}$ 107-9°. n_D^{26} 1.5979.

Schmidt, M. *et al*, *Chem. Ber.*, 1968, **101**, 281 (*synth*)
Vahrenkamp, H., *J. Organomet. Chem.*, 1971, **28**, 181 (*ir, raman*)
Cragg, R.H. *et al*, *J. Inorg. Nucl. Chem.*, 1973, **35**, 3685 (*synth, props, ms, ir, nmr*)
Nöth, H. *et al*, *Z. Anorg. Allg. Chem.*, 1975, **418**, 97 (*props*)
Odom, J.D. *et al*, *J. Organomet. Chem.*, 1979, **173**, 15 (*nmr*)
Goetze, R. *et al*, *Chem. Ber.*, 1981, **114**, 1884 (*nmr*)

$C_8H_{12}BBr_4O_4^{\ominus}$ B-00533

Bis[2,3-dibromo-1,4-butanediolato(2−)-*O,O'*]borate(1−), 10CI

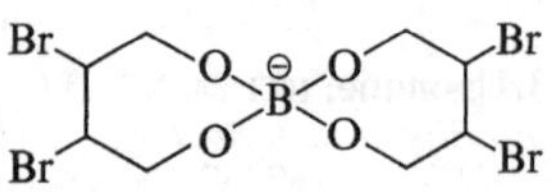

M 502.606 (ion)

Guanidinium salt: [74211-02-2]. *Hydrogen bis[2,3-dibromo-1,4-butanediolato(2−)-O,O′]borate(1−) compd. with guanidine (1:1), 10CI. Guanidinium bis(2,3-dibromo-1,4-butanediolato)borate.*
$C_9H_{18}BBr_4N_3O_4$ M 562.685
Flame retardant for urethane polymers, epoxy resins and polyesters.

Eur. Pat., 5 497, (*1979*); *CA*, **93**, 46174h (*use*)
Austrian Pat., 357 336, (*1980*); *CA*, **93**, 151192k (*use*)
Austrian Pat., 358 063, (*1980*); *CA*, **94**, 103330c

$C_8H_{12}BN$ B-00534

(Diethylamino)diethynylborane

N,N-*Diethyl-1,1-diethynylboranamine, 10CI*
[62705-36-6]

$Et_2NB(C{\equiv}CH)_2$

M 133.000
Air- and moisture-sensitive liq. Bp_{18} 54-6°.

Wrackmeyer, B. *et al*, *Chem. Ber.*, 1977, **110**, 1086 (*nmr, pmr*)
Berger, H.O. *et al*, *Chem. Ber.*, 1980, **113**, 1235 (*synth*)

$C_8H_{12}BN$ B-00535

Dimethylaminodi-1-propynylborane

N,N-*Dimethyl-1,1-di-1-propynylboranamine, 10CI*
[62654-64-2]

$(H_3CC{\equiv}C)_2BNMe_2$

M 133.000
Air- and moisture-sensitive solid. Mp 64-6°. $Bp_{0.1}$ 29°.

Nöth, H. *et al*, *Chem. Ber.*, 1977, **110**, 1086 (*synth, nmr*)
Nöth, H. *et al*, *Chem. Ber.*, 1980, **113**, 1235 (*nmr*)

$C_8H_{12}BN$ B-00536

Dimethyl(phenylamino)borane

*1,1-Dimethyl-*N*-phenylboranamine, 10CI, 9CI. Anilinodimethylborane, 8CI.* B*-Dimethyl-*N*-phenylborazene*
[4011-26-1]

Me_2BNHPh

M 133.000
Colourless liq., darkens in air. Bp 184° (175°) (extrap.); Bp_{12} 77°.

Wiberg, E. *et al*, *Z. Anorg. Allg. Chem.*, 1948, **257**, 138 (*synth*)
Becher, H.J., *Z. Anorg. Allg. Chem.*, 1957, **289**, 262 (*synth, ir*)
Nöth, H. *et al*, *Z. Naturforsch., B*, 1961, **166**, 470 (*synth*)
Beck, W. *et al*, *Chem. Ber.*, 1972, **105**, 2883 (*nmr*)
Nöth, H. *et al*, *Chem. Ber.*, 1977, **110**, 3172 (*nmr*)

$C_8H_{12}BN_4^{\ominus}$ B-00537

Dimethylbis(1*H*-pyrazolato-*N'*)borate(1−), 9CI

Dimethylbis(1-pyrazolyl)borate

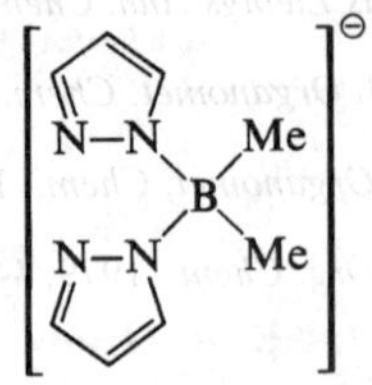

M 175.020 (ion)

Na salt: [56953-94-7].
$C_8H_{12}BN_4Na$ M 198.009
Used in metal complex formn.

Breakell, K.R. *et al*, *J. Chem. Soc., Dalton Trans.*, 1975, 749 (*synth*)

$C_8H_{13}BN_2$ B-00538

Bis(methylamino)phenylborane, 8CI

N,N′*-Dimethyl-1-phenylboranediamine. Bis(methylamino)phenylborine*
[19125-66-7]

$PhB(NHMe)_2$

M 148.014
Wrongly named as *N,N*-dimethyl-1-phenylboranediamine, 10CI, 9CI. Liq., sensitive to moist air. d_4^{20} 0.965. Bp_{16} 106-7°. n_D^{20} 1.5275.

Burch, J.E. *et al*, *J. Chem. Soc.*, 1962, 2200.
Burch, J.E. *et al*, *Spectrochim. Acta*, 1962, **18**, 1403 (*ir, struct*)
Barfield, P.A. *et al*, *J. Chem. Soc. (A)*, 1968, 554.
Beck, W. *et al*, *Chem. Ber.*, 1972, **105**, 2883 (*nmr*)
Niedenzu, K. *et al*, *Z. Anorg. Allg. Chem.*, 1978, **439**, 112 (*nmr*)
Goetze, R. *et al*, *Chem. Ber.*, 1981, **114**, 1884 (*nmr*)

$C_8H_{14}BBr$ B-00539

9-Bromo-9-borabicyclo[3.3.1]nonane, 10CI

B*-Bromo-9-BBN*
[22086-45-9]

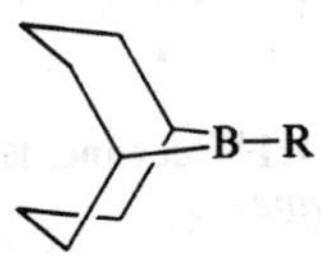

R = Br

M 200.913
Reagent for cleavage of ethers. Colourless viscous liq., air- and moisture-sensitive. Bp_{10} 90-2°, $Bp_{1.0}$ 64-5°.

Köster, R. *et al*, *Justus Liebigs Ann. Chem.*, 1968, **719**, 169 (*synth*)
Brown, H.C. *et al*, *Tetrahedron Lett.*, 1977, 3007 (*synth*)
Bhatt, M.V., *J. Organomet. Chem.*, 1978, **156**, 221 (*synth, nmr*)
Brown, H.C. *et al*, *J. Org. Chem.*, 1979, **44**, 2422 (*synth*)
Brown, H.C. *et al*, *J. Organomet. Chem.*, 1979, **168**, 281 (*synth, nmr*)
Brown, H.C. *et al*, *J. Org. Chem.*, 1981, **46**, 3978 (*nmr*)

$C_8H_{14}BCl$ B-00540

9-Chloro-9-borabicyclo[3.3.1]nonane, 10CI

B*-Chloro-9-BBN*
[22086-34-6]
As 9-Bromo-9-borabicyclo[3.3.1]nonane, B-00539 with

R = Cl

M 156.462

Air- and moisture-sensitive solid. Mp 25-8°, 37°. $Bp_{1.0}$ 54-6°, $Bp_{0.3}$ 37-40°, 41-3°.

Köster, R. *et al*, *Justus Liebigs Ann. Chem.*, 1969, **719**, 169 (*synth*)
Kramer, G.W. *et al*, *J. Organomet. Chem.*, 1974, **73**, 1 (*synth, nmr, pmr*)
Brown, H.C. *et al*, *J. Organomet. Chem.*, 1979, **168**, 281 (*synth, pmr*)
Brown, H.C. *et al*, *J. Org. Chem.*, 1979, **44**, 2422; 1980, **45**, 846 (*synth, nmr*)

$C_8H_{14}BN$ B-00541

Borane *N,N*-dimethylaniline

(N,N-*Dimethylbenzenamine*)*trihydroboron*, 10CI. N,N-*Dimethylbenzenamine compd. with borane*(*1:1*), 9CI. N,N-*Dimethylaniline compd. with borine*

[1769-74-0]

$PhNMe_2$—BH_3

M 135.015

Tested as rocket fuel.

Mikheeva, V.I. *et al*, *Zh. Neorg. Khim.*, 1957, **2**, 604.
B.P., 889 599, (*1962*); *CA*, **61**, P9198d
Fr. Pat., 1 350 849, (*1964*); *CA*, **60**, P14327f
Ashby, E.C., *J. Organomet. Chem.*, 1965, **3**, 371.
Beachley, O.T. *et al*, *Inorg. Chem.*, 1973, **12**, 1128.

$C_8H_{14}BNO$ B-00542

1-Pyrrolylborane tetrahydrofuran

*Dihydro-1*H-*pyrrol-1-yl*(*tetrahydrofuran*)*boron*, 10CI

[58178-51-1]

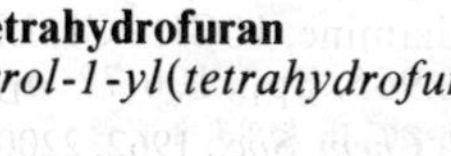

M 151.015

Prepd. from pyrrole and borane-THF. Cannot be isol. on evap. of solv.

Ioffe, S.L. *et al*, *Khim. Geterotsikl. Soedin.*, 1971, **7**, 1056; *CA*, **76**, 25012n (*synth, ir, nmr*)
Gyori, B. *et al*, *Acta Chim. Acad. Sci. Hung.*, 1975, **86**, 235 (*synth*)
Añez, M. *et al*, *Synthesis*, 1981, 214.

$C_8H_{14}B_2$ B-00543

2,6-Diboratricyclo[3.3.1.1^3,7^]decane, 10CI

2,6-Diboraadamantane

[69961-38-2]

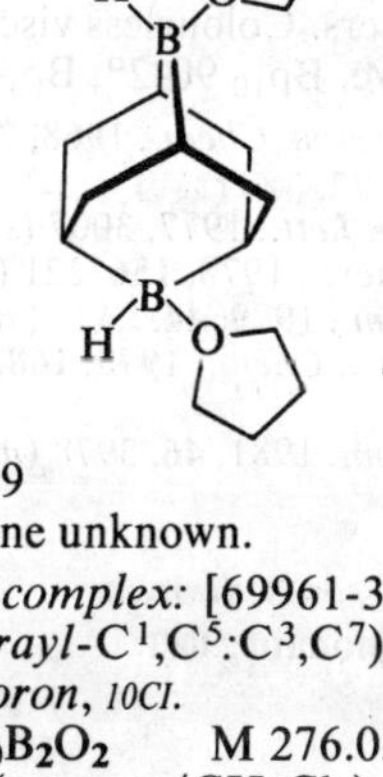

THF complex

M 131.819

Free borane unknown.

Bis-THF complex: [69961-39-3]. [μ-(*1,3,5,7-Cyclooctanetetrayl*-C^1,C^5:C^3,C^7)]*dihydrobis*(*tetrahydrofuran*)*diboron*, 10CI.

$C_{16}H_{30}B_2O_2$ M 276.032

Cryst. (pentane/CH_2Cl_2). Prob. air-sensitive. Mp 145-7°. Unusually strong THF complex; other complexes also reported.

Kulkarni, S.U. *et al*, *J. Org. Chem.*, 1979, **44**, 1747 (*synth, struct, props, nmr*)

$C_8H_{15}B$ B-00544

3-Borabicyclo[3.3.1]nonane, 10CI

3-BBN

[4632-82-0]

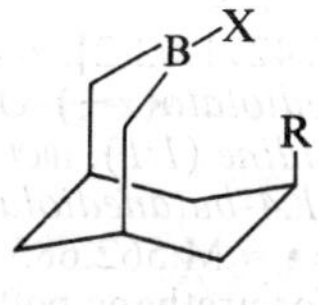

X = R = H

M 122.017

Probably exists as dimer in uncomplexed state. Air- and moisture-sensitive.

Bubnov, Yu.N. *et al*, *Izv. Akad. Nauk SSSR, Ser. Khim.*, 1968, 824 (*synth*)

$C_8H_{15}B$ B-00545

9-Borabicyclo[3.3.1]nonane, 10CI, 9CI

9-BBN

[280-64-8]

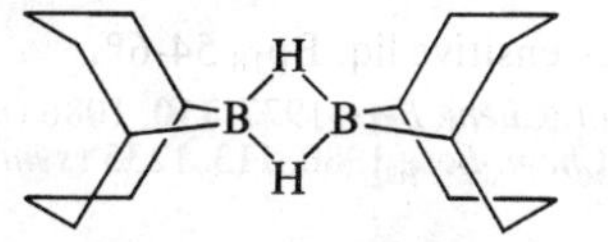

Dimer

M 122.017

Dimeric both as solid and in soln. in most solvs. Unknown as the monomer except in form of complexes.

Dimer: [70658-61-6].

$C_{16}H_{30}B_2$ M 244.033

Important selective hydroborating agent. Cryst. (1,2-dimethoxyethane); remarkably stable to air though dec. slowly. Sol. THF (0.29M at 25°), 1,2-dimethoxyethane (0.07M at 25°). Mp 153-5° (142°). ir ν_{max} 1570 cm^{-}.

▷Flammable

THF complex: [76422-63-4]. *9-Borabicyclo*[*3.3.1*]*nonane tetrahydrofuran*.

$C_{12}H_{23}BO$ M 194.123

Present to a significant extent in THF solns. of 9-BBN dimer. Hydroborating agent. Unstable, found only in soln.

Köster, R. *et al*, *Angew. Chem.*, 1960, **72**, 626 (*synth*)
Brown, H.C. *et al*, *J. Am. Chem. Soc.*, 1974, **96**, 7765 (*synth, ir, pmr, ms*)
Brown, H.C., *Organic Synthesis via Boranes*, 1975, Wiley, N.Y., 191 (*synth*)
Brown, H.C. *et al*, *Heterocycles*, 1977, **7**, 453 (*rev*)
Contreras, R. *et al*, *Z. Naturforsch., B*, 1980, **35**, 1236 (*nmr*)
Wang, K.K. *et al*, *J. Org. Chem.*, 1980, **45**, 1748, 5303 (*props, cmr*)
Soderquist, J.A. *et al*, *J. Org. Chem.*, 1981, **46**, 4599 (*synth*)
Fieser, M. *et al*, *Reagents for Organic Synthesis*, Wiley, 1967-83, **8**, 47.

$C_8H_{16}B^{\ominus}$ B-00546

9-Boratabicyclo[3.3.1]nonane, 10CI, 9CI
9-BBN hydride
[25301-61-5]

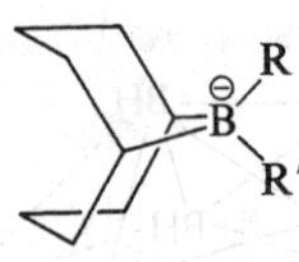

R = R′ = H

M 123.024 (ion)
Li salt: [76448-08-3].
$C_8H_{16}BLi$ M 129.965
Air- and moisture-sensitive.
Na salt: [76448-07-2].
$C_8H_{16}BNa$ M 146.014
Air- and moisture-sensitive.
K salt: [76448-06-1].
$C_8H_{16}BK$ M 162.123
Air- and moisture-sensitive.

Pinazzi, C.P. *et al*, *Eur. Polym. J.*, 1976, **12**, 83.
Boudjouk, P. *et al*, *J. Org. Chem.*, 1978, **43**, 3979.
Hubbard, J.L. *et al*, *J. Organomet. Chem.*, 1978, **156**, 81 (*deriv, nmr*)
Moreau, R. *et al*, *C.R. Hebd. Seances Acad. Sci.*, 1978, **287**, 39.
Brown, H.C. *et al*, *J. Org. Chem.*, 1981, **46**, 2712, 4541 (*synth, ir, nmr*)

$C_8H_{16}BN$ B-00547

1-Aza-6-borabicyclo[4.4.0]decane
9-Aza-10-boradecahydronaphthalene
[18903-54-3]

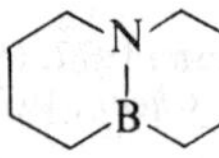

M 137.031
Oil. Bp_{27} 81-4°. n_D^{24} 1.4837.

Dewar, M.J.S. *et al*, *J. Am. Chem. Soc.*, 1968, **90**, 2137 (*synth*)
Davis, F.A. *et al*, *J. Am. Chem. Soc.*, 1969, **91**, 2094 (*nmr, ms*)

$C_8H_{16}BN_3$ B-00548

Bis(dimethylamino)-1-pyrrolylborane
N,N,N′,N′-*Tetramethyl-1-(1*H*-pyrrol-1-yl)boranediamine, 10CI, 9CI*
[42051-53-6]

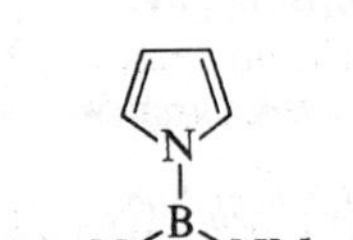

M 165.045
Liq., sensitive to moisture. Bp_1 42-3°.

Nöth, H. *et al*, *Chem. Ber.*, 1973, **106**, 1145 (*nmr, props*)
Ebraheem, K.A.K., *Org. Magn. Reson.*, 1977, **10**, 258 (*props*)
Jallali-Heravi, M. *et al*, *J. Mol. Struct.*, 1979, **55**, 113 (*props*)

$C_8H_{17}B$ B-00549

2-Butenyldiethylborane, 10CI
Crotyldiethylborane
[67682-90-0]

$Et_2BCH_2CH{=}CHCH_3$

M 124.032
Undergoes ready allyl rearr., which causes equilibration of *E*- and *Z*-isomers. Used in the synth. of 2-alkoxy-1,4-pentadienes. Liq., sensitive to air and moisture. Sol. nonhydroxylic solvs. Bp_{50} 62-4°. n_D^{20} 1.4278.

Köster, R. *et al*, *Justus Liebigs Ann. Chem.*, 1976, 1116 (*synth*)
Mikhailov, B.M. *et al*, *J. Organomet. Chem.*, 1978, **154**, 113 (*synth*)
Mikhailov, B.M. *et al*, *Izv. Akad. Nauk SSSR, Ser. Khim.*, 1978, 1586; *Bull. Acad. Sci. USSR, Div. Chem. Sci.*, 1385 (*synth, nmr*)
Bubnov, Yu.N. *et al*, *Synthesis*, 1980, 902 (*props*)

$C_8H_{17}B$ B-00550

Diethyl(2-methyl-2-propenyl)borane, 10CI
(2-Methyl-2-propenyl)diethylborane. Diethyl(methallyl)borane
[67741-07-5]

$H_2C{=}C(CH_3)CH_2BEt_2$

M 124.032
Prepd. from alkoxydiethylborane and a methallyl-metal deriv. Synthetic intermed. Air- and moisture sensitive liq. Sol. THF. Bp_{45} 56°. n_D^{20} 1.4235. Stable up to 110° under N_2 atm.

Mikhailov, B.M. *et al*, *Izv. Akad. Nauk SSSR, Ser. Khim.*, 1978, 1594 (*synth*)
Mikhailov, B.M. *et al*, *J. Organomet. Chem.*, 1978, **154**, 113 (*synth*)
Pelter, A. *et al*, *Comprehensive Organic Chemistry*, (*Barton, D.H.R. et al*, Ed.), Pergamon, Oxford, 1979, **3**, 695 (*rev*)
Bubnov, Y.N. *et al*, *Synthesis*, 1980, 902 (*use*)

$C_8H_{17}B$ B-00551

3,6-Dimethylborepane, 10CI, 9CI, 8CI
[39546-63-9]

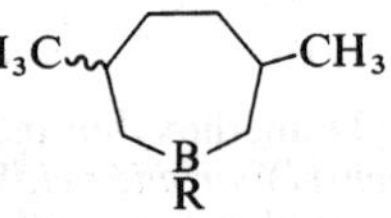

R = H

M 124.032
Dimeric. Has not been isol. pure, but has been fully identified spectroscopically and as its derivs. Hydroborating agent.
B-*methoxy:* [32503-24-5]. *1-Methoxy-3,6-dimethylborepane.*
$C_9H_{19}BO$ M 154.059
Liq., sensitive to air and moisture. Bp_{20} 76-9°.
2-Dimethylaminoethanol deriv.: [32503-13-2]. *2-[(3,6-Dimethyl-1-borepanyl)oxy]*-N,N-*dimethylethanamine.*
$C_{12}H_{26}BNO$ M 211.154
Liq. Bp_1 104-5°. n_D^{20} 1.4825.

Brown, H.C. *et al*, *J. Am. Chem. Soc.*, 1971, **93**, 3777 (*use*)
Zweifel, G. *et al*, *Synthesis*, 1972, 557 (*use*)
Negishi, E. *et al*, *J. Am. Chem. Soc.*, 1973, **95**, 6757 (*synth, pmr, ir, deriv*)
Zweifel, G. *et al*, *Synthesis*, 1973, 37 (*props*)

$C_8H_{17}B$ B-00552

Octyldichloroborane
Dichlorooctylboron. Dichlorooctylborane

$H_3C(CH_2)_7BCl_2$

M 124.032
Me_2S complex: [72205-94-8]. *Dichlorooctyl[thiobis[methane]]boron, 10CI.*
$C_{10}H_{23}BCl_2S$ M 257.068

Prepd. by hydroboration of 1-octene with $HBCl_2$-SMe_2. Air- and moisture-sensitive liq. Bp_2 65-7°.

Brown, H.C. *et al*, *J. Org. Chem.*, 1980, **45**, 384 (*synth*)

$C_8H_{17}BClNO$ B-00553

Diethylaminochloro(1-butenyloxy)borane

1-(1-Butenyloxy)-1-chloro-N,N-*diethylboranamine*, *10CI*

[71466-96-1]

$H_3CCH_2CH{=}CHOBClNEt_2$

M 189.491

Moisture-sensitive.

Sugasawa, T. *et al*, *Synth. Commun.*, 1979, **9**, 583 (*synth, use*)

$C_8H_{17}BCl_2$ B-00554

Dichlorooctylborane, 10CI

Dichlorooctylborine. Octyldichloroborane

[63348-82-3]

$H_3C(CH_2)_7BCl_2$

M 194.938

Liq., sensitive to air and moisture. Sol. pentane and other inert solvs. Bp_{19} 92-4°. ^{11}B nmr δ 64.0 ppm.

U.S.P., 2 930 809, (*1960*); *CA*, **54**, P19484h
U.S.P., 3 133 961, (*1964*); *CA*, **61**, P5690d
Gerrard, W. *et al*, *J. Chem. Soc.*, 1964, 740 (*synth*)
Brown, H.C. *et al*, *J. Org. Chem.*, 1977, **42**, 2533 (*synth*)
Brown, H.C. *et al*, *J. Org. Chem.*, 1980, **45**, 384 (*synth, derivs*)

$C_8H_{17}BO_2$ B-00555

(2-Methylcyclopentyl)dimethoxyborane

(2-Methylcyclopentyl)boronic acid dimethyl ester, *10CI*, *9CI*. *Dimethyl 2-methylcyclopentylboronate. Dimethyl 2-methylcyclopentaneboronate. Dimethoxy(2-methylcyclopentyl)borane*

[37981-95-6]

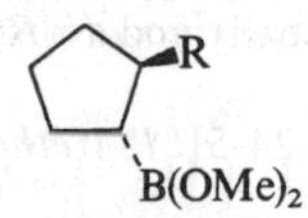

R = CH_3

M 156.031

(***1RS,2RS)-form*** [56118-57-1]

(±)-trans-*form*

Synthetic intermed. Air- and moisture-sensitive liq. Sol. THF. Bp_{15} 60-1°. n_D^{20} 1.4290.

Brown, H.C. *et al*, *J. Am. Chem. Soc.*, 1972, **94**, 5893 (*synth*)
Brown, H.C. *et al*, *J. Am. Chem. Soc.*, 1975, **97**, 2791 (*synth, pmr, ir*)
Brown, H.C. *et al*, *J. Organomet. Chem.*, 1977, **135**, C10.

$C_8H_{17}B_9Ru$ B-00556

(η^6-Benzene)[(7,8,9,10,11-η)-undecahydro-7,8-dicarbaundecaborato(2–)]ruthenium

3,1,2-(η^6-Benzene)-$RuC_2B_9H_{11}$

[82704-20-9]

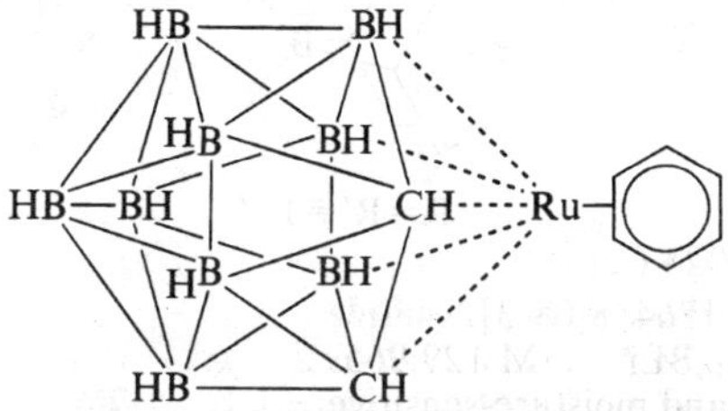

M 311.582

Canary-yellow needles (Me_2CO/hexane). Mp 308-9° dec.

Hanusa, T. *et al*, *Polyhedron*, 1982, **1**, 77 (*synth, pmr, ir, nmr, ms*)

$C_8H_{18}BBr$ B-00557

Bromobis(1-methylpropyl)borane, 10CI

Bromodi-sec-*butylborane*

[13317-63-0]

$[H_3CCH_2CH(CH_3)]_2BBr$

M 204.944

Synthetic intermed. Air- and moisture-sensitive liq. Sol. THF. d^{25} 1.070. Bp_5 50.0-50.1°. n_D^{25} 1.444.

McCusker, P.A. *et al*, *J. Inorg. Nucl. Chem.*, 1966, **28**, 2261 (*synth*)
Brown, H.C. *et al*, *Synthesis*, 1977, 695 (*synth*)
Brown, H.C. *et al*, *J. Org. Chem.*, 1979, **44**, 2417 (*synth*)

$C_8H_{18}BBr$ B-00558

Bromodibutylborane, 10CI, 9CI, 8CI

Bromodibutylborine

[5674-70-4]

$(H_3CCH_2CH_2CH_2)_2BBr$

M 204.944

Synthetic intermed.; catalyst in polymerisation of vinyl compds. Colourless liq., hydrol. by water, oxid. by air. Sol. THF. Bp_4 44°.

Johnson, J.R. *et al*, *J. Am. Chem. Soc.*, 1938, **60**, 115 (*synth, props*)
Skinner, H.A. *et al*, *J. Chem. Soc.*, 1953, 3378 (*synth*)
McCusker, P.A. *et al*, *J. Inorg. Nucl. Chem.*, 1966, **28**, 2261 (*synth*)
Laurent, J.P. *et al*, *C.R. Hebd. Seances Acad. Sci.*, 1968, **267**, 1535 (*synth*)
Köster, R. *et al*, *Justus Liebigs Ann. Chem.*, 1969, **719**, 169 (*synth*)
Pelter, A. *et al*, *J. Chem. Soc., Chem. Commun.*, 1975, 531 (*synth*)
Brown, H.C. *et al*, *Synthesis*, 1977, 695 (*synth*)

$C_8H_{18}BBr_2N$ B-00559

Dibromo(dibutylamino)borane

1,1-Dibromo-N,N-*dibutylboranamine*, *10CI*, *9CI*

[54336-83-3]

$Br_2BN(CH_2CH_2CH_2CH_3)_2$

M 298.855

Moisture-sensitive liq. d_4^{18} 1.373. Bp_1 84°. n_D^{18} 1.4875.

Spitsyn, V.I. *et al*, *Izv. Akad. Nauk SSSR, Ser. Khim.*, 1974, 2626; *Bull. Acad. Sci. USSR, Div. Chem. Sci.*, 2535 (*synth*)
Bryukhova, E.V. *et al*, *Izv. Akad. Nauk SSSR, Ser. Khim.*, 1976, 1614; *Bull. Acad. Sci. USSR, Div. Chem. Sci.*, 1529 (*nqr*)
Sheikh, S.U. *et al*, *Bull. Chem. Soc. Jpn.*, 1976, **49**, 2879 (*synth*)
Balabanova, E.A. *et al*, *Izv. Akad. Nauk SSSR, Ser. Khim.*, 1978, 2638; *Bull. Acad. Sci. USSR, Div. Chem. Sci.*, 2360 (*nmr*)

$C_8H_{18}BBr_3S_2$ B-00560

Tribromoborane dibutyl disulfide

Tribromo(dibutyl disulfide-S)boron, 10CI. *Tribromoborane compd. with dibutyl disulfide(1:1). Boron tribromide dibutyl disulfide*

[69246-02-2]

$(H_3CCH_2CH_2CH_2)_2S_2—BBr_3$

M 428.872

Ganyushin, L.A. *et al*, *Zh. Obshch. Khim.*, 1978, **48**, 2478; 1979, **49**, 2090.

$C_8H_{18}BCl$ B-00561

Chlorobis(2-methylpropyl)borane, 10CI, 9CI

Chlorodiisobutylborane, 8CI. *Chlorodiisobutylborine*

[13317-64-1]

$[(H_3C)_2CHCH_2]_2BCl$

M 160.493

Catalyst for polymerisation of acrylonitrile and other vinyl compds.; synthetic intermed. Air- and moisture-sensitive liq. Bp_{747} 160-1°. n_D^{20} 1.4230.

Booth, R.B. *et al*, *J. Am. Chem. Soc.*, 1952, **74**, 1415 (*synth*)
McCusker, P.A. *et al*, *J. Am. Chem. Soc.*, 1957, **79**, 5192 (*synth*)
Mikhailov, B.M., *Izv. Akad. Nauk SSSR, Ser. Khim.*, 1962, 413.
McCusker, P.A. *et al*, *J. Inorg. Nucl. Chem.*, 1966, **28**, 2261.
Pasto, D.J. *et al*, *J. Am. Chem. Soc.*, 1969, **89**, 295 (*synth*, *nmr*)
Brown, H.C. *et al*, *J. Org. Chem.*, 1977, **42**, 2533; 1979, **44**, 2417 (*synth*)

$C_8H_{18}BCl$ B-00562

Dibutylchloroborane, 10CI, 9CI, 8CI

Dibutylchloroborine

[1730-69-4]

$(H_3CCH_2CH_2CH_2)_2BCl$

M 160.493

Catalyst for butadiene oligomerisation; also used in polyester manufacture. Clear colourless liq., easily hydrolysed, sensitive to air. Sol. non-hydroxylic solvs. Bp 173°, Bp_{12} 58-60°. $\Delta H_{Hydrolysis}$ −18.54 kcal mol^{-1}. ΔH_f −31.4 kcal mol^{-1}.

▷Spontaneously flammable in air

Booth, R.B. *et al*, *J. Am. Chem. Soc.*, 1952, **74**, 1415 (*synth*, *props*)
Buls, V.W. *et al*, *J. Am. Chem. Soc.*, 1957, **79**, 337 (*synth*)
Mostovoi, N.V. *et al*, *Izv. Akad. Nauk SSSR, Ser. Khim.*, 1966, 90 (*synth*, *props*)
Köster, R. *et al*, *Justus Liebigs Ann. Chem.*, 1969, **719**, 169 (*synth*)
Brown, H.C. *et al*, *J. Org. Chem.*, 1977, **42**, 2534; 1979, **44**, 2417 (*synth*)
Sax, N.I., *Dangerous Properties of Industrial Materials*, 5th Ed., Van Nostrand-Reinhold, 1979, 551.

$C_8H_{18}BCl_2N$ B-00563

Dichloro(dibutylamino)borane, 8CI

N,N-*Dibutyl-1,1-dichloroboranamine*, 10CI, 9CI

[18379-71-0]

$(H_3CCH_2CH_2CH_2)_2NBCl_2$

M 209.953

Moisture-sensitive.

Gros, G. *et al*, *C.R. Hebd. Seances Acad. Sci.*, 1967, **265**, 1065.
Gallais, F. *et al*, *J. Chim. Phys. Physicochim. Biol.*, 1971, **68**, 247.
Sheikh, S.U., *Bull. Chem. Soc. Jpn.*, 1976, **49**, 2879.
Sheikh, S.U. *et al*, *Pak. J. Sci. Ind. Res.*, 1980, **23**, 21.

$C_8H_{18}BF$ B-00564

Dibutylfluoroborane, 10CI, 9CI, 8CI

Dibutylfluoroborine. Fluorodibutylborane

[589-69-5]

$(H_3CCH_2CH_2CH_2)_2BF$

M 144.039

Colourless mobile liq., sensitive to moisture. Sol. non-hydroxylic solvs. d_4^{24} 0.783. Bp_{20} 48°, 49-52°.

▷Inflames in air

Buls, V.W. *et al*, *J. Am. Chem. Soc.*, 1957, **79**, 337 (*synth*)
DeWitt, E.L. *et al*, *J. Org. Chem.*, 1961, **26**, 4156 (*synth*)
Tuchagues, J.P. *et al*, *Bull. Soc. Chim. Fr.*, 1967, 4160 (*synth*)
Laurent, J.P., *C.R. Hebd. Seances Acad. Sci.*, 1968, **267**, 789 (*props*)
Köster, R., *Justus Liebigs Ann. Chem.*, 1968, **719**, 169 (*synth*)
Costes, J.P. *et al*, *J. Organomet. Chem.*, 1979, **175**, 257 (*props*)

$C_8H_{18}BF_3O$ B-00565

Trifluoroborane dibutyl ether

Trifluoro[1,1′-oxybis[butane]]boron, 10CI. *1,1′-Oxybis-[butane] compd. with trifluoroborane(1:1)*, 9CI. *Butyl ether compd. with boron fluoride(BF_3)(1:1)*, 8CI

[593-04-4]

$(H_3CCH_2CH_2CH_2)_2O—BF_3$

M 198.035

Catalyst in polymerisation reactions. Moisture-sensitive liq. Fp −30°.

Palko, A.A. *et al*, *J. Chem. Phys.*, 1967, **46**, 2297.
Topchii, V.A. *et al*, *Zh. Obshch. Khim.*, 1969, **39**, 2689.
Fratiello, A. *et al*, *J. Org. Chem.*, 1972, **37**, 2237 (*nmr*)
Voloshchuk, A.M. *et al*, *Kernenergie*, 1974, **17**, 121.
Borovikov, Y.Y. *et al*, *Zh. Obshch. Khim.*, 1980, **50**, 2531.
Chmiel-Pela, J., *CA*, 1981, **95**, 189369 (*rev*)

$C_8H_{18}B_2$ B-00566

1,6-Diboracyclodecane

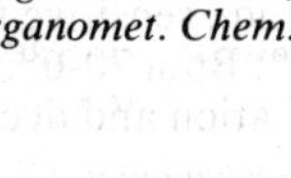

M 135.850

Originally thought to be the simple H-bridged dimer of Borolane, B-00227 . Colourless liq. Insol. H_2O, misc. most org. solvs. Bp_{16} 88-9°, Bp_5 59-60°. n_D^{20} 1.4886, 1.4895. Thermally stable; stable to air, water and AcOH at r.t.

Köster, R. *et al*, *Angew. Chem., Int. Ed. Engl.*, 1963, **2**, 219 (*synth*)
Breuer, E. *et al*, *J. Am. Chem. Soc.*, 1969, **91**, 4164 (*struct*)
Inorg. Synth., 1979, **19**, 239 (*synth, struct, ir, nmr*)

$C_8H_{19}B$ B-00567

Dibutylborane, 10CI, 9CI, 8CI
Dibutylborine
[34266-14-3]

$(H_3CCH_2CH_2CH_2)_2BH$

M 126.048
Dimeric in the uncomplexed state.
Dimer:
$C_{16}H_{38}B_2$ M 252.096
Sol. Et_2O, THF. Readily disproportionates; more stable as Et_2O or Py complexes.

Auten, R.W. *et al*, *J. Am. Chem. Soc.*, 1952, **74**, 3398 (*synth*)
Brown, H.C. *et al*, *J. Organomet. Chem.*, 1971, **32**, C1 (*synth*)
Gupta, S.K., *J. Organomet. Chem.*, 1978, **156**, 95 (*synth*)
Yaniguchi, H., *Nippon Kagaku Kaishi*, 1979, 906; *CA*, **91**, 210504x (*rev*)

$C_8H_{19}BN_2$ B-00568

Bis(dimethylamino)(1-methyl-2-propenyl)borane
N,N,N′,N′-*Tetramethyl-1-(1-methyl-2-propenyl)boranediamine*, 10CI. *1-Buten-3-ylbis(dimethylamino)borane*
[51783-26-7]

$H_2C{=}CHCH(CH_3)B(NMe_2)_2$

M 154.062
Synthetic intermed. Liq., sensitive to air and moisture. Sol. THF. Bp_{48} 89°, Bp_{30} 70-6°. Stable at 200°; at 225° some isomerisation and dec., even under N_2.

Hancock, K.G. *et al*, *J. Organomet. Chem.*, 1974, **64**, C29 (*synth*)
Hoffmann, R.W. *et al*, *Angew. Chem., Int. Ed. Engl.*, 1980, **19**, 218 (*props*)
Hoffmann, R.W. *et al*, *J. Organomet. Chem.*, 1980, **195**, 137 (*synth, pmr*)
Hoffmann, R.W. *et al*, *J. Org. Chem.*, 1981, **46**, 1309 (*props*)

$C_8H_{19}BN_2$ B-00569

2-Butenylbis(dimethylamino)borane
1-(2-Butenyl)-N,N,N′,N′-*tetramethylboranediamine*, 10CI, 9CI
[51783-27-8]

$H_3CCH{=}CHCH_2B(NMe_2)_2$

M 154.062
The *Z* and *E* isomers react diastereospecifically with carbonyl compds.
(*E*)-*form* [69611-00-3]
Liq., sensitive to air and moisture. Bp_{100} 98.6°.
(*Z*)-*form* [69610-99-7]
Liq., sensitive to air and moisture. Bp_{100} 100.8-101.4°.

Hancock, K.G. *et al*, *J. Organomet. Chem.*, 1974, **64**, C29 (*synth*)
Hoffmann, R.W. *et al*, *Angew. Chem., Int. Ed. Engl.*, 1979, **18**, 306; 1980, **19**, 218 (*synth, isom*)
Hoffmann, R.W. *et al*, *J. Org. Chem.*, 1981, **46**, 1309 (*synth, pmr, cmr, isom*)
Hoffmann, R.W. *et al*, *Chem. Ber.*, 1981, **114**, 2802 (*props*)

$C_8H_{19}BO$ B-00570

Butoxydiethylborane
Diethylborinic acid butyl ester, 10CI, 9CI. *Diethyl(butoxy)borane*
[55848-21-0]

$Et_2BO(CH_2)_3CH_3$

M 142.048
Used in synth. of 2-Methylallyl(diethyl)borane and other diethylboryl compds. Air- and moisture-sensitive liq. Sol. nonhydroxylic solvs. Bp_{15} 48-9°.

Köster, R. *et al*, *Justus Liebigs Ann. Chem.*, 1975, 352 (*pmr, synth*)
Mikhailov, B.M. *et al*, *Izv. Akad. Nauk SSSR, Ser. Khim.*, 1978, 1594; *Bull. Acad. Sci. USSR, Div. Chem. Sci.*, **27**, 1392 (*use*)
Mikhailov, B.M. *et al*, *J. Organomet. Chem.*, 1978, **154**, 113 (*props, use*)

$C_8H_{19}BO$ B-00571

Dibutyl(hydroxy)borane
Dibutylborinic acid, 10CI, 9CI
[1189-31-7]

$(H_3CCH_2CH_2CH_2)_2BOH$

M 142.048
Liq. $Bp_{0.5}$ 84°.

Letsinger, R.L. *et al*, *J. Am. Chem. Soc.*, 1954, **76**, 4174 (*synth*)
Brown, H.C. *et al*, *J. Am. Chem. Soc.*, 1971, **93**, 2802; 1972, **94**, 2112; 1976, **96**, 1789, 1798 (*synth*)
Galiullina, R.F. *et al*, *Zh. Obshch. Khim.*, 1980, **50**, 1657.

$C_8H_{19}BO$ B-00572

Ethoxydipropylborane
Dipropylborinic acid ethyl ester, 10CI
[53678-60-7]

$(H_3CCH_2CH_2)_2BOEt$

M 142.048
Air- and moisture-sensitive liq. Sol. nonhydroxylic solvs. Bp_{721} 148-51°, Bp_9 37°. n_D^{20} 1.4024.

Vetter, H.J., *Z. Anorg. Allg. Chem.*, 1964, **330**, 233 (*synth*)
Vetter, H.J., *Z. Naturforsch., B*, 1964, **19**, 72 (*synth*)
Airiyan, L.Sh. *et al*, *Arm. Khim. Zh.*, 1979, **32**, 725; *CA*, **92**, 181248e (*synth, props*)

$C_8H_{19}BO_2$ B-00573

Dihydroxyoctylborane
Octylboronic acid, 10CI. *1-Octaneboronic acid*, 8CI
[28741-08-4]

$H_3C(CH_2)_7B(OH)_2$

M 158.047
Bifunctional reversible inhibitor of many enzyme systems. Solid, somewhat sensitive to air. Mp 79°.

Torssell, K., *Acta Chem. Scand.*, 1957, **11**, 404 (*synth*)
Köster, R., *Justus Liebigs Ann. Chem.*, 1958, **618**, 31 (*synth*)
Gerrard, W. *et al*, *J. Chem. Soc.*, 1964, 740.
Antonov, V.K., *Mol. Biol.*, 1970, **4**, 558 (*use*)
Jarv, J., *Bioorg. Khim.*, 1978, **4**, 1364 (*use*)
Wiecko, J. *et al*, *J. Am. Chem. Soc.*, 1979, **101**, 979 (*ms*)

$C_8H_{19}BO_2$ B-00574

Hexyldimethoxyborane

Hexylboronic acid dimethyl ester, 10CI. *1-Hexaneboronic acid dimethyl ester. Dimethyl hexylboronate*

[2344-23-2]

$H_3C(CH_2)_5B(OMe)_2$

M 158.047

Synthetic intermed. Air- and moisture-sensitive liq. Sol. non-hydroxylic solvs. d^{20} 0.841. Bp_{10} 59-61°. n_D^{20} 1.4082.

Mikhailov, B.M., *Izv. Akad. Nauk SSSR, Ser. Khim.*, 1962, 1756 (*synth*)
Mikhailov, B.M. *et al*, *Zh. Obshch. Khim.*, 1965, **35**, 925, 1073 (*synth*)
Brown, H.C. *et al*, *J. Am. Chem. Soc.*, 1977, **99**, 7097 (*synth*)
Brown, H.C. *et al*, *J. Org. Chem.*, 1980, **45**, 384 (*props*)

$C_8H_{20}B^{\ominus}$ B-00575

Tetraethylborate(1−), 10CI, 9CI, 8CI

[44772-63-6]

$Et_4B^{\ominus}$

M 127.056 (ion)

Li salt: [15003-13-1].
$C_8H_{20}BLi$ M 133.997
Battery electrolyte. Air- and moisture-sensitive.

Na salt: [15523-24-7].
$C_8H_{20}BNa$ M 150.046
White cryst. (toluene); sensitive to air, light, acid; highly hygroscopic. Sol. H_2O, Et_2O, THF diglyme, insol. pentane, sl. sol. C_6H_6. Mp 145°. Aqueous solns. alkaline.

Thompson, R.J. *et al*, *Inorg. Chem.*, 1965, **4**, 1464 (*props*)
Binger, P. *et al*, *Justus Liebigs Ann. Chem.*, 1968, **717**, 21 (*use*)
Nöth, H. *et al*, *J. Organomet. Chem.*, 1968, **12**, 23 (*nmr*)
Inorg. Synth., 1974, **15**, 136 (*synth, use*)
Horowitz, H.H. *et al*, *Proc. Electrochem. Soc.*, 1981, **81**, 131 (*use*)

$C_8H_{20}BBrN_2$ B-00576

Bromobis(diethylamino)borane

*1-Bromo-*N,N,N′,N′*-tetraethylboranediamine*, 10CI, 9CI

[38762-31-1]

$BrB(NEt_2)_2$

M 234.973

Probably liq., sensitive to moisture.

Beck, W. *et al*, *Chem. Ber.*, 1972, **105**, 2883 (*nmr*)
Davidson, G. *et al*, *CA*, 1977, **87**, 175182j (*raman, ir*)
Maringgele, W. *et al*, *Z. Anorg. Allg. Chem.*, 1978, **445**, 107 (*props*)
Bessler, E. *et al*, *Z. Anorg. Allg. Chem.*, 1981, **480**, 117 (*props*)
Davidson, G. *et al*, *J. Chem. Soc., Dalton Trans.*, 1981, 306 (*ir, raman*)

$C_8H_{20}BClN_2$ B-00577

Chlorobis(diethylamino)borane, 8CI

*1-Chloro-*N,N,N′,N′*-tetraethylboranediamine*, 10CI, 9CI. *Chlorobis(diethylamino)borine. Bisdiethylaminoboron chloride*

[868-25-7]

$ClB(NEt_2)_2$

M 190.522

Liq., sensitive to moisture. Bp_{16} 83-7°. n_D^{20} 1.4549.

Gerrard, W. *et al*, *J. Chem. Soc.*, 1957, 381 (*synth*)
Bellamy, L.J. *et al*, *J. Chem. Soc.*, 1958, 2412 (*ir*)
Panidi, I.S., *Dokl. Akad. Nauk SSSR, Ser. Sci. Khim.*, 1965, **41**, 226 (*synth, props*)
Panidi, I.S., *CA*, 1967, **66**, 18485j (*synth*)
Davis, F.A. *et al*, *J. Org. Chem.*, 1971, **36**, 1300 (*nmr*)
Beck, W. *et al*, *Chem. Ber.*, 1972, **105**, 2883 (*nmr*)

$C_8H_{20}BN$ B-00578

Aminodibutylborane, 8CI

1,1-Dibutylboranamine, 10CI, 9CI. *Aminodibutylborine*

[1767-37-9]

$(H_3CCH_2CH_2CH_2)_2BNH_2$

M 141.063

Air- and moisture-sensitive liq. Bp_8 56-8°. n_D^{20} 1.4278.

Mikhailhov, B.M. *et al*, *Zh. Obshch. Khim.*, 1962, **32**, 1969 (*synth*)
Mikhailhov, B.M. *et al*, *Dokl. Akad. Nauk SSSR, Ser. Sci. Khim.*, 1970, **194**, 595 (*nmr*)
Beck, W. *et al*, *Chem. Ber.*, 1972, **105**, 2883 (*nmr*)
Dorokhov, V.A. *et al*, *Izv. Akad. Nauk SSSR, Ser. Khim.*, 1976, 1432; *Bull. Acad. Sci. USSR, Div. Chem. Sci.*, 1375.
Dorokhov, V.A. *et al*, *Izv. Akad. Nauk SSSR, Ser. Khim.*, 1977, 1921; *Bull. Acad. Sci. USSR, Div. Chem. Sci.*, 1785.
Dorokhov, V.A. *et al*, *Dokl. Akad. Nauk SSSR, Ser. Sci. Khim.*, 1981, **258**, 351.

$C_8H_{20}BN$ B-00579

Diethylaminodiethylborane, 8CI

N,N,*1,1-Tetraethylboranamine*, 10CI, 9CI

[4023-39-6]

Et_2BNEt_2

M 141.063

Catalyst for polymerisation of propene. Air- and moisture-sensitive liq. Bp 154° (161°). n_D^{20} 1.4245.

Niedenzu, K. *et al*, *J. Am. Chem. Soc.*, 1960, **82**, 4223 (*synth*)
Wyman, G.M. *et al*, *J. Chem. Soc.*, 1962, 4068 (*ir*)
Nöth, H. *et al*, *Chem. Ber.*, 1966, **99**, 1049 (*nmr*)
Beck, W. *et al*, *Chem. Ber.*, 1972, **105**, 2883 (*nmr*)
Binger, P. *et al*, *J. Organomet. Chem.*, 1974, **73**, 205 (*use*)
Gupta, S.K. *et al*, *J. Organomet. Chem.*, 1978, **156**, 95 (*synth*)
Nöth, H. *et al*, *Chem. Ber.*, 1981, **114**, 1150 (*cmr*)

$C_8H_{20}BN$ B-00580

Dimethylaminodipropylborane

N,N*-Dimethyl-1,1-dipropylboranamine*, 10CI, 9CI

[13113-78-5]

$(H_3CCH_2CH_2)_2BNMe_2$

M 141.063

Air- and moisture-sensitive liq. Bp 136-8°, Bp_{12} 57-9°.

Nöth, H. *et al*, *Z. Naturforsch., B*, 1961, **16**, 618 (*synth*)
Nöth, H. *et al*, *Chem. Ber.*, 1964, **97**, 709 (*synth*)
Eubanks, I.D. *et al*, *J. Am. Chem. Soc.*, 1966, **88**, 2425 (*uv*)
Costes, J.P. *et al*, *J. Organomet. Chem.*, 1979, **175**, 257 (*pmr*)
Emerick, D.P. *et al*, *Z. Anorg. Allg. Chem.*, 1980, **464**, 44 (*nmr*)

$C_8H_{20}B_2N_4$ B-00581

Tetramethylethylenediamine bis(cyanoborane)

*Bis(cyano-*C*)tetrahydro(μ-*N,N,N′,N′*-tetramethyl-1,2-ethanediamine-*N:N′*)diboron*, 10CI. *Boranecarbonitrile compd. with* N,N,N′,N′*-tetramethyl-1,2-ethanediamine(2:1)*, 9CI

[30353-64-1]

$NCBH_2—Me_2NCH_2CH_2NMe_2—H_2BCN$

M 193.893

White powder. Mp 169-71°.

▷ED1090000.

Kelly, H.C. *et al*, *J. Chem. Soc., Chem. Commun.*, 1970, 1619 (*synth*)
Weidig, C. *et al*, *Inorg. Chem.*, 1974, **13**, 1763 (*synth, ir, pmr*)
Spielvogel, B.F. *et al*, *J. Inorg. Nucl. Chem.*, 1979, **41**, 1223 (*pharmacol*)
Hall, I.H. *et al*, *J. Pharm. Sci.*, 1980, **69**, 1025; 1981, **70**, 339 (*pharmacol*)

$C_8H_{20}B_2O$ B-00582

Oxybis[diethylborane], 8CI

Diethylborinic acid anhydride, 10CI, 9CI. *Bis(diethylboron)oxide. Diethylboric oxide. Tetraethyldiboroxane*

[7318-84-5]

Et_2BOBEt_2

M 153.865

Air-sensitive liq. Sol. nonhydroxylic solvs. Bp 143°, Bp_{11} 47°.

Meerwein, H. *et al*, *J. Prakt. Chem.*, 1936, **147**, 251 (*synth*)
Rothstein, E. *et al*, *J. Chem. Soc.*, 1952, 2987.
Nöth, H. *et al*, *Chem. Ber.*, 1966, **99**, 1049 (*nmr*)
Nöth, H. *et al*, *J. Organomet. Chem.*, 1968, **12**, 23 (*pmr*)
Köster, R. *et al*, *Justus Liebigs Ann. Chem.*, 1974, 54 (*synth, ms, nmr*)
Shitov, O.P. *et al*, *Izv. Akad. Nauk SSSR, Ser. Khim.*, 1974, 2782; *Bull. Acad. Sci. USSR, Div. Chem. Sci.*, **23**, 2684 (*props*)
Biffar, W. *et al*, *Chem. Ber.*, 1980, **113**, 333 (*nmr*)

$C_8H_{22}B_2N_2O_4$ B-00583

***N,N,N′,N′*-Tetramethylethylenediamine bis(carboxyborane)**

Boranecarboxylic acid compd. with N,N,N′,N′*-tetramethyl-1,2-ethanediamine(2:1)*, 10CI

[72695-30-8]

$HOOCBH_2–NMe_2CH_2CH_2NMe_2–BH_2COOH$

M 231.893

Prepd. from $TMEDA(BH_2CN)_2$. Shows anticholesteraemic and antihypolipidaemic activity. Air-sensitive white cryst. (H_2O). Sol. $CHCl_3$. Mp 270° dec.

Pelter, A. *et al*, *Compr. Org. Chem.*, (*Barton, D.H.R. et al*, Ed.), 1979, **3**, 695 (*rev*)
Spielvogel, B.F. *et al*, *J. Inorg. Nucl. Chem.*, 1979, **41**, 1223 (*synth*)
Hall, I.H. *et al*, *J. Pharm. Sci.*, 1980, **69**, 1025; 1981, **70**, 339 (*pharmacol*)

$C_8H_{22}B_8Fe$ B-00584

Bis[(2,3,4,6-η)-pentahydro-4,5-dimethyl-4,5-dicarbahexaborato(1–)-H^{μ}]iron

Bis(2,3-dimethyl-2,3-dicarbahexaboranyl)iron dihydride. $[2,3\text{-}[C_2B_4H_4(CH_3)_2]]_2FeH_2$

[58846-86-9]

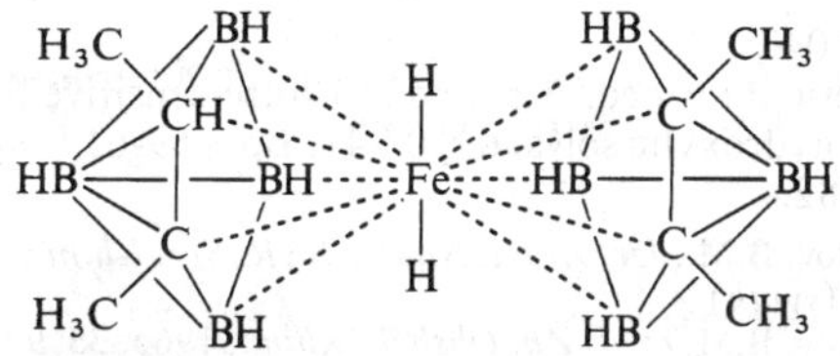

M 260.589

H-Fe-H atoms not located. Extremely air-sensitive deep-red cryst. Bp 0° *in vacuo* subl.

Maxwell, W.M. *et al*, *Inorg. Chem.*, 1976, **15**, 1343 (*synth, nmr, pmr, ms, ir*)
Pipal, J.R. *et al*, *Inorg. Chem.*, 1979, **18**, 263 (*synth, struct*)

$C_8H_{24}B_{10}$ B-00585

1-Hexyl-1,2-dicarbadodecaborane(12), 10CI

*1-Hexyl-*o*-carborane*

[20740-05-0]

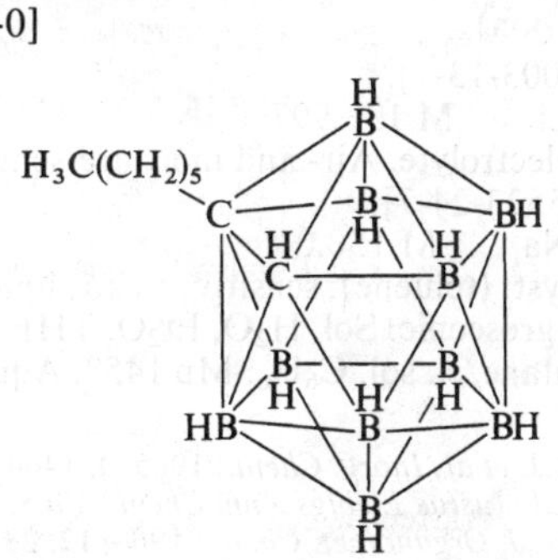

M 228.378

closo-. Rocket propellant burn rate modifier. Liq. Bp_{760} 326.7°, $Bp_{0.5}$ 102°. n_D^{25} 1.5211.

▷HS9350000.

Fein, M.M. *et al*, *Inorg. Chem.*, 1963, **2**, 1115 (*synth*)
Heying, T.L. *et al*, *Inorg. Chem.*, 1963, **2**, 1089 (*synth*)

$C_8H_{30}B_{18}Fe^{\ominus}$ B-00586

Bis[(7,8,9,10,11-η)-1,2,3,4,5,6,9,10,11-nonahydro-7,8-dimethyl-7,8-dicarbaundecaborato(2–)]ferrate(1–), 10CI

Bis[undecahydro-7,8-dimethyl-π-7,8-dicarbaundecaborato(2–)]ferrate(1–), 8CI

[11070-28-3]

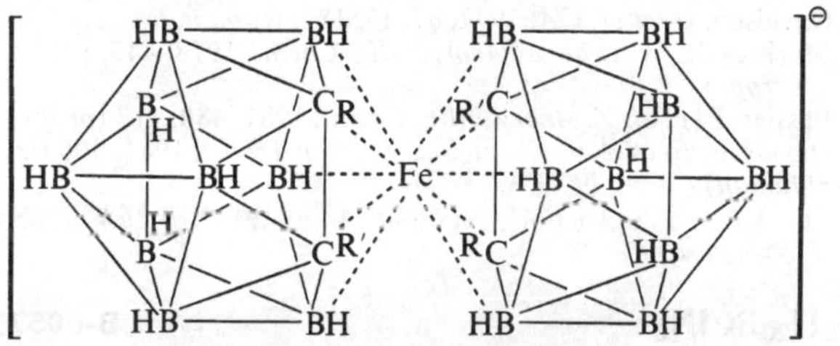

R = R′ = CH_3

M 376.752 (ion)

Paramagnetic. Reduced to a deep-lavender-blue ferrate (2–) species.

Tetramethylammonium salt: [12556-51-3].
$C_{12}H_{42}B_{18}FeN$ M 450.898

Red cryst. (Me_2CO aq.). Mp 247-9°.

Gray, H.B. *et al*, *Inorg. Chem.*, 1971, **10**, 1559.
Ammeter, J.H. *et al*, *J. Magn. Reson.*, 1978, **30**, 299 (*esr*)

$C_8H_{32}B_{20}Ti^{\ominus\ominus}$ B-00587

Bis[η^6-decahydro-*C,C'*-dimethyldicarbadodecaborato(2−)]-titanate(2−), 9CI
*4,4'-Bis(decahydro-1,6-dimethyl-1,6-dicarba-4-titana-*closo-*tridecaborate)(2−)*

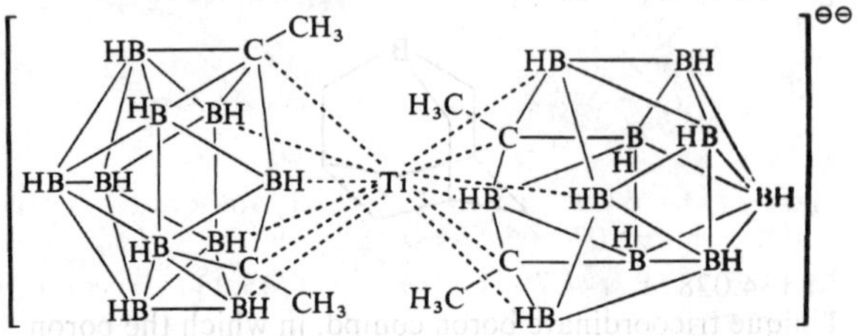

M 392.421 (ion)
Bis(tetramethylammonium) salt: [55266-92-7].
$C_{16}H_{56}B_{20}N_2Ti$ M 540.712
Red cryst. + $2Me_2CO$ (Me_2CO/EtOH), air-stable in solid state.

Salentine, C.G. *et al*, *J. Am. Chem. Soc.*, 1975, **97**, 426 (*synth, pmr, nmr, props*)
Yo, F.Y. *et al*, *J. Am. Chem. Soc.*, 1975, **97**, 428 (*cryst struct*)

$C_9H_9BFeO_4$ B-00588

Tricarbonyl[(3,4,5,6-η)-1,2-dihydro-1-methoxyborin]iron, 10CI
[72401-05-9]

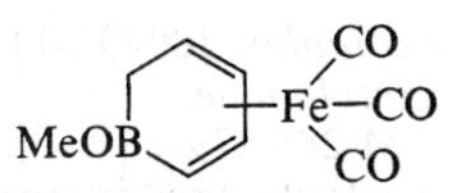

M 247.825
Precursor for bis(borabenzene) iron. Air-sensitive yellow oil.

Ashe, A.J. *et al*, *J. Am. Chem. Soc.*, 1979, **101**, 7066 (*synth, nmr, pmr, ms*)

$C_9H_{10}BN_6^{\ominus}$ B-00589

Hydrotris(1*H*-pyrazolato-*N*[1])borate(1−), 10CI, 9CI
[46755-84-4]

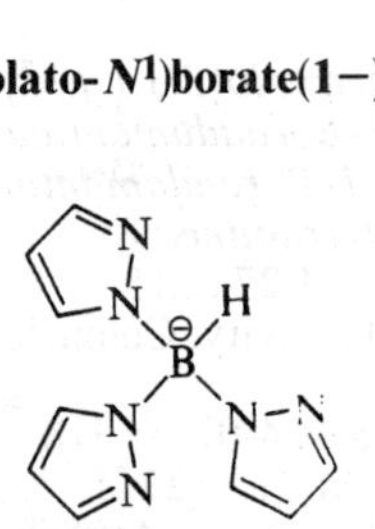

M 213.028 (ion)
Used in metal complex formn.
Li salt: [18583-61-4]. Needles (toluene). Mp 264-5°. Bp_2 240° subl.
Na salt: [18583-62-5].
$C_9H_{10}BN_6Na$ M 236.018
Cryst. (1,2-dimethoxyethane or by subl.). Moisture-sensitive. Mp 222-3°. Bp_2 210-5° subl.
K salt: [18583-60-3].
$C_9H_{10}BKN_6$ M 252.127
Fine needles (anisole). Mp 188-9°.

Trofimenko, S., *J. Am. Chem. Soc.*, 1967, **89**, 3170 (*synth*)
Inorg. Synth., 1970, **12**, 102 (*synth*)
Trofimenko, S., *Chem. Rev.*, 1972, **72**, 497 (*rev*)
Bruno, G. *et al*, *Inorg. Chim. Acta*, 1981, **48**, 61 (*struct, pe*)
Fragala, I. *et al*, *CA*, 1981, **94**, 217059b (*uv, pe*)

$C_9H_{11}BCl_2$ B-00590

Dichloro(2,4,6-trimethylphenyl)borane, 10CI
Dichloromesitylborine. Dichloromesitylborane. Mesityldichloroborane
[69464-76-2]

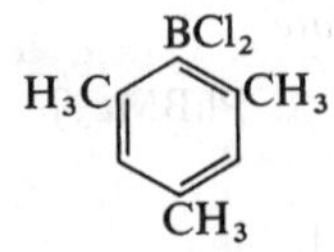

M 200.902
Fuming liq., rapidly hydrol. by H_2O. Sol. toluene.

Muetterties, E.L., *J. Am. Chem. Soc.*, 1959, **81**, 2597 (*synth*)
Hakan, G.K., *Acta Chem. Scand., Ser. B*, 1978, **32**, 765 (*use*)
Paetzold, P. *et al*, *Chem. Ber.*, 1979, **112**, 3811 (*use*)

$C_9H_{12}B_2FeO_3S$ B-00591

Tricarbonyl[(1,2,3,4,5-η)-3,4-diethyl-2,5-dihydro-1,2,5-thiadiborole]iron, 10CI
[63339-02-6]

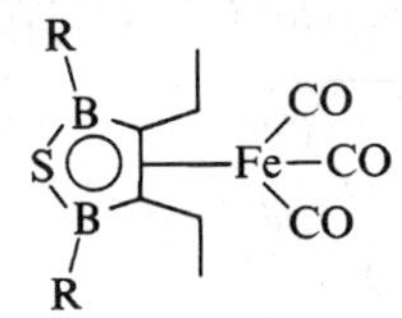

R = H

M 277.719
Yellow liq. Bp_1 49°.

Siebert, W. *et al*, *J. Organomet. Chem.*, 1977, **131**, 1 (*synth, pmr, nmr, ir, ms*)

$C_9H_{12}B_4FeO_2$ B-00592

Dicarbonyl(η^5-2,4-cyclopentadien-1-yl)[(2,3-η)-heptahydro-4,5-dicarbahexaborato(1−)]iron
[38999-56-3]

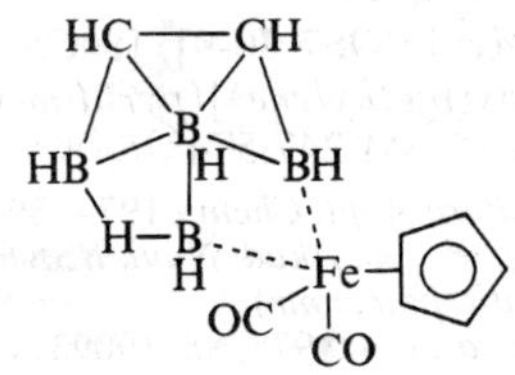

M 251.280
Intermed. for synthesis of (η^5-2,4-Cyclopentadien-1-yl)-[(2,3,4,5,6-η)-heptahydro-2,5-dicarbahexaborato(1−)-H^{μ}]iron, B-00485 . Bright-yellow.

Sneddon, L.G. *et al*, *J. Am. Chem. Soc.*, 1972, **94**, 7161; 1973, **95**, 6623 (*synth, ir, ms, pmr, nmr*)

$C_9H_{13}BO_2$ B-00593

Dihydroxy(3-phenylpropyl)borane
(3-Phenylpropyl)boronic acid, 10CI, 9CI
[36329-85-8]

$PhCH_2CH_2CH_2B(OH)_2$

M 164.011
Bifunctional reversible inhibitor of many enzyme systems.

▷Toxic

Rotonova, T.V., *Bioorg. Khim.*, 1976, **2**, 837 (*use, pharmacol*)
Carter, D.B., *Biochim. Biophys. Acta*, 1977, **484**, 103 (*use, pharmacol*)
Rotonova, T.V., *Bioorg. Khim.*, 1978, **4**, 694 (*use, pharmacol*)

$C_9H_{14}B^{\ominus}$ B-00594

Trimethylphenylborate(1−), 10CI
Phenyltrimethylborate

$PhBMe_3^{\ominus}$

M 133.020 (ion)
Li salt: [78885-81-1].
$C_9H_{14}BLi$ M 139.961
Used in battery electrolytes.

Horowitz, H.H. *et al*, *Proc. Electrochem. Soc.*, 1981, **81**, 131 (*props, use*)
Klemann, L.P. *et al*, *Proc. Electrochem. Soc.*, 1981, **81**, 179 (*props*)

$C_9H_{14}BCl$ B-00595

4-Chloro-1-boraadamantane
4-Chloro-1-boratricyclo[3.3.1.1^{3,7}]decane, 10CI, 9CI

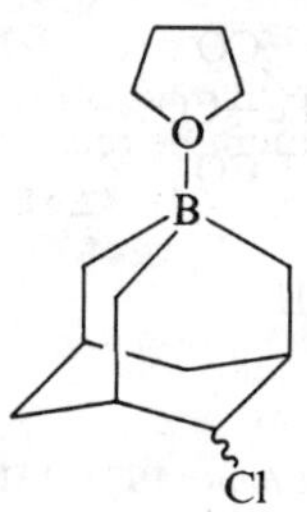

THF complex

M 168.473
Known only in the form of its complexes.
THF complex: [61392-02-7]. [*2-Chloro-1,3,5-cyclohexanetriyltris(methylene)*](*tetrahydrofuran*)*boron*, 10CI.
$C_{13}H_{22}BClO$ M 240.579
White cryst. Mp 47-9°.
Pyridine complex: [55967-70-9]. [*2-Chloro-1,3,5-cyclohexanetriyltris(methylene)*](*pyridine*)*boron*, 10CI.
$C_{14}H_{19}BClN$ M 247.574

Mikhailov, B.M., *Pure Appl. Chem.*, 1974, **39**, 505 (*rev*)
Mikhailov, B.M. *et al*, *Izv. Akad. Nauk SSSR, Ser. Khim.*, 1976, 2056 (*synth, pmr, nmr*)
Mikhailov, B.M. *et al*, *CA*, 1978, **88**, 190937y (*synth*)

$C_9H_{14}BF_3O_3S$ B-00596

9-[[(Trifluoromethyl)sulfonyl]oxy]-9-borabicyclo[3.3.1]nonane, 10CI
9-Borabicyclo[3.3.1]nonane trifluoromethanesulfonate
9-BBN triflate
[62731-43-5]
As 9-Bromo-9-borabicyclo[3.3.1]nonane, B-00539 with

$R = —OSO_2CF_3$

M 270.073
Reagent for generation of highly stereoselective boron enolates. Liq., moisture-sensitive. $Bp_{0.03}$ 38°.

Inoue, T. *et al*, *Chem. Lett.*, 1977, 153, 1215.
Hirama, M. *et al*, *Tetrahedron Lett.*, 1979, 3937.
Yamamota, Y. *et al*, *Yuki. Gosei. Kagaku Kyokaishi*, 1979, **37**, 1008; *CA*, **92**, R139884j (*synth, use*)
Inoue, T. *et al*, *Bull. Chem. Soc. Jpn.*, 1980, **53**, 174 (*synth, ir*)

$C_9H_{15}B$ B-00597

1-Boratricyclo[3.3.1.1^{3,7}]decane, 10CI, 9CI
1,3,5-Cyclohexanetriyltris[methylene]boron, 10CI. *1-Boraadamantane*
[38673-97-1]

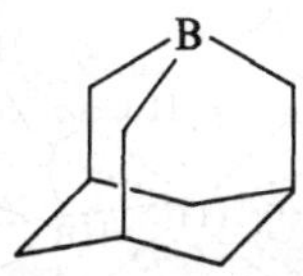

M 134.028
Unique tricoordinate boron compd. in which the boron atom exists in pyramidal and not trigonal state. Forms complexes with amines and related ligands readily. Colourless cryst. by subl. Thermally unstable, fumes in air. Mp 80°. High reactivity by comparison with simple trialkylboranes.
NMe_3 complex: [69874-56-2]. *1-Boraadamantane trimethylamine.*
$C_{12}H_{24}BN$ M 193.138
Cryst. Shows antiviral props. against influenza virus A. Can be vacuum-distilled without dissociation.
Octadecylamine complex: [69492-12-2]. *1-Boraadamantane octadecanamine. 1-Boraadamantane stearylamine.*
$C_{27}H_{54}BN$ M 403.540
Oil.
Adamantylamine complex: [69492-11-1]. *1-Boraadamantane 1-adamantylamine.*
$C_{19}H_{32}BN$ M 285.279
Shows antiviral activity; used in treatment of influenza A infections in chickens. Cryst. (EtOH). Mp 190-2°.
(1-Adamantylethyl)amine complex: [69874-58-4] *1-Boratricyclo[3.3.1.1^{3,7}]decane compd. with α-methyltricyclo[3.3.1.1^{3,7}]decane-1-methanamine(1:1)*, 10CI. *1-Boraadamantane 1-adamantylethylamine.*
$C_{21}H_{36}BN$ M 313.332
Heat-stabiliser for diene rubbers. Shows virucidal activity.
1,3,5,7-Tetraazatricyclo[3.3.1.1^{3,7}]decane complex: [61632-80-2]. *1-Boraadamantane 1,3,5,7-tetraazaadamantane. 1-Boraadamantane hexamethylenetetramine.*
$C_{15}H_{27}BN_4$ M 274.215
Shows virucidal activity. Colourless cryst. (hexane). Mp 275-8°.
Py complex: [37863-04-0].
$C_{14}H_{20}BN$ M 213.129
Shows virucidal activity. Monoclinic cryst. (EtOH). Mp 160-2°.
2-Methylpyridine complex: [69464-63-7]. *1-Boraadamantane 2-methylpyridine. 1-Boraadamantane α-picoline.*
$C_{15}H_{22}BN$ M 227.156
Cryst. (EtOH). Mp 135-8°.
3-Methylpyridine complex: [69492-10-0]. *1-Boraadamantane 3-methylpyridine. 1-Boraadamantane β-picoline.* Cryst. (EtOH). Mp 127-9°.
THF complex: [61357-36-6].
$C_{13}H_{23}BO$ M 206.134
Thermally unstable colourless cryst. (hexane). Mp 84-8° (79-83°).

Et_2O complex: [55969-64-1].
$C_{13}H_{25}BO$ M 208.150
Useful for synth. of other 1-boraadamantane complexes with stronger ligands. Shows antiviral activity. Colourless, viscous liq., partially dissociated at 25°.
$EtNH_2$ complex: [61632-81-3]. *[1,3,5-Cyclohexanetriyltris[methylene]](ethanamine)boron*, 10CI.
$C_{11}H_{22}BN$ M 179.112
Colourless cryst. (pentane or MeOH aq.). Mp 148-52°.

Mikhailov, B.M. *et al*, *Izv. Akad. Nauk SSSR, Ser. Khim.*, 1972, 1672; 1973, 2165 (*synth, props*)
Mikhailov, B.M. *et al*, *Izv. Akad. Nauk SSSR, Ser. Khim.*, 1976, 2302; 1979, 2544; *Bull. Acad. Sci. USSR, Div. Chem. Sci.*, **25**, 2148; **28**, 2361 (*synth, pmr, nmr*)
Mikhailov, B.M. *et al*, *Dokl. Akad. Nauk SSSR, Ser. Sci. Khim.*, 1978, **243**, 929 (*synth*)
Mikhailov, B.M. *et al*, *Khim.-Farm. Zh.*, 1979, **13**, 35; *CA*, **90**, 168661y (*synth, use*)
Buyanovskaya, A.G. *et al*, *Zh. Anal. Khim.*, 1980, **35**, 2106; *CA*, **94**, 113985j.
Mikhailov, B.M. *et al*, *Izv. Akad. Nauk SSSR, Ser. Khim.*, 1980, 1106, 1621; *Bull. Acad. Sci. USSR, Div. Chem. Sci.*, **29**, 813, 1164 (*props, cmr*)
Mikhailov, B.M., *Pure Appl. Chem.*, 1980, **52**, 691 (*rev, synth, nmr, props*)
U.S.S.R. Pat., 726 128, (*1980*); *CA*, **93**, 96561c (*use*)
Erokhina, L.M. *et al*, *CA*, 1981, **95**, 125883h (*use*)
Gurskii, M.E. *et al*, *Izv. Akad. Nauk SSSR, Ser. Khim.*, 1981, 394; *Bull. Acad. Sci. USSR, Div. Chem. Sci.*, **30**, 312 (*props*)
Vorontsova, L.G. *et al*, *Izv. Akad. Nauk SSSR, Ser. Khim.*, 1981, 595; *Bull. Acad. Sci. USSR, Div. Chem. Sci.*, **30**, 438 (*synth, struct*)

$C_9H_{15}B$ B-00598

2-Boratricyclo[3.3.1.1^{3,7}]decane, 10CI
2-Boraadamantane
[56701-60-1]

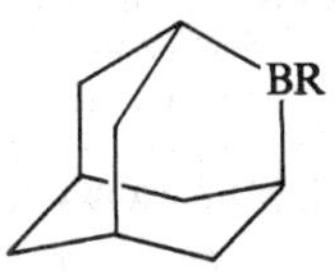

R = H

M 134.028
Dimeric (BH-bridged). Forms complexes with amines.
Dimer:
$C_{18}H_{30}B_2$ M 268.055
Solid. Mp 193-8°.

Mikhailov, B.M. *et al*, *Izv. Akad. Nauk SSSR, Ser. Khim.*, 1977, 894; *Bull. Acad. Sci. USSR, Div. Chem. Sci.*, **26**, 818 (*synth, ir, pmr, nmr*)
Mikhailov, B.M., *Pure Appl. Chem.*, 1980, **52**, 691 (*synth*)

$C_9H_{15}B$ B-00599

Tri-2-propenylborane, 10CI, 9CI
Triallylborane, 8CI. *Triallylborine. Triallylboron*
[688-61-9]

$$(H_2C{=}CHCH_2)_3B$$

M 134.028
Prepd. from allylmagnesium bromide and $BF_3{\cdot}Et_2O$ under N_2. Synthetic intermed. used for transfer of an allyl group. Air- and moisture-sensitive liq. Sol. Et_2O, other non-hydroxylic solvs. d^{20} 0.718. Bp_{15} 62-5°. Undergoes allylic rearrangement at 25°; at 135° the pmr shows only two signals, at δ 5.80 (1*H*, quintet) and δ 3.5 (4*H*, doublet).
▷Ignites in air
Py complex: [14535-62-7]. *Pyridine triallylborane.*
$C_{14}H_{20}BN$ M 213.129
Can be distilled. Bp_4 116-8°.

Bogdanov, V.S. *et al*, *Izv. Akad. Nauk SSSR, Ser. Khim.*, 1971, 1363 (*nmr*)
Bogdanov, V.S. *et al*, *Zh. Obshch. Khim.*, 1971, **41**, 1533 (*derivs*)
Mikhailov, B.M., *Organomet. Chem. Rev., Sect. A*, 1972, **8**, 1 (*rev*)
Mikhailov, B.M. *et al*, *Izv. Akad. Nauk SSSR, Ser. Khim.*, 1974, 1137; *Bull. Acad. Sci. USSR, Div. Chem Sci.*, 1079 (*rev*)
Mikhailov, B.M. *et al*, *Zh. Obshch. Khim.*, 1974, **44**, 1878 (*derivs*)
Brueser, W. *et al*, *Z. Anorg. Allg. Chem.*, 1976, **421**, 89 (*ir, raman*)
Schroeder, S. *et al*, *Z. Anorg. Allg. Chem.*, 1977, **428**, 225 (*synth*)

$C_9H_{15}BN^{\ominus}$ B-00600

9-Cyano-9-hydro-9-boratabicyclo[3.3.1]nonane, 10CI
9-Cyano-9-hydrido-9-borabicyclo[3.3.1]nonane. B-Cyano-9-BBN hydride
As 9-Boratabicyclo[3.3.1]nonane, B-00546 with

R = CN, R′ = H

M 148.034 (ion)
Na salt: [60645-80-9].
$C_9H_{15}BNNa$ M 171.024
Reducing agent. Tacky, air- and moisture-sensitive solid. Difficult to purify, cannot be recryst.

Hutchins, R.O. *et al*, *J. Org. Chem.*, 1977, **42**, 82 (*synth, ir*)

$C_9H_{15}BO$ B-00601

4-Oxa-3-boratricyclo[4.3.1.1^{3,8}]undecane
4-Oxa-3-borahomoadamantane

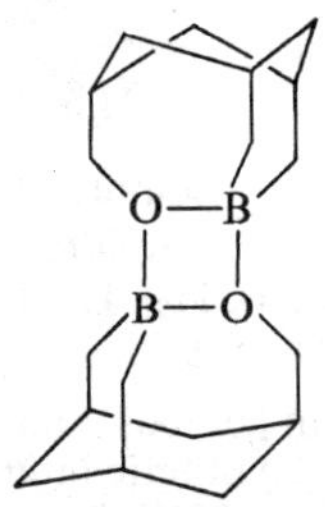

Dimer

M 150.027
Dimeric.
Dimer: [72930-44-0]. *Bis[μ-[3,5-dimethylcyclohexanemethanolato(3−)]]diboron*, 10CI.
$C_{18}H_{30}B_2O_2$ M 300.054
White cryst. stable in air. Mp >360°. Converts to polymeric mass on heating.

Mikhailov, B.M. *et al*, *Izv. Akad. Nauk SSSR, Ser. Khim.*, 1979, 2729; *Bull. Acad. Sci. USSR, Div. Chem. Sci.*, 2537 (*synth*)
Gurskii, M.E. *et al*, *J. Org. Chem.*, 1980, **199**, 171 (*nmr*)
Mikhailov, B.M. *et al*, *Izv. Akad. Nauk SSSR, Ser. Khim.*, 1980, 1106; *Bull. Acad. Sci. USSR, Div. Chem. Sci.*, 813 (*synth, nmr, pmr*)
Gurskii, M.E. *et al*, *Izv. Akad. Nauk SSSR, Ser. Khim.*, 1981, 341; *Bull. Acad. Sci. USSR, Div. Chem. Sci.*, 264 (*nmr*)

$C_9H_{15}BO_3$ B-00602

Tri(2-propenyloxy)borane
Boric acid (H_3BO_3) tri-2-propenyl ester, 10CI, 9CI. *Boric acid (H_3BO_3) triallyl ester*, 8CI. *Triallyl borate. Allyl borate. Triallyl orthoborate. Tris(allyloxy)borane. Triallyloxyborane*

[1693-71-6]

$(H_2C{=}CHCH_2O)_3B$

M 182.026

Crosslinking agent for unsatd. polyesters in presence of H_2O. Moisture-sensitive liq. Sol. non-hydroxylic solvs. d_4^{20} 0.926, d^{10} 1.04. Bp_{630} 163-5°, Bp_{15} 76°, Bp_7 42°. n_D^{21} 1.4276.

▷ED4810000.

Haider, S.Z. *et al*, *J. Appl. Chem.*, 1954, **4**, 93 (*props*)
Gerrard, W. *et al*, *J. Chem. Soc.*, 1956, 3285 (*synth*)
Brown, H.C. *et al*, *J. Chem. Soc.*, 1956, **78**, 3613 (*props*)
Bogdanov, V.S. *et al*, *CA*, 1968, **68**, 100409n (*nmr*)
Pola, M. *et al*, *Collect. Czech. Chem. Commun.*, 1979, **44**, 3688 (*ir, props*)
U.S.P., 4 283 533, (*1981*); *CA*, **95**, 171468z (*use*)

$C_9H_{16}BBr$ B-00603

3-Bromo-7-methyl-3-borabicyclo[3.3.1]nonane, 10CI

3-Bromo-7-methyl-3-BBN

[74789-85-8]

As 3-Borabicyclo[3.3.1]nonane, B-00544 with

X = Br, R = CH_3

M 214.939

Air- and moisture-sensitive liq. Sol. CH_2Cl_2. $Bp_{8.5}$ 91-2°. n_D^{20} 1.5175.

Gurskii, M.E. *et al*, *J. Organomet. Chem.*, 1980, **199**, 171 (*cmr*)
Mikhailov, B.M. *et al*, *Izv. Akad. Nauk SSSR, Ser. Khim.*, 1980, 1106; *Bull. Acad. Sci. USSR, Div. Chem. Sci.*, 813 (*synth*)
Mikhailov, B.M. *et al*, *Izv. Akad. Nauk SSSR, Ser. Khim.*, 1981, 341; *Bull. Acad. Sci. USSR, Div. Chem. Sci.*, 264 (*cmr*)

$C_9H_{17}B$ B-00604

Di-2-propenylpropylborane, 10CI, 9CI

Propyldi-2-propenylborane. Diallylpropylborane. Propyldiallylborane

[56347-95-6]

$H_3CCH_2CH_2B(CH_2CH{=}CH_2)_2$

M 136.043

Prepd. from allylaluminum compds. and propyldialkoxyboranes. Synthetic intermed. Air- and moisture-sensitive liq. Sol. THF. Bp_8 37-8°. n_D^{20} 1.4385.

Mikhailov, B.M. *et al*, *Izv. Akad. Nauk SSSR, Ser. Khim.*, 1975, 979; 1978, 1586 (*synth*)
Mikhailov, B.M. *et al*, *J. Organomet. Chem.*, 1978, **154**, 113 (*synth*)
Pelter, A. *et al*, *Comprehensive Organic Chemistry*, (*Barton, D.H.R., et al*, Ed.), Pergamon, Oxford, 1979, **3**, 695 (*rev*)

$C_9H_{17}B$ B-00605

9-Methyl-9-borabicyclo[3.3.1]nonane, 10CI

B-*Methyl-9-BBN*

[23418-81-7]

As 9-Bromo-9-borabicyclo[3.3.1]nonane, B-00539 with

R = Me

M 136.043

Oil, sensitive to air. $Bp_{1.0}$ 38-40°.

Yamamoto, Y. *et al*, *Chem. Lett.*, 1975, 1199 (*nmr*)
Brown, H.C. *et al*, *Tetrahedron Lett.*, 1977, 3007.
Wrackmeyer, B., *Z. Naturforsch., B*, 1980, **35**, 439 (*nmr*)
Contreras, R. *et al*, *Z. Naturforsch., B*, 1980, **35**, 1236 (*nmr*)

Whiteley, C.G., *J. Chem. Soc., Chem. Commun.*, 1981, 5 (*synth*)
Whiteley, C.G. *et al*, *S. Afr. J. Chem.*, 1982, **35**, 9 (*synth*)

$C_9H_{17}BClNO$ B-00606

(Diethylamino)chloro(cyclopentenyloxy)borane

*1-Chloro-1-(1-cyclopenten-1-yloxy)-*N,N*-diethylboranamine, 10CI*

[71466-95-0]

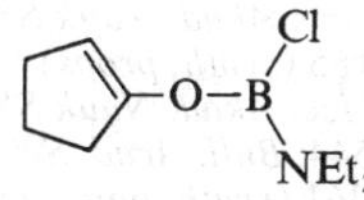

M 201.502

Moisture-sensitive.

Sugasawa, T. *et al*, *Synth. Commun.*, 1979, **9**, 583.

$C_9H_{17}BO$ B-00607

3-Methoxy-3-borabicyclo[3.3.1]nonane, 10CI

MBBN

[17428-04-5]

As 3-Borabicyclo[3.3.1]nonane, B-00544 with

X = OMe, R = H

M 152.043

Important parent compd. with many known derivs. having substituents at the 7α-position. Air- and moisture-sensitive. Sol. CH_2Cl_2.

Mikhailov, B.M. *et al*, *Izv. Akad. Nauk SSSR, Ser. Khim.*, 1967, 2290 (*synth*)
Gurskii, M.E. *et al*, *J. Organomet. Chem.*, 1980, **199**, 171 (*cmr*)
Mikhailov, B.M. *et al*, *Izv. Akad. Nauk SSSR, Ser. Khim.*, 1981, 341; *Bull. Acad. Sci. USSR, Div. Chem. Sci.*, 264 (*cmr*)
Gurskii, M.E. *et al*, *Izv. Akad. Nauk SSSR, Ser. Khim.*, 1981, 394; *Bull. Acad. Sci. USSR, Div. Sci. Chem.*, 312.

$C_9H_{17}BO$ B-00608

9-Methoxy-9-borabicyclo[3.3.1]nonane, 10CI, 9CI

B-*Methoxy-9-*BBN

[38050-71-4]

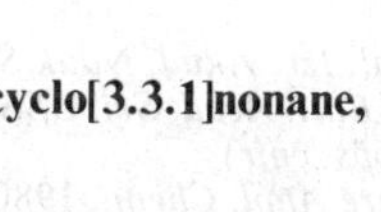

M 152.043

Liq., moisture- and air-sensitive. Bp_2 48-50°, $Bp_{0.08}$ 30-3°.

Brown, H.C. *et al*, *J. Am. Chem. Soc.*, 1972, **94**, 3561 (*synth, pmr*)
Kramer, G.W. *et al*, *J. Organomet. Chem.*, 1974, **73**, 1 (*synth, nmr*)
Biffar, W. *et al*, *Chem. Ber.*, 1980, **113**, 333 (*nmr*)
Brown, H.C. *et al*, *J. Org. Chem.*, 1980, **45**, 846 (*nmr*)

$C_9H_{17}B_2CoF_4N_4O_5$ B-00609

Aqua[bis[μ-[(2,3-butanedione dioximato)(2−)-O:O′]][tetrafluorodiborato(2−)-N,N′,N″,N‴]methylcobalt, 10CI

[Tetrafluorobis[dimethylglyoximato(2−)]diborato]-aquomethylcobalt

[53729-64-9]

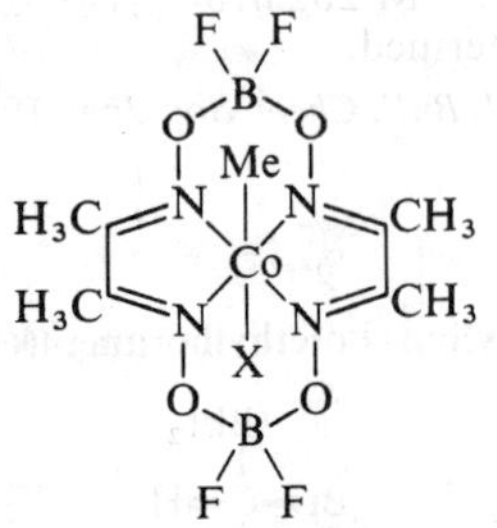

X = OH_2

M 417.804

Vitamin B_{12} analogue. Yellow cryst. (MeOH aq.). Sol. H_2O. Co−Me bond cleaved by light, OH^- and Hg^{2+}. H_2O ligand easily replaced by Lewis bases, e.g. Py.

Schrauzer, G.N. *et al*, *J. Am. Chem. Soc.*, 1966, **88**, 3738 (*synth*)
Adin, A. *et al*, *J. Chem. Soc., Chem. Commun.*, 1971, 653.
Cartano, A.V. *et al*, *Bioinorg. Chem.*, 1977, **7**, 351 (*synth*)
Brown, K.L., *J. Am. Chem. Soc.*, 1979, **101**, 6600 (*pmr, props*)

$C_9H_{18}B^{\ominus}$ B-00610

9-Methyl-9-boratabicyclo[3.3.1]nonane, 10CI

B-*Methyl-9-BBN hydride*

As 9-Boratabicyclo[3.3.1]nonane, B-00546 with

R = Me, R′ = H

M 137.051 (ion)

Li salt: [60824-29-5].
$C_9H_{18}BLi$ M 143.992
Air- and moisture-sensitive.

Krishnamurthy, S. *et al*, *J. Org. Chem.*, 1976, **41**, 3064.
U.S.P., 4 082 810, (*1978*); *CA*, **89**, 109936y (*synth*)
Hubbard, J.L. *et al*, *J. Organomet. Chem.*, 1978, **156**, 81 (*nmr*)
Brown, H.C. *et al*, *J. Organomet. Chem.*, 1980, **188**, 1 (*synth, nmr*)

$C_9H_{18}BNO_3$ B-00611

[1,1′,1″-Nitrilotris[2-propanolato](3−)-N,O,O′,O″]boron, 10CI, 9CI

[1,1′,1″-Nitrilotri-2-propanolato(3−)]boron, 8CI. 1,1′,1″-Nitrilotri-2-propanol cyclic borate. Triisopropanolamine borate

[27664-53-5]

As [[2,2′,2″-Nitrilotris[ethanolato]](3−)-*N,O,O′,O″*]-boron, B-00397 with

R = CH_3

M 199.056

Dipole moment 6.69D.

Onak, T.P. *et al*, *J. Phys. Chem.*, 1963, **67**, 1741 (*nmr*)
Pugh, J.M. *et al*, *Aust. J. Chem.*, 1963, **16**, 204.
Ger. Pat., 1 908 844, (*1969*); *CA*, **72**, 4052c (*use*)
Lukevics, E. *et al*, *Latv. PSR Zinat. Akad. Vestis. Kim. Ser.*, 1975, 483.
Eur. Pat., 6 634, (*1980*); *CA*, **93**, 113942x (*use*)
Shatz, V.D. *et al*, *J. Chromatogr.*, 1980, **200**, 105 (*chromatog*)

$C_9H_{18}BNO_3$ B-00612

[3,3′,3″-Nitrilotris[1-propanolato](3−)-N,O,O′,O″]boron, 10CI, 9CI

[3,3′,3″-Nitrilotri-1-propanolato(3−)]boron, 8CI. 3,3′,3″-Nitrilotri-1-propanol cyclic borate. Tripropanolamine borate

[27664-60-4]

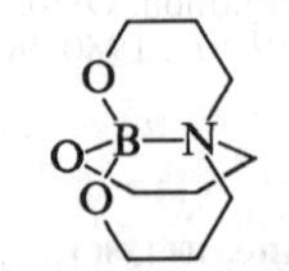

M 199.056

Onak, T.P. *et al*, *J. Phys. Chem.*, 1963, **67**, 1741 (*nmr*)
Ger. Pat., 1 908 844, (*1969*); *CA*, **72**, 4052c (*use*)
Lukevics, E. *et al*, *Latv. PSR Zinat. Akad. Vestis. Kim. Ser.*, 1975, 483.
Shatz, V.D. *et al*, *J. Chromatogr.*, 1980, **200**, 105 (*chromatog*)

$C_9H_{18}BN_2^{\ominus}$ B-00613

Triethyl(1H-imidazolato-N[1])borate(1−), 10CI

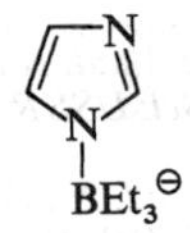

M 165.065 (ion)

Li salt: [69206-21-9].
$C_9H_{18}BLiN_2$ M 172.006
Shows electrical resistivity; used in battery electrolytes.

U.S.P., 4 117 213, (*1978*); *CA*, **90**, 124645s (*props, use*)
U.S.P., 4 104 450, (*1978*); *CA*, **90**, 124646t (*use*)
Ger. Pat., 2 833 943, (*1979*); *CA*, **91**, 57173s (*synth, props*)

$C_9H_{18}B_2O_6$ B-00614

Trimethylene borate

1,3-Propanediol cyclic ester with boric acid (H_3BO_3) trimethylene ester, 8CI. 2,2′-[(1,3-Propanediyl)bis(oxy)-]bis(1,3,2-dioxaborinane)

[20905-35-5]

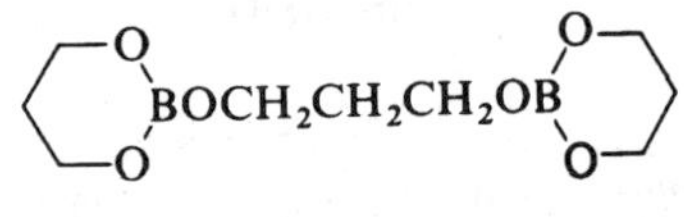

M 243.858

Liq., sensitive to moisture. $Bp_{0.05}$ 125°.

Finch, A. *et al*, *J. Org. Chem.*, 1961, **26**, 3250 (*synth*)

$C_9H_{19}B$ B-00615

Diethyl(3-methyl-2-butenyl)borane, 10CI, 9CI

3-Methyl-2-butenyldiethylborane. Diethylisopentenylborane

[36934-17-5]

$(H_3C)_2C{=}CHCH_2BEt_2$

M 138.059

Although the compd. should undergo allyl rearrangement readily, this isomer is much more stable than its alternative, so predominates. Air-sensitive liq. Sol. THF. Bp_7 40°. n_D^{20} 1.4367. Stable below 100°; does not easily symmetrise.

Mikhailov, B.M. *et al*, *Dokl. Akad. Nauk SSSR, Ser. Sci. Khim.*, 1972, **204**, 612 (*synth*)
Mikhailov, B.M. *et al*, *Zh. Obshch. Khim.*, 1974, **44**, 1878 (*nmr*)
Mikhailov, B.M. *et al*, *Izv. Akad. Nauk SSSR, Ser. Khim.*, 1978, 1586 (*synth*)
Mikhailov, B.M. *et al*, *J. Organomet. Chem.*, 1978, **154**, 113 (*synth, pmr*)
Pelter, A. *et al*, *Comprehensive Organic Chemistry*, (*Barton, D.H.R. et al*, Ed.), Pergamon, Oxford, 1979, **3**, 695 (*rev*)
Bubnov, Y.N. *et al*, *Synthesis*, 1980, 902 (*use*)

$C_9H_{19}B$ B-00616

2-Propenyldipropylborane, 10CI, 9CI

Dipropyl-2-propenylborane. Allyldipropylborane

[51851-75-3]

$H_2C{=}CHCH_2B(CH_2CH_2CH_3)_2$

M 138.059

Synthetic intermed., often generated *in situ*. Air- and moisture-sensitive liq. Sol. THF. Bp_{12} 48-50°. n_D^{20} 1.4267. ^{11}Bnmr δ 83.5 ppm.

Kacheishvili, G.E. *et al*, *Zh. Obshch. Khim.*, 1973, **43**, 2402 (*synth*)
Mikhailov, B.M. *et al*, *Izv. Akad. Nauk SSSR, Ser. Khim.*, 1975, 979; *Bull. Acad. Sci. USSR, Div. Chem. Sci.*, 897 (*synth*)
Bubnov, Y.N. *et al*, *Izv. Akad. Nauk SSSR, Ser. Khim.*, 1976, 2842; *Bull. Acad. Sci. USSR, Div. Chem. Sci.*, 2653 (*props*)
Mikahilov, B.M. *et al*, *Izv. Akad. Nauk SSSR, Ser. Khim.*, 1978, 1586; *Bull. Acad. Sci. USSR, Div. Chem. Sci.*, 1392 (*synth*)
Mikhailov, B.M. *et al*, *J. Organomet. Chem.*, 1978, **154**, 113 (*synth*)
Pelter, A. *et al*, *Compr. Org. Chem.*, (*Barton, D.H.R. et al*, Ed.), 1979, **3**, 695 (*rev*)

$C_9H_{19}BBr_2S$ B-00617

Dibromo(2-methylcyclohexyl)borane dimethyl sulfide

Dibromo(2-methylcyclohexyl)[thiobis[methane]]boron, 10CI. *2-Methylcyclohexyldibromoborane dimethyl sulfide*

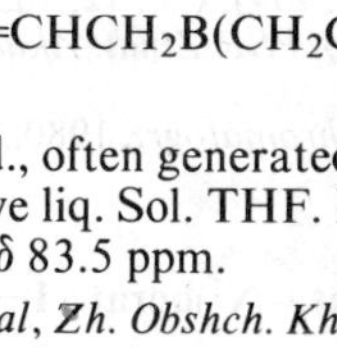

M 329.927

(*1RS,2SR*)-**form** [64770-12-3]

(+)-trans-*form*

Synthetic intermediate. Air- and moisture-sensitive liq. $Bp_{0.5}$ 68-9°.

Brown, H.C. *et al*, *J. Am. Chem. Soc.*, 1977, **99**, 7097 (*synth*)

$C_9H_{19}BO_2$ B-00618

(2-Ethylcyclopentyl)dimethoxyborane

(2-Ethylcyclopentyl)boronic acid dimethyl ester, 10CI, 9CI. *Dimethyl 2-ethylcyclopentylboronate*

As (2-Methylcyclopentyl)dimethoxyborane, B-00555 with

R = CH_2CH_3

M 170.058

***trans*-form** [60134-95-4]

Synthetic intermed., precursor for prostaglandin model systems. Colourless liq., sensitive to air and moisture. Bp_1 40°.

Evans, D.A. *et al*, *J. Org. Chem.*, 1976, **41**, 3947 (*synth, ir, nmr*)
Evans, D.A. *et al*, *Tetrahedron Lett.*, 1976, 1427 (*props*)

$C_9H_{20}B^{\ominus}$ B-00619

1-Hexenyltrimethylborate(1−), 10CI

$H_3C(CH_2)_3CH{=}CHBMe_3^{\ominus}$

M 139.067 (ion)

Cu(I) salt: [75213-13-7].

$C_9H_{20}BCu$ M 202.613

Synthetic intermed.

Miyaura, N. *et al*, *Bull. Chem. Soc. Jpn.*, 1980, **53**, 1471 (*synth, props*)

$C_9H_{20}BBr$ B-00620

(1-Bromo-2-methylbutyl)diethylborane, 10CI

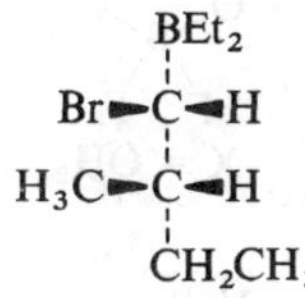

M 218.971

(*1RS,2SR*)-**form** [69798-76-1]

Air- and moisture-sensitive. Sol. THF.

Midland, M.M. *et al*, *J. Am. Chem. Soc.*, 1979, **101**, 248.

$C_9H_{20}BIO_2$ B-00621

Dibutoxy(iodomethyl)borane

(Iodomethyl)boronic acid dibutyl ester, 10CI. *Iodomethaneboronic acid dibutyl ester*, 8CI. *Dibutyl iodomethylboronate*

[13251-29-1]

$ICH_2B(OCH_2CH_2CH_2CH_3)_2$

M 297.970

Synthetic intermed. Air- and moisture-sensitive liq. $Bp_{0.2}$ 78-80°.

Matteson, D.S. *et al*, *J. Organomet. Chem.*, 1966, **6**, 100 (*synth, nmr*)
Matteson, D.S. *et al*, *J. Org. Chem.*, 1968, **33**, 3055 (*synth, nmr*)
Lindquist, R.N. *et al*, *J. Am. Chem. Soc.*, 1977, **99**, 6435 (*props*)
Matteson, D.S. *et al*, *J. Organomet. Chem.*, 1979, **170**, 259 (*synth, props*)

$C_9H_{21}B$ B-00622

Triisopropylborane, 8CI

Tris(1-methylethyl)borane, 10CI, 9CI

[1776-66-5]

$B[CH(CH_3)_2]_3$

M 140.075

Synthetic intermed. and catalyst. Air-sensitive liq. distillable *in vacuo* below 100°. Sol. Et_2O, THF.

Love, P., *J. Phys. Chem.*, 1963, **39**, 3044 (*nqr*)
McCusker, P.A. *et al*, *J. Org. Chem.*, 1963, **28**, 2889 (*synth*)
Ital. Pat., 696 030, (*1965*); *CA*, **66**, 19019x (*derivs*)
Weigert, F.J. *et al*, *Inorg. Chem.*, 1973, **12**, 313 (*nmr*)
Mikhailov, B.M. *et al*, *Zh. Obshch. Khim.*, 1976, **46**, 1057 (*derivs*)
Vanchagova, V.K. *et al*, *Zh. Obshch. Khim.*, 1976, **46**, 989 (*props*)

$C_9H_{21}B$ B-00623

Tripropylborane, 10CI, 9CI

[1116-61-6]

$B(CH_2CH_2CH_3)_3$

M 140.075

Liq., sensitive to air. Mp −65°. Bp 156°, 160°, 164°.

▷ED2450000.

Rosenblum, L, *J. Org. Chem.*, 1960, **25**, 1652 (*synth*)
Leleu, J., *Cah. Notes Doc.*, 1977, **88**, 361; *CA*, **88**, 26973m (*rev, haz*)
Postnikova, T.K. *et al*, *CA*, 1978, **90**, 5469 (*ms*)
Spoormaker, T., *Recl. Trav. Chim. Pays-Bas*, 1979, **98**, 380 (*nmr*)
Yamamoto, Y. *et al*, *Tetrahedron Lett.*, 1980, **21**, 3599 (*cmr*)

$C_9H_{21}BO$ B-00624

Dibutylmethoxyborane

Dibutylborinic acid methyl ester, 10CI, 9CI, 8CI. *Methyl dibutylborinate*

[2344-21-0]

$$(H_3CCH_2CH_2CH_2)_2BOMe$$

M 156.074

Air- and moisture-sensitive liq. d_{20} 0.785. Bp_{42} 90-1°, Bp_5 56-8°. n_D^{20} 1.4180, 1.4144.

Mikhailov, B.M. *et al*, *Dokl. Akad. Nauk SSSR, Ser. Sci. Khim.*, 1961, **136**, 82[illegible] (*synth, props*)
Mikhailov, B.M. *et al*, *Izv. Akad. Nauk SSSR, Ser. Khim.*, 1962, 1756 (*synth*)
Brown, H.C. *et al*, *J. Am. Chem. Soc.*, 1972, **94**, 2112 (*synth*)
Brown, H.C. *et al*, *J. Am. Chem. Soc.*, 1976, **98**, 1785, 1798 (*synth*)
Mikhailov, B.M. *et al*, *J. Organomet. Chem.*, 1978, **154**, 113 (*nmr, synth*)
Bagdasaryan, G.B. *et al*, *Arm. Khim. Zh.*, 1979, **32**, 303; *CA*, **92**, 6578r (*synth, props*)

$C_9H_{21}BO$ B-00625

Methoxybis(1-methylpropyl)borane

Bis(1-methylpropyl)borinic acid methyl ester, 10CI, 9CI. *Di-*sec*-butylborinic acid methyl ester*, 8CI. *Di-*sec*-butylmethoxyborane. Methyl di-*sec*-butylborinate*

[32705-45-6]

$$[H_3CCH_2CH(CH_3)]_2BOMe$$

M 156.074

Used in synth. of di-*sec*-butyl ketone *via* reaction with dichloromethyl methyl ether. Air- and moisture-sensitive liq.

Brown, H.C. *et al*, *J. Am. Chem. Soc.*, 1972, **94**, 2112 (*synth*)
Carlson, B.A. *et al*, *J. Am. Chem. Soc.*, 1973, **95**, 6876 (*use*)
Brown, H.C. *et al*, *J. Am. Chem. Soc.*, 1976, **98**, 1785 (*synth*)
Brown, H.C. *et al*, *J. Org. Chem.*, 1977, **42**, 2533; 1979, **44**, 2417 (*synth*)
Brown, H.C. *et al*, *J. Organomet. Chem.*, 1977, **131**, 163 (*props*)

$C_9H_{21}BO$ B-00626

Methoxybis(2-methylpropyl)borane

Bis(2-methylpropyl)borinic acid methyl ester, 10CI, 9CI. *Diisobutylborinic acid methyl ester*, 8CI. *Diisobutylmethoxyborane. Methyl diisobutylborinate*

[17832-17-6]

$$[(H_3C)_2CHCH_2]_2BOMe$$

M 156.074

Used in synth. of diisobutyl ketone *via* reaction with dichloromethyl methyl ether. Air- and moisture-sensitive liq.

McCusker, P.A. *et al*, *J. Am. Chem. Soc.*, 1962, **84**, 4362 (*props*)
Brown, H.C. *et al*, *J. Am. Chem. Soc.*, 1972, **94**, 2112 (*synth*)
Carlson, B.A. *et al*, *J. Am. Chem. Soc.*, 1973, **95**, 6876 (*use*)
Brown, H.C. *et al*, *J. Am. Chem. Soc.*, 1976, **98**, 1785 (*synth*)
Brown, H.C. *et al*, *J. Organomet. Chem.*, 1977, **131**, 163 (*props*)
Gupta, S.K., *J. Organomet. Chem.*, 1978, **156**, 95 (*props*)

$C_9H_{21}BO$ B-00627

Propoxydipropylborane

Dipropylborinic acid propyl ester, 10CI

[72168-54-8]

$$(H_3CCH_2CH_2)_2BOCH_2CH_2CH_3$$

M 156.074

Air- and moisture-sensitive liq. Sol. nonhydroxylic solvs.

Bagdasaryan, G.B. *et al*, *Arm. Khim. Zh.*, 1979, **32**, 192; *CA*, **92**, 6577q.
Bagdasaryan, G.B. *et al*, *Arm. Khim. Zh.*, 1980, **33**, 744; *CA*, **94**, 156990z.

$C_9H_{21}BO_3$ B-00628

Triisopropoxyborane

Boric acid (H_3BO_3) tris(1-methylethyl) ester, 10CI, 9CI. *Triisopropyl borate. Isopropyl borate*

[5419-55-6]

$$B[OCH(CH_3)_2]_3$$

M 188.073

Catalyst, vulcanizing agent, insect sterilant. Moisture-sensitive liq. Sol. non-hydroxylic solvs. d_4^{28} 0.815. Bp 139-40°, Bp_{76} 75°. n_D^{25} 1.3750.

▷ED5950000.

Brown, H.C. *et al*, *J. Am. Chem. Soc.*, 1956, **78**, 3613 (*synth, props*)
Steinberg, H. *et al*, *Ind. Eng. Chem.*, 1957, **49**, 174 (*synth, props*)
Christopher, P.M. *et al*, *J. Chem. Eng. Data*, 1969, **14**, 437; 1979, **24**, 67 (*props*)
Brown, C.A., *J. Am. Chem. Soc.*, 1973, **95**, 4100 (*use*)
Jaworski, K. *et al*, *Przem. Chem.*, 1981, **60**, 18; *CA*, **95**, 42269m (*synth*)

$C_9H_{21}BO_3$ B-00629

Tripropoxyborane

Boric acid (H_3BO_3) tripropyl ester, 10CI, 9CI, 8CI. *Tripropyl borate. Propyl borate. Tripropyl orthoborate. Tris(propoxy)borane*

[688-71-1]

$$(H_3CCH_2CH_2O)_3B$$

M 188.073

Catalyst for crosslinking of epoxy resins; used as heat stabilizer; insect sterilant. Moisture-sensitive liq. Sol. non-hydroxylic solvs. d_4^{20} 0.858, 0.861. Bp 177°, Bp_{12} 68.5-69°. n_D^{20} 1.3948, 1.3969.

▷ED5785000.

Gerrard, W. *et al*, *J. Chem. Soc.*, 1955, 3084 (*synth, props*)
O'Brien, K.G., *Aust. J. Chem.*, 1957, **10**, 91 (*synth, props*)
Fallon, P.J. *et al*, *Int. J. Mass Spectrom. Ion Phys.*, 1968, **1**, 133 (*ms*)
Christopher, P.M. *et al*, *J. Chem. Eng. Data*, 1969, **14**, 437 (*props*)
Maijs, L. *et al*, *CA*, 1973, **79**, 131073 (*pmr*)
Pola, M. *et al*, *Collect. Czech. Chem. Commun.*, 1979, **44**, 3688 (*ir, props*)
Jaworski, K. *et al*, *Przem. Chem.*, 1981, **60**, 18; *CA*, **95**, 42269m (*synth*)
Sax, N.I., *Dangerous Properties of Industrial Materials*, 5th Ed., Van Nostrand-Reinhold, 1979, 1069.

$C_9H_{21}BS$ **B-00630**

Dibutyl(methylthio)borane

Dibutylthioborinic acid methyl ester, 10CI, 9CI, 8CI. *Methylthiodibutylborane*

[20524-64-5]

$(H_3CCH_2CH_2CH_2)_2BSMe$

M 172.135

Used in synth. of dibutylhalogenoboranes. Colourless liq., air- and moisture-sensitive. $Bp_{1.1}$ 43-4°, Bp_1 27°. m/e 115 for characteristic ion $RB{=}SMe^{\oplus}$.

Schmidt, M. *et al*, *Z. Naturforsch.*, *B*, 1968, **23**, 1258 (*synth*)
Siebert, W. *et al*, *J. Organomet. Chem.*, 1970, **22**, 511 (*synth*, *props*)
Pelter, A. *et al*, *J. Chem. Soc., Chem. Commun.*, 1975, 531 (*use*)
Siebert, W. *et al*, *J. Organomet. Chem.*, 1975, **90**, 13 (*props*)
Pelter, A. *et al*, *J. Chem. Soc., Dalton Trans.*, 1976, 2087 (*use*, *synth*, *ms*)

$C_9H_{21}BS$ **B-00631**

Diethyl(pentylthio)borane

Diethylthioborinic acid pentyl ester, 10CI. *Amylthio(diethyl)borane*

[67682-92-2]

$Et_2BS(CH_2)_4CH_3$

M 172.135

Used in synth. of allyldiethylboranes. Air- and moisture-sensitive liq. Sol. nonhydroxylic solvs. Bp_4 69-70°.

Mikhailov, B.M. *et al*, *Izv. Akad. Nauk SSSR, Ser. Khim.*, 1978, 1594; *Bull. Acad. Sci. USSR, Div. Chem. Sci.*, **27**, 1392 (*synth*, *props*)
Mikhailov, B.M. *et al*, *J. Organomet. Chem.*, 1978, **154**, 113 (*synth*)

$C_9H_{22}B^{\ominus}$ **B-00632**

Triisopropylhydroborate

Hydrotris(1-methylethyl)borate(1–), 10CI

$[H_3CCH(CH_3)]_3BH^{\ominus}$

M 141.083 (ion)

K salt: [63706-88-7].

$C_9H_{22}BK$ M 180.181

Air- and moisture-sensitive.

Strouf, O. *et al*, *Acta Chem. Scand., Ser. A*, 1977, **31**, 391 (*props*)
Brown, C.A. *et al*, *J. Organomet. Chem.*, 1978, **156**, 111 (*synth*)

$C_9H_{22}BN$ **B-00633**

(3-Aminopropyl)dipropylborane

[61018-71-1]

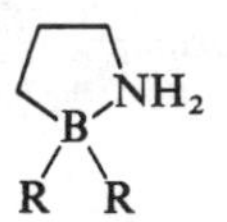

$R = CH_2CH_2CH_3$

M 155.090

Wrongly indexed as the 2-isomer in 10CI. Liq. (?). Dist. under Ar.

Gal'chenko, G.L. *et al*, *Tr. Khim. Khim. Tekhnol.*, 1975, 17; *CA*, **86**, 9289j (*props*)
Zolotarev, B.M. *et al*, *Izv. Akad. Nauk SSSR, Ser. Khim.*, 1979, 2226; *Bull. Acad. Sci. USSR, Div. Chem. Sci.*, 2048 (*synth*, *ms*)

$C_9H_{22}BO_3^{\ominus}$ **B-00634**

Triisopropoxyhydroborate

Hydrotris(2-propanolato)borate(1–), 10CI, 9CI. *Hydrotriisopropoxyborate. Triisopropoxyborohydride. Triisopropoxyhydridoborate*

$[(H_3C)_2CHO]_3BH^{\ominus}$

M 189.081 (ion)

K salt: [42278-67-1]. *Feeble hydride.*

$C_9H_{22}BKO_3$ M 228.179

Catalyst for carbonylation of organoboranes. Mild reducing agent, comparable in props. to $NaBH_4$, but sol. in THF. Moisture-sensitive. Sol. THF. Commercially available as THF soln.

▷Flash pt. of THF soln. –15°

Brown, C.A. *et al*, *J. Chem. Soc., Chem. Commun.*, 1973, 391 (*synth*)
Brown, H.C., *CA*, 1974, **80**, 37205y.
Brown, C.A. *et al*, *J. Am. Chem. Soc.*, 1979, **101**, 3964 (*props*, *use*)
Brown, H.C. *et al*, *Synthesis*, 1979, 701 (*use*)
Steinmetz, G.R. *et al*, *J. Am. Chem. Soc.*, 1981, **103**, 1278 (*use*)

$C_9H_{22}BO_6^{\ominus}$ **B-00635**

Hydrotris(2-methoxyethoxy)borate

Hydrotris(2-methoxyethanolato-O′-borate)(1–), 10CI. *Monohydridotris(2-methoxyethoxy)borate*

$(MeOCH_2CH_2O)_3BH^{\ominus}$

M 237.079 (ion)

Na salt: [57972-25-5].

$C_9H_{22}BNaO_6$ M 260.069

Reducing agent for aldehydes, ketones, acyl chlorides. Moisture-sensitive. Sol. arom. hydrocarbons, ethers.

Kritz, O. *et al*, *Collect. Czech. Chem. Commun.*, 1977, **42**, 421 (*use*)
Czech. Pat., 174 515, (*1978*); *CA*, **90**, 39423y (*use*)
Kritz, O. *et al*, *Collect. Czech. Chem. Commun.*, 1978, **43**, 3130 (*struct*, *ir*, *nmr*)

$C_9H_{24}B_3N_3$ **B-00636**

2,4,6-Triethyl-1,3,5-trimethylborazine, 10CI, 9CI, 8CI

B,B,B-*Triethyl*-N,N,N-*trimethylborazole*

[715-62-8]

As 2,4,6-Tribromoborazine, B-00015 with

X = Et, Y = Me

M 206.739

Liq., sl. sensitive to air. Fp 1-2°. $Bp_{0.65}$ 82°, 98°. $n_D^{22.5}$ 1.4791.

Ryschkewitsch, G.E. *et al*, *J. Am. Chem. Soc.*, 1958, **80**, 4515 (*synth*, *props*)
Watanabe, H. *et al*, *J. Am. Chem. Soc.*, 1960, **82**, 2428 (*props*, *synth*)
Phillips, C.S.G. *et al*, *Z. Anal. Chem.*, 1963, **197**, 202 (*synth*, *chromatog*)
Cros, G. *et al*, *Bull. Soc. Chim. Fr.*, 1969, 2601 (*nmr*)
Gallais, F. *et al*, *J. Chim. Phys. Phys. Chim. Biol.*, 1969, **66**, 122; *CA*, **70**, 101171g (*props*)
Wrackmeyer, B. *et al*, *Chem. Ber.*, 1976, **109**, 3480 (*nmr*)

$C_9H_{24}B_3N_3$ B-00637

1,3,5-Tripropylborazine, 9CI

Hexahydro-1,3,5-tripropyl-1,3,5-triazatriborine. N,N′-,N″-*Tripropylborazole*

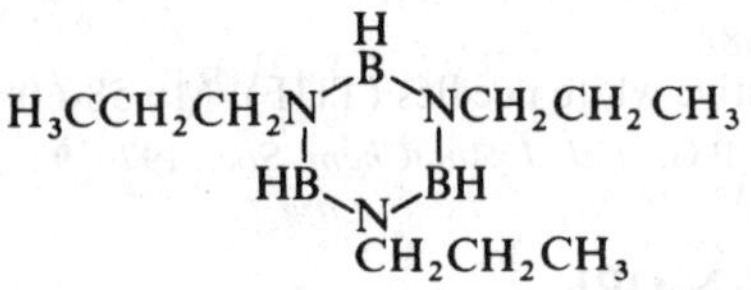

M 206.739

Bp 203°.

Hough, W.V. *et al*, *J. Am. Chem. Soc.*, 1955, **77**, 864 (*synth*)
Powell, P. *et al*, *J. Chem. Soc. (A)*, 1971, 2951 (*ms*)

$C_9H_{27}BO_3Sn_3$ B-00638

Tris[(trimethylstannyl)oxy]borane, 9CI

Tris(trimethylstannyl)borate

[51823-76-8]

$(Me_3SnO)_3B$

M 550.191

Colourless liq. Bp_{14} 130-2°.

Mehrotra, S.K. *et al*, *J. Organomet. Chem.*, 1974, **65**, 361.
Mehrotra, S.K. *et al*, *Indian J. Chem., Sect. A*, 1976, **14**, 137.

$C_9H_{27}B_3N_6$ B-00639

2,4,6-Tris(dimethylamino)-1,3,5-trimethylborazine, 8CI

N,N,N′,N′,N″,N″,*1,3,5-Nonamethyl-2,4,6-borazinetriamine,* 10CI

[16912-59-7]

As 2,4,6-Tribromoborazine, B-00015 with

X = NMe_2, Y = Me

M 251.783

Solid. Mp 66°. Bp_3 122-4°. V. slow dec. at 360° *in vacuo*.

Niedenzu, K. *et al*, *Chem. Ber.*, 1961, **94**, 671 (*synth*)
Gutmann, V. *et al*, *Monatsh. Chem.*, 1963, **94**, 1071 (*synth, ir*)
Toeniskoetter, R.H. *et al*, *Inorg. Chem.*, 1963, **2**, 29 (*props*)
Meller, A. *et al*, *Monatsh. Chem.*, 1967, **98**, 1358 (*ir*)
Wrackmeyer, B. *et al*, *Chem. Ber.*, 1976, **109**, 3480 (*nmr*)
Nöth, H. *et al*, *Chem. Ber.*, 1981, **114**, 1150 (*nmr, cmr*)

$C_{10}H_9BCl_2Fe$ B-00640

Dichloroborylferrocene, 9CI, 8CI

Dichloroferrocenylborane. Ferrocenyldichloroborane

[32007-87-7]

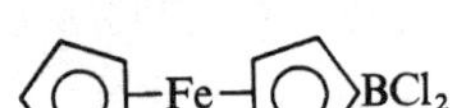

M 266.744

Red-orange solid. Sol. hexane. Mp 33°.

Py complex: [31811-20-8]. *Dichloroferrocenyl(pyridine)-boron,* 10CI.

$C_{15}H_{14}BCl_2FeN$ M 345.845

Yellow-orange air-sensitive solid by subl. Sol. hexane. Mp 162-5°.

Kotz, J.C. *et al*, *J. Am. Chem. Soc.*, 1968, **90**, 4503 (*synth, pmr*)
Kotz, J.C. *et al*, *Inorg. Chem.*, 1970, **9**, 1661 (*synth, ir, pmr, uv, nmr*)
Ruf, W. *et al*, *J. Organomet. Chem.*, 1974, **64**, C45 (*synth, pmr*)
Renk, T. *et al*, *J. Organomet. Chem.*, 1976, **120**, 1.

$C_{10}H_9BF_2O_2$ B-00641

Difluoro(1-phenyl-1,3-butanedionato-*O,O′*)boron, 10CI

Benzoylacetonatodifluoroborane. Benzoylacetonatoboron difluoride. Boron benzoylacetone difluoride

[14871-98-8]

As Difluoro(2,4-pentanedionato-*O,O′*)boron, B-00320 with

R = Ph

M 209.987

Cryst. (C_6H_6, $CHCl_3$ or EtOAc). Mp 155°.

▷ED7650000.

Morgan, G.T. *et al*, *J. Chem. Soc.*, 1924, **125**, 1963 (*synth*)
Durden, J.A. *et al*, *J. Org. Chem.*, 1965, **30**, 1684 (*synth, ir*)
Smith, J.A.S. *et al*, *J. Chem. Soc. (A)*, 1966, 1749 (*nmr*)
Trestianu, A. *et al*, *Tetrahedron*, 1968, **24**, 2499 (*pmr*)
Hanson, A.W. *et al*, *Acta Crystallogr., Sect. B*, 1972, **28**, 1961 (*cryst struct*)
Neplyuev, V.M. *et al*, *Zh. Org. Khim.*, 1980, **16**, 2588 (*synth*)

$C_{10}H_9BO$ B-00642

3-Hydroxy-3*H*-3-benzoborepin, 9CI, 8CI

*3(3*H*)-Benzoborepin-3-ol*

[31569-68-3]

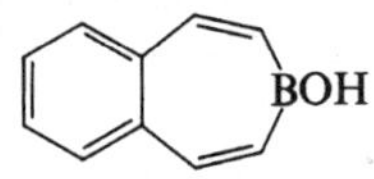

M 155.991

Mp ca. 110°. Sensitive to air.

Axelrad, G. *et al*, *J. Chem. Soc., Chem. Commun.*, 1971, 291 (*synth, ir, ms, uv, pmr*)

$C_{10}H_9BO_2$ B-00643

Dihydroxy(1-naphthalenyl)borane

(1-Naphthalenyl)boronic acid, 10CI, 9CI, 8CI. *α-Naphthaleneboronic acid. 1-(Dihydroxyboryl)naphthalene*

[13922-41-3]

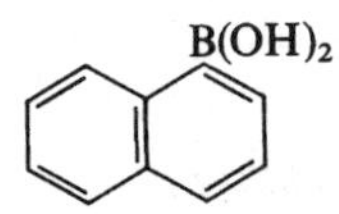

M 171.990

Used for detection of bifunctional compds. by glc analysis. Cryst. (C_6H_6). Sol. Et_2O. Mp 225-7°.

Mikhailov, B.M. *et al*, *Izv. Akad. Nauk SSSR, Ser. Khim.*, 1960, 1590.
Nencetti, G. *et al*, *Ann. Chim. (Paris)*, 1963, **53**, 1875; *CA*, **60**, 12037e
Bruce, R.L. *et al*, *Aust. J. Chem.*, 1971, **24**, 1085 (*props, use*)
Kliegel, W. *et al*, *Chem. Ber.*, 1977, **110**, 1623.
Poole, C.F. *et al*, *J. Chromatogr.*, 1978, **158**, 33 (*use*)
Youngless, T.L. *et al*, *J. Mass Spectrom. Ion Phys.*, 1980, **34**, 9 (*props*)

The first digit of the Entry number defines the Supplement in which the Entry is found. 0 indicates the Main Work

$C_{10}H_{10}BCl_2N_6ORh$ B-00644

Carbonyldichloro[hydrotris(1*H*-pyrazolato-*N'*)borato(1−)]-rhodium

Carbonyldichloro[tris(pyrazolyl)borato]rhodium

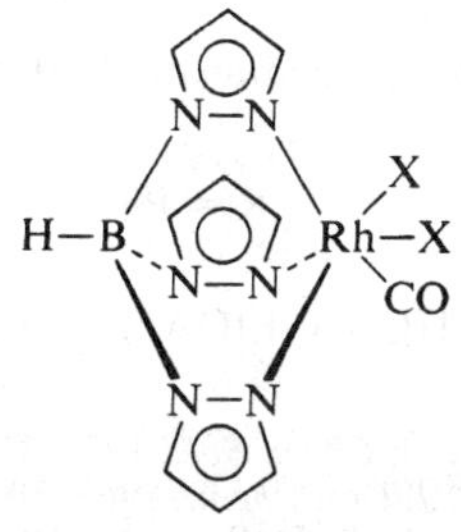

X = Cl

M 414.850

Struct. by analogy with Carbonyl[hydrotris(1*H*-pyrazolato-*N'*)borato(1−)]diiodorhodium, B-00648 . Loses CO slowly in solid state and rapidly on attempted recryst. ir ν_{CO} 2129 cm^{-1} (KBr).

Cocivera, M. *et al*, *Organometallics*, 1982, **1**, 1125 (*synth, ir*)

$C_{10}H_{10}BCuN_6O$ B-00645

Carbonyl[hydrotris(1*H*-pyrazolato-*N*¹)borato(1−)-*N*²,*N*²'-,*N*²'']copper, 9CI

[52374-64-8]

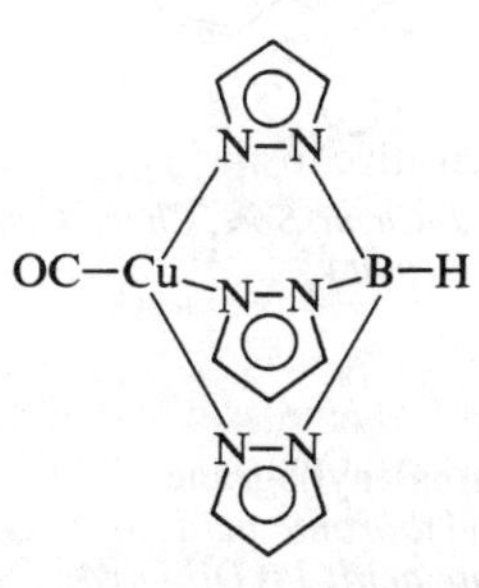

M 304.585

Cryst. (pet. ether). Sol. Me_2CO, THF. Mp 164-6°. Stable under CO atm. Rapid oxid. in soln.

Bruce, M.I. *et al*, *J. Chem. Soc., Dalton Trans.*, 1973, 2433 (*synth*)

Churchill, M.R. *et al*, *Inorg. Chem.*, 1975, **14**, 2051 (*cryst struct*)

Inorg. Synth., 1982, **21**, 108 (*synth*)

$C_{10}H_{10}BF_3Pb$ B-00646

Bis(cyclopentadienyl)lead boron trifluoride

Plumbocene compd. with trifluoroborane(1:1), *10CI*

[50647-92-2]

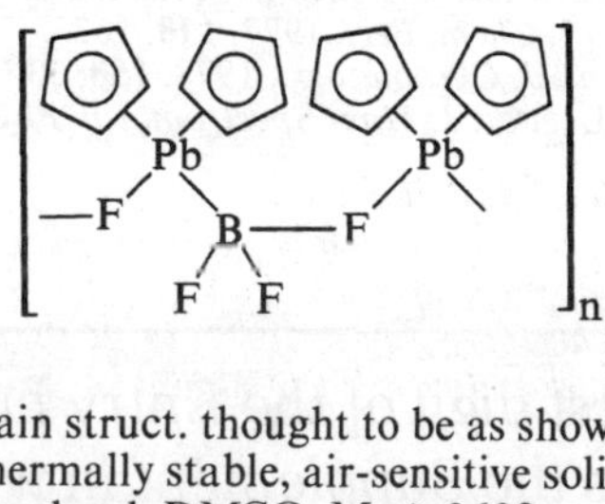

M 405.194

Polymer chain struct. thought to be as shown. Pale-yellow, thermally stable, air-sensitive solid. Insol. most org. solvs., sl. sol. DMSO. Mp >360°.

Holliday, A.K. *et al*, *J. Chem. Soc., Dalton Trans.*, 1979, 228 (*synth, pmr*)

$C_{10}H_{10}BF_3Sn$ B-00647

Di-π-cyclopentadienyl(trifluoroboryl)tin, 8CI

Bis(cyclopentadienyl)tin boron trifluoride

$(C_5H_5)_2SnBF_3$

M 316.684

Air-senstive white needles (THF). Mp 58-60°.

Harrison, P.G. *et al*, *J. Am. Chem. Soc.*, 1970, **92**, 2577.

$C_{10}H_{10}BI_2N_6ORh$ B-00648

Carbonyl[hydrotris(1*H*-pyrazolato-*N'*)borato(1−)]diiodo-rhodium, 9CI

Carbonyldiiodo[tris(pyrazolyl)borato]rhodium

[52504-76-4]

As Carbonyldichloro[hydrotris(1*H*-pyrazolato-*N'*)-borato(1−)]rhodium, B-00644 with

X = I

M 597.753

Deep-red cryst. (CH_2Cl_2/hexane). ir ν_{CO} 2095 cm^{-1} (KBr).

Cocivera, M. *et al*, *Organometallics*, 1982, **1**, 1125 (*synth, pmr, ir, struct*)

$C_{10}H_{11}BFeO_2$ B-00649

Ferroceneboronic acid

Boronoferrocene, *9CI*. *Ferrocenylboronic acid*

[12152-94-2]

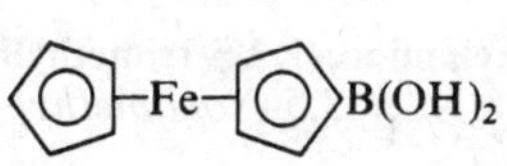

M 229.853

Yellow cryst. Mp 143-8° (sealed tube). Rapidly hydrol. by hot H_2O giving ferrocene.

Nesmeyanov, A.N. *et al*, *Dokl. Akad. Nauk SSSR*, 1959, **126**, 1004 (*synth*)

Shechter, H. *et al*, *J. Org. Chem.*, 1961, **26**, 1034 (*synth*)

Nesmeyanov, A.N. *et al*, *Izv. Akad. Nauk SSSR, Otd. Khim. Nauk*, 1968, 2371 (*synth*)

Org. Reactions, 1969, **17**, 1 (*synth*)

Epton, R. *et al*, *J. Organomet. Chem.*, 1978, **150**, 93, 101 (*synth, ir, pmr, ms*)

$C_{10}H_{11}BO_2Ru$ B-00650

Ruthenoceneboronic acid

Boronoruthenocene, *9CI*

[38816-64-7]

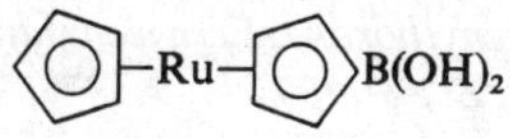

M 275.076

White cryst. Sol. Et_2O.

Nesmeyanov, A.N. *et al*, *Izv. Akad. Nauk SSSR, Ser. Khim.*, 1972, 1823 (*synth*)

$C_{10}H_{12}B_2Fe$ B-00651

Bis(η^6-boratabenzene)iron, 10CI

[68344-23-0]

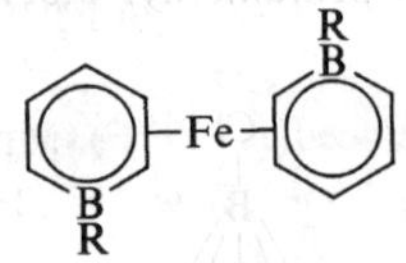

R = H

M 209.672

Red octahedral cryst. Bp 165° subl. Undergoes nucleophilic substitution at B and electrophilic substitution at C.

Ashe, A.J. *et al*, *J. Am. Chem. Soc.*, 1979, **101**, 7066 (*synth, ms, pmr, nmr*)

$C_{10}H_{12}B_2FeO_4$ B-00652

1,1′-Ferrocenediboronic acid

Ferrocenylene-1,1′-diboronic acid

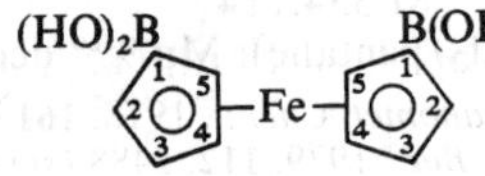

M 273.669

Yellow cryst. (H_2O). Mp 180° dec., 220° dec. Unstable in hot water.

Nesmeyanov, A.N. *et al*, *Chem. Ber.*, 1960, **93**, 2717 (*synth, use*)
Nesmeyanov, A.N. *et al*, *Izv. Akad. Nauk SSSR, Ser. Khim.*, 1968, 2371 (*synth*)
Org. React., 1969, **17**, 1 (*synth*)

$C_{10}H_{13}B$ B-00653

1-Phenylborolane, 10CI, 9CI

1-Phenylboracyclopentane. Phenylcyclotetramethyleneborin

[1075-17-8]

As Borolane, B-00227 with

R = Ph

M 144.023

Air-sensitive liq. Sol. nonhydroxylic solvs. d_{20} 0.924. Mp −19°. Bp_{11} 85-7°, Bp_1 50-1°. n_D^{20} 1.5278.

▷Spont. inflammable

Torssell, K. *et al*, *Acta Chem. Scand.*, 1954, **8**, 1779 (*synth*)
Mikhailov, B.M. *et al*, *Dokl. Akad. Nauk SSSR, Ser. Sci. Khim.*, 1960, **133**, 119 (*synth*)
Greenwood, N.N. *et al*, *J. Chem. Soc.*, 1965, 448 (*synth, ir*)
Wrackmeyer, B. *et al*, *Chem. Ber.*, 1976, **109**, 1075 (*nmr*)
Schmidt, A. *et al*, *Z. Naturforsch., B*, 1978, **33**, 855.
Odom, J.D. *et al*, *J. Organomet. Chem.*, 1979, **173**, 15 (*nmr*)

$C_{10}H_{13}BN_2$ B-00654

Cyanoborane tetrahydroisoquinoline

(Cyano-C)dihydro(1,2,3,4-tetrahydroisoquinoline)boron, 10CI, 9CI. Boranecarbonitrile, compd. with 1,2,3,4-tetrahydroisoquinoline(1:1)

[73109-50-9]

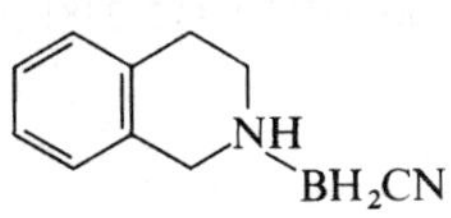

M 172.036

Singh, H. *et al*, *Indian J. Chem., Sect. B*, 1979, **18**, 312 (*synth*)

$C_{10}H_{14}BNO_2$ B-00655

2-Phenyl-6-aza-1,3-dioxa-2-boracyclooctane

[[2,2′-Iminobis(ethanolato)](2−)-N,O,O′]phenylboron, 10CI. B-Phenyldiptychboroxazolidine

[56296-15-2]

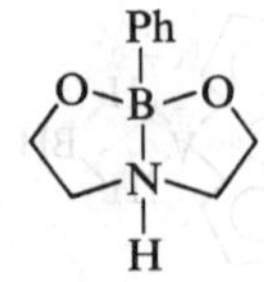

M 191.036

Colourless cryst. (EtOH/Me_2CO). Stable in air. Associated in soln., MW(Rast) >600. Mp 214-5° (209.5-210°). $Bp_{0.04}$ 180° subl.

Letsinger, R.L. *et al*, *J. Am. Chem. Soc.*, 1955, **77**, 2491 (*synth*)
Musgrave, O.C. *et al*, *Chem. Ind. (London)*, 1955, 1552 (*synth, ir*)
Rettig, S.J. *et al*, *Can. J. Chem.*, 1975, **53**, 1393 (*struct*)
Roscoe, C.W. *et al*, *J. Pharm. Sci.*, 1977, **66**, 1505.
Mazeika, I. *et al*, *Zh. Obshch. Khim.*, 1979, **49**, 2528 (*ms*)
Csuk, R. *et al*, *Monatsh. Chem.*, 1981, **112**, 879.

$C_{10}H_{14}BNb$ B-00656

Bis(η^5-2,4-cyclopentadien-1-yl)[tetrahydroborato(1−)-*H,H′*-]niobium, 9CI

Bis(cyclopentadienyl)niobium borohydride

[37298-41-2]

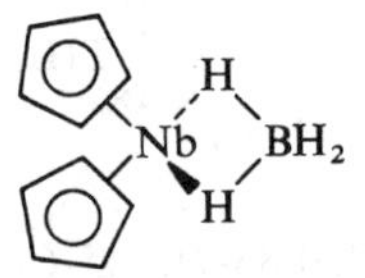

M 237.937

Dark-green solid (toluene), highly air-sensitive. Sol. Et_2O, aromatic hydrocarbons. Bp 80° subl. (part. dec.).

▷Pyrophoric in air

Kirillova, N.I. *et al*, *Zh. Strukt. Khim.*, 1974, **15**, 718 (*struct*)
Inorg. Synth., 1976, **16**, 107 (*synth, ir*)

$C_{10}H_{14}BTi$ B-00657

Bis(η^5-2,4-cyclopentadien-1-yl)[tetrahydroborato(1−)*H,H′*-]titanium, 10CI

Bis(η^5-cyclopentadienyl)titanium tetrahydridoborate. Titanocene borohydride

[12772-20-2]

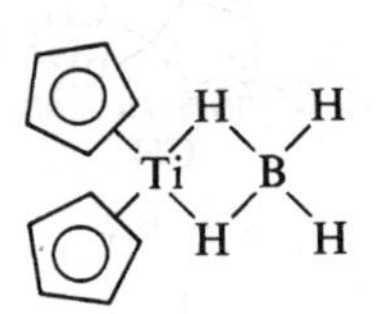

M 192.911

Purple needles. Subl. *in vacuo*.

Nöth, H. *et al*, *Naturwissenschaften*, 1960, **47**, 57 (*esr*)
Davies, N. *et al*, *J. Chem. Soc. (A)*, 1969, 2601 (*ir*)
Marks, T.J. *et al*, *Inorg. Chem.*, 1972, **11**, 2540 (*synth, ir*)
Melmed, K.M. *et al*, *Inorg. Chem.*, 1973, **12**, 232 (*cryst struct*)
Inorg. Synth., 1977, **17**, 91 (*synth*)

$C_{10}H_{14}BV$ B-00658

Bis(η^5-2,4-cyclopentadien-1-yl)[tetrahydroborato(1−)-*H,H'*-]vanadium, 9CI

Bis(cyclopentadienyl)vanadium borohydride

[55744-43-9]

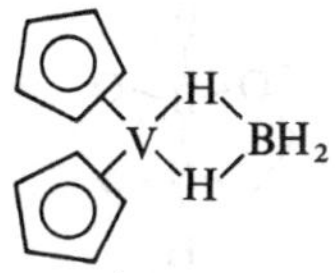

M 195.972

Deep-violet cryst. by subl. Sol. aromatic hydrocarbons. $Bp_{0.001}$ 55° subl. Dec. slowly at r.t. under N_2. Dec. >80°. Diamagnetic.

▷Pyrophoric

Marks, T.J. *et al*, *J. Am. Chem. Soc.*, 1975, **97**, 1439 (*synth, ir, pmr*)

$C_{10}H_{15}B$ B-00659

9-Ethynyl-9-borabicyclo[3.3.1]nonane, 10CI

B-*Ethynyl-9-BBN*

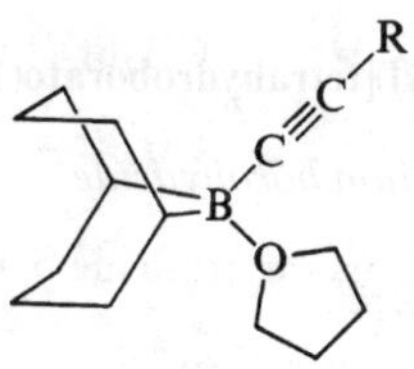

THF complex

R = H

M 146.039

THF complex: [63220-83-7]. *1,5-Cyclooctanediylethynyl(tetrahydrofuran)boron, 10CI.*

$C_{14}H_{23}BO$ M 218.145

White cryst. solid, dec. on warming from −78°.

Brown, H.C. *et al*, *J. Organomet. Chem.*, 1977, **131**, 163 (*synth, nmr*)

$C_{10}H_{15}BBr_2$ B-00660

Dibromo(1,2,3,4,5-pentamethyl-2,4-cyclopentadien-1-yl)borane, 10CI

5-Dibromoboryl-1,2,3,4,5-pentamethyl-1,3-cyclopentadiene

[71245-46-0]

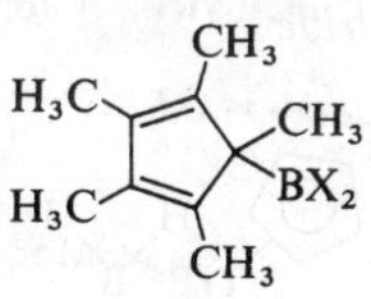

X = Br

M 305.847

Shows fluxional behavior in nmr spectrum. Prepd. from the reaction of trimethyl(pentamethylcyclopentadienyl)germane with dibromoborane. Sol. CH_2Cl_2. $Bp_{0.4}$ 60°.

Jutzi, P. *et al*, *Chem. Ber.*, 1979, **112**, 2481, 2488 (*synth, nmr, props*)

$C_{10}H_{15}BCl^{\oplus}$ B-00661

Chloro[(1,2,3,4,5-η)-1,2,3,4,5-pentamethyl-2,4-cyclopentadien-1-yl]boron(1+)

1-Chloro-2,3,4,5,6-pentamethyl-2,3,4,5,6-pentacarbahexaborane(6)

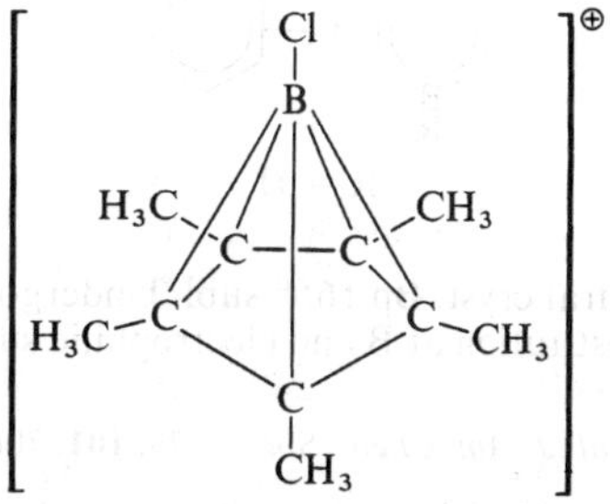

M 181.492 (ion)

nido-. Stable deriv. of one boron nidocarborane, isoelectronic with B_6H_{10}. Sol. CH_2Cl_2.

Tetrachloroborate: [71237-75-7].

$C_{10}H_{15}B_2Cl_5$ M 334.114

Needles (CH_2Cl_2/pentane). Mp 97° dec.

Jutzi, P. *et al*, *J. Organomet. Chem.*, 1978, **161**, C5 (*synth, nmr*)
Jutzi, P. *et al*, *Chem. Ber.*, 1979, **112**, 2488 (*synth, nmr, ms*)

$C_{10}H_{15}BClN$ B-00662

Chloro(diethylamino)phenylborane, 8CI

*1-Chloro-*N,N*-diethyl-1-phenylboranamine, 10CI, 9CI*

[33441-65-5]

$PhBClNEt_2$

M 195.498

Air- and moisture-sensitive liq. $Bp_{0.1}$ 50°.

Sheikh, S.U., *Bull. Chem. Soc. Jpn.*, 1976, **49**, 2879.
Cragg, R.H. *et al*, *J. Organomet. Chem.*, 1981, **217**, 1; 1982, **232**, 201 (*synth, pmr, nmr*)
Meller, A. *et al*, *Z. Naturforsch., B*, 1981, **36**, 1411.

$C_{10}H_{15}BCl_2$ B-00663

Dichloro(1,2,3,4,5-pentamethyl-2,4-cyclopentadien-1-yl)borane, 10CI

(1,2,3,4,5-Pentamethyl-2,4-cyclopentadien-1-yl)dichloroborane

[60828-63-9]

As Dibromo(1,2,3,4,5-pentamethyl-2,4-cyclopentadien-1-yl)borane, B-00660 with

X = Cl

M 216.945

Undergoes rapid degenerate sigmatropic rearr. Air- and moisture-sensitive colourless liq. Sol. CH_2Cl_2. $Bp_{0.5}$ 55°.

▷Ignites in air

Py complex: [71245-48-2]. *Dichloro(1,2,3,4,5-pentamethyl-2,4-cyclopentadien-1-yl)(pyridine)boron, 10CI.*

$C_{15}H_{20}BCl_2N$ M 296.046

Solid. Mp 125° dec.

Jutzi, P. *et al*, *Chem. Ber.*, 1979, **112**, 2481 (*synth, nmr*)

$C_{10}H_{15}BFeINO_2$ B-00664

[(3,4,5-η)-1-*tert*-Butyl-2,3-dihydro-2-methyl-1*H*-1,2-azaborol-3-yl-*B*[2],*N*[1]]iodoiron

*Dicarbonyl[(3,4,5-η)-1-(1,1-dimethylethyl)-2,3-dihydro-2-methyl-1*H*-1,2-azaborol-3-yl-*B[2],N[1]*]iodoiron*, 10CI

[75334-22-4]

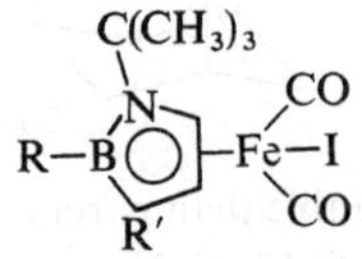

R = Me,R′ = H

M 374.796

Red-brown cryst. (pet. ether). Mp 87-8°.

Schulze, J. *et al*, *J. Organomet. Chem.*, 1980, **193**, 83 (*synth, ms, ir, pmr, nmr*)

$C_{10}H_{15}BI_2$ B-00665

Diiodo(1,2,3,4,5-pentamethyl-2,4-cyclopentadien-1-yl)borane, 9CI

(1,2,3,4,5-Pentamethyl-2,4-cyclopentadien-1-yl)diiodoborane. 5-Diiodoboryl-1,2,3,4,5-pentamethyl-1,3-cyclopentadiene

[71245-47-1]

As Dibromo(1,2,3,4,5-pentamethyl-2,4-cyclopentadien-1-yl)borane, B-00660

X = I

M 399.848

Fluxional molecule under normal conditions. Prepd. *in situ* from BI_3 and trimethyl(pentamethylcyclopentadienyl)germane. Air- and moisture-sensitive. Sol. CH_2Cl_2. pmr $\delta(CH_3)$ 1.72(s) ppm.

Jutzi, P. *et al*, *Chem. Ber.*, 1979, **112**, 2481, 2488 (*synth, use*)
Pelter, A. *et al*, *Comprehensive Organic Chemistry*, (*Barton, D.H.R. et al*, Ed.), Pergamon, Oxford, 1979, **3**, 695 (*rev*)

$C_{10}H_{15}BS_2$ B-00666

Bis(ethylthio)phenylborane

Phenyldithioboronic acid diethyl ester, 10CI, 9CI. *Dithiobenzeneboronic acid diethyl ester*, 8CI. *Bisethylthio phenylboronate. Diethyl phenyldithioboronate*

[1870-68-4]

$PhB(SEt)_2$

M 210.159

Liq., sensitive to moist air. Sol. non-hydroxylic solvs. $Bp_{1.5}$ 104-5° ($Bp_{0.7}$ 77-8°). n_D^{23} 1.5538, n_D^{19} 1.5779.

Abel, E.W. *et al*, *J. Chem. Soc.*, 1965, 3045 (*synth, props*)
Lalancette, J.M. *et al*, *Tetrahedron Lett.*, 1965, 2547 (*ir, synth, props*)
Cragg, R.H. *et al*, *J. Chem. Soc. (A)*, 1967, 947 (*props, synth, ir*)
Cragg, R.H. *et al*, *J. Inorg. Nucl. Chem.*, 1973, **35**, 3685 (*ir, ms, props, nmr*)
Niedenzu, K. *et al*, *Synth. React. Inorg. Metal-Org. Chem.*, 1980, **10**, 373 (*synth, nmr*)

$C_{10}H_{15}BZr$ B-00667

Bis(η^5-2,4-cyclopentadien-1-yl)hydro[tetrahydroborato(1−)-*H,H′*]zirconium

Bis(η-cyclopentadienyl)hydrido(tetrahydridoborato)zirconium

[12116-93-7]

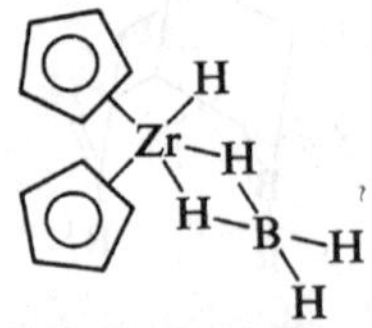

M 237.259

White solid. Subl. *in vacuo*.

James, B.D. *et al*, *Inorg. Chem.*, 1967, **6**, 1979 (*synth, ir, ms, nmr*)
Marks, T.J. *et al*, *Inorg. Chem.*, 1972, **11**, 2540 (*ir, raman*)
Marks, T.J. *et al*, *J. Am. Chem. Soc.*, 1975, **97**, 3397 (*nmr*)

$C_{10}H_{16}BBr_4O_4^{\ominus}$ B-00668

Bis[2,2-bis(bromomethyl)-1,3-propanediolato(2−)-*O,O′*]borate(1−), 10CI

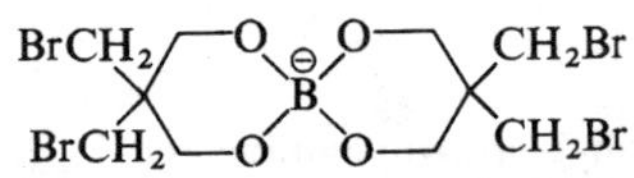

M 530.660 (ion)

Salts used as flame retardants for urethane polymers, epoxy resins, and polyesters.

Na salt: [74211-05-5].
$C_{10}H_{16}BBr_4NaO_4$ M 553.650
No phys. props. abstracted.

Ca salt: [74210-95-0].
$C_{20}H_{32}B_2Br_8CaO_8$ M 1101.400
No phys. props. abstracted.

K salt: [76772-21-9].
$C_{10}H_{16}BBr_4KO_4$ M 569.758
No phys. props. abstracted.

Guanidinium salt: [74211-04-4].
$C_{11}H_{22}BBr_4N_3O_4$ M 590.739
No phys. props. abstracted.

European Pat., 5 497, (*1979*); *CA*, **93**, 46174h (*use*)
Austrian Pat., 357 336, (*1980*); *CA*, **93**, 151192k (*use*)
Austrian Pat., 358 063, (*1980*); *CA*, **94**, 103330c

$C_{10}H_{16}BClO$ B-00669

6-Chloro-3-methoxy-7-methylene-3-borabicyclo[3.3.1]nonane, 10CI

6-Chloro-7-methylene-3-MBBN

[37490-52-1]

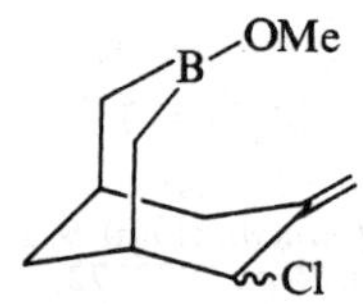

M 198.499

Air- and moisture-sensitive.

Mikhailov, B.M. *et al*, *Dokl. Akad. Nauk SSSR, Ser. Sci. Khim.*, 1972, **203**, 837 (*props*)
Mikhailov, B.M., *Pure Appl. Chem.*, 1974, **39**, 505 (*rev*)
Mikhailov, B.M. *et al*, *Izv. Akad. Nauk SSSR, Ser. Khim.*, 1976, 2056; *Bull. Acad. Sci. USSR, Div. Chem. Sci.*, 1927 (*use*)

$C_{10}H_{16}BClO$ B-00670

7-(Chloromethyl)-3-methoxy-3-borabicyclo[3.3.1]non-6-ene, 10CI, 9CI

[37490-50-9]

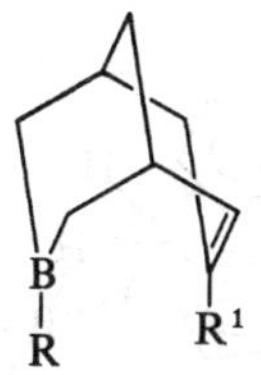

R = OMe, R[1] = CH_2Cl

M 198.499

Air- and moisture-sensitive.

Mikhailov, B.M. *et al*, *Dokl. Akad. Nauk SSSR*, *Ser. Sci. Khim.*, 1972, **203**, 837.
Mikhailov, B.M. *et al*, *Zh. Obshch. Khim.*, 1973, **43**, 1102, 1108.
Mikhailov, B.M. *et al*, *Izv. Akad. Nauk SSSR*, *Ser. Khim.*, 1976, 2056; *Bull. Acad. Sci. USSR*, *Div. Chem. Sci.*, 1927 (*synth*)
Mikhailov, B.M. *et al*, *CA*, 1978, **88**, 190937y (*synth*)

$C_{10}H_{16}BN$ B-00671

(Diethylamino)di-1-propynylborane, 8CI

N,N-*Diethyl-1,1-di-1-propynylboranamine*, 10CI, 9CI

[5271-93-2]

$Et_2NB(C{\equiv}CCH_3)_2$

M 161.053

Air- and moisture-sensitive liq. Bp_1 90°, $Bp_{0.001}$ 44-6°, 49°.

Soulié, J. *et al*, *Bull. Soc. Chim. Fr.*, 1966, 3850 (*ir, uv, synth*)
Berger, H.O. *et al*, *Z. Naturforsch.*, *B*, 1975, **30**, 641.
Wrackmeyer, B. *et al*, *Chem. Ber.*, 1977, **110**, 1086 (*nmr, pmr*)
Berger, H.O. *et al*, *Chem. Ber.*, 1980, **113**, 1235 (*synth*)

$C_{10}H_{16}BN_4^{\ominus}$ B-00672

Bis(3,5-dimethyl-1*H*-pyrazolato-*N'*)dihydroborate(1−), 9CI

Dihydrobis(3,5-dimethyl-1-pyrazolyl)borate

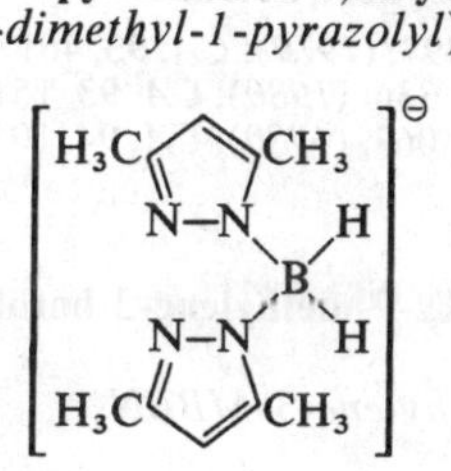

M 203.073 (ion)

K salt: [53241-68-2].

$C_{10}H_{16}BKN_4$ M 242.172

Cryst. (toluene).

Trofimenko, S., *Inorg. Chem.*, 1970, **9**, 2493 (*synth*)
Trofimenko, S., *Chem. Rev.*, 1972, **72**, 497 (*rev*)
McCurdy, W.H., *Inorg. Chem.*, 1975, **14**, 2292 (*pmr*)

$C_{10}H_{17}B$ B-00673

3-Boratricyclo[4.3.1.1^{3,8}]undecane, 10CI

3-Borahomoadamantane

[75612-44-1]

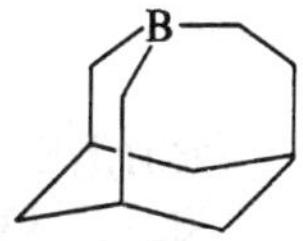

M 148.054

Complexes more stable than parent compd.

Me_3N complex: [75612-45-2].

$C_{13}H_{26}BN$ M 207.165

Cryst. (EtOH). Mp 150-6° (impure).

THF complex: [75612-46-3].

$C_{14}H_{25}BO$ M 220.161

Cryst. (2-methylbutane). Mp 55-70° (impure).

Mikhailov, B.M. *et al*, *Izv. Akad. Nauk SSSR*, *Ser. Khim.*, 1980, 1621; *Bull. Acad. Sci. USSR*, *Div. Chem. Sci.*, **29**, 1164 (*synth, pmr, derivs*)
Mikhailov, B.M., *Pure Appl. Chem.*, 1980, **52**, 691 (*synth*)

$C_{10}H_{17}B$ B-00674

3-Methyl-7-methylene-3-borabicyclo[3.3.1]nonane, 10CI

7-Methylene-3-methyl-3-borabicyclo[3.3.1]nonane

[76170-55-3]

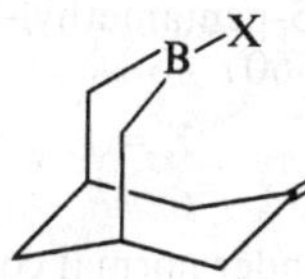

X = Me

M 148.054

Low melting, air-sensitive cryst. (pentane). Sol. ethers. Bp_{21} 77-80°. Predominately in the chair-chair conformn.

Gurskii, M.E. *et al*, *Izv. Akad. Nauk SSSR*, *Ser. Khim.*, 1980, 2188 (*synth*)
Mikhailov, B.M. *et al*, *J. Organomet. Chem.*, 1980, **201**, 81 (*ir, cmr, pmr, synth*)

$C_{10}H_{17}BN_2$ B-00675

Bis(dimethylamino)phenylborane, 8CI

N,N,N′,N′-*Tetramethyl-1-phenylboranediamine*, 10CI, 9CI

[1201-45-2]

$PhB(NMe_2)_2$

M 176.068

Liq., sensitive to moisture. Sol. Et_2O. d_4^{19} 0.924. Bp_9 91-2°, $Bp_{2.5}$ 61°. n_D^{20} 1.5183.

Niedenzu, K. *et al*, *Inorg. Chem.*, 1962, **1**, 738 (*synth*)
Ruff, J.K., *J. Org. Chem.*, 1962, **27**, 1020 (*synth, nmr, ir*)
Nöth, H. *et al*, *Chem. Ber.*, 1965, **98**, 962 (*ir, synth*)
Barfield, P.A. *et al*, *J. Chem. Soc. (A)*, 1968, 554 (*props*)
Cragg, R.H. *et al*, *J. Chem. Soc.*, *Chem. Commun.*, 1972, 206 (*ms*)
Gragg, B.R. *et al*, *J. Organomet. Chem.*, 1977, **132**, 29 (*nmr*)
Odom, J.D. *et al*, *J. Organomet. Chem.*, 1979, **173**, 15 (*nmr*)

$C_{10}H_{17}BN_2$ B-00676

Diethyl(2-pyridylmethylamino)borane

*Diethyl(2-pyridinemethanaminato-*N[1],N[2]*)boron*, 10CI

[78577-62-5]

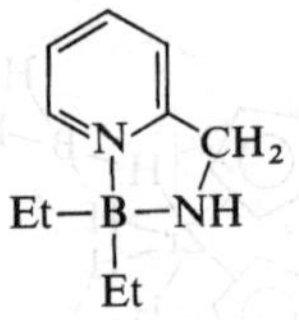

M 176.068

Exists in intramol. coord. form as shown.

Niedenzu, K. *et al*, *Z. Anorg. Allg. Chem.*, 1981, **473**, 139 (*synth*)

$C_{10}H_{17}BO$ B-00677

2-Methoxy-2-boratricyclo[3.3.1.1^3,7^]decane, 10CI, 9CI

2-Methoxy-2-boraadamantane

[56348-11-9]

As 2-Boratricyclo[3.3.1.$1^{3,7}$]decane, B-00598 with

R = OMe

M 164.054

Liq., sensitive to air and moisture. Bp_2 53-5° (55-60°). n_D^{20} 1.4960, 1.4965.

Mikhailov, B.M. *et al*, *Izv. Akad. Nauk SSSR, Ser. Khim.*, 1975, 980; *Bull. Acad. Sci. USSR, Div. Chem. Sci.*, **24**, 899 (*synth*)

Mikhailov, B.M. *et al*, *Izv. Akad. Nauk SSSR, Ser. Khim.*, 1977, 894; *Bull. Acad. Sci. USSR, Div. Chem. Sci.*, **26**, 818 (*synth, props*)

Shchegoleva, T.A. *et al*, *Izv. Akad. Nauk SSSR, Ser. Khim.*, 1980, 2169.

$C_{10}H_{17}BO$ B-00678

3-Methoxy-7-methylene-3-borabicyclo[3.3.1]nonane, 10CI

7-Methylene-3-methoxy-3-borabicyclo[3.3.1]nonane. 7-Methylene-3-MBBN

[39238-98-7]

As 3-Methyl-7-methylene-3-borabicyclo[3.3.1]nonane, B-00674 with

X = OMe

M 164.054

Air- and moisture-sensitive. Sol. Et_3N, CH_2Cl_2.

Mikhailov, B.M. *et al*, *Dokl. Akad. Nauk SSSR, Ser. Sci. Khim.*, 1972, **206**, 125 (*synth*)

Mikhailov, B.M. *et al*, *Pure Appl. Chem.*, 1974, **39**, 505 (*rev*)

Mikhailov, B.M. *et al*, *Izv. Akad. Nauk SSSR, Ser. Khim.*, 1979, 2544; *Bull. Acad. Sci. USSR, Div. Chem. Sci.*, 2361 (*synth*)

Gurskii, M.E. *et al*, *J. Organomet. Chem.*, 1980, **199**, 171 (*cmr, pmr*)

Mikhailov, B.M. *et al*, *Izv. Akad. Nauk SSSR, Ser. Khim.*, 1981, 394; *Bull. Acad. Sci. USSR, Div. Chem. Sci.*, 312.

$C_{10}H_{17}BO_2$ B-00679

9-Acetoxy-9-borabicyclo[3.3.1]nonane

9-(Acetyloxy)-9-borabicyclo[3.3.1]nonane, 10CI. *B-Acetoxy-9-*BBN

[62015-69-4]

As 9-Bromo-9-borabicyclo[3.3.1]nonane, B-00539 with

R = —OAc

M 180.053

Not obt. pure. Solid by subl. Mp >100°. Part. dec. during subl.

Kramer, G.W. *et al*, *J. Org. Chem.*, 1977, **42**, 2292.

Kramer, G.W. *et al*, *J. Organomet. Chem.*, 1977, **132**, 9.

Brown, H.C. *et al*, *J. Org. Chem.*, 1981, **46**, 3978.

$C_{10}H_{18}BBr$ B-00680

Bromodicyclopentylborane, 10CI, 9CI

Dicyclopentylbromoborane

[57476-09-2]

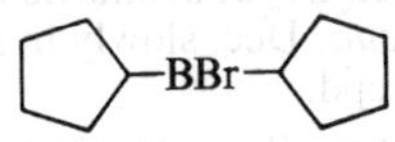

M 228.966

Synthetic intermed. Air- and moisture-sensitive liq. Sol. THF. Bp_1 68-70°.

Pelter, A. *et al*, *J. Chem. Soc., Chem. Commun.*, 1975, 531, 532 (*synth, use*)

Pelter, A. *et al*, *J. Chem. Soc., Dalton Trans.*, 1976, 2087 (*synth*)

Maddocks, P.J. *et al*, *J. Chem. Soc., Perkin Trans. 1*, 1981, 653 (*use*)

$C_{10}H_{18}BCl$ B-00681

Chlorodicyclopentylborane, 10CI, 9CI

Dicyclopentylchloroborane

[36140-18-8]

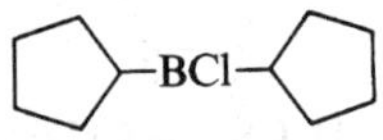

M 184.515

Synthetic intermed. Liq., air- and moisture-sensitive. V. sol. non-hydroxylic solvs. $Bp_{1.2}$ 69-70°.

Brown, H.C. *et al*, *J. Am. Chem. Soc.*, 1972, **94**, 2112; 1976, **98**, 1785 (*synth*)

Yamamoto, Y. *et al*, *Bull. Chem. Soc. Jpn.*, 1975, **48**, 3682 (*props*)

Brown, H.C. *et al*, *J. Org. Chem.*, 1977, **42**, 2533; 1979, **44**, 2417 (*synth, use*)

$C_{10}H_{18}BNO$ B-00682

Borane ephedrine

*Trihydro[α-[1-(methylamino)ethyl]benzenemethanol-*N*]boron*, 10CI. *α-[1-(Methylamino)ethyl]benzenemethanol compd. with borane(1:1). 1,*N*-Dimethyl-2-hydroxy-2-phenylethylamine borane*

[73798-15-9]

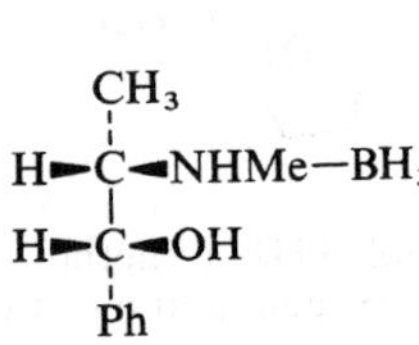

M 179.068

Chiral reducing complex.

Johnson, C.R. *et al*, *Tetrahedron Lett.*, 1979, 4713.

Grundon, M.F. *et al*, *J. Chem. Soc.*, *Perkin Trans. 1*, 1981, 231 (*synth*)
Mancilla, T. *et al*, *Tetrahedron Lett.*, 1982, **23**, 1561 (*ir*, *raman*, *nmr*)

$C_{10}H_{18}B_2Hf$ B-00683

Bis(η^5-2,4-cyclopentadien-1-yl)bis[tetrahydroborato(1−)-*H,H′*]hafnium, 9CI
Hafnocene bis(borohydride)
[56420-26-9]

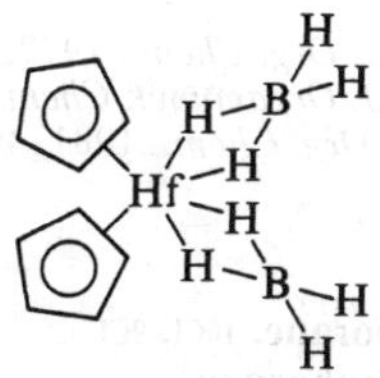

M 338.362
Colourless solid. Sol. Et_2O, aromatic hydrocarbons. Bp 110° subl. *in vacuo*. Dec. slowly in air. Deutero analogue also prepd.

Davies, N. *et al*, *J. Chem. Soc. (A)*, 1969, 2601 (*synth*, *ir*, *nmr*)
Marks, T.J. *et al*, *J. Am. Chem. Soc.*, 1975, **97**, 3397 (*nmr*)

$C_{10}H_{18}B_2N_4$ B-00684

Bis[(3,5-dimethyl-1-pyrazolyl)borane]
*Bis[μ-(3,5-dimethyl-1*H*-pyrazolato-*N^1:N^2*)]tetrahydrodiboron*, 10CI. *1,3,5,7-Tetramethyl-4*H*,4*H*,8*H*,8*H*-7*a*,8*a*-diaza-3*a*,4*a*-diazonia-4,8-diborata-*s*-indacene*, 8CI
[16998-92-8]

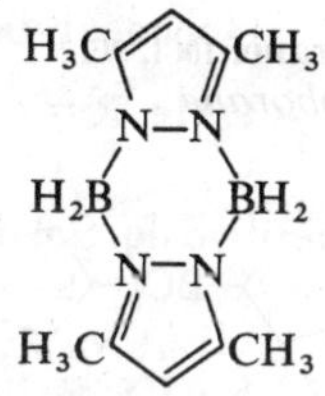

M 215.899
Cryst. Mp 184-5°.

Trofimenko, S., *J. Am. Chem. Soc.*, 1967, **89**, 3165 (*synth*, *nmr*)
Fr. Pat., 1 466 118, (*1967*); *CA*, **67**, P43814u (*synth*)
May, C.E. *et al*, *Z. Naturforsch.*, *B*, 1976, **31**, 1662 (*ms*)
May, C.E. *et al*, *Z. Naturforsch.*, *B*, 1978, **33**, 220 (*cmr*)

$C_{10}H_{18}B_2U$ B-00685

Bis(η^5-2,4-cyclopentadien-1-yl)bis[tetrahydroborato(1−)-*H,H′*]uranium, 10CI
Bis(cyclopentadienyl)uranium bis(tetrahydroborate)
[65888-45-1]

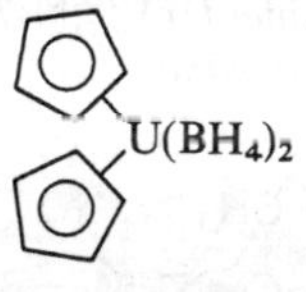

M 397.901
Red-brown solid. Sol. THF, dimethoxyethane. Subl. at 60° *in vacuo*. V. oxygen- and moisture-sensitive.

Zanella, P. *et al*, *J. Organomet. Chem.*, 1977, **142**, C21 (*synth*, *struct*, *ir*, *nmr*)

$C_{10}H_{18}B_2Zr$ B-00686

Bis(η^5-2,4-cyclopentadien-1-yl)bis[tetrahydroborato(1−)-*H,H′*]zirconium, 10CI
Zirconocene bis(borohydride)
[12083-77-1]

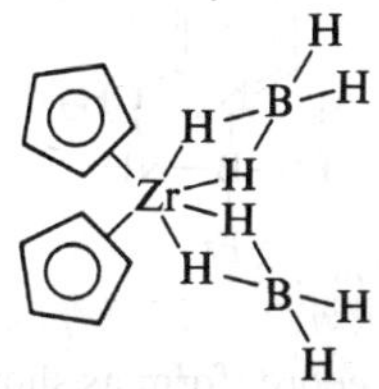

M 251.092
Pale-yellow solid by subl. Mp 155° dec. Bp 110-5° subl. *in vacuo*. Hydrolysed only slowly by water. BD_4 analogue also made.

Nanda, R.K. *et al*, *Inorg. Chem.*, 1964, **3**, 1798 (*synth*, *ir*)
Davies, N. *et al*, *J. Chem. Soc. (A)*, 1969, 2601 (*ir*, *nmr*)
Marks, T.J. *et al*, *J. Am. Chem. Soc.*, 1975, **97**, 3397 (*ir*, *nmr*)
Smith, B.E. *et al*, *J. Inorg. Nucl. Chem.*, 1976, **38**, 1973 (*synth*, *ir*, *raman*)

$C_{10}H_{19}B$ B-00687

Butyldi-2-propenylborane, 10CI, 9CI
Diallylbutylborane. Butyldiallylborane
[56347-96-7]

$$H_3CCH_2CH_2CH_2B(CH_2CH{=}CH_2)_2$$

M 150.070
Synthetic intermed. Liq., sensitive to air and moisture. Sol. nonhydroxylic solvs. Bp_4 33-5°. n_D^{20} 1.4390.

Mikhailov, B.M. *et al*, *Izv. Akad. Nauk SSSR, Ser. Khim.*, 1975, 979; *Bull. Acad. Sci. USSR, Div. Chem. Sci.*, 897 (*synth*)
Mikhailov, B.M. *et al*, *Izv. Akad. Nauk SSSR, Ser. Khim.*, 1978, 1586; *Bull. Acad. Sci. USSR, Div. Chem. Sci.*, 1385 (*synth*, *pmr*)
Mikhailov, B.M. *et al*, *J. Organomet. Chem*, 1978, **154**, 113 (*synth*)
Bubnov, Yu.N. *et al*, *Synthesis*, 1980, 902 (*props*)

$C_{10}H_{19}B$ B-00688

3,7-Dimethyl-3-borabicyclo[3.3.1]nonane, 10CI
3,7-Dimethyl-3-BBN
[76062-77-6]
As 3-Borabicyclo[3.3.1]nonane, B-00544 with

X = R = CH_3

M 150.070
Air-sensitive liq. Sol. CH_2Cl_2. Bp_{20} 80-9°.

Gurskii, M.E. *et al*, *J. Organomet. Chem.*, 1980, **199**, 171 (*nmr*)
Mikhailov, B.M. *et al*, *Izv. Akad. Nauk SSSR, Ser. Khim.*, 1981, 341; *Bull. Acad. Sci. USSR, Div. Chem. Sci.*, 264 (*nmr*)
Gurskii, M.E. *et al*, *Izv. Akad. Nauk SSSR, Ser. Khim.*, 1981, 394; *Bull. Acad. Sci. USSR, Div. Chem. Sci.*, 312.

$C_{10}H_{19}B$ B-00689

9-Ethyl-9-borabicyclo[3.3.1]nonane, 9CI
B-*Ethyl-9-*BBN
[52102-17-7]

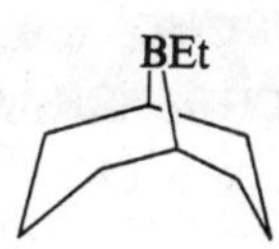

M 150.070

Reagent for transfer of ethyl group. Liq., sensitive to air. Bp_{18} 88-91°, $Bp_{0.2}$ 32-4°.

Koster, R. *et al*, *Justus Liebigs Ann. Chem.*, 1964, **672**, 1 (*synth*)
Kramer, G.W. *et al*, *J. Organomet. Chem.*, 1974, **73**, 1 (*synth*)
Brown, C.A., *J. Org. Chem.*, 1975, **40**, 3154 (*synth*)
Midland, M.M. *et al*, *J. Organomet. Chem.*, 1978, **156**, 203 (*use*)
Brown, H.C. *et al*, *J. Org. Chem.*, 1980, **45**, 846 (*nmr*)

$C_{10}H_{19}B$ B-00690

Monoisopinocampheylborane

(2,6,6-Trimethylbicyclo[3.3.1]hept-3-yl)borane, 9CI. *3-Pinanylborane*, 8CI. *Dihydro(2,6,6-trimethylbicyclo[3.1.1]hept-3-ylborane. Isopinocampheylborane*

[28875-08-3]

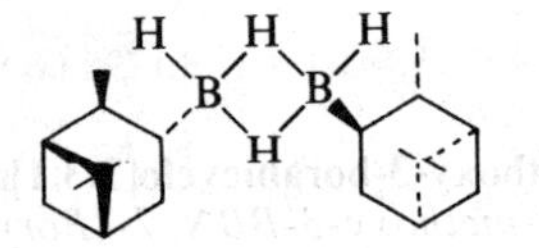

Dimer

M 150.070

Dimeric in uncomplexed state; avail. in both enantiomeric forms.

Dimer: [64350-68-1]. *Di-μ-hydrodihydrobis(2,6,6-trimethylbicyclo[3.3.1]hept-3-yl)diboron*, 10CI.
$C_{20}H_{38}B_2$ M 300.140
Hydroborating agent for converting hindered alkenes into opt. active alcohols. Air- and moisture-sensitive distillable liq. Sol. THF. $Bp_{0.05}$ 56-7°.
Disproportionates slowly at 25°.

NEt_3 complex: [64065-16-3]. (N,N-*Diethylethanamine)-dihydro(2,6,6-trimethylbicyclo[3.1.1]hept-3-yl)boron*, 10CI. *Monoisopinocampheylborane triethylamine. 3-Pinanylborane triethylamine.*
$C_{16}H_{34}BN$ M 251.261
Useful means of storage of parent compd. Air-sensitive viscous liq.

Tetramethylethylenediamine complex: [68297-73-4].
$C_{16}H_{35}BN_2$ M 266.276
Useful means of storage of parent compd. Treatment with BF_3 ppts. TMED.2BF_3 leaving free monoisopinocamphenylborane. Air-stable cryst. (pentane), stable for several weeks as cryst. or in THF soln. Mp 113-5°.

Brown, H.C. *et al*, *J. Am. Chem. Soc.*, 1977, **99**, 5514 (*synth*)
Brown, H.C. *et al*, *J. Organomet. Chem.*, 1977, **135**, C10 (*synth*)
Brown, H.C. *et al*, *J. Org. Chem.*, 1978, **43**, 4395 (*synth, use*)
Brown, H.C. *et al*, *Synthesis*, 1978, 146 (*synth*)
Singaram, B. *et al*, *J. Organomet. Chem.*, 1978, **156**, C1 (*synth*)
Brown, H.C. *et al*, *Inorg. Chem.*, 1979, **18**, 51 (*synth, nmr*)
Mandal, A.K. *et al*, *J. Org. Chem.*, 1980, **45**, 3543 (*use*)
Bell, T.W., *J. Am. Chem. Soc.*, 1981, **103**, 1163 (*use*)
Brown, H.C. *et al*, *J. Org. Chem.*, 1981, **46**, 2712, 5047 (*use*)
Fieser, M. *et al*, *Reagents for Organic Synthesis*, Wiley, 1967-83, **8**, 267.

$C_{10}H_{19}BCl_2$ B-00691

(1-Butyl-1-hexenyl)dichloroborane, 10CI

(5-Decenyl)dichloroborane

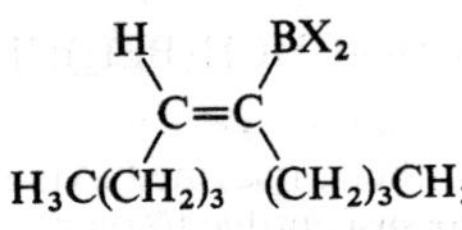

X = Cl

M 220.976

(Z)-form [64285-25-2]
Synthetic intermed. Colourless liq. Sol. CCl_4. Bp_2 64-70°. Air- and moisture-sensitive; dec. on storage.

Brown, H.C. *et al*, *J. Am. Chem. Soc.*, 1976, **98**, 1798 (*synth*)
Yatagai, H. *et al*, *J. Org. Chem.*, 1979, **44**, 2566 (*pmr*)

$C_{10}H_{19}BN^{\ominus}$ B-00692

Triethyl-1*H*-pyrrol-1-ylborate(1−), 10CI

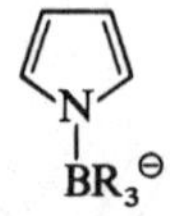

R = Et

M 164.077 (ion)

Li salt: [69350-02-3].
$C_{10}H_{19}BLiN$ M 171.018
Shows electrical resistivity; used in battery electrolytes.

Na salt: [69350-01-2].
$C_{10}H_{19}BNNa$ M 187.067
Shows electrical resistivity; used in battery electrolytes.

U.S.P., 4 104 450, (*1978*); *CA*, **90**, 124646t (*use*)
U.S.P., 4 117 213, (*1978*); *CA*, **90**, 124645s (*synth, props, use*)
Ger. Pat., 2 833 943, (*1979*); *CA*, **91**, 57173s (*synth, props*)
Newman, G.H. *et al*, *J. Electrochem. Soc.*, 1980, **127**, 2097 (*props, use*)

$C_{10}H_{19}BO$ B-00693

3-Methoxy-7-methyl-3-borabicyclo[3.3.1]nonane, 10CI

7-Methyl-3-methoxy-3-borabicyclo[3.3.1]nonane. 7-Methyl-3-MBBN

[17428-08-9]

As 3-Borabicyclo[3.3.1]nonane, B-00544 with

X = OMe, R = CH_3

M 166.070

Air- and moisture-sensitive liq. Sol. CH_2Cl_2. Bp_2 49.5-50.5°.

Mikhailov, B.M. *et al*, *Izv. Akad. Nauk SSSR, Ser. Khim.*, 1976, 2302; *Bull. Acad. Sci. USSR, Div. Chem. Sci.*, 2148 (*synth*)
Gurskii, M.E. *et al*, *J. Organomet. Chem.*, 1980, **199**, 171 (*cmr, pmr*)
Mikhailov, B.M. *et al*, *Izv. Akad. Nauk SSSR, Ser. Khim.*, 1981, 341; *Bull. Acad. Sci. USSR, Div. Chem. Sci.*, 264 (*cmr, pmr*)
Gurskii, M.E. *et al*, *Izv. Akad. Nauk SSSR, Ser. Khim.*, 1981, 394; *Bull. Acad. Sci. USSR, Div. Chem. Sci.*, 312.

$C_{10}H_{20}BBrS$ B-00694

9-Bromo-9-borabicyclo[3.3.1]nonane dimethyl sulfide

Bromo-1,5-cyclooctanediyl[thiobis[methane]]boron, 10CI. *9-Bromo-9-borabicyclo[3.3.1]nonane compd. with thiobis[methane](1:1)*. B-*Bromo-9-*BBN*-dimethyl sulfide*

[70160-59-7]

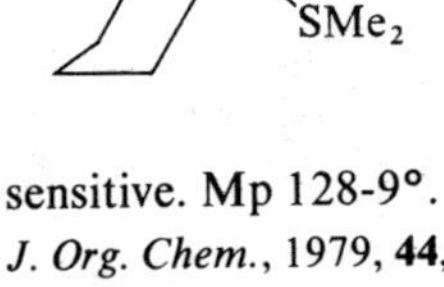

M 263.042

Cryst., moisture sensitive. Mp 128-9°.

Brown, H.C. *et al*, *J. Org. Chem.*, 1979, **44**, 2422 (*synth, pmr, nmr*)

Brown, H.C. *et al*, *J. Organomet. Chem.*, 1979, **168**, 281 (*synth, pmr, nmr*)

$C_{10}H_{20}BClS$ B-00695

9-Chloro-9-borabicyclo[3.3.1]nonane dimethyl sulfide

Chloro-1,5-cyclooctanediyl[thiobis[methane]]boron, *10CI*. *9-Chloro-9-borabicyclo[3.3.1]nonane compd. with thiobis[methane](1:1)*. B-*Chloro-9-*BBN-*dimethyl sulfide*

[70160-58-6]

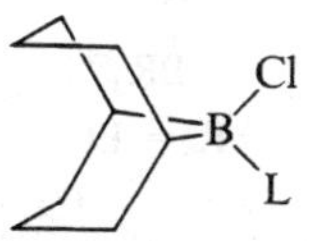

L = Me_2S

M 218.591

Cryst. (hexane); moisture- and air-sensitive. Mp 104-5°.

Brown, H.C. *et al*, *J. Org. Chem.*, 1979, **44**, 2422 (*synth, nmr*)
Brown, H.C. *et al*, *J. Organomet. Chem.*, 1979, **168**, 281 (*synth, nmr*)

$C_{10}H_{20}BIS$ B-00696

9-Iodo-9-borabicyclo[3.3.1]nonane dimethyl sulfide

1,5-Cyclooctanediyliodo[thiobis[methane]]boron, *10CI*. *9-Iodo-9-borabicyclo[3.3.1]nonane, compd. with thiobis[methane]*. B-*Iodo-9-BBN dimethyl sulfide*. B-*Iodo-9-BBN-methyl sulfide*

[70160-56-4]

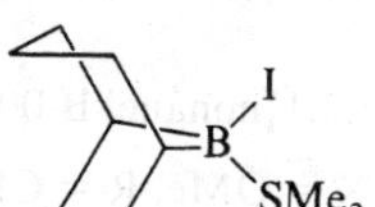

M 310.043

Cryst. (hexane), sensitive to moisture. Mp 133-4°.

Brown, H.C. *et al*, *J. Org. Chem.*, 1979, **44**, 2422 (*nmr*)
Brown, H.C. *et al*, *J. Organomet. Chem.*, 1979, **168**, 281 (*synth, nmr*)

$C_{10}H_{20}BN$ B-00697

9-Dimethylamino-9-borabicyclo[3.3.1]nonane

N,N-*Dimethyl-9-borabicyclo[3.3.1]nonan-9-amine*, *10CI*. B-(N,N-*Dimethylamino)-9-BBN*

[63366-67-6]

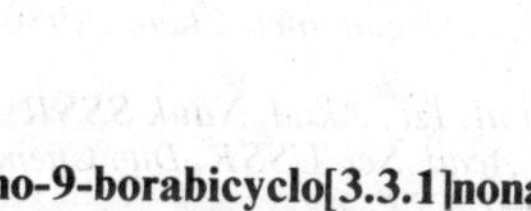

M 165.085

Air- and moisture-sensitive oil. Sol. cyclopentane, $CHCl_3$, CCl_4. Bp_7 88°.

Kramer, G.W. *et al*, *J. Organomet. Chem.*, 1977, **132**, 9 (*synth, nmr*)
Nelsen, S.F. *et al*, *J. Org. Chem.*, 1980, **45**, 2116 (*pe, ir, nmr, synth*)

$C_{10}H_{20}BNO$ B-00698

9-(2-Aminoethoxy)-9-borabicyclo[3.3.1]nonane

*(2-Aminoethanolato-*N,O*)(1,5-cyclooctanediyl)boron*, *10CI*. *2-(9-Borabicyclo[3.3.1]non-9-yloxy)ethanamine*, *10CI*. B-*Hydroxy-9-*BBN *ethanolamine ester*

[62059-31-8]

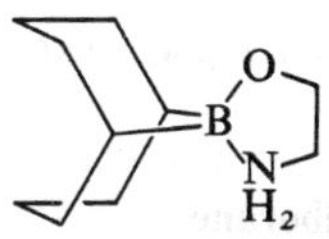

M 181.084

White solid, stable in air. Insol. pentane. Mp 202-203.5°.

Kramer, G.W. *et al*, *J. Org. Chem.*, 1977, **42**, 2292 (*synth, pmr, nmr*)
Brown, H.C. *et al*, *J. Org. Chem.*, 1977, **42**, 4169 (*synth*)

$C_{10}H_{20}B_2O$ B-00699

7-(Borylmethyl)-3-methoxy-3-borabicyclo[3.3.1]nonane, 10CI

7-(Borylmethyl)-3-methoxy-3-BBN. *7-(Borylmethyl)-3-MBBN*

[66352-26-9]

As 3-Borabicyclo[3.3.1]nonane, B-00544 with

X = OMe, R = $-CH_2BH_2$

M 177.887

Prepd. *in situ* by hydroboration of 3-Methoxy-7-methylene-3-borabicyclo[3.3.1]nonane, B-00678 . Intermed. in the synth. of 1-boraadamantane. Air- and moisture-sensitive.

Mikhailov, B.M. *et al*, *CA*, 1975, **88**, 190937y.
Mikhailov, B.M. *et al*, *Dokl. Akad. Nauk SSSR, Ser. Sci. Khim.*, 1978, **243**, 929.

$C_{10}H_{21}B$ B-00700

2-Butenyldipropylborane, 10CI

Crotyldipropylborane

[67682-91-1]

$(H_3CCH_2CH_2)_2BCH_2CH{=}CHCH_3$

M 152.086

Undergoes ready allyl rearr. causing equilibration of (*E*) and (*Z*)-isomers. Used in synth. of 2-alkoxy-1,4-pentadienes. Colourless liq.; air- and moisture-sensitive. Sol. nonhydroxylic solvs. Bp_7 51-2°. n_D^{20} 1.4317.

Mikhailov, B.M. *et al*, *Izv. Akad. Nauk SSSR, Ser. Khim.*, 1978, 1586; *Bull. Acad. Sci. USSR, Div. Chem. Sci.*, 1385 (*synth, nmr*)
Mikhailov, B.M. *et al*, *J. Organomet. Chem.*, 1978, **154**, 113 (*synth, nmr*)
Bubnov, Yu.N. *et al*, *Synthesis*, 1980, 902 (*props*)

$C_{10}H_{21}B$ B-00701

Diethyl(1-ethyl-2-butenyl)borane, 10CI, 9CI

1-Ethyl-2-butenyldiethylborane

[76284-46-3]

$H_3CCH{=}CHCH(BEt_2)CH_2CH_3$

M 152.086

Prepd. *in situ*. Synthetic intermed. Air-sensitive. Sol. THF. Should be susceptible to ready isomerisation *via* allyl rearrangement.

Pelter, A. *et al*, *Comprehensive Organic Chemistry*, (*Barton, D.H.R. et al*, Ed.), Pergamon, Oxford, 1979, **3**, 695 (*rev*)

Yamaguchi, M. *et al*, *Chem. Lett.*, 1980, 993 (*use*)

$C_{10}H_{21}B$ B-00702

(2-Methyl-2-propenyl)dipropylborane, 10CI

Dipropyl(2-methyl-2-propenyl)borane. Methallyldipropylborane

[67570-19-8]

$$H_2C{=}C(CH_3)CH_2B(CH_2CH_2CH_3)_2$$

M 152.086

Prepd. by exchange reaction between $(C_3H_7)_2BOC_6H_{13}$ and $[H_2C{=}C(CH_3)CH_2]_3B$. Synthetic intermed. Air- and moisture-sensitive liq. Sol. THF. Bp_9 58-9°. n_D^{20} 1.4315.

Mikhailov, B.M. *et al*, *Izv. Akad. Nauk SSSR, Ser. Khim.*, 1978, 1594; *Bull. Acad. Sci. USSR, Div. Chem. Sci.*, 1392 (*synth*)
Mikhailov, B.M. *et al*, *J. Organomet. Chem.*, 1978, **154**, 113, 131 (*synth, use*)
Pelter, A. *et al*, *Compr. Org. Chem.*, (Barton, D.H.R. *et al*, Ed.), 1979, **3**, 695 (*rev*)
Bubnov, Y.N. *et al*, *Synthesis*, 1980, 902 (*use*)

$C_{10}H_{21}BO$ B-00703

Diethyl(1,1-dimethyl-3-buten-1-yloxy)borane

Diethylborinic acid 1,1-dimethyl-3-butenyl ester, 10CI, 9CI

[55951-89-8]

$$Et_2BOC(CH_3)_2CH_2CH{=}CH_2$$

M 168.085

Prod. of reaction of allyldiethylborane with Me_2CO. Intermed. in prepn. of corresponding alcohol, which is obt. upon hydrolysis. Air- and moisture-sensitive liq. Sol. nonhydroxylic solvs. Bp_{30} 77-8°. $n_D^{23.5}$ 1.4193.

Mikhailov, B.M. *et al*, *Izv. Akad. Nauk SSSR, Ser. Khim.*, 1975, 483; *Bull. Acad. Sci. USSR, Div. Chem. Sci.*, **24**, 416 (*synth*)
Mikhailov, B.M. *et al*, *Izv. Akad. Nauk SSSR, Ser. Khim.*, 1978, 1892; *Bull. Acad. Sci. USSR, Div. Chem. Sci.*, **27**, 1663 (*synth, ir, pmr*)
Mikhailov, B.M. *et al*, *J. Organomet. Chem.*, 1978, **154**, 131 (*use*)

$C_{10}H_{21}BO$ B-00704

Diethyl(1-hexen-4-yloxy)borane

Diethylborinic acid 1-ethyl-3-butenyl ester, 10CI, 9CI

[55951-88-7]

$$Et_2BOCH(CH_2CH_3)CH_2CH{=}CH_2$$

M 168.085

Prod. of reaction of allyldiethylborane with propanal. Intermed. in prepn. of corresponding alcohol, which is obt. upon hydrolysis. Air- and moisture-sensitive liq. Sol. nonhydroxylic solvs. Bp_{20} 68-70°. $n_D^{23.5}$ 1.4165.

Mikhailov, B.M. *et al*, *Izv. Akad. Nauk SSSR, Ser. Khim.*, 1975, 483; *Bull. Acad. Sci. USSR, Div. Chem. Sci.*, **24**, 416 (*synth*)
Mikhailov, B.M. *et al*, *Izv. Akad. Nauk SSSR, Ser. Khim.*, 1978, 1892; *Bull. Acad. Sci. USSR, Div. Chem. Sci.*, **27**, 1663 (*synth, ir, pmr*)
Mikhailov, B.M. *et al*, *J. Organomet. Chem.*, 1978, **154**, 131 (*use*)

$C_{10}H_{21}BS$ B-00705

9-Borabicyclo[3.3.1]nonane dimethyl sulfide

1,5-Cyclooctanediylhydro[thiobis[methane]]boron, 10CI. 9-Borabicyclo[3.3.1]nonane, compd. with thiobis[methane](1:1). 9-BBN dimethyl sulfide

[64045-91-6]

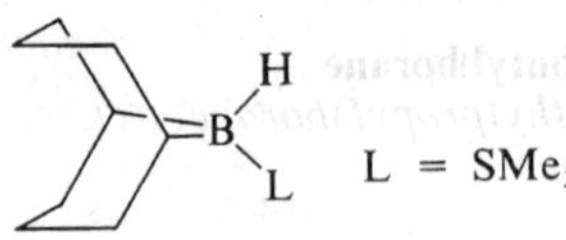

M 184.146

Useful monofunctional hydroborating agent; weakly complexed.

Brown, H.C. *et al*, *Inorg. Chem.*, 1977, **16**, 3090 (*synth, nmr*)
Soderquist, J.A. *et al*, *J. Org. Chem.*, 1981, **46**, 4599 (*use*)

$C_{10}H_{22}BBr$ B-00706

Bromodipentylborane, 10CI, 9CI

Dipentylbromoborane

[13317-62-9]

$$[H_3C(CH_2)_4]_2BBr$$

M 232.998

Precursor for synth. of unsymmetrical trialkylboranes. Air- and moisture-sensitive colourless liq. $Bp_{0.9}$ 58°.

McCusker, P.A. *et al*, *J. Inorg. Nucl. Chem.*, 1966, **28**, 2261 (*synth*)
Günter, S. *et al*, *Chem. Ber.*, 1967, **100**, 2899.
Pelter, A. *et al*, *J. Chem. Soc., Chem. Commun.*, 1975, 531, 532 (*synth, use*)
Pelter, A. *et al*, *J. Chem. Soc., Dalton Trans.*, 1976, 2087 (*synth*)
Maddocks, P.J. *et al*, *J. Chem. Soc., Perkin Trans. 1*, 1981, 653 (*use*)

$C_{10}H_{22}BN$ B-00707

[3-(Cyclohexylmethylamino)propyl]borane

[3-(Cyclohexylmethylamino)propyl)-C,N]dihydroboron, 10CI. N-Methyl-N-cyclohexyl-1,2-azaborolidine

[78207-46-2]

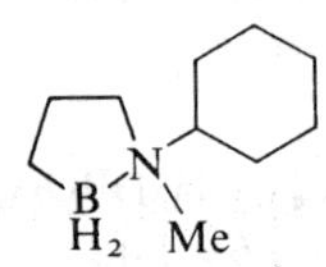

M 167.101

Baboulène, M. *et al*, *Bull. Soc. Chim. Fr., Part II*, 1980, 565 (*synth, ir, nmr*)

$C_{10}H_{22}BN_3$ B-00708

[*N*-(1-Iminoethyl)ethanimidamidato-*N′*,*N^N*]dipropylboron, 10CI, 9CI

[34629-68-0]

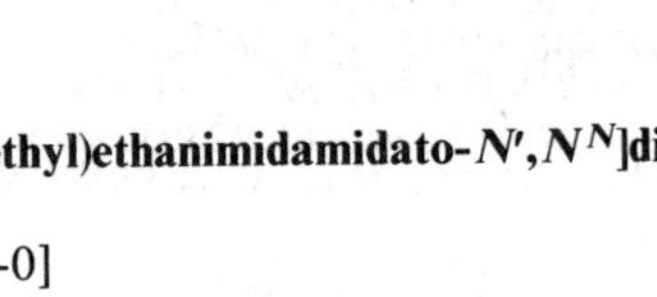

M 195.114

Dorokhov, V.A. *et al*, *Zh. Obshch. Khim.*, 1971, **41**, 2702.
Mikhailov, B.M. *et al*, *Dokl. Akad. Nauk SSSR, Ser. Sci. Khim.*, 1971, **199**, 1328.
Mikhailov, B.M. *et al*, *Zh. Obshch. Khim.*, 1974, **44**, 1888.

Zolotarev, B.M. *et al*, *Izv. Akad. Nauk SSSR, Ser. Khim.*, 1978, 1312; *Bull. Acad. Sci. USSR, Div. Chem. Sci.*, 1140 (*ms*)
Dorokhov, V.A. *et al*, *Dokl. Akad. Nauk SSSR, Ser. Sci. Khim.*, 1981, **258**, 351 (*synth*)

$C_{10}H_{23}B$ B-00709

Bis(3-methyl-2-butyl)borane

Bis(1,2-dimethylpropyl)borane, 10CI, 9CI. *Disiamylborane*

[6838-83-1]

$(H_3C)_2CH(CH_3)CHBHCH(CH_3)CH(CH_3)_2$

M 154.102

Usually exists as a bridged dimer. Selective reducing agent for carbonyl functions, unhindered alcohols and alkynes. Mp 35-40°. Air- and moisture-sensitive.

Brown, H.C. *et al*, *J. Am. Chem. Soc.*, 1961, **83**, 1241 (*synth, use*)
Holan, G. *et al*, *Tetrahedron Lett.*, 1973, 673 (*use*)
Brown, H.C. *et al*, *J. Org. Chem.*, 1974, **39**, 1631 (*use*)
Negishi, E. *et al*, *Tetrahedron Lett.*, 1977, 411 (*use*)
Hubbard, J.C. *et al*, *J. Organomet. Chem.*, 1978, **156**, 81 (*synth, nmr, use*)
Fieser, M. *et al*, *Reagents for Organic Synthesis*, Wiley, 1967-83, **8**, 41 (*use*)

$C_{10}H_{23}BI_2S$ B-00710

Diiodooctylborane dimethyl sulfide

Diiodooctyl[thiobis[methane]]boron, 10CI. *Diiodooctylborane compd. with thiobis[methane](1:1)*

[72206-01-0]

$H_3C(CH_2)_7BI_2—SMe_2$

M 439.971

Air- and moisture-sensitive.

Brown, H.C. *et al*, *J. Org. Chem.*, 1980, **45**, 384 (*synth*)

$C_{10}H_{23}BO$ B-00711

Butoxydipropylborane

Dipropylborinic acid butyl ester, 10CI. *Butyloxydipropylborane*

[3578-40-3]

$(H_3CCH_2CH_2)_2BO(CH_2)_3CH_3$

M 170.101

Air- and moisture-sensitive liq. Sol. nonhydroxylic solvs. Bp_7 65-8°, $Bp_{0.8}$ 34°. n_D^{20} 1.4132.

Mikhailov, B.M. *et al*, *Dokl. Akad. Nauk SSSR, Ser. Sci. Khim.*, 1961, **136**, 828 (*synth*)
Ger. Pat., 1 097 446, (*1961*); *CA*, **56**, 1478a
Laurent, J.P., *Bull. Soc. Chim. Fr.*, 1963, 558.
Nikitina, A.N. *et al*, *CA*, 1964, **61**, 12803b (*uv*)
Mikhailov, B.M. *et al*, *Izv. Akad. Nauk SSSR, Ser. Khim.*, 1965, 1487.

$C_{10}H_{23}BO$ B-00712

Dibutylethoxyborane

Dibutylborinic acid ethyl ester, 10CI, 9CI, 8CI

[32429-86-0]

$(H_3CCH_2CH_2CH_2)_2BOEt$

M 170.101

Air- and moisture-sensitive liq. Bp_{20} 86-8°.

Laurent, J.P. *et al*, *Bull. Soc. Chim. Fr.*, 1963, 558 (*synth*)

Davies, A.G. *et al*, *J. Chem. Soc. (B)*, 1971, 698.
Mikhailov, B.M. *et al*, *Izv. Akad. Nauk SSSR, Ser. Khim.*, 1977, 2633; *Bull. Acad. Sci. USSR, Div. Chem. Sci.*, **26**, 2437 (*props*)
Bagdasaryan, G.B. *et al*, *Arm. Khim. Zh.*, 1979, **32**, 203; *CA*, **92**, 6578 (*props, synth*)
Airiyan, L.Sh. *et al*, *Arm. Khim. Zh.*, 1979, **32**, 725; *CA*, **92**, 181248e (*synth, props*)
Pola, M. *et al*, *Collect. Czech. Chem. Commun.*, 1979, **44**, 3688 (*ir*)

$C_{10}H_{23}BO$ B-00713

Diethyl(hexyloxy)borane

Diethylborinic acid hexyl ester, 10CI

[67682-98-8]

$Et_2BO(CH_2)_5CH_3$

M 170.101

Air- and moisture-sensitive liq. Sol. nonhydroxylic solvs. Bp_3 45-6°. n_D^{20} 1.4160.

Mikhailov, B.M. *et al*, *Izv. Akad. Nauk SSSR, Ser. Khim.*, 1978, 1586; *Bull. Acad. Sci. USSR, Div. Chem. Sci.*, **27**, 1385 (*synth*)
Mikhailov, B.M. *et al*, *Izv. Akad. Nauk SSSR, Ser. Khim.*, 1978, 1594; *Bull. Acad. Sci. USSR, Div. Chem. Sci.*, **27**, 1392.
Mikhailov, B.M. *et al*, *J. Organomet. Chem.*, 1978, **154**, 113.

$C_{10}H_{23}BO_2$ B-00714

Dibutoxyethylborane

Ethylboronic acid dibutyl ester, 10CI. *Ethaneboronic acid dibutyl ester*, 8CI. *Ethyldibutoxyborane. Dibutyl ethaneboronate*

[3027-60-9]

$EtB(OCH_2CH_2CH_2CH_3)_2$

M 186.101

Air- and moisture-sensitive liq. Sol. non-hydroxylic solvs. d^{20} 0.822. Bp_{14} 86-8°. n_D^{20} 1.4060, 1.4140.

Mikhailov, B.M. *et al*, *Izv. Akad. Nauk SSSR, Ser. Khim.*, 1957, 1080 (*synth, props*)
Gallais, F. *et al*, *C.R. Hebd. Seances Acad. Sci.*, 1964, **259**, 4262 (*props*)
Baudet, J. *et al*, *C.R. Hebd. Seances Acad. Sci.*, 1967, **264**, 1090 (*props*)
Sterlin, R.N. *et al*, *CA*, 1968, **68**, 13025r (*props*)
Mikhailov, B.M. *et al*, *Izv. Akad. Nauk SSSR, Ser. Khim.*, 1978, 1586; *Bull. Acad. Sci. USSR, Div. Chem. Sci.*, **27**, 1385 (*synth, props*)
Mikhailov, B.M. *et al*, *J. Organomet. Chem.*, 1978, **154**, 113 (*synth, props*)

$C_{10}H_{23}BO_2$ B-00715

Di-*tert*-butoxyethylborane

Ethylboronic acid bis(1,1-dimethylethyl) ester, 10CI. *Ethaneboronic acid di-tert-butyl ester*, 8CI. *Ethylboronic acid di-*tert*-butyl ester. Di-*tert*-butyl ethaneboronate. Ethylbis-*tert-*butoxyborane*

[7318-83-4]

$EtB[OC(CH_3)_3]_2$

M 186.101

Air- and moisture-sensitive liq. Sol. THF. Bp_9 43° (Bp_8 46-8°). n_D^{20} 1.3968.

Nöth, H. *et al*, *Chem. Ber.*, 1966, **99**, 1049 (*nmr*)
Nöth, H. *et al*, *J. Organomet. Chem.*, 1968, **12**, 23 (*pmr, synth*)
Fenzl, W. *et al*, *Justus Liebigs Ann. Chem.*, 1978, 1030 (*synth, pmr, ms*)
Biffar, W. *et al*, *Chem. Ber.*, 1980, **113**, 333 (*nmr*)

$C_{10}H_{23}BS$ B-00716

(Butylthio)dipropylborane

Dipropylthioborinic acid butyl ester, 10CI, 9CI

[2938-91-2]

$(H_3CCH_2CH_2)_2BS(CH_2)_3CH_3$

M 186.162

Air- and moisture-sensitive liq. Sol. nonhydroxylic solvs. $Bp_{11.5}$ 98.5-102°, Bp_3 70-5°. n_D^{20} 1.4580 (1.4528).

Mikhailov, B.M. *et al*, *Izv. Akad. Nauk SSSR, Ser. Khim.*, 1960, 1307 (*synth*)
Mikhailov, B.M. *et al*, *Dokl. Akad. Nauk SSSR, Ser. Sci. Khim.*, 1961, **136**, 828 (*synth, props*)
Laurent, J.P., *Bull. Soc. Chim. Fr.*, 1963, 558 (*props*)
Mikhailov, B.M. *et al*, *Izv. Akad. Nauk SSSR, Ser. Khim.*, 1963, 641 (*synth, props*)
Bagdasaryan, G.B. *et al*, *Arm. Khim. Zh.*, 1979, **32**, 203; *CA*, **92**, 6578r (*synth, props*)

$C_{10}H_{23}BS$ B-00717

Diethyl(hexylthio)borane

Diethylthioborinic acid hexyl ester, 10CI, 9CI. *Hexylthio(diethyl)borane*

[55951-86-5]

$Et_2BS(CH_2)_5CH_3$

M 186.162

Used in synth. of allyldiethylboranes. Air- and moisture-sensitive liq. Sol. nonhydroxylic solvs. Bp_1 61-2°. n_D^{20} 1.4605.

Mikhailov, B.M. *et al*, *Izv. Akad. Nauk SSSR, Ser. Khim.*, 1975, 483; *Bull. Acad. Sci. USSR, Div. Chem. Sci.*, **24**, 416.
Mikhailov, B.M. *et al*, *Izv. Akad. Nauk SSSR, Ser. Khim.*, 1978, 1594; *Bull. Acad. Sci. USSR, Div. Chem. Sci.*, **27**, 1392 (*synth, nmr, props*)
Mikhailov, B.M. *et al*, *J. Organomet. Chem.*, 1978, **154**, 113 (*synth, nmr, props*)

$C_{10}H_{24}B^{\ominus}$ B-00718

Methyltripropylborate(1−), 10CI, 9CI

$MeB(CH_2CH_2CH_3)_3^{\ominus}$

M 155.110 (ion)

Cu(I) salt: [59598-77-5].
$C_{10}H_{24}BCu$ M 218.656
Formed *in situ* by adding CuI to the Li salt. Synth. intermed. Air-sensitive. Sol. THF. Not fully characterised.

Miyaura, N. *et al*, *Synthesis*, 1976, 618 (*props*)
Miyaura, N. *et al*, *Tetrahedron Lett.*, 1976, 255; 1977, 173, 3369 (*synth, use*)
Miyaura, N. *et al*, *Bull. Chem. Soc. Jpn.*, 1977, **50**, 2199.
Yamada, K. *et al*, *Bull. Chem. Soc. Jpn.*, 1979, **52**, 275 (*synth, props*)

$C_{10}H_{24}BN$ B-00719

Dibutyl(dimethylamino)borane, 8CI

1,1-Dibutyl-N,N-dimethylboranamine, 10CI, 9CI

[7289-86-3]

$(H_3CCH_2CH_2CH_2)_2BNMe_2$

M 169.116

Iodine-coloured liq., air- and moisture-sensitive. Bp_{10} 78-80°. Reported colour prob. due to impurities.

Ruff, J.K., *J. Org. Chem.*, 1962, **27**, 1020 (*synth, ir, nmr*)
Nöth, H. *et al*, *Chem. Ber.*, 1966, **99**, 1049 (*nmr*)
Schmidt, M. *et al*, *Z. Naturforsch., B*, 1968, **23**, 1258 (*synth*)
Nöth, H. *et al*, *Chem. Ber.*, 1969, **102**, 161.
Beck, W. *et al*, *Chem. Ber.*, 1972, **105**, 2883 (*nmr*)
Costes, J.P. *et al*, *J. Organomet. Chem.*, 1979, **175**, 257 (*nmr*)

$C_{10}H_{24}BN$ B-00720

Di-*tert*-butyl(dimethylamino)borane

1,1-Bis(1,1-dimethylethyl)-N,N-dimethylboranamine, 10CI

[76826-48-7]

$[(H_3C)_3C]_2BNMe_2$

M 169.116

Biffar, W. *et al*, *Chem. Ber.*, 1981, **114**, 49 (*nmr*)
Nöth, H. *et al*, *Chem. Ber.*, 1981, **114**, 1150 (*nmr*)
Klusik, H. *et al*, *J. Organomet. Chem.*, 1982, **232**, C21.

$C_{11}H_{10}BBrN_6O_2Ru$ B-00721

Bromodicarbonyl[hydrotris(pyrazolato)borato(1−)]ruthenium, 8CI

[33849-21-7]

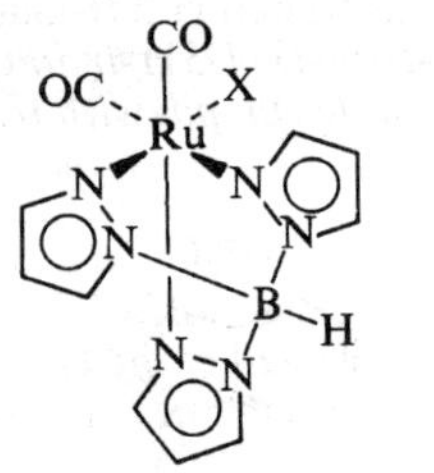

X = Br

M 450.023

Yellow microcryst. (MeOH). Mp 256°.

Bruce, M.I. *et al*, *J. Organomet. Chem.*, 1971, **31**, 269 (*synth, pmr, ir*)

$C_{11}H_{10}BClN_6O_2Ru$ B-00722

Dicarbonylchloro[hydrotris(pyrazolato)borato(1−)]ruthenium, 8CI

[33849-22-8]

As Bromodicarbonyl[hydrotris(pyrazolato)borato(1−)]-ruthenium, B-00721 with

X = Cl

M 405.572

White cryst. (MeOH). Mp 261-2°.

Bruce, M.I. *et al*, *J. Organomet. Chem.*, 1971, **31**, 269 (*synth, pmr, ir*)

$C_{11}H_{10}BIN_6O_2Ru$ B-00723

Dicarbonyl[hydrotris(pyrazolato)borato(1−)]iodoruthenium, 8CI

[33849-20-6]

As Bromodicarbonyl[hydrotris(pyrazolato)borato(1−)]-ruthenium, B-00721 with

X = I

M 497.024

Bright-yellow cryst. (MeOH). Mp 243°.

Bruce, M.I. *et al*, *J. Organomet. Chem.*, 1971, **31**, 269 (*synth, pmr, ir*)

$C_{11}H_{13}BCl_2N_7Rh$ B-00724

(Acetonitrile)dichloro[hydrotris(1*H*-pyrazolato-N^1)borato(1−)-N^2,$N^{2'}$,$N^{2''}$]rhodium, 10CI

(Hydrotrispyrazolylborato)dichloro(acetonitrile)rhodium

[73137-32-3]

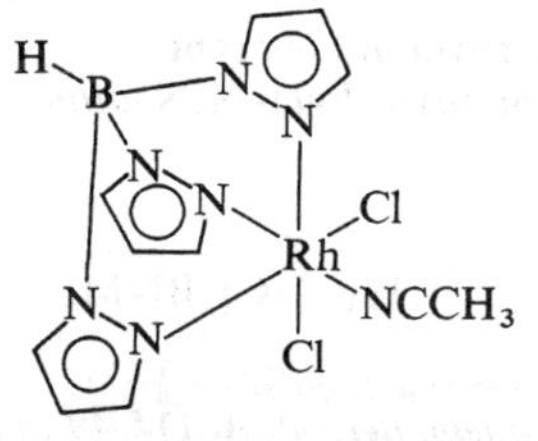

M 427.892

Starting material for hydrotris(1-pyrazolyl)borato-rhodium complexes. Dull-gold prisms (MeCN). Mp >360°.

May, S. *et al*, *Inorg. Chem.*, 1980, **19**, 1582 (*synth, pmr*)

$C_{11}H_{13}BF_3N$ B-00725

Trifluoroborane 2,3,3-trimethyl-3*H*-indole

*Trifluoro(2,3,3-trimethyl-3*H*-indole)boron, 10CI. 2,3,3-Trimethyl-3*H*-indole compd. with trifluoroborane(1:1)*

[69975-70-8]

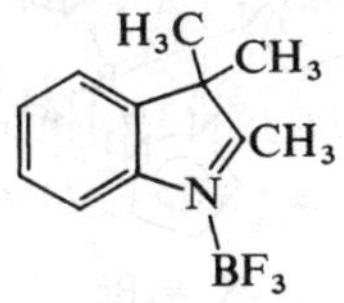

M 227.036

Mushkalo, N.N. *et al*, *Ukr. Khim. Zh. (Russ. Ed.)*, 1979, **45**, 79, 231 (*synth, nmr*)

$C_{11}H_{13}BFe$ B-00726

(η^5-2,4-Cyclopentadien-1-yl)[(1,2,3,4,5,6-η)-1-methylboratabenzene]iron, 10CI

[65777-53-9]

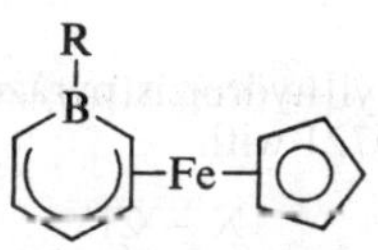

R = Me

M 211.881

Bright-red cryst. (pentane). Mp 56-7°. Subl. *in vacuo*.

Herberich, G.E. *et al*, *J. Organomet. Chem.*, 1978, **144**, C1 (*synth, ms, pmr, nmr*)

$C_{11}H_{15}B$ B-00727

3-Methyl-1-phenylborolane, 10CI, 9CI, 8CI

1-Phenyl-3-methylborolane. 3-Methyl-1-phenylboracyclopentane

[33673-18-6]

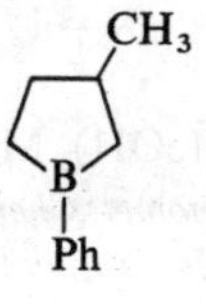

M 158.050

Air-sensitive liq. d_{20} 0.908. Bp_2 61-4°. n_D^{20} 1.5205.

Mikhailov, B.M. *et al*, *Dokl. Akad. Nauk SSSR, Ser. Sci. Khim.*, 1960, **133**, 119 (*synth*)
Wrackmeyer, B. *et al*, *J. Organomet. Chem.*, 1976, **108**, C21 (*props*)
Odom, J.D. *et al*, *J. Organomet. Chem.*, 1979, **173**, 15 (*nmr*)
Contreras, R. *et al*, *Z. Naturforsch., B*, 1980, **35**, 1229 (*nmr*)

$C_{11}H_{15}B$ B-00728

1-Phenylborinane, 10CI, 9CI

1-Phenyl-1-boracyclohexane

[59073-88-0]

As Borinane, B-00327 with

R = Ph

M 158.050

Characterized spectroscopically. Sensitive to air.

Wrackmeyer, B. *et al*, *Chem. Ber.*, 1976, **109**, 1075.
Wrackmeyer, B. *et al*, *J. Organomet. Chem.*, 1976, **108**, C21.
Odom, J.D. *et al*, *J. Organomet. Chem.*, 1979, **173**, 15 (*cmr, nmr*)
Braeuchle, C. *et al*, *Chem. Phys.*, 1980, **53**, 373 (*props*)

$C_{11}H_{15}B_2Co$ B-00729

(η^5-2,4-Cyclopentadien-1-yl)[(2,3,5,6-η)-1,4-dihydro-1,4-dimethyl-1,4-diborin]cobalt, 10CI

Cyclopentadienyl(1,4-dimethyl-1,4-dibora-2,5-cyclohexadiene)cobalt

[75039-09-7]

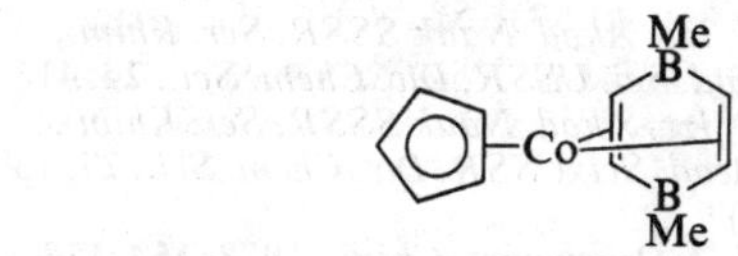

M 227.793

Red cryst. (pentane). Mp 122°. Air stable; thermally stable to 300°.

Herberich, G.E. *et al*, *J. Organomet. Chem.*, 1980, **192**, 421 (*synth, struct*)

$C_{11}H_{16}BNO_2$ B-00730

[[2,2′-(Methylimino)bis[ethanolato]](2−)-*N,O,O′*]phenylboron, 10CI

[66035-56-1]

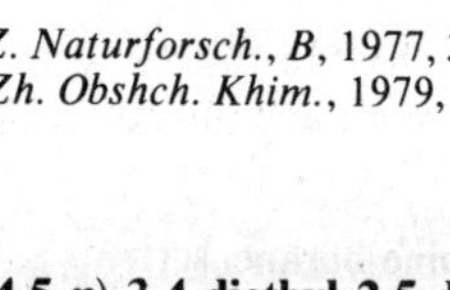

M 205.063

Gerwarth, U.W., *Z. Naturforsch., B*, 1977, **32**, 1408.
Mazeika, I. *et al*, *Zh. Obshch. Khim.*, 1979, **49**, 2528 (*ms*)

$C_{11}H_{16}B_2FeO_3S$ B-00731

Tricarbonyl[(1,2,3,4,5-η)-3,4-diethyl-2,5-dihydro-2,5-dimethyl-1,2,5-thiadiborole]iron, 10CI

[54688-79-8]

As Tricarbonyl[(1,2,3,4,5-η)-3,4-diethyl-2,5-dihydro-1,2,5-thiadiborole]iron, B-00591 with

R = Me

M 305.773

Yellow-orange cryst. Mp 23-5°. $Bp_{0.02}$ 35°.

Siebert, W. *et al*, *J. Organomet. Chem.*, 1977, **131**, 1 (*synth, pmr, nmr, ir, ms*)

$C_{11}H_{19}B$ B-00732

4,6-Dimethyl-1-boratricyclo[3.3.1.1^{3,7}]decane

4,6-Dimethyl-1-boraadamantane

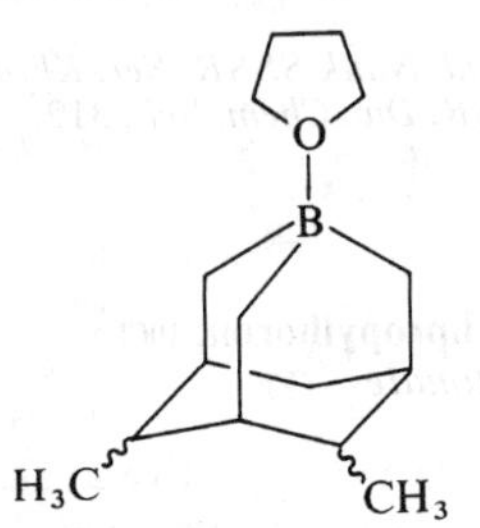

THF complex

M 162.081

Known only in the form of its complexes.

THF complex: [73207-64-4]. [(*2,4-Dimethyl-1,3,5-cyclohexanetriyl*)*tris*(*methylene*)](*tetrahydrofuran*)*boron*, 10CI.

$C_{15}H_{27}BO$ M 234.188

Cryst. (2-methylbutane). Mp 53-56.5°.

Py complex: [73207-65-5]. [(*2,4-Dimethyl-1,3,5-cyclohexanetriyl*)*tris*(*methylene*)](*pyridine*)*boron*, 10CI. *Pyridine compd. with 4,6-dimethyl-1-boratricyclo*[*3.3.1.1*^{3,7}]*decane*(*1:1*).

$C_{16}H_{24}BN$ M 241.182

Cryst. (hexane). Mp 90.5-91.5°.

Mikhailov, B.M. *et al*, *Izv. Akad. Nauk SSSR, Ser. Khim.*, 1979, 2551; *Bull. Acad. Sci. USSR, Div. Chem. Sci.*, 2368 (*synth, pmr*)

$C_{11}H_{19}B$ B-00733

3,5-Dimethyl-1-boratricyclo[3.3.1.1^{3,7}]decane, 10CI, 9CI

3,5-Dimethyl-1-boraadamantane

[55967-67-4]

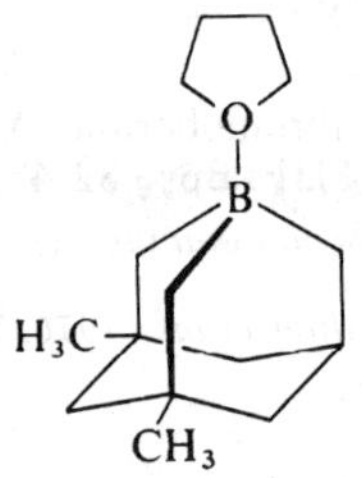

THF complex

M 162.081

THF complex: [72890-35-8]. [(*1,3-Dimethyl-1,3,5-cyclohexanetriyl*)*tris*(*methylene*)](*tetrahydrofuran*)*boron*, 10CI.

$C_{15}H_{27}BO$ M 234.188

Liq., unstable to air and moisture. Bp_1 72-4°. n_D^{20} 1.5087.

Py complex: [55967-68-5]. [(*1,3-Dimethyl-1,3,5-cyclohexanetriyl*)*tris*(*methylene*)](*pyridine*)*boron*, 10CI. *Pyridine compd. with 3,5-dimethyl-1-boratricyclo*[*3.3.1.1*^{3,7}]*decane*, 9CI, 8CI.

$C_{16}H_{24}BN$ M 241.182

Cryst. (hexane). Mp 54-6°. Bp_1 162-4°.

Et_2O complex: [55967-74-3]. [(*1,3-Dimethyl-1,3,5-cyclohexanetriyl*)*tris*(*methylene*)][*1,1'-oxybis*[*ethane*]-]*boron*, 10CI. *3,5-Dimethyl-1-boratricyclo*[*3.3.1.1*^{3,7}]-*decane compd. with 1,1'-oxybis*[*ethane*](*1:1*), 9CI.

$C_{15}H_{29}BO$ M 236.204

Colourless liq. d_4^{20} 0.92. Fairly stable but partially dissociates at r.t.

Mikhailov, B.M., *Pure Appl. Chem.*, 1974, **39**, 505 (*synth, nmr*)

Mikhailov, B.M. *et al*, *Izv. Akad. Nauk SSSR, Ser. Khim.*, 1976, 2302; *Bull. Acad. Sci. USSR, Div. Chem. Sci.*, 2148 (*synth*)

Mikhailov, B.M. *et al*, *Izv. Akad. Nauk SSSR, Ser. Khim.*, 1979, 2724; *Bull. Acad. Sci. USSR, Div. Chem. Sci.*, 2533 (*synth, cmr, pmr*)

Mikhailov, B.M. *et al*, *J. Organomet. Chem.*, 1980, **201**, 81.

$C_{11}H_{19}B$ B-00734

4,4-Dimethyl-1-boratricyclo[3.3.1.1^{3,7}]decane, 10CI

4,4-Dimethyl-1-boraadamantane

[72646-18-5]

M 162.081

Liq., sensitive to air. Bp_1 73-8°. More stable as complexes with Py and Et_2O.

Mikhailov, B.M. *et al*, *Izv. Akad. Nauk SSSR, Ser. Khim.*, 1979, 2340; *Bull. Acad. Sci. USSR, Div. Chem. Sci.*, **28**, 2161 (*synth, nmr*)

$C_{11}H_{19}B$ B-00735

9-(2-Propenyl)-9-borabicyclo[3.3.1]nonane, 10CI, 9CI

B-*Allyl-9-borabicyclo*[*3.3.1*]*nonane*. B-*Allyl-9-BBN*

[53317-08-1]

As 9-Bromo-9-borabicyclo[3.3.1]nonane, B-00539 with

$$R = —CH_2CH{=}CH_2$$

M 162.081

Fluxional molecule. Synthetic intermed. Clear liq., sensitive to air and moisture. $Bp_{0.05}$ 41.5-42°.

Kramer, G.W. *et al*, *J. Organomet. Chem.*, 1974, **73**, 1.

Kramer, G.W. *et al*, *J. Org. Chem.*, 1977, **42**, 2292 (*nmr*)

Kramer, G.W. *et al*, *J. Organomet. Chem.*, 1977, **132**, 9 (*synth, pmr, nmr*)

Brown, H.C. *et al*, *J. Organomet. Chem.*, 1980, **188**, 1.

Yamamoto, Y. *et al*, *Tetrahedron Lett.*, 1980, **21**, 3599 (*nmr*)

$C_{11}H_{19}BO_2$ B-00736

3-Methoxy-7-(methoxymethyl)-3-borabicyclo[3.3.1]non-6-ene, 10CI

[69464-55-7]

As 7-(Chloromethyl)-3-methoxy-3-borabicyclo[3.3.1]-non-6-ene, B-00670 with

$$R = OMe, R' = CH_2OMe$$

M 194.080

Air- and moisture-sensitive liq. Bp_1 67-8°. $n_D^{18.5}$ 1.4860.

Mikhailov, B.M. *et al*, *Dokl. Akad. Nauk SSSR, Ser. Sci. Khim.*, 1978, **243**, 929 (*synth*)

Mikhailov, B.M. *et al*, *Izv. Akad. Nauk SSSR, Ser. Khim.*, 1979, 2541, 2544; *Bull. Acad. Sci. USSR, Div. Chem. Sci.*, 2358, 2361 (*synth, ir, pmr*)

Mikhailov, B.M. *et al*, *Izv. Akad. Nauk SSSR, Ser. Khim.*, 1980, 1881; *Bull. Acad. Sci. USSR, Div. Chem. Sci.*, 1355 (*nmr, pmr*)

$C_{11}H_{21}B$ B-00737

9-Isopropyl-9-borabicyclo[3.3.1]nonane

9-(1-Methylethyl)-9-borabicyclo[*3.3.1*]*nonane*, 10CI, 9CI.

B-(*2-Propyl*)*-9-BBN*

[51475-51-5]

As 9-Bromo-9-borabicyclo[3.3.1]nonane, B-00539 with

$$R = -CH(CH_3)_2$$

M 164.097

Synthetic intermed. Oil, sensitive to air. $Bp_{0.25}$ 35.5-37.0°, $Bp_{0.05}$ 28-30°.

Brown, H.C. *et al*, *J. Am. Chem. Soc.*, 1974, **96**, 311.
Kramer, G.W. *et al*, *J. Organomet. Chem.*, 1974, **73**, 1 (*synth, pmr, nmr*)
Brown, H.C. *et al*, *Tetrahedron Lett.*, 1977, 3007.
Katz, J.J. *et al*, *Bull. Soc. Chim. Fr.*, 1977, 683.

$C_{11}H_{21}BO$ B-00738

Dicyclopentylmethoxyborane

Dicyclopentylborinic acid methyl ester, *10CI, 9CI*. *Methyl dicyclopentylborinate*

[36140-24-6]

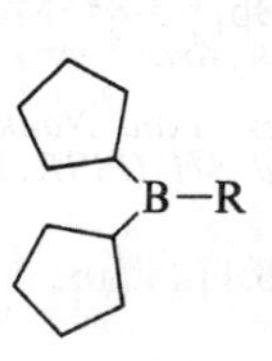

R = OMe

M 180.096

Used in synth. of dicyclopentyl ketone *via* reaction with dichloromethyl methyl ether. Air- and moisture-sensitive liq. Sol. THF. Bp_3 90-2°, Bp_2 82-4°. n_D^{20} 1.4705.

Brown, H.C. *et al*, *J. Am. Chem. Soc.*, 1972, **94**, 2112 (*synth*)
Carlson, B.A. *et al*, *J. Am. Chem. Soc.*, 1973, **95**, 6876 (*use*)
Brown, H.C. *et al*, *J. Am. Chem. Soc.*, 1976, **98**, 1785 (*synth, props*)
Brown, H.C. *et al*, *J. Org. Chem.*, 1979, **44**, 2417 (*synth*)

$C_{11}H_{21}BO$ B-00739

7-(Methoxymethyl)-3-methyl-3-borabicyclo[3.3.1]nonane, 10CI

[[5-(Methoxymethyl)-1,3-cyclohexanediyl]bis(methylene)-C,C′,O]methylboron, *10CI*

[76062-79-8]

As 3-Borabicyclo[3.3.1]nonane, B-00544 with

$$X = Me, R = CH_2OMe$$

M 180.096

endo-form [76070-07-0]

Intramolecularly coordinated. Air-sensitive. Sol. $CHCl_3$, CH_2Cl_2. Bp_2 62-3°.

Gurskii, M.E. *et al*, *J. Organomet. Chem.*, 1980, **199**, 171 (*cmr, synth*)
Gurskii, M.E. *et al*, *Izv. Akad. Nauk SSSR, Ser. Khim.*, 1981, 341; *Bull. Acad. Sci. USSR, Div. Chem. Sci.*, 264 (*nmr*)
Mikhailov, B.M. *et al*, *Izv. Akad. Nauk SSSR, Ser. Khim.*, 1981, 394; *Bull. Acad. Sci. USSR, Div. Chem. Sci.*, 312 (*synth*)

$C_{11}H_{21}BO_2$ B-00740

3-Methoxy-7-(methoxymethyl)-3-borabicyclo[3.3.1]nonane, 10CI

7-(Methoxymethyl)-3-methoxy-3-borabicyclo[3.3.1]-nonane. *7-(Methoxymethyl)-3-MBBN*

[74789-87-0]

As 3-Borabicyclo[3.3.1]nonane, B-00544 with

$$X = OMe, R = -CH_2OMe$$

M 196.096

Air- and moisture-sensitive. Sol. $CHCl_3$. Bp_3 85-6°.

Gurskii, M.E. *et al*, *J. Organomet. Chem.*, 1980, **199**, 171 (*nmr, synth*)
Mikhailov, B.M. *et al*, *Izv. Akad. Nauk SSSR, Ser. Khim.*, 1980, 1106; *Bull. Acad. Sci. USSR, Div. Chem. Sci.*, 813 (*synth, pmr*)
Mikhailov, B.M. *et al*, *Izv. Akad. Nauk SSSR, Ser. Khim.*, 1981, 341; *Bull. Acad. Sci. USSR, Div. Chem. Sci.*, 264 (*nmr, synth*)
Gurskii, M.E. *et al*, *Izv. Akad. Nauk SSSR, Ser. Khim.*, 1981, 394; *Bull. Acad. Sci. USSR, Div. Chem. Sci.*, 312.

$C_{11}H_{21}BO_2$ B-00741

(2,4-Pentanedionato-*O,O′*)dipropylboron, 10CI

Dipropylboryl acetylacetonate

[42562-18-5]

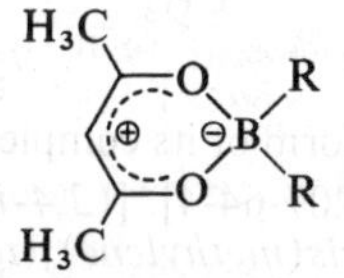

$R = CH_2CH_2CH_3$

M 196.096

Dorokhov, V.A. *et al*, *Zh. Obshch. Khim.*, 1973, **43**, 1115 (*synth*)
Mikhailov, B.M. *et al*, *Zh. Obshch. Khim.*, 1973, **43**, 862.
Zolotarev, B.M. *et al*, *Izv. Akad. Nauk SSSR, Ser. Khim.*, 1979, 80; *Bull. Acad. Sci. USSR, Div. Chem. Sci.*, 70 (*ms*)
Costes, J.P. *et al*, *Synth. React. Inorg. Metal-Org. Chem.*, 1981, **11**, 383.

$C_{11}H_{21}BS$ B-00742

Dicyclopentyl(methylthio)borane

Dicyclopentylthioborinic acid methyl ester, *10CI, 9CI*

[57476-23-0]

As Dicyclopentylmethoxyborane, B-00738 with

$$R = MeS-$$

M 196.157

Used in synth. of dicyclopentylbromoborane. Air- and moisture-sensitive liq. Sol. THF. $Bp_{1.3}$ 82-4°. m/e 127.

Pelter, A. *et al*, *J. Chem. Soc., Chem. Commun.*, 1975, 531 (*synth, use*)
Pelter, A. *et al*, *J. Chem. Soc., Dalton Trans.*, 1976, 2087 (*synth, props*)

$C_{11}H_{23}B$ B-00743

Dibutyl-2-propenylborane, 10CI, 9CI

Allyldibutylborane

[51851-76-4]

$$(H_3CCH_2CH_2CH_2)_2BCH_2CH{=}CH_2$$

M 166.113

Colourless mobile liq., stable under N_2 but readily hydrolysed and sensitive to air. Sol. non-hydroxylic solvs. Bp_{756} 114-5°. n_D^{20} 1.4152, n_D^{22} 1.4381.

Mikhailov, B.M. *et al*, *Izv. Akad. Nauk SSSR, Ser. Khim.*, 1975, 493, 979, 1586, 1594; *Bull. Acad. Sci. USSR, Div. Chem. Sci.*, 416, 897, 1385, 1392 (*synth, props*)
Mikhailov, B.M. *et al*, *J. Organomet. Chem.*, 1978, **154**, 113, 131 (*synth, use*)

$C_{11}H_{23}B$ B-00744

(3-Methyl-2-butenyl)dipropylborane, 10CI
Dipropyl(3-methyl-2-butenyl)borane.
Prenyldipropylborane
[67741-08-6]

$$(H_3C)_2C{=}CHCH_2B(CH_2CH_2CH_3)_2$$

M 166.113
Prepd. from 1-Bromo-3-methyl-2-butene, magnesium and phenoxydipropylborane. Synthetic intermed. Air- and moisture-sensitive liq. Sol. Et_2O. Bp_1 37-8°. n_D^{20} 1.4370.

Mikhailov, B.M. *et al*, *Izv. Akad. Nauk SSSR, Ser. Khim.*, 1978, 1586, 1892; *Bull. Acad. Sci. USSR, Div. Chem. Sci.*, 1392, 1663 (*synth*)
Mikhailov, B.M. *et al*, *J. Organomet. Chem.*, 1978, **154**, 113 (*synth, pmr*)
Pelter, A. *et al*, *Compr. Org. Chem.*, (*Barton, D.H.R. et al*, Ed.), 1979, **3**, 695 (*rev*)
Bubnov, Y.N. *et al*, *Synthesis*, 1980, 904 (*use*)
Mikhailov, B.M. *et al*, *Synthesis*, 1980, 902 (*use*)

$C_{11}H_{23}B_9Fe$ B-00745

[(1,2,3,4,5,6-η)-Methylbenzene][(7,8,9,10,11-η)-1,2,3,4,5,6,8,10,11-nonahydro-7,9-dimethyl-7,9-dicarbaundecaborato(2−)]iron
[closo-*1-(η^6-Toluene)-2,4-dimethyl-1,2,4-$FeC_2B_9H_9$*]
[80464-85-3]

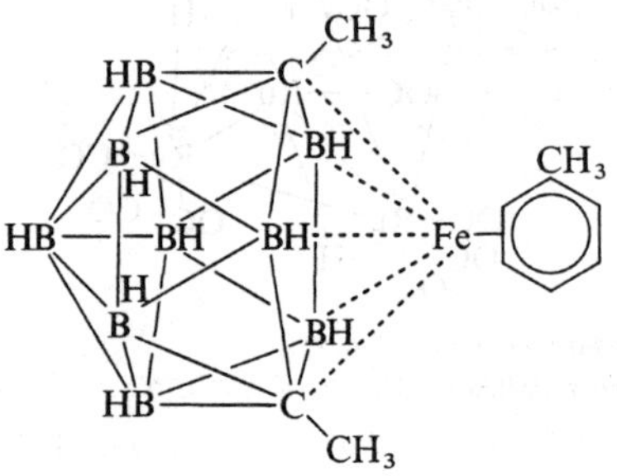

M 308.440
Red cryst. Mp 198°.

Garcia, M.P. *et al*, *J. Chem. Soc., Chem. Commun.*, 1981, 871 (*synth, struct, ir, nmr, cmr*)

$C_{11}H_{23}B_9Fe$ B-00746

[(1,2,3,4,5,6-η)-1,3,5-Trimethylbenzene][(7,8,9,10,11-η)-undecahydro-7,8-dicarbaundecaborato(2−)]iron
3,1,2-(η^6-Mesitylene)-$FeC_2B_9H_{11}$
[82733-98-0]

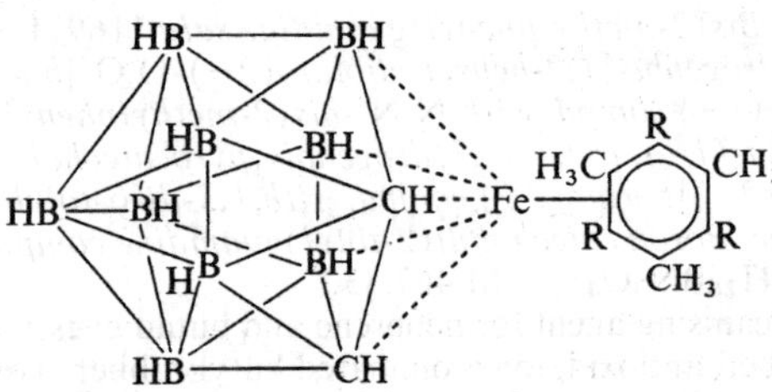

R = H

M 308.440
Orange cryst. (CH_2Cl_2/hexane). Mp 249-51° dec.

Hanusa, T. *et al*, *Polyhedron*, 1982, **1**, 77 (*synth, struct, pmr, ir, nmr*)

$C_{11}H_{24}BN$ B-00747

9-Borabicyclo[3.3.1]nonane isopropylamine
1,5-Cyclooctanediylhydro(2-propanamine)boron, 10CI.
2-Propanamine, compd. with 9-borabicyclo[3.3.1]nonane(1:1). 9-BBN isopropylamine complex
[70338-03-3]
As 9-Borabicyclo[3.3.1]nonane dimethyl sulfide, B-00705 with

$$L = H_2NCH(CH_3)_2$$

M 181.127
Completely associated in cyclohexane soln.

Brown, H.C. *et al*, *Recl. Trav. Chim. Pays-Bas*, 1979, **98**, 117.
Brown, H.C. *et al*, *J. Org. Chem.*, 1980, **45**, 1748 (*cmr, props*)

$C_{11}H_{24}BN$ B-00748

9-Borabicyclo[3.3.1]nonane propylamine
1,5-Cyclooctanediylhydro(1-propanamine)boron, 10CI.
1-Propanamine, compd. with 9-borabicyclo[3.3.1]nonane(1:1). 9-BBN propylamine complex
[70338-02-2]
As 9-Borabicyclo[3.3.1]nonane dimethyl sulfide, B-00705 with

$$L = H_2NCH_2CH_2CH_3$$

M 181.127
Completely associated in cyclohexane soln.

Brown, H.C. *et al*, *Recl. Trav. Chim. Pays-Bas*, 1979, **98**, 117.
Brown, H.C. *et al*, *J. Org. Chem.*, 1980, **45**, 1748 (*cmr, props*)

$C_{11}H_{24}BN$ B-00749

9-Borabicyclo[3.3.1]nonane trimethylamine
*1,5-Cyclooctanediyl(*N,N*-dimethylmethanamine)hydroboron.* N,N*-Dimethylmethanamine, compd. with 9-borabicyclo[3.3.1]nonane. 9-BBN trimethylamine complex*
[64070-34-4]
As 9-Borabicyclo[3.3.1]nonane dimethyl sulfide, B-00705 with

$$L = Me_3N$$

M 181.127
Colourless cryst. (pentane); unstable in air. Mp 74-5°.

Brown, H.C. *et al*, *Inorg. Chem.*, 1977, **16**, 3090 (*nmr*)

$C_{11}H_{25}BN^{\ominus}$ B-00750

(2,2-Dimethyl-1-propaniminato)triethylborate(1−), 10CI

$$(H_3C)_3CCH{=}NBEt_3^{\ominus}$$

M 182.135 (ion)
Li salt: [61867-25-2].
$C_{11}H_{25}BLiN$ M 189.076
No phys. props. reported.
Na salt: [61867-22-9].
$C_{11}H_{25}BNNa$ M 205.125
Solid. Mp 82°.

Hoberg, H. *et al*, *J. Organomet. Chem.*, 1976, **118**, C3, C6 (*props, ir*)

$C_{11}H_{25}BO$ B-00751

Dibutylpropoxyborane
Dibutylborinic acid propyl ester, 10CI

[63787-76-8]

$(H_3CCH_2CH_2CH_2)_2BOCH_2CH_2CH_3$

M 184.128

Air- and moisture-sensitive liq.

Bagdasaryan, G.B. *et al*, *Arm. Khim. Zh.*, 1977, **30**, 399; *CA*, **87**, 135509q.
Bagdasaryan, G.B. *et al*, *Arm. Khim. Zh.*, 1979, **32**, 192; *CA*, **92**, 6577q.
Airiyan, L.Sh. *et al*, *Arm. Khim. Zh.*, 1979, **32**, 725; *CA*, **92**, 181248e (*synth*, *props*)

$C_{11}H_{26}BN$ B-00752

(3-Aminopropyl)dibutylborane

*(3-Aminopropyl-*C,N*)dibutylboron*, 10CI

[50589-33-8]

As (3-Aminopropyl)dipropylborane, B-00633 with

$R = (CH_2)_3CH_3$

M 183.143

Dorokhov, V.A. *et al*, *Zh. Obshch. Khim.*, 1973, **43**, 1955 (*synth*)
Bochkareva, M.N. *et al*, *Zh. Obshch. Khim.*, 1975, **45**, 780.
Gal'chenko, G.L. *et al*, *Tr. Khim. Khim. Tekhnol.*, 1975, 17; *CA*, **86**, 9289j.
Bureiko, S.F. *et al*, *Kinet. Katal.*, 1979, **20**, 1414; *CA*, **92**, 146003z.
Zolotarev, B.M. *et al*, *Izv. Akad. Nauk SSSR, Ser. Khim.*, 1979, 2226; *Bull. Acad. Sci. USSR, Div. Chem. Sci.*, 2048 (*ms*)

$C_{11}H_{26}BN$ B-00753

(3-Dimethylaminopropyl)dipropylborane

*[3-(Dimethylamino)propyl-*C,N*]dipropylboron*, 10CI

[72910-22-6]

As (3-Dimethylaminopropyl)borane, B-00342 with

$R = CH_2CH_2CH_3$

M 183.143

Zolotarev, B.M. *et al*, *Izv. Akad. Nauk SSSR, Ser. Khim.*, 1979, 2226; *Bull. Acad. Sci. USSR, Div. Chem. Sci.*, 2048 (*ms*)

$C_{12}BBr_8O_4^{\ominus}$ B-00754

Bis[3,4,5,6-tetrabromo-1,2-benzenediolato(2−)-*O*,*O'*]borate(1−), 10CI

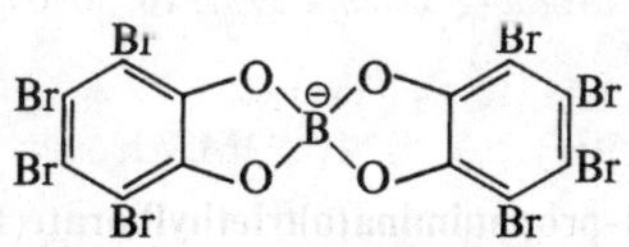

M 858.172 (ion)

Tetraethylammonium salt: [74222-94-9].

$C_{20}H_{20}BBr_8NO_4$ M 988.424

Flame retardant for urethane polymers, epoxy resins and polyesters.

Eur. Pat., 5 497, (*1979*); *CA*, **93**, 46174h (*use*, *synth*)
Austrian Pat., 357 336, (*1980*); *CA*, **93**, 151192k (*use*)
Austrian Pat., 358 063, (*1980*); *CA*, **94**, 103330c (*use*)

$C_{12}H_3BFe_4O_{12}$ B-00755

[(μ_4-(Boryl-*B*:*B*:*H*:*H'*)]dodecacarbonyl-μ-hydrotetrairon

[80572-82-3]

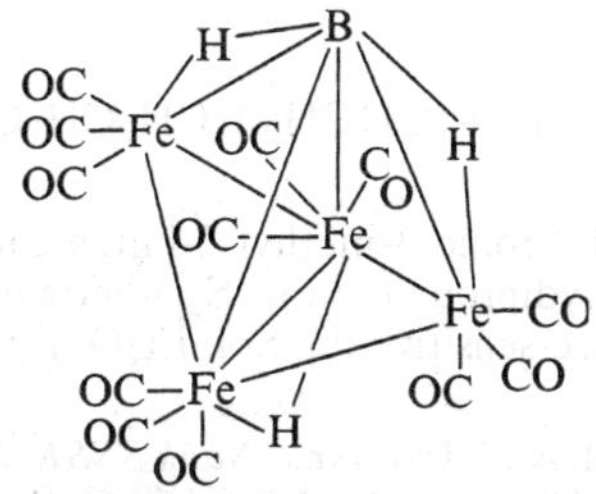

M 573.347

Air-sensitive dark-brown cryst.

Wong, K.W. *et al*, *J. Am. Chem. Soc.*, 1982, **104**, 1111 (*synth*, *pmr*, *nmr*)
Fehlner, T.P. *et al*, *Organometallics*, 1983, **2**, 825 (*synth*, *struct*)

$C_{12}H_3BO_{12}Ru_4$ B-00756

[μ_4-(Boryl-*B*:*B*:*H*:*H'*)]dodecacarbonyl-μ-hydrotetraruthenium, 10CI

[63374-84-5]

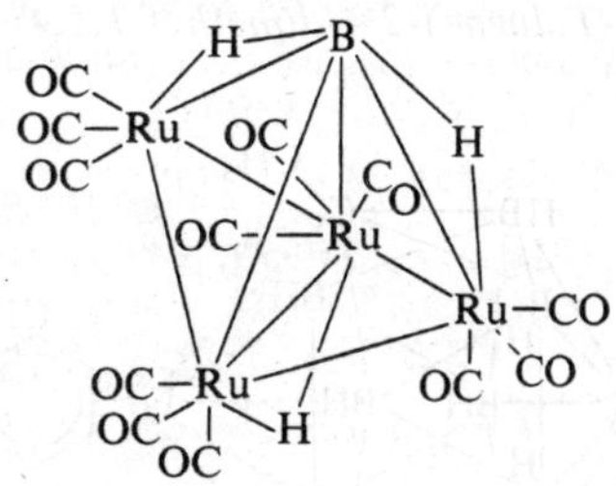

M 754.239

Air-sensitive yellow solid.

Eady, C.R. *et al*, *J. Chem. Soc., Dalton Trans.*, 1977, 477 (*synth*, *ms*, *ir*, *pmr*)

$C_{12}H_8BO_4^{\ominus}$ B-00757

Bis[1,2-benzenediolato(2−)-*O*,*O'*]borate(1−), 10CI, 9CI

Bis[pyrocatecholato(2−)]borate(1−), 8CI

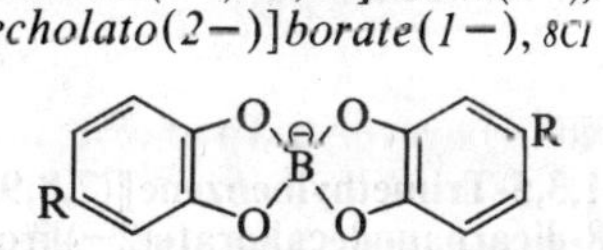

R = H

M 227.003 (ion)

N,N'-*Bis(2-methylphenyl)guanidine salt:* [16971-82-7]. *Hydrogenbis[1,2-benzenediolato(2−)-O,O']borate(1−) compd. with N,N'-bis(2-methylphenyl)guanidine(1:1)*, 10CI, 9CI. *Hydrogenbis[pyrocatecholato(2−)]borate(1−) compd. with 1,3-di-o-tolylguanidine. Boron catechol di(2-tolyl)guanidine complex.*

$C_{27}H_{26}BN_3O_4$ M 467.330

Vulcanising agent for neoprene and butadienestyrene rubber; antioxid. for brominated butyl rubber. Toxic to *Tetrahymena pyriformis*.

Epstein, S.S. *et al*, *J. Protozool.*, 1967, **14**, 238 (*tox*)
Morozova, N.N. *et al*, *CA*, 1972, **76**, 126551t (*synth*)
Timar, J. *et al*, *Rubber Chem. Technol.*, 1979, **52**, 319 (*use*)
Japan. Pat., 79 154 446, (*1979*); *CA*, **92**, 130387s (*use*)

$C_{12}H_{10}BBr$ B-00758

Bromodiphenylborane, 10CI, 9CI

Bromodiphenylborine

[5123-17-1]

Ph_2BBr

M 244.925

Synthetic intermed. Viscous liq., air- and moisture-sensitive. $Bp_{0.4}$ 105-10°. Comly. available.

Lappert, M.F. *et al*, *J. Organomet. Chem.*, 1964, **1**, 384 (*synth*)
Bowie, R.A. *et al*, *J. Chem. Soc.* (*C*), 1966, 566.
Finch, A. *et al*, *J. Chem. Soc., Chem. Commun.*, 1967, 1054.
Finch, A. *et al*, *Trans. Faraday Soc.*, 1967, **63**, 1880.
Schmidbach, C.D. *et al*, *Inorg. Chem.*, 1969, **8**, 1414.
Nahm, F.C. *et al*, *J. Organomet. Chem.*, 1972, **35**, 9 (*ir*)
Gragg, B.R. *et al*, *J. Organomet. Chem.*, 1977, **132**, 29 (*cmr*)

$C_{12}H_{10}BCl$ B-00759

Chlorodiphenylborane, 10CI, 9CI

Chlorodiphenylborine. Diphenylchloroborane

[3677-81-4]

Ph_2BCl

M 200.474

Catalyst for alkylbromide reaction with K_2CO_3. Colourless fuming liq., sensitive to air and moisture. d^{20} 1.116. $Bp_{0.1}$ 98°. n_D^{20} 1.6118.

Abel, E.W. *et al*, *J. Chem. Soc.*, 1956, 4697 (*synth, props*)
Niedenzu, K. *et al*, *Inorg. Chem.*, 1962, **1**, 738 (*synth*)
Nöth, H. *et al*, *Chem. Ber.*, 1966, **99**, 1049 (*nmr*)
Inorg. Synth., 1971, **10**, 32; 1974, **15**, 149 (*synth*)
Nahm, F.C. *et al*, *J. Organomet. Chem.*, 1972, **35**, 9 (*ir, nmr*)
Wrackmeyer, B. *et al*, *Chem. Ber.*, 1976, **109**, 1075 (*pmr*)
Gragg, B.R. *et al*, *J. Organomet. Chem.*, 1977, **132**, 29 (*nmr*)

$C_{12}H_{10}BCl_2N$ B-00760

Dichloro(diphenylamino)borane, 8CI

*1,1-Dichloro-*N,N*-diphenylboranamine*, 10CI, 9CI. *Dichloro(diphenylamino)borine*

[1139-65-7]

Cl_2BNPh_2

M 249.934

White solid readily hydrolysed and oxidized in moist air. Mp 78°.

Becher, H.J., *Z. Anorg. Allg. Chem.*, 1957, **289**, 262 (*synth, ir, raman*)
Becher, H.J. *et al*, *Chem. Ber.*, 1965, **98**, 526 (*synth*)
Casanova, J. *et al*, *Inorg. Chem.*, 1974, **13**, 2783 (*ir, nmr*)
Zettler, F. *et al*, *Chem. Ber.*, 1975, **108**, 2269 (*struct*)
Ger. Pat., 2 615 706, (*1975*); *CA*, **86**, 29473t (*synth*)
Baudler, M. *et al*, *Z. Anorg. Allg. Chem.*, 1981, **474**, 18 (*synth*)

$C_{12}H_{11}BO$ B-00761

Hydroxydiphenylborane

Diphenylborinic acid, 10CI, 9CI. *Diphenylboric acid*

[2622-89-1]

Ph_2BOH

M 182.028

Used as additive in paper chromatog. Mp 57.5°. Bp_{20} 150-5° (Bp 210-3° as anhydride). Slowly dec. in air. May be kept for up to three years as ethanolamine deriv.

Coates, G.E. *et al*, *J. Chem. Soc.*, 1961, 4909 (*synth*)
B.P., 906 145, (*1962*); *CA*, **58**, 2471 (*synth*)
Spielvogel, B.F. *et al*, *J. Am. Chem. Soc.*, 1975, **97**, 1609 (*nmr*)
Wrackmeyer, B., *Z. Naturforsch.*, *B*, 1980, **35**, 439 (*nmr*)
Sax, N.I., *Dangerous Properties of Industrial Materials*, 5th Ed., Van Nostrand-Reinhold, 1979, 622.

$C_{12}H_{12}BN_8^{\ominus}$ B-00762

Tetrakis(1*H*-imidazolato-*N*¹)borate(1−), 10CI

Tetrakis(1-imidazolyl)borate

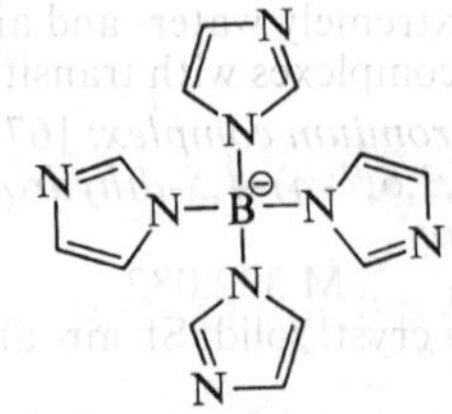

M 279.090 (ion)

Na salt: [68146-65-6].

$C_{12}H_{12}BN_8Na$ M 302.080

Used in detn. of deuterium by gravimetry. Yellow-white cryst. (EtOH). Mp 360° dec.

Tetramethylammonium salt: [68146-67-8].

$C_{16}H_{24}BN_9$ M 353.236

Cryst. (EtOH or Me_2CO/C_6H_6). Mp 200° dec.

Chao, S. *et al*, *Anal. Chim. Acta*, 1978, **100**, 457 (*synth, use, uv, ir, nmr*)
Pacey, G. *et al*, *Talanta*, 1980, **27**, 1013 (*nmr*)
Hockman, D.C. *et al*, *Anal. Lett., Part A*, **14**, 719 (*nmr*)

$C_{12}H_{12}BN_8^{\ominus}$ B-00763

Tetrakis(1*H*-pyrazolato-*N*²)borate(1−), 10CI, 9CI

Tetrakis(1-pyrazolyl)borate

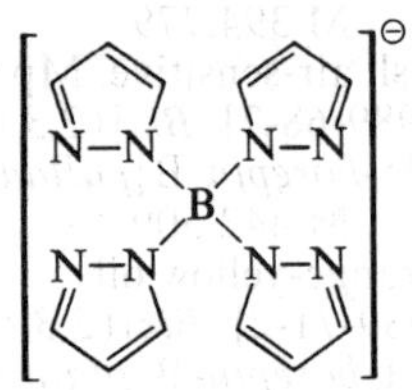

M 279.090 (ion)

Used in formn. of metal complexes.

Free acid: [17967-50-9]. Solid (CH_2Cl_2). Mp 70-1°.

K salt: [14782-58-2].

$C_{12}H_{12}BKN_8$ M 318.189

Cryst. (EtOH). Sol. DMF, DMSO, spar. sol. H_2O. Mp 253-4°.

Na salt: [14695-83-1].

$C_{12}H_{12}BNaN_8$ M 302.080

Cryst. (EtOH). Mp 238-9°.

Li salt: [14728-62-2].

$C_{12}H_{12}BLiN_8$ M 286.031

Cryst. (H_2O). Mp 394-5°. Sublimes at 300°/2 mm.

Trofimenko, S., *J. Am. Chem. Soc.*, 1967, **89**, 3170 (*synth*)
Inorg. Synth., 1970, **12**, 103 (*synth*)
Trofimenko, S., *Chem. Rev.*, 1972, **72**, 497 (*rev*)
Onishi, M. *et al*, *Bull. Chem. Soc. Jpn.*, 1980, **53**, 961, 1459, 2540 (*use*)

$C_{12}H_{13}B$ **B-00764**

4,5-Dihydro-1-phenylborepin, 10CI, 9CI

[59269-92-0]

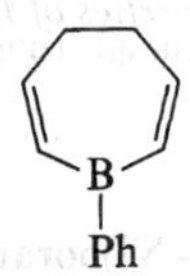

M 168.045

Colourless liq., extremely water- and air-sensitive. $Bp_{0.002}$ 52-6°. Forms complexes with transition metals.

Tetracarbonylchromium complex: [67158-11-6]. *Tetracarbonyl[(1,2,3,6,7-η)-4,5-dihydro-1-phenyl-1H-borepin]chromium.*

$C_{16}H_{13}BCrO_4$ M 332.082

Orange-yellow cryst. solid. Sl. air- and water-sensitive. Mp 71°.

Cyclopentadienylcobalt complex: [77070-13-4]. *(η^5-2,4-Cyclopentadien-1-yl)[(2,3,6,7-η)-4,5-dihydro-1-phenyl-1H-borepin-B]cobalt.*

$C_{17}H_{18}BCo$ M 292.072

Tricarbonyliron complex: see Tricarbonyl[(1,2,3,6,7-η)-4,5-dihydro-1-phenyl-1H-borepin]iron, B-00945

Carbonylcyclopentadienylmanganese complex: see Carbonyl(η^5-2,4-cyclopentadien-1-yl)[(1,2,3,6,7-η)-4,5-dihydro-1-phenyl-1H-borepin]manganese, B-01068

Tetracarbonylmolybdenum complex: [67158-12-7]. *Tetracarbonyl[(1,2,3,6,7-η)-4,5-dihydro-1-phenyl-1H-borepin]molybdenum.*

$C_{16}H_{13}BMoO_4$ M 376.026

Orange-yellow cryst. Sl. air- and water-sensitive. Mp 94-5°.

Ni complex: [74989-65-4]. *Bis[(2,3,6,7-η)-4,5-dihydro-1-phenyl-1H-borepin-B]nickel.*

$C_{24}H_{26}B_2Ni$ M 394.779

Orange cryst., sl. air-sensitive. Mp 95-6°.

Pd complex: [74989-68-7]. *Bis[(2,3,6,7-η)-4,5-dihydro-1-phenyl-1H-borepin-B]palladium.*

$C_{24}H_{26}B_2Pd$ M 442.509

Air-sensitive orange-yellow oil.

Pt complex: [74989-71-2]. *Bis[(2,3,6,7-η)-4,5-dihydro-1-phenyl-1H-borepin-B]platinum.*

$C_{24}H_{26}B_2Pt$ M 531.169

Air-sensitive yellow oil.

Tetracarbonyltungsten complex: [67158-13-8]. *Tetracarbonyl[(1,2,3,6,7-η)-4,5-dihydro-1-phenyl-1H-borepin]tungsten.*

$C_{16}H_{13}BO_4W$ M 463.936

Orange-yellow cryst., sl. air-sensitive. Mp 105°.

Herberich, G.E. *et al*, *Chem. Ber.*, 1977, **110**, 760 (*nmr, pmr, deriv, synth, ms*)

Koelle, U. *et al*, *J. Organomet. Chem.*, 1978, **152**, 7 (*derivs, synth, ms, props, ir, nmr*)

Herberich, G.E. *et al*, *J. Organomet. Chem.*, 1980, **191**, 27 (*derivs*)

Koelle, U., *Inorg. Chim. Acta*, 1981, **47**, 13 (*props, deriv*)

$C_{12}H_{13}BCoF_3N_4$ **B-00765**

(η^5-2,4-Cyclopentadien-1-yl)[dihydrobis(1*H*-pyrazolato-N^1)borato(1–)-N^2,$N^{2'}$](trifluoromethyl)cobalt, 9CI

[52470-75-4]

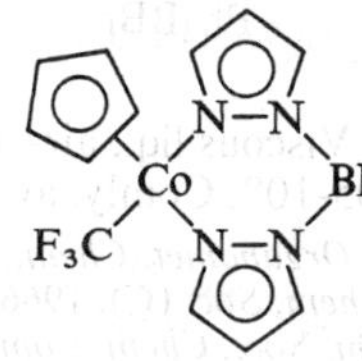

M 340.000

Air-stable red cryst. (hexane/Et_2O). Mp 147°. ^{19}F nmr δ 11.9s (Et_2O).

King, R.B. *et al*, *J. Am. Chem. Soc.*, 1974, **96**, 1334 (*synth, nmr*)

$C_{12}H_{14}B_2Fe_2N_2O_4$ **B-00766**

Di-μ-carbonyldicarbonylbis[(3,4,5-η)-2,3-dihydro-2-methyl-1*H*-1,2-azaborol-3-yl-B^2,N^1]diiron, 10CI

[78534-54-0]

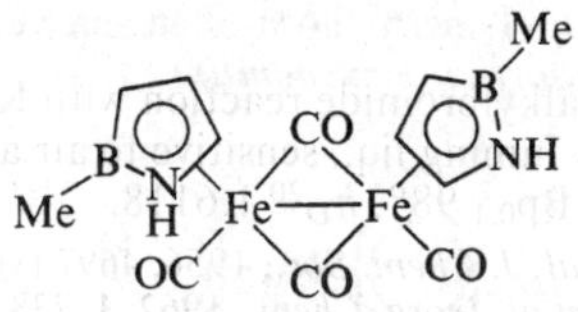

M 383.568

Dark-red cryst. (toluene). Mp 140° dec. $Bp_{0.01}$ 80° subl.

Schulze, J. *et al*, *Chem. Ber.*, 1981, **114**, 1297 (*synth, ir, ms, pmr, nmr*)

$C_{12}H_{15}BFe$ **B-00767**

Dimethylborylferrocene

Ferrocenyldimethylboron, 10CI. *(Dimethylboryl)ferrocene. Ferrocenyldimethylborane*

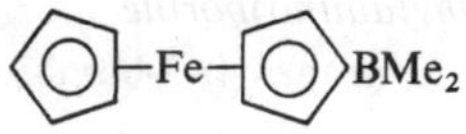

M 225.908

Py complex: [61649-81-8].

$C_{17}H_{20}BFeN$ M 305.009

Solid. Mp 113-5°.

Renk, T. *et al*, *J. Organomet. Chem.*, 1976, **120**, 1 (*synth*)

$C_{12}H_{16}B_2Co$ **B-00768**

Bis(1-methylborinato)cobalt

1-Methylboratabenzene cobalt complex, 10CI, 9CI. *Bis[(1,2,3,4,5,6-η)-1-methylboratabenzene]cobalt*

[36534-27-7]

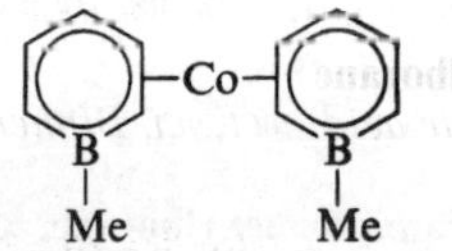

M 240.812

Solid. Mp 98°.

Herberich, G.E. *et al*, *J. Chem. Soc., Chem. Commun.*, 1971, 1328 (*synth, props*)

Huttner, G. *et al*, *Chem. Ber.*, 1972, **105**, 3424 (*struct*)

Koelle, U., *J. Organomet. Chem.*, 1978, **152**, 225 (*props*)

Clack, D.W. *et al*, *Inorg. Chem.*, 1979, **18**, 513 (*props*)
Herberich, G.E. *et al*, *J. Organomet. Chem.*, 1980, **197**, 1.

$C_{12}H_{16}B_2Fe$ B-00769

Bis[(1,2,3,4,5,6-η)-1-methylboratabenzene]iron, 9CI

Bis(1-methylborinato)iron

[54853-80-4]

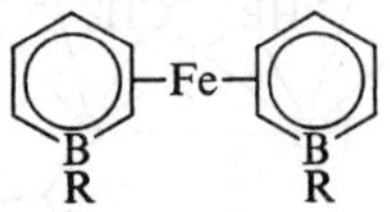

R = Me

M 237.725

Orange air-stable cryst. (pentane). Mp 44-5°, 71.2-71.7°.

Herberich, G. *et al*, *Chem. Ber.*, 1974, **107**, 3780 (*synth, pmr, ir, ms*)
Ashe, A.J. *et al*, *J. Am. Chem. Soc.*, 1975, **97**, 6865 (*synth, mössbauer, nmr, uv, ms*)

$C_{12}H_{16}B_2Ru$ B-00770

Bis[(1,2,3,4,5,6-η)-1-methylboratabenzene]ruthenium, 9CI

Bis(1-methylborinato)ruthenium

[60552-37-6]

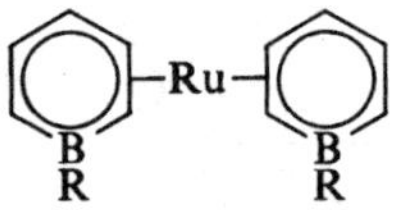

R = Me

M 282.948

Air stable, pale-yellow cryst. (pentane). Mp 64-5°. $Bp_{0.01}$ 60° subl.

Herberich, G.E. *et al*, *Chem. Ber.*, 1976, **109**, 2382 (*synth, pmr, nmr, ms*)

$C_{12}H_{18}BCl$ B-00771

7-(Chloromethyl)-3-(2-propenyl)-3-borabicyclo[3.3.1]non-6-ene, 10CI, 9CI

3-Allyl-7-(chloromethyl)-3-borabicyclo[3.3.1]non-6-ene

[43184-99-2]

As 7-(Chloromethyl)-3-methoxy-3-borabicyclo[3.3.1]-non-6-ene, B-00670 with

$$R = —CH_2CH{=}CH_2,\ R' = CH_2Cl$$

M 208.537

Air- and moisture-sensitive.

Mikhailov, B.M. *et al*, *Zh. Obshch. Khim.*, 1973, **43**, 1102, 1108.
Mikhailov, B.M. *et al*, *Izv. Akad. Nauk SSSR, Ser. Khim.*, 1976, 2056; *Bull. Acad. Sci. USSR, Div. Chem. Sci.*, 1927 (*pmr, synth*)
Mikhailov, B.M. *et al*, *CA*, 1978, **88**, 190937y (*synth*)

$C_{12}H_{19}BClN$ B-00772

Chloro(diisopropylamino)phenylborane, 8CI

1-Chloro-N,N-bis(1-methylethyl)-1-phenylboranamine, 10CI, 9CI

[33441-68-8]

$$PhBClN[CH(CH_3)_2]_2$$

M 223.552

Air- and moisture-sensitive liq. Bp_4 94°, $Bp_{0.1}$ 64°.

Niedenzu, K. *et al*, *Chem. Ber.*, 1963, **96**, 2653 (*synth*)
Imbery, D. *et al*, *Org. Magn. Reson.*, 1970, **2**, 271 (*nmr*)
Brown, C. *et al*, *J. Organomet. Chem.*, 1978, **149**, C34 (*nmr*)
Cragg, R.H. *et al*, *J. Organomet. Chem.*, 1981, **217**, 1; 1982, **232**, 201 (*synth, pmr, nmr*)

$C_{12}H_{19}BClN$ B-00773

Chloro(dipropylamino)phenylborane, 8CI

1-Chloro-1-phenyl-N,N-dipropylboranamine, 10CI, 9CI

[33441-66-6]

$$PhBClN(CH_2CH_2CH_3)_2$$

M 223.552

Air- and moisture-sensitive.

Brown, C. *et al*, *J. Organomet. Chem.*, 1978, **149**, C34; 1981, **220**, C25 (*nmr*)
Imbery, D. *et al*, *Org. Magn. Reson.*, 1979, **2**, 271 (*nmr*)
Cragg, R.H. *et al*, *J. Organomet. Chem.*, 1981, **217**, 1.

$C_{12}H_{20}BBr_3$ B-00774

Bromobis(2-bromo-1-hexenyl)borane, 10CI

$$[H_3C(CH_2)_3CBr{=}CH]_2BBr$$

M 414.812

Formed by bromoboronation of 1-hexyne. Air- and moisture-sensitive.

(E,Z)-form [63011-71-2]

Can undergo acetolysis with retention of configuration.

Blackborrow, R.J., *J. Organomet. Chem.*, 1977, **128**, 161 (*synth*)

$C_{12}H_{20}BCl_2N$ B-00775

Phenyldichloroborane triethylamine

Dichloro(N,N-diethylethanamine)phenylboron, 10CI. Dichlorophenylborane N,N-diethylethanamine, 9CI, 8CI. Dichlorophenylborane triethylamine

[17520-47-7]

$$Et_3N\text{-}BPhCl_2$$

M 260.013

Solid. Mp 88-90°.

Currell, B.R. *et al*, *J. Organomet. Chem.*, 1967, **8**, 411 (*synth*)
Sheikh, S.U., *J. Math. Sci.*, 1974, **1**, 131 (*ms*)
Sheikh, S.U., *Bull. Chem. Soc. Jpn.*, 1976, **49**, 2879.
Sheikh, S.U., *Pak. J. Sci. Ind. Res.*, 1976, **19**, 48.
Sheikh, S.U., *Therm. Anal.*, 1977, 509.
Sheikh, S.U., *J. Therm. Anal.*, 1980, **18**, 299.

$C_{12}H_{20}BCl_3$ B-00776

Chlorobis(2-chloro-l-hexenyl)borane, 10CI

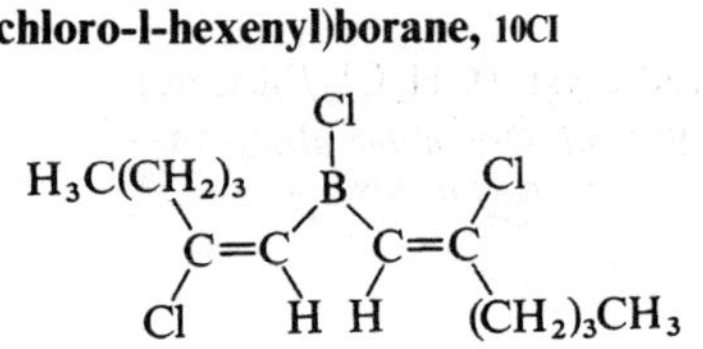

M 281.459

(E,Z)-form [63011-73-4]

Prod. of addn. of BCl_3 to 1-hexyne. Dec. on heating.

Blackborow, R.J., *J. Organomet. Chem.*, 1977, **128**, 161 (*synth, pmr*)

$C_{12}H_{20}BN_3O_4$ **B-00777**
Diethyl(ethyl nitrocarbamato)borane pyridine
Diethyl(ethyl nitrocarbamato)(pyridine)boron, *10CI*
[62261-12-5]

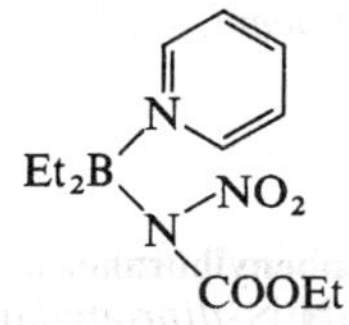

M 281.118
Air- and moisture-sensitive. Insol. hexane, Et_2O. Mp 75-80°.

Ioffe, S.L. *et al*, *Izv. Akad. Nauk SSSR, Ser. Khim.*, 1976, 2547; *Bull. Acad. Sci. USSR, Div. Chem. Sci.*, 2371 (*ir, nmr, synth*)
Ioffe, S.L. *et al*, *CA*, 1977, **87**, 23357g.

$C_{12}H_{20}B_4Fe$ **B-00778**
(η^6-Benzene)[(2,3,4,5,6-η)-2,2-diethyl-1,4,5,6-tetrahydro-2,3-dicarbahexaborato(2−)]iron
[84582-99-0]

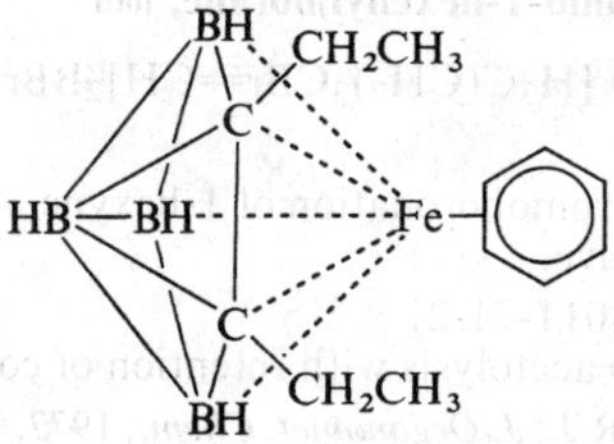

M 263.377
Golden-yellow air-stable cryst. (C_6H_6/hexane). Mp 156-7°.

Swisher, R.G. *et al*, *Organometallics*, 1983, **2**, 506 (*synth, pmr, cmr, nmr, ms, ir, struct*)

$C_{12}H_{20}B_4Fe$ **B-00779**
[(1,2,3,4,5,6-η)-1,3,5-Cyclooctatriene][(2,3,4,5,6-η)-2,3-dimethyl-1,4,5,6-tetrahydro-2,3-dicarbahexaborato(2−)]iron
[84583-04-0]

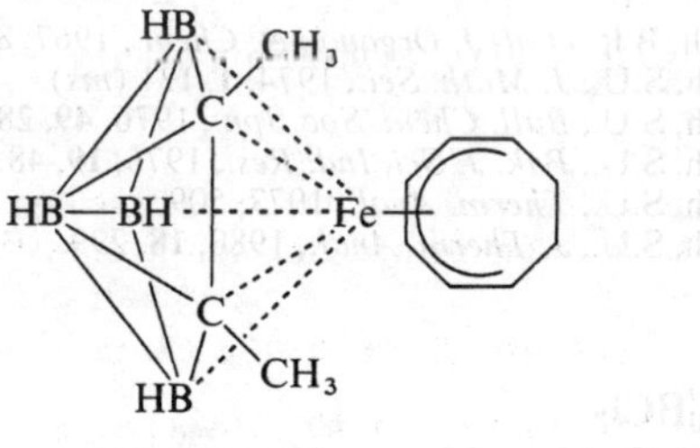

M 263.377
Air-stable red cryst. (CH_2Cl_2/hexane).

Maynard, R.B. *et al*, *Organometallics*, 1983, **2**, 500 (*synth, struct, pmr, ir, nmr, cmr, ms*)

$C_{12}H_{20}B_{10}Fe$ **B-00780**
(1,2-Dicarbadodecaboran(12)-1-yl)ferrocene, 9CI
[11064-94-1]

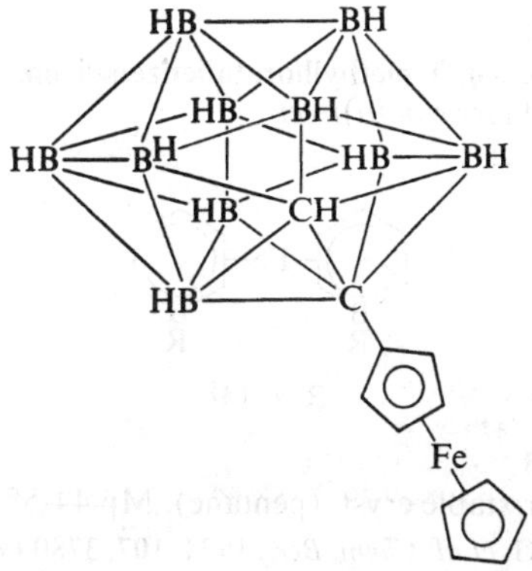

M 328.237
Cryst. (hexane). Mp 127-127.5°.

Zakharkin, L.I. *et al*, *Zh. Obshch. Khim.*, 1970, **40**, 2246 (*synth*)

$C_{12}H_{20}B_{10}Fe$ **B-00781**
(1,2-Dicarbadodecaboran(12)-3-yl)ferrocene, 9CI
(o-*Carboran-3-yl)ferrocene*
[37298-49-0]

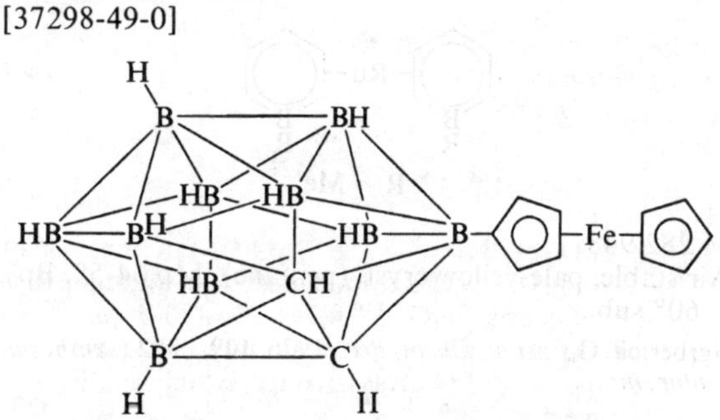

M 328.237
Cryst. (hexane). Mp 154-5°.

Krieg, V. *et al*, *Z. Anorg. Allg. Chem.*, 1969, 44 (*synth*)
Zakharkin, L.I. *et al*, *J. Organomet. Chem.*, 1972, **40**, 15 (*synth, pmr, ir*)

$C_{12}H_{21}B$ **B-00782**
9-(2-Butenyl)-9-borabicyclo[3.3.1]nonane, 10CI
B-*Crotyl-9-BBN*
[61967-28-0]
As 9-Bromo-9-borabicyclo[3.3.1]nonane, B-00539 with

$$R = —CH_2CH{=}CHCH_3$$

M 176.108
Clear liq., air- and moisture-sensitive. $Bp_{0.05}$ 55-6°.
(*E*)-*form* [76371-55-6]
Characterised spectroscopically.
(*Z*)-*form* [76371-56-7]
Characterised spectroscopically.

Kramer, G.W. *et al*, *J. Org. Chem.*, 1977, **42**, 2292.
Kramer, G.W. *et al*, *J. Organomet. Chem.*, 1977, **132**, 9 (*synth, nmr, pmr, isom*)
Brown, H.C. *et al*, *J. Organomet. Chem.*, 1980, **188**, 1.
Yatagai, H. *et al*, *J. Am. Chem. Soc.*, 1980, **102**, 4548.
Yamamoto, Y. *et al*, *Tetrahedron Lett.*, 1980, **21**, 3599 (*isom, nmr*)
Yamamoto, Y. *et al*, *J. Am. Chem. Soc.*, 1981, **103**, 1969.

$C_{12}H_{21}B$ B-00783

Perhydro-9*b*-boraphenylene

*Dodecahydro-9*b*-boraphenylene*, 10CI, 9CI

[16664-33-8]

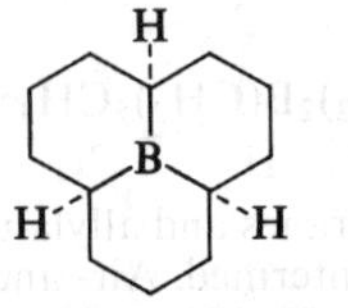

(3aα,6aα,9aα)-*form*

M 176.108

Sensitive to air.

(3aα,6aα,9aα)-form [1130-59-2]

cis,cis,cis-*form*

d_4^{35} 0.92. Mp 31°. Bp_{10} 113.5°.

(3aα,6aα,9aβ)-form [2938-53-6]

cis,cis,trans-*form*

Liq. d_4^{35} 0.94. Bp_{10} 115°.

Rotermund, G.W. *et al*, *Justus Liebigs Ann. Chem.*, 1965, **686**, 153 (*synth*, *ir*)
Brown, H.C. *et al*, *J. Am. Chem. Soc.*, 1969, **91**, 1226 (*struct*, *pmr*)
Pelter, A. *et al*, *J. Chem. Soc., Chem. Commun.*, 1978, 805 (*use*)
Brown, H.C. *et al*, *J. Organomet. Chem.*, 1980, **188**, 1 (*use*)

$C_{12}H_{21}B$ B-00784

Tri-2-butenylborane, 10CI, 9CI, 8CI

Tricrotylborane

[17304-01-7]

$(H_3CCH{=}CHCH_2)_3B$

M 176.108

Prepd. from crotylbromide and $Et_2O{\cdot}BF_3$ in presence of $HgCl_2$ and Al. Synthetic intermed. Air- and moisture-sensitive liq. Sol. Et_2O. d^{20} 0.786. Bp_2 70-2°. n_D^{20} 1.4682. As normally prepd. contains a mixt. of *trans* (70%) and *cis* (30%) crotyl groups, which readily interconvert at 150°.

Mikhailov, B.M. *et al*, *Izv. Akad. Nauk SSSR, Ser. Khim.*, 1967, 1477 (*synth*)
Mikhailov, B.M., *Organomet. Chem. Rev., Sect. A*, 1972, **8**, 1 (*rev*)
Stefani, A., *Helv. Chim. Acta*, 1973, **56**, 1192 (*use*)
Mikhailov, B.M. *et al*, *Zh. Obshch. Khim.*, 1975, **45**, 51 (*props*)
Brueser, W. *et al*, *Z. Anorg. Allg. Chem.*, 1976, **421**, 89 (*ir*)
Schroeder, S. *et al*, *Z. Anorg. Allg. Chem.*, 1977, **428**, 225 (*synth*)
Mikhailov, B.M. *et al*, *Izv. Akad. Nauk SSSR, Ser. Khim.*, 1979, 2551; *Bull. Acad. Sci. USSR, Div. Chem. Sci.*, 2368 (*use*)

$C_{12}H_{21}B$ B-00785

Tris(2-methyl-2-propenyl)borane, 10CI, 9CI

Tris(2-methylallyl)borane, 8CI. *Trimethallylborane*

[32784-73-9]

$[H_2C{=}C(CH_3)CH_2]_3B$

M 176.108

Synthetic intermed., reacts readily with aldehydes, ketones and activated alkynes. Air- and moisture-sensitive liq. Sol. non-hydroxylic solvs. Undergoes allylic rearrangement readily even at 25°, resulting in pmr signal coalescence at elevated temp.

Mikhailov, B.M., *Organomet. Chem. Rev., Sect. A*, 1972, **8**, 1 (*rev*)
Stefani, A. *et al*, *Helv. Chim. Acta*, 1972, **55**, 1110 (*nmr*)
Mikhailov, B.M. *et al*, *Zh. Obshch. Khim.*, 1974, **44**, 1878 (*nmr*)
Lapkin, I.I. *et al*, *Zh. Obshch. Khim.*, 1975, **45**, 59 (*ir*, *raman*, *deriv*)
Mikhailov, B.M. *et al*, *Zh. Obshch. Khim.*, 1975, **45**, 324 (*props*, *deriv*)
Brueser, W. *et al*, *Z. Anorg. Allg. Chem.*, 1976, **421**, 89 (*ir*, *raman*)
Schroeder, S. *et al*, *Z. Anorg. Allg. Chem.*, 1977, **428**, 225 (*synth*)

$C_{12}H_{22}B^{\ominus}$ B-00786

Dodecahydro-9*b*-borataphenylene, 10CI

1,5,9-Cyclododecanetriylhydroborate(1−), 9CI. *Perhydro-9*b*-boraphenalyl hydride*. *Perhydro-9*b*-boraphenylene hydride*

[67392-85-2]

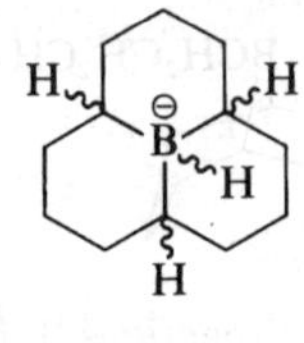

M 177.116 (ion)

Li salt: [36005-35-3]. *Lithium 1,5,9-cyclododecanetriylhydroborate(1−)*, 9CI.

$C_{12}H_{22}BLi$ M 184.057

Hindered reducing agent. Air- and moisture-sensitive.

Na salt: [67276-08-8].

$C_{12}H_{22}BNa$ M 200.106

Obt. in soln.

K salt: [67993-15-1].

$C_{12}H_{22}BK$ M 216.214

Obt. in soln.

Corey, E.J. *et al*, *J. Am. Chem. Soc.*, 1971, **93**, 7319 (*use*)
Brown, H.C. *et al*, *J. Am. Chem. Soc.*, 1978, **100**, 3343 (*ir*, *nmr*)
U.S.P., 4 082 810, (*1978*); *CA*, **89**, 109936y (*synth*)
Brown, C.A. *et al*, *J. Organomet. Chem.*, 1978, **156**, 111 (*ir*, *nmr*)
Brown, H.C. *et al*, *J. Org. Chem.*, 1979, **44**, 5004.
Brown, H.C. *et al*, *J. Organomet. Chem.*, 1980, **188**, 1 (*synth*, *nmr*, *ir*)

$C_{12}H_{22}BCl$ B-00787

Chlorodicyclohexylborane, 10CI, 9CI

Chlorodicyclohexylborine. *Dicyclohexylchloroborane*

[36140-19-9]

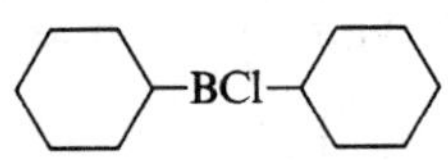

M 212.569

Synthetic intermed. Colourless fuming liq., sensitive to both air and moisture. Sol. non-hydroxylic solvs. Bp_{12} 82.6°.

Hartmann, H. *et al*, *Z. Anorg. Allg. Chem.*, 1959, **299**, 174 (*synth*)
B.P., 873 267, (*1961*); *CA*, **57**, 11104d (*synth*)
U.S.P., 3 133 961, (*1964*); *CA*, **61**, 5690d (*synth*)
Brown, H.C. *et al*, *J. Am. Chem. Soc.*, 1972, **94**, 2114 (*props*)
Paget, W.E. *et al*, *J. Chem. Soc., Chem. Commun.*, 1980, 1169 (*synth*)

$C_{12}H_{22}BCl$ B-00788

Chlorodi-1-hexenylborane, 10CI, 9CI

Di-1-hexenylchloroborane

$H_3C(CH_2)_3CH{=}CHBClCH{=}CH(CH_2)_3CH_3$

M 212.569

(E,E)-form [41929-38-8]

Obt. *in situ* by hydroboration of 1-hexyne with BH_2Cl. Synth. intermed., e.g. for synth. of dodecadienes. Air- and moisture-sensitive liq. Sol. nonhydroxylic solvs. $Bp_{0.05}$ 79-81°.

Brown, H.C. *et al, J. Org. Chem.*, 1973, **38**, 1617 (*synth*)
Brown, H.C. *et al, J. Am. Chem. Soc.*, 1976, **98**, 1785 (*synth*)
Yamamoto, Y. *et al, J. Am. Chem. Soc.*, 1977, **99**, 5652 (*props*)
Yatagai, H. *et al, J. Org. Chem.*, 1979, **44**, 2566 (*pmr*)
Yatagai, H. *et al, J. Org. Chem.*, 1980, **45**, 1640 (*synth, props*)

$C_{12}H_{23}B$ B-00789

9-Butyl-9-borabicyclo[3.3.1]nonane, 9CI

B-Butyl-9-BBN

[23532-74-3]

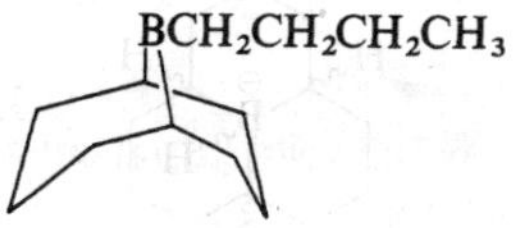

M 178.124

Used in hydroboration reactions. Liq., sensitive to air. $Bp_{0.5}$ 58-9°.

Brown, H.C. *et al, J. Am. Chem. Soc.*, 1969, **91**, 4304 (*synth*)
Kramer, G.W. *et al, J. Organomet. Chem.*, 1974, **73**, 1 (*synth*)
Yamamoto, Y. *et al, Chem. Lett.*, 1975, 1199; *CA*, **84**, 16629 (*nmr*)
Midland, M.M. *et al, J. Organomet. Chem.*, 1977, **134**, C17 (*use*)
Yamamoto, Y. *et al, Tetrahedron Lett.*, 1980, **21**, 3599 (*cmr*)
Yatagai, H., *J. Org. Chem.*, 1980, **45**, 1640 (*use*)

$C_{12}H_{23}B$ B-00790

9-*tert*-Butyl-9-borabicyclo[3.3.1]nonane

9-(1,1-Dimethylethyl)-9-borabicyclo[3.3.1]nonane, 10CI. B-tert-*Butyl-9-BBN*

[42928-43-8]

As 9-Bromo-9-borabicyclo[3.3.1]nonane, B-00539 with

$R = —C(CH_3)_3$

M 178.124

Liq. $Bp_{0.2}$ 27-8°.

Brown, H.C. *et al, J. Org. Chem.*, 1973, **38**, 3968.
Brown, H.C. *et al, J. Am. Chem. Soc.*, 1974, **96**, 311 (*synth*)
Kramer, G.W. *et al, J. Organomet. Chem.*, 1974, **73**, 1 (*synth*)
Katz, J-J. *et al, Bull. Soc. Chim. Fr.*, 1977, 683.
Brown, H.C. *et al, J. Org. Chem.*, 1980, **45**, 846 (*nmr*)

$C_{12}H_{23}B$ B-00791

Dicyclohexylborane

[1568-65-6]

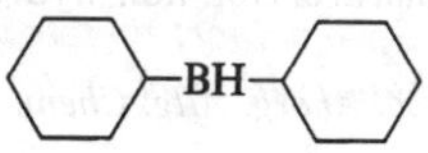

M 178.124

Selective hydroborating reagent. Cryst. from tetrahydrofuran as the bridged dimer. Air- and moisture-sensitive.

Zweifel, G. *et al, J. Am. Chem. Soc.*, 1963, **85**, 2072 (*synth*)
Rao, V.V.R. *et al, Tetrahedron*, 1978, **34**, 2223 (*use*)
Soderquist, J.A. *et al, J. Organomet. Chem.*, 1978, **156**, C12 (*use*)
Zweifel, G. *et al, J. Organomet. Chem.*, 1978, **156**, 159 (*synth, use*)
Midland, M.M. *et al, J. Org. Chem.*, 1980, **45**, 747 (*use*)
Sonnet, P. *et al, J. Chem. Ecol.*, 1980, **6**, 221 (*use*)
Fieser, M. *et al, Reagents for Organic Synthesis*, Wiley, 1967-83, **8**, 162 (*use*)

$C_{12}H_{23}B$ B-00792

Hexyldi-2-propenylborane, 10CI, 9CI

Di-2-propenylhexylborane. Diallylhexylborane. Hexyldiallylborane

[56347-97-8]

$(H_2C{=}CHCH_2)_2B(CH_2)_5CH_3$

M 178.124

Prepd. from hexyldialkoxyboranes and allylaluminum sesquibromide. Synthetic intermed. Air- and moisture-sensitive liq. Sol. THF. Bp_2 58-60°. n_D^{20} 1.4475, ^{11}Bnmr δ 81.2 ppm.

Mikhailov, B.M. *et al, Izv. Akad. Nauk SSSR, Ser. Khim.*, 1975, 979; *Bull. Acad. Sci. USSR, Div. Chem. Sci.*, 897 (*synth*)
Mikhailov, B.M. *et al, Izv. Akad. Nauk SSSR, Ser. Khim.*, 1978, 1586, 1892; *Bull. Acad. Sci. USSR, Div. Chem. Sci.*, 1392, 1663 (*synth*)
Mikhailov, B.M. *et al, J. Organomet. Chem.*, 1978, **154**, 113 (*synth*)
Pelter, A. *et al, Compr. Org. Chem.*, (*Barton, D.H.R. et al*, Ed.), 1979, **3**, 695 (*rev*)

$C_{12}H_{23}B$ B-00793

9-(1-Methylpropyl)-9-borabicyclo[3.3.1]nonane, 10CI, 9CI

9-sec-Butyl-9-borabicyclo[3.3.1]nonane. B-sec-*Butyl-9-BBN*

[53317-06-9]

As 9-Bromo-9-borabicyclo[3.3.1]nonane, B-00539 with

$R = —CH(CH_3)CH_2CH_3$

M 178.124

Reagent for transfer of *sec*-butyl group. Sensitive to air. $Bp_{0.2}$ 56-8°.

Kramer, G.W. *et al, J. Organomet. Chem.*, 1974, **73**, 1.
Levy, A.B., *J. Org. Chem.*, 1978, **43**, 4684.
Levy, A.B., *Tetrahedron Lett.*, 1979, 4021.
Brown, H.C. *et al, J. Org. Chem.*, 1980, **45**, 846 (*nmr*)
Slayden, S.W., *J. Org. Chem.*, 1981, **46**, 2311.
Whiteley, C.G., *S. Afr. J. Chem.*, 1982, **35**, 9 (*synth*)

$C_{12}H_{23}B$ B-00794

9-(2-Methylpropyl)-9-borabicyclo[3.3.1]nonane, 10CI

9-Isobutyl-9-borabicyclo[3.3.1]nonane. B-Isobutyl-9-BBN

[63942-77-8]

As 9-Bromo-9-borabicyclo[3.3.1]nonane, B-00539 with

$R = —CH_2CH(CH_3)_2$

M 178.124

Reagent for transfer of an isobutyl group. Sensitive to air.

Midland, M.M. *et al, J. Organomet. Chem.*, 1977, **134**, C17; 1978, **156**, 203.
Levy, A.B., *J. Org. Chem.*, 1978, **43**, 4684.
Levy, A.B., *Tetrahedron Lett.*, 1979, 4021.
Slayden, S.W., *J. Org. Chem.*, 1981, **46**, 2311; 1982, **47**, 2753.

$C_{12}H_{23}BO_2Si_2$ B-00795

Bis(trimethylsilyloxy)phenylborane

Phenylboronic acid bis(trimethylsilyl) ester, 10CI. Benzeneboronic acid bis(trimethylsilyl) ester. Phenylbis(trimethylsilyloxy)borane

[7560-51-2]

$PhB(OSiMe_3)_2$

M 266.294

Air- and moisture-sensitive liq. Sol. non-hydroxylic solvs. d^{20} 0.913. Bp_{13} 114°. n_D^{20} 1.4560.

Gridina, V.F. *et al*, *Zh. Obshch. Khim.*, 1966, **36**, 1283 (*synth, props, ir*)
Van den Heuvel, W.J.A. *et al*, *Org. Mass Spectrom.*, 1977, **12**, 724 (*ms*)
Wannagat, U. *et al*, *Z. Naturforsch., B*, 1978, **33**, 475 (*synth*)

$C_{12}H_{24}B^{\ominus}$ B-00796

9-Butyl-9-boratabicyclo[3.3.1]nonane, 10CI
B-*butyl-9-BBN hydride*

As 9-Boratabicyclo[3.3.1]nonane, B-00546 with

$$R = H_3C(CH_2)_3-,\ R' = H$$

M 179.132 (ion)

Li salt: [67813-46-1].
$C_{12}H_{24}BLi$ M 186.073
Air- and moisture-sensitive.

Hubbard, J.L. *et al*, *J. Organomet. Chem.*, 1978, **156**, 81 (*nmr, synth*)
Brown, H.C. *et al*, *J. Org. Chem.*, 1979, **44**, 467 (*use*)

$C_{12}H_{24}B^{\ominus}$ B-00797

Dicyclohexyldihydroborate(1−), 10CI
Dicyclohexylborohydride

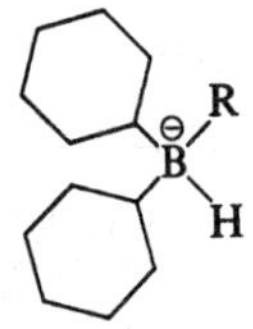

R = H

M 179.132 (ion)

Li salt: [67813-27-8].
$C_{12}H_{24}BLi$ M 186.073
Air- and moisture-sensitive. Stable for at least 3 months in dry, air-free THF soln.

Hubbard, J.L. *et al*, *J. Organomet. Chem.*, 1978, **156**, 81 (*nmr, synth*)
Brown, H.C. *et al*, *J. Org. Chem.*, 1981, **46**, 2712.
Brown, H.C. *et al*, *J. Org. Chem.*, 1981, **46**, 4541 (*synth, ir, nmr*)

$C_{12}H_{24}BCl$ B-00798

Chloro(1-ethyl-1-butenyl)(1,1,2-trimethylpropyl)borane, 10CI
Chloro(3-hexen-3-yl)(2,3-dimethyl-2-butyl)borane. Chloro(3-hexen-3-yl)thexylborane. Thexyl(3-hexen-3-yl)chloroborane

$$H_3CCH_2CH{=}C(CH_2CH_3)BClC(CH_3)_2CH(CH_3)_2$$

M 214.585

(*Z*)-*form* [76173-98-3]
Formed *in situ* by reacn. of thexylchloroborane with 3-hexene. Synthetic intermed. Prob. oil, sensitive to air and moisture. Sol. THF with slow reacn.

Zweifel, G. *et al*, *J. Org. Chem.*, 1981, **46**, 829 (*synth*)

$C_{12}H_{24}BCl$ B-00799

Chloro-1-hexenyl(1,1,2-trimethylpropyl)borane, 10CI
Chloro(1-hexenyl)(2,3-dimethyl-2-butyl)borane. Chloro(1-hexenyl)thexylborane. Thexyl(1-hexenyl)-chloroborane

$$H_3C(CH_2)_3CH{=}CHBClC(CH_3)_2CH(CH_3)_2$$

M 214.585

(*E*)-*form* [76173-94-9]
Formed *in situ* by reacn. of thexylchloroborane with 1-hexyne. Synthetic intermed. Prob oil. Air- and moisture-sensitive. Sol. non-hydroxylic solvs. Wrongly reported in *CA* as the (*Z*)-*form*.

Zweifel, G. *et al*, *J. Org. Chem.*, 1981, **46**, 829 (*synth, use*)

$C_{12}H_{24}BClO$ B-00800

9-Chloro-9-borabicyclo[3.3.1]nonane diethyl ether
Chloro-1,5-cyclooctanediyl[1,1'-oxybis[ethane]]boron, 10CI. B-*Chloro-9-BBN-diethyl ether. 9-Chloro-9-borabicyclo[3.3.1]nonane compd. with 1,1'-oxybis[ethane](1:1)*
[70160-57-5]

As 9-Chloro-9-borabicyclo[3.3.1]nonane dimethyl sulfide, B-00695 with

$$L = Et_2O$$

M 230.584

Air- and moisture-sensitive.

Brown, H.C. *et al*, *J. Org. Chem.*, 1979, **44**, 2422.

$C_{12}H_{24}B_2$ B-00801

1,1'-(1,4-Butanediyl)bis[borolane], 10CI, 9CI
1,1'-Tetramethylenediborolane, 8CI. *1,1'-Tetramethylenebisborolane. 1,4-Bis(1-boracyclopentyl)butane. 1,4-Bis(1-borolyl)butane*
[15909-61-2]

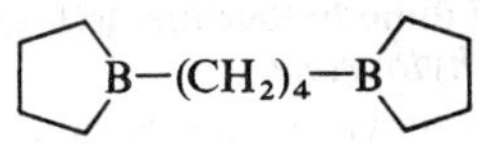

M 189.942

Air-sensitive liq. Sol. THF. Bp_9 113-5°, $Bp_{0.5}$ 83-4°.

▷Spont. inflammable at r.t.

Köster, R., *Angew. Chem.*, 1959, **71**, 520 (*synth*)
Zakharkin, L.I. *et al*, *Zh. Obshch. Khim.*, 1962, **32**, 1421 (*synth*)
Mikhailov, B.M. *et al*, *Dokl. Akad. Nauk SSSR, Ser. Sci. Khim.*, 1964, **155**, 141.
Brown, H.C. *et al*, *J. Am. Chem. Soc.*, 1970, **92**, 2460; 1971, **93**, 3400 (*props*)
Nöth, H. *et al*, *Chem. Ber.*, 1974, **107**, 3089 (*nmr*)

$C_{12}H_{24}B_2Cl_2S_2$ B-00802

2,6-Dichloro-2,6-diboraadamantane bis(dimethyl sulfide)
Dichloro[μ-(1,3,5,7-cyclooctanetetrayl-C^1,C^5:C^3,C^7*)-]bis[thiobis[methane]]diboron,* 10CI
[69961-35-9]

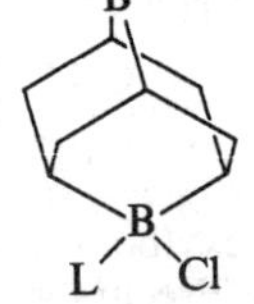

$L = Me_2S$

M 324.968

Cryst. (CH_2Cl_2/pentane). Mp 167-8°.

Kulkarni, S.U. *et al*, *J. Org. Chem.*, 1979, **44**, 1747 (*synth, nmr*)

$C_{12}H_{25}B$ B-00803

2-Butenyldibutylborane, 10CI

Dibutylcrotylborane

[67570-20-1]

$(H_3CCH_2CH_2CH_2)_2BCH_2CH{=}CHCH_3$

M 180.140

Fluxional mol. at ambient temps. Used in allylboration of aldehydes, ketones, and esters. Liq., stable up to 110° under N_2 atm., but sensitive to air and moisture. Sol. THF. Bp_2 58-9°. n_D^{20} 1.4382.

Mikhailov, B.M. *et al*, *Izv. Akad. Nauk SSSR, Ser. Khim.*, 1978, 1586, 1892; *Bull. Acad. Sci. USSR, Div. Chem. Sci.*, 1385, 1663 (*props*)
Mikhailov, B.M. *et al*, *J. Organomet. Chem.*, 1978, **154**, 113, 131 (*synth, props, nmr*)
Bubnov, Y.N. *et al*, *Synthesis*, 1980, 902, 904 (*props*)

$C_{12}H_{25}B$ B-00804

1-Butyl-3,6-dimethylborepane, 10CI

[76885-32-0]

As 3,6-Dimethylborepane, B-00551 with

$R = H_3C(CH_2)_3–$

M 180.140

Synthetic intermed. No phys. data recorded.

Yamamoto, Y. *et al*, *J. Chem. Soc., Chem. Commun.*, 1980, 1072 (*props*)

$C_{12}H_{25}B$ B-00805

(1-Butyl-1-hexenyl)dimethylborane, 10CI

5-Decenyldimethylborane

As (1-Butyl-1-hexenyl)dichloroborane, B-00691 with

X = Me

M 180.140

(Z)-form [64285-24-1]

Air-sensitive liq. Sol. THF. Bp_1 43-5°.

Brown, H.C. *et al*, *J. Am. Chem. Soc.*, 1976, **98**, 1798 (*synth, pmr, props*)
Yamamoto, Y. *et al*, *J. Am. Chem. Soc.*, 1977, **99**, 5652 (*synth*)
Yatagai, H. *et al*, *J. Org. Chem.*, 1979, **44**, 2566 (*pmr*)

$C_{12}H_{25}B$ B-00806

Cyclohexyl(1,1,2-trimethylpropyl)borane, 10CI, 9CI

Cyclohexyl(2,3-dimethyl-2-butyl)borane. Cyclohexylthexylborane. Thexylcyclohexylborane

[35331-48-7]

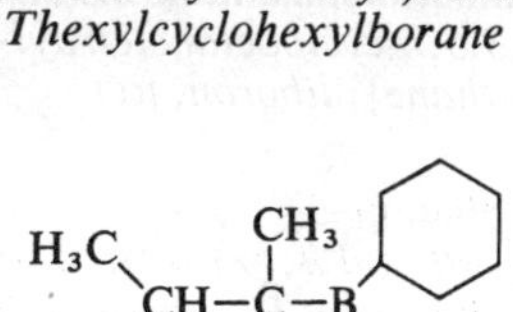

M 180.140

Prepd. *in situ* from thexylborane and cyclohexene. Synthetic intermed. Precursor to mixed ketones *via* reaction with an alkene and then carbonylation or cyanidation. Air- and moisture-sensitive. Sol. THF.

Negishi, E. *et al*, *Synthesis*, 1974, 77 (*rev*)
Brown, H.C. *et al*, *J. Am. Chem. Soc.*, 1975, **97**, 2791, 2799 (*synth*)
Pelter, A. *et al*, *Tetrahedron Lett.*, 1978, 4715 (*synth, use*)
Pelter, A. *et al*, *Comprehensive Organic Chemistry*, (*Barton, D.H.R. et al*, Ed.), Pergamon, Oxford, 1979, **3**, 695 (*rev*)

$C_{12}H_{25}BO_{10}$ B-00807

Hydrogen [bis[3,3′-oxybis[1,2-propanediolato]](2−)-*O*¹,*O*²]-borate(1−)

[58748-90-6]

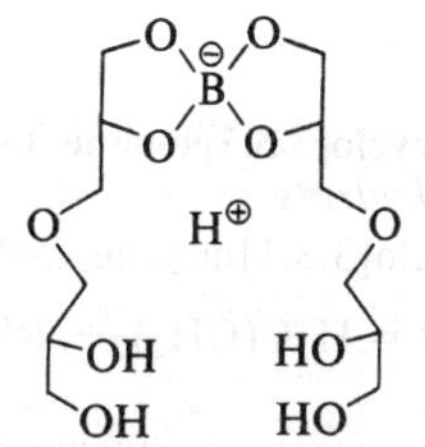

M 340.134

Not properly characterised; probably a mixt. of several components in equilibrium. 10CI/9CI name appears incorrect. Used in brake fluids for automobiles.

Japan. Pat., 75 157 430, (*1975*); *CA*, **85**, 23014b
Ger. Pat., 2 831 538, (*1980*); *CA*, **93**, 134858q (*use*)
U.S.P., 4 192 759, (*1980*); *CA*, **94**, 86976y (*use*)

$C_{12}H_{26}BBr$ B-00808

Bromodihexylborane, 10CI, 9CI

Dihexylbromoborane

[57476-26-3]

$[H_3C(CH_2)_5]_2BBr$

M 261.051

Synthetic intermed. Colourless liq., fumes in air. Sensitive to air and moisture. Sol. THF. $Bp_{0.8}$ 78°.

Pelter, A. *et al*, *J. Chem. Soc., Dalton Trans.*, 1976, 2087 (*synth*)
Fletcher, A.S. *et al*, *J. Chem. Soc., Chem. Commun.*, 1979, 573 (*use*)
Maddocks, P.J. *et al*, *J. Chem. Soc., Perkin Trans. 1*, 1981, 653 (*use*)

$C_{12}H_{26}BN$ B-00809

9-Borabicyclo[3.3.1]nonane *tert*-butylamine

1,5-Cyclooctanediylhydro(2-methyl-2-propanamine)-boron, 10CI, 9CI. *2-Methyl-2-propanamine, compd. with 9-borabicyclo[3.3.1]nonane(1:1). 9-BBN* tert-*butylamine complex*

[70338-04-4]

As 9-Borabicyclo[3.3.1]nonane dimethyl sulfide, B-00705 with

$L = H_2NC(CH_3)_3$

M 195.154

Completely associated in cyclohexane soln.

Brown, H.C. *et al*, *Recl. Trav. Chim. Pays-Bas*, 1978, **98**, 117 (*synth, props*)

$C_{12}H_{26}BN$ B-00810

9-Borabicyclo[3.3.1]nonane diethylamine

*1,5-Cyclooctanediylhydro(*N*-ethylethanamine)boron*, 10CI, 9CI

[70338-05-5]

As 9-Borabicyclo[3.3.1]nonane dimethyl sulfide, B-00705 with

L = Et_2NH

M 195.154

Totally associated in cyclohexane soln.

Brown, H.C. *et al*, *Recl. Trav. Chim. Pays-Bas*, 1979, **98**, 117 (*synth, props*)
Brown, H.C. *et al*, *J. Org. Chem.*, 1980, **45**, 1748 (*cmr, props*)

$C_{12}H_{27}B$ B-00811

Bis(1,1,2-trimethylpropyl)borane, 10CI, 9CI

Bis(2,3-dimethyl-2-butyl)borane. Dithexylborane

[20622-63-3]

$[(H_3C)_2CHC(CH_3)_2]_2BH$

M 182.155

Unusually does not form a bridged dimer. Characterised spectroscopically. Air- and moisture-sensitive. Sol. THF.

Brown, H.C. *et al*, *J. Am. Chem. Soc.*, 1962, **84**, 1478 (*synth*)
Negishi, E. *et al*, *J. Am. Chem. Soc.*, 1972, **94**, 4025 (*ir*)
Fussstetter, H. *et al*, *Chem. Ber.*, 1977, **110**, 3172 (*nmr*)

$C_{12}H_{27}B$ B-00812

Tri-*tert*-butylborane

Tris(1,1-dimethylethyl)borane, 10CI, 9CI. *Tris*(tert-*butyl*)*borane*, 8CI

[1113-42-4]

$B[C(CH_3)_3]_3$

M 182.155

Many earlier references to this compound are incorrect, the actual compound obt. being an isomer. Only references after 1977 should be consulted. Air-sensitive liq. Sol. THF. Bp_{12} 71°.

Radical anion (1−): [81174-43-8]. Formed by reduction with Na/K alloy in diglyme/THF/dicyclohexyl-18-crown-6. Sol. THF/diglyme.

Nöth, H. *et al*, *J. Organomet. Chem.*, 1977, **142**, 281 (*nmr, ir, synth*)
Berndt, A. *et al*, *J. Organomet. Chem.*, 1981, **222**, C25 (*deriv*)

$C_{12}H_{27}B$ B-00813

Tributylborane, 10CI, 9CI

[122-56-5]

$B(CH_2CH_2CH_2CH_3)_3$

M 182.155

Additive used in polym. of alkenes. Reagent for transfer of a butyl group. Liq. Bp_{20} 108-10°, Bp_9 90-1°.

▷ED1850000.

Johnson, J.R. *et al*, *J. Am. Chem. Soc.*, 1938, **60**, 115 (*synth*)
Kuimova, M.E. *et al*, *Izv. Akad. Nauk SSSR, Ser. Khim.*, 1979, 314 (*nmr*)
Pelter, A. *et al*, *Comprehensive Org. Chem.*, (*Barton, D.H.R. et al*, Ed.), 1979, Pergamon, Oxford, **3**, 792 (*rev*)
Brown, H.C. *et al*, *J. Organomet. Chem.*, 1980, **188**, 1 (*use*)

$C_{12}H_{27}B$ B-00814

Tris(1-methylpropyl)borane, 10CI, 9CI

Tri-sec-*butyl borane*, 8CI. *Tri-2-butylborane*

[1113-78-6]

$[H_3CCH_2CH(CH_3)]_3B$

M 182.155

Prepd. by hydroboration of 2-butene. Synthetic intermed., generally prepd. *in situ*, for transfer of *sec*-butyl group. Air-sensitive liq. Sol THF, other non-hydroxylic solvs. $Bp_{2.5}$ 60°.

Hennion, G.F. *et al*, *J. Am. Chem. Soc.*, 1957, **79**, 5190 (*synth*)
McCusker, P.A. *et al*, *J. Org. Chem.*, 1963, **28**, 2889 (*synth, props*)
Wahab, M.A. *et al*, *Rev. Roum. Phys.*, 1964, **9**, 599; *CA*, **62**, 12614d (*spectra*)
Yamamoto, Y. *et al*, *Chem. Lett.*, 1975, 1199 (*nmr*)
Suzuki, A. *et al*, *Synthesis*, 1976, 687 (*synth, use*)
Reetz, M.T. *et al*, *Angew. Chem., Int. Ed. Engl.*, 1977, **89**, 46 (*use*)
Takahashi, Y. *et al*, *Chem. Lett.*, 1977, 999 (*synth*)
Pelter, A. *et al*, *Compr. Org. Chem.*, (*Barton, D.H.R. et al*, Ed.), 1979, **3**, 695 (*rev*)
Brown, H.C. *et al*, *J. Organomet. Chem.*, 1980, **188**, 1 (*use*)

$C_{12}H_{27}B$ B-00815

Tris(2-methylpropyl)borane, 10CI, 9CI

Triisobutylborane, 8CI

[1116-39-8]

$[(H_3C)_2CHCH_2]_3B$

M 182.155

Prepd. by hydroboration of isobutene. Synthetic intermed.; catalyst. Air-sensitive liq. Sol. Et_2O, THF, other non-hydroxylic solvs. Bp 188.5°, Bp_{20} 86°.

Krause, E. *et al*, *Chem. Ber.*, 1921, **54**, 2784 (*synth*)
Hurd, D.T. *et al*, *J. Am. Chem. Soc.*, 1948, **70**, 2053 (*synth*)
Nobuyuki, A., *Bull. Chem. Soc. Jpn.*, 1958, **31**, 540 (*synth, use*)
Ashby, E.C., *J. Am. Chem. Soc.*, 1959, **81**, 4791 (*synth*)
Hare, D.G. *et al*, *Chem. Ind. (London)*, 1959, 1519 (*spectra*)
McCusker, P.A. *et al*, *J. Org. Chem.*, 1963, **28**, 2889 (*synth*)
Vanchagova, V.K. *et al*, *Zh. Obshch. Khim.*, 1976, **46**, 989.
Kuimova, M.E. *et al*, *Izv. Akad. Nauk SSSR, Ser. Khim.*, 1979, 314; *Bull. Acad. Sci. USSR, Div. Chem. Sci.*, 288 (*nmr*)

$C_{12}H_{27}BBr_3N$ B-00816

Tribromoborane tributylamine

Tribromo(N,N-*dibutyl-1-butanamine*)*boron*, 10CI. N,N-*Dibutyl-1-butanamine compd. with tribromoborane(1:1). Boron tribromide tributylamine*

[54706-18-2]

$(H_3CCH_2CH_2CH_2)_3N—BBr_3$

M 435.874

Sheikh, S.U., *J. Math. Sci.*, 1974, **1**, 131 (*ms*)
Sheikh, S.U., *Pak. J. Sci. Ind. Res.*, 1976, **19**, 48.
Sheikh, S.U., *J. Therm. Anal.*, 1980, **18**, 299 (*synth*)

$C_{12}H_{27}BCl_3N$ B-00817

Trichloroborane tributylamine

Trichloro(N,N-*dibutyl-1-butanamine*)*boron*, 10CI. N,N-*Dibutyl-1-butanamine compd. with trichloroborane(1:1). Boron trichloride tributylamine*

[54706-17-1]

$(H_3CCH_2CH_2CH_2)_3N—BCl_3$

M 302.521

Sheikh, S.U., *J. Math. Sci.*, 1974, **1**, 131 (*ms*)
Sheikh, S.U., *Pak. J. Sci. Ind. Res.*, 1976, **19**, 48; 1980, **23**, 21.
Sheikh, S.U., *J. Therm. Anal.*, 1980, **18**, 299 (*synth*)

$C_{12}H_{27}BO$ **B-00818**

Ethoxydipentylborane

Dipentylborinic acid ethyl ester, 10CI. *Dipentyl(ethoxy)-borane*

[65344-93-6]

$[H_3C(CH_2)_4]_2BOEt$

M 198.155

Air- and moisture-sensitive liq. Sol. nonhydroxylic solvs.

Mikhailov, B.M. *et al*, *Izv. Akad. Nauk SSSR, Ser. Khim.*, 1977, 2633; *Bull. Acad. Sci. USSR, Div. Chem. Sci.*, **26**, 2437 (*synth*)
Mikhailov, B.M. *et al*, *Izv. Akad. Nauk SSSR, Ser. Khim.*, 1979, 855; *Bull. Acad. Sci. USSR, Div. Chem. Sci.*, **28**, 799.

$C_{12}H_{27}BO$ **B-00819**

(Hexyloxy)dipropylborane

Dipropylborinic acid hexyl ester, 10CI

[2117-96-6]

$(H_3CCH_2CH_2)_2BO(CH_2)_5CH_3$

M 198.155

Used in synth. of allyl(dialkyl)boranes. Air- and moisture-sensitive liq. Sol. nonhydroxylic solvs. d_{20} 0.786. Bp_3 63-4°, Bp_{12} 100-2°. n_D^{20} 1.4220.

Mikhailov, B.M. *et al*, *Izv. Akad. Nauk SSSR, Ser. Khim.*, 1962, 628 (*synth*)
Mikhailov, B.M. *et al*, *Zh. Obshch. Khim.*, 1965, **35**, 925.
Mikhailov, B.M. *et al*, *Izv. Akad. Nauk SSSR, Ser. Khim.*, 1978, 1586, 1594; *Bull. Acad. Sci. USSR, Div. Chem. Sci.*, **27**, 1385, 1392 (*synth, props, use*)
Mikhailov, B.M. *et al*, *J. Organomet. Chem.*, 1978, **154**, 113 (*synth, props*)

$C_{12}H_{27}BO_3$ **B-00820**

Tributoxyborane

Boric acid (H_3BO_3) tributyl ester, 10CI, 9CI, 8CI. *Tributylborate. Tributyl orthoborate. Butyl borate*

[688-74-4]

$(H_3CCH_2CH_2CH_2O)_3B$

M 230.154

Used in manuf. of oxalate esters; oxidation catalyst, gelling agent, stabilizer, vulcanizing agent; also used to treat glass for fluorescent lamps. Moisture-sensitive liq. d_4^{28} 0.847. Bp 229-31°, Bp_8 105°. n_D^{28} 1.4089.

▷ED4900000.

Bean, F.R. *et al*, *J. Am. Chem. Soc.*, 1932, **54**, 4415 (*synth, props*)
Scattergood, A. *et al*, *J. Am. Chem. Soc.*, 1945, **67**, 2150 (*synth, props*)
Lappert, M.F., *J. Chem. Soc.*, 1953, 667 (*synth*)
Haiden, S.Z. *et al*, *J. Appl. Chem. Biotechnol.*, 1954, **4**, 93 (*synth*)
Org. Synth., Coll. Vol., **2**, 106 (*synth, props*)
Fallon, P.J. *et al*, *Int. J. Mass Spectrom. Ion Phys.*, 1968, **1**, 133 (*ms*)
Maijs, L. *et al*, *CA*, 1973, **79**, 131073e (*pmr*)
Pola, M. *et al*, *Collect. Czech. Chem. Commun.*, 1979, **44**, 3688 (*ir, props*)
Draney, D. *et al*, *J. Am. Chem. Soc.*, 1981, **103**, 1041 (*nmr*)

$C_{12}H_{27}BO_3$ **B-00821**

Tri-*tert*-butoxyborane

Boric acid (H_3BO_3) tris(1,1-dimethylethyl)ester, 10CI, 9CI. *Boric acid (H_3BO_3) tri-tert-butyl ester*, 8CI. *Tri-*tert-*butyl borate.* tert-*Butyl borate.* tert-*Butyl orthoborate. Tris*(tert-*butyloxy)borane*

[7397-43-5]

$B[OC(CH_3)_3]_3$

M 230.154

Catalyst for polymn. of olefins and for crosslinking of epoxy resins; heat stabilizer; insect sterilant. Moisture-sensitive liq. Sol. non-hydroxylic solvs. Mp 18-9°. Bp 175°, Bp_{74} 101°, Bp_{12} 60°. n_D^{20} 1.3872.

Lappert, M.F., *J. Chem. Soc.*, 1953, 667 (*synth*)
George, P.D. *et al*, *J. Am. Chem. Soc.*, 1955, **77**, 1900 (*synth, ir, props*)
Brown, H.C. *et al*, *J. Am. Chem. Soc.*, 1956, **78**, 3613 (*synth, props*)
O'Brien, K.G., *Aust. J. Chem.*, 1957, **10**, 91 (*props*)
Davis, F.A. *et al*, *J. Org. Chem.*, 1971, **36**, 1300 (*nmr*)
Kronawitter, I. *et al*, *Chem. Ber.*, 1972, **105**, 2423 (*synth, nmr, ir*)
Brown, C.A. *et al*, *J. Org. Chem.*, 1978, **43**, 2731 (*synth*)
Pola, M. *et al*, *Collect. Czech. Chem. Commun.*, 1979, **44**, 3688 (*ir, props*)

$C_{12}H_{27}BO_3$ **B-00822**

Tris(2-butoxy)borane

Boric acid (H_3BO_3) tris(1-methylpropyl) ester, 10CI, 9CI. *Boric acid (H_3BO_3) tri-sec-butyl ester*, 8CI. sec-*Butyl borate. Tri-1-methylpropyl borate. Tri-sec-butoxyborane. Tri-2-butyl borate*

[22238-17-1]

$[H_3CCH_2CH(CH_3)O]_3B$

M 230.154

Catalyst for oxidation of cyclododecatriene. Moisture-sensitive liq. d_4^{25} 0.826. Bp 195.4-195.8°, Bp_{10} 92°. n_D^{25} 1.3960.

▷ED4905000.

Scattergood, A. *et al*, *J. Am. Chem. Soc.*, 1945, **67**, 2150 (*props*)
Gerrard, W. *et al*, *J. Chem. Soc.*, 1951, 2545 (*props*)
Aroney, M. *et al*, *J. Chem. Soc.*, 1961, 4140 (*struct, props*)
Davis, F.A. *et al*, *J. Org. Chem.*, 1971, **36**, 1300 (*nmr*)
Kobayashi, Y., *Tetrahedron Lett.*, 1972, 5093 (*use*)
Brown, C.A. *et al*, *J. Org. Chem.*, 1978, **43**, 2731 (*props*)
Eur. Pat., 12 198, (*1980*); *CA*, **94**, 47554q (*use*)
Sax, N.I., *Dangerous Properties of Industrial Materials*, 5th Ed., Van Nostrand-Reinhold, 1979, 1041.

$C_{12}H_{27}BO_3$ **B-00823**

Tris(2-methylpropoxy)borane

Boric acid (H_3BO_3) tris(2-methylpropyl) ester, 10CI, 9CI. *Boric acid (H_3BO_3) triisobutyl ester*, 8CI. *Triisobutoxyborane. Isobutyl borate. Tris(2-methylpropyl)borate. Triisobutyl borate*

[13195-76-1]

$[(H_3C)_2CHCH_2O]_3B$

M 230.154

Insect sterilant; moisture stabilizer for urethane polymer adhesives. Moisture-sensitive liq. d_4^{25} 0.840. Bp 207-9°, Bp_{12} 96°. n_D^{25} 1.4014.

▷ED5250000.

Gerrard, W. *et al*, *J. Chem. Soc.*, 1951, 2545 (*props*)
Lappert, M.F., *J. Chem. Soc.*, 1953, 667 (*synth*)
Aroney, M. *et al*, *J. Chem. Soc.*, 1961, 4140 (*struct, props*)
Christopher, P.M. *et al*, *J. Chem. Eng. Data*, 1967, **12**, 333 (*props*)
Davis, F.A. *et al*, *J. Org. Chem.*, 1971, **36**, 1300 (*props, nmr*)
Christopher, P.M. *et al*, *J. Chem. Eng. Data*, 1979, **24**, 67 (*props*)

Deshpande, P.B. *et al*, *J. Chem. Eng. Data*, 1979, **24**, 263 (*props*)
Jaworski, K. *et al*, *Przem. Chem.*, 1981, **60**, 18; *CA*, **94**, 47554q (*synth, props*)
Sax, N.I., *Dangerous Properties of Industrial Materials*, 5th Ed., Van Nostrand-Reinhold, 1979, 1056.

$C_{12}H_{27}BS$ B-00824

Dibutyl(butylthio)borane

Dibutylthioborinic acid butyl ester, 10CI, 9CI, 8CI

[2938-93-4]

$$(H_3CCH_2CH_2CH_2)_2BSCH_2CH_2CH_2CH_3$$

M 214.215

Air- and moisture-sensitive liq. d_{20} 0.837. Bp_8 117.5-118.5°. n_D^{20} 1.4548.

Mikhailov, B.M. *et al*, *Izv. Akad. Nauk SSSR, Ser. Khim.*, 1959, 172 (*synth*)
Laurent, J.P., *Bull. Soc. Chim. Fr.*, 1962, 558.
Mikhailov, B.M. *et al*, *Izv. Akad. Nauk SSSR, Ser. Khim.*, 1976, 1823; *Bull. Acad. Sci. USSR, Div. Chem. Sci.*, **25**, 1717 (*use*)
Dorokhov, V.A. *et al*, *Izv. Akad. Nauk SSSR, Ser. Khim.*, 1977, 1843; *Bull. Acad. Sci. USSR, Div. Chem. Sci.*, **26**, 1704 (*props*)
Bagdasaryan, G.B. *et al*, *Arm. Khim. Zh.*, 1979, **32**, 303; *CA*, **92**, 6578r (*synth, props*)
Dorokhov, V.A. *et al*, *Izv. Akad. Nauk SSSR, Ser. Khim.*, 1979, 1340; *Bull. Acad. Sci. USSR, Div. Chem. Sci.*, **28**, 1253 (*use*)

$C_{12}H_{27}B_3Cl_3N_3$ B-00825

1,3,5-Tributyl-2,4,6-trichloroborazine, 10CI, 9CI, 8CI

1,3,5-Tributyl-2,4,6-trichloroborazole

[3982-02-3]

As 2,4,6-Tribromoborazine, B-00015 with

$$X = Cl, Y = -(CH_2)_3CH_3$$

M 352.154

Solid at r.t., sensitive to moisture. Mp 30°. $Bp_{0.5}$ 115-20° ($Bp_{0.2}$ 118°).

Turner, H.S. *et al*, *Chem. Ind. (London)*, 1958, 526 (*synth*)
Butcher, I.M. *et al*, *J. Inorg. Nucl. Chem.*, 1965, **27**, 817 (*props*)
Turner, H.S. *et al*, *J. Chem. Soc.*, 1965, 6421 (*synth, props*)
Pasdeloup, M. *et al*, *J. Chim. Phys. Phys. Chim. Biol.*, 1972, **69**, 1022; *CA*, **77**, 87302m.
Sheikh, S.U., *CA*, 1981, **94**, 75999g.

$C_{12}H_{27}B_3O_3$ B-00826

Tributylboroxin, 10CI, 9CI, 8CI

[7359-98-0]

As Boroxin, B-00019 with

$$R = -(CH_2)_3CH_3$$

M 251.774

Used in production of epoxide resins. Moisture-sensitive liq., disproportionates above 200°. d^{25} 0.872. Fp −50°. Bp 259° (extrapolated), Bp_{16} 134°. n_D^{25} 1.4193.

Carrazzoni, N.E. *et al*, *Anales. Assoc. Quim. Arg.*, 1961, **49**, 46 (*ir*)
McCusker, P.A. *et al*, *J. Org. Chem.*, 1964, **29**, 2093 (*synth*)
Brooks, C.J.W. *et al*, *Org. Mass Spectrom.*, 1970, **3**, 231 (*glc, ms*)
Laurent, J.P. *et al*, *Bull. Soc. Chim. Fr.*, 1970, **3**, 836 (*pmr, nmr*)
Davies, A.G. *et al*, *Acc. Chem. Res.*, 1972. **5**, 387 (*props*)
Tsvetkov, V.G. *et al*, *Zh. Obshch. Khim.*, 1981, **51**, 269.

$C_{12}H_{27}B_3O_6$ B-00827

Tributoxyboroxin, 10CI, 9CI, 8CI

Butyl metaborate

[101-36-0]

As Boroxin, B-00019 with

$$R = -O(CH_2)_3CH_3$$

M 299.772

Used in production of heat resistant epoxide resins and for removal of nitrogen from petrol hydrocarbons. Liq., dec. on distillation, hydrolysed by water, sensitive to air. d^{20} 1.014. Bp 280° dec. n_D^{25} 1.420.

Lappert, M.F. *et al*, *J. Chem. Soc.*, 1958, 2790 (*synth*)
Aubrey, D.W. *et al*, *J. Chem. Soc.*, 1960, 1931 (*ir*)
Laurent, J.P. *et al*, *Bull. Soc. Chim. Fr.*, 1966, **3**, 908.
Settepani, J.A. *et al*, *J. Econ. Entomol.*, 1969, **62**, 375 (*use*)
Dzhemilev, U.M. *et al*, *Izv. Akad. Nauk SSSR, Ser. Khim.*, 1979, 553; *Bull. Acad. Sci. USSR, Div. Chem. Sci.*, 509 (*use*)

$C_{12}H_{28}B^{\ominus}$ B-00828

Hydrotris(1-methylpropyl)borate(1−), 10CI, 9CI

Tri-sec-*butylhydroborate. Tri*-sec-*butylborohydride. Selectride*

$$[H_3CCH_2CH(CH_3)]_3BH^{\ominus}$$

M 183.163 (ion)

Li salt: [38721-52-7]. *L-Selectride.*

$C_{12}H_{28}BLi$ M 190.104

Often prepd. *in situ*. Available commercially. Stereoselective reducing agent for cyclic and bicyclic ketones. Air- and moisture-sensitive. Sol. THF, Et_2O.

▷Pyrophoric. Flash point of THF soln. −17°

K salt: [54575-49-4]. *K-Selectride.*

$C_{12}H_{28}BK$ M 222.262

Pure compd. not isol. Available commercially as THF soln. Stereoselective reducing agent. Colourless to yellow soln. in THF, stable indefinitely under N_2 but v. sensitive to air and moisture. Sol. THF.

▷Flammable. Flash pt. of soln. −17°. Irritant, may cause burns

Brown, H.C. *et al*, *J. Am. Chem. Soc.*, 1972, **94**, 7159 (*synth*)
Brown, C.A., *J. Am. Chem. Soc.*, 1973, **95**, 4100 (*synth, use*)
Ganem, B. *et al*, *J. Org. Chem.*, 1975, **40**, 146, 2846 (*use, props*)
Strouf, O. *et al*, *Acta Chem. Scand., Ser. A*, 1977, **31**, 391 (*props*)
Inorg. Synth., 1977, **17**, 26 (*synth, ir*)
Dauben, W.G. *et al*, *Tetrahedron Lett.*, 1978, 4487 (*props*)
U.S.P., 4 082 810, (*1978*); *CA*, **89**, 109936y (*synth*)
Brown, H.C. *et al*, *J. Organomet. Chem.*, 1980, **188**, 1 (*synth, nmr*)
Yamamoto, Y. *et al*, *CA*, 1980, **92**, 139884j (*rev*)

$C_{12}H_{28}B^{\ominus}$ B-00829

Hydrotris(2-methylpropyl)borate(1−), 10CI

Triisobutylhydroborate. Triisobutylborohydride

$$[(H_3C)_2CHCH_2]_3BH^{\ominus}$$

M 183.163 (ion)

Li salt: [63717-73-7].

$C_{12}H_{28}BLi$ M 190.104

Reducing agent. Air- and moisture-sensitive. Sol. THF. Commercially available as THF soln.

▷Flash point of soln. −15°
K salt: [67940-40-3].
$C_{12}H_{28}BK$ M 222.262
Generally prepd. *in situ*; no phys. props. reported. Reducing agent. Air- and moisture-sensitive. Sol. THF.

Strouf, O. *et al*, *Acta Chem. Scand., Ser. A*, 1977, **31**, 391 (*props*)
Brown, C.A., *J. Organomet. Chem.*, 1978, **156**, C17.
Brown, H.C. *et al*, *J. Am. Chem. Soc.*, 1978, **100**, 3343 (*synth, ir, nmr*)
Brown, C.A. *et al*, *J. Am. Chem. Soc.*, 1979, **101**, 3964 (*synth, nmr*)
Brown, H.C. *et al*, *J. Org. Chem.*, 1979, **44**, 5004 (*nmr, ir, synth*)
Brown, H.C. *et al*, *J. Organomet. Chem.*, 1979, **166**, 271; 1980, **188**, 1 (*synth, nmr, ir*)

$C_{12}H_{28}B^{\ominus}$ B-00830

Tri-*tert*-butylhydroborate
Tris(1,1-dimethylethyl)hydroborate(1−), 10CI. *Tri-*tert*-butylborohydride*

$[(H_3C)_3C]_3BH^{\ominus}$

M 183.163 (ion)
Li salt: [63717-92-0].
$C_{12}H_{28}BLi$ M 190.104
Characterised spectroscopically.

Strouf, O. *et al*, *Acta Chem. Scand., Ser. A*, 1977, **31**, 391 (*props*)
Nöth, H. *et al*, *J. Organomet. Chem.*, 1977, **142**, 281 (*synth, nmr*)

$C_{12}H_{28}B^{\ominus}$ B-00831

Tributylhydroborate(1−), 10CI
Tributylborohydride

$(H_3CCH_2CH_2CH_2)_3BH^{\ominus}$

M 183.163 (ion)
Li salt: [67335-72-2].
$C_{12}H_{28}BLi$ M 190.104
Selective reducing agent. Air- and moisture-sensitive; stable in THF soln. under N_2. Sol. THF.
K salt: [67940-39-0].
$C_{12}H_{28}BK$ M 222.262
Selective reducing agent. Air- and moisture-sensitive. Sol. THF.

Brown, C.A. *et al*, *J. Organomet. Chem.*, 1978, **156**, C17, 111.
Brown, H.C. *et al*, *J. Am. Chem. Soc.*, 1978, **100**, 3343; 1979, **44**, 5004 (*ir, nmr, synth*)
Brown, C.A. *et al*, *J. Am. Chem. Soc.*, 1979, **101**, 3964 (*synth, nmr*)
Brown, H.C. *et al*, *J. Organomet. Chem.*, 1979, **166**, 281; 1980, **188**, 1 (*synth, nmr*)

$C_{12}H_{28}BN$ B-00832

Cyclohexylborane triethylamine
Cyclohexyl(N,N-diethylethanamine)dihydroboron, 10CI. *N,N-Diethylethanamine, compd. with cyclohexylborane(1:1)*, 9CI. *Monocyclohexylborane triethylamine*
[56118-66-2]

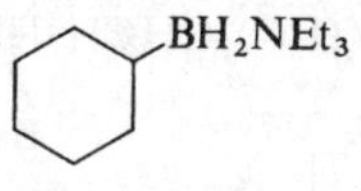

M 197.170
Colourless viscous liq.

Brown, H.C. *et al*, *J. Am. Chem. Soc.*, 1975, **97**, 2791 (*synth*)
Brown, H.C. *et al*, *J. Organomet. Chem.*, 1977, **135**, C10 (*synth*)

$C_{12}H_{28}BN$ B-00833

Dibutyl(diethylamino)borane, 8CI
1,1-Dibutyl-N,N-diethylboranamine, 10CI, 9CI. *Dibutyl(diethylamino)borine*
[4268-15-9]

$(H_3CCH_2CH_2CH_2)_2BNEt_2$

M 197.170
Air- and moisture-sensitive liq. d_{20} 0.757. $Bp_{2.0}$ 75-6°, $Bp_{1.5}$ 52-5°. n_D^{20} 1.4358.

Mikhailov, B.M. *et al*, *Dokl. Akad. Nauk SSSR, Ser. Sci. Khim.*, 1961, **136**, 356 (*synth*)
Mikhailov, B.M. *et al*, *Izv. Akad. Nauk SSSR, Ser. Khim.*, 1972, 2372.
Mikhailov, B.M. *et al*, *Izv. Akad. Nauk SSSR, Ser. Khim.*, 1976, 870; *Bull. Acad. Sci. USSR, Div. Chem. Sci.*, 848, 1924, 2053.
Gupta, S.K., *J. Organomet. Chem.*, 1978, **156**, 95.
Bagdasaryan, G.B. *et al*, *Arm. Khim. Zh.*, 1979, **32**, 203; *CA*, **92**, 6578r.

$C_{12}H_{28}BN$ B-00834

(2-Methylcyclopentyl)borane
Dihydro(2-methylcyclopentyl)boron

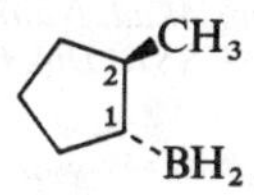

M 197.170
(1RS,2SR)-form
(±)-trans-*form*
Et₃N complex: [56118-61-7].
$C_{12}H_{28}BN$ M 197.170
Useful store for the otherwise unstable 2-methylcyclopentylborane. Liq. Bp_{15} 60-1°. n_D^{20} 1.429.

Brown, H.C. *et al*, *J. Am. Chem. Soc.*, 1975, **97**, 2791 (*synth, ir*)
Brown, H.C. *et al*, *J. Organomet. Chem.*, 1977, **135**, C10 (*synth*)

$C_{12}H_{28}B_2O$ B-00835

Oxybis[dipropylborane], 10CI, 9CI
Bis(dipropylboron)oxide. Tetrapropyldiboron oxide
[52102-14-4]

$(H_3CCH_2CH_2)_2BOB(CH_2CH_2CH_3)_2$

M 209.973
Prepd. by hydrol. of $(H_3CCH_2CH_2)_3B$ in presence of $(H_3C)_3CCOOBEt_2$. Air- and moisture-sensitive liq. Sol. THF.

Köster, R. *et al*, *Justus Liebigs Ann. Chem.*, 1974, 54 (*synth*)
Dorokhov, V.A. *et al*, *Izv. Akad. Nauk SSSR, Ser. Khim.*, 1977, 1919; *Bull. Acad. Sci. USSR, Div. Chem. Sci.*, 1783 (*use*)
Mikhailov, B.M. *et al*, *Izv. Akad. Nauk SSSR, Ser. Khim.*, 1978, 1586; *Bull. Acad. Sci. USSR, Div. Chem Sci.*, 1392 (*synth*)
Mikhailov, B.M. *et al*, *J. Organomet. Chem.*, 1978, **154**, 113 (*synth*)
Pelter, A. *et al*, *Compr. Org. Chem.*, (*Barton, D.H.R. et al*, Ed.), 1979, **3**, 695 (*rev*)
Paetzold, P. *et al*, *Z. Naturforsch., B*, 1980, **35**, 568 (*use*)

$C_{12}H_{30}BN$ B-00836

Tri-*tert*-butylborane ammonia
Amminetris(1,1-dimethylethyl)boron, 10CI. *Tris(1,1-dimethylethyl)borane, ammoniate*

[65568-78-7]

$[(H_3C)_3C]_3B—NH_3$

M 199.186
Mp 91-3°.

Nöth, H. *et al*, *J. Organomet. Chem.*, 1977, **142**, 281 (*synth*, *pmr*, *cmr*)

$C_{12}H_{30}BN$ B-00837

Tripropylborane trimethylamine
(N,N-*Dimethylmethanamine*)*tripropylboron*, *10CI*.
N,N-*Dimethylmethanamine*, *compd. with tripropylborane*(*1:1*)
[62685-91-0]

$(H_3CCH_2CH_2)_3B—NMe_3$

M 199.186

Costes, J.P. *et al*, *J. Chim. Phys. Phys.-Chim. Biol.*, 1976, **73**, 16.

$C_{12}H_{30}BN_3$ B-00838

Tris(diethylamino)borane, 8CI
Hexaethylboranetriamine, *10CI*, *9CI*.
Trisdiethylaminoboron
[867-97-0]

$B(NEt_2)_3$

M 227.199
Chemosterilant against screwworm flies. Liq., sensitive to moisture. Sol. pentane. d_{20} 0.826. Bp 220°, $Bp_{0.4}$ 50-3°. n_D^{20} 1.4450.

Gerrard, W. *et al*, *J. Chem. Soc.*, 1957, 381 (*synth*, *props*)
Dornow, A. *et al*, *Z. Anorg. Allg. Chem.*, 1958, **294**, 81.
Beck, W. *et al*, *Chem. Ber.*, 1972, **105**, 2883 (*nmr*)
Gal'chenko, G.L. *et al*, *Zh. Fiz. Khim.*, 1973, **47**, 1606; *CA*, **79**, 91431m (*synth*, *props*)
Inorg. Synth., 1977, **17**, 159 (*synth*)
Davidson, G. *et al*, *J. Chem. Soc., Dalton Trans.*, 1981, 306 (*ir*, *raman*)

$C_{12}H_{30}B_3N_3$ B-00839

Hexaethylborazine, 9CI
Hexaethylborazole
[839-71-4]

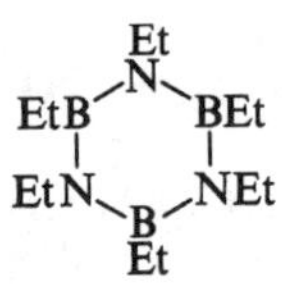

M 248.819
Mp 88°. $Bp_{0.2}$ 98°.

Butcher, I.M. *et al*, *J. Chem. Soc.*, 1964, 4528 (*synth*)
Viswamitra, M.A. *et al*, *Z. Kristallogr., Kristallgeom., Kristallphys., Kristallchem.*, 1965, **121**, 472 (*cryst struct*)
Cros, G. *et al*, *Bull. Soc. Chim. Fr.*, 1969, 2601 (*pmr*)

$C_{12}H_{32}B_2N_4O_2$ B-00840

Tetramethylethylenediamine bis[(ethylcarbamoyl)borane]
Bis[(*ethylamino*)*carbonyl*]*tetrahydro*[μ-(N,N,N′,N′-*tetramethyl-1,2-ethanediamine*-N:N′)]*diboron*, *10CI*.
N-*Ethylboranecarboxamide compd. with* N,N,N′,N′-*tetramethyl-1,2-ethanediamine*(*2:1*). N,N,N′,N′-*Tetramethylethylenediaminebis*(N-*ethylamido*)*borane*
[72695-29-5]

$EtNHCOBH_2—Me_2NCH_2CH_2NMe_2—BH_2CONHEt$

M 286.030
White solid. Sol. CH_2Cl_2. Mp 103-10°.

Spielvogel, B.F. *et al*, *J. Inorg. Nucl. Chem.*, 1979, **41**, 1223 (*synth*)
Hall, I.H. *et al*, *J. Pharm. Sci.*, 1980, **69**, 1025; 1981, **70**, 339 (*pharmacol*)

$C_{12}H_{42}B_8P_2Pt$ B-00841

[η^3-Dodecahydrooctaborato(2−)]bis(triethylphosphine)platinum, 9CI
[*Octaborane*(*12*)]*bis*(*triethylphosphine*)*platinum*, *8CI*
[12563-57-4]

$[Pt(B_8H_{12})(PEt_3)_2]$

M 529.971
Photographic fogging agent.

Kane, A.R. *et al*, *J. Am. Chem. Soc.*, 1970, **92**, 2571.
Riggs, W.M., *Anal. Chem.*, 1972, **44**, 830.

$C_{12}H_{42}B_{10}P_2Pt$ B-00842

[(5,6,7,8,9,10-η)Dodecahydrodecaborato(2−)]bis(triethylphosphine)platinum, 9CI
[12375-22-3]

$[Pt(B_{10}H_{12})(PEt_3)_2]$

M 551.591
Possible developing agent for $AgNO_3$-based invisible inks. Yellow cryst. (MeCN), dec. >180°.

Klanberg, F. *et al*, *Inorg. Chem.*, 1968, **7**, 2072 (*synth*, *ir*, *nmr*)
Riggs, W.M., *Anal. Chem.*, 1972, **44**, 830 (*pe*)

$C_{13}H_{10}BCoO_2$ B-00843

Dicarbonyl[(1,2,3,4,5,6-η)-1-phenylboratabenzene]cobalt, 9CI
[53852-42-9]

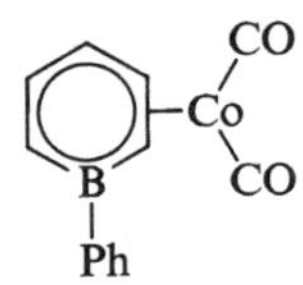

M 267.964
Cryst. (pentane). ir ν_{co} 2043, 1990 cm^{-1} (hexane). ^{11}B nmr −20.5 ppm ($BF_3 \cdot Et_2O$ ref.).

Herberich, G.E. *et al*, *Z. Naturforsch., B*, 1974, **29**, 439 (*synth*, *ir*, *nmr*, *ms*)

$C_{13}H_{12}BBrClN$ B-00844

[Methyl(4-bromophenyl)amino]chloro(phenyl)borane
N-(*4-Bromophenyl*)-*1-chloro*-N-*methyl-1-phenylboranamine*, *10CI*
[66338-33-8]

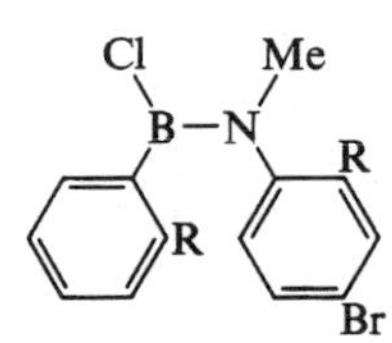

R = H

M 308.412
Air- and moisture-sensitive.

Ziegler, M.L. *et al*, *Z. Naturforsch., B*, 1978, **33**, 200 (*struct*)
Friebolin, H. *et al*, *Org. Magn. Reson.*, 1977, **10**, 157 (*synth, nmr*)

$C_{13}H_{12}BN$ B-00845

10-Methyl-10,9-borazarophenanthrene

5,6-Dihydro-6-methyldibenz[c,e][*1,2*]*azaborine*, 9CI

[15813-13-5]

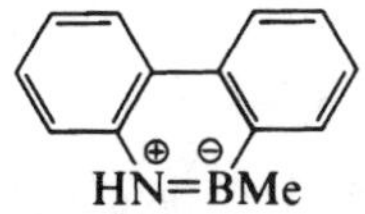

M 193.055

Cryst. (pet. ether). Mp 103-4°. λ_{max} 220 (log ϵ 4.203), 224 (4.195), 230 (4.191), 236 (4.191), 240 (4.203), 246 (4.211), 250 (4.205), 298 (3.725), 312 (3.944) and 324 (3.995) nm.

Dewar, M.J.S. *et al*, *J. Chem. Soc.*, 1958, 3073; *J. Am. Chem. Soc.*, 1968, **90**, 706.
Org. Synth., 1966, **46**, 65.

$C_{13}H_{13}BFeN_6O_3$ B-00846

Acetyldicarbonyl[hydrotris(1*H*-pyrazolato-*N*[1])borato(1−)-*N*[2],*N*[2′],*N*[2″]]iron, 9CI

[43070-44-6]

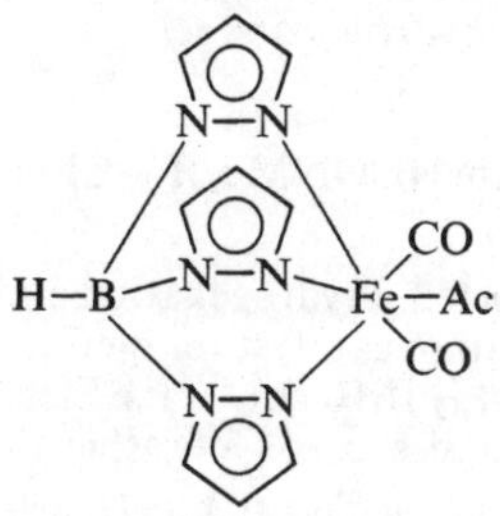

M 367.941

Bright-yellow cryst. (CH_2Cl_2/hexane). Decarbonylates to the Fe-Me compd. in soln.

Cotton, F.A. *et al*, *Inorg. Chim. Acta*, 1973, **7**, 161 (*synth, ir, pmr, struct*)

$C_{13}H_{13}BO$ B-00847

Methoxydiphenylborane

Diphenylborinic acid methyl ester, 10CI, 9CI, 8CI. *Methyl diphenylborinate*

[13471-36-8]

Ph_2BOMe

M 196.055

Liq., moisture-sensitive, but stable indefinitely when stored under inert atm. Sol. nonhydroxylic solvs. Bp_{10} 132°.

Aubrey, D.W. *et al*, *Proc. Chem. Soc. London*, 1960, 148 (*props*)
Hofmeister, H.K. *et al*, *J. Inorg. Nucl. Chem.*, 1964, **26**, 1209 (*pmr*)
Jefferson, R. *et al*, *J. Chem. Soc. (A)*, 1966, 1584.
Jacob, P., *J. Organomet. Chem.*, 1978, **156**, 101 (*synth, pmr*)
Paetzold, P. *et al*, *Chem. Ber.*, 1979, **112**, 663.

$C_{13}H_{15}B$ B-00848

4,4-Dimethyl-1-phenyl-1-bora-2,5-cyclohexadiene

1,4-Dihydro-4,4-dimethyl-1-phenylborin, 10CI

[62496-51-9]

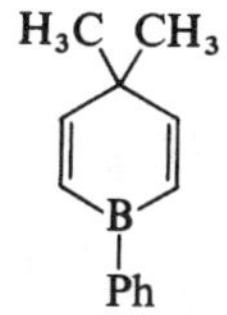

M 182.072

Characterized as its transition metal complex derivs. Air-sensitive.

Cyclopentadienylcobalt complex: [62479-24-7]. (η^5-*2,4-Cyclopentadien-1-yl*)[(*1,2,3,5,6-η*)-*1,4-dihydro-4,4-dimethyl-1-phenylborin*]*cobalt*.
$C_{18}H_{20}BCo$ M 306.099
Red cryst. (pentane). Mp 60°.

Tricarbonyliron complex: see Tricarbonyl[(*2,3,5,6-η*)-*1,4-dihydro-4,4-dimethyl-1-phenylborin*]*iron*, *B-00987*

Ni complex: [74982-02-8]. *Bis*[(*2,3,5,6-η*)-*1,4-dihydro-4,4-dimethyl-1-phenylborin*-B]*nickel*.
$C_{26}H_{30}B_2Ni$ M 422.833
Yellow-orange needles. Mp 183-4°.

Pd complex: [74989-66-5]. *Bis*[(*2,3,5,6-η*)-*1,4-dihydro-4,4-dimethyl-1-phenylborin*-B]*palladium*.
$C_{26}H_{30}B_2Pd$ M 470.563
Yellow-orange cryst. Mp 138-9°.

Pt complex: [74989-69-8]. *Bis*[(*2,3,5,6-η*)-*1,4-dihydro-4,4-dimethyl-1-phenylborin*-B]*platinum*.
$C_{26}H_{30}B_2Pt$ M 559.223
Yellow needles. Mp 173-4°. $Bp_{0.000001}$ 120° subl.

Herberich, G.E. *et al*, *Chem. Ber.*, 1977, **110**, 760 (*synth, nmr, pmr, props*)
Herberich, G.E. *et al*, *Chem. Ber.*, 1977, **110**, 1167 (*props*)
Herberich, G.E. *et al*, *J. Organomet. Chem.*, 1980, **191**, 27 (*props, pmr, nmr*)

Having problems with locating a compound? Have you checked the indexes?

$C_{13}H_{18}BN_6Rh$ B-00850

Bis(ethene)[hydrotris(1*H*-pyrazolato-*N*)borato(1–)-*N*[2],*N*[2']]-rhodium, 9CI

Bis(ethylene)[tris(pyrazolyl)borato]rhodium

[52668-39-0]

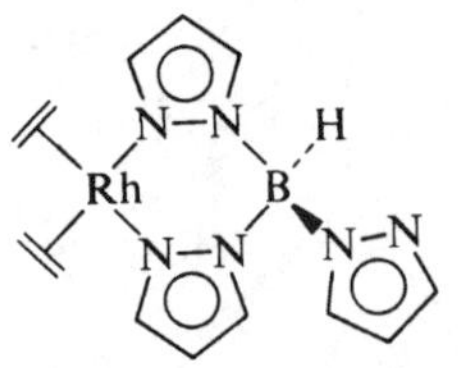

M 372.041

Pyrazolyl groups are fluxional and magnetically equivalent. Ethylene groups are also fluxional. Yellow, air-stable cryst. Dec. >140°, or on attempted subl.

Trofimenko, S. *et al*, *J. Am. Chem. Soc.*, 1969, **91**, 588 (*synth*, *pmr*)
O'Sullivan, D. *et al*, *J. Organomet. Chem.*, 1974, **65**, C47 (*struct*)

$C_{13}H_{19}BClN$ B-00851

9-Chloro-9-borabicyclo[3.3.1]nonane pyridine

Chloro-1,5-cyclooctanediyl(pyridine)boron, 10CI. *Pyridine compd. with 9-chloro-9-borabicyclo[3.3.1]-nonane(1:1)*, 8CI

[22086-36-8]

As 9-Chloro-9-borabicyclo[3.3.1]nonane dimethyl sulfide, B-00695 with

L = Py

M 235.563

Cryst. Mp 139-40°.

Köster, R. *et al*, *Justus Liebigs Ann. Chem.*, 1968, **719**, 169 (*synth*)
Brown, H.C. *et al*, *J. Org. Chem.*, 1980, **45**, 846 (*cmr*, *nmr*)

$C_{13}H_{20}BCl$ B-00852

9-(5-Chloro-1-pentynyl)-9-borabicyclo[3.3.1]nonane, 10CI

[62459-80-7]

As 9-Bromo-9-borabicyclo[3.3.1]nonane, B-00539 with

$R = —C{\equiv}CCH_2CH_2CH_2Cl$

M 222.564

Sensitive to air.

Sinclair, J.A. *et al*, *J. Am. Chem. Soc.*, 1977, **99**, 954.
Molander, G.A. *et al*, *J. Org. Chem.*, 1977, **42**, 3106.

$C_{13}H_{20}BN$ B-00853

9-Borabicyclo[3.3.1]nonane pyridine

1,5-Cyclooctanediylhydro(pyridine)boron, 10CI. *Pyridine, compd. with 9-borabicyclo[3.3.1]nonane(1:1)*.
9-BBN pyridine complex

[64045-95-0]

As 9-Borabicyclo[3.3.1]nonane dimethyl sulfide, B-00705 with

L = pyridine

M 201.118

Useful selective reducing agent for aldehydes in presence of ketones or other groups. Cryst. (Et_2O/pentane); stable for several hours in air with no change in Mp. Mp 75.5-76°.

Brown, H.C. *et al*, *Inorg. Chem.*, 1977, **16**, 3090 (*synth*)
Brown, H.C. *et al*, *J. Org. Chem.*, 1977, **42**, 4169 (*use*)
Brown, H.C. *et al*, *Recl. Trav. Chim. Pays-Bas*, 1979, **98**, 117.
Brown, H.C. *et al*, *J. Org. Chem.*, 1980, **45**, 846; 1748 (*props*, *nmr*)
Soderquist, J.A. *et al*, *J. Org. Chem.*, 1981, **46**, 4599.

$C_{13}H_{21}B$ B-00854

9-(1-Cyclopenten-1-yl)-9-borabicyclo[3.3.1]nonane, 10CI

[74142-80-6]

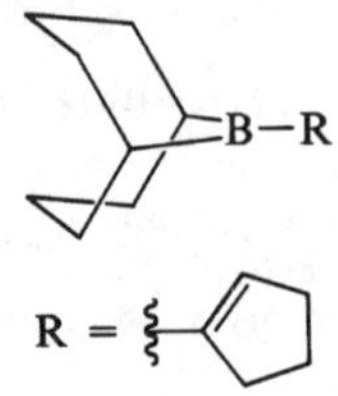

M 188.119

Synthetic intermed. Sensitive to air.

Miyaura, N. *et al*, *Tetrahedron Lett.*, 1980, **21**, 537.

$C_{13}H_{21}BCl_2N_3ORh$ B-00855

Dichloro(*N,N*-dimethylformamide-*O*)bis(pyridine)[tetrahydroborato(1–)-*H*]rhodium, 10CI

Dichloro(N,N-*dimethylformamide*)*bis*(*pyridine*)*rhodium*(*1+*) *tetrahydroborate*, 8CI

[72930-45-1]

$Rh(HCONMe_2)Py_2Cl_2(BH_4)$

M 419.950

Struct. not reported. Hydrogenation and hydroformylation catalyst for olefins. Dark-red cryst. ($CHCl_3$). ir ν_{CO} 1610 cm^{-1}. 1:1 Electrolyte in DMF with proposed dissoc. of $Cl^{\ominus}$ rather than $BH_4^{\ominus}$.

Abley, P. *et al*, *J. Chem. Soc.* (*C*), 1971, 840 (*synth*, *ir*, *pmr*, *mv*, *ms*)
U.S.P., 4 178 312, (*1979*); *CA*, **92**, 110538p (*use*)

$C_{13}H_{21}BN_2S$ B-00856

(*N,N*-Diethyl-*N'*-phenylthioureato)dimethylborane

(N,N-*Diethyl*-N'-*phenylthioureato*-N',S)*dimethylboron*, 10CI

[73768-06-6]

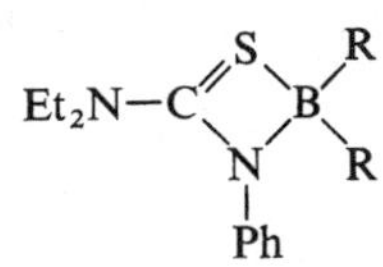

R = Me

M 248.192

Formed from the reaction of $Et_2NCSNPhSiMe_3$ with Me_2BCl.

Marringgle, W., *Z. Naturforsch.*, *B*, 1980, **35**, 164 (*synth*)

$C_{13}H_{21}BO$ B-00857

1,5-Diallyl-3-(methoxymethyl)-1-bora-2-cyclohexene

1,2,3,4-Tetrahydro-5-(methoxymethyl)-1,3-di-2-propenylborin, 10CI

[69594-46-3]

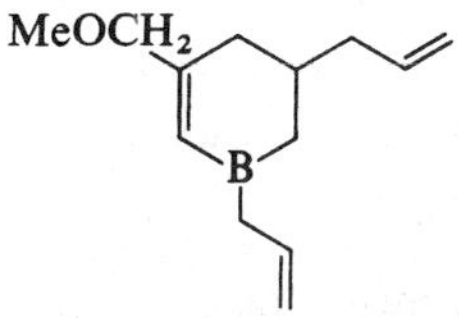

M 204.118

Prepd. by allylboronation of 3-methoxypropyne. Air-sensitive. Not isol. pure; part. cyclises on dist.

Mikhailov, B.M. *et al*, *Dokl. Akad. Nauk SSSR, Ser. Sci. Khim.*, 1978, **243**, 929 (*props*)
Mikhailov, B.M. *et al*, *Izv. Akad. Nauk SSSR, Ser. Khim.*, 1979, 2541; *Bull. Acad. Sci. USSR, Div. Chem. Sci.*, **28**, 2358 (*synth*)

$C_{13}H_{21}BO$ B-00858

1,5-Diallyl-3-methoxy-2-methyl-1-bora-2-cyclohexene

1,2,3,4-Tetrahydro-5-methoxy-6-methyl-1,3-di-2-propenylborin, 10CI

[72997-22-9]

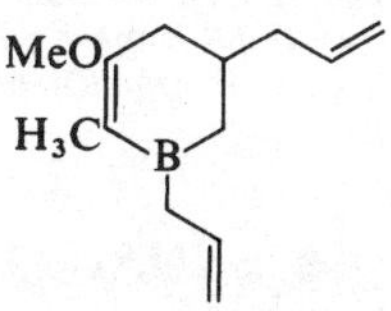

M 204.118

Prepd. by allylboronation of 1-methoxypropyne. Air-sensitive liq. $Bp_{0.05}$ 75-8°. n_D^{20} 1.5015.

Mikhailov, B.M. *et al*, *Izv. Akad. Nauk SSSR, Ser. Khim.*, 1979, 2335; *Bull. Acad. Sci. USSR, Div. Chem. Sci.*, **28**, 2156 (*synth, ir, pmr*)

$C_{13}H_{21}BO$ B-00859

7-(Methoxymethyl)-3-(2-propenyl)-3-borabicyclo[3.3.1]non-6-ene, 10CI

3-Allyl-7-(methoxymethyl)-3-borabicyclo[3.3.1]non-6-ene

[69464-54-6]

As 7-(Chloromethyl)-3-methoxy-3-borabicyclo[3.3.1]-non-6-ene, B-00670 with

$$R = —CH_2CH{=}CH_2,\ R' = —CH_2OMe$$

M 204.118

Air- and moisture-sensitive liq. Bp_1 76-7°. n_D^{19} 1.4490.

Mikhailov, B.M. *et al*, *Dokl. Akad. Nauk SSSR, Ser. Sci. Khim.*, 1978, **243**, 929 (*synth*)
Mikhailov, B.M. *et al*, *Izv. Akad. Nauk SSSR, Ser. Khim.*, 1979, 2541; *Bull. Acad. Sci. USSR, Div. Chem. Sci.*, 2358 (*synth, ir, pmr*)
Mikhailov, B.M. *et al*, *Izv. Akad. Nauk SSSR, Ser. Khim.*, 1980, 1881; *Bull. Acad. Sci. USSR, Div. Chem. Sci.*, 1355 (*pmr*)

$C_{13}H_{21}BO_2$ B-00860

3-Borabicyclo[3.3.1]non-3-yl acetylacetonate

[*1,3-Cyclohexanediylbis(methylene)*](*2,4-pentanedionato-O,O′*)*boron*, 10CI, 9CI

[76069-95-9]

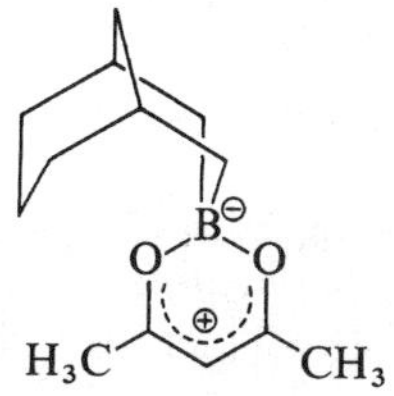

M 220.118

Bright-yellow cryst. Mp 65-75°.

Gursky, M.E. *et al*, *J. Organomet. Chem.*, 1980, **199**, 171 (*pmr, cmr*)
Gurskii, M.E. *et al*, *Izv. Akad. Nauk SSSR, Ser. Khim.*, 1981, 341; *Bull. Acad. Sci. USSR, Div. Chem. Sci.*, 264 (*pmr*)
Gurskii, M.E. *et al*, *Izv. Akad. Nauk SSSR, Ser. Khim.*, 1981, 394; *Bull. Acad. Sci. USSR, Div. Chem. Sci.*, 312 (*synth, ir, pmr*)

$C_{13}H_{22}BFeNO_3Si$ B-00861

Tricarbonyl[(3,4-η)-4,5-diethyl-1,2,2,3-tetramethyl-1-aza-2-sila-5-boracyclopent-3-ene-B^5,N^1]iron

Tricarbonyl[(2,3,4,5-η)-4,5-diethyl-1,2,2,3-tetramethyl-Δ³-1,2,5-azasilaboroline]iron

[81952-05-8]

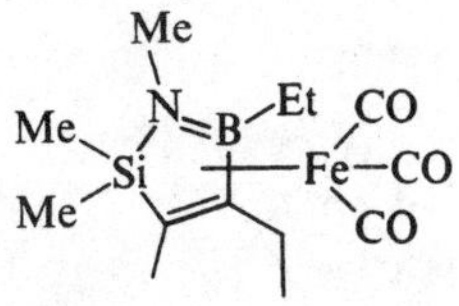

M 335.064

Orange-red cryst. (pentane). Mp 189-90° dec. $Bp_{0.001}$ 60° subl.

Köster, R. *et al*, *Chem. Ber.*, 1982, **115**, 738 (*synth, struct, ir, ms, pmr, cmr, nmr*)

$C_{13}H_{23}B$ B-00862

9-Cyclopentyl-9-borabicyclo[3.3.1]nonane, 10CI

B-Cyclopentyl-9-BBN

[49623-12-3]

As 9-(1-Cyclopenten-1-yl)-9-borabicyclo[3.3.1]nonane, B-00854 with

R = cyclopentyl

M 190.135

Useful reagent for transfer of a cyclopentyl group. Liq., sensitive to air. $Bp_{0.2}$ 68-70°.

Kramer, G.W. *et al*, *J. Organomet. Chem.*, 1974, **73**, 1 (*synth*)
Levy, A.B., *J. Org. Chem.*, 1978, **43**, 4684.
Midland, M.M. *et al*, *J. Organomet. Chem.*, 1978, **156**, 203.
Brown, H.C. *et al*, *Synthesis*, 1979, 701.
Levy, A.B., *Tetrahedron Lett.*, 1979, 4021.
Brown, H.C. *et al*, *J. Org. Chem.*, 1980, **45**, 4067.

$C_{13}H_{23}B$ B-00863

9-(1-Pentenyl)-9-borabicyclo[3.3.1]nonane, 10CI

B-1-Pentenyl-9-BBN

[77161-71-8]

As 9-(1-Cyclopenten-1-yl)-9-borabicyclo[3.3.1]nonane, B-00854 with

$$R = —CH{=}CHCH_2CH_2CH_3$$

M 190.135

(E)-form [69322-44-7]

Formed by hydroboration of 1-pentyne with 9-BBN. Reagent for transfer of 1-pentenyl group, sometimes stereospecifically. Air-sensitive.

Brown, H.C. *et al*, *J. Am. Chem. Soc.*, 1979, **101**, 96.
Levy, A.B. *et al*, *Synthesis*, 1980, 945.

$C_{13}H_{24}B^{\ominus}$ B-00864

9-Cyclopentyl-9-boratabicyclo[3.3.1]nonane, 10CI

B-Cyclopentyl-9-BBN hydride

As 9-Boratabicyclo[3.3.1]nonane, B-00546 with

R = cyclopentyl, R′ = H

M 191.143 (ion)

Li salt: [69593-80-2].
$C_{13}H_{24}BLi$ M 198.084
Air- and moisture-sensitive.

Brown, H.C. *et al*, *J. Org. Chem.*, 1979, **44**, 467.
Brown, H.C. *et al*, *J. Organomet. Chem.*, 1980, **188**, 1 (*synth, nmr*)

$C_{13}H_{24}BCl_2N$ B-00865

(1,2,3,4,5-Pentamethyl-2,4-cyclopentadien-1-yl)dichloroborane trimethylamine

Dichloro(N,N-*dimethylmethanamine*)(*1,2,3,4,5-pentamethyl-2,4-cyclopentadien-1-yl*)*boron*, 10CI. *1-Dichloroboryl-1,2,3,4,5-pentamethyl-2,4-cyclopentadiene trimethylamine*

[60828-65-1]

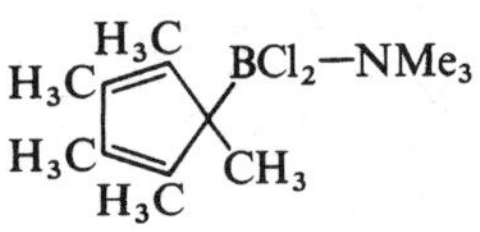

M 276.055

Cryst. (methylcyclohexane). Mp 135° dec.

Jutzi, P. *et al*, *Angew. Chem.*, 1977, **89**, 44 (*pmr, ms, synth*)
Jutzi, P. *et al*, *Chem. Ber.*, 1979, **112**, 2481 (*synth, ir, nmr, pmr*)

$C_{13}H_{24}BFeNO_2Si$ B-00866

Dicarbonyl[(3,4,5-η)-1-(1,1-dimethylethyl)-2,3-dihydro-2-methyl-1*H*-1,2-azaborol-3-yl-*B*[2],*N*[1]](trimethylsilyl)iron, 10CI

(*1*-tert-*Butyl-2,3-dihydro-2-methyl-1*H-*azaborol-3-yl*)*dicarbonyl*(*trimethylsilyl*)*iron*

[75325-35-8]

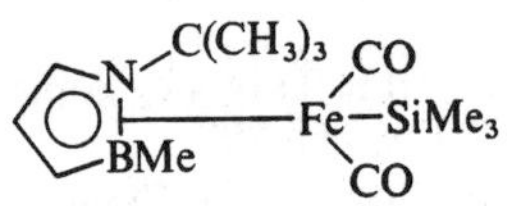

M 321.081

Waxy yellow solid.

Schulze, J. *et al*, *J. Organomet. Chem.*, 1980, **193**, 83 (*synth, ms, ir, pmr, nmr*)

$C_{13}H_{24}BO^{\ominus}$ B-00867

2-Furanyltripropylborate(1−), 10CI

2-Furyltripropylborate

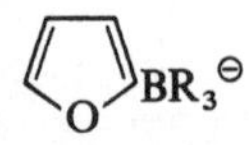

R = $H_3CCH_2CH_2$—

M 207.142 (ion)

Li salt: [70415-39-3].
$C_{13}H_{24}BLiO$ M 214.083
Synthetic intermed., prepd. *in situ*.

Akimoto, I. *et al*, *Synthesis*, 1979, 146 (*synth*)

$C_{13}H_{25}B$ B-00868

3-*tert*-Butyl-7-methyl-3-borabicyclo[3.3.1]nonane

3-(1,1-Dimethylethyl)-7-methyl-3-borabicyclo[3.3.1]nonane, 10CI. *3-*tert*-Butyl-7-methyl-3-BBN*

[76062-78-7]

As 3-Borabicyclo[3.3.1]nonane, B-00544 with

X = $-C(CH_3)_3$, R = CH_3

M 192.151

Air-sensitive liq. Sol. CH_2Cl_2. Bp_1 81-3°.

Gurskii, M.E. *et al*, *J. Organomet. Chem.*, 1980, **199**, 171 (*cmr*)
Mikhailov, B.M. *et al*, *Izv. Akad. Nauk SSSR, Ser. Khim.*, 1981, 341; *Bull. Acad. Sci. USSR, Div. Chem. Sci.*, 264 (*cmr*)
Gurskii, M.E. *et al*, *Izv. Akad. Nauk SSSR, Ser. Khim.*, 1981, 394; *Bull. Acad. Sci. USSR, Div. Chem. Sci.*, 312 (*synth*)

$C_{13}H_{25}B$ B-00869

9-(1,2-Dimethylpropyl)-9-borabicyclo[3.3.1]nonane, 10CI

B-Siamyl-9-borabicyclo[3.3.1]nonane. B-Siamyl-9-BBN

[63942-78-9]

As 9-(1-Cyclopenten-1-yl)-9-borabicyclo[3.3.1]nonane, B-00854 with

R = $-CH(CH_3)CH(CH_3)_2$

M 192.151

Useful as a mild reducing agent for aldehydes. Sensitive to air.

Midland, M.M. *et al*, *J. Organomet. Chem.*, 1977, **134**, C17.
Midland, M.M. *et al*, *J. Org. Chem.*, 1978, **43**, 1470.
Midland, M.M. *et al*, *J. Organomet. Chem.*, 1978, **156**, 203.
Midland, M.M. *et al*, *J. Organomet. Chem.*, 1979, **182**, C53.
Brown, H.C. *et al*, *J. Org. Chem.*, 1980, **45**, 846 (*nmr*)

$C_{13}H_{25}BO$ B-00870

Dicyclohexyl(methoxy)borane

Dicyclohexylborinic acid methyl ester, 10CI, 9CI, 8CI. *Methyl dicyclohexylborinate*

[32705-46-7]

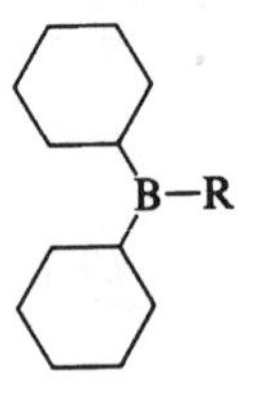

R = OMe

M 208.150

Used in synth. of dicyclohexylketone by reaction with dichloromethyl methyl ether. Liq., can be dist. under N_2, but sensitive to air and moisture. $Bp_{0.005}$ 75-80°.

Brown, H.C. *et al*, *J. Am. Chem. Soc.*, 1971, **93**, 2802 (*synth*)
Carlson, B.A. *et al*, *J. Am. Chem. Soc.*, 1973, **95**, 6876 (*use*)
Carlson, B.A. *et al*, *J. Organomet. Chem.*, 1974, **67**, C39 (*use*)
Kramer, G.W. *et al*, *J. Organomet. Chem.*, 1974, **73**, 1 (*synth*, *pmr*)
Brown, H.C. *et al*, *J. Org. Chem.*, 1977, **42**, 1392 (*synth*, *props*)
Brown, H.C. *et al*, *J. Organomet. Chem.*, 1977, **131**, 163 (*synth*, *nmr*)

$C_{13}H_{25}BS$ B-00871

Dicyclohexyl(methylthio)borane

Dicyclohexylthioborinic acid methyl ester, 10CI, 9CI

[57476-24-1]

As Dicyclohexyl(methoxy)borane, B-00870 with

R = SMe

M 224.211

Prepd. *in situ* from tricyclohexylborane. Used in synth. of dicyclohexylbromoborane and of unsymmetrical conjugated diynes. Air- and moisture-sensitive liq. Bp_1 109-10°, $Bp_{0.1}$ 96°. n_D^{20} 1.518, m/e 141.

Mikhailov, B.M. *et al*, *Izv. Akad. Nauk SSSR, Ser. Khim.*, 1963, 646 (*props*)
Pelter, A. *et al*, *J. Chem. Soc., Chem. Commun.*, 1975, 531 (*use*)
Pelter, A. *et al*, *J. Chem. Soc., Dalton Trans.*, 1976, 2087 (*synth*, *ms*, *props*)
Pelter, A. *et al*, *Tetrahedron Lett.*, 1976, 4385 (*use*)

$C_{13}H_{26}B^{\ominus}$ B-00872

Dicyclohexylhydromethylborate(1−), 10CI

Dicyclohexylmethylhydroborate

As Dicyclohexyldihydroborate(1−), B-00797 with

R = Me

M 193.158 (ion)

Li salt: [61721-94-6].

$C_{13}H_{26}BLi$ M 200.099

Minor product from reaction of methyllithium with dicyclohexylborane, which gives a mixture of the Li salts of dicyclohexyldihydro- and dicyclohexyldimethylborates. Stereoselective reducing agent.

U.S.P., 3 974 183, (*1976*); *CA*, **86**, 72219b (*use*)
Hubbard, J.L. *et al*, *J. Organomet. Chem.*, 1978, **156**, 81 (*nmr*)

$C_{13}H_{26}BO^{\ominus}$ B-00873

(Ethoxyethynyl)tripropylborate(1−), 10CI

(*Ethoxyethynyl*)*tripropylboranate*

$$(H_3CCH_2CH_2)_3BC\equiv COEt^{\ominus}$$

M 209.158 (ion)

Li salt: [65467-78-9].

$C_{13}H_{26}BLiO$ M 216.099

Prepd. *in situ* from ethoxyethynyllithium and tripropylborane. Synthetic intermed.

Mikhailov, B.M. *et al*, *Izv. Akad. Nauk SSSR, Ser. Khim.*; 1977, 2633; *Bull. Acad. Sci. USSR, Div. Chem. Sci.*, **26**, 2437 (*props*)
Mikhailov, B.M. *et al*, *Izv. Akad. Nauk SSSR, Ser. Khim.*, 1979, 855; *Bull. Acad. Sci. USSR, Div. Chem. Sci.*, **28**, 799 (*synth*)

$C_{13}H_{27}BO$ B-00874

Dibutyl(2-penten-3-yloxy)borane

Dibutylborinic acid 1-ethyl-1-propenyl ester, 10CI

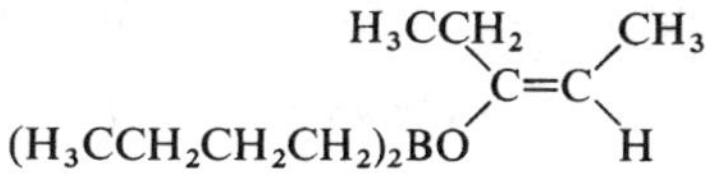

M 210.166

(*Z*)-**form** [78024-10-9]

Prepd. *in situ*. Intermed. in stereoselective aldol condensations. Air- and moisture-sensitive. Characterised spectroscopically.

Evans, D.A. *et al*, *J. Am. Chem. Soc.*, 1981, **103**, 3099 (*use*, *synth*, *pmr*)
Evans, D.A. *et al*, *Pure Appl. Chem.*, 1981, **53**, 1109 (*use*, *synth*, *pmr*)

$C_{13}H_{27}BSi$ B-00875

9-[1-(Trimethylsilyl)ethyl]-9-borabicyclo[3.3.1]nonane, 10CI

[72610-06-1]

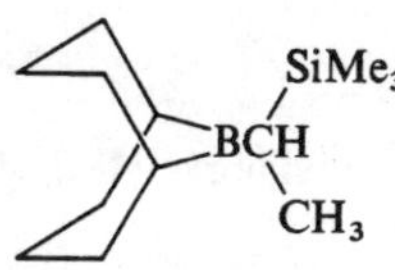

M 222.252

Bp_1 88-92°.

Brown, H.C. *et al*, *J. Org. Chem.*, 1980, **45**, 846 (*nmr*)
Soderquist, J.A. *et al*, *J. Org. Chem.*, 1980, **45**, 3571 (*nmr*, *synth*, *ms*, *pmr*)

$C_{13}H_{27}BSi$ B-00876

9-(2-Trimethylsilylethyl)-9-borabicyclo[3.3.1]nonane, 10CI

[72610-05-0]

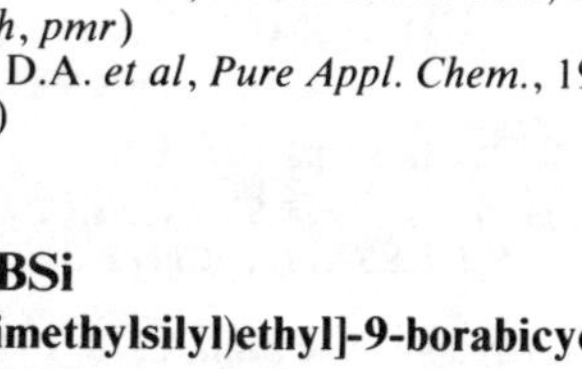

Py complex

M 222.252

Formed by hydroboration of vinyltrimethylsilane with 9-BBN.

Py complex: [72638-39-2]. *1,5-Cyclooctanediyl(pyridine)[2-(trimethylsilyl)ethyl]boron*, 10CI.

$C_{18}H_{32}BNSi$ M 301.353

Ramsay, B.G. *et al*, *J. Organomet. Chem.*, 1977, **141**, 257 (*uv*, *nmr*, *struct*)
Brown, H.C. *et al*, *J. Org. Chem.*, 1980, **45**, 846 (*nmr*)
Soderquist, J.A. *et al*, *J. Org. Chem.*, 1980, **45**, 3571 (*synth*)

$C_{13}H_{29}BO_2$ B-00877

Dibutoxypentylborane

Pentylboronic acid dibutyl ester, 10CI. *1-Pentaneboronic acid dibutyl ester*, 8CI

[3027-57-4]

$$H_3C(CH_2)_4B(OCH_2CH_2CH_2CH_3)_2$$

M 228.181

Air- and moisture-sensitive liq. Sol. non-hydroxylic solvs. d^{20} 0.829. $Bp_{1.5}$ 144°. n_D^{25} 1.4167.

Laurent, J.P. *et al*, *C.R. Hebd. Seances Acad. Sci.*, 1961, **253**, 1812 (*synth, props*)
Gallais, F. *et al*, *C.R. Hebd. Seances Acad. Sci.*, 1964, **259**, 4262 (*props*)
Baudet, J. *et al*, *C.R. Hebd. Seances Acad. Sci.*, 1967, **264**, 1090 (*props*)
Tsvetkov, V.G. *et al*, *Zh. Obshch. Khim.*, 1979, **49**, 1852, 2406 (*props*)

$C_{13}H_{29}BS$ B-00878

Dibutyl(pentylthio)borane

Dibutylthioborinic acid pentyl ester, 10CI

[67682-93-3]

$(H_3CCH_2CH_2CH_2)_2BS(CH_2)_4CH_3$

M 228.242

Air- and moisture-sensitive liq. Bp_4 103-4°.

Mikhailov, B.M. *et al*, *Izv. Akad. Nauk SSSR, Ser. Khim.*, 1978, 1594; *Bull. Acad. Sci. USSR, Div. Chem. Sci.*, **27**, 1392 (*synth, props, nmr*)
Mikhailov, B.M. *et al*, *J. Organomet. Chem.*, 1978, **154**, 113 (*synth, props, nmr*)

$C_{13}H_{30}B^{\ominus}$ B-00879

Methyltris(2-methylpropyl)borate(1−), 10CI, 9CI

Triisobutylmethylborate

$[(H_3C)_2CHCH_2]_3BMe^{\ominus}$

M 197.190 (ion)

Cu(I) salt: [59560-22-4].

$C_{13}H_{30}BCu$ M 260.736

Formed *in situ* by adding CuI to the Li salt. Synth. intermed. Air-sensitive. Sol. THF. Not fully characterised.

Miyaura, N. *et al*, *Tetrahedron Lett.*, 1976, 255; 1977, 173, 3369 (*synth, use*)
Yamada, K. *et al*, *Bull. Chem. Soc. Jpn.*, 1979, **52**, 275 (*synth, props*)

$C_{13}H_{30}B^{\ominus}$ B-00880

Tributylmethylborate(1−), 10CI, 9CI

Methyltributylborate

$(H_3CCH_2CH_2CH_2)_3BMe^{\ominus}$

M 197.190 (ion)

Cu (I) salt: [59560-21-3].

$C_{13}H_{30}BCu$ M 260.736

Prepd. *in situ* from Li salt and CuI. Synth. intermed. Air-sensitive. Sol. THF. Not fully characterised.

Maruyama, K. *et al*, *J. Am. Chem. Soc.*, 1977, **99**, 8068 (*props*)
Miyaura, N. *et al*, *Bull. Chem. Soc. Jpn.*, 1977, **50**, 2199 (*synth, props*)
Miyaura, N. *et al*, *Tetrahedron Lett.*, 1977, 3369 (*synth, props*)
Yamada, K. *et al*, *Bull. Chem. Soc. Jpn.*, 1979, **52**, 275.

$C_{13}H_{30}BN$ B-00881

Dibutyl[3-(dimethylamino)propyl-*C,N*]boron, 10CI

*3-(Dibutylboryl)-*N,N*-dimethylpropylamine. Dibutyl[3-(dimethylamino)propyl]borane*

[72910-21-5]

$Me_2NCH_2CH_2CH_2B(CH_2CH_2CH_2CH_3)_2$

M 211.197

Exists as intramol. coord. form. Liq. d_{20} 0.838. Bp_2 98-100°.

Mostovoi, N.V. *et al*, *Izv. Akad. Nauk SSSR, Ser. Khim.*, 1966, 90 (*synth*)
Zolotarev, B.M. *et al*, *Izv. Akad. Nauk SSSR, Ser. Khim.*, 1979, 2226; *Bull. Acad. Sci. USSR, Div. Chem. Sci.*, 2048 (*ms*)

$C_{13}H_{30}BN$ B-00882

Dibutyl[3-(ethylamino)propyl-*C,N*]boron, 10CI

Dibutyl[3-(ethylamino)propyl]borane

[72909-88-7]

$(H_3CCH_2CH_2CH_2)_2BCH_2CH_2CH_2NHEt$

M 211.197

Exists as intramol. coord. form.

Zolotarev, B.M., *Izv. Akad. Nauk SSSR, Ser. Khim.*, 1979, 2226; *Bull. Acad. Sci. USSR, Chem. Div. Sci.*, 2048 (*ms*)

$C_{14}H_8BO_6^{\ominus}$ B-00883

Bis[2-hydroxybenzoato(2−)-O^1,O^2]borate(1−)

Bis(salicylato)borate. Borodisalicylate

[38403-08-6]

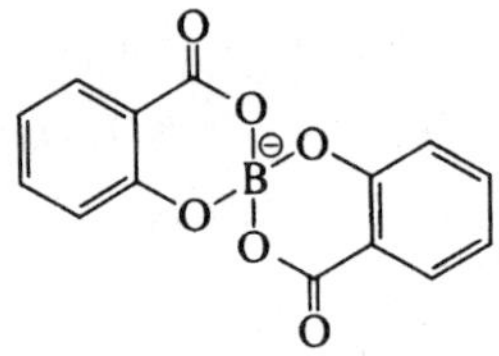

M 283.024 (ion)

Anion formed in solns. of salicylic acid in presence of trihydroxyborane (boric acid). Known with many different counterions.

Na salt: [22364-90-5].

$C_{14}H_8BNaO_6$ M 306.013

Used in manuf. of org. solvent-soluble dye salts. Opaque cryst.

K salt: [72556-01-5].

$C_{14}H_8BKO_6$ M 322.122

Rectangular platelets or microscopic leaflets.

Zn salt, decahydrate: [72556-00-4].

$C_{28}H_{36}B_2O_{22}Zn$ M 811.579

Solid.

NH_4 salt: [19443-40-4].

$C_{14}H_{12}BNO_6$ M 301.062

Amorph. white solid (EtOH). Mp 283-7°.

Meulenhoff, J., *Recl. Trav. Chim. Pays-Bas*, 1925, **44**, 161.
Meulenhoff, J., *Z. Anorg. Allg. Chem.*, 1925, **142**, 373.
Larrsson, R. *et al*, *Acta Chem. Scand.*, 1972, **26**, 1503 (*ir*)
Bassett, J. *et al*, *J. Inorg. Nucl. Chem.*, 1978, **40**, 987 (*synth, props*)
Kostromina, N.A. *et al*, *Koord. Khim.*, 1980, **6**, 501; *CA*, **93**, 87473y (*pmr*)
Mardanenko, V.K. *et al*, *Koord. Khim.*, 1980, **6**, 1193; *CA*, **93**, 178691j (*synth, ir, props*)
Mardanenko, V.K. *et al*, *CA*, 1980, **93**, 214717h.

$C_{14}H_{10}BN_2O_6^{\ominus}$ B-00884

Bis[4-amino-2-hydroxybenzoato(2−)-O^1,O^2]borate(1−), 10CI

*Bis(*p*-aminosalicylato)borate*

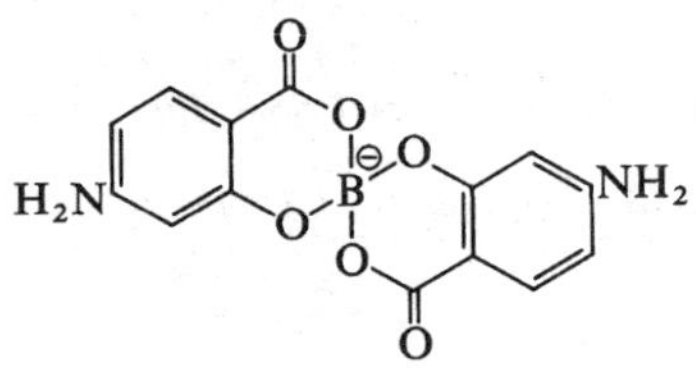

M 313.053 (ion)

H deriv.: [64012-96-0].
$C_{14}H_{11}BN_2O_6$ M 314.061
Colourless prismatic cryst. V. sol. $PhNH_2$, DMF, DMSO, H_2O, pract. insol. toluene, C_6H_6, AcOH. Mp 283°.
Na salt: [67481-59-8].
$C_{14}H_{10}BN_2NaO_6$ M 336.043
Rb salt: [67481-61-2].
$C_{14}H_{10}BN_2O_6Rb$ M 398.521
Triclinic cryst. + $1H_2O$.

Svarcs, E. *et al*, *CA*, 1977, **87**, 110732u.
Mardananeko, V.K., *CA*, 1978, **88**, 142344f; **89**, 122223h; 1979, **90**, 80145x; 1981, **94**, 21248t (*synth, ir, props*)
Kostromina, N.A. *et al*, *Koord. Khim.*, 1980, **6**, 501 (*pmr*)
Zviedre, I. *et al*, *CA*, 1980, **93**, 177693z, 105188r (*cryst struct*)

$C_{14}H_{11}BO_2$ B-00885

Dihydroxy(9-phenanthrenyl)borane
9-Phenanthrenylboronic acid, 10CI, 9CI. *9-(Dihydroxyboryl)phenanthrene*
[68572-87-2]

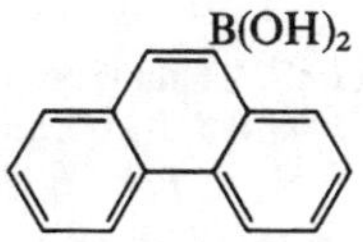

M 222.050
Selective fluorescent reagent for hplc of bifunctional compds.

Poole, C.F., *J. High Resolut. Chromatogr. Chromatogr. Commun.*, 1978, **1**, 96 (*use*)

$C_{14}H_{15}B$ B-00886

Ethyldiphenylborane, 10CI, 9CI
Diphenylethylborane
[40218-53-9]

Ph_2BEt

M 194.083
Prepd. from diphenylborane (itself prepd. *in situ*) and ethylene. Synthetic intermed. Sl. yellow liq., sensitive to air. Sol. THF, pentane, CCl_4. $Bp_{0.08}$ 85-7°.
Py complex: Ethyldiphenylborane pyridine.
$C_{19}H_{20}BN$ M 273.184
Mp 128°.

Soulié, J. *et al*, *Bull. Soc. Chim. Fr.*, 1966, 1981 (*deriv*)
Cabiddu, S. *et al*, *Gazz. Chim. Ital.*, 1972, **102**, 555 (*synth*)
Jacob, P., *J. Organomet. Chem.*, 1978, **156**, 101 (*synth, pmr*)
Pelter, A. *et al*, *Comprehensive Organic Chemistry*, (Barton, *D.H.R. et al*, Ed.), Pergamon, Oxford, 1979, **3**, 695 (*rev*)

$C_{14}H_{15}BCoN_6^{\oplus}$ B-00887

(η^5-2,4-Cyclopentadien-1-yl)[hydrotris(1-pyrazoyl)borato]-cobalt(1+), 9CI

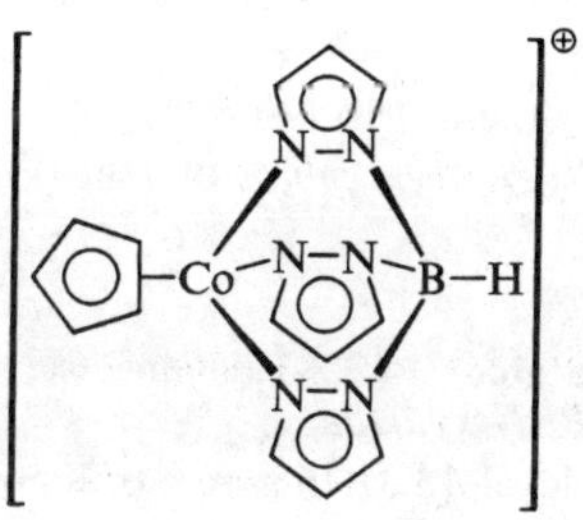

M 337.056 (ion)

Iodide: [51839-61-3].
$C_{14}H_{15}BCoIN_6$ M 463.960
No phys. props. reported.

O'Sullivan, D.J. *et al*, *J. Organomet. Chem.*, 1973, **57**, C58 (*synth*)

$C_{14}H_{15}BO$ B-00888

Ethoxydiphenylborane
Diphenylborinic acid ethyl ester, 10CI, 9CI. *Ethyl diphenylborinate*
[43185-52-0]

Ph_2BOEt

M 210.082
Fungicide, fuel additive; used in synth. of diphenylborane (prepd. *in situ*). Air- and moisture-sensitive liq. Sol. nonhydroxylic solvs. Bp 100°.

Hawthorne, M.F., *J. Am. Chem. Soc.*, 1958, **80**, 4293 (*synth*)
U.S.P., 3 027 396, (*1962*); *CA*, **57**, 11231i
Dorokhov, V.A. *et al*, *Izv. Akad. Nauk SSSR, Ser. Khim.*, 1977, 1587; *Bull. Acad. Sci. USSR, Div. Chem. Sci.*, **26**, 1458 (*props*)
Jacob, P., *J. Organomet. Chem.*, 1978, **156**, 101 (*use*)
Paetzold, P. *et al*, *Chem. Ber.*, 1979, **112**, 663.

$C_{14}H_{15}BO$ B-00889

Hydroxybis(4-methylphenyl)borane
Bis(4-methylphenyl)borinic acid, 10CI. *Di-*p*-tolylborinic acid*
[66117-64-4]

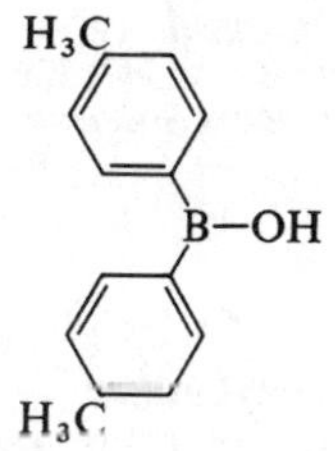

M 210.082
Solid, somewhat unstable to air, and readily dehydrated. Mp 105°.

Mikhailov, B.M. *et al*, *Izv. Akad. Nauk SSSR, Ser. Khim.*, 1958, 419 (*synth*)
Coates, G.E. *et al*, *J. Chem. Soc.*, 1961, 4909 (*synth*)
Thorpe, F.G. *et al*, *J. Organomet. Chem.*, 1978, **145**, C26.
Stegmann, H.B. *et al*, *Tetrahedron Lett.*, 1979, 3689 (*props*)

$C_{14}H_{16}B^{\ominus}$ B-00890

Dimethyldiphenylborate(1−), 10CI

$Me_2BPh_2^{\ominus}$

M 195.090 (ion)
Li salt: [68132-95-6].
$C_{14}H_{16}BLi$ M 202.031
Potential battery electrolyte.

Horowitz, H.H. *et al*, *Proc. Electrochem. Soc.*, 1981, **81-4**, 131 (*props*)
Klemann, L.P. *et al*, *Proc. Electrochem. Soc.*, 1981, **81-4**, 179 (*use*)

$C_{14}H_{16}BN$ B-00891

Dimethylaminodiphenylborane
N,N-*Dimethyl-1,1-diphenylboranamine*, 10CI, 9CI
[2237-45-8]

Ph_2BNMe_2

M 209.097

Colourless liq., stable in air. d_{20} 0.999. $Bp_{2.5}$ 121-2°, Bp_1 103-5°. n_D^{15} 1.5718.

Coates, G.E. *et al*, *J. Chem. Soc.*, 1961, 1000 (*synth, ir*)
Ruff, J.K., *J. Org. Chem.*, 1962, **27**, 1020 (*synth, ir, nmr*)
Becher, H.J. *et al*, *Adv. Chem. Ser.*, 1964, **42**, 71 (*ir*)
Becher, H.J. *et al*, *Chem. Ber.*, 1965, **98**, 2159 (*pmr*)
Wrackmeyer, B. *et al*, *Chem. Ber.*, 1976, **109**, 1075 (*nmr*)
Brown, C. *et al*, *J. Organomet. Chem.*, 1981, **217**, 139 (*synth, cmr*)

$C_{14}H_{16}BNO$ B-00892

(2-Aminoethoxy)diphenylborane

*(2-Aminoethanolato-*N,O*)diphenylboron*, 10CI. *Diphenylborinic acid 2-aminoethylester. Aminoethyl diphenylborinate. Hydroxydiphenylborane ethanolamine*

[15614-89-8]

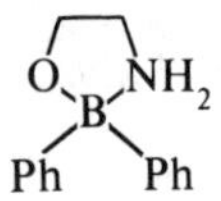

M 225.097

Intramolecularly coordinated cyclic molecule. Shows bactericidal, fungicidal and antiprotozoal props. Cryst. (EtOH aq.). Mp 190-1° (186-8°). Air-stable.

Letsinger, R.L. *et al*, *J. Am. Chem. Soc.*, 1955, **77**, 2491 (*synth*)
Mikhailov, B.M. *et al*, *Izv. Akad. Nauk SSSR, Ser. Khim.*, 1957, 989 (*synth*)
Bellamy, L.J. *et al*, *J. Chem. Soc.*, 1958, 2412 (*ir*)
Rettig, S.J. *et al*, *Can. J. Chem.*, 1976, **54**, 3130 (*cryst struct*)
Kliegel, W. *et al*, *Chem. Ber.*, 1977, **110**, 2067.
Hohaus, E. *et al*, *Int. J. Mass Spectrom. Ion Phys.*, 1979, **31**, 113 (*ms*)

$C_{14}H_{16}BNO_3$ B-00893

(2-Aminoethoxy)diphenoxyborane

Boric acid (H_3BO_3) 2-aminoethyl diphenyl ester, 10CI, 9CI

[51685-18-8]

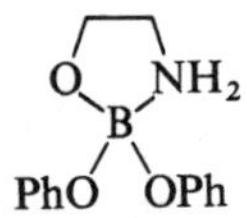

M 257.095

Intramolecularly coordinated. Nmr shift reagent for flavonoids; spray reagent for sugar detection after chromatog.

Niemann, G.J., *J. Chromatogr.*, 1979, **170**, 227 (*use*)
Homberg, H. *et al*, *Phytochemistry*, 1980, **19**, 2443 (*use*)

$C_{14}H_{18}BN_3$ B-00894

Diethyl(*N*-2-pyridinyl-2-pyridinaminato-*N*^{N},*N*′)boron, 10CI

[64800-08-4]

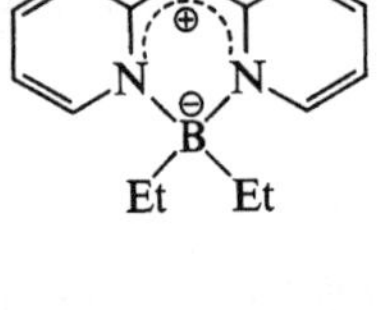

M 239.126

Yellow cryst., stable in air. Mp 77-9°. Bp_1 140-5°.

Dorokhov, V.A. *et al*, *Izv. Akad. Nauk SSSR, Ser. Khim.*, 1977, 1921; *Bull. Acad. Sci. USSR, Div. Chem. Sci.*, 1785 (*synth, nmr, ms, ir*)
Zolotarev, B.M. *et al*, *Izv. Akad. Nauk SSSR, Ser. Khim.*, 1978, 1312; *Bull. Acad. Sci. USSR, Div. Chem. Sci.*, 1140 (*ms*)
Lagutkin, N.A. *et al*, *Khim.-Farm. Zh.*, 1982, **16**, 695.

$C_{14}H_{18}B_2FeO$ B-00895

[(1,2,3,4,5,6-η)-2-Acetyl-1-methylboratabenzene][(1,2,3,4,5,6-η)-1-methylboratabenzene]iron, 9CI

[57692-67-8]

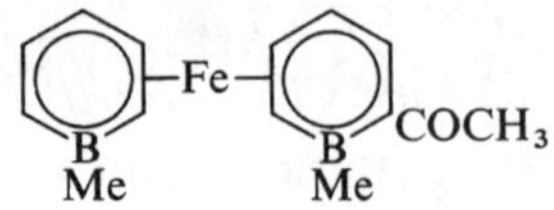

M 279.763

Ashe, A.J. *et al*, *J. Am. Chem. Soc.*, 1975, **97**, 6865 (*synth, pmr, ms, ir*)

$C_{14}H_{19}B$ B-00896

9-Phenyl-9-borabicyclo[3.3.1]nonane, 10CI, 9CI, 8CI

B-*Phenyl-9-borabicyclo[3.3.1]nonane.* B-*Phenyl-9-BBN*

[23418-91-9]

As 9-(1-Cyclopenten-1-yl)-9-borabicyclo[3.3.1]nonane, B-00854 with

R = Ph

M 198.114

Sensitive to air. $Bp_{0.4}$ 90° ($Bp_{0.3}$ 105°).

Piperidine complex: [65481-79-0].
$C_{19}H_{30}BN$ M 283.263

Brown, H.C. *et al*, *J. Am. Chem. Soc.*, 1969, **91**, 4304 (*synth*)
Kramer, G.W. *et al*, *J. Organomet. Chem.*, 1974, **73**, 1.
Negishi, E. *et al*, *J. Org. Chem.*, 1975, **40**, 814.
Ramsey, B.G. *et al*, *J. Organomet. Chem.*, 1977, **141**, 257 (*uv, nmr, struct, deriv*)
Whitely, C.G., *J. Chem. Soc., Chem. Commun.*, 1981, 5 (*synth*)
Whitely, C.G., *S. Afr. J. Chem.*, 1982, **35**, 9 (*synth*)

$C_{14}H_{19}BCl_2N_7Rh$ B-00897

Dichloro[hydrotris(1*H*-pyrazolato-*N*^{1})borato(1−)-*N*^{2},*N*$^{2'}$,*N*$^{2''}$](1-isocyanobutane)rhodium, 10CI

*(Butyl isonitrile)dichloro[hydrotris(1*H*-pyrazolato-*N^1*-borato]rhodium*

[73117-45-0]

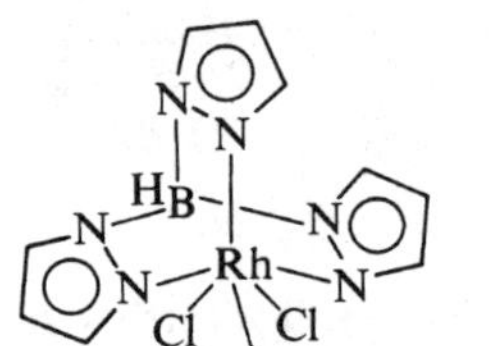

M 469.973

Bright yellow prisms. Mp 148-50°.

May, S. *et al*, *Inorg. Chem.*, 1980, **19**, 1582 (*synth, ir, pmr*)

$C_{14}H_{20}B^{\ominus}$ B-00898

9-Phenyl-9-boratabicyclo[3.3.1]nonane, 10CI

B-*Phenyl-9-BBN hydride*

As 9-Boratabicyclo[3.3.1]nonane, B-00546 with

R = Ph, R′ = H

M 199.122 (ion)

Li salt: [61722-00-7].
$C_{14}H_{20}BLi$ M 206.063
Air- and moisture-sensitive.

Hubbard, J.L. *et al*, *J. Organomet. Chem.*, 1978, **156**, 81 (*nmr*)
Brown, H.C. *et al*, *J. Organomet. Chem.*, 1980, **188**, 1 (*synth, nmr*)

$C_{14}H_{20}BCl$ B-00899

Chloro(2-phenylethenyl)(1,1,2-trimethylpropyl)borane, 10CI
Chloro(2,3-dimethyl-2-butyl)(2-phenylethenyl)borane. Chloro(2-phenylethenyl)thexylborane. Thexyl(2-phenylethenyl)chloroborane. Chlorostyrylthexylborane

$PhCH{=}CHBClC(CH_3)_2CH(CH_3)_2$

M 234.575

(*E*)-*form* [76173-97-2]
Formed *in situ* from thexylchloroborane and PhC≡CH. Synthetic intermed. Air- and moisture-sensitive. Sol. non-hydroxylic solvs. Wrongly reported in *CA* as the *Z-form*.

Zweifel, G. *et al*, *J. Org. Chem.*, 1981, **46**, 829 (*synth, use*)

$C_{14}H_{20}B_2CoF_4N_5O_4$ B-00900

[Bis[μ-[(2,3-butanedione dioximato)(2−)-*O:O'*]]tetrafluorodiborato(2−)-*N,N',N'',N'''*]methyl(pyridine)cobalt, 9CI
[Bis[dimethylglyoximato(2−)]tetrafluorodiborato]-methyl(pyridine)cobalt
[28627-87-4]
As Aqua[bis[μ-[(2,3-butanedione dioximato)(2−)-*O:O'*-]]][tetrafluorodiborato(2−)-*N,N',N'',N'''*]methylcobalt, B-00609 with

X = Py

M 478.890
Vitamin B_{12} analogue. Yellow cryst. (Me_2CO). Light-sensitive in soln.
▷Explodes at 245°

Schrauzer, G.N. *et al*, *J. Am. Chem. Soc.*, 1966, **88**, 3738 (*synth, props*)
Giannotti, C. *et al*, *J. Organomet. Chem.*, 1975, **91**, 357 (*props*)

$C_{14}H_{21}BN^{\ominus}$ B-00901

Triethyl-1*H*-indol-1-ylborate(1−), 10CI

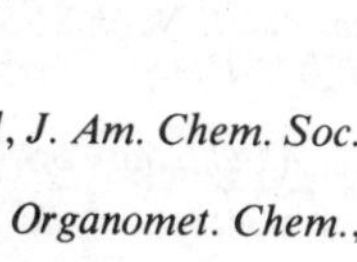

M 214.137 (ion)

Li salt: [69289-42-5].
$C_{14}H_{21}BLiN$ M 221.078
Shows electrical resistivity; used in battery electrolytes.

U.S.P., 4 117 213, (*1978*); *CA*, **90**, 124645s (*props, use*)
U.S.P., 4 104 450, (*1978*); *CA*, **90**, 124646t (*use*)
Ger. Pat., 2 833 943, (*1979*); *CA*, **91**, 57173s (*synth, props*)

$C_{14}H_{21}BO_2$ B-00902

7-Methylene-3-borabicyclo[3.3.1]nonyl acetylacetonate
[(5-Methylene-1,3-cyclohexanediyl)bis[methylene]]-(2,4-pentanedionato-O,O')boron, 10CI
[76069-96-0]

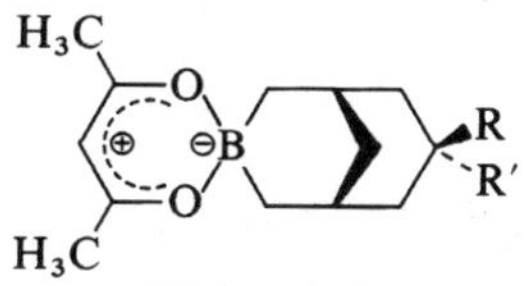

RR' = CH_2

M 232.129
Bright-yellow cryst. (hexane). Mp 54-60°.

Gurskii, M.E. *et al*, *J. Organomet. Chem.*, 1980, **199**, 171 (*cmr, synth*)
Gurskii, M.E. *et al*, *Izv. Akad. Nauk SSSR, Ser. Khim.*, 1981, 341; *Bull. Acad. Sci. USSR, Div. Chem. Sci.*, 264 (*cmr, struct*)
Gurskii, M.E. *et al*, *Izv. Akad. Nauk SSSR, Ser. Khim.*, 1981, 394; *Bull. Acad. Sci. USSR, Div. Chem. Sci.*, 312 (*synth, ir, pmr*)

$C_{14}H_{22}BN$ B-00903

9-Borabicyclo[3.3.1]nonane 2-methylpyridine
1,5-Cyclooctanediylhydro(2-methylpyridine)boron, 10CI, 9CI. *2-Methylpyridine, compd. with 9-borabicyclo[3.3.1]-nonane(1:1). 9-BBN α-picoline complex*
[70338-10-2]
As 9-Borabicyclo[3.3.1]nonane dimethyl sulfide, B-00705 with

L = 2-methylpyridine

M 215.145
ca. 91% associated as a 0.13 molar soln. in cyclohexane.

Brown, H.C. *et al*, *Recl. Trav. Chim. Pays-Bas*, 1979, **98**, 117 (*synth*)
Brown, H.C. *et al*, *J. Org. Chem.*, 1980, **45**, 1748 (*cmr, props*)

$C_{14}H_{22}BN$ B-00904

9-Borabicyclo[3.3.1]nonane 4-methylpyridine
1,5-Cyclooctanediylhydro(4-methylpyridine)boron, 10CI, 9CI. *4-Methylpyridine, compd. with 9-borabicyclo[3.3.1]-nonane(1:1). 9-BBN γ-picoline complex*
[64045-94-9]
As 9-Borabicyclo[3.3.1]nonane dimethyl sulfide, B-00705 with

L = 4-methylpyridine

M 215.145
Shows selective reducing props. Cryst. ($CHCl_3$/pentane); stable for several hr. in air. Mp 116-7°.

Brown, H.C. *et al*, *Inorg. Chem.*, 1977, **16**, 3090 (*nmr, props*)

$C_{14}H_{23}B$ B-00905

Bis(bicyclo[2.2.1]hept-2-yl)borane, *10CI*
Di-2-norbornylborane, 8CI

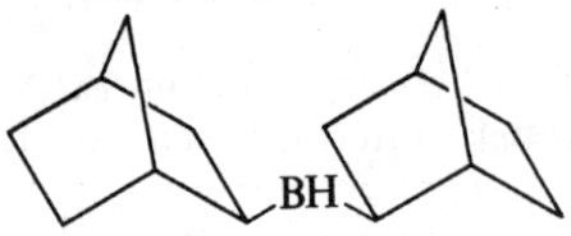

M 202.146
Uncomplexed material would exist as the H-bridged dimer.

exo,exo-form [34239-26-4]
Intermed. in hydroboration of norbornene with BH_3. Air- and moisture-sensitive. Sol. THF.

Brown, H.C. *et al*, *J. Organomet. Chem.*, 1971, **32**, C1 (*synth*)
Taniguchi, H., *Bull. Chem. Soc. Jpn.*, 1979, **52**, 2946.
Taniguchi, H., *Nippon Kagaku Kaishi*, 1979, **10**, 906; *CA*, **91**, 210504x.

$C_{14}H_{23}B$ B-00906

9-(1-Cyclohexen-1-yl)-9-borabicyclo[3.3.1]nonane, *10CI*
[74142-79-3]

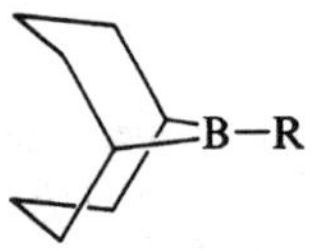

R = 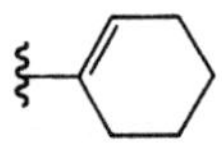

M 202.146
Synthetic intermed. Sensitive to air.

Miyaura, N. *et al*, *Tetrahedron Lett.*, 1980, **21**, 537.

$C_{14}H_{23}B$ B-00907

9-(3,3-Dimethyl-1-butynyl)-9-borabicyclo[3.3.1]nonane, *10CI*
9-(tert-Butylethynyl)-9-BBN
[62276-26-0]
As 9-(1-Cyclohexen-1-yl)-9-borabicyclo[3.3.1]nonane, B-00906 with

$$R = —C{\equiv}CC(CH_3)_3$$

M 202.146
Synthetic intermed. Sensitive to air.
THF complex: [63220-87-1].
$C_{18}H_{31}BO$ M 274.252
Cryst. (pentane), air-sensitive but stable indefinitely under N_2.

Molander, G.A. *et al*, *J. Org. Chem.*, 1977, **42**, 3106.
Brown, H.C. *et al*, *J. Organomet. Chem.*, 1977, **131**, 163 (*synth, nmr*)
Sinclair, J.A. *et al*, *J. Am. Chem. Soc.*, 1977, **99**, 954 (*synth*)
Molander, G.A. *et al*, *Synthesis*, 1979, 104.

$C_{14}H_{23}B$ B-00908

9-(1-Hexynyl)-9-borabicyclo[3.3.1]nonane, *10CI*
B-(1-Hexynyl)-9-BBN
[62459-81-8]
As 9-(1-Cyclohexen-1-yl)-9-borabicyclo[3.3.1]nonane, B-00906 with

$$R = —C{\equiv}C(CH_2)_3CH_3$$

M 202.146
Synthetic intermed. Sensitive to air.
THF complex: [63220-86-0].
$C_{18}H_{31}BO$ M 274.252
Cryst. (pentane), air-sensitive but stable indefinitely under N_2.

Brown, H.C. *et al*, *J. Organomet. Chem.*, 1977, **131**, 163 (*synth, nmr*)
Molander, G.A. *et al*, *J. Org. Chem.*, 1977, **42**, 3106 (*synth*)
Sinclair, J.A. *et al*, *J. Am. Chem. Soc.*, 1977, **99**, 954 (*synth*)
Molander, G.A. *et al*, *Synthesis*, 1979, 104.
Whiteley, C.G. *et al*, *S. Afr. J. Chem.*, 1982, **35**, 9 (*synth*)

$C_{14}H_{23}BClN$ B-00909

Chloro(dibutylamino)phenylborane, *8CI*
N,N-Dibutyl-1-chloro-1-phenylboranamine, 10CI
[33441-67-7]

$$PhBClN(CH_2CH_2CH_2CH_3)_2$$

M 251.605
Air- and moisture-sensitive liq. $Bp_{0.1}$ 75°.

Imberry, D. *et al*, *Org. Magn. Reson.*, 1970, **2**, 271 (*nmr*)
Sheikh, S.U., *Bull. Chem. Soc. Jpn.*, 1976, **49**, 2879.
Cragg, R.H. *et al*, *J. Organomet. Chem.*, 1981, **217**, 1.
Brown, C. *et al*, *J. Organomet. Chem.*, 1981, **220**, C25 (*cmr*)
Cragg, R.H. *et al*, *J. Organomet. Chem.*, 1982, **232**, 201 (*pmr, nmr, synth*)

$C_{14}H_{23}BO_2$ B-00910

7-Methyl-3-borabicyclo[3.3.1]nonyl acetylacetonate
[(5-Methyl-1,3-cyclohexanediyl)bis[methylene]](2,4-pentanedionato-O,O′)boron, 10CI
[76069-98-2]
As 7-Methylene-3-borabicyclo[3.3.1]nonyl acetylacetonate, B-00902 with

$$R = H, R' = CH_3$$

M 234.145
Bright-yellow cryst. solid (hexane). Mp 104-8°.

Gurskii, M.E. *et al*, *J. Organomet. Chem.*, 1980, **199**, 171 (*cmr, synth*)
Gurskii, M.E. *et al*, *Izv. Akad. Nauk SSSR, Ser. Khim.*, 1981, 341; *Bull. Acad. Sci. USSR, Div. Chem. Sci.*, 264 (*cmr, struct*)
Gurskii, M.E. *et al*, *Izv. Akad. Nauk SSSR, Ser. Khim.*, 1981, 394; *Bull. Acad. Sci. USSR, Div. Chem. Sci.*, 312 (*synth, pmr, ir*)

$C_{14}H_{24}BNO_2$ B-00911

2-Adamantyl-6-aza-1,3-dioxa-2-boracyclooctane
[[2,2′-Iminobis(ethanolato)](2−)-N,O,O′]-tricyclo[3.3.1.1^{3,7}]dec-1-ylboron, 10CI. 1-Adamantylboric acid diethanolamine ester
[79217-35-9]

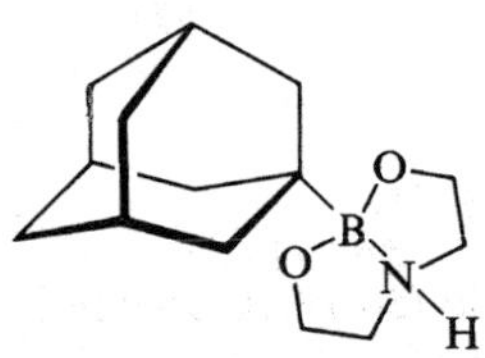

M 249.159
Fine, colourless cryst. Stable in air. Mp 288.5-290°.

Shchegoleva, T.A. *et al*, *Izv. Akad. Nauk SSSR, Ser. Khim.*, 1981, 1098; *Bull. Acad. Sci. USSR, Div. Chem. Sci.*, 858 (*synth*)

$C_{14}H_{24}B_2Br_2N_4$ B-00912

Bis[(4-bromo-1-pyrazolyl)diethylborane]

*Bis[μ-(4-bromo-1*H*-pyrazolato-N[1]:N[2])]tetraethyldiboron*, 10CI. *2,6-Dibromo-4,4,8,8-tetraethyl-4*H*,4*H*,8*H*,8*H*-7*a*,8*a*-diaza-3*a*,4*a*-diazonia-4,8-diborata-*s*-indacene*, 8CI

[14695-71-7]

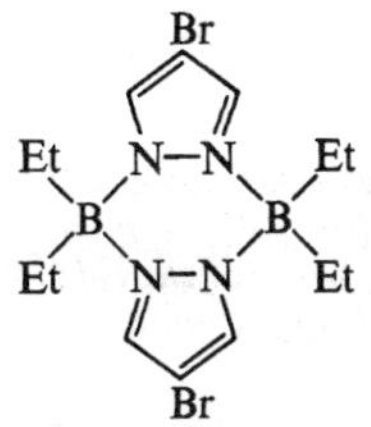

M 429.798

Cryst. (EtOH). Mp 127-8°.

Trofimenko, S., *J. Am. Chem. Soc.*, 1966, **88**, 1842 (*synth*)
Trofimenko, S., *J. Am. Chem. Soc.*, 1967, **89**, 3165 (*synth, pmr, nmr*)
Holt, E.M. *et al*, *Acta Crystallogr., Sect. B*, 1977, **33**, 1986 (*cryst struct*)
May, C.E. *et al*, *Z. Naturforsch., B*, 1978, **33**, 220 (*uv, nmr*)

$C_{14}H_{25}B$ B-00913

9-Cyclohexyl-9-borabicyclo[3.3.1]nonane, 10CI

B-*Cyclohexyl-9-BBN*

[53535-83-4]

As 9-(1-Cyclohexen-1-yl)-9-borabicyclo[3.3.1]nonane, B-00906 with

R = cyclohexyl

M 204.162

Useful reagent for transfer of a cyclohexyl group. Sensitive to air.

Prager, R.H. *et al*, *Aust. J. Chem.*, 1974, **27**, 1457.
Levy, A.B., *J. Org. Chem.*, 1978, **43**, 4684.
Midland, M.M. *et al*, *J. Organomet. Chem.*, 1978, **156**, 203.
Brown, H.C. *et al*, *J. Org. Chem.*, 1980, **45**, 4067.
Levy, A.B. *et al*, *Synthesis*, 1980, 945.
Slayden, S.W., *J. Org. Chem.*, 1981, **46**, 2311 (*synth*)

$C_{14}H_{25}B$ B-00914

9-(Cyclopentylmethyl)-9-borabicyclo[3.3.1]nonane, 10CI

B-*(Cyclopentylmethyl)-9-BBN*

[74763-89-6]

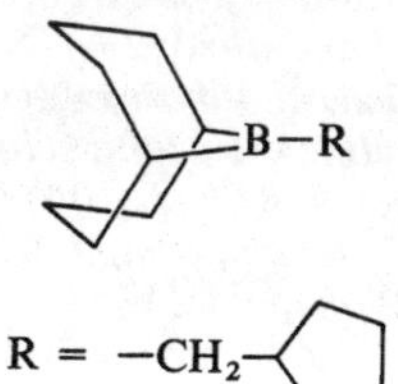

M 204.162

Sensitive to air.

Brown, H.C. *et al*, *J. Org. Chem.*, 1980, **45**, 4067 (*synth*)
Brown, H.C. *et al*, *J. Org. Chem.*, 1981, **46**, 647.

$C_{14}H_{25}B$ B-00915

9-(3,3-Dimethyl-1-butenyl)-9-borabicyclo[3.3.1]nonane, 10CI

*9-(*tert*-Butylethenyl)-9-BBN*

[79073-86-2]

As 9-(Cyclopentylmethyl)-9-borabicyclo[3.3.1]nonane, B-00914 with

$R = —CH{=}CHC(CH_3)_3$

M 204.162

Formed *in situ* by hydroboration of *tert*-butylethyne with 9-BBN. Synthetic intermed. Sensitive to air.

(E)-form [69322-47-0]

Brown, H.C. *et al*, *J. Am. Chem. Soc.*, 1979, **101**, 96 (*synth*)
Singaram, B. *et al*, *Heterocycles*, 1981, **15**, 231.
Brown, H.C. *et al*, *J. Org. Chem.*, 1981, **46**, 645.
Hartner, F.W. *et al*, *J. Am. Chem. Soc.*, 1981, **103**, 4979.

$C_{14}H_{25}B$ B-00916

9-(1-Ethyl-1-butenyl)-9-borabicyclo[3.3.1]nonane, 10CI

9-(3-Hexen-3-yl)-9-borabicyclo[3.3.1]nonane. B-*(3-Hexen-3-yl)-9-BBN*

As 9-(Cyclopentylmethyl)-9-borabicyclo[3.3.1]nonane, B-00914 with

$R = —C(CH_2CH_3){=}CHCH_2CH_3$

M 204.162

(E)-form [76036-41-4]

Air-sensitive.

(Z)-form [69322-51-6]

Prepd. by hydroboration of 3-hexyne with 9-BBN. Synthetic reagent, generally prepd. *in situ*. Air-sensitive.

Brown, H.C. *et al*, *J. Am. Chem. Soc.*, 1979, **101**, 96.
Brown, H.C. *et al*, *J. Org. Chem.*, 1981, **46**, 645.

$C_{14}H_{25}B$ B-00917

9-(1-Hexenyl)-9-borabicyclo[3.3.1]nonane, 10CI

B-*(1-Hexenyl)-9-BBN*

[74142-78-2]

As 9-(Cyclopentylmethyl)-9-borabicyclo[3.3.1]nonane, B-00914 with

$R = —CH{=}CH(CH_2)_3CH_3$

M 204.162

(E)-form [69322-45-8]

Synthetic reagent, generally prepd. *in situ* by hydroboration of 1-hexyne with 9-BBN. Air-sensitive. $Bp_{0.01}$ 52-5°.

(Z)-form [67826-85-1]

Air-sensitive.

Campbell, J.B. *et al*, *J. Organomet. Chem.*, 1978, **156**, 71; 1980, **45**, 549.
Brown, H.C. *et al*, *J. Am. Chem. Soc.*, 1979, **101**, 96.
Yatagai, H. *et al*, *J. Org. Chem.*, 1979, **44**, 2566 (*pmr*)
Miyaura, N. *et al*, *Tetrahedron Lett.*, 1980, **21**, 537.
Singaram, B. *et al*, *Heterocycles*, 1981, **15**, 231.

$C_{14}H_{25}B$ B-00918

9-(4-Hexenyl)-9-borabicyclo[3.3.1]nonane, 10CI

B-*(4-Hexenyl)-9-BBN*

[69362-75-0]

As 9-(Cyclopentylmethyl)-9-borabicyclo[3.3.1]nonane, B-00914 with

$R = —(CH_2)_3CH{=}CHCH_3$

M 204.162

Sensitive to air.

Hara, S. *et al*, *Tetrahedron Lett.*, 1978, 2891.

$C_{14}H_{25}B$ B-00919

9-(2-Methylcyclopentyl)-9-borabicyclo[3.3.1]nonane, 10CI

B-(2-Methylcyclopentyl)-9-BBN

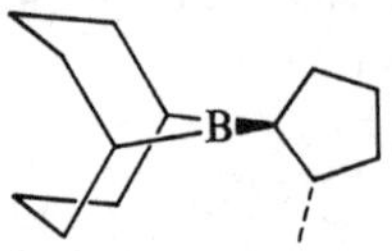

M 204.162

Reagent for transfer of a 2-methylcyclopentyl group.

trans-form [63942-79-0]

Generally prepd. *in situ* by hydroboration of 1-methylcyclopentene with 9-BBN. Synthetic intermed. Air-sensitive.

Midland, M.M. *et al*, *J. Organomet. Chem.*, 1977, **134**, C17; 1978, **156**, 203; 1979, **182**, C53.
Brown, H.C. *et al*, *Synthesis*, 1979, 701.
Brown, H.C. *et al*, *J. Org. Chem.*, 1980, **45**, 4067.
Midland, M.M. *et al*, *J. Am. Chem. Soc.*, 1982, **104**, 528 (*synth*)

$C_{14}H_{25}B_3Fe$ B-00920

[(1,2,3,4,5,6-η)-1,3,5-Cyclooctatriene][(1,2,3,4,5-η)-pentahydro-1,2-diethyl-1,2-dicarbapentaborato(2−)]iron

[84582-97-8]

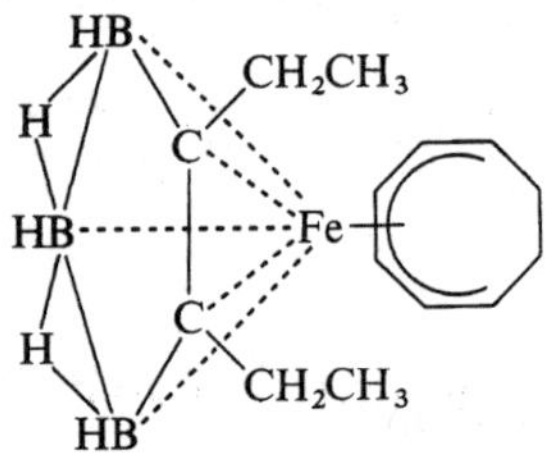

M 281.629

Red oil.

Swisher, R. *et al*, *Organometallics*, 1983, **2**, 506 (*synth, pmr, cmr, nmr, ms*)

$C_{14}H_{27}B$ B-00921

9-(3-Hexyl)-9-borabicyclo[3.3.1]nonane

9-(1-Ethylbutyl)-9-borabicyclo[3.3.1]nonane, 10CI. *B-(3-Hexyl)-9-BBN*

[78964-99-5]

As 9-(Cyclopentylmethyl)-9-borabicyclo[3.3.1]nonane, B-00914 with

$$R = —CH(CH_2CH_3)CH_2CH_2CH_3$$

M 206.177

Prepd. *in situ* by hydroboration of 3-hexene with 9-BBN. Synthetic reagent for transfer of a 3-hexyl group. Sensitive to air.

Brown, H.C. *et al*, *J. Org. Chem.*, 1981, **46**, 4313.

$C_{14}H_{27}B$ B-00922

9-Hexyl-9-borabicyclo[3.3.1]nonane, 10CI, 9CI

B-Hexyl-9-BBN

[42371-64-2]

As 9-(Cyclopentylmethyl)-9-borabicyclo[3.3.1]nonane, B-00914 with

$$R = —(CH_2)_5CH_3$$

M 206.177

Reagent for transfer of a hexyl group. Sensitive to air.

Naruse, M. *et al*, *Tetrahedron*, 1974, **30**, 2159.
Brown, H.C. *et al*, *Synthesis*, 1979, 701.
Ncube, S. *et al*, *Tetrahedron Lett.*, 1979, 1893.
Marinelli, E.R. *et al*, *Tetrahedron Lett.*, 1979, 2313.
Bestmann, H.J. *et al*, *Angew. Chem., Int. Ed. Engl.*, 1981, **20**, 1038 (*synth*)

$C_{14}H_{27}B$ B-00923

***B*-Thexyl-9-borabicyclo[3.3.1]nonane**

9-(1,1,2-Trimethylpropyl)-9-borabicyclo[3.3.1]nonane, 10CI, 9CI. *9-(2,3-Dimethyl-2-butyl)-9-borabicyclo[3.3.1]-nonane. B-Thexyl-9-BBN*

[17304-18-6]

As 9-(Cyclopentylmethyl)-9-borabicyclo[3.3.1]nonane, B-00914 with

$$R = —C(CH_3)_2CH(CH_3)_2$$

M 206.177

Liq., sensitive to air. $Bp_{0.4}$ 93-4°.

Brown, H.C. *et al*, *J. Am. Chem. Soc.*, 1972, **94**, 3567 (*glc*)
Midland, M.M. *et al*, *J. Organomet. Chem.*, 1977, **134**, C17; 1978, **156**, 203.
Brown, C.A. *et al*, *J. Am. Chem. Soc.*, 1979, **101**, 3964.
Brown, H.C. *et al*, *J. Organomet. Chem.*, 1980, **188**, 1 (*nmr*)

$C_{14}H_{27}BN_2$ B-00924

Bis(dimethylamino)(1,2,3,4,5-pentamethyl-2,4-cyclopentadienyl)borane

N,N,N′,N′-Tetramethyl-1-(1,2,3,4,5-pentamethyl-2,4-cyclopentadien-1-yl)boranediamine, 10CI

[60828-64-0]

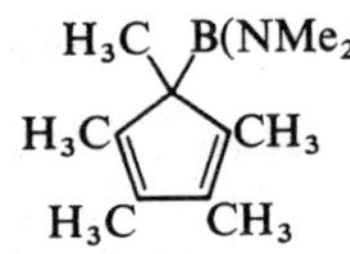

M 234.191

Static struct. at 40°, unlike many cyclopentadienylboron compds. Colourless viscous liq., sensitive to air and moisture. Sol. non-hydroxylic solvs. $Bp_{0.05}$ 74°.

Jutzi, P. *et al*, *Angew. Chem., Int. Ed. Engl.*, 1977, **16**, 41 (*synth, pmr, ms*)
Jutzi, P. *et al*, *Chem. Ber.*, 1979, **112**, 2481 (*synth, struct, nmr*)

$C_{14}H_{27}BN_2$ B-00925

9-Borabicyclo[3.3.1]nonane 1,4-diazabicyclo[2.2.2]octane

1,5-Cyclooctanediyl(1,4-diazabicyclo[2.2.2]octane-N¹)hydroboron, 10CI. *9-BBN-DABCO. 1,4-Diazabicyclo[2.2.2]octane, compd. with 9-borabicyclo[3.3.1]nonane(1:1)*

[70338-09-9]

As 9-Borabicyclo[3.3.1]nonane dimethyl sulfide, B-00705 with

L = 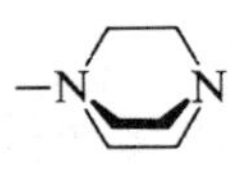

M 234.191

Brown, H.C. *et al*, *Recl. Trav. Chim. Pays-Bas*, 1979, **98**, 117.

$C_{14}H_{27}BO$ **B-00926**

Dibutyl(1-cyclohexenyloxy)borane

Dibutylborinic acid 1-cyclohexen-1-yl ester, 10CI, 8CI. 1-Cyclohexenyl dibutylborinate

[17814-54-9]

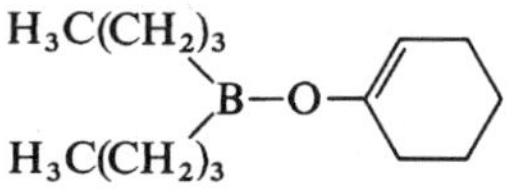

M 222.177

Air- and moisture-sensitive liq. d^{20} 0.862. Bp_1 73-4°. n_D^{20} 1.4558.

Ponomarev, S.V. *et al, Zh. Obshch. Khim.*, 1967, **37**, 2204 (*synth, ir, nmr*)
Evans, D.A. *et al, J. Am. Chem. Soc.*, 1979, **101**, 6120 (*props*)
Evans, D.A. *et al, J. Am. Chem. Soc.*, 1981, **103**, 3099 (*props*)

$C_{14}H_{28}B^{\ominus}$ **B-00927**

Tributylethynylborate(1−), 10CI, 9CI

$(H_3CCH_2CH_2CH_2)_3BC{\equiv}CH^{\ominus}$

M 207.185 (ion)

Li salt: [57088-67-2].
$C_{14}H_{28}BLi$ M 214.126
Synth. intermed. Air- and acid-sensitive. Sol. non-hydroxylic solvs.

Brown, H.C. *et al, J. Am. Chem. Soc.*, 1975, **97**, 5017 (*use*)
Midland, M.M. *et al, J. Org. Chem.*, 1975, **40**, 2845 (*use*)
Utimoto, K. *et al, Tetrahedron Lett.*, 1976, 3969 (*use*)

$C_{14}H_{28}B^{\ominus}$ **B-00928**

9-(1,1,2-Trimethylpropyl)-9-boratabicyclo[3.3.1]nonane, 10CI

B-*Thexyl-9-BBN hydride*

As 9-Boratabicyclo[3.3.1]nonane, B-00546 with

$R = -C(CH_3)_2CH(CH_3)_2$, $R' = H$

M 207.185 (ion)

Li salt: [74365-24-5].
$C_{14}H_{28}BLi$ M 214.126
Air- and moisture-sensitive.

Brown, H.C. *et al, J. Organomet. Chem.*, 1980, **188**, 1 (*synth, nmr*)

$C_{14}H_{29}B_9Fe$ **B-00929**

[(1,2,3,4,5,6-η)-Hexamethylbenzene][(7,8,9,10,11-η)-undecahydro-7,8-dicarbaundecaborato(2−)]iron

3,1,2-(η^6-Hexamethylbenzene)-$FeC_2B_9H_{11}$

[82704-19-6]

As [(1,2,3,4,5,6-η)-1,3,5-Trimethylbenzene][(7,8,9,10,11-η)-undecahydro-7,8-dicarbaundecaborato(2−)]iron, B-00746 with

$R = CH_3$

M 350.520

Reddish-orange cryst. (C_6H_6/heptane). Mp 299-302° dec.

Hanusa, T.P. *et al, Polyhedron*, 1981, **1**, 77 (*synth, pmr, ir, nmr*)

$C_{14}H_{30}BCl$ **B-00930**

Chlorooctyl(1,1,2-trimethylpropyl)borane, 10CI

Chloro(2,3-dimethyl-2-butyl)octylborane. Thexyloctylchloroborane. Chlorooctylthexylborane

[75052-81-2]

$(H_3C)_2CHC(CH_3)_2BCl(CH_2)_7CH_3$

M 244.654

Prepd. *in situ* by reaction of thexylchloroborane with 1-octene. Synthetic intermed. Prob. oil, sensitive to air and moisture. Sol. non-hydroxylic solvs.

Zweifel, G. *et al, J. Am. Chem. Soc.*, 1980, **102**, 5919 (*synth, use*)

$C_{14}H_{30}BN$ **B-00931**

9-Borabicyclo[3.3.1]nonane triethylamine

1,5-Cyclooctanediyl(N,N-*diethylethanamine)hydroboron, 10CI.* N,N-*Diethylethanamine, compd. with 9-borabicyclo[3.3.1]nonane(1:1). 9-BBN triethylamine complex*

[64045-92-7]

As 9-Borabicyclo[3.3.1]nonane dimethyl sulfide, B-00705 with

$L = Et_3N$

M 223.208

Partially dissociated in soln.

Brown, H.C. *et al, Inorg. Chem.*, 1977, **16**, 3090 (*nmr*)

$C_{14}H_{31}B$ **B-00933**

Dibutyl(1,1,2-trimethylpropyl)borane, 10CI, 9CI

Dibutyl(2,3-dimethyl-2-butyl)borane. Dibutylthexylborane. Thexyldibutylborane

[42928-38-1]

$(H_3C)_2CHC(CH_3)_2B(CH_2CH_2CH_2CH_3)_2$

M 210.209

Prepd. *in situ* by the hydroboration of 1-butene with thexylborane. Synthetic intermed. Prob. oil, sensitive to air. Sol. THF.

Brown, H.C. *et al, J. Org. Chem.*, 1973, **38**, 3968 (*synth, use*)
Miyaura, N. *et al, Tetrahedron Lett.*, 1974, 2961 (*synth*)
Negishi, E. *et al, Synthesis*, 1974, 77 (*synth*)
Pelter, A. *et al, Tetrahedron Lett.*, 1975, 1633 (*use*)
Slayden, S.W. *et al, J. Org. Chem.*, 1981, **46**, 2311 (*synth, use*)

$C_{14}H_{31}BN_2$ **B-00934**

9-Borabicyclo[3.3.1]nonane tetramethylethylenediamine

1,5-Cyclooctanediylhydro(N,N,N′,N′-*tetramethyl-1,2-ethanediamine-*N*)boron, 10CI.* N,N,N′,N′-*Tetramethyl-1,2-ethanediamine, compd. with 9-borabicyclo[3.3.1]nonane(1:1). 9-BBN TMEDA complex*

[64045-93-8]

As 9-Borabicyclo[3.3.1]nonane dimethyl sulfide, B-00705 with

$L = Me_2NCH_2CH_2NMe_2$

M 238.222

Mobile equilibrium between complexed and uncomplexed species in soln.

Brown, H.C. *et al*, *Inorg. Chem.*, 1977, **16**, 3090 (*synth, nmr*)

$C_{14}H_{31}BO$ B-00935

Butoxydipentylborane

Dipentylborinic acid butyl ester, 10CI, 8CI. *Dipentyl(butoxy)borane*

[3027-55-2]

$[H_3C(CH_2)_4]_2BOCH_2CH_2CH_2CH_3$

M 226.208

Air- and moisture-sensitive. Sol. nonhydroxylic solvs.

Laurent, J.-P., *Bull. Soc. Chim. Fr.*, 1963, 558 (*synth*)
Baudet, J. *et al*, *C.R. Hebd. Seances Acad. Sci.*, 1967, **264B**, 1090 (*props*)
Tsvetkov, V.G. *et al*, *Zh. Obshch. Khim.*, 1979, **49**, 2406 (*props*)
Alyasov, V.N. *et al*, *Zh. Obshch. Khim.*, 1980, **50**, 362.

$C_{14}H_{31}BO$ B-00936

Dibutyl(hexyloxy)borane

Dibutylborinic acid hexyl ester, 10CI

[67682-99-9]

$(H_3CCH_2CH_2CH_2)_2BO(CH_2)_5CH_3$

M 226.208

Air- and moisture-sensitive liq. Bp_3 88-90°. n_D^{20} 1.4281.

Mikhailov, B.M. *et al*, *Izv. Akad. Nauk SSSR, Ser. Khim.*, 1978, 1586; *Bull. Acad. Sci. USSR, Div. Chem. Sci.*, **27**, 1385 (*synth, nmr*)
Mikhailov, B.M. *et al*, *Izv. Akad. Nauk SSSR, Ser. Khim.*, 1978, 1594; *Bull. Acad. Sci. USSR, Div. Chem. Sci.*, **27**, 1392 (*props*)
Mikhailov, B.M. *et al*, *J. Organomet. Chem.*, 1978, **154**, 113 (*synth, nmr*)

$C_{14}H_{31}BO_2$ B-00937

Dibutoxyhexylborane

Hexylboronic acid dibutyl ester, 10CI. *1-Hexaneboronic acid dibutyl ester*, 8CI. *Dibutyl hexaneboronate*

[3089-96-1]

$H_3C(CH_2)_5B(OCH_2CH_2CH_2CH_3)_2$

M 242.208

Used to stabilize polyacrylonitrile. Air- and moisture-sensitive liq. Sol. non-hydroxylic solvs. d_{20} 0.828. Bp_{21} 145°, Bp_3 98-100°. n_D^{20} 1.4231.

Laurent, J.P., *C.R. Hebd. Seances Acad. Sci.*, 1961, **253**, 1812 (*synth, props*)
Gallais, F. *et al*, *C.R. Hebd. Seances Acad. Sci.*, 1964, **259**, 4262 (*props*)
Japan. Pat., 65 9 629, (*1965*); *CA*, **63**, 18376f
Baudet, J. *et al*, *C.R. Hebd. Seances Acad. Sci.*, 1967, **264**, 1090 (*props*)
Mikhailov, B.M. *et al*, *Izv. Akad. Nauk SSSR, Ser. Khim.*, 1978, 1586; *Bull. Acad. Sci. USSR, Div. Chem. Sci.*, **27**, 1385 (*synth, nmr*)
Mikhailov, B.M. *et al*, *J. Organomet. Chem.*, 1978, **154**, 113 (*synth, props, nmr*)

$C_{14}H_{31}BS$ B-00938

Dibutyl(hexylthio)borane

Dibutylthioborinic acid hexyl ester, 10CI

[67682-97-7]

$(H_3CCH_2CH_2CH_2)_2BS(CH_2)_5CH_3$

M 242.269

Air- and moisture-sensitive liq. Bp_2 127-8°. n_D^{20} 1.4618.

Mikhailov, B.M. *et al*, *Izv. Akad. Nauk SSSR, Ser. Khim.*, 1978, 1594; *Bull. Acad. Sci. USSR, Div. Chem. Sci.*, **27**, 1392 (*synth, props, nmr*)
Mikhailov, B.M. *et al*, *J. Organomet. Chem.*, 1978, **154**, 113 (*synth, nmr*)

$C_{14}H_{33}BN_2$ B-00939

2,4,4-Trimethyl-3-pentylborane tetramethylethylenediamine

[*2,2-Dimethyl-1-(1-methylethyl)propyl*]*dihydro*(N,N,N',N'-*tetramethyl-1,2-ethanediamine*-N)*boron*, 10CI. N,N,N',N'-*Tetramethyl-1,2-ethanediamine, compd. with* [*2,2-dimethyl-1-(1-methylethyl)propyl*]-*borane(1:1)*. *DIB-borane TMEDA*

[67826-88-4]

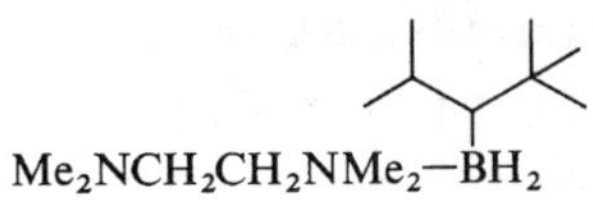

M 240.238

Convenient storage for the useful monoalkylborane, DIB-borane. Stable cryst. (pentane). Mp 92-5°. Dec. slowly in THF soln.

Singaram, B. *et al*, *J. Organomet. Chem.*, 1978, **156**, C1 (*synth*)
Brown, H.C. *et al*, *Inorg. Chem.*, 1979, **18**, 51 (*synth, pmr, nmr*)

$C_{14}H_{34}BN_3$ B-00940

Dimethylaminobis(diisopropylamino)borane

N,N-*Dimethyl*-N',N',N'',N''-*tetrakis(1-methylethyl)-boranetriamine*, 10CI. *Bis(diisopropylamino)-(dimethylamino)borane*

[61134-33-6]

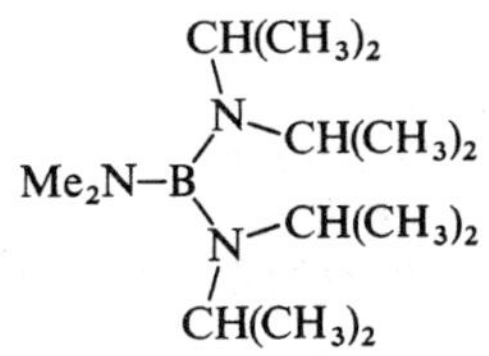

M 255.253

Characterised spectroscopically. Probably liq., sensitive to moisture.

Nöth, H. *et al*, *Chem. Ber.*, 1968, **101**, 969.
Nöth, H. *et al*, *J. Organomet. Chem.*, 1968, **12**, 23.
Curry, K.K. *et al*, *J. Am. Chem. Soc.*, 1976, **98**, 8262; 1978, **100**, 1442 (*nmr, props*)

$C_{14}H_{39}B_7P_2Ru$ B-00941

[η^6-Nonahydrodicarbanonaborato(2−)-*B,H*]bis(triethylphosphine)ruthenium, 10CI

hyper-closo-*6,6-Bis(triethylphosphine)-6,2,3-RuC$_2$B$_7$H$_9$*

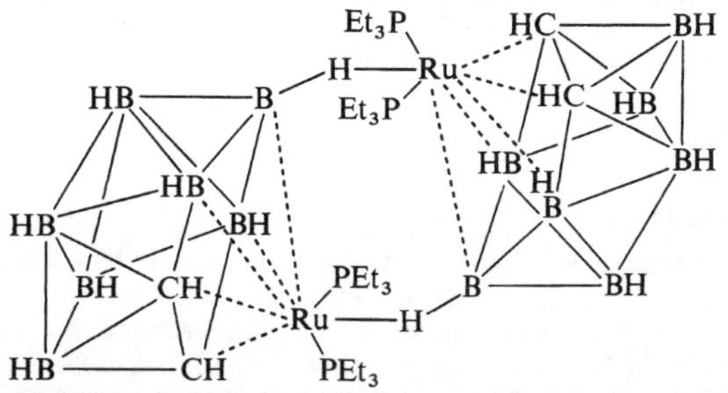

M 446.150

Dimeric.

Dimer: [76798-39-5]. *Bis[μ-[η^6-nonahydrodicarbanonaborato(2−)-B,H]]tetrakis(triethylphosphine)diruthenium,* 10CI.

$C_{28}H_{78}B_{14}P_4Ru_2$ M 892.299

Large blue-black or small bright-green cryst. (toluene/pentane).

Jung, C.W. *et al*, *J. Am. Chem. Soc.*, 1981, **103**, 810 (*synth, nmr, pmr, ir*)

$C_{15}H_{10}BF_6N$ B-00942

1,1-Diphenyl-*N*-[2,2,2-trifluoro-1-(trifluoromethyl)ethylidene]boranamine, 10CI, 9CI

Diphenyl[[trifluoro-1-(trifluoromethyl)ethylidene]amino]borane, 8CI

[29215-41-6]

$Ph_2B—N{=}C(CF_3)_2$

M 329.051

Cryst. (C_6H_6), moisture-sensitive. Bp_{22} 110°.

Niedenzu, K. *et al*, *Tetrahedron Lett.*, 1970, **28**, 2441 (*ms, ir, synth*)
Gragg, B.R. *et al*, *J. Organomet. Chem.*, 1977, **132**, 29 (*cmr*)
Emerick, D.P. *et al*, *J. Organomet. Chem.*, 1978, **153**, 9 (*ir, nmr*)

$C_{15}H_{11}BF_2O_2$ B-00943

(1,3-Diphenyl-1,3-propanedionato-*O,O'*)difluoroboron, 10CI, 9CI

Difluoroboryl 1,3-diphenyl-1,3-propanedionate. Difluoro(1,3-diphenyl-1,3-propanedionato)boron

[14947-61-6]

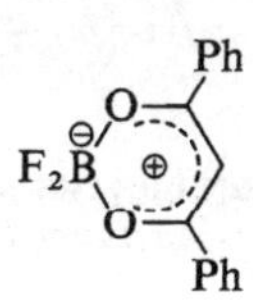

M 272.058

Light-yellow solid by subl. Mp 197° (193-4°)

▷ED7560000.

Toporcer, R.E. *et al*, *Inorg. Chem.*, 1965, **4**, 1649 (*synth, ir, nmr*)
Smith, J.A.S. *et al*, *J. Chem. Soc. (A)*, 1966, 1749 (*nmr*)
Dessy, R.E. *et al*, *J. Am. Chem. Soc.*, 1968, **90**, 2005.
Trestianu, A. *et al*, *Tetrahedron*, 1968, **24**, 2499 (*pmr*)
Brown, N.M.D., *J. Chem. Soc. (A)*, 1969, 526 (*synth, ir, uv, nmr*)
Neplyuev, V.M. *et al*, *Zh. Org. Khim.*, 1980, **16**, 2558 (*synth, spectra*)

$C_{15}H_{12}BMnO_3$ B-00944

Tricarbonyl[(1,2,3,4,5,6-η)-4-methyl-1-phenylboratabenzene]manganese, 10CI

[62980-54-5]

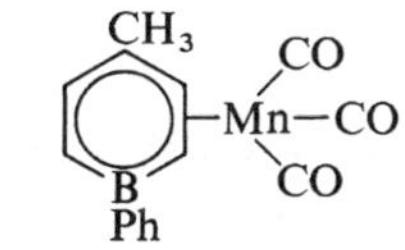

M 306.006

Yellow cryst. Mp 73-5°.

Herberich, G.E. *et al*, *Chem. Ber.*, 1977, **110**, 1167 (*synth, nmr*)

$C_{15}H_{13}BFeO_3$ B-00945

Tricarbonyl[(1,2,3,6,7-η)-4,5-dihydro-1-phenyl-1*H*-borepin]iron

[62479-26-9]

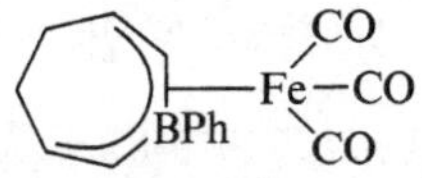

M 307.923

Bright-yellow air-sensitive cryst. (pentane). Mp 64.5-65.5°.

Herberich, G.E. *et al*, *Chem. Ber.*, 1977, **110**, 760 (*synth, pmr, nmr, ir, ms*)

$C_{15}H_{13}BFeO_3$ B-00946

Tricarbonyl[(1,2,3,4,5-η)-2-ethyl-1-phenyl-1*H*-borole]iron, 10CI

[60840-45-1]

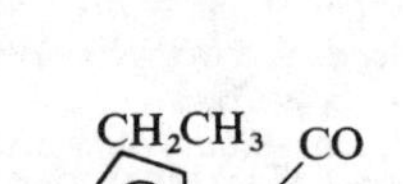

M 307.923

Air- and light-sensitive amber coloured oil. Mp >−30°.

Herberich, G.E. *et al*, *Angew. Chem., Int. Ed. Engl.*, 1977, **16**, 43 (*synth, ms, ir, pmr*)

$C_{15}H_{15}BFeO_3Si$ B-00947

Tricarbonyl[(2,3,4,5,6-η)-1,1-dimethyl-4-phenyl-1-sila-4-boracyclohexa-2,5-diene]iron, 10CI

[62479-25-8]

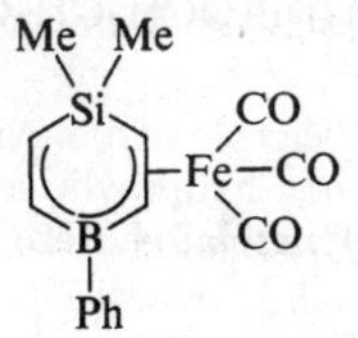

M 338.024

Pale-yellow cryst. (Et_2O). Mp 99°.

Herberich, G.E. *et al*, *Chem. Ber.*, 1977, **110**, 760 (*synth, struct, pmr, nmr, ir, ms*)

$C_{15}H_{16}BBrClN$ B-00948

Methyl(2-methyl-4-bromophenyl)aminochloro(2-methylphenyl)borane

N-*(4-Bromo-2-methylphenyl)-1-chloro-*N*-methyl-1-(2-methylphenyl)boranamine,* 10CI

[66338-32-7]

As [Methyl(4-bromophenyl)amino]chloro(phenyl)borane, B-00844 with

R = CH_3

M 336.465
Air- and moisture-sensitive.

Ziegler, M.L. *et al*, *Z. Naturforsch.*, *B*, 1978, **33**, 200 (*struct*)
Friebolin, H. *et al*, *Org. Magn. Reson.*, 1977, **10**, 157 (*synth, nmr*)

$C_{15}H_{16}BN_6Ru^{\oplus}$ B-00949

(η^6-Benzene)[hydrotris(1*H*-pyrazolato-*N'*)borato(1−)-N^2,$N^{2'}$,$N^{2''}$]ruthenium(1+), 9CI

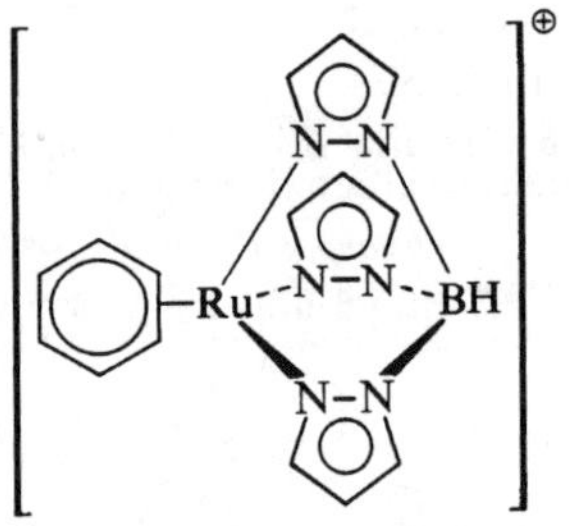

M 392.212 (ion)
Hexafluorophosphate: [52015-87-9].
$C_{15}H_{16}BF_6N_6PRu$ M 537.176
Yellow cryst.

O'Sullivan, D.J. *et al*, *J. Organomet. Chem.*, 1973, **57**, C58 (*synth*)

$C_{15}H_{17}BMoN_6O_2$ B-00950

Dicarbonyl[hydrotris(1*H*-pyrazolato-N^1)borato(1−)-N^2,N^2,N^2][(1,2,3-η-2-methyl-2-propenyl]molybdenum, 9CI
[38834-21-8]

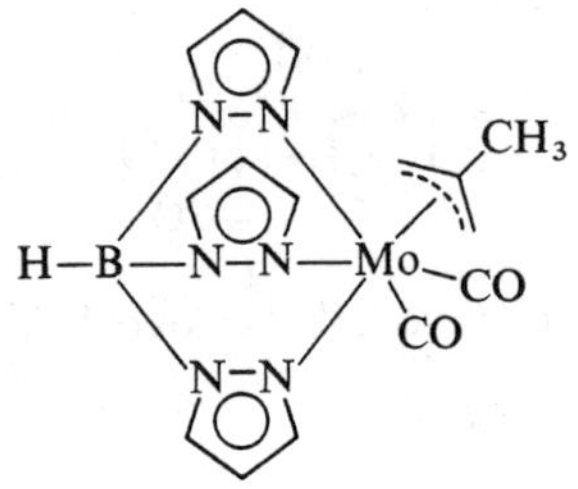

M 420.088
Yellow cryst.

Trofimenko, S. *et al*, *J. Am. Chem. Soc.*, 1969, **91**, 588, 3183 (*synth, nmr*)
Meekin, P. *et al*, *J. Am. Chem. Soc.*, 1972, **94**, 5677 (*nmr*)
Holt, E.M. *et al*, *J. Chem. Soc., Dalton Trans.*, 1973, 2444 (*cryst struct*)

$C_{15}H_{17}BN_6O_2W$ B-00951

Dicarbonyl[hydrotris(pyrazolato)borato(1−)](η^3-2-methylpropenyl)tungsten
Dicarbonyl[hydrotris(pyrazolato)borato(1−)](2-methyl-π-allyl)tungsten, 8CI
[12242-63-6]

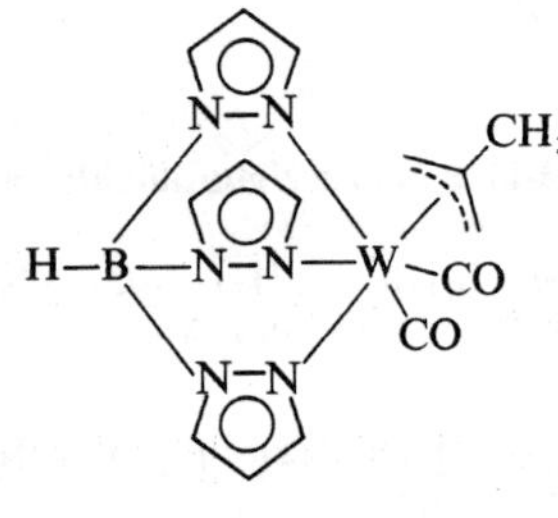

M 507.998
Yellow air-stable cryst.

Trofimenko, S., *J. Am. Chem. Soc.*, 1967, **89**, 3904; 1969, **91**, 558 (*synth, props*)

$C_{15}H_{18}BNO$ B-00952

(3-Aminopropoxy)diphenylborane
Diphenylborinic acid 3-aminopropyl ester, 10CI. B,B-*Diphenyltetrahydroboroxazine*
[1475-88-3]

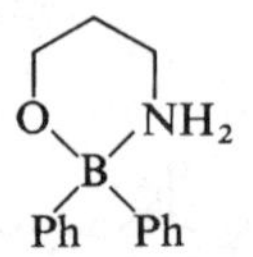

M 239.123
Exists in intramolecularly coordinated cyclic form, as shown. Solid, stable in air. Sol. EtOAc, Me_2CO, MeCN, insol. C_6H_6, Et_2O, H_2O. Mp 190-2°.

Weidmann, H. *et al*, *Justus Liebigs Ann. Chem.*, 1958, **619**, 28 (*synth, props*)
Zimmerman, H.K. *et al*, *Justus Liebigs Ann. Chem.*, 1962, **655**, 54 (*props*)
Zimmerman, H.K., *Adv. Chem. Ser.*, 1964, **42**, 23.
Hohaus, E. *et al*, *Int. J. Mass Spectrom. Ion Phys.*, 1979, **31**, 113 (*ms*)
Mazeika, I. *et al*, *Zh. Obshch. Khim.*, 1979, **49**, 2528 (*ms*)

$C_{15}H_{21}B$ B-00953

9-(4-Methylphenyl)-9-borabicyclo[3.3.1]nonane, 10CI, 9CI
9-p-Tolyl-9-borabicyclo[3.3.1]nonane, 8CI. B-p-*Tolyl-9-BBN*
[23418-90-8]

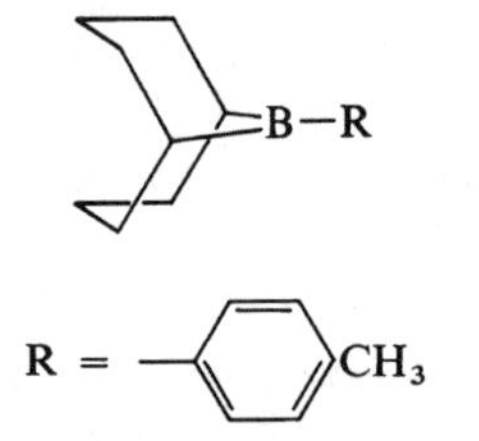

M 212.141
Colourless liq., sensitive to air. $Bp_{5.5}$ 155-60°, $Bp_{0.05}$ 72°.

Brown, H.C. *et al*, *J. Am. Chem. Soc.*, 1969, **91**, 4304 (*synth*)
Kramer, G.W. *et al*, *J. Organomet. Chem.*, 1974, **73**, 1.
Ramsey, B.G. *et al*, *J. Organomet. Chem.*, 1977, **141**, 257 (*nmr, uv, synth*)
Brown, H.C. *et al*, *J. Organomet. Chem.*, 1980, **188**, 1 (*nmr*)

$C_{15}H_{21}BO$ B-00954

3-Methoxy-7-phenyl-3-borabicyclo[3.3.1]nonane, 10CI
7-Phenyl-3-methoxy-3-borabicyclo[3.3.1]nonane. 7-Phenyl-3-MBBN
[33586-48-0]
As 3-Borabicyclo[3.3.1]nonane, B-00544 with

X = OMe, R = Ph

M 228.140
Exists in double chair form. Air- and moisture-sensitive. Sol. $CHCl_3$, CH_2Cl_2.

Mikhailov, B.M. *et al*, *Izv. Akad. Nauk SSSR, Ser. Khim.*, 1971, 1244 (*synth*)
Gurskii, M.E. *et al*, *J. Organomet. Chem.*, 1980, **199**, 171 (*nmr*)
Gurskii, M.E. *et al*, *Izv. Akad. Nauk SSSR, Ser. Khim.*, 1981, 341; *Bull. Acad. Sci. USSR, Div. Chem. Sci.*, 264 (*cmr*)

Mikhailov, B.M. *et al*, *Izv. Akad. Nauk SSSR, Ser. Khim.*, 1981, 394; *Bull. Acad. Sci. USSR, Div. Chem. Sci.*, 312 (*props*)

$C_{15}H_{22}BN_6^{\ominus}$ B-00955

Tris(3,5-dimethyl-1*H*-pyrazolato-*N'*)hydroborate(1–), 10CI, 9CI

Hydrotris(3,5-dimethyl-1-pyrazolyl)borate

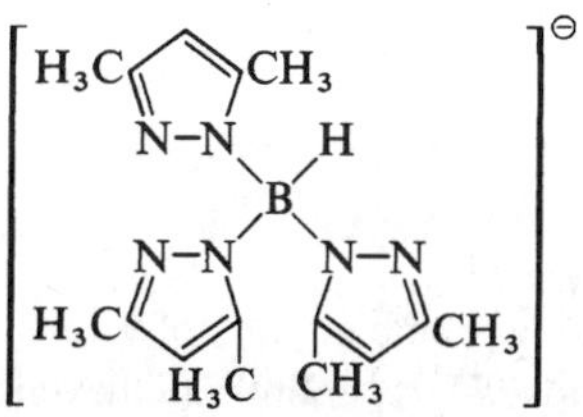

M 297.189 (ion)

Used in metal complex formn.

K salt: [17567-17-8].
$C_{15}H_{22}BKN_6$ M 336.287
Mp 298-300° dec. Sublimes at 280°/1 mm.

Li salt: [17857-04-4].
$C_{15}H_{22}BLiN_6$ M 304.130
Cryst. (xylene). Mp 304-5°. Sublimes *in vacuo*.

Trofimenko, S., *J. Am. Chem. Soc.*, 1967, **89**, 6288 (*synth*)
Trofimenko, S., *Chem. Rev.*, 1972, **72**, 497 (*rev*)

$C_{15}H_{24}BN$ B-00956

9-Borabicyclo[3.3.1]nonane 2,6-dimethylpyridine

1,5-Cyclooctanediyl(2,6-dimethylpyridine)hydroboron, 10CI. 2,6-Dimethylpyridine, compd. with 9-borabicyclo[3.3.1]nonane(1:1). 9-BBN 2,6-dimethylpyridine

[70338-07-7]

As 9-Borabicyclo[3.3.1]nonane dimethyl sulfide, B-00705 with

L = 2,6 dimethylpyridine

M 229.171

Complex shown to be completely dissociated into its components in cyclohexane soln.

Brown, H.C. *et al*, *Recl. Trav. Chim. Pays-Bas*, 1979, **98**, 117 (*synth, props*)

$C_{15}H_{24}BN$ B-00957

9-Borabicyclo[3.3.1]nonane 2-ethylpyridine

1,5-Cyclooctanediyl(2-ethylpyridine)hydroboron, 10CI, 9CI. 2-Ethylpyridine, compd. with 9-borabicyclo[3.3.1]-nonane(1:1). 9-BBN 2-ethylpyridine complex

[70338-11-3]

As 9-Borabicyclo[3.3.1]nonane dimethyl sulfide, B-00705 with

L = 2-Ethylpyridine

M 229.171

ca. 59% associated as a 0.13 molar soln. in cyclohexane.

Brown, H.C. *et al*, *Recl. Trav. Chim. Pays-Bas*, 1979, **98**, 117 (*props*)
Brown, H.C. *et al*, *J. Org. Chem.*, 1980, **45**, 1748 (*cmr, props*)

$C_{15}H_{25}B$ B-00958

9-Bicyclo[2.2.1]hept-2-yl-9-borabicyclo[3.3.1]nonane, 10CI

9-(2-Norbornyl)-9-borabicyclo[3.3.1]nonane

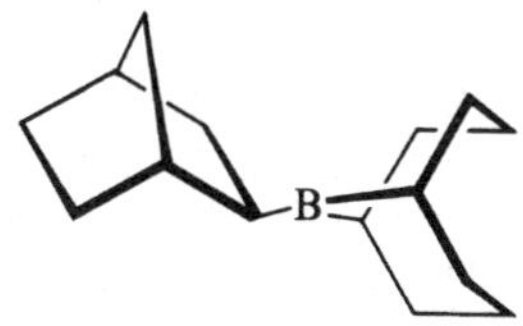

M 216.173

exo-form [67753-91-7]

Reagent for norbornyl group transfers.

Midland, M.M. *et al*, *J. Organomet. Chem.*, 1978, **156**, 203.
Brown, H.C. *et al*, *Synthesis*, 1979, 701.
Brown, H.C. *et al*, *J. Org. Chem.*, 1980, **45**, 4067 (*synth*)

$C_{15}H_{25}BClN_3O$ B-00959

Dibutyl(5-chloro-2-imino-*N*-methyl-1(2*H*)-pyridinecarboxamidato-N^{N1},N^2)boron

[71231-69-1]

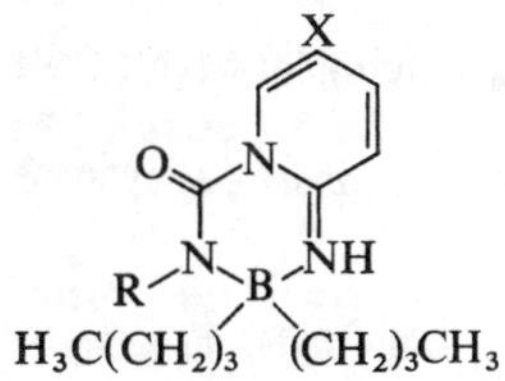

R = Me, X = Cl

M 309.645

Exists as intramol. coord. complex as shown. Possesses some antiviral activity. Cryst. (EtOH). Mp 124-5° dec.

Mikhailov, B.M. *et al*, *Izv. Akad. Nauk SSSR, Ser. Khim.*, 1979, 1085; *Bull. Acad. Sci. USSR, Div. Chem. Sci.*, 1014 (*synth, nmr, ir*)
Mikhailov, B.M. *et al*, *Khim.-Farm. Zh.*, 1982, **16**, 695.

$C_{15}H_{25}BN_2$ B-00960

(*N,N'*-Dimethylbenzenecarboximidamidato-*N,N'*)diisopropylboron, 10CI

(N,N'*-Dimethylbenzenecarboximidamidato*-N,N')-*bis(1-methylethyl)boron, 10CI*

[69138-06-3]

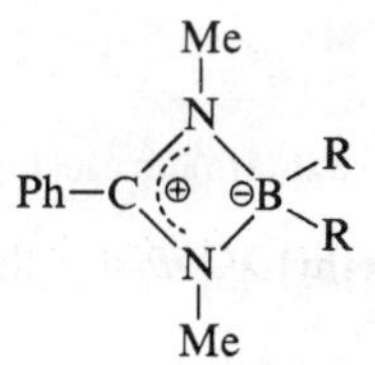

R = $-CH(CH_3)_2$

M 244.186

Liq. Bp_1 78-83°. ^{11}B nmr –11.0 ppm ($CHCl_3$, 60°).

Mikhailov, B.M. *et al*, *Izv. Akad. Nauk SSSR, Ser. Khim.*, 1978, 2578; *Bull. Acad. Sci. USSR, Div. Chem. Sci.*, 2304 (*synth, pmr, nmr*)

$C_{15}H_{25}BN_2$ B-00961

(*N,N'*-Dimethylbenzenecarboximidamidato-*N,N'*)dipropylboron, 10CI

[69138-07-4]

As (*N*,*N′*-Dimethylbenzenecarboximidamidato-*N*,*N′*)diisopropylboron, B-00960 with

R = $CH_2CH_2CH_3$

M 244.186
Liq. Bp_1 82-3°. ^{11}B nmr −9, −48 ppm (C_6H_6, 80°).

Mikhailov, B.M. *et al*, *Izv. Akad. Nauk SSSR, Ser. Khim.*, 1978, 2578; *Bull. Acad. Sci. USSR, Div. Chem. Sci.*, 2304 (*synth, nmr, pmr*)
Dorokov, V.A. *et al*, *Dokl. Akad. Nauk SSSR, Ser. Sci. Khim.*, 1979, **245**, 121 (*synth*)

$C_{15}H_{25}BN_2S$ B-00962

Diethyl(*N*,*N*-diethyl-*N′*-phenylthioureato)borane
(N,N-*Diethyl*-N′-*phenylthioureato*-N′,S)*diethylboron*, *10CI*
[73768-04-4]
As (*N*,*N*-Diethyl-*N′*-phenylthioureato)dimethylborane, B-00856 with

R = Et

M 276.246
Formed from the reaction of $Et_2NCSNPhSiMe_2$ with Et_2BCl.

Marringgle, W., *Z. Naturforsch.*, *B*, 1980, **35**, 164 (*synth*)

$C_{15}H_{25}BO_3$ B-00963

7-Methoxymethyl-3-borabicyclo[3.3.1]nonyl acetylacetonate
[[5-(*Methoxymethyl*)-*1*,*3-cyclohexanediyl*]*bis*[*methylene*]](*2*,*4-pentanedionato*-O,O′)*boron*
[76069-97-1]
As 7-Methylene-3-borabicyclo[3.3.1]nonyl acetylacetonate, B-00902 with

R, R′ = H, $-CH_2OMe$

M 264.171
Bright-yellow cryst. (hexane). Mp 95-6°.

Gurskii, M.E. *et al*, *J. Organomet. Chem.*, 1980, **199**, 171 (*cmr, synth*)
Gurskii, M.E. *et al*, *Izv. Akad. Nauk SSSR, Ser. Khim.*, 1981, 341; *Bull. Acad. Sci. USSR, Div. Chem. Sci.*, 264 (*cmr, struct*)
Gurskii, M.E. *et al*, *Izv. Akad. Nauk SSSR, Ser. Khim.*, 1981, 394; *Bull. Acad. Sci. USSR, Div. Chem. Sci.*, 312 (*synth, ir, pmr*)

$C_{15}H_{26}BN_3O$ B-00964

Dibutyl(2-imino-*N*-methyl-1(2*H*)-pyridinecarboxamidato-N^{N1},N^2)boron, 10CI
[70469-68-0]
As Dibutyl(5-chloro-2-imino-*N*-methyl-1(2*H*)-pyridinecarboxamidato-N^{N1},N^2)boron, B-00959 with

R = Me, X = H

M 275.200
Exists as an intramol. coord. complex. Shows some antiviral activity. Cryst. (hexane); stable in air or EtOH. Sol. THF, Et_2O, $CHCl_3$. Mp 148-9°.

Dorokhov, V.A. *et al*, *Dokl. Akad. Nauk SSSR, Ser. Sci. Khim.*, 1979, **245**, 121 (*synth*)
Dorokhov, V.A. *et al*, *Izv. Akad. Nauk SSSR, Ser. Khim.*, 1979, 1085; *Bull. Acad. Sci. USSR, Div. Chem. Sci.*, 1014 (*synth, ir, nmr*)
Lagutkin, N.A. *et al*, *Khim.Farm. Zh.*, 1982, **16**, 695.

$C_{15}H_{27}B$ B-00965

9-[(2-Methylcyclopentyl)methyl]-9-borabicyclo[3.3.1]nonane, 10CI
B-[(*2-Methylcyclopentyl*)*methyl*]-*9-BBN*

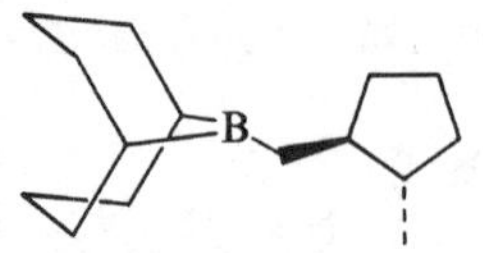

M 218.188
trans-form [74763-94-3]
Formed *in situ* by homologation of 9-(2-Methylcyclopentyl)-9-borabicyclo[3.3.1]nonane, B-00919 . Air-sensitive.

Brown, H.C. *et al*, *J. Org. Chem.*, 1980, **45**, 4067; 1981, **46**, 647.

$C_{15}H_{27}B$ B-00966

Tricyclopentylborane, 10CI, 9CI, 8CI
[23985-40-2]

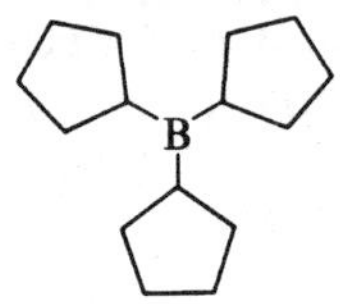

M 218.188
Prepd. by hydroboration of cyclopentene. Synthetic intermed., generally prepd. *in situ*, for transfer of a cyclopentyl group. Liq., sensitive to air. Sol. THF. Bp_2 118-9°.

Brown, H.C. *et al*, *J. Am. Chem. Soc.*, 1962, **84**, 1478 (*synth*)
Takahashi, Y. *et al*, *Chem. Lett.*, 1977, 999 (*use*)
Tokuda, M. *et al*, *J. Chem. Soc., Chem. Commun.*, 1977, 690 (*use*)
Akimoto, I. *et al*, *Synthesis*, 1979, 146 (*use*)
Mikhailov, B.M. *et al*, *Izv. Akad. Nauk SSSR, Ser. Khim.*, 1979, 855; *Bull. Acad. Sci. USSR, Div. Chem. Sci.*, 799 (*props*)
Sotoyama, T. *et al*, *Bull. Chem. Soc. Jpn.*, 1979, **52**, 1865 (*use*)

$C_{15}H_{27}BN_2O$ B-00967

Dipropyl[1-(2-pyridinylamino)-1-butanolato]boron, 10CI
Dipropylboryl-[*2*-(*α-hydroxybutyl*)*aminopyridinate*]
[64167-75-5]

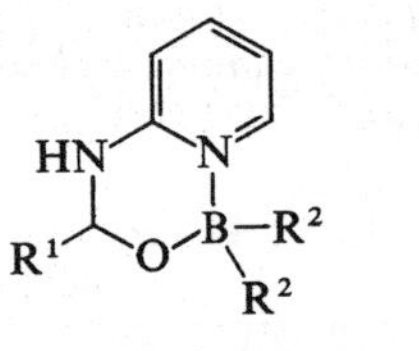

$R^1 = R^2 = CH_2CH_2CH_3$

M 262.201
Cryst., thermally and chemically unstable. Mp >85° dec.

Dorokov, V.A. *et al*, *Izv. Akad. Nauk SSSR, Ser. Khim.*, 1977, 1587; *Bull. Acad. Sci. USSR, Div. Chem. Sci.*, 1458 (*synth, nmr, ir, ms*)
Zolotarev, B.M. *et al*, *Izv. Akad. Nauk SSSR, Ser. Khim.*, 1979, 80; *Bull. Acad. Sci. USSR, Div. Chem. Sci.*, 70 (*ms*)

$C_{15}H_{27}BO_3$ B-00968

3-Methoxy-7-[[(tetrahydro-2*H*-pyran-2-yl)oxy]methyl]-3-borabicyclo[3.3.1]nonane, 10CI
*7-[[(Tetrahydro-2*H*-pyran-2-yl)oxy]methyl]-3-methoxy-3-borabicyclo[3.3.1]nonane. 7-[[(Tetrahydro-2*H*-pyran-2-yl)oxy]methyl]-3-MBBN*
[72890-45-0]

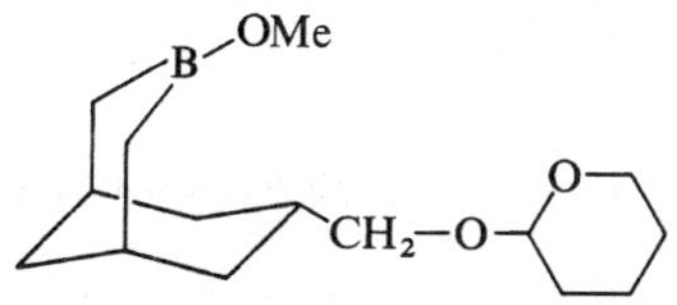

M 266.187
Air- and moisture-sensitive. $Bp_{1.0}$ 113.5-114°. n_D^{20} 1.4944.

Mikhailov, B.M. *et al*, *Izv. Akad. Nauk SSSR, Ser. Khim.*, 1979, 2729; *Bull. Acad. Sci. USSR, Div. Chem. Sci.*, 2537 (*synth, pmr*)
Gurskii, M.E. *et al*, *Izv. Akad. Nauk SSSR, Ser. Khim.*, 1981, 394; *Bull. Acad. Sci. USSR, Div. Chem. Sci.*, 312.

$C_{15}H_{28}B^{\ominus}$ B-00969

9-Butyl-9-(2-propenyl)-9-boratabicyclo[3.3.1]nonane, 10CI
9-Butyl-9-allyl-9-boratabicyclo[3.3.1]nonane
As 9-Boratabicyclo[3.3.1]nonane, B-00546 with

$$R = H_3C(CH_2)_3—, R' = H_2C{=}CHCH_2—$$

M 219.196 (ion)
Li salt: [68046-28-6].
$C_{15}H_{28}BLi$ M 226.137
Air- and (probably) moisture-sensitive.

Yamamoto, Y. *et al*, *J. Am. Chem. Soc.*, 1978, **100**, 6282.
Yamamoto, Y. *et al*, *Tetrahedron Lett.*, 1980, **21**, 3599 (*cmr*)

$C_{15}H_{28}B^{\ominus}$ B-00970

Tricyclopentylhydroborate(1–), 10CI
Tricyclopentylborohydride

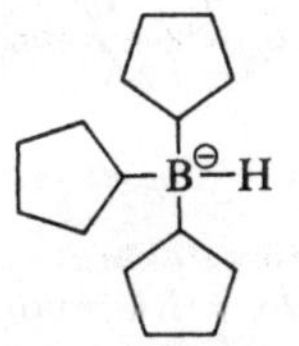

M 219.196 (ion)
Li salt: [61721-97-9].
$C_{15}H_{28}BLi$ M 226.137
Not isolated in pure, solvent-free state. Prepd. *in situ* as THF soln. Stereoselective reducing agent. Air- and moisture-sensitive.
Na salt: [67276-05-5].
$C_{15}H_{28}BNa$ M 242.186
Prepd. *in situ* as THF soln. Stereoselective reducing agent. Air- and moisture-sensitive. Sol. THF.
K salt: [63706-76-3].
$C_{15}H_{28}BK$ M 258.295
Prepd. *in situ* as THF soln. Stereoselective reducing agent. Air- and moisture-sensitive. Sol. THF. Indefinitely stable under dry N_2.

U.S.P., 3 974 183, (*1976*); *CA*, **86**, 72219b (*synth, use*)
Strouf, O. *et al*, *Acta Chem. Scand., Ser. A*, 1977, **31**, 391 (*props*)
Brown, C.A. *et al*, *J. Organomet. Chem.*, 1978, **156**, 111 (*synth, ir, nmr*)
Brown, H.C. *et al*, *J. Am. Chem. Soc.*, 1978, **100**, 3343 (*synth, ir, nmr*)
Brown, H.C. *et al*, *J. Org. Chem.*, 1979, **44**, 467, 5004 (*nmr, ir, use*)
Brown, H.C. *et al*, *J. Organomet. Chem.*, 1979, **166**, 271; 1980, **188**, 1 (*synth, nmr, ir*)

$C_{15}H_{29}BN_2$ B-00971

(1-Methyl-2(1*H*)-pyridinimine-*N*²)tripropylboron, 10CI
*1-Methyl-2(1*H*)-pyridinimine compd. with tripropylborane(1:1),* 9CI. *2-Imino-1-methyl-1,2-dihydropyridine tripropylborane*
[60334-09-0]

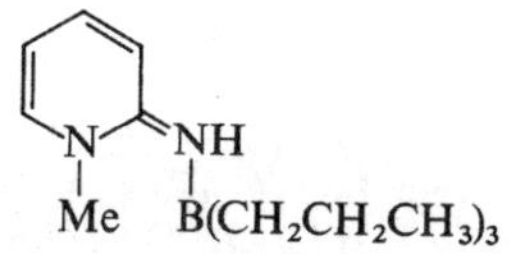

M 248.218

Dorokhov, V.A. *et al*, *Zh. Obshch. Khim.*, 1976, **46**, 1057.
Dorokhov, V.A. *et al*, *Izv. Akad. Nauk SSSR, Ser. Khim.*, 1977, 1593; *Bull. Acad. Sci. USSR, Div. Chem. Sci.*, 1464.

$C_{15}H_{29}BSn$ B-00972

9-[1-(Trimethylstannyl)-2-butenyl]-9-borabicyclo[3.3.1]nonane, 10CI
α-Stannylcrotyl-9-borabicyclo[3.3.1]nonane
[79154-10-2]

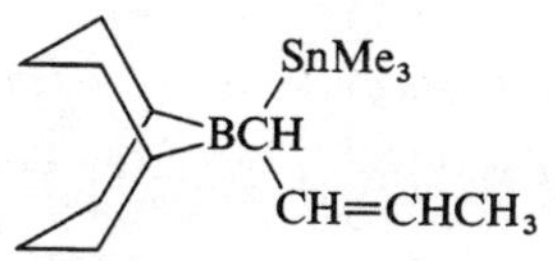

M 338.894
Prepd. *in situ*. Reagent for the stereoregulated synth. of acrylic systems. Air-sensitive.

Yamamoto, Y. *et al*, *J. Am. Chem. Soc.*, 1981, **103**, 3229 (*use*)

$C_{15}H_{30}BCl$ B-00973

(5-Chloro-1-pentenyl)bis(1,2-dimethylpropyl)borane, 10CI
(5-Chloro-1-pentenyl)bis(3-methyl-2-butyl)borane. (5-Chloro-1-pentenyl)disiamylborane
[66315-09-1]

$$[(H_3C)_2CHCH(CH_3)]_2BCH{—}CHCH_2CH_2CH_2Cl$$

M 256.665
Prepd. *in situ* from 5-chloro-1-pentyne with disiamylborane. Synthetic intermed. Prob. oil, sensitive to air. Sol. non-hydroxylic solvs.

Yatagai, H. *et al*, *J. Chem. Soc., Chem. Commun.*, 1977, 852 (*synth, use*)
Campbell, J.B. *et al*, *J. Organomet. Chem.*, 1978, **156**, 71 (*nmr, synth*)
Yatagai, H. *et al*, *Bull. Chem. Soc. Jpn.*, 1980, **53**, 1670 (*synth, use*)

$C_{15}H_{30}BN$ B-00974

1-Boratricyclo[3.3.1.1^{3,7}]decane triethylamine
[1,3,5-Cyclohexanetriyltris(methylene)](N,N-diethylethanamine)boron, 10CI. *N,N-Diethylethanamine compd. with 1-boratricyclo[3.3.1.1^{3,7}]decane(1:1),* 9CI. *1-Boraadamantane triethylamine*
[55967-65-2]
Et_3N complex of B-00597

M 235.219

Easily exchanges Et_3N for other ligands, so useful for synth. of other complexes. Colourless cryst. (isopentane or hexane); complex not dissociated at 200°. Mp 72-5°.

Mikhailov, B.M., *Pure Appl. Chem.*, 1974, **39**, 505.
Mikhailov, B.M. *et al*, *Izv. Akad. Nauk SSSR, Ser. Khim.*, 1976, 2302; *Bull. Acad. Sci. USSR, Div. Chem. Sci.*, 2148 (*synth*)
Mikhailov, B.M. *et al*, *Dokl. Akad. Nauk SSSR, Ser. Sci. Khim.*, 1978, **243**, 929.
Mikhailov, B.M. *et al*, *Izv. Akad. Nauk SSSR, Ser. Khim.*, 1979, 2544; *Bull. Acad. Sci. USSR, Div. Chem. Sci.*, 2361 (*synth*)

$C_{15}H_{31}BOS$ B-00975

Dibutyl(1-*tert*-butylthio-1-propenyloxy)borane

Dibutylborinic acid 1-[(1,1-dimethylethyl)thio]-1-propenyl ester, 10CI

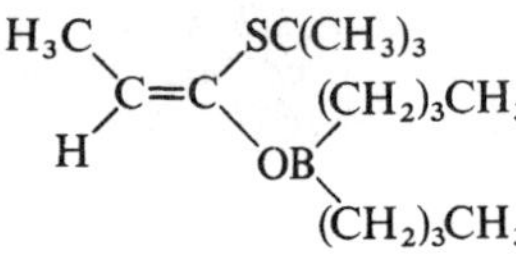

M 270.279

Prepd. *in situ* from *S-tert* butyl propanothioate and dibutylboryl triflate. Intermed. in stereoselective aldol condensations. Air- and moisture-sensitive.

(Z)-form [78024-07-4]

Major stereoisomer prod.

Evans, D.A. *et al*, *J. Am. Chem. Soc.*, 1981, **103**, 3099 (*use, synth*)
Evans, D.A. *et al*, *Pure Appl. Chem.*, 1981, **53**, 1109 (*use, synth*)

$C_{15}H_{33}B$ B-00976

Tripentylborane, 10CI, 9CI, 8CI

[1883-38-1]

$[H_3C(CH_2)_4]_3B$

M 224.236

Prepd. from 1-pentene and borane in THF. This method gives a product contining ca. 6% of 2-pentyl groups. Synthetic intermed., used for transfer of 1-pentyl group; usually prepd. *in situ*. Air-sensitive liq. Sol. THF and nonpolar solvs.

Van der Werf, C.A. *et al*, *CA*, 1957, **51**, 2626 (*props*)
Kate, S. *et al*, *Bull. Chem. Soc. Jpn.*, 1963, **36**, 868 (*synth*)
Laurent, J.P., *Bull. Soc. Chim. Fr.*, 1963, 558 (*synth*)
Mikhailov, B.M. *et al*, *Izv. Akad. Nauk SSSR, Ser. Khim.*, 1977, 2633; *Bull. Acad. Sci. USSR, Div. Chem. Sci.*, 2437 (*use*)
Nikolaev, N.I., *CA*, 1979, **90**, 71606w.
Tsvetkov, V.G. *et al*, *Zh. Obshch. Khim.*, 1979, **49**, 2406 (*props*)

$C_{15}H_{33}B$ B-00977

Tris(1,2-dimethylpropyl)borane, 10CI, 9CI, 8CI

Tris(3-methyl-2-butyl)borane. Trisiamylborane

[32327-52-9]

$[(H_3C)_2CHCH(CH_3)]_3B$

M 224.236

Prepd. *in situ* from borane dimethyl sulfide and 2-methyl-2-butene in THF. Synthetic intermed. Air-sensitive liq. Sol. THF, CH_2Cl_2, pentane.

Brown, H.C. *et al*, *J. Am. Chem. Soc.*, 1962, **84**, 1478 (*synth*)
Brown, H.C. *et al*, *J. Chem. Soc., Dalton Trans.*, 1971, 699 (*props*)
Arase, A. *et al*, *Bull. Chem. Soc. Jpn.*, 1974, **47**, 2511 (*use*)
Brown, H.C. *et al*, *J. Am. Chem. Soc.*, 1978, **100**, 3343 (*props*)
Brown, H.C. *et al*, *J. Organomet. Chem.*, 1980, **188**, 1 (*use*)
Brown, H.C. *et al*, *J. Org. Chem.*, 1981, **46**, 2712 (*synth, nmr*)

$C_{15}H_{33}BO_2$ B-00978

Dihexyloxypropylborane

Propylboronic acid dihexyl ester, 10CI. *Propylbis(hexyloxy)borane. 1-Propaneboronic acid dihexyl ester. Dihexyl propaneboronate*

[67683-00-5]

$H_3CCH_2CH_2B[O(CH_2)_5CH_3]_2$

M 256.235

Air- and moisture-sensitive liq. Sol. non-hydroxylic solvs. Bp_3 95-6°. n_D^{20} 1.4240.

Mikhailov, B.M. *et al*, *Izv. Akad. Nauk SSSR, Ser. Khim.*, 1978, 1586; *Bull. Acad. Sci. USSR, Div. Chem. Sci.*, **27**, 1385 (*synth, props*)
Mikhailov, B.M. *et al*, *J. Organomet. Chem.*, 1978, **154**, 113 (*synth, props*)

$C_{15}H_{33}BO_3$ B-00979

Tripentyloxyborane

Boric acid (H_3BO_3) tripentyl ester, 10CI, 9CI, 8CI. *Pentyl borate. Tripentyl borate. Tripentyl orthoborate. Tris(pentoxy)borane. Amyl borate*

[621-78-3]

$[H_3C(CH_2)_4O]_3B$

M 272.234

Catalyst for oxidn. of cycloalkanes, corrosion inhibitor. Moisture-sensitive liq. Sol. non-hydroxylic solvs. d_4^{27}0.852. Bp 274.5-276.1°, Bp_{16} 146-8°. n_D^{27} 1.4197.

▷ED5650000.

Scattergood, A. *et al*, *J. Am. Chem. Soc.*, 1945, **67**, 2150 (*synth, props*)
Org. Synth., Coll. Vol., **2**, 106 (*synth*)
O'Brien, K.G., *Aust. J. Chem.*, 1957, **10**, 91 (*props*)
Davis, F.A. *et al*, *J. Org. Chem.*, 1971, **36**, 1300 (*nmr*)
Hammerum, S. *et al*, *Org. Mass Spectrom.*, 1974, **8**, 217 (*ms*)
Itskovich, V.A. *et al*, *CA*, 1978, **89**, 196672a (*use*)

$C_{15}H_{34}B^{\ominus}$ B-00980

Tris(1,2-dimethylpropyl)hydroborate(1−), 10CI

Trisiamylborohydride. Trisiamylhydroborate

$[(H_3C)_2CHCH(CH_3)]_3BH^{\ominus}$

M 225.244 (ion)

Li salt: [60217-34-7]. *Lithium triisoamylborohydride.*
$C_{15}H_{34}BLi$ M 232.185
Highly stereoselective reducing agent. Air- and moisture-sensitive. Sol. THF as THF complex.

Li salt, THF complex: [68004-72-8].
$C_{19}H_{42}BLiO$ M 304.291

K salt: [67966-25-0].
$C_{15}H_{34}BK$ M 264.342
Highly stereoselective reducing agent. Air- and moisture-sensitive.

Brown, C.A. *et al*, *J. Organomet. Chem.*, 1978, **156**, 111 (*synth*)
Dauben, W.G. *et al*, *Tetrahedron Lett.*, 1978, 4487 (*props*)
Krishnamurthy, S., *J. Organomet. Chem.*, 1978, **156**, 171 (*props*)
Brown, C.A. *et al*, *J. Am. Chem. Soc.*, 1979, **101**, 3964.
Brown, H.C. *et al*, *J. Org. Chem.*, 1979, **44**, 5004 (*ir, nmr*)
Brown, H.C. *et al*, *J. Organomet. Chem.*, 1979, **166**, 271; 1980, **188**, 1 (*synth, nmr, ir*)
Fieser, M. *et al*, *Reagents for Organic Synthesis*, Wiley, 1967-83, **7**, 216.

$C_{15}H_{36}BN$ **B-00981**

Tributylborane trimethylamine

Tributyl(N,N*-dimethylmethanamine)boron*, 10CI. N,N-*Dimethylmethanamine compd. with tributylborane(1:1)*

[62685-92-1]

$Me_3N—B(CH_2CH_2CH_2CH_3)_3$

M 241.266

Costes, J.P. *et al*, *J. Chim. Phys. Phys. Chim. Biol.*, 1976, **73**, 16.

$C_{16}H_8BBr_4S_4^{\ominus}$ **B-00982**

Tetrakis(5-bromo-2-thienyl)borate(1–), 10CI

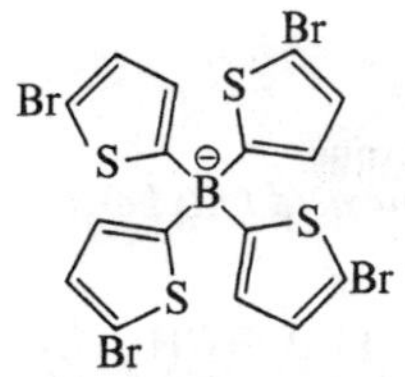

M 658.905 (ion)

Cs salt: [70177-29-6].

$C_{16}H_8BBr_4CsS_4$ M 791.811

Analytical reagent. No phys. props. reported.

Pacey, G.E. *et al*, *Anal. Chim. Acta*, 1979, **105**, 353 (*synth*)
Pacey, G. *et al*, *Talanta*, 1980, **27**, 1013 (*nmr*)

$C_{16}H_{12}BS_4^{\ominus}$ **B-00983**

Tetra-2-thienylborate(1–), 10CI

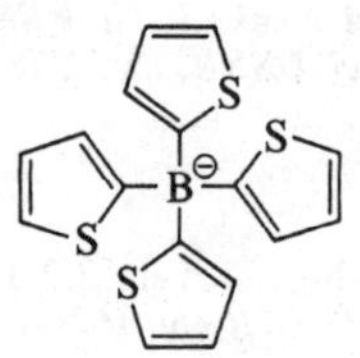

M 343.321 (ion)

Cs salt: [70177-26-3].

$C_{16}H_{12}BCsS_4$ M 476.226

Nmr shift reagent. Cryst. (Me_2CO). Mp >300°.

Pacey, G.E. *et al*, *Anal. Chim. Acta*, 1979, **105**, 353 (*use, synth, props*)
Pacey, G.E. *et al*, *Talanta*, 1980, **27**, 1013 (*nmr, use*)

$C_{16}H_{13}B$ **B-00984**

3-Phenyl-3*H*-benzoborepin, 9CI

3-Phenyl[3]benzoborepin

[16703-13-2]

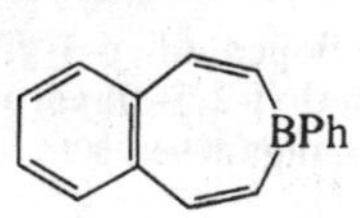

M 216.089

Cryst. Mp 134-7°. λ_{max} 232 (log ε 4.50), 285 (4.68), 301 (4.45), 322 (3.98), 336 (4.22) and 352 nm (4.21) (cyclohexane). Stable in air for several days. Half-life 7 days in cyclohexane.

Leusink, A.J. *et al*, *Tetrahedron Lett.*, 1967, 1263 (*synth, uv, pmr*)
Allinger, N.L. *et al*, *J. Am. Chem. Soc.*, 1975, **97**, 752 (*struct*)
Jeffries, A.T. *et al*, *J. Phys. Chem.*, 1976, **80**, 287 (*uv, struct*)

$C_{16}H_{15}BCo^{\oplus}$ **B-00985**

(η^5-2,4-Cyclopentadien-1-yl)[(1,2,3,4,5,6-η)-1-phenylboratabenzene]cobalt(1+), 10CI

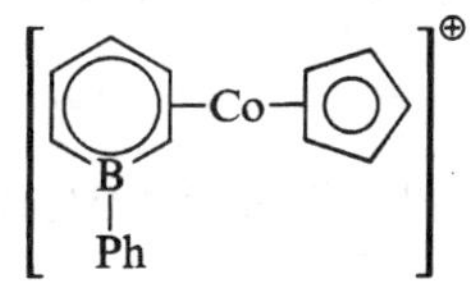

M 277.038 (ion)

Hexafluorophosphate: [51151-11-2].

$C_{16}H_{15}BCoF_6P$ M 422.002

Orange-red cryst. (Me_2CO aq.). Mp 139° dec. ^{11}B nmr δ –23.3 ($(CD_3)_2CO$, reference $BF_3{\cdot}Et_2O$). Reacts with nucleophiles to give stable adducts. Undergoes ring-contraction with amines, OH^- and CN^-.

Herberich, G.E. *et al*, *Chem. Ber.*, 1975, **108**, 3413 (*synth, ms*)
Herberich, G.E. *et al*, *Chem. Ber.*, 1979, **112**, 607 (*synth, nmr*)

$C_{16}H_{15}BFe$ **B-00986**

(η^5-2,4-Cyclopentadien-1-yl)[(1,2,3,4,5,6-η)-1-phenylboratabenzene]iron, 10CI

[65774-56-3]

As (η^5-2,4-Cyclopentadien-1-yl)[(1,2,3,4,5,6-η)-1-methylboratabenzene]iron, B-00726 with

R = Ph

M 273.952

Red cryst. Mp 77.5-78.5°.

Herberich, G.E. *et al*, *J. Organomet. Chem.*, 1978, **144**, C1 (*synth, ms, pmr, nmr*)

$C_{16}H_{15}BFeO_3$ **B-00987**

Tricarbonyl[(2,3,5,6-η)-1,4-dihydro-4,4-dimethyl-1-phenylborin]iron, 10CI

Tricarbonyl(4,4-dimethyl-1-phenyl-1-bora-2,5-cyclohexadiene)iron

[62980-53-4]

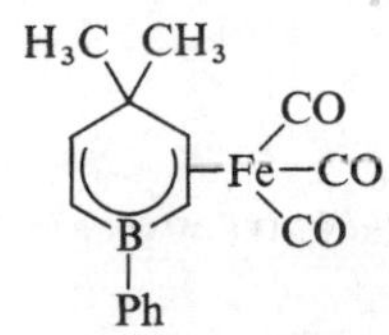

M 321.950

Golden-yellow cryst. (pentane). Mp 85.5°, dec. >200°. $Bp_{0.001}$ 60° subl.

Herberich, G.E. *et al*, *Chem. Ber.*, 1977, **110**, 760 (*synth, pmr, nmr, ir, ms*)

$C_{16}H_{16}B_2Fe_2O_4$ **B-00988**

Di-μ-carbonyldicarbonylbis[(1,2,3,4,5,6-η)-1-methylboratabenzene]diiron, 9CI

Di-μ-carbonyldicarbonylbis(1-methylborinato)diiron

[54853-78-0]

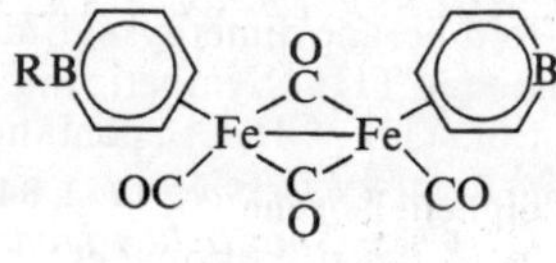

R = Me

M 405.614

Violet plates. Mp 130° dec. Subl. at 80° *in vacuo*.

Herberich, G. *et al*, *Chem. Ber.*, 1974, **107**, 3780 (*synth, pmr, ir, ms*)
Huttner, G. *et al*, *Chem. Ber.*, 1974, **107**, 3786 (*struct*)

$C_{16}H_{19}B$ B-00989

9-(Phenylethynyl)-9-borabicyclo[3.3.1]nonane, 10CI

B-Phenylethynyl-9-BBN

[67553-90-6]

As 9-Ethynyl-9-borabicyclo[3.3.1]nonane, B-00659 with

R = Ph

M 222.136

THF complex: [63220-90-6]. *1,5-Cyclooctanediyl(phenylethynyl)(tetrahydrofuran)boron, 10CI.*
$C_{20}H_{27}BO$ M 294.243
Cryst., stable for >1 yr. at 20° under N_2, but sensitive to air. Sol. pentane.

Sinclair, J.A. *et al*, *J. Am. Chem. Soc.*, 1977, **99**, 954.
Molander, G.A. *et al*, *J. Org. Chem.*, 1977, **42**, 3106 (*synth*)
Brown, H.C. *et al*, *J. Organomet. Chem.*, 1977, **131**, 163 (*synth, nmr*)

$C_{16}H_{19}BFeINO_2$ B-00990

(1-*tert*-Butyl-3-methyl-2-phenyl-1*H*-1,2-azaborol-3-yl-*B*^{2},*N*^{1})dicarbonyliodoiron

*Dicarbonyl[(3,4,5-η)-1-(1,1-dimethylethyl)-2,3-dihydro-3-methyl-2-phenyl-1*H*-1,2-azaborol-3-yl-*B^2,N^1]*-iodoiron, 10CI*

[75059-85-7]

As [(3,4,5-η)-1-*tert*-Butyl-2,3-dihydro-2-methyl-1*H*-1,2-azaborol-3-yl-*B*^{2},*N*^{1}]iodoiron, B-00664 with

R = Ph, R′ = CH_3

M 450.893

Brown plates (pet. ether). Mp 95-6°.

Schulze, J. *et al*, *Chem. Ber.*, 1980, **113**, 2348 (*synth, ir, pmr, nmr, cryst struct*)

$C_{16}H_{19}BO$ B-00991

Butoxydiphenylborane

Diphenylborinic acid butyl ester, 10CI, 9CI, 8CI. Butyl diphenylborinate

[15323-04-3]

$Ph_2BO(CH_2)_3CH_3$

M 238.136

Stabilizer for vinyl chloride. Air- and moisture-sensitive. Sol. nonhydroxylic solvs.

U.S.P., 2 862 952, (*1958*); *CA*, **53**, 9149a (*synth*)
U.S.P., 2 884 441, (*1959*); *CA*, **54**, 298g (*synth*)
Mikhailov, B.M. *et al*, *Izv. Akad. Nauk SSSR, Ser. Khim.*, 1962, 413 (*props*)
Dorokhov, V.A. *et al*, *Zh. Obshch. Khim.*, 1975, **45**, 1772.
Dorokhov, V.A. *et al*, *Izv. Akad. Nauk SSSR, Ser. Khim.*, 1979, 174; *Bull. Acad. Sci. USSR, Div. Chem. Sci.*, **28**, 163.
Dorokhov, V.A. *et al*, *Izv. Akad. Nauk SSSR, Ser. Khim.*, 1981, 869; *Bull. Acad. Sci. USSR, Div. Chem. Sci.*, **30**, 649.

$C_{16}H_{19}BO$ B-00992

(2-Methylpropoxy)diphenylborane

Diphenylborinic acid 2-methylpropyl ester, 10CI. Diphenylborinic acid isobutyl ester, 8CI. Isobutyl diphenylborinate. Isobutyloxydiphenylborane

[23147-97-9]

$Ph_2BOCH_2CH(CH_3)_2$

M 238.136

Air- and moisture-sensitive liq. Used for derivatising nucleosides. Sol. nonhydroxylic solvs. Bp 142-5°.

Mikhailov, B.M. *et al*, *Izv. Akad. Nauk SSSR, Ser. Khim.*, 1956, 322; *Bull. Acad. Sci. USSR, Div. Chem. Sci.*, 1956, 311 (*synth*)
Yurkevich, A.M. *et al*, *Zh. Obshch. Khim.*, 1968, **38**, 2115 (*use*)
Arbuzova, B.A. *et al*, *Izv. Akad. Nauk SSSR, Ser. Khim.*, 1979, 2349; *Bull. Acad. Sci. USSR, Div. Chem. Sci.*, **28**, 2170 (*props*)
Arbuzov, B.A. *et al*, *Izv. Akad. Nauk SSSR, Ser. Khim.*, 1980, 954.

$C_{16}H_{19}BS$ B-00993

(Butylthio)diphenylborane

Diphenylthioborinic acid butyl ester, 10CI

[61209-21-0]

$Ph_2BS(CH_2)_3CH_3$

M 254.196

Used in synth. of diaryl ketiminoboranes. Air- and moisture-sensitive. Sol. nonhydroxylic solvs.

Mikhailov, B.M. *et al*, *Izv. Akad. Nauk SSSR, Ser. Khim.*, 1976, 1820, 1823; *Bull. Acad. Sci. USSR, Div. Chem. Sci.*, **25**, 1715, 1717 (*use*)
Dorokhov, V.A. *et al*, *Izv. Akad. Nauk SSSR, Ser. Khim.*, 1978, 253; *Bull. Acad. Sci. USSR, Div. Chem. Sci.*, **27**, 224.
Mikhailov, B.M. *et al*, *Izv. Akad. Nauk SSSR, Ser. Khim.*, 1978, 1385; *Bull. Acad. Sci. USSR, Div. Chem. Sci.*, **27**, 1205.
Dorokhov, V.A. *et al*, *Izv. Akad. Nauk SSSR, Ser. Khim.*, 1979, 1340; *Bull. Acad. Sci. USSR, Div. Chem. Sci.*, **28**, 1253.

$C_{16}H_{21}B$ B-00994

9-(2-Phenylethenyl)-9-borabicyclo[3.3.1]nonane, 10CI

*B-(2-Phenylethenyl)-9-BBN. B-Styryl-9-*BBN

As 9-(4-Methylphenyl)-9-borabicyclo[3.3.1]nonane, B-00953 with

R = —CH=CHPh

M 224.152

(*E*)-*form* [69322-49-2]
Formed by hydroboration of phenylacetylene with 9-BBN. Reagent for phenylethenyl group transfer. Air-sensitive.

Brown, H.C. *et al*, *J. Am. Chem. Soc.*, 1979, **101**, 96.

$C_{16}H_{21}B_2FeMnO_3S$ B-00995

Tricarbonyl[(η^5-2,4-cyclopentadien-1-yl)iron][μ-[(1,2,3,4,5-η:1,2,3,4,5-η)-3,4-diethyl-2,5-dihydro-2,5-dimethyl-1,2,5-thiadiborole]]manganese, 10CI

[66921-71-9]

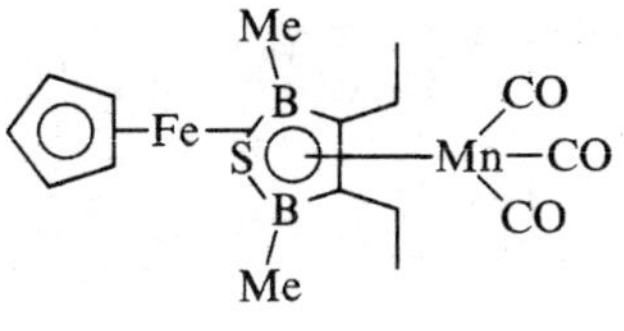

M 425.805

Green air-stable cryst. Mp 194°.

Siebert, W. *et al*, *Angew. Chem., Int. Ed. Engl.*, 1978, **17**, 527 (*synth, pmr, nmr, ir, ms*)

$C_{16}H_{22}BNO_2$ B-00996

3-(2-Aminophenoxy)-7-(methoxymethyl)-3-borabicyclo[3.3.1]-6-nonene

*(2-Aminophenolato-*N,O*)[[5-(methoxymethyl)-4-cyclohexene-1,3-diyl]bis(methylene)]boron*, 10CI

[76014-64-7]

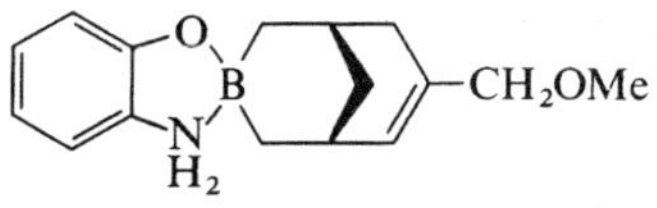

M 271.165

Pale-yellow cryst. V. sensitive to moisture and air.

Mikhailov, B.M. *et al*, *Izv. Akad. Nauk SSSR, Ser. Khim.*, 1980, 1881; *Bull. Acad. Sci. USSR, Div. Chem. Sci.*, 1355 (*synth, ir, pmr*)

$C_{16}H_{22}BN_8Tl$ B-00997

Diethyl[tetrakis(1*H*-pyrazolato-N^1)borato(1–)-N^2,$N^{2'}$]thallium

[74882-66-9]

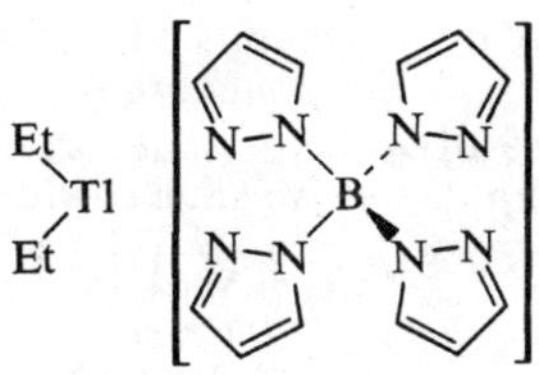

M 541.596

Monomeric fluxional and having low conductivity in soln. Sol. common org. solvs. Mp 194°.

▷Toxic

Onishi, M. *et al*, *Bull. Chem. Soc. Jpn.*, 1980, **53**, 1459 (*synth, ms, ir, pmr*)

$C_{16}H_{23}B$ B-00998

9-(2-Phenylethyl)-9-borabicyclo[3.3.1]nonane, 10CI

B-*(2-Phenylethyl)-9-BBN*

[67753-90-6]

As 9-(4-Methylphenyl)-9-borabicyclo[3.3.1]nonane, B-00953 with

R = —CH_2CH_2Ph

M 226.168

Pale-yellow oil giving thick white fumes when exposed to air. $Bp_{0.35}$ 119-20°.

Midland, M.M. *et al*, *J. Organomet. Chem.*, 1978, **156**, 203.
Vishwakarma, L.C. *et al*, *J. Org. Chem.*, 1980, **45**, 5306 (*synth, nmr*)

$C_{16}H_{25}B$ B-00999

9-(Cyclohexylethynyl)-9-borabicyclo[3.3.1]nonane, 10CI

[62459-79-4]

As 9-(4-Methylphenyl)-9-borabicyclo[3.3.1]nonane, B-00953 with

R = —C≡C—cyclohexyl

M 228.184

Synthetic intermed. Sensitive to air.

Molander, G.A. *et al*, *J. Org. Chem.*, 1977, **42**, 3106.
Sinclair, J.A. *et al*, *J. Am. Chem. Soc.*, 1977, **99**, 954 (*synth*)

$C_{16}H_{26}BCl_2N_6ORh$ B-01000

Dichloro(methanol)[tris(3,5-dimethyl-1*H*-pyrazolato-N^1)-hydroborato(1–)-N^2,$N^{2'}$,$N^{2''}$]rhodium

Hydrotris(3,5-dimethylpyrazolyl)borato]dichloro-(methanol)rhodium

[73117-95-0]

M 503.043

Incorrectly named in 10CI. Starting material for hydrotris(3,5-dimethylpyrazolyl)borato-rhodium complexes. Gold prisms (MeOH). Mp 296-8°.

May, S. *et al*, *Inorg. Chem.*, 1980, **19**, 1582 (*synth, pmr*)

$C_{16}H_{26}BN$ B-01001

9-Borabicyclo[3.3.1]nonane 2-isopropylpyridine

1,5-Cyclooctanediylhydro[2-(1-methylethyl)pyridine]-boron, 10CI. *2-(1-Methylethyl)pyridine, compd. with 9-borabicyclo[3.3.1]nonane(1:1)*

[70338-12-4]

As 9-Borabicyclo[3.3.1]nonane dimethyl sulfide, B-00705 with

L = 2-Isopropylpyridine

M 243.198

ca. 26% associated at 0.13 molar in cyclohexane.

Brown, H.C. *et al*, *Recl. Trav. Chim. Pays-Bas*, 1979, **98**, 117 (*synth, props*)
Brown, H.C. *et al*, *J. Org. Chem.*, 1980, **45**, 1748 (*cmr, props*)

$C_{16}H_{27}B$ B-01002

9-(Bicyclo[2.2.1]hept-2-ylmethyl)-9-borabicyclo[3.3.1]nonane, 10CI

9-(2-Norbornylmethyl)-9-borabicyclo[3.3.1]nonane

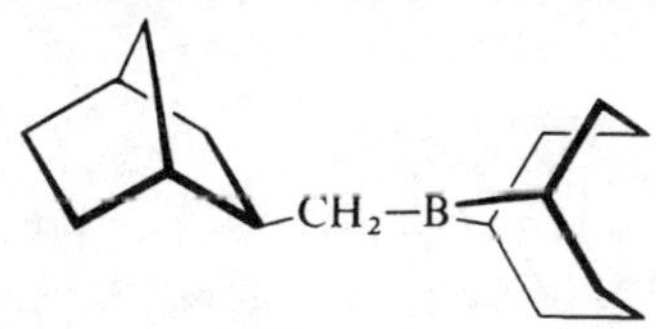

exo-form

M 230.199

exo-form [74763-92-1]

No phys. props. reported.

endo-form [74763-91-0]

No phys. props. reported.

Brown, H.C. *et al*, *J. Org. Chem.*, 1980, **45**, 4067 (*synth*)

$C_{16}H_{27}B$ B-01003

9-(1-Octynyl)-9-borabicyclo[3.3.1]nonane

B-*(1-Octynyl)-9-BBN*

As 9-Ethynyl-9-borabicyclo[3.3.1]nonane, B-00659 with

R = -$(CH_2)_5CH_3$

M 230.199

THF complex: [63220-89-3]. *1,5-Cyclooctanediyl-1-octynyl(tetrahydrofuran)boron*, 10CI.
$C_{20}H_{35}BO$ M 302.306
Cryst., stable for >1 yr. at 20° under N_2, but sensitive to air. Sol. pentane.

Brown, H.C. *et al*, *J. Organomet. Chem.*, 1977, **131**, 163 (*synth*)

$C_{16}H_{27}B$ **B-01004**

7-Pentyl-3-(2-propenyl)-3-borabicyclo[3.3.1]non-6-ene, 10CI

3-Allyl-7-amyl-3-borabicyclo[3.3.1]non-6-ene

[67683-02-7]

As 7-(Chloromethyl)-3-methoxy-3-borabicyclo[3.3.1]-non-6-ene, B-00670 with

$$R = —CH_2CH{=}CH_2,\ R' = -(CH_2)_4CH_3$$

M 230.199

Air- and moisture-sensitive liq. Bp_3 124-5°.

Bubnov, Yu.N. *et al*, *Zh. Obshch. Khim.*, 1970, **40**, 1316 (*synth*)
Mikhailov, B.M. *et al*, *Izv. Akad. Nauk SSSR, Ser. Khim.*, 1978, 1594; *Bull. Acad. Sci. USSR, Div. Chem. Sci.*, 1392.
Mikhailov, B.M. *et al*, *J. Organomet. Chem.*, 1978, **154**, 113.

$C_{16}H_{29}B$ **B-01005**

9-Cyclooctyl-9-borabicyclo[3.3.1]nonane, 10CI

B-*Cyclooctyl-9-BBN*

[67753-92-8]

As 9-(4-Methylphenyl)-9-borabicyclo[3.3.1]nonane, B-00953 with

$$R = \text{cyclooctyl}$$

M 232.215

Synthetic intermed. Sensitive to air.

Midland, M.M. *et al*, *J. Organomet. Chem.*, 1978, **156**, 203 (*synth*)
Brown, H.C. *et al*, *J. Org. Chem.*, 1980, **45**, 4067.

$C_{16}H_{29}B$ **B-01006**

9-[3-Methyl-1-(1-methylethyl)-2-butenyl]-9-borabicyclo[3.3.1]nonane, 10CI

B-*(3,3-Dimethyl-1-isopropylallyl)-9-BBN*

[61967-14-4]

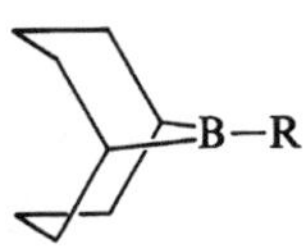

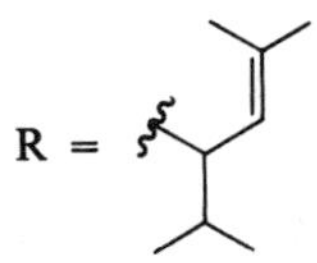

M 232.215

Synthetic intermed. Colourless oil, sensitive to air and somewhat so to moisture. $Bp_{0.001}$ 84-6°.

Kramer, G.W. *et al*, *J. Org. Chem.*, 1977, **42**, 2292.
Kramer, G.W. *et al*, *J. Organomet. Chem.*, 1977, **132**, 9 (*nmr, pmr, synth*)

$C_{16}H_{29}B$ **B-01007**

9-(1-Octenyl)-9-borabicyclo[3.3.1]nonane, 10CI

B-*Octenyl-9-borabicyclo[3.3.1]nonane.* B-*Octenyl-9-BBN*

[74142-77-1]

As 9-(4-Methylphenyl)-9-borabicyclo[3.3.1]nonane, B-00953 with

$$R = —CH{=}CH(CH_2)_5CH_3$$

M 232.215

Reagent for transfer of 1-octenyl group, often stereospecifically.

(E)-form [73062-42-7]

Air-sensitive.

(Z)-form [75579-48-5]

Formed by hydroboration of 1-octyne with 9-BBN. Air-sensitive.

Miyaura, N. *et al*, *Tetrahedron Lett.*, 1980, **21**, 537.
Yatagai, H., *Bull. Chem. Soc. Jpn.*, 1980, **53**, 1670.
Yatagai, H., *J. Org. Chem.*, 1980, **45**, 1640.

$C_{16}H_{30}BNO_3$ **B-01008**

Dibutyl[3-(1-hydroxyethenyl)-4-(1-methylethyl)-2-oxazolidinonato-O^2,O^3]boron, 10CI

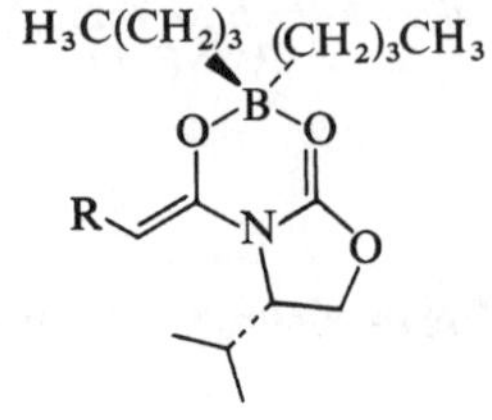

R = H

M 295.228

(S)-form [79616-69-6]

Intermed. in enantioselective aldol condensations.

Evans, D.A. *et al*, *Pure Appl. Chem.*, 1981, **53**, 1109 (*synth*)

$C_{16}H_{30}BO^{\ominus}$ **B-01009**

2-Furanyltris(1-methylpropyl)borate(1−), 10CI

*2-Furyltri-*sec*-butylborate*

As 2-Furanyltripropylborate(1−), B-00867 with

$$R = —CH(CH_3)CH_2CH_3$$

M 249.222 (ion)

Li salt: [70469-69-1].

$C_{16}H_{30}BLiO$ M 256.163

Synthetic intermed., prepd. *in situ*.

Akimoto, I. *et al*, *Synthesis*, 1979, 146 (*synth*)

$C_{16}H_{30}BO^{\ominus}$ **B-01010**

2-Furanyltris(2-methylpropyl)borate(1−), 10CI

2-Furyltriisobutylborate

As 2-Furanyltripropylborate(1−), B-00867 with

$$R = -CH_2CH(CH_3)_2$$

M 249.222 (ion)

Li salt: [70440-86-7].

$C_{16}H_{30}BLiO$ M 256.163

Synthetic intermed., prepd. *in situ*.

Akimoto, I. *et al*, *Synthesis*, 1979, 146 (*synth*)

$C_{16}H_{30}B_{18}Fe^{\ominus}$ **B-01011**

Bis[(7,8,9,10,11-η)-1,2,3,4,5,6,8,9,10,11-decahydro-7-phenyl-7,8-dicarbaundecaborato(2−)]ferrate(1−), 10CI

Bis[decahydro-7-phenyl-π-7,8-dicarbaundecaborato-(2−)]ferrate(1−)

As Bis[(7,8,9,10,11-η)-1,2,3,4,5,6,9,10,11-nonahydro-7,8-dimethyl-7,8-dicarbaundecaborato(2−)]ferrate(1−), B-00586 with

$$R = Ph,\ R' = H$$

M 472.840 (ion)
Paramagnetic. Reduced to a deep-blue ferrate (2−) species.
Tetramethylammonium salt: [12302-58-8]. Red cryst. Mp >300°.

Hawthorne, M.F. *et al*, *J. Am. Chem. Soc.*, 1968, **90**, 879 (*synth*)
Ammeter, J.H. *et al*, *J. Magn. Reson.*, 1978, **30**, 299 (*esr*)

$C_{16}H_{31}B$ B-01012

Dicyclopentyl(1,1,2-trimethylpropyl)borane, 10CI, 9CI
Dicyclopentyl(2,3-dimethyl-2-butyl)borane. Dicyclopentylthexylborane. Thexyldicyclopentylborane
[58879-31-5]
As Dicyclopentylmethoxyborane, B-00738 with

$$R = -C(CH_3)_2CH(CH_3)_2$$

M 234.231
Prepd. *in situ* by hydroboration of cyclopentene with thexylborane. Synthetic intermed., e.g. for synth. of dicyclopentyl ketone. Air-sensitive liq. Sol. THF.

Hughes, R.J. *et al*, *J. Chem. Soc., Perkin Trans. 1*, 1977, 1172 (*use*)
Hughes, R.J. *et al*, *Tetrahedron Lett.*, 1979, 1895 (*use*)
Pelter, A. *et al*, *Comprehensive Organic Chemistry*, (*Barton, D.H.R. et al*, Ed.), Pergamon, Oxford, 1979, **3**, 695 (*rev*)

$C_{16}H_{31}B$ B-01013

4,8-Dimethyl-2-(1,1,2-trimethylpropyl)-2-borabicyclo[3.3.1]-nonane, 10CI
B-*Thexyl-4,8-dimethyl-2-borabicyclo[3.3.1]nonane. Thexyllimonylborane*
[31753-03-4]

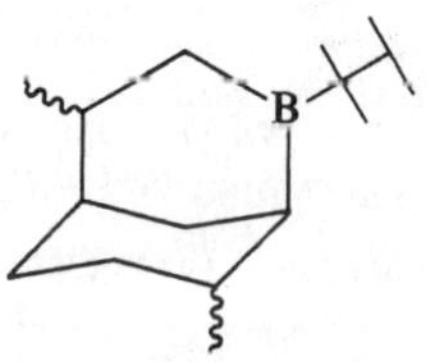

M 234.231
Prepd. by hydroboration of limonene with thexylborane. Hydride adduct used for stereoselective synth. of alcohols and selective reduction of prostaglandin intermeds. Air-sensitive liq. Sol. THF. $Bp_{0.4}$ 88°.

Corey, E.J. *et al*, *J. Am. Chem. Soc.*, 1971, **93**, 1491 (*use*)
Brown, H.C. *et al*, *J. Am. Chem. Soc.*, 1972, **94**, 3567 (*synth*)
Brown, H.C. *et al*, *Tetrahedron*, 1975, **31**, 925 (*pmr, synth, use*)
Brown, H.C. *et al*, *J. Organomet. Chem.*, 1980, **188**, 1 (*nmr, use*)

$C_{16}H_{31}B$ B-01014

9-Octyl-9-borabicyclo[3.3.1]nonane, 10CI, 9CI, 8CI
B-*Octyl-9-BBN*
[30089-00-0]
As 9-[3-Methyl-1-(1-methylethyl)-2-butenyl]-9-borabicyclo[3.3.1]nonane, B-01006 with

$$R = -(CH_2)_7CH_3$$

M 234.231
Formed by hydroboration of 1-octene with 9-BBN. Reagent for transfer of 1-octyl group. Sensitive to air.

Midland, M.M. *et al*, *J. Organomet. Chem.*, 1977, **134**, C17; 1978, **156**, 203.
Jigajinni, V.B. *et al*, *Tetrahedron Lett.*, 1978, 181.
Brown, H.C. *et al*, *J. Org. Chem.*, 1980, **45**, 4067.
Jigajinni, V.B. *et al*, *J. Chem. Res. Synop.*, 1981, 376.
Midland, M.M. *et al*, *J. Am. Chem. Soc.*, 1982, **104**, 525.

$C_{16}H_{32}B^{\ominus}$ B-01015

9,9-Dibutyl-9-boratabicyclo[3.3.1]nonane, 10CI
Dibutyl-1,5-cyclooctanediylborate(1−), 9CI. *9,9-Dibutyl-9-borabicyclo[3.3.1]nonanate*
As 9-Boratabicyclo[3.3.1]nonane, B-00546 with

$$R = R' = H_3C(CH_2)_3-$$

M 235.239 (ion)
Li salt: [56050-29-4].
$C_{16}H_{32}BLi$ M 242.180
Selective reducing agent for *tert*-alkyl, benzyl and allyl halides to give hydrocarbons.

Yamamoto, Y. *et al*, *Chem. Lett.*, 1975, 1199 (*nmr*)
Yamamoto, Y. *et al*, *J. Am. Chem. Soc.*, 1975, **97**, 2558; 1976, **98**, 1965 (*use*)
Yamamoto, Y. *et al*, *J. Chem. Soc., Chem. Commun.*, 1976, 672 (*use*)
Hubbard, J.L. *et al*, *J. Organomet. Chem.*, 1978, **156**, 81 (*synth, nmr*)
Yamamoto, Y. *et al*, *Tetrahedron Lett.*, 1980, **21**, 3599 (*cmr*)

$C_{16}H_{32}B^{\ominus}$ B-01016

4,8-Dimethyl-2-(1,1,2-trimethylpropyl)-2-boratabicyclo[3.3.1]nonane, 10CI
[(6-Methyl-1,3-cyclohexanediyl)(1-methyl-1,2-ethanediyl)](1,1,2-trimethylpropyl)borate(1−), 9CI. *2-(2,3-Dimethyl-2-butyl)-4,8-dimethyl-2-boratabicyclo[3.3.1]-nonane.* B-*Thexyl-4,8-dimethyl-2-boratabicyclo[3.3.1]-nonane*

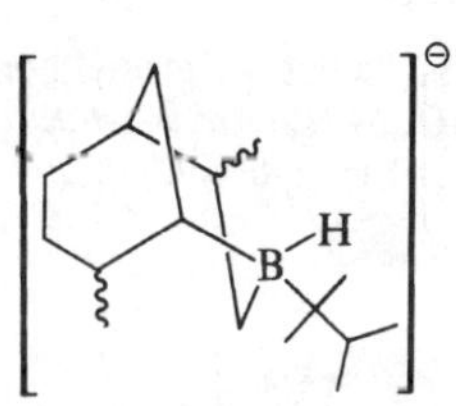

M 235.239 (ion)
Li salt: [38218-68-7].
$C_{16}H_{32}BLi$ M 242.180
Hindered, stereoselective reducing agent. Air- and moisture-sensitive. Sol. THF.

Ger. Pat., 2 207 987, (*1972*); *CA*, **77**, 152342r
Corey, E.J. *et al*, *J. Am. Chem. Soc.*, 1972, **94**, 8616.
U.S.P., 3 974 183, (*1976*); *CA*, **86**, 72219b
Brown, H.C. *et al*, *J. Organomet. Chem.*, 1980, **188**, 1.

$C_{16}H_{32}BBr$ B-01017

(1-Bromo-1-hexenyl)bis(1,2-dimethylpropyl)borane, 10CI
Disiamyl(1-bromo-1-hexenyl)borane
[67532-02-9]

$$[(H_3C)_2CHCH(CH_3)]_2BCBr{=}CH(CH_2)_3CH_3$$

M 315.143
Synthetic intermed. Sensitive to air. Sol. THF.

Masuda, Y. *et al*, *Bull. Chem. Soc. Jpn.*, 1980, **53**, 1652.

$C_{16}H_{32}BO^{\ominus}$ B-01018

Tributyl(ethoxyethynyl)borate(1–), 10CI

$[H_3C(CH_2)_3]_3BC{\equiv}COEt^{\ominus}$

M 251.238 (ion)

Li salt: [63964-04-5].

$C_{16}H_{32}BLiO$ M 258.179

Prepd. *in situ* from tributylborane and 2-ethoxyethynyllithium. Synth. intermed. Air- and acid-sensitive. Sol. THF.

Mikhailov, B.M. *et al*, *Izv. Akad. Nauk SSSR, Ser. Khim.*, 1977, 1456, 2633; 1979, 855; *Bull. Acad. Sci. USSR, Div. Chem. Sci.*, *1977, 1342, 2437; 1979, 799.*

$C_{16}H_{33}B$ B-01019

Bis(1,2-dimethylpropyl)-1-hexenylborane, 10CI, 9CI

1-Hexenyldisiamylborane. Bis(3-methyl-2-butyl)-1-hexenylborane

[55168-92-8]

$[(H_3C)_2CHCH(CH_3)]_2BCH{=}CH(CH_2)_3CH_3$

M 236.247

Prepd. *in situ* from disiamylborane and 1-hexyne (gives *trans*-isomer). Synthetic intermed. Sensitive to air. Sol. THF.

Brown, H.C., *Boranes in Organic Chemistry*, Cornell, Ithaca, N.Y., 1972.
Brown, H.C., *Organic Synthesis via Boranes*, Wiley, N.Y., 1975.
Negishi, E. *et al*, *J. Organomet. Chem.*, 1975, **92**, C4 (*synth*)
Yamamoto, Y. *et al*, *J. Org. Chem.*, 1975, **40**, 3434 (*cmr*)
Yamamoto, Y. *et al*, *J. Am. Chem. Soc.*, 1977, **99**, 5652 (*props*)
Yatagai, H. *et al*, *J. Org. Chem.*, 1979, **44**, 2566 (*pmr*)
Yatagai, H., *Bull. Chem. Soc. Jpn.*, 1980, **53**, 1670 (*synth*)

$C_{16}H_{33}B$ B-01020

Butylcyclohexyl(1,1,2-trimethylpropyl)borane, 10CI

Butylcyclohexyl(2,3-dimethyl-2-butyl)borane. Butylcyclohexylthexylborane. Thexylcyclohexylbutylborane

[77123-50-3]

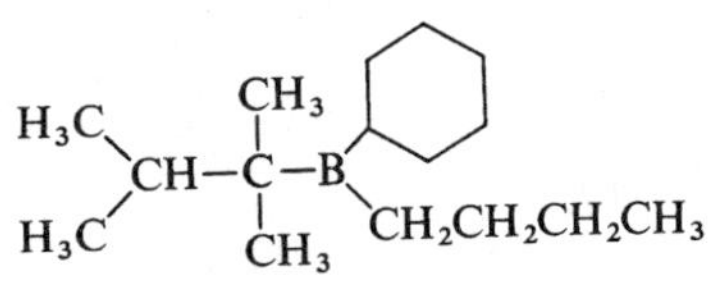

M 236.247

Synth. intermed., prepd. *in situ* by sequential hydroboration reactions. Oil, sensitive to air but stable under N_2 for many hours. Sol. THF.

Slayden, S.W., *J. Org. Chem.*, 1981, **46**, 2311 (*synth, props*)

$C_{16}H_{33}B$ B-01021

3,6-Dimethyl-1-octylborepane, 10CI

[66951-21-1]

As 3,6-Dimethylborepane, B-00551 with

$R = H_3C(CH_2)_7{-}$

M 236.247

Synth. intermed. for transfer of 1-octyl group. No phys. props. reported. Sol. THF.

Jigajinni, V.B. *et al*, *Tetrahedron Lett.*, 1978, 181 (*use*)

$C_{16}H_{35}B$ B-01022

Butylhexyl(1,1,2-trimethylpropyl)borane, 10CI

Butylhexyl(2,3-dimethyl-2-butyl)borane. Butylhexylthexylborane. Thexylbutylhexylborane

[75052-82-3]

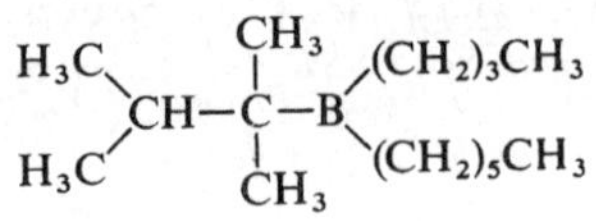

M 238.263

Prepd. *in situ* from thexylchloroborane and 1-hexene, then butyllithium. Synthetic intermed. Oil, sensitive to air; stable to disproportionation over prolonged periods. Sol. THF.

Zweifel, G. *et al*, *J. Am. Chem. Soc.*, 1980, **102**, 5919 (*synth*)

$C_{16}H_{36}B^{\ominus}$ B-01023

Tetrabutylborate(1–), 10CI, 9CI, 8CI

[24651-47-6]

$(H_3CCH_2CH_2CH_2)_4B^{\ominus}$

M 239.270 (ion)

Li salt: [15243-31-9].

$C_{16}H_{36}BLi$ M 246.211

Used in battery electrolytes; catalyst for butadiene polym. Air- and acid-sensitive solid. Insol. hexane, pentane, sol. H_2O. Mp 205° dec.

Tetrabutylammonium salt: [23231-91-6].

$C_{32}H_{72}BN$ M 481.738

Crosslinking agent for epoxy resins.

Jäger, H. *et al*, *Chem. Ber.*, 1962, **95**, 345 (*use*)
Damico, R.A. *et al*, *J. Org. Chem.*, 1964, **29**, 1971 (*synth, ir*)
Jenkins, D.K. *et al*, *Chem. Ind. (London)*, 1966, 1887 (*synth, ir*)
Dixon, C. *et al*, *Eur. Polym. J.*, 1972, **8**, 13 (*use*)
Grindley, T. *et al*, *J. Chem. Phys.*, 1972, **56**, 3602 (*props*)
Takezawa, S. *et al*, *J. Phys. Chem.*, 1973, **77**, 2133 (*props*)
Yamamoto, Y. *et al*, *Chem. Lett.*, 1975, 1199 (*pmr, cmr*)
Japan. Pat., 76 08 337, (*1976*); *CA*, **84**, 151708x (*use*)
Negishi, E. *et al*, *J. Chem. Soc., Perkin Trans. 2*, 1978, 1225 (*synth, use*)
Gill, D.S., *J. Solution Chem.*, 1979, **8**, 691 (*props*)

$C_{16}H_{36}BClN_2$ B-01024

Chlorobis(dibutylamino)borane, 8CI

N,N,N′,N′-*Tetrabutyl-1-chloroboranediamine*, 10CI

[18379-69-6]

$ClB[N(CH_2CH_2CH_2CH_3)_2]_2$

M 302.737

Used in synthesis of corresponding monomeric cyanoborane. Probably liq., sensitive to moisture.

Gros, G. *et al*, *C.R. Hebd. Seances Acad. Sci.*, 1967, **265**, 1065 (*props*)
Gallais, F. *et al*, *J. Chim. Phys. Phys. Chim. Biol.*, 1971, **68**, 247 (*props*)
Meller, A. *et al*, *J. Organomet. Chem.*, 1977, **141**, 249 (*props*)
Meller, A. *et al*, *Monatsh. Chem.*, 1977, **108**, 1279 (*props*)

$C_{16}H_{36}B_2O$ B-01025

Oxybis[dibutylborane], 10CI, 8CI

Bis(dibutylboron)oxide

[3027-53-0]

$(H_3CCH_2CH_2CH_2)_2BOB(CH_2CH_2CH_2CH_3)_2$

M 266.080

Air- and moisture-sensitive liq. Sol. THF. Bp_{10} 110-2°. Heat of formn. (liq.); ΔH_f^0 −212.8 Kcal mol^{-1}.

Nöth, H. *et al*, *Chem. Ber.*, 1964, **97**, 1464 (*synth*)
Tuchagues, J.P. *et al*, *Bull. Soc. Chim. Fr.*, 1968, 2009 (*use*)
Brown, H.C. *et al*, *J. Am. Chem. Soc.*, 1971, **93**, 2802 (*synth*)
Davies, A.G. *et al*, *J. Chem. Soc. (B)*, 1971, 698 (*props*)
Dorokhov, V.A. *et al*, *Izv. Akad. Nauk SSSR, Ser. Khim.*, 1977, 1919; *Bull. Acad. Sci. USSR, Div. Chem. Sci.*, 1977, 1783 (*use*)
Pola, M. *et al*, *Collect. Czech. Chem. Commun.*, 1979, **44**, 3688 (*ir*)

$C_{17}H_{12}BNOS_2$ B-01026

(8-Quinolinolato-*N*[1],*O*[8])di-2-thienylboron, 10CI

[74344-31-3]

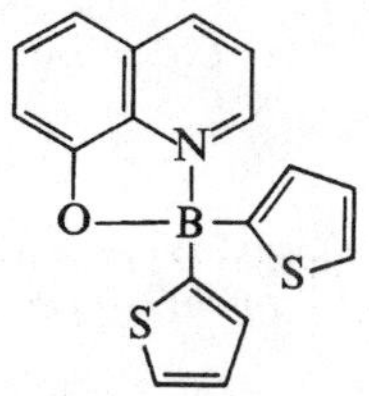

M 321.218

Antibacterial agent.

Bailey, P.J. *et al*, *Antimicrob. Agents. Chemother.*, 1980, **17**, 549 (*use*)

$C_{17}H_{16}BN$ B-01027

Diphenylborane pyridine

Hydrodiphenyl(pyridine)boron, 10CI. Pyridine, compd. with diphenylborane(1:1). Pyridine diphenylborane

[67975-72-8]

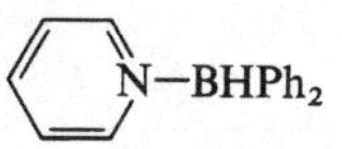

M 245.130

Large colourless cryst. (C_6H_6/hexane), reported to be air stable. Mp 107-8°.

Hawthorne, M.F., *J. Am. Chem. Soc.*, 1958, **80**, 4293 (*synth*)
Jacob, P., *J. Organomet. Chem.*, 1978, **156**, 101 (*synth*)

$C_{17}H_{16}BRu^{\oplus}$ B-01028

(η^6-Benzene)[(1,2,3,4,5,6-η)-1-phenylboratabenzene]ruthenium(1+), 10CI

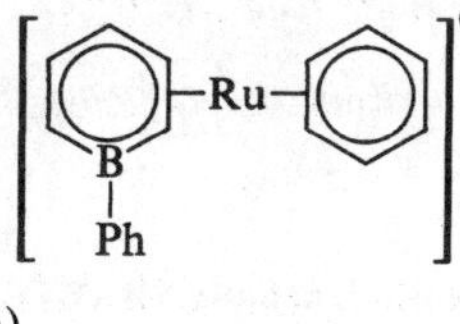

M 332.193 (ion)

Hexafluorophosphate: [69891-00-5].
$C_{17}H_{16}BF_6PRu$ M 477.158
Pearl-coloured plates. Mp 171-171.5°. Dec. >240°.

Herberich, G.E. *et al*, *Chem. Ber.*, 1979, **112**, 607 (*synth, pmr, nmr*)

$C_{17}H_{17}BN_2$ B-01029

Diphenylborane 2-aminopyridine

Hydrodiphenyl(2-pyridinamine-N′)boron, 10CI. 2-Pyridinamine compd. with diphenylborane(1:1)

[70262-74-7]

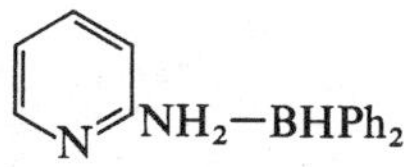

M 260.145

Palmer, D.W. *et al*, *Synth. React. Inorg. Met.-Org. Chem.*, 1979, **9**, 13.

$C_{17}H_{17}BO_2$ B-01030

(2,4-Pentanedionato-*O,O′*)diphenylboron, 10CI

Diphenylboryl acetylacetonate

[19426-80-3]

As (2,4-Pentanedionato-*O,O′*)dipropylboron, B-00741 with

R = Ph

M 264.130

Cryst. (C_6H_6). Dipole moment 4.90D.

Trestianu, H. *et al*, *Tetrahedron*, 1968, **24**, 2499 (*nmr*)
Hammel, J.C. *et al*, *J. Chem. Soc. (A)*, 1969, 1461, 2883 (*nmr, ir*)
Balaban, A.T. *et al*, *Tetrahedron*, 1977, **33**, 3265.
Kopteva, T.S. *et al*, *Zh. Obshch. Khim.*, 1978, **48**, 1587 (*ir*)
Zolotarev, B.M. *et al*, *Izv. Akad. Nauk SSSR, Ser. Khim.*, 1979, 80; *Bull. Acad. Sci. USSR, Div. Chem. Sci.*, 70 (*ms*)

$C_{17}H_{19}BN_2O_2$ B-01031

[2-[1-(4,5-Dihydro-1*H*-imidazol-2-yl)-2-phenylethyl]phenyl]-dihydroxyborane

*[2-[1-(4,5-Dihydro-1*H*-imidazol-2-yl)-2-phenylethyl]-phenyl]boronic acid, 10CI.* (N^3-B)-*2-[1-(o-Dihydroxyborylphenyl)-2-phenylethyl]-2-imidazoline*

[65734-27-2]

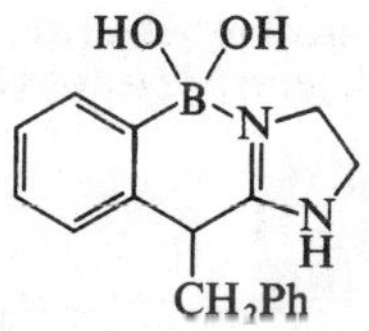

M 294.159

Isol. only as the hydrochloride, but has been produced in soln.

B,HCl:
$C_{17}H_{20}B_2ClN_2O_2$ M 341.430
Used for determination of tartaric acid. White cryst. solid (MeCN aq.). Mod. sol. H_2O. indefinite Mp and dec. pt.

Kankare, J.J. *et al*, *Acta Chem. Scand., Ser. B*, 1977, **31**, 914 (*synth*)
Kankare, J.J. *et al*, *Anal. Chim. Acta*, 1978, **99**, 151 (*use*)

$C_{17}H_{21}BN_2O$ B-01032

Hydroxy[2-[[1-(phenylamino)ethylidene]amino]phenyl]propylboron, 10CI

[69934-19-6]

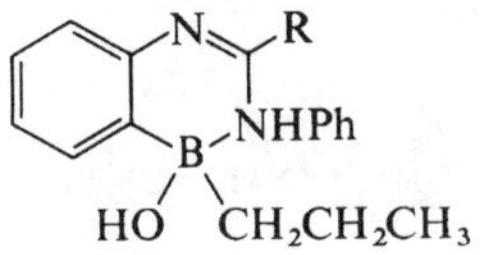

R = CH_3

M 280.176

Colourless cryst. Sl. sol. org. solvs. Mp 184-98°.

Dorokhov, V.A. *et al*, *Izv. Akad. Nauk SSSR, Ser. Khim.*, 1979, 174; *Bull. Acad. Sci. USSR, Div. Chem. Sci.*, 163 (*synth, ir*)

Dorokhov, V.A. *et al*, *Izv. Akad. Nauk SSSR, Ser. Khim.*, 1979, 411; *Bull. Acad. Sci. USSR, Div. Chem. Sci.*, 380 (*ir, tautom*)

$C_{17}H_{22}BN_6Rh$ B-01033

[(1,2,5,6-η)-1,5-Cyclooctadiene][hydrotris(1*H*-pyrazolato-*N*¹)borato(1−)-*N*²,*N*²′]rhodium, 9CI

[52668-35-6]

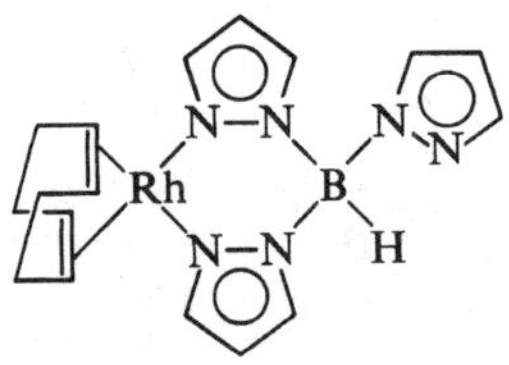

M 424.117

Yellow cryst. (CH_2Cl_2/Et_2O). Mp 167°. Stereochem. non-rigid at r.t.

O'Sullivan, D.J. *et al*, *J. Organomet. Chem.*, 1974, **65**, C47 (*synth*)

King, R.B. *et al*, *J. Organomet. Chem.*, 1974, **73**, 115 (*synth, nmr*)

$C_{17}H_{23}BF_2N_2O$ B-01034

[4-Ethyl-2-[3-ethyl-5-methoxy-4-methyl-2*H*-pyrrol-2-ylidene)methyl]-3,5-dimethyl-1*H*-pyrrolato-*N*¹,*N*²]difluoroboron, 10CI

[66585-84-0]

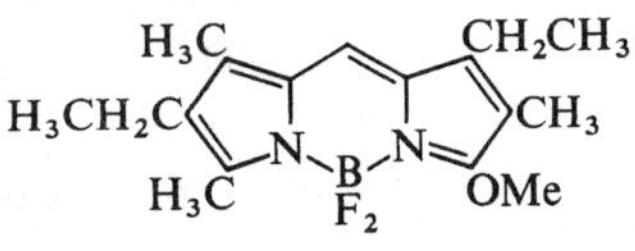

M 320.188

Falk, H. *et al*, *Monatsh. Chem.*, 1978, **109**, 183, 883.

$C_{17}H_{24}BNO_2S$ B-01035

[α-Ethyl-α-[(*N*-methyl-*S*-phenylsulfonimidoyl)methyl]benzenemethanol-*N*]trihydroboron, 10CI

α-Ethyl-α-[(N-methyl-S-phenylsulfonimidoyl)methyl-]benzenemethanol, compd. with borane(1:1)

[73808-90-9]

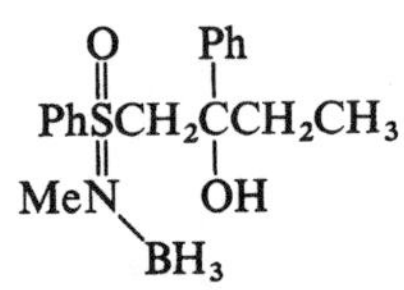

M 317.252

Known in two diastereoisomeric forms. Asymmetric reducing agent. Prepared *in situ* at low temp.

Johnson, C.R. *et al*, *Tetrahedron Lett.*, 1979, 4713.

$C_{17}H_{27}BN_2O$ B-01036

Dibutyl[*N*-(1-iminoethyl)benzamidato-*N*^N^,*O*]boron, 10CI

[43163-49-1]

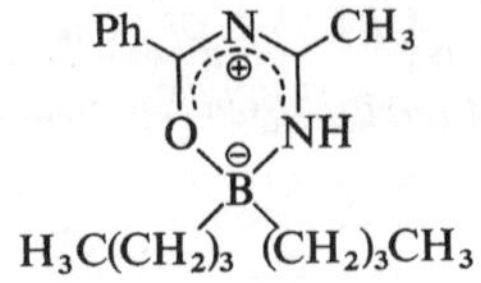

M 286.223

Dorokhov, V.A. *et al*, *Zh. Obshch. Khim.*, 1973, **43**, 1115 (*synth*)

Zolotarev, B.M. *et al*, *Izv. Akad. Nauk SSSR, Ser. Khim.*, 1979, 80; *Bull. Acad. Sci. USSR, Div. Chem. Sci.*, 70 (*ms*)

$C_{17}H_{27}BO$ B-01037

Dibutyl(1-phenyl-1-propenyloxy)borane

Dibutylborinic acid 1-phenyl-1-propenyl ester, 10CI

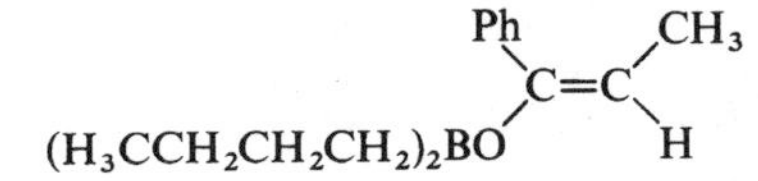

M 258.210

(*Z*)-***form*** [78024-09-6]

Prepd. *in situ*. Intermed. in stereoselective aldol condensations. Air- and moisture-sensitive. Characterised spectroscopically.

Evans, D.A. *et al*, *J. Am. Chem. Soc.*, 1981, **103**, 3099 (*use, synth, pmr*)

Evans, D.A. *et al*, *Pure Appl. Chem.*, 1981, **53**, 1109 (*use, synth, pmr*)

$C_{17}H_{28}BN$ B-01038

1,1-Dibutyl-*N*-(1-phenylpropylidene)boranamine, 10CI, 9CI

[54948-31-1]

$$H_3CCH_2CPh{=}NB(CH_2CH_2CH_2CH_3)_2$$

M 257.225

Air- and moisture-sensitive liq. d_4^{20} 0.883. $Bp_{2.0}$ 118-21°. n_D^{20} 1.5080.

Mikhailov, B.M. *et al*, *Izv. Akad. Nauk SSSR, Ser. Khim.*, 1974, 2389; *Bull. Acad. Sci. USSR, Div. Chem. Sci.*, 2307 (*synth, ir, nmr*)

Mikhailov, B.M. *et al*, *Izv. Akad. Nauk SSSR, Ser. Khim.*, 1976, 2053, 2756; *Bull. Acad. Sci. USSR, Div. Chem. Sci.*, 1924, 2566.

Mikhailov, B.M. *et al*, *Khim. Elementoorg. Soedin*, 1976, 3.

$C_{17}H_{28}BN_3O$ B-01039

Dipropylboryl-(*N*-methylcarbamoyl)-*N*,*N*′-dimethylbenzamidinate

[N,N′-Dimethyl-N-[methylamino(carbonyl)]benzenecarboximidamidato-N^N^,N′]dipropylboron, 10CI

[69109-12-2]

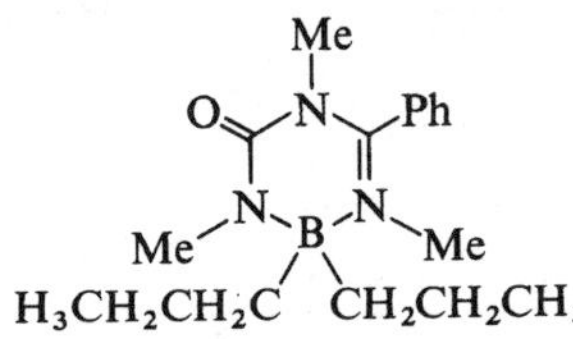

M 301.238

Cryst. Mp 111-3°.

Mikhailov, B.M. *et al*, *Izv. Akad. Nauk SSSR, Ser. Khim.*, 1978, 2578; *Bull. Acad. Sci. USSR, Div. Chem. Sci.*, 2304 (*synth, ms, ir, nmr*)

$C_{17}H_{29}B$ B-01040

Dicyclohexyl(3-methyl-1,3-butadienyl)borane, 10CI, 9CI

3-Methyl-1,3-butadienyldicyclohexylborane

[31569-53-6]

As Dicyclohexyl(methoxy)borane, B-00870 with

$$R = H_2C{=}C(CH_3)CH{=}CH{-}$$

M 244.226

Characterised spectroscopically. Sensitive to air. Sol. THF.

Zweifel, G. *et al*, *J. Am. Chem. Soc.*, 1971, **93**, 1308, 3395 (*synth, use*)

Zweifel, G. *et al*, *J. Organomet. Chem.*, 1976, **117**, 303 (*ir, nmr, uv*)

Zweifel, G. *et al*, *J. Am. Chem. Soc.*, 1977, **99**, 5192 (*use*)

Pelter, A. *et al*, *Comprehensive Organic Chemistry*, (*Barton, D.H.R. et al*, Ed.), Pergamon, Oxford, 1979, **3**, 695 (*rev*)

$C_{17}H_{29}BN_2$ B-01041

Dibutyl(*N,N'*-dimethylbenzenecarboximidamidato-*N,N'*)-boron, 10CI

N-*Dibutylboryl*-N,N'-*dimethylphenylamidine*

[69138-08-5]

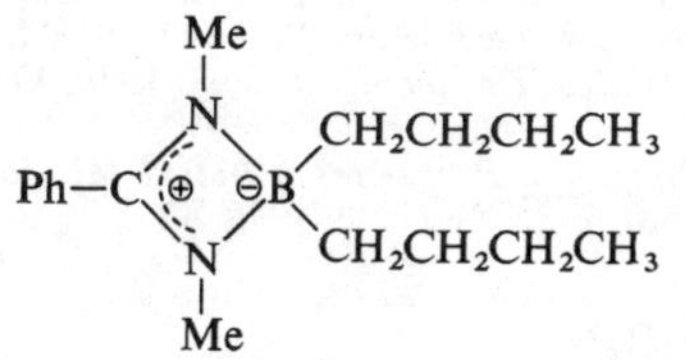

M 272.240

Exists in intramol. coord. form, as shown. Air-sensitive liq. Bp_1 93-5°.

Mikhailov, B.M. *et al*, *Izv. Akad. Nauk SSSR, Ser. Khim.*, 1978, 2578; *Bull. Acad. Sci. USSR, Div. Chem. Sci.*, 2304 (*pmr, nmr, ir, synth*)

$C_{17}H_{29}BN_2$ B-01042

(*N,N*-Dimethylbenzenecarboximidamidato-*N,N'*)bis(2-methylpropyl)boron, 10CI

(N,N-*Dimethylbenzenecarboximidamidato*)*diisobutylborane*

[69202-83-1]

As (*N,N'*-Dimethylbenzenecarboximidamidato-*N,N'*)-diisopropylboron, B-00960 with

$$R = -CH_2CH(CH_3)_2$$

M 272.240

Liq. Bp_1 89-90°. ^{11}B nmr −11, −49 ppm (hexane, 60°).

Mikhailov, B.M. *et al*, *Izv. Akad. Nauk SSSR, Ser. Khim.*, 1978, 2578; *Bull. Acad. Sci. USSR, Div. Chem. Sci.*, 2304 (*synth, nmr, pmr*)

$C_{17}H_{29}BO$ B-01043

9-[(1-Cyclohexyl-1-propenyl)oxy]-9-borabicyclo[3.3.1]nonane, 10CI

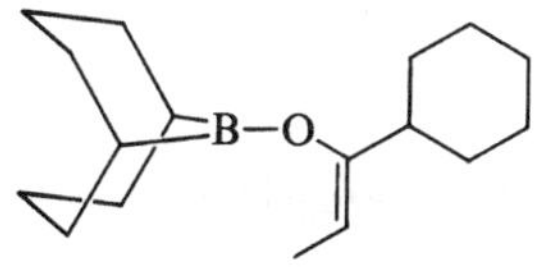

M 260.226

(*Z*)-*form* [72449-20-8]

Useful in stereoselective aldol condensations. Air- and moisture-sensitive.

Van Horn, D.E. *et al*, *Tetrahedron Lett.*, 1979, 2229 (*synth*)

$C_{17}H_{31}BOS$ B-01044

Dicyclopentyl(1-*tert*-butylthio-1-propenyloxy)borane

Dicyclopentylborinic acid 1-[(1,1-dimethylethyl)thio]-1-propenyl ester, 10CI

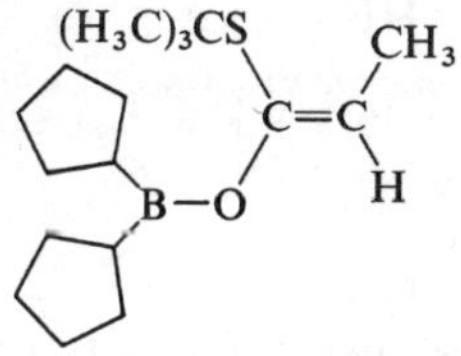

M 294.301

(*Z*)-*form* [72251-55-9]

Prepd. *in situ*. Intermed. in stereoselective synth. of β-hydroxy-α-methylcarboxylic acid thiol esters. Air- and moisture-sensitive. Sol. THF.

Hirama, M. *et al*, *Tetrahedron Lett.*, 1979, 2225 (*synth, use*)

Evans, D.A. *et al*, *J. Am. Chem. Soc.*, 1981, **103**, 3099 (*props*)

$C_{17}H_{32}BNO_3$ B-01045

Dibutyl[1-(4-isopropyl-2-oxo-3-oxazolidinyl)propen-1-yloxy]borane

Dibutylborinic acid 1-[4-(1-methylethyl)-2-oxo-3-oxazolidinyl]-1-propenyl ester, 10CI

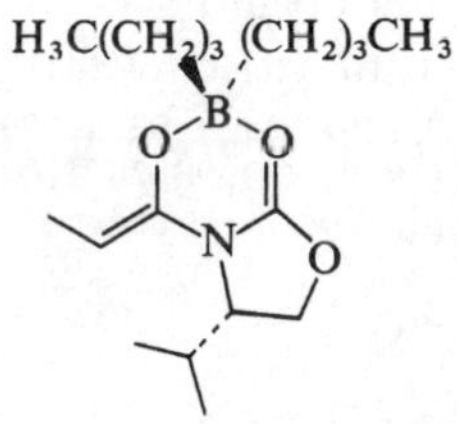

M 309.255

Prepd. *in situ*.

(*S*)-*form* [79563-39-6]

Intermed. in stereoselective aldol condensations. No phys. props. reported.

Evans, D.A. *et al*, *Pure Appl. Chem.*, 1981, **53**, 1109 (*use, synth*)

$C_{17}H_{32}BNO_3S$ B-01046

Dibutyl[3-[1-hydroxy-2-(methylthio)ethenyl]-4-(1-methylethyl)-2-oxazolidinonato-O^2,O^3]boron, 10CI

As Dibutyl[3-(1-hydroxyethenyl)-4-(1-methylethyl)-2-oxazolidinonato-O^2,O^3]boron, B-01008 with

$$R = MeS$$

M 341.315

(S)-form [79593-16-1]

Intermed. in enantioselective aldol condensations.

Evans, D.A. *et al*, *Pure Appl. Chem.*, 1981, **53**, 1109 (*synth*)

$C_{17}H_{33}B$ B-01047

Bis(2-methylcyclohexyl)propylborane, 10CI

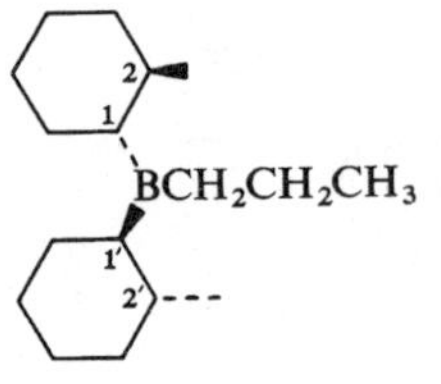

M 248.258

(1RS,1′RS,2RS,2′RS)-form [73959-25-8]

(±)-trans,trans-*form*

Formed *in situ* by reaction of bis(2-methylcyclohexyl)-borane with 1-propene. Synthetic intermed. Air-sensitive. Sol. THF.

Pelter, A. *et al*, *Comprehensive Organic Chemistry*, (*Barton, D.H.R. et al*, Ed.), 1979, Pergamon, Oxford, **3**, 695 (*rev*)
Takahashi, Y. *et al*, *Chem. Lett.*, 1980, 461 (*synth*)

$C_{18}BF_{16}^{\ominus}$ B-01048

Fluorotris(pentafluorophenyl)borate(1−), 10CI

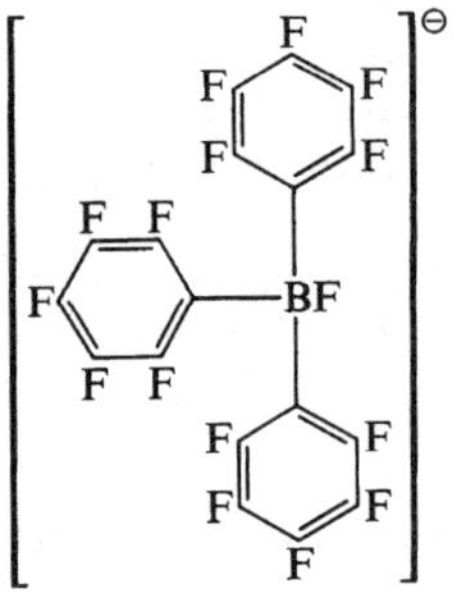

M 530.982 (ion)

Li salt: [70021-19-1].

$C_{18}BF_{16}Li$ M 537.923

Component of battery electrolytes.

Ger. Pat., 2 901 367, (*1979*); *CA*, **92**, 61799p (*use*)
U.S.P., 4 139 681, (*1979*); *CA*, **90**, 207210x (*use*)
Horowitz, H.H. *et al*, *Proc. Electrochem. Soc.*, 1981, **81-4**, 131 (*props*)

$C_{18}H_{13}BN_2O_2$ B-01049

[[2,2′-Azobis(phenolato)](2−)-*N,O,O′*]phenylboron, 10CI

2,2′-Dihydroxyazobenzene, cyclic ester with phenylboronic acid

[74300-72-4]

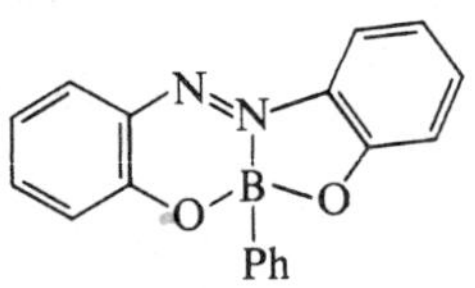

M 300.123

Dark-red cryst. (C_6H_6/pet. ether). Mp 128°.

Hohaus, E. *et al*, *Z. Naturforsch., B*, 1980, **35**, 316 (*ms*)
Hohaus, E. *et al*, *Z. Naturforsch., B*, 1980, **35**, 319 (*synth, nmr, ir*)

$C_{18}H_{13}BN_2O_3$ B-01050

[4-[(2-Hydroxyphenyl)azo]-1,3-benzenediolato(2−)]phenylboron, 10CI

[74302-87-7]

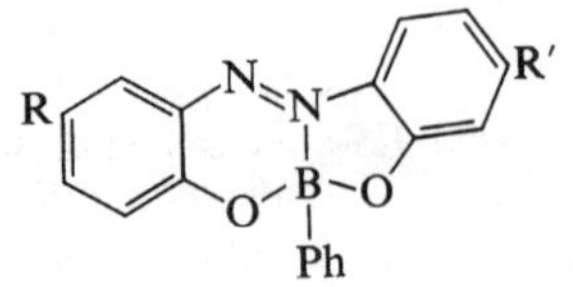

R = H,R′ = OH

M 316.122

Dark-red cryst. (pet. ether/C_6H_6). Mp 128°.

Hohaus, E. *et al*, *Z. Naturforsch., B*, 1980, **35**, 316, 319 (*ms, synth*)

$C_{18}H_{15}B$ B-01051

Triphenylborane, 10CI, 9CI

Triphenylborine

[960-71-4]

BPh_3

M 242.127

Cryst. (Et_2O). Mp 142°. Bp_{15} 245-50°.

Krause, E. *et al*, *Ber.*, 1922, **55**, 1261 (*synth*)
Zettler, F. *et al*, *J. Organomet. Chem.*, 1974, **72**, 157 (*struct*)
Wahab, M.A. *et al*, *Egypt. J. Chem.*, 1976, **19**, 357 (*synth*)
Smith, J.H. *et al*, *Inorg. Chem.*, 1977, **16**, 20 (*uv*)
Nöth, H. *et al*, *NMR Spectroscopy of Boron Compounds* (*NMR Basic Principles and Progess*, Part 14), 1978, Springer (*nmr*)
Odom, J.D. *et al*, *J. Organomet. Chem.*, 1979, **173**, 15 (*cmr, nmr*)
Sax, N.I., *Dangerous Properties of Industrial Materials*, 5th Ed., Van Nostrand-Reinhold, 1979, 429.

$C_{18}H_{15}BBr_3OP$ B-01052

Tribromoborane triphenylphosphine oxide

Tribromo(triphenylphosphine oxide-O)boron, 10CI. Triphenylphosphine oxide compd. with tribromoborane(1:1), 9CI

[67535-58-4]

Ph_3PO—BBr_3

M 528.812

Hoste, S.C.F. *et al*, *Phys. Scr.*, 1977, **16**, 367 (*pe, cmr*)
Hoste, S. *et al*, *J. Electron Spectrosc. Relat. Phenom.*, 1979, **17**, 191 (*pe*)

$C_{18}H_{15}BBr_3P$ B-01053

Tribromoborane triphenylphosphine

Tribromo(triphenylphosphine)boron, 10CI. Triphenylphosphine compd. with tribromoborane(1:1), 9CI

[3426-98-0]

Ph_3P—BBr_3

M 512.812

Moisture-sensitive solid. Mp 194-5°.

Gee, W. *et al*, *J. Chem. Soc.*, 1964, 4180 (*synth*)
Muylle, E. *et al*, *Spectrochim. Acta, Part A*, 1975, **31**, 1039; 1976, **32**, 1149 (*nmr*)
Romm, I.P. *et al*, *Zh. Obshch. Khim.*, 1976, **46**, 2445.
Hoste, S.C.F. *et al*, *Phys. Scr.*, 1977, **16**, 367 (*pe, cmr*)
Hoste, S. *et al*, *J. Electron. Spectrosc. Relat. Phenom.*, 1979, **17**, 191 (*pe*)

$C_{18}H_{15}BCl_3OP$ B-01054

Trichloroborane triphenylphosphine oxide

Trichloro(triphenylphosphine oxide-O)boron, 10CI. *Triphenylphosphine oxide compd. with trichloroborane(1:1)*

[67535-57-3]

$Ph_3PO—BCl_3$

M 395.459

Hoste, S.C.F. *et al*, *Phys. Scr.*, 1977, **16**, 367 (*nmr*)
Hoste, S. *et al*, *J. Electron Spectrosc. Relat. Phenom.*, 1979, **17**, 191 (*pe*)

$C_{18}H_{15}BCl_3P$ B-01055

Trichloroborane triphenylphosphine

Trichloro(triphenylphosphine)boron, 10CI. *Triphenylphosphine compd. with trichloroborane(1:1)*, 9CI. *Triphenylphosphine compd. with boron chloride(BCl_3)(1:1)*, 8CI

[3230-03-3]

$Ph_3P—BCl_3$

M 379.459

Cryst. which fume in air, readily hydrol. by moisture, dec. slowly. Mp 220°, 230°.

Phillips, G.M. *et al*, *J. Chem. Soc.*, 1945, 146.
Muylle, E. *et al*, *Spectrochim. Acta, Part A*, 1975, **31**, 1039; 1976, **32**, 1149 (*nmr*)
Plummer, J.F. *et al*, *Inorg. Chem.*, 1975, **14**, 1505 (*ir*)
Hoste, S.C.F. *et al*, *Phys. Scr.*, 1977, **16**, 367 (*pe, cmr*)
Symons, M.C.R. *et al*, *J. Chem. Res. (S)*, 1978, 122
Iijima, K. *et al*, *Bull. Chem. Soc. Jpn.*, 1979, **52**, 3204 (*struct*)

$C_{18}H_{15}BF_3OP$ B-01056

Trifluoroborane triphenylphosphine oxide

Trifluoro(triphenylphosphine oxide-O)boron, 10CI. *Triphenylphosphine oxide compd. with trifluoroborane(1:1)*, 9CI

[850-13-5]

$Ph_3PO—BF_3$

M 346.095

Fireproofing agent for thermoplastics. White cryst. Mp 239°.

Frazer, M.J. *et al*, *J. Inorg. Nucl. Chem.*, 1963, **25**, 637 (*synth*)
U.S.P., 3 660 350, (*1972*); *CA*, **77**, 35740h
Hoste, S.C.F. *et al*, *Phys. Scr.*, 1977, **16**, 367 (*pe, nmr*)
Hoste, S. *et al*, *J. Electron. Spectrosc. Relat. Phenom.*, 1979, **17**, 191 (*pe*)

$C_{18}H_{15}BF_3P$ B-01057

Trifluoroborane triphenylphosphine

Trifluoro(triphenylphosphine)boron, 10CI. *Triphenylphosphine compd. with trifluoroborane(1:1)*, 9CI. *Triphenylphosphine compd. with boron fluoride(BF_3)(1:1)*, 8CI

[850-07-7]

$Ph_3P—BF_3$

M 330.095

Curing agent for epoxy resins. Yellow foam, sensitive to and liquifies in air. Insol. nonpolar solvs. Mp 128-30°.

▷Ignited by conc. HNO_3

Fluck, E. *et al*, *Pure Appl. Chem.*, 1975, **44**, 373 (*pe*)
Muylle, E. *et al*, *Spectrochim. Acta, Part A*, 1975, **31**, 1039.
U.S.P., 3 910 878, (*1975*); *CA*, **84**, 59735f (*synth*)
Muylle, E. *et al*, *Spectrochim, Acta, Part A*, 1976, **32**, 1149 (*nmr*)
Hoste, S.C.F. *et al*, *Phys. Scr.*, 1977, **16**, 367 (*nmr, pe*)
Hoste, S. *et al*, *J. Electron. Spectrosc. Relat. Phenom.*, 1979, **17**, 191 (*pe*)

$C_{18}H_{15}BI_3P$ B-01058

Triiodoborane triphenylphosphine

Triiodo(triphenylphosphine)boron, 10CI. *Triphenylphosphine compd. with boron iodide*

[3053-69-8]

$Ph_3P—BI_3$

M 653.814

Ivory-coloured cryst. ($CHCl_3$). Mp 168-9°, >220° dec.

Muetterties, E.L., *J. Inorg. Nucl. Chem.*, 1960, **15**, 182 (*synth*)
Gee, W. *et al*, *J. Chem. Soc.*, 1964, 4180.
Hoste, S.C.F. *et al*, *Phys. Scr.*, 1977, **16**, 367 (*cmr, pe*)
Hoste, S. *et al*, *J. Electron Spectrosc. Relat. Phenom.*, 1979, **17**, 191 (*pe*)

$C_{18}H_{15}BO_2$ B-01059

Diphenoxyphenylborane

Phenylboronic acid diphenyl ester, 10CI, 9CI. *Benzeneboronic acid diphenyl ester*, 8CI. *Diphenyl phenylboronate. Phenyldiphenoxyborane*

[2938-25-2]

$PhB(OPh)_2$

M 274.125

Herbicide. Moisture-sensitive solid. Sol. non-hydroxylic solvs. Mp 98-100°, 115-7°. $Bp_{0.5}$ 140-50°.

Dandegaonker, S.H. *et al*, *J. Chem. Soc.*, 1957, 2872 (*synth, props*)
U.S.P., 3 062 859, (*1960*); *CA*, **58**, 7973g (*use*)
Nencetti, G. *et al*, *Ann. Chim. (Rome)*, 1963, **53**, 1875 (*synth, props*)
Faniran, J.A. *et al*, *Can. J. Chem.*, 1968, **46**, 2089 (*ir*)
Wieber, M., *Z. Anorg. Allg. Chem.*, 1974, **403**, 107 (*synth*)
Batrakov, S.G. *et al*, *CA*, 1981, **95**, 183274w (*props*)

$C_{18}H_{15}BO_3$ B-01060

Triphenoxyborane

Boric acid (H_3BO_3) triphenyl ester, 10CI, 9CI, 8CI. *Phenyl borate. Triphenyl borate. Triphenyl orthoborate. Tris(phenoxy)borane*

[1095-03-0]

$(PhO)_3B$

M 290.125

Catalyst for polymn. reactions, heat stabilizer, insect sterilant. Cryst. (CH_2Cl_2), slowly hydrol. by moisture. Sol. non-hydroxylic solvs. Mp 89-93°. $Bp_{0.05}$ 155-8°.

▷Highly toxic orally. Emits highly toxic fumes on heating. ED5775000.

Colclough, T. *et al*, *J. Chem. Soc.*, 1955, 907 (*synth*)
Frazer, M.J. *et al*, *J. Chem. Soc.*, 1955, 2959 (*props*)
Washburn, R.M., *Adv. Chem. Ser.*, 1959, **23**, 129 (*synth, props, use*)
Shurvell, H.F. *et al*, *Can. J. Chem.*, 1968, **46**, 2081 (*ir*)
Pola, M. *et al*, *Collect. Czech. Chem. Commun.*, 1979, **44**, 3688 (*ir, props*)
Ger. Pat., 2 951 673, (*1980*); *CA*, **93**, 240269h (*use*)

Sax, N.I., *Dangerous Properties of Industrial Materials*, 5th Ed., Van Nostrand-Reinhold, 1979, 1067.

$C_{18}H_{15}B_3Cl_3N_3$ B-01061

2,4,6-Trichloro-1,3,5-triphenylborazine, 10CI, 9CI, 8CI

B,B,B-*Trichloro*-N,N,N-*triphenylborazole*

[981-87-3]

As 2,4,6-Tribromoborazine, B-00015 with

X = Cl, Y = Ph

M 412.126

Used to fireproof cotton textiles; curing agent for epoxy resins. Moisture-sensitive monoclinic cryst. (C_6H_6). Mp 273-5° (270-2°). Must be stored in desiccator or under dry C_6H_6 or toluene.

▷ED3800000.

Groszos, S.J. *et al*, *J. Am. Chem. Soc.*, 1958, **80**, 1357 (*synth, props*)
Hohnstedt, L.F. *et al*, *J. Am. Chem. Soc.*, 1960, **82**, 89 (*synth, props*)
Snedden, W., *CA*, 1963, **59**, 2304f (*ms*)
Haworth, D.T. *et al*, *J. Chromatogr.*, 1967, **27**, 302 (*chromatog*)
Wrackmeyer, B. *et al*, *Chem. Ber.*, 1976, **109**, 3480 (*nmr*)
Schwarz, W. *et al*, *Cryst. Struct. Commun.*, 1977, **6**, 431 (*struct*)

$C_{18}H_{15}B_3O_3$ B-01062

Triphenylboroxin, 10CI, 9CI, 8CI

Phenyl metaborate. Benzeneboronic anhydride. 1,3,5-Triphenyl-1,3,5-tribora-2,4,6-trioxane

[3262-89-3]

As Boroxin, B-00019 with

R = Ph

M 311.745

Lubricant. Moisture-sensitive solid. Mp 214-8°.

Molinari, M.A. *et al*, *Anales. Assoc. Quim. Arg.*, 1961, **49**, 46.
Nöth, H. *et al*, *Chem. Ber.*, 1964, **97**, 1464 (*synth*)
Aroney, M.J. *et al*, *J. Chem. Soc. (A)*, 1966, **11**, 1066 (*struct*)
Santucci, L. *et al*, *J. Chem. Soc. (A)*, 1969, **3**, 392 (*uv*)
Brookes, C.J.W. *et al*, *Org. Mass. Spectrom.*, 1970, **3**, 231 (*glc, ms*)
Griffiths, C.J., *Carbohydr. Res.*, 1980, **81**, 7 (*use*)

$C_{18}H_{15}B_3O_6$ B-01063

Triphenoxyboroxin, 10CI, 9CI, 8CI

Phenoxy metaborate

[7187-84-0]

As Boroxin, B-00019 with

R = OPh

M 359.743

Used as chemosterilant for flies, fire extinguisher, plasticizer and welding flux. Moisture-sensitive solid. Mp 98-101° (75-8°).

U.S.P., 3 093 675, (*1960*); *CA*, **59**, 13882e (*synth, use*)
Settepani, J.A. *et al*, *J. Econ. Entomol.*, 1969, **62**, 375 (*use*)
Ger. Pat., 2 214 821, (*1972*); *CA*, **80**, 84153q (*use*)
U.S.P., 3 864 380, (*1974*); *CA*, **82**, 155421e (*use*)
Shelden, R.A., *J. Catal.*, 1974, **34**, 242 (*use*)
Dzhemilev, U.M., *Izv. Akad. Nauk SSSR, Ser. Khim.*, 1979, **3**, 553 (*use*)

$C_{18}H_{15}B_3S_3$ B-01064

Triphenylborthiin, 10CI, 9CI, 8CI

Triphenylborsulfole

[1870-69-5]

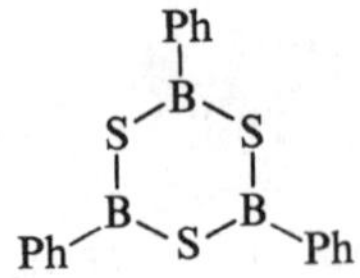

M 359.927

Moisture-sensitive solid. Mp 230-5°.

Abel, E.W. *et al*, *J. Chem. Soc.*, 1965, 3045 (*synth*)
Cragg, R.H. *et al*, *J. Chem. Soc. (A)*, 1967, 947 (*synth, rev, ir*)
Schmidt, M. *et al*, *Z. Naturforsch., B*, 1970, **25**, 1344 (*synth*)
Nöth, H. *et al*, *J. Organomet. Chem.*, 1977, **134**, 297.

$C_{18}H_{16}B^{\ominus}$ B-01065

Hydrotriphenylborate(1−), 10CI

Triphenylhydroborate. Triphenylborohydride

$Ph_3BH^{\ominus}$

M 243.134 (ion)

Li salt: [63717-74-8].

$C_{18}H_{16}BLi$ M 250.075

Reducing agent. Air- and moisture-sensitive.

Strouf, O. *et al*, *Acta Chem. Scand., Ser. A*, 1977, **31**, 391 (*props*)
Brown, H.C. *et al*, *J. Organomet. Chem.*, 1980, **188**, 1 (*synth, nmr*)

$C_{18}H_{17}BIP$ B-01066

Iodoborane triphenylphosphine

Dihydroiodo(triphenylphosphine)boron, 10CI. *Triphenylphosphine iodoborane*

[59499-21-7]

$IBH_2—PPh_3$

M 402.021

Cryst. solid, air-stable for up to 15 hr.

Denniston, M.L. *et al*, *J. Inorg. Nucl. Chem.*, 1976, **38**, 379 (*synth*)
Pennington, B.T. *et al*, *J. Inorg. Nucl. Chem.*, 1979, **40**, 389 (*synth, nmr*)

$C_{18}H_{18}BCo$ B-01067

[(2,3,5,6-η)Bicyclo[2.2.1]hepta-2,5-diene][(1,2,3,4,5,6-η-)-1-phenylboratabenzene]cobalt, 10CI

η^4-Norbornadiene(η^6-1-phenylboratabenzene)cobalt

[69021-84-7]

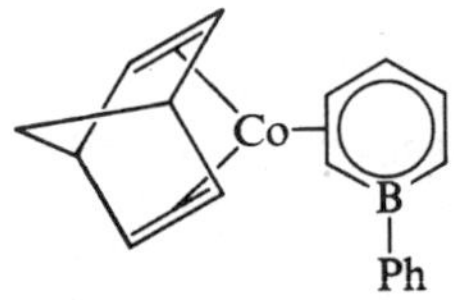

M 304.083

Red-brown cryst. Mp 189-90°. ^{11}B nmr δ +21.2 (C_6D_6, reference BF_3Et_2O).

Herberich, G.E. *et al*, *J. Organomet. Chem.*, 1978, **160**, 17 (*synth, ms, nmr*)

$C_{18}H_{18}BMnO$ B-01068

Carbonyl(η^5-2,4-cyclopentadien-1-yl)[(1,2,3,6,7-η)-4,5-dihydro-1-phenyl-1*H*-borepin]manganese

[67158-14-9]

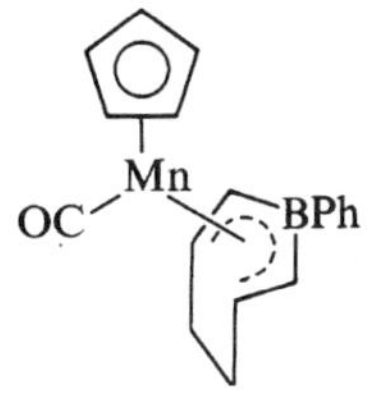

M 316.088

Violet oil, characterised by ir and nmr.

Koelle, U. *et al*, *J. Organomet. Chem.*, 1978, **152**, 7 (*synth, ir, nmr*)

$C_{18}H_{18}BN$ B-01069

Borane triphenylamine

(N,N-*Diphenylbenzenamine*)*trihydroboron*, 10CI. N,N-*Diphenylbenzenamine, compd. with borane*(*1:1*). *Triphenylamine borane*

[69809-42-3]

$Ph_3N—BH_3$

M 259.157

Japan. Pat., 78 131 844, (*1978*); *CA*, **91**, P81565s
Japan. Pat., 78 131 845, (*1978*); *CA*, **91**, P81566t

$C_{18}H_{18}BNO$ B-01070

Triphenylborane hydroxylamine

(*Hydroxylamine*-N)*triphenylboron*, 10CI. *Hydroxylamine, compd. with triphenylborane*(*1:1*)

[69001-94-1]

$Ph_3B—NH_2OH$

M 275.156

Lapkin, I.I. *et al*, *Zh. Obshch. Khim.*, 1978, **48**, 2215 (*synth*)

$C_{18}H_{18}BP$ B-01071

Borane triphenylphosphine

Trihydro(*triphenylphosphine*)*boron*, 10CI. *Triphenylphosphine compd. with* BH_3. *Triphenylphosphine borane*

[2049-55-0]

$Ph_3P—BH_3$

M 276.124

Extremely stable borane complex. Used in heat developable photoimaging processes. Comly. available. Colourless needles (C_6H_6). Mp 187-8°.

Heal, H.G. *J. Inorg. Nucl. Chem.*, 1961, **16**, 208 (*synth*)
Berclaz, T. *et al*, *Mol. Phys.*, 1976, **32**, 815.
Durand, M. *et al*, *J. Chem. Soc., Dalton Trans.*, 1977, 57 (*pmr, nmr*)
Van Paasschen, J.M. *et al*, *Inorg. Chem.*, 1978, **17**, 3302.
Hassler, K. *et al*, *Z. Anorg. Allg. Chem.*, 1979, **448**, 161 (*ir, raman*)

$C_{18}H_{18}B_3N_3$ B-01072

2,4,6-Triphenylborazine, 10CI, 9CI, 8CI

B,B,B-*Triphenylborazole*

[976-28-3]

As Borazine, B-00020 with

R = Ph, R′ = H

M 308.790

Insect sterilant; component of flame-retarding polymers. Cryst. (hexane). Sol. most org. solvs. Mp 183-5°. V. high thermal stability; dec. >370°. Slowly hydrol. by water.

Niedenzu, K. *et al*, *Angew. Chem.*, 1959, **71**, 651 (*synth*)
Gerrard, W. *et al*, *J. Chem. Soc.*, 1961, 3153 (*ir*)
Hawthorne, M.F. *et al*, *J. Am. Chem. Soc.*, 1961, **83**, 833 (*synth, ir, nmr*)
Santucci, L., *J. Chem. Soc.* (*A*), 1969, 392 (*uv*)
Nöth, H. *et al*, *Chem. Ber.*, 1976, **109**, 3480 (*nmr*)
Odom, J.D. *et al*, *J. Organomet. Chem.*, 1979, **173**, 15 (*nmr*)

$C_{18}H_{22}BF$ B-01073

Fluorobis(2,4,6-trimethylphenyl)borane, 10CI, 9CI

Fluorodimesitylborane. Fluorodimesitylborine. Dimesitylfluoroborane

[436-59-9]

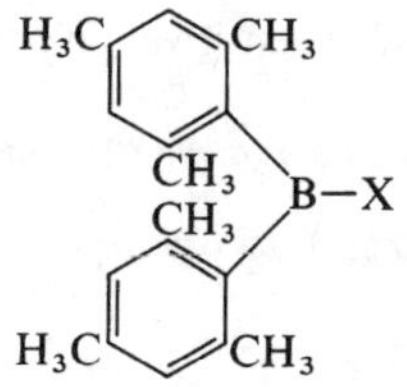

X = F

M 268.180

Prepd. from mesitylmagnesium bromide and $Et_2O·BF_3$ at r.t. Used in synth. of other dimesitylboryl compds. Air- and moisture-sensitive orange solid. Sol. Et_2O. Mp 75.5-76°. $Bp_{0.5}$ 127-9°.

Brown, H.C. *et al*, *J. Am. Chem. Soc.*, 1957, **79**, 2302 (*synth*)
Leffler, J.E. *et al*, *Chem. Ind.* (*London*), 1961, 512 (*use*)
Hooz, J. *et al*, *J. Am. Chem. Soc.*, 1974, **96**, 274 (*props*)
Brown, N.M.D. *et al*, *J. Organomet. Chem.*, 1980, **193**, 271 (*nmr*)
Davidson, F. *et al*, *J. Organomet. Chem.*, 1981, **204**, 147 (*use*)
Lukehart, C.M. *et al*, *J. Am. Chem. Soc.*, 1981, **103**, 4166 (*use*)

$C_{18}H_{22}B_2Cl_2N_2$ B-01074

2,6-Dichloro-2,6-diboraadamantane bis(pyridine)

Dichloro[μ-(*1,3,5,7-cyclooctanetetrayl*-C^1,C^5:C^3,C^7)-]*bis*(*pyridine*)*diboron*, 10CI

[69961-36-0]

As 2,6 Dichloro-2,6-diboraadamantane bis(dimethyl sulfide), B-00802 with

L = Py

M 358.911

Kulkarni, S.U. *et al*, *J. Org. Chem.*, 1979, **44**, 1747 (*synth*)

$C_{18}H_{23}BO$ B-01075

Hydroxybis(2,4,6-trimethylphenyl)borane

Bis(*2,4,6-trimethylphenyl*)*borinic acid*, 10CI, 9CI. *Dimesitylborinic acid*, 8CI. *Hydroxydimesitylborane*

[20631-84-9]

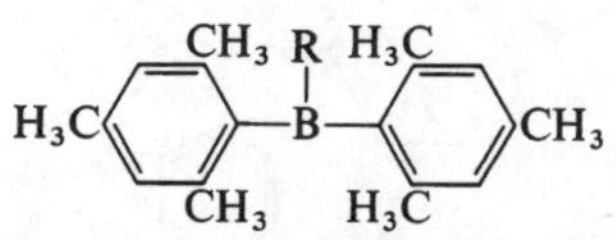

R = OH

M 266.189

White cryst. (pet. ether), stable in air. Mp 140-1°.

Brown, H.C. *et al*, *J. Am. Chem. Soc.*, 1957, **79**, 2302 (*synth*)
Lapkin, I.I. *et al*, *Zh. Obshch. Khim.*, 1962, **32**, 1967 (*synth*)
Lapkin, I.I. *et al*, *CA*, 1968, **69**, 47731d (*ir*)
Lapkin, I.I. *et al*, *CA*, 1968, **69**, 47839v (*uv*)
Brown, N.M.D. *et al*, *J. Organomet. Chem.*, 1980, **193**, 271 (*cmr, nmr*)

$C_{18}H_{24}BN$ B-01076

Aminobis(2,4,6-trimethylphenyl)borane

1,1-Bis(2,4,6-trimethylphenyl)boranamine, 10CI. *Aminodimesitylborane*

[75018-24-5]

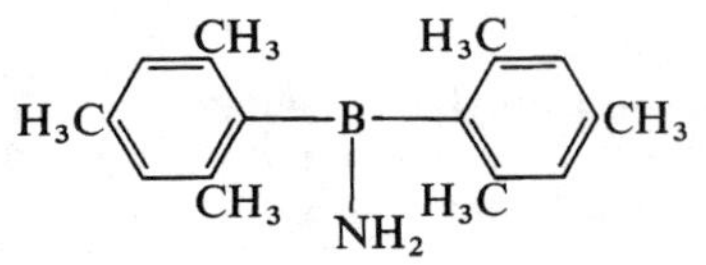

M 265.204

White solid, moisture-sensitive. Sol. most non-hydroxylic org. solvs. Mp 158-62°.

Brown, N.M.D. *et al*, *J. Organomet. Chem.*, 1980, **192**, 133 (*synth, pmr, nmr, ir*)
Davidson, F. *et al*, *Org. Mass Spectrom.*, 1981, **16**, 467 (*ms*)

$C_{18}H_{26}BN_3$ B-01077

Dibutyl(*N*-2-pyridinyl-2-pyridinaminato-N^N,N^1)boron, 10CI

Dibutylboryl-[2-(2-pyridyl)aminopyridinate]

[64800-07-3]

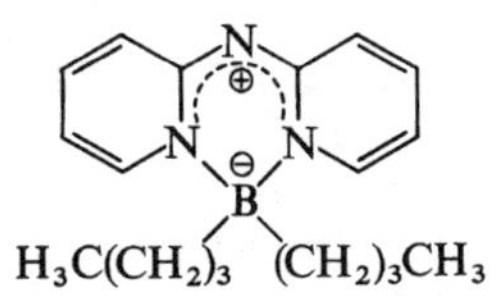

M 295.234

Shows some antiviral activity. Yellow cryst. (EtOH); stable in air. Mp 93-5°. Bp_1 150-5°.

Dorokhov, V.A. *et al*, *Izv. Akad. Nauk SSSR, Ser. Khim.*, 1977, 1921; *Bull. Acad. Sci. USSR, Div. Chem. Sci.*, 1785 (*synth, ir, ms, nmr*)
Zolotarev, B.M. *et al*, *Izv. Akad. Nauk SSSR, Ser. Khim.*, 1978, 1312; *Bull. Acad. Sci. USSR, Div. Chem. Sci.*, 1140 (*ms*)
Dorokhov, V.A. *et al*, *Izv. Akad. Nauk SSSR, Ser. Khim.*, 1981, 1371; *Bull. Acad. Sci. USSR, Div. Chem. Sci.*, 1097 (*synth, nmr*)
Lagutkin, N.A. *et al*, *Khim.-Farm. Zh.*, 1982, **16**, 695.

$C_{18}H_{28}BNO$ B-01078

9-Methyl-9-borabicyclo[3.3.1]nonane 4-dimethylaminobenzaldehyde

1,5-Cyclooctanediyl[4-(dimethylamino)benzaldehyde-N]methylboron, 10CI. *4-(Dimethylamino)benzaldehyde, compd. with 9-methyl-9-borabicyclo[3.3.1]nonane*. *B-Methyl-9-BBN* p*-dimethylaminobenzaldehyde*

[70719-24-3]

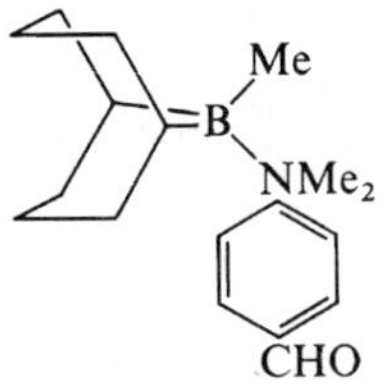

M 285.235

Midland, M.M. *et al*, *J. Am. Chem. Soc.*, 1979, **101**, 2352.

$C_{18}H_{28}B_2Fe$ B-01079

Bis(1-*tert*-butylborabenzene)iron

Bis[(1,2,3,4,5,6-η)-1-(1,1-dimethylethyl)boratabenzene]iron, 9CI

[57692-68-9]

As Bis(η^6-boratabenzene)iron, B-00651 with

$$R = (H_3C)_3C-$$

M 321.886

Red-orange solid. Mp 144-5°.

Ashe, A.J. *et al*, *J. Am. Chem. Soc.*, 1975, **97**, 6865 (*synth, pmr, nmr, uv, ms, mössbauer, pe*)

$C_{18}H_{29}B$ B-01080

Bis(1,2-dimethylpropyl)(2-phenylethenyl)borane, 10CI

2-Phenylethenyldisiamylborane. *Bis(3-methyl-2-butyl)-(2-phenylethenyl)borane*. *Disiamylstyrylborane*

[66315-07-9]

$$[(H_3C)_2CHCH(CH_3)]_2BCH{=}CHPh$$

M 256.237

Prepd. *in situ* by reaction of disiamylborane with phenylacetylene (gives *trans*-isomer). Sol. THF.

Miyaura, N. *et al*, *J. Chem. Soc., Chem. Commun.*, 1979, **19**, 866 (*synth*)
Miyaura, N. *et al*, *Tetrahedron Lett.*, 1979, 3437 (*synth*)
Pelter, A. *et al*, *Comprehensive Organic Chemistry*, 1979, (*Barton, D.H.R. et al*, Ed.), Pergamon, Oxford, **3**, 695 (*rev*)
Yatagai, H., *Bull. Chem. Soc. Jpn.*, 1980, **53**, 1670.
Miyaura, N. *et al*, *J. Organomet. Chem.*, 1982, **233**, C13.

$C_{18}H_{30}BN$ B-01081

1,1-Dibutyl-*N*-(1-phenylbutylidene)boranamine, 10CI, 9CI

[54948-32-2]

$$H_3CCH_2CH_2CPh{=}NB(CH_2CH_2CH_2CH_3)_2$$

M 271.252

Air- and moisture-sensitive liq. d_4^{20} 0.866. $Bp_{1.0}$ 108-10°. n_D^{20} 1.4949.

Mikhailov, B.M. *et al*, *Izv. Akad. Nauk SSSR, Ser. Khim.*, 1974, 2389; *Bull. Acad. Sci. USSR, Div. Chem. Sci.*, 2307 (*synth, ir, nmr*)
Mikhailov, B.M. *et al*, *Izv. Akad. Nauk SSSR, Ser. Khim.*, 1976, 2053, 2756; *Bull. Acad. Sci. USSR, Div. Chem. Sci.*, 1924, 2566.
Mikhailov, B.M. *et al*, *Khim. Elementoorg. Soedin*, 1976, 3.

$C_{18}H_{31}B^{\ominus}$ B-01082

9-Isopinocampheyl-9-boratabicyclo[3.3.1]nonane

[69102-10-9]

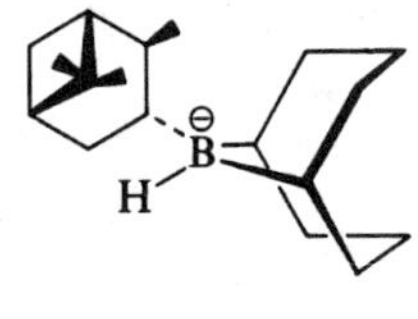

M 258.253 (ion)

H deriv.: [69102-10-9]. *1,5-Cyclooctanediylhydro(2,6,6-trimethylbicyclo[3.1.1]hept-3-yl)-borate(1−)*, 10CI. *B-Isopinocampheyl-9-borabicyclo[3.3.1]nonyl hydride*. *9-Hydro-9-(3-pinanyl)-9-boratabicyclo[3.3.1]nonane*.

$C_{18}H_{32}B$ M 259.261
Chiral reducing agent.
Li salt: [64081-12-5].
$C_{18}H_{31}BLi$ M 265.194
Chiral reducing agent for ketones.

Krishnamurphy, S. *et al*, *J. Org. Chem.*, 1977, **42**, 2534 (*use*)
Brown, H.C., *CA*, 1980, **92**, 68730t.
Brown, H.C. *et al*, *J. Organomet. Chem.*, 1980, **188**, 1 (*synth, nmr*)

$C_{18}H_{31}B$ B-01083

9-(2,6,6-Trimethylbicyclo[3.1.1]hept-3-yl)-9-borabicyclo[3.3.1]nonane, 10CI
B-3-Pinanyl-9-borabicyclo[3.3.1]nonane. B-Isopinocampheyl-9-BBN. Alpine-borane
[73624-47-2]

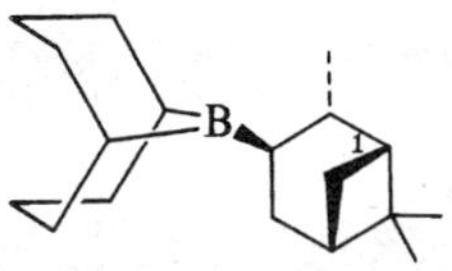

(1′*R*)-*form*

M 258.253
Useful chiral reducing agent for aldehydes and unhindered ketones. Commercially available.
(1′R)-form [64106-79-2]
Liq., air-sensitive. Fp −17°. $[\alpha]_D^{20}$ −3.0° (neat) (coml. sample).
(1′S)-form [76695-88-0]
Air-sensitive liq. Fp −17°. $[\alpha]_D^{22}$ +1.2° (neat) (coml. sample).

Midland, M.M. *et al*, *J. Organomet. Chem.*, 1979, **182**, C53.
Yamamota, Y. *et al*, *Yuki Gosei Kagaku Kyokaishi*, 1979, **37**, 1008; *CA*, **92** R139884j (*synth, use, props*)
Midland, M.M. *et al*, *J. Org. Chem.*, 1981, **46**, 4107.
Brown, H.C. *et al*, *J. Org. Chem.*, 1982, **47**, 1606.
Midland, M.M. *et al*, *J. Am. Chem. Soc.*, 1982, **104**, 525, 528 (*synth, props*)

$C_{18}H_{32}B^{\ominus}$ B-01084

Tributylphenylborate(1−), 10CI, 9CI
Phenyltributylborate

$(H_3CCH_2CH_2CH_2)_3BPh^{\ominus}$

M 259.261 (ion)
Li salt: [54926-82-8].
$C_{18}H_{32}BLi$ M 266.202
Prepd. *in situ* from tributylborane and PhLi. Air- and acid-sensitive. Sol. THF.

Negishi, E. *et al*, *J. Chem. Soc., Chem. Commun.*, 1974, 860.
Negishi, E. *et al*, *J. Chem. Soc., Perkin Trans. 2*, 1978, 1225 (*ir, nmr, synth*)

$C_{18}H_{32}BBr$ B-01085

1-Bromo-1-hexenyl)dicyclohexylborane, 10CI
[67532-01-8]

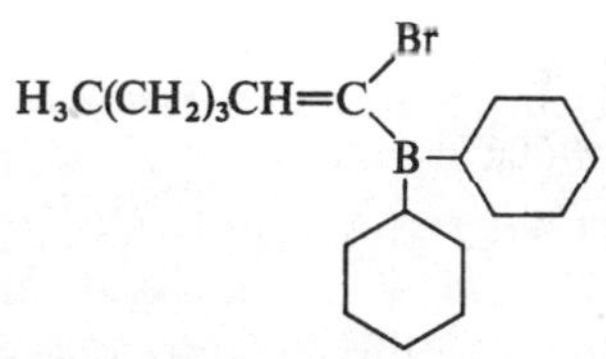

M 339.165
Synthetic intermed. Sensitive to air.

Campbell, J.B. *et al*, *J. Organomet. Chem.*, 1978, **156**, 71 (*synth*)

Masuda, Y. *et al*, *Bull. Chem. Soc. Jpn.*, 1980, **53**, 1652.

$C_{18}H_{32}BCl_2N$ B-01086

Phenyldichloroborane tributylamine
Dichloro(N,N-dibutyl-1-butanamine)phenylboron, 10CI.
Dichlorophenylborane tributylamine
[61133-58-2]

$(H_3CCH_2CH_2CH_2)_3N—BPhCl_2$

M 344.174
Formed in thermogravimetric analysis of appropriate tetraalkylammonium tetrahaloborate. Solid.

Sheikh, S.U., *Bull. Chem. Soc. Jpn.*, 1976, **49**, 2879.
Sheikh, S.U., *Therm. Anal.*, 1977, 509.
Sheikh, S.U., *J. Therm. Anal.*, 1980, **18**, 299 (*synth*)
Sheikh, S.U. *et al*, *Pak. J. Sci. Ind. Res.*, 1980, **23**, 21, 23.

$C_{18}H_{32}BNO_2S$ B-01087

Trihydro[5-methyl-2-(1-methylethyl)-1-[(*N*-methyl-*S*-phenylsulfonimidoyl)methyl]cyclohexanol-*N*]boron, 10CI
5-Methyl-2-(1-methylethyl)-1-[(N-methyl-S-phenylsulfonimidoyl)methyl], compd. with borane(1:1)
[73808-91-0]

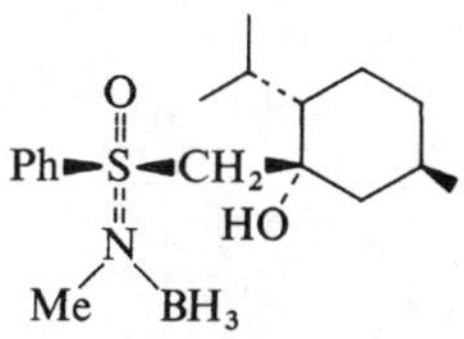

M 337.326
Chiral reducing agent, prepd. *in situ*.

Johnson, C.R. *et al*, *Tetrahedron Lett.*, 1979, 4713.

$C_{18}H_{33}B$ B-01088

9-(1-Butyl-1-hexenyl)-9-borabicyclo[3.3.1]nonane, 10CI
9-(5-Decen-5-yl)-9-BBN
[69322-52-7]

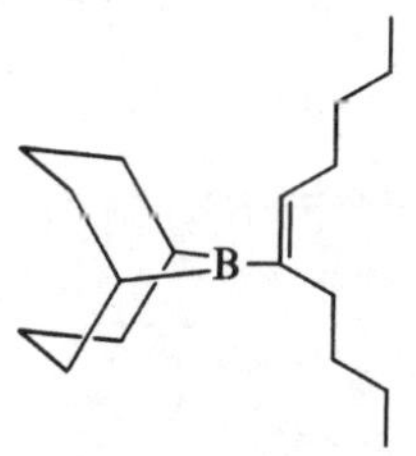

M 260.269
Synthetic intermed. Sensitive to air.

Brown, H.C. *et al*, *J. Am. Chem. Soc.*, 1979, **101**, 96.
Yatagai, H. *et al*, *J. Org. Chem.*, 1979, **44**, 2566 (*pmr*)
Yatagai, H., *Bull. Chem. Soc. Jpn.*, 1980, **53**, 1670.

The symbol ▷ in Entries highlights hazard or toxicity information

$C_{18}H_{33}B$ B-01089

[2-(1-Cyclohexen-1-yl)ethenyl]bis(1,2-dimethylpropyl)borane, 10CI

[2-(1-Cyclohexenyl)ethenyl]bis(3-methyl-2-butyl)borane. [2-(1-Cyclohexenyl)ethenyl]disiamylborane

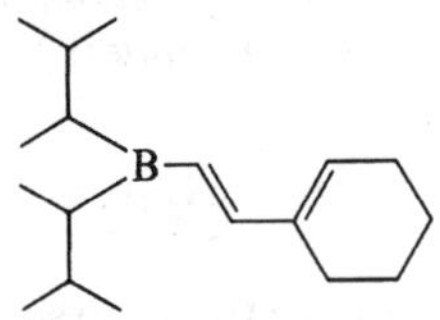

M 260.269

(E)-form [64106-86-1]

Prepd. *in situ* from disiamylborane and cyclohexenylacetylene. Synthetic intermed. Prob. oil, air-sensitive.

Zweifel, G. *et al, J. Am. Chem. Soc.*, 1977, **99**, 5192 (*synth, use*)

$C_{18}H_{33}B$ B-01090

Dicyclohexyl-1-hexenylborane, 10CI

1-Hexenyldicyclohexylborane

[62594-04-1]

As Dicyclohexyl(methoxy)borane, B-00870 with

$$X = -CH{=}CH(CH_2)_3CH_3$$

M 260.269

Liq., sensitive to air. Sol. THF.

(E)-form [37609-12-4]

Prepd. and used *in situ* by hydroboration of 1-hexyne with dicyclohexylborane. Synthetic intermed.

(Z)-form [56962-83-5]

Negishi, E. *et al, J. Organomet. Chem.*, 1975, **92**, C4 (*synth, nmr*)
Campbell, J.B. *et al, J. Organomet. Chem.*, 1978, **156**, 71 (*synth, nmr*)
Zweifel, G. *et al, J. Organomet. Chem.*, 1978, **156**, 159 (*synth*)
Yatagai, H. *et al, J. Org. Chem.*, 1979, **44**, 2566 (*nmr*)
Brown, H.C. *et al, J. Org. Chem.*, 1980, **45**, 550 (*use*)
Campbell, J.B. *et al, J. Org. Chem.*, 1980, **45**, 549 (*use*)

$C_{18}H_{33}B$ B-01091

Tricyclohexylborane, 10CI, 9CI, 8CI

[1088-01-3]

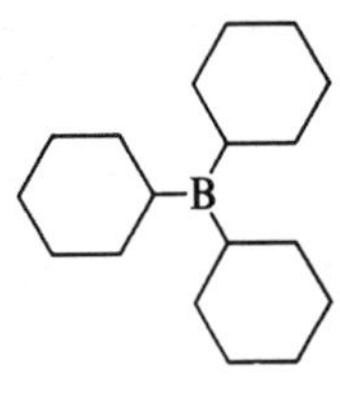

M 260.269

Prepd. by hydroboration of cyclohexene. Synthetic intermed., generally prepd. *in situ*. Air-sensitive solid. Sol. THF. Mp 118°. $Bp_{0.6}$ 125-30°.

Köster, R. *et al, Justus Liebigs Ann. Chem.*, 1958, **618**, 31 (*synth*)
Köster, R. *et al, Justus Liebigs Ann. Chem.*, 1965, **682**, 21 (*props*)
Finch, A.P.J. *et al, Trans Faraday Soc.*, 1967, **63**, 1880 (*props*)
Brown, H.C. *et al, J. Org. Chem.*, 1973, **38**, 2422; 1981, **46**, 2712 (*synth*)
Manzer, L.E. *et al, Inorg. Chem.*, 1976, **15**, 3114 (*nmr*)

$C_{18}H_{33}B$ B-01092

Tris(2-methylcyclopentyl)borane, 10CI, 9CI

[51134-66-8]

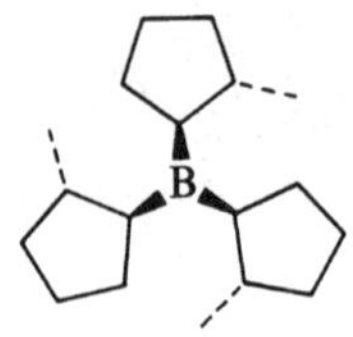

M 260.269

trans,trans,trans-form [40911-83-9]

Prepd. by hydroboration of 1-methylcyclopentene with borane-THF. Reagent for transfer of 2-methylcyclopentyl group. Air-sensitive. Sol. THF, other non-hydroxylic solvs.

Leung, T. *et al, J. Am. Chem. Soc.*, 1974, **96**, 5620 (*use*)
Utimoto, K. *et al, J. Am. Chem. Soc.*, 1974, **96**, 5601 (*use*)
Zweifel, G. *et al, Synthesis*, 1975, 376 (*use*)
Pelter, A. *et al, J. Chem. Soc., Dalton Trans.*, 1976, 2087 (*use*)
Zweifel, G. *et al, J. Am. Chem. Soc.*, 1978, **100**, 5561 (*use*)
Brown, C.A. *et al, J. Am. Chem. Soc.*, 1979, **101**, 3964 (*use*)

$C_{18}H_{33}BO_3$ B-01093

Tricyclohexyloxyborane

Boric acid (H_3BO_3) tricyclohexyl ester, 10CI, 9CI, 8CI. Tris(cyclohexyloxy)borane. Tricyclohexyl borate. Tricyclohexyl orthoborate. Cyclohexyl borate

[2467-16-5]

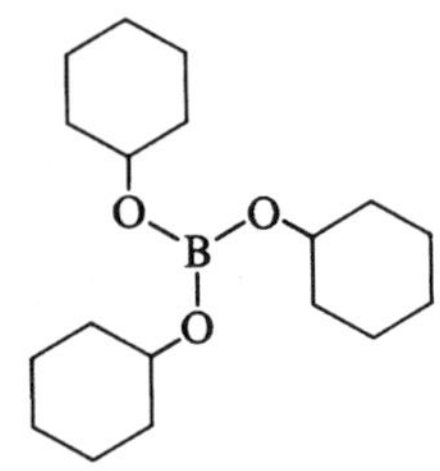

M 308.267

Catalyst for oxidn. of cycloalkanes, gelling agent, sterilant for control of *Cochliomyia hominivorax*. Solid, moisture-sensitive. Sol. non-hydroxylic solvs. Mp 56.4°. Bp_{750} 323-6°, Bp_{17} 201-3°.

Kustov, V.K. *et al, Izv. Akad. Nauk SSSR, Ser. Khim.*, 1956, 733 (*synth*)
O'Brien, K.G., *Aust. J. Chem.*, 1957, **10**, 91 (*synth, props*)
Strauss, I. *et al, CA*, 1971, **74**, 69142t (*props*)
Maijs, L. *et al, CA*, 1973, **79**, 131073e (*pmr*)
Itskovich, V.A. *et al, CA*, 1978, **89**, 196672a (*use*)
Japan. Pat., 79 24 826, (*1979*); *CA*, **91**, 123403v (*synth*)

$C_{18}H_{33}BS$ B-01094

7-Pentyl-3-pentylthio-3-borabicyclo[3.3.1]non-6-ene, 10CI

7-Amyl-3-amylthio-3-borabicyclo[3.3.1]non-6-ene

[67682-94-4]

As 7-(Chloromethyl)-3-methoxy-3-borabicyclo[3.3.1]-non-6-ene, B-00670 with

$$R = H_3C(CH_2)_4S—,\ R' = —(CH_2)_4CH_3$$

M 292.329

Air- and moisture-sensitive liq. $Bp_{2.5}$ 174-6°. n_D^{20} 1.5021.

Mikhailov, B.M. *et al, Izv. Akad. Nauk SSSR, Ser. Khim.*, 1978, 1594; *Bull. Acad. Sci. USSR, Div. Chem. Sci.*, 1392 (*synth, nmr*)

Mikhailov, B.M. *et al*, *J. Organomet. Chem.*, 1978, **154**, 113 (*nmr, ir*)

$C_{18}H_{33}B_3O_6$ B-01095

Tris(cyclohexyloxy)boroxin, 10CI, 9CI, 8CI

Cyclohexyl metaborate trimer. CHMB

[1172-69-6]

As Boroxin, B-00019 with

R = cyclohexyl

M 377.885

Used to induce dec. of alkyl hydroperoxides. Cryst. (C_6H_6/pet. ether); moisture-sensitive. Mp 166-7° (215-25°).

O'Connor, G.L. *et al*, *J. Am. Chem. Soc.*, 1955, **77**, 1578 (*synth*)
O'Brien, K.G. *et al*, *Aust. J. Chem.*, 1963, **5**, 915 (*synth*)
Steinberg, H., *Organoboron Chemistry*, 1964, Wiley, N.Y., Vol. 1.
Wolf, P.E. *et al*, *J. Org. Chem.*, 1975, **40**, 1875 (*synth, use*)
Dzhemilev, U.M. *et al*, *Izv. Akad. Nauk SSSR, Ser. Khim.*, 1979, 553; *Bull. Acad. Sci. USSR, Div. Chem. Sci.*, 509 (*use*)
Won, P.L., *Neftekhimiya*, 1980, **204**, 568; *CA*, **93**, 23844t.

$C_{18}H_{34}B^{\ominus}$ B-01096

Hydrotris(2-methylcyclopentyl)borate(1−), 10CI, 9CI

Tris(2-methylcyclopentyl)borohydride

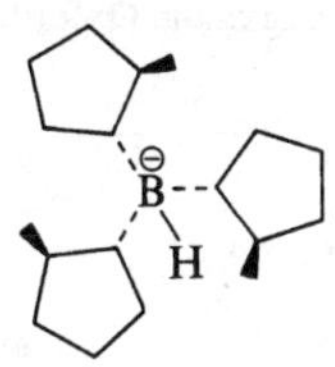

M 261.277 (ion)

trans,trans,trans-form

Li salt: [60284-40-4].

$C_{18}H_{34}BLi$ M 268.218

Stereoselective reducing agent. Air and moisture-sensitive.

Krishnamurthy, S. *et al*, *J. Am. Chem. Soc.*, 1976, **98**, 3383 (*synth, use*)
Hutchins, R.O., *J. Org. Chem.*, 1977, **42**, 920 (*props, use*)
Brown, H.C. *et al*, *J. Org. Chem.*, 1979, **44**, 5004 (*use*)
Brown, H.C. *et al*, *J. Organomet. Chem.*, 1979, **166**, 271.
Brown, H.C. *et al*, *J. Organomet. Chem.*, 1980, **188**, 1 (*synth, nmr*)

$C_{18}H_{34}B^{\ominus}$ B-01097

Tricyclohexylhydroborate(1−), 10CI

Tricyclohexylborohydride

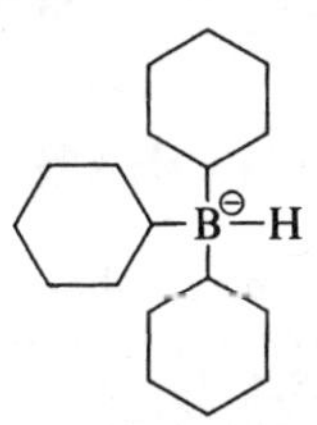

M 261.277 (ion)

Li salt: [69786-43-2].

$C_{18}H_{34}BLi$ M 268.218

Not isolated in pure, solvent-free state. Prepd. *in situ*. Stereoselective reducing agent. Air- and moisture-sensitive. Sol. THF.

Na salt: [67276-06-6].

$C_{18}H_{34}BNa$ M 284.266

Prepd. *in situ*. Stereoselective reducing agent. Air- and moisture-sensitive. Sol. THF. Stable under dry N_2.

K salt: [63802-48-2].

$C_{18}H_{34}BK$ M 300.375

Prepd. *in situ*. Stereoselective reducing agent. Air- and moisture-sensitive. Sol. THF. Indefinitely stable under dry N_2.

Strouf, O. *et al*, *Acta Chem. Scand., Ser. A*, 1977, **31**, 391 (*props*)
Brown, C.A. *et al*, *J. Organomet. Chem.*, 1978, **56**, 111 (*synth, ir, nmr*)
Brown, H.C. *et al*, *J. Am. Chem. Soc.*, 1978, **100**, 3343 (*synth, ir, nmr*)
Brown, H.C. *et al*, *J. Org. Chem.*, 1979, **44**, 467, 5004 (*use, nmr, synth*)
Brown, H.C. *et al*, *J. Organomet. Chem.*, 1979, **166**, 271 (*synth, props*)
Brown, H.C. *et al*, *J. Organomet. Chem.*, 1980, **188**, 1 (*synth, nmr, ir*)

$C_{18}H_{35}B$ B-01098

Dicyclohexyl(1-ethylbutyl)borane, 10CI

Dicyclohexyl(3-hexyl)borane. (1-Ethylbutyl)dicyclohexylborane

[72487-19-5]

As Dicyclohexyl(methoxy)borane, B-00870 with

R = $H_3CCH_2CH_2CH(CH_2CH_3)$–

M 262.285

Prepd. *in situ* from dicyclohexylborane dimer and 3-hexene in diglyme at 0°. 3-Hexyl group transfer agent. Air-sensitive. Sol. THF.

Taniguchi, H. *et al*, *Bull. Chem. Soc. Jpn.*, 1979, **52**, 2942 (*synth, use*)
Brown, H.C. *et al*, *J. Org. Chem.*, 1981, **46**, 4313 (*synth*)

$C_{18}H_{35}B$ B-01099

Dicyclohexylhexylborane, 10CI, 9CI

Hexyldicyclohexylborane

[52251-52-2]

As Dicyclohexyl(methoxy)borane, B-00870 with

R = $-(CH_2)_5CH_3$

M 262.285

Prepd. *in situ* by hydroboration of 1-hexene with dicyclohexylborane. Synthetic intermed. for transfer of a hexyl group. Air-sensitive. Sol. THF.

Larock, R.C. *et al*, *J. Organomet. Chem.*, 1974, **67**, 353 (*use*)
Masuda, Y. *et al*, *Bull. Chem. Soc. Jpn.*, 1978, **51**, 901 (*use*)
Taniguchi, H. *et al*, *Bull. Chem. Soc. Jpn.*, 1979, **52**, 2942 (*use*)
Pelter, A. *et al*, *Comprehensive Organic Chemistry*, (Barton, *D.H.R. et al*, Ed.), Pergamon, Oxford, 1979, **3**, 695 (*rev*)
Brown, H.C. *et al*, *J. Org. Chem.*, 1981, **46**, 4313 (*synth*)

$C_{18}H_{35}BO_3$ B-01100

Dibutoxy(3,7-dimethyl-1,6-octadien-3-yloxy)borane

Boric acid (H_3BO_3) dibutyl 1-ethenyl-1,5-dimethyl-4-hexenyl ester, 10CI

[76184-42-4]

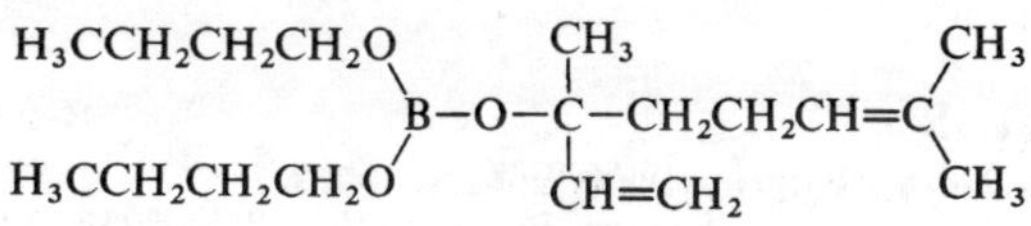

M 310.283

Moisture-sensitive.

Eur. Pat., 12 198, (*1980*); *CA*, **94**, 47554q (*synth, props*)
U.S.P., 4 254 291, (*1981*); *CA*, **95**, 81277g (*props*)

$C_{18}H_{36}B^{\ominus}$ B-01101

Tributyl-1-hexynylborate(1−), 10CI, 9CI

$$(H_3CCH_2CH_2CH_2)_3BC{\equiv}C(CH_2)_3CH_3^{\ominus}$$

M 263.292 (ion)

Li salt: [42126-47-6].
$C_{18}H_{36}BLi$ M 270.233
Prepd. *in situ* from tributylborane and hexynyllithium. Synth. intermed. Prob. oil. Air- and acid-sensitive. Sol. THF, glyme, diglyme.

Suzuki, A. *et al*, *J. Am. Chem. Soc.*, 1973, **95**, 3080 (*use*)
Pelter, A. *et al*, *J. Chem. Soc., Perkin Trans. 1*, 1976, 2419, 2435 (*use*)
Mikhailov, B.M. *et al*, *Izv. Akad. Nauk SSSR, Ser. Khim.*, 1977, 2633; *Bull. Acad. Sci. USSR, Div. Chem. Sci.*, **26**, 2437 (*synth*)
Negishi, E. *et al*, *J. Chem. Soc., Perkin Trans. 2*, 1978, 1225 (*nmr, ir*)

$C_{18}H_{36}BBr$ B-01102

(1-Bromo-1-hexenyl)bis(2-methylpentyl)borane, 10CI

[67532-04-1]

$$[H_3CCH_2CH_2CH(CH_3)CH_2]_2BCBr{=}CH(CH_2)_3CH_3$$

M 343.196

Synthetic intermed. Sensitive to air. Sol. THF.

Masuda, Y. *et al*, *Bull. Chem. Soc. Jpn.*, 1980, **53**, 1652.

$C_{18}H_{36}BBr$ B-01103

(1-Bromo-1-hexenyl)dihexylborane, 10CI

[67532-03-0]

$$[H_3C(CH_2)_5]_2BCBr{=}CH(CH_2)_3CH_3$$

M 343.196

Synthetic intermed. Sensitive to air. Sol. THF.

Masuda, Y. *et al*, *Bull. Chem. Soc. Jpn.*, 1980, **53**, 1652.

$C_{18}H_{37}B$ B-01104

Bis(1,2-dimethylpropyl)-1-octenylborane, 10CI, 9CI

1-Octenyldisiamylborane. Bis(3-methyl-2-butyl)-1-octenylborane

[41285-63-6]

$$[(H_3C)_2CHCH(CH_3)]_2BCH{=}CH(CH_2)_5CH_3$$

M 264.300

Prepd. *in situ* from disiamylborane and 1-octyne (gives *trans*-isomer). Synthetic intermed. Air-sensitive. Sol. THF.

Kow, R. *et al*, *J. Am. Chem. Soc.*, 1973, **95**, 2715 (*synth*)
Miyaura, N., *et al*, *Tetrahedron Lett.*, 1979, 3437; 1980, 2865 (*synth*)
Miyaura, N. *et al*, *Chem. Lett.*, 1981, 879 (*synth*)

$C_{18}H_{39}B$ B-01105

Bis(1,2-dimethylpropyl)octylborane, 10CI, 9CI

Disiamyl-1-octylborane. Bis(3-methyl-2-butyl)octylborane

[32327-50-7]

$$[(H_3C)_2CHCH(CH_3)]_2B(CH_2)_7CH_3$$

M 266.316

Prepd. *in situ* by hydroboration of 1-octene with disiamylborane. Synthetic intermed. Sensitive to air. Sol. THF and other inert org. solvs.

Bigley, D.B. *et al*, *J. Inorg. Nucl. Chem.*, 1971, **33**, 1157 (*props*)
Org. Synth., 1973, **53**, 77 (*synth*)
Brown, H.C. *et al*, *Synthesis*, 1980, **2**, 151 (*synth*)

$C_{18}H_{39}B$ B-01106

Dihexyl(1,1,2-trimethylpropyl)borane, 10CI, 9CI

Dihexyl(2,3-dimethyl-2-butyl)borane. Thexyldihexylborane. Dihexylthexylborane

[30038-49-4]

$$(H_3C)_2CHC(CH_3)_2B[(CH_2)_5CH_3]_2$$

M 266.316

Prepd. *in situ* by reaction of 1-hexene with thexylborane. Synth. intermed., e.g. for prepn. of 7-tridecanone. Air-sensitive liq. Sol. THF.

Pelter, A. *et al*, *J. Chem. Soc., Perkin Trans. 1*, 1975, 145 (*use*)
Pelter, A. *et al*, *Tetrahedron Lett.*, 1975, 1633 (*use*)
Arase, A. *et al*, *Chem. Lett.*, 1976, 1115 (*synth*)
Pelter, A. *et al*, *Comprehensive Organic Chemistry*, (*Barton, D.H.R. et al*, Ed.), Pergamon, Oxford, 1979, **3**, 695 (*rev*)

$C_{18}H_{39}B$ B-01107

Trihexylborane, 10CI, 9CI, 8CI

[1188-92-7]

$$[H_3C(CH_2)_5]_3B$$

M 266.316

Generally prepd. *in situ* by hydroboration of 1-hexene. Synthetic intermed. used for transfer of 1-hexyl group. Air-sensitive liq. Sol. THF and other relatively nonpolar solvs. d^{20} 0.774. Bp_8 112°. n_D^{20} 1.4412.

McCusker, P.A. *et al*, *J. Am. Chem. Soc.*, 1955, **77**, 4253 (*synth*)
Brown, H.C. *et al*, *J. Org. Chem.*, 1957, **22**, 1137 (*synth*)
Timofeeva, G.A., *Izv. Akad. Nauk SSSR, Ser. Khim.*, 1967, 1152 (*props*)
Gal'chenko, G.L. *et al*, *Zh. Fiz. Khim.*, 1967, **41**, 2181 (*props*)
Org. Synth., 1973, **53**, 77 (*synth*)
Kabalka, G.W. *et al*, *J. Chem. Educ.*, 1976, **53**, 549 (*rev*)

$C_{18}H_{39}B$ B-01108

Tris(1-ethylbutyl)borane, 10CI, 9CI, 8CI

Tri-3-hexylborane

[1883-34-7]

$$B[CH(CH_2CH_3)CH_2CH_2CH_3]_3$$

M 266.316

Prepd. by hydroboration of 3-hexene. Synthetic intermed. Air-sensitive oil. Sol. C_6H_6, THF and other non-polar solvs. $Bp_{0.5}$ 79-80°.

Fr. Pat., 1 393 282, (*1965*); *CA*, **63**, 4331h (*synth*)
Bigley, D.B. *et al*, *J. Chem. Soc.* (*B*), 1970, 1811 (*use*)
Leung, T. *et al*, *J. Am. Chem. Soc.*, 1974, **96**, 5620 (*use*)
Zweifel, G. *et al*, *Synthesis*, 1975, 376 (*use*)
Larson, G.L. *et al*, *J. Organomet. Chem.*, 1980, **198**, 15 (*use*)
Brown, H.C. *et al*, *J. Org. Chem.*, 1981, **46**, 4313 (*props*)

$C_{18}H_{39}B$ B-01109

Tris(2-methylpentyl)borane, 10CI, 9CI, 8CI

[1188-50-7]

$[H_3CCH_2CH_2CH(CH_3)CH_2]_3B$

M 266.316

Prepd. by hydroboration of 2-methyl-1-pentene. Synthetic intermed., usually prepd. *in situ*, for transfer of a 2-methylpentyl group. Air-sensitive liq. Sol. pentane, THF.

Brown, H.C. *et al*, *J. Am. Chem. Soc.*, 1971, **93**, 1506 (*props*)
Negishi, E. *et al*, *J. Org. Chem.*, 1975, **40**, 814 (*props*)
Zweifel, G. *et al*, *Synthesis*, 1975, 376 (*use*)
Kabalka, G.W. *et al*, *J. Organomet. Chem.*, 1977, **125**, 273 (*props*)
Marinelli, E.R. *et al*, *Tetrahedron Lett.*, 1979, 2313 (*use*)
Kabalka, G.W. *et al*, *J. Org. Chem.*, 1981, **46**, 4296 (*synth*)

$C_{18}H_{39}BO_3$ B-01110

Trihexyloxyborane

Boric acid (H_3BO_3) trihexyl ester, 10CI, 9CI. *1-Hexanol triester with boric acid (H_3BO_3). Trihexyl borate. Trihexyl orthoborate. Hexyl borate*

[5337-36-0]

$[H_3C(CH_2)_5O]_3B$

M 314.314

Gelling agent for liq. hydrocarbons, secondary plasticizer for vinyl chloride polymers; used in pharmaceutical aerosols; insect sterilant. Moisture-sensitive liq. d_4^{28}0.847. Bp 310-1°, Bp_{15} 182.2-183°. n_D^{20} 1.4250.

▷ED5120000.

Scattergood, A. *et al*, *J. Am. Chem. Soc.*, 1945, **67**, 2150 (*synth, props*)
Steinberg, H. *et al*, *Ind. Eng. Chem.*, 1957, **49**, 174 (*synth, props*)
Strauss, I. *et al*, *CA*, 1972, **77**, 145170e (*props*)
Brown, H.C. *et al*, *Synthesis*, 1979, 702 (*props*)

$C_{18}H_{42}B_2N_2$ B-01111

Bis(2-methylcyclopentylborane) tetramethylethylenediamine

Tetrahydrobis(2-methylcyclopentyl)[μ-(N,N,N′,N′-tetramethyl-1,2-ethanediamine-N:N′)]diboron, 10CI

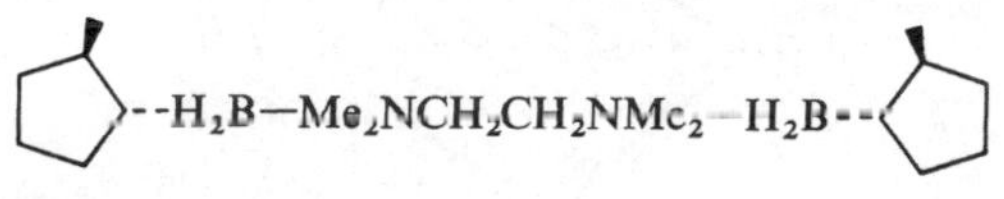

M 308.163

Convenient way of storing the otherwise unstable monoalkylborane.

trans-form [69319-82-0]

N,N,N′,N′-*Tetramethyl-1,2-ethanediamine compd. with* trans-*(2-methylcyclopentyl)borane(1:2)*

Solid. Mp 123-4°.

Brown, H.C. *et al*, *J. Org. Chem.*, 1979, **44**, 465 (*nmr, synth*)

$C_{19}H_3BF_{15}^{\ominus}$ B-01112

Methyltris(pentafluorophenyl)borate(1−), 10CI

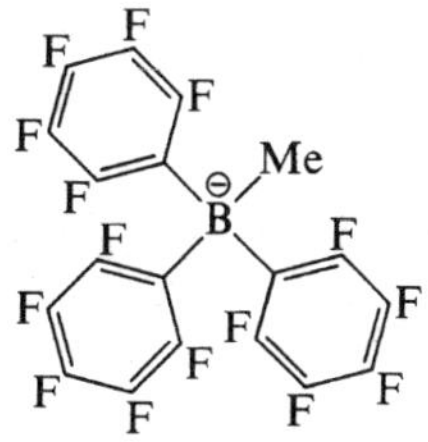

M 527.019 (ion)

Li salt: [70083-57-7].

$C_{19}H_3BF_{15}Li$ M 533.960

Used in battery electrolytes.

U.S.P., 4 139 681, (*1979*); *CA*, **90**, 207210 (*synth, props*)
Ger. Pat., 2 901 367, (*1979*); *CA*, **92**, 61799p (*synth*)
Horowitz, H.H. *et al*, *Proc. Electrochem. Soc.*, 1981, **81**, 131.

$C_{19}H_{15}BF_3^{\ominus}$ B-01113

Tris(4-fluorophenyl)methylborate(1−), 10CI

M 311.133 (ion)

Li salt: [78955-00-7].

$C_{19}H_{15}BF_3Li$ M 318.074

Used in battery electrolytes.

Horowitz, H.H. *et al*, *Proc. Electrochem. Soc.*, 1981, **81**, 131 (*use, props*)
Klemann, L.P. *et al*, *Proc. Electrochem. Soc.*, 1981, **81**, 179 (*props*)

$C_{19}H_{15}BFeO_2$ B-01114

Dicarbonyl(η⁵-2,4-cyclopentadien-1-yl)diphenylboraneiron

Dicarbonyl-π-cyclopentadienyl(diphenylboryl)iron, 8CI. *Dicarbonyl(η^5-cyclopentadienyl)irondiphenylborane. (η^5-Cyclopentadienyl)dicarbonyl(diphenylboryl)iron*

[12277-78-0]

As Dicarbonyl(η^5-2,4-cyclopentadien-1-yl)-dichloroboraneiron, B-00462 with

X = Ph

M 341.983

Beige solid. Mp 113-6°.

Schmid, G. *et al*, *Chem. Ber.*, 1967, **100**, 2899 (*synth*)

$C_{19}H_{15}BN^{\ominus}$ B-01115

(Cyano-*C*)triphenylborate(1−), 10CI, 9CI

Triphenylcyanoborate

[47107-21-1]

$[Ph_3BCN]^{\ominus}$

M 268.144 (ion)

Na salt: [14568-16-2]. *Cesignost.*

$C_{19}H_{15}BNNa$ M 291.134

Used to separate Cs-137 and Be-7 in radioactive fallout and to detect Cs-137 in feed. Used in solvent extraction of alkali met

als. Monohydrate. Commercially available.
Cs salt: [52337-56-1].
$C_{19}H_{15}BCsN$ M 401.050
Nmr shift reagent. Analytical reagent for mechanistic studies.

Lee, D.A., *J. Inorg. Nucl. Chem.*, 1972, **34**, 2895 (*use*)
Bunge, M. *et al*, *Eur. Polym. J.*, 1977, **13**, 283 (*use*)
Pacey, G. *et al*, *Talanta*, 1980, **27**, 1013 (*use*)
Selig, W., *Microchem. J.*, 1980, **25**, 200.
Stankovic, S. *et al*, *CA*, 1980, **92**, 36885f (*use*)

$C_{19}H_{15}BN_2O_2$ B-01116

[2-[(2-Hydroxyphenyl)azo]-4-methylphenolato(2−)]phenylboron, 10CI
[74302-86-6]
As [4-[(2-Hydroxyphenyl)azo]-1,3-benzenediolato(2−)]-phenylboron, B-01050 with

$$R = CH_3, R' = H$$

M 314.150
Dark-red cryst. (pet. ether/C_6H_6). Mp 126°.

Hohaus, E. *et al*, *Z. Naturforsch.*, *B*, 1980, **35**, 316, 319 (*ms, synth*)

$C_{19}H_{17}BNP$ B-01117

Cyanoborane triphenylphosphine
(Cyano-C)dihydro(triphenylphosphine)boron, 10CI. *Boranecarbonitrile, compd. with triphenylphosphine(1:1)*
[67374-60-1]

$$Ph_3P—BH_2CN$$

M 301.134
White solid. Mp 172-4°.

Martin, D.R. *et al*, *J. Inorg. Nucl. Chem.*, 1978, **40**, 9 (*synth, nmr, ir*)
Wisian-Nielson, P. *et al*, *J. Inorg. Nucl. Chem.*, 1981, **43**, 457 (*props*)

$C_{19}H_{18}B^{\ominus}$ B-01118

Methyltriphenylborate(1−), 10CI

$$MeBPh_3^{\ominus}$$

M 257.161 (ion)
Li salt: [78885-82-2].
$C_{19}H_{18}BLi$ M 264.102
Used in battery electrolytes. Air- and moisture-sensitive.

Horowitz, H.H. *et al*, *Proc. Electrochem. Soc.*, 1981, **81**, 131.
Klemann, L.P. *et al*, *Proc. Electrochem. Soc.*, 1981, **81**, 179.

$C_{19}H_{19}BN_2O$ B-01119

Diphenyl[1-(2-pyridinylamino)ethanolato]boron, 10CI, 9CI
Diphenylboryl-[2-(α-hydroxyethyl)aminopyridinate]
[61300-57-0]
As Dipropyl[1-(2-pyridinylamino)-1-butanolato]boron, B-00967 with

$$R^1 = CH_3, R^2 = Ph$$

M 302.182
Cryst., mod. stable in air. Mp 165-8° dec.

Gragg, B.R. *et al*, *J. Organomet. Chem.*, 1976, **116**, 135.
Dorokov, V.A. *et al*, *Izv. Akad. Nauk SSSR, Ser. Khim.*, 1977, 1587; *Bull. Acad. Sci. USSR, Div. Chem. Sci.*, 1458 (*synth, ir, nmr, pmr*)

$C_{19}H_{20}BN_3$ B-01120

Triphenylborane guanidine
(Guanidine-N″)triphenylboron, 10CI. *Guanidine compd. with triphenylborane(1:1)*
[67007-35-6]

$$HN{=}C(NH_2)NH_2—BPh_3$$

M 301.197

Yuzhakova, G.A. *et al*, *Zh. Obshch. Khim.*, 1978, **48**, 811 (*synth*)

$C_{19}H_{22}BCo$ B-01121

[(1,2,5,6-η)-1,5-Cyclooctadiene][(1,2,3,4,5,6-η)-1-phenylboratabenzene]cobalt, 10CI
[69021-83-6]

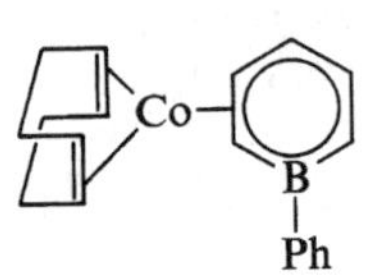

M 320.126
Yellow-brown cryst. Mp 168-9°. ^{11}B nmr δ +20.9 (C_6D_6, reference $BF_3{\cdot}Et_2O$).

Herberich, G.E. *et al*, *J. Organomet. Chem.*, 1978, **160**, 17 (*synth, ms, nmr*)

$C_{19}H_{24}BF_6IrN_6O_5$ B-01122

Aquabis(trifluoroacetato-*O*)[tris(3,5-dimethyl-1*H*-pyrazolato-*N*¹)hydroborato(1−)-*N*², *N*²′, *N*²″]iridium, 10CI
[Hydrotris(3,5-dimethylpyrazolyl)borato]bis(trifluoroacetato)aquoiridium
[73117-36-9]

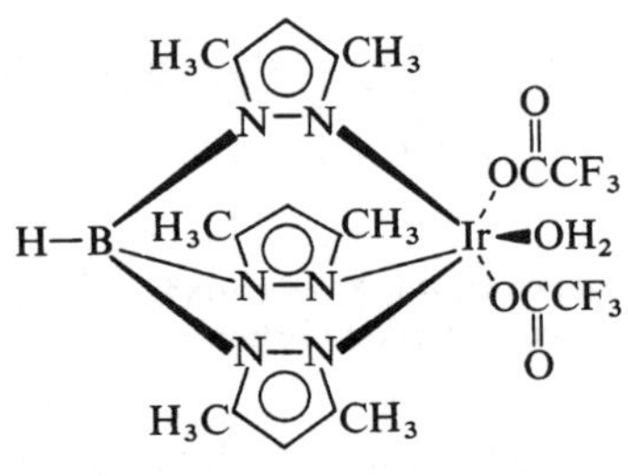

M 733.456
Pale-yellow needles (toluene). Mp 166-7°.

May, S. *et al*, *Inorg. Chem.*, 1980, **19**, 1582 (*synth, pmr*)

$C_{19}H_{24}BF_6N_6O_5Rh$ **B-01123**

Aquabis(trifluoroacetato-*O*)[tris(3,5-dimethyl-1*H*-pyrazolato-*N*¹)hydroborato(1−)-*N*²,*N*²′,*N*²″]rhodium, 10CI

[Hydrotris(3,5-dimethylpyrazolyl)borato]bis(trifluoroacetato)aquorhodium

[73117-64-3]

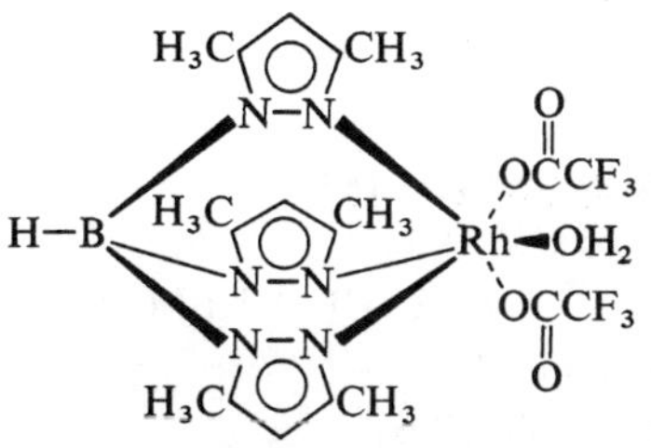

M 644.142

Pale-yellow plates (CH_2Cl_2/hexane). Mp 231-3°.

May, S. *et al*, *Inorg. Chem.*, 1980, **19**, 1582 (*synth, pmr*)

$C_{19}H_{25}B$ **B-01124**

Methylbis(2,4,6-trimethylphenyl)borane, 10CI

Dimesitylmethylborane. Methyldimesitylborane

[74178-56-6]

As Fluorobis(2,4,6-trimethylphenyl)borane, B-01073 with

X = Me

M 264.217

Prepd. from Fluorobis(2,4,6-trimethylphenyl)borane, B-01073 and methylmagnesium bromide. Synthetic intermed. Air-sensitive. Sol. THF, $CHCl_3$. Mp 78°. pmr 2.21 (s), 2.27 (s), 6.80 (s), 1.50 (s) ppm ($CDCl_3$).

Finocchiaro, P. *et al*, *J. Am. Chem. Soc.*, 1973, **95**, 7029 (*synth, pmr*)
Pelter, A. *et al*, *Compr. Org. Chem.*, (*Barton, D.H.R. et al*, Ed.), 1979, **3**, 695 (*rev*)
Brown, N.M.D. *et al*, *J. Organomet. Chem.*, 1980, **185**, 277 (*nmr*)
Wilson, J.W. *et al*, *J. Organomet. Chem.*, 1980, **186**, 297 (*synth*)
Pelter, A. *et al*, *Tetrahedron Lett.*, 1983, **24**, 625, 627 (*use*)

$C_{19}H_{25}BN_6Rh^{\oplus}$ **B-01125**

[Hydrotris(1*H*-pyrazolato-*N*¹)borato(1−)-*N*²,*N*²′,*N*²″](η⁵-pentamethylcyclopentadienyl)rhodium(1+)

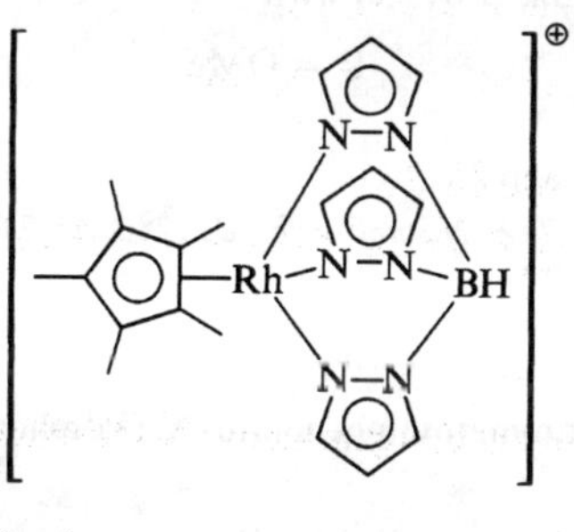

M 451.162 (ion)

Hexafluorophosphate: [52015-82-4].
$C_{19}H_{25}BF_6N_6PRh$ M 596.126
Yellow, air-stable cryst. (CH_2Cl_2/Et_2O). Mp 220-7°.

Restivo, R.J. *et al*, *Inorg. Chem.*, 1975, **14**, 3046 (*synth, ir, pmr, cryst struct*)

$C_{19}H_{25}BO$ **B-01126**

Methoxybis(2,4,6-trimethylphenyl)borane

Bis(2,4,6-trimethylphenyl)borinic acid methyl ester, 10CI, 9CI. *Dimesitylborinic acid methyl ester. Dimesitylmethoxyborane*

[50517-71-0]

As Hydroxybis(2,4,6-trimethylphenyl)borane, B-01075 with

R = OMe

M 280.216

Colourless cryst. solid, stable in air. Sol. most org. solvs. Mp 75-6°. Slowly oxid. in soln.

Finocchiaro, P. *et al*, *J. Am. Chem. Soc.*, 1973, **95**, 7029 (*nmr*)
Brown, N.M.D. *et al*, *J. Organomet. Chem.*, 1980, **193**, 271 (*synth, nmr, cmr*)
Brown, N.M.D. *et al*, *J. Organomet. Chem.*, 1981, **210**, 1 (*nmr*)

$C_{19}H_{25}BS$ **B-01127**

(Methylthio)bis(2,4,6-trimethylphenyl)borane

Thiobis(2,4,6-trimethylphenyl)borinic acid methyl ester, 10CI. *Dimesityl(methylthio)borane. (Methylthio)-dimesitylborane*

[77657-15-9]

As Hydroxybis(2,4,6-trimethylphenyl)borane, B-01075 with

R = −SMe

M 296.277

Moisture-sensitive white cryst. (pentane or pentane/Et_2O). Mp 99-102°.

Davidson, F. *et al*, *J. Organomet. Chem.*, 1981, **204**, 147 (*synth, pmr, cmr, nmr*)
Brown, N.M.D. *et al*, *J. Organomet. Chem.*, 1981, **210**, 1 (*cmr*)

$C_{19}H_{26}BN$ **B-01128**

(Methylamino)bis(2,4,6-trimethylphenyl)borane

N-Methyl-1,1-bis(2,4,6-trimethylphenyl)boranamine, 10CI. *Dimesityl(methylamino)borane*

[75018-25-6]

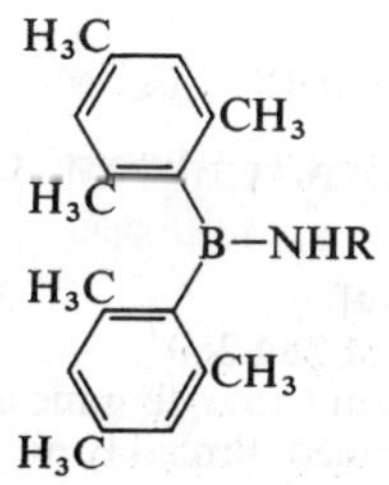

R = Me

M 279.231

Cryst. (hexane), moisture-sensitive. Sol. most org. solvs. Mp 131-4°.

Brown, N.M.D. *et al*, *J. Organomet. Chem.*, 1980, **192**, 133 (*synth, nmr, ir*)
Brown, N.M.D. *et al*, *J. Organomet. Chem.*, 1981, **210**, 1 (*cmr*)

$C_{19}H_{33}BS$ **B-01129**

Dibutyl[1-(phenylthio)pentyl]borane, 10CI, 9CI

[67808-82-6]

$(H_3CCH_2CH_2CH_2)_2BCH(SPh)CH_2CH_2CH_2CH_3$

M 304.340

Synthetic intermed. Sensitive to air. Sol. THF.

Mendoza, A. *et al*, *J. Organomet. Chem.*, 1978, **156**, 149 (*synth*)
Mendoza, A. *et al*, *J. Chem. Soc., Chem. Commun.*, 1978, 357.

$C_{19}H_{35}BO$ B-01130

3-(Hexyloxy)-7-pentyl-3-borabicyclo[3.3.1]non-6-ene, 10CI
7-Amyl-3-hexyloxy-3-borabicyclo[3.3.1]non-6-ene
[67682-96-6]
As 7-(Chloromethyl)-3-methoxy-3-borabicyclo[3.3.1]-non-6-ene, B-00670 with

$$R = —O(CH_2)_5CH_3,\ R' = —(CH_2)_4CH_3$$

M 290.295
Liq., air- and moisture-sensitive. Bp_2 159-60°. n_D^{20} 1.4792.

Mikhailov, B.M. *et al*, *Izv. Akad. Nauk SSSR, Ser. Khim.*, 1978, 1594; *Bull. Acad. Sci. USSR, Div. Chem. Sci.*, 1392 (*synth, nmr*)
Mikhailov, B.M. *et al*, *J. Organomet. Chem.*, 1978, **154**, 113 (*nmr*)

$C_{19}H_{35}BS$ B-01131

3-(Hexylthio)-7-pentyl-3-borabicyclo[3.3.1]non-6-ene, 10CI
7-Amyl-3-hexylthio-3-borabicyclo[3.3.1]non-6-ene
[67682-95-5]
As 7-(Chloromethyl)-3-methoxy-3-borabicyclo[3.3.1]-non-6-ene, B-00670 with

$$R = —S(CH_2)_5CH_3,\ R' = —(CH_2)_4CH_3$$

M 306.356
Air- and moisture-sensitive liq. Bp_2 198-200°. n_D^{20} 1.5002.

Mikhailov, B.M. *et al*, *Izv. Akad. Nauk SSSR, Ser. Khim.*, 1978, 1594; *Bull. Acad. Sci. USSR, Div. Chem. Sci.*, 1392 (*synth, nmr*)
Mikhailov, B.M. *et al*, *J. Organomet. Chem.*, 1978, **154**, 113 (*nmr*)

$C_{19}H_{38}B^{\ominus}$ B-01132

Tributyl-1-heptynylborate(1−), 10CI, 9CI

$$(H_3CCH_2CH_2CH_2)_3BC{\equiv}C(CH_2)_4CH_3^{\ominus}$$

M 277.319 (ion)
Li salt: [43128-72-9].
$C_{19}H_{38}BLi$ M 284.260
Prepd. *in situ* from tributylborane and heptynyllithium. Synth. intermed. Probably oil. Air- and acid-sensitive. Sol. THF, glyme, diglyme.

Naruse, M. *et al*, *Tetrahedron Lett.*, 1973, 795, 2741 (*use*)
Naruse, M. *et al*, *Tetrahedron*, 1974, **30**, 835 (*use*)
Pelter, A. *et al*, *J. Chem. Soc., Perkin Trans. 1*, 1976, 2419 (*props, synth*)
Hara, S. *et al*, *Chem. Lett.*, 1980, 221 (*use*)

$C_{19}H_{38}BO^{\ominus}$ B-01133

(Ethoxyethynyl)tripentylborate(1−), 10CI
(Ethoxyethynyl)tripentylboranate

$$[H_3C(CH_2)_4]_3BC{\equiv}COEt^{\ominus}$$

M 293.319 (ion)
Li salt: [65684-82-4].
$C_{19}H_{38}BLiO$ M 300.260
Prepd. *in situ* from ethoxyethynyllithium and tripentylborane. Synthetic intermed.

Mikhailov, B.M. *et al*, *Izv. Akad. Nauk SSSR, Ser. Khim.*, 1977, 2633; *Bull. Acad. Sci. USSR, Div. Chem. Sci.*, **26**, 2437 (*props*)
Mikhailov, B.M. *et al*, *Izv. Akad. Nauk SSSR, Ser. Khim.*, 1979, 855; *Bull. Acad. Sci. USSR, Div. Chem. Sci.*, **28**, 799 (*synth*)

$C_{19}H_{42}B^{\ominus}$ B-01134

Trihexylmethylborate(1−), 10CI, 9CI
Methyltrihexylborate

$$[H_3C(CH_2)_5]_3BMe^{\ominus}$$

M 281.351 (ion)
Cu(I) salt: [59560-23-5].
$C_{19}H_{42}BCu$ M 344.897
Prepd. *in situ*. Used in synth. of hexyl aryl ketones and other compds.

Miyaura, N. *et al*, *Bull. Chem. Soc. Jpn.*, 1977, **50**, 2199 (*synth, props*)
Miyaura, N. *et al*, *Tetrahedron Lett.*, 1977, 173 (*use*)
Yamada, K. *et al*, *Bull. Chem. Soc. Jpn.*, 1979, **52**, 275 (*props*)

$C_{20}H_{15}BN_2O_3$ B-01135

[2-[(2-Hydroxy-5-methylphenyl)azo]benzoato(2−)]phenylboron, 10CI
[74300-64-4]

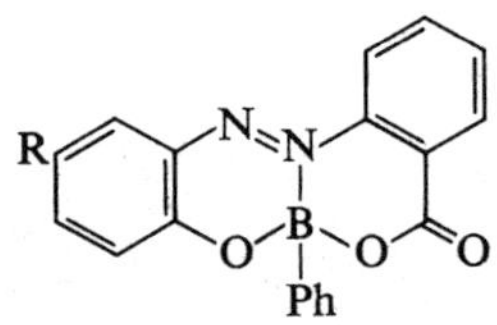

R = CH_3

M 342.160
Red-violet to brown-violet cryst. (C_6H_6). Mp 196°.

Hohaus, E. *et al*, *Z. Naturforsch., B*, 1980, **35**, 316, 319 (*ms, synth, nmr, ir*)

$C_{20}H_{15}BN_2O_4$ B-01136

[2-[(2-Hydroxy-5-methoxyphenyl)azo]benzoato(2−)]phenylboron, 10CI
[74300-65-5]
As [2-[(2-Hydroxy-5-methylphenyl)azo]benzoato(2−)]-phenylboron, B-01135 with

R = OMe

M 358.160
Violet cryst. Mp 251°.

Hohaus, E. *et al*, *Z. Naturforsch., B*, 1980, **35**, 316, 319 (*ms, synth, nmr, ir*)

$C_{20}H_{18}BNO$ B-01137

[2-[(Methylimino)methyl]phenolato-*N,O*]diphenylboron, 10CI, 9CI
[54721-42-5]

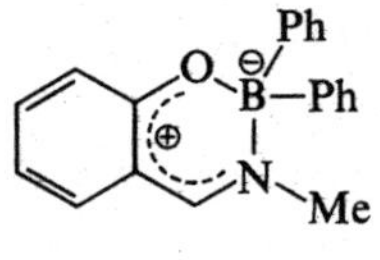

M 299.178
Solid. Mp 179-80°.

Hohaus, E. *et al*, *Z. Naturforsch.*, *B*, 1974, **29**, 663; 1976, **31**, 324.
Bregadze, V.I. *et al*, *J. Organomet. Chem.*, 1980, **192**, 1 (*uv*, *ir*, *struct*)
Hohaus, E., *Monatsh. Chem.*, 1980, **111**, 863.

$C_{20}H_{19}BFe_2O$ B-01138

Diferrocenylhydroxyborane

Diferrocenylborinic acid, *8CI*. *(Borinic acid)diferrocene*

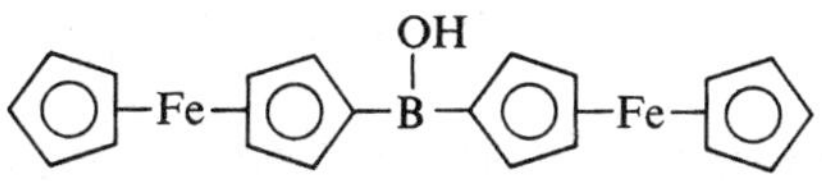

M 397.874

Air-stable orange-red cryst. Mp 159-61°.

Post, E.W. *et al*, *Inorg. Chem.*, 1970, **9**, 1670 (*synth*, *ms*, *pmr*, *ir*)

$C_{20}H_{24}BO_4^{\ominus}$ B-01139

Bis(4-*tert*-butyl-1,2-benzenediolato)borate

Bis[4-(1,1-dimethylethyl)-1,2-benzenediolato(2−)-O,O′]borate(1−), *10CI*

As Bis[1,2-benzenediolato(2−)-*O*,*O′*]borate(1−), B-00757 with

R = $(H_3C)_3C$–

M 339.217 (ion)

Na salt: [67494-93-3].

$C_{20}H_{24}BO_4Na$ M 362.207

Isocyanate polymerisation catalyst.

Ger. Pat., 2 754 011, (*1978*); *CA*, **89**, 111249v (*use*)
U.S.P., 4 151 334, (*1979*); *CA*, **91**, 21447v (*use*)

$C_{20}H_{26}AgB_8P$ B-01140

Triphenylphosphine[(7,8,9-η)undecahydro-5,6-dicarbadecaborato(1−)]silver

[74354-46-4]

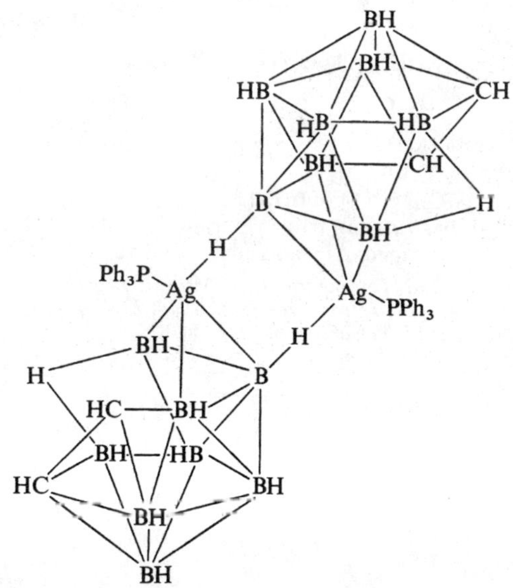

M 491.747

Dimeric in solid state with two *arachno*-AgC_2B_8 cages linked by a pair of Ag—H—B bridges.

Dimer: 11-Bis[triphenylphosphine-11-argenta-5,6-dicarbundecaborane(11)].

$C_{40}H_{52}Ag_2B_{16}P_2$ M 983.494

Cream cryst. ($CHCl_3$/hexane). Sol. $CHCl_3$.

Colquhoun, H.M. *et al*, *J. Chem. Soc., Chem. Commun.*, 1980, 192 (*synth*, *nmr*, *cryst struct*)

$C_{20}H_{26}B_8CuP$ B-01141

(Triphenylphosphine)[(7,8,9-η)-undecahydro-5,6-dicarbadecaborato(1−)]copper, 10CI

[74354-44-2]

$(Ph_3P)Cu(C_2B_8H_{11})$

M 447.425

Struct. not known. Bright-yellow cryst. (Me_2CO). Sol. Me_2CO.

Colquhoun, H.M. *et al*, *J. Chem. Soc., Chem. Commun.*, 1980, 192 (*synth*, *nmr*)

$C_{20}H_{27}B$ B-01142

Ethylbis(2,4,6-trimethylphenyl)borane, 10CI, 9CI

Ethyldimesitylborane

[50517-74-3]

As Hydroxybis(2,4,6-trimethylphenyl)borane, B-01075 with

R = Et

M 278.243

Prepd. from EtMgBr in Et_2O and fluorodimesitylborane. White needles, sensitive to air. Sol. CS_2. Mp 77-8°.

Finocchiaro, P. *et al*, *J. Am. Chem. Soc.*, 1973, **95**, 7029 (*synth*)
Pelter, A. *et al*, *Comprehensive Organic Chemistry*, (Barton, D.H.R. *et al*, Ed.), Pergamon, Oxford, 1979, **3**, 695 (*rev*)
Brown, N.M.D. *et al*, *J. Organomet. Chem.*, 1980, **185**, 277; 1981, **209**, 1 (*nmr*)

$C_{20}H_{27}BClN_3O$ B-01143

Dibutyl(5-chloro-2-imino-*N*-phenyl-1(2*H*)-pyridinecarboxamidato-N^{N^1},N^2)boron, 10CI

[71231-70-4]

As Dibutyl(5-chloro-2-imino-*N*-methyl-1(2*H*)-pyridinecarboxamidato-N^{N^1},N^2)boron, B-00959 with

R = Ph, X = Cl

M 371.716

Exists as intramol. coord. complex. Possesses some antiviral activity. Cryst. (EtOH).

Mikhailov, B.M. *et al*, *Izv. Akad. Nauk SSSR, Ser. Khim.*, 1979, 1085; *Bull. Acad. Sci. USSR, Div. Chem. Sci.*, 1014 (*synth*, *ir*, *nmr*)
Mikhailov, B.M. *et al*, *Khim.-Farm. Zh.*, 1982, **16**, 695.

$C_{20}H_{27}BS$ B-01144

(Ethylthio)bis(2,4,6-trimethylphenyl)borane

Thiobis(2,4,6-trimethylphenyl)borinic acid ethyl ester, *10CI*. *(Ethylthio)dimesitylborane*

[77657-16-0]

As Hydroxybis(2,4,6-trimethylphenyl)borane, B-01075 with

R = −SEt

M 310.303

Moisture-sensitive white cryst. (pentane or pentane/Et_2O). Mp 102-4°.

Davidson, F. *et al*, *J. Organomet. Chem.*, 1981, **204**, 147 (*synth*, *pmr*, *cmr*, *nmr*)
Brown, N.M.D. *et al*, *J. Organomet. Chem.*, 1981, **210**, 1 (*cmr*)

$C_{20}H_{28}BNO_4$ B-01145

[1-[(3,4-Dimethoxyphenyl)methyl]-1,2,3,4-tetrahydro-6-methoxy-2-methyl-7-isoquinolinol-N^2]trihydroboron, 10CI
[61659-89-0]

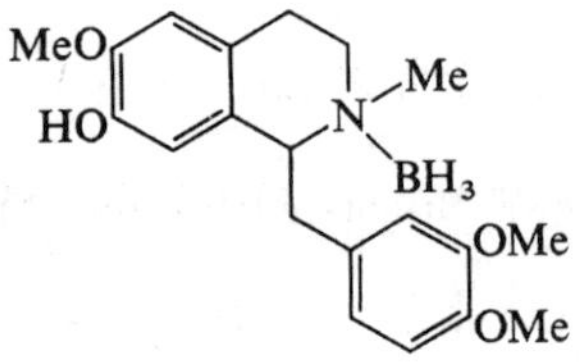

M 357.256
Colourless amorph. solid.

Kupchan, S.M. *et al*, *J. Org. Chem.*, 1978, **43**, 4076 (*synth*, *ms*)

$C_{20}H_{28}BN_3O$ B-01146

Dibutyl(*N*-phenyl-*N'*-2-pyridinylureato-*N*,$N^{N'}$)boron, 10CI
[71521-36-3]

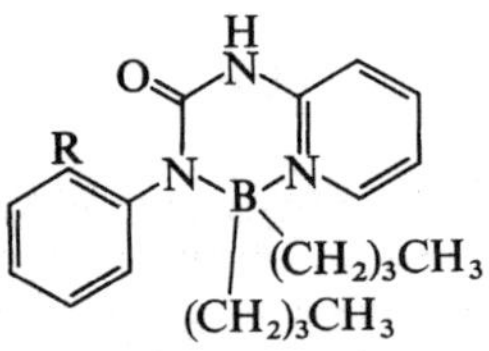

R = H

M 337.271
Dimeric in C_6H_6. Shows some antiviral activity. Cryst. (heptane). Mp 143-6°.

Dorokhov, V.A. *et al*, *Izv. Akad. Nauk SSSR, Ser. Khim.*, 1979, 1340; *Bull. Acad. Sci. USSR, Div. Chem. Sci.*, 1253 (*synth*, *nmr*)
Tolstikov, G.A. *et al*, *CA*, 1981, **95**, P114910p.
Lagutkin, N.A. *et al*, *Khim.-Farm. Zh.*, 1982, **16**, 695.

$C_{20}H_{28}B_2N_2S_2$ B-01147

Tetramethylbis[*N*-(*N*-phenylethanethioamidato-*N*:*S*)]diboron, 10CI
[63835-74-5]

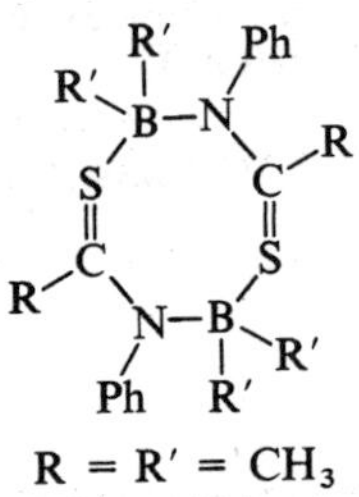

R = R' = CH_3

M 382.195
Liq. Sol. CH_2Cl_2. $Bp_{0.05}$ 82°.

Maringgele, W. *et al*, *J. Organomet. Chem.*, 1977, **141**, 241 (*synth*, *nmr*, *ms*)
Maringgele, W. *et al*, *Monatsh. Chem.*, 1977, **108**, 751.

$C_{20}H_{29}B$ B-01148

Dicyclohexyl(2-phenylethenyl)borane, 10CI
2-Phenylethenyldicyclohexylborane.
Dicyclohexylstyrylborane

As Dicyclohexyl(methoxy)borane, B-00870 with

R = —CH=CHPh

M 280.259

(*E*)-***form*** [62072-25-7]
Prepd. *in situ* from dicyclohexylborane and phenylacetylene in THF. Synthetic intermed. for transfer of a phenylethenyl group. Air-sensitive. Sol. THF.

Zweifel, G., *J. Am. Chem. Soc.*, 1971, **93**, 3395 (*synth*)
Zweifel, G. *et al*, *J. Organomet. Chem.*, 1976, **117**, 303 (*ir*, *nmr*, *uv*)
Brown, H.C. *et al*, *J. Org. Chem.*, 1980, **45**, 550 (*use*)
Yatagai, H., *Bull. Chem. Soc. Jpn.*, 1980, **53**, 1670 (*use*)

$C_{20}H_{30}B_2Fe_2N_2O_4$ B-01149

Di-μ-carbonyldicarbonylbis[(3,4,5-η)-1-(1,1-dimethylethyl)-2,3-dihydro-2-methyl-1*H*-1,2-azaborol-3-yl-B^2,N^1]-diiron, 10CI
*Tetracarbonylbis(η^5-1-*tert*-butyl-2-methyl-1,2-azaborolinyl)diiron*

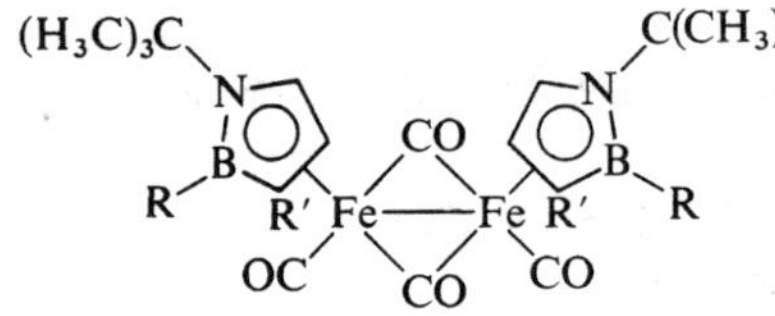

cis-form

R = Me, R' = H

M 495.782
Two isomers which interconvert in soln.

cis-form [75325-34-7]
Characterised spectroscopically.

trans-form [75364-55-5]
Dark-red cryst. (toluene/pet. ether).

Schulze, J. *et al*, *J. Organomet. Chem.*, 1980, **193**, 83 (*synth*, *ms*, *ir*, *pmr*, *nmr*)

$C_{20}H_{32}B^{\ominus}$ B-01150

Tributyl(phenylethynyl)borate(1−), 10CI, 9CI

$(H_3CCH_2CH_2CH_2)_3BC{\equiv}CPh^{\ominus}$

M 283.283 (ion)

Li salt: [42033-93-2].
$C_{20}H_{32}BLi$ M 290.224
Prepd. *in situ* from tributylborane and PhC≡CLi. Synth. intermed. Air- and acid-sensitive. Sol. THF.

Naruse, M. *et al*, *Tetrahedron Lett.*, 1973, 795 (*use*)
Miyaura, N. *et al*, *Tetrahedron Lett.*, 1974, 2961 (*props*)
Naruse, M. *et al*, *Tetrahedron*, 1974, **30**, 835, 2159 (*use*)
Pelter, A. *et al*, *J. Chem. Soc., Perkin Trans. 1*, 1976, 2435 (*props*)
Hara, S. *et al*, *Chem. Lett.*, 1980, 221 (*synth*)

$C_{20}H_{35}B$ B-01151

Diisopinocampheylborane

Bis(2,6,6-trimethylbicyclo[3.1.1]hept-3-yl)borane, 10CI, 9CI. *Di-3-pinanylborane*, 8CI

[1091-56-1]

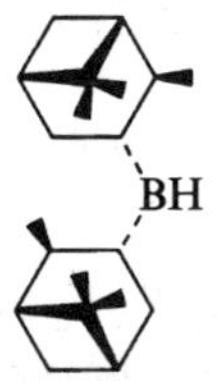

(+)-*form*

M 286.307

Exists as dimer in uncomplexed state. Generally prepd. *in situ* by hydroboration of α-pinene.

(+)-***form*** [21947-87-5]

Prepd. from (−) α-pinene. Chiral hydroborating and reducing agent.

Dimer:

$C_{40}H_{70}B_2$ M 572.613

Air- and moisture-sensitive cryst. solid. Sol. THF.

(−)-***form*** [21932-54-7]

Obt. from (+) α-pinene. Chiral hydroborating and reducing agent.

Dimer: Air- and moisture-sensitive cryst. solid. Sol. THF.

Brown, H.C. *et al*, *J. Am. Chem. Soc.*, 1964, **86**, 393 (*synth, struct*)
Lee, J.B. *et al*, *Tetrahedron*, 1967, **23**, 2789.
Grundson, M. *et al*, *J. Chem. Soc. (C)*, 1971, 2560 (*synth, use*)
Herz, J. *et al*, *J. Chem. Soc. (C)*, 1971, 3504 (*synth*)
Crawford, R.J. *et al*, *Can. J. Chem.*, 1973, **51**, 3718 (*use*)
Brown, H.C. *et al*, *J. Org. Chem.*, 1978, **43**, 4395; 1981, **46**, 4541 (*use*)
Ruethmann, A. *et al*, *Helv. Chim. Acta*, 1980, **63**, 1456 (*use*)

$C_{20}H_{35}BO_2$ B-01152

9-(10-Methoxycarbonyl-1-decenyl)-9-borabicyclo[3.3.1]nonane

11-(9-Borabicyclo[3.3.1]non-9-yl)-10-undecenoic acid methyl ester, 10CI

[73062-43-8]

As 9-Bromo-9-borabicyclo[3.3.1]nonane, B-00539 with

$$R = OCH{=}CH(CH_2)_8COOMe$$

M 318.305

Yatagai, H., *J. Org. Chem.*, 1980, **45**, 1640 (*synth*)

$C_{20}H_{37}B$ B-01153

Dicyclohexyl-1-octenylborane, 10CI

1-Octenyldicyclohexylborane

As Dicyclohexyl(methoxy)borane, B-00870 with

$$R = {-}CH{=}CH(CH_2)_5CH_3$$

M 288.322

(***E***)-***form*** [62072-20-2]

Prepd. *in situ* from dicyclohexylborane and 1-octyne in THF. Synthetic intermed. for transfer of an octenyl group. Air-sensitive. Sol. THF.

Zweifel, G., *J. Am. Chem. Soc.*, 1971, **93**, 3395 (*synth, nmr*)
Zweifel, G. *et al*, *J. Organomet. Chem.*, 1976, **117**, 303 (*uv, pmr, ir*)
Negishi, E. *et al*, *J. Chem. Soc., Perkin Trans. 2*, 1978, 1225 (*use*)
Pelter, A. *et al*, *Comprehensive Organic Chemistry*, (*Barton, D.H.R. et al*, Ed.), Pergamon, Oxford, 1979, **3**, 695 (*rev*)
Singaram, B. *et al*, *Heterocycles*, 1981, **15**, 231 (*use*)

$C_{20}H_{38}BCl$ B-01154

Bis(1-butyl-1-hexenyl)chloroborane, 10CI

Chlorobis(5-decenyl)borane

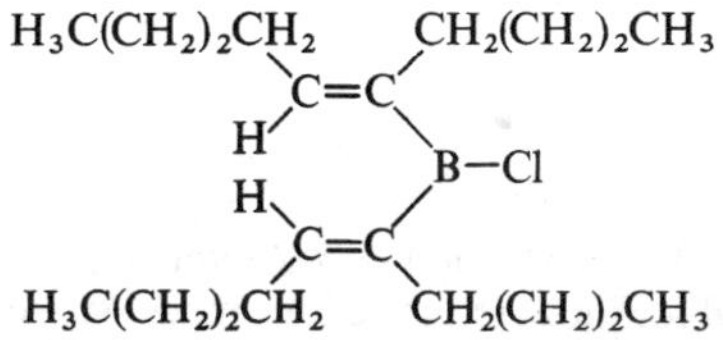

M 324.783

(***Z,Z***)-***form*** [64323-54-2]

Air- and moisture-sensitive liq. $Bp_{0.005}$ 110-5°.

Yamamoto, Y. *et al*, *J. Am. Chem. Soc.*, 1977, **99**, 5652 (*synth, pmr*)
Yamamoto, Y. *et al*, *J. Chem. Soc., Chem. Commun.*, 1978, **16**, 702 (*synth*)
Yatagai, H. *et al*, *J. Org. Chem.*, 1979, **44**, 2566 (*pmr*)

$C_{20}H_{40}B^{\ominus}$ B-01155

Dibutyl-1-hexynyl(1,1,2-trimethylpropyl)borate(1−), 10CI, 9CI

Dibutylhexynylthexylborate. Thexyldibutylhexynylborate

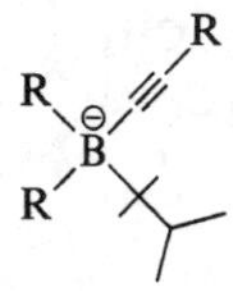

$$R = {-}(CH_2)_3CH_3$$

M 291.346 (ion)

Li salt: [56556-79-7].

$C_{20}H_{40}BLi$ M 298.287

Synthetic intermediate, prepared *in situ*. No phys. props. reported.

Pelter, A. *et al*, *Tetrahedron Lett.*, 1975, 1633 (*synth, props*)
Pelter, A. *et al*, *J. Chem. Soc., Perkin Trans. 1*, 1976, 2419, 2435 (*synth, props*)

$C_{20}H_{40}B^{\ominus}$ B-01156

Ethynyltrihexylborate(1−), 10CI

$$[H_3C(CH_2)_5]_3BC{\equiv}CH^{\ominus}$$

M 291.346 (ion)

Li salt: [62349-87-5].

$C_{20}H_{40}BLi$ M 298.287

Synthetic intermed.

Brown, H.C. *et al*, *J. Am. Chem. Soc.*, 1975, **97**, 5017 (*synth*)
Utimoto, K. *et al*, *Tetrahedron Lett.*, 1976, 3969 (*synth, props*)

$C_{20}H_{41}B$ B-01157

(1-Butyl-1-hexenyl)bis(1,2-dimethylpropyl)borane, *10CI*

5-Decenylbis(3-methyl-2-butyl)borane. (5-Decenyl)-disiamylborane

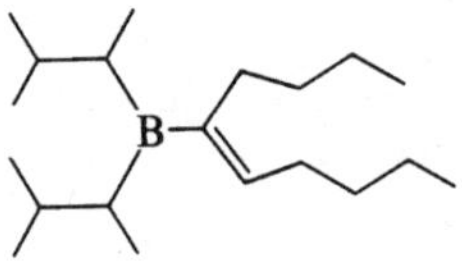

M 292.354

(Z)-form [69082-89-9]

Formed by hydroboration of 5-decyne with disiamylborane. Synthetic intermed. Air-sensitive oil. Sol. THF.

Yatagai, H. *et al, J. Chem. Soc., Chem. Commun.*, 1978, 702 (*synth, props*)
Yatagai, H. *et al, J. Org. Chem.*, 1979, **44**, 2566 (*pmr*)
Yatagai, H., *Bull. Chem. Soc. Jpn.*, 1980, **53**, 1670 (*synth, props*)

$C_{20}H_{41}B$ B-01158

Cyclohexyloctyl(1,1,2-trimethylpropyl)borane, *10CI*

Cyclohexyl(2,3-dimethyl-2-butyl)octylborane. Cyclohexyloctylthexylborane. Thexylcyclohexyloctylborane

[62594-03-0]

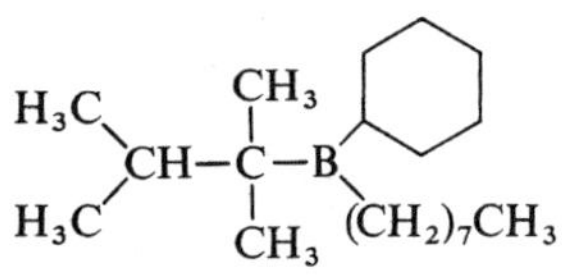

M 292.354

Prepd. from thexylborane-PhNEt$_2$ and cyclohexene then 1-octene. Synthetic intermed., e.g. for prepn. of 1-cyclohexylnonanone. Oil, sensitive to air. Stable to disproportionation for considerable periods. Sol. THF.

Negishi, E. *et al, Synthesis*, 1974, 77 (*rev*)
Kabalka, G.W. *et al, J. Organomet. Chem.*, 1977, **125**, 273.
Pelter, A. *et al, Tetrahedron Lett.*, 1978, 4715 (*synth, use*)
Pelter, A. *et al, Comprehensive Organic Chemistry*, (*Barton, D.H.R. et al*, Ed.), Pergamon, Oxford, 1979, **3**, 695 (*rev*)

$C_{20}H_{43}BO_{10}$ B-01159

(2-Butoxyethoxy)bis(1,4,7,10-tetraoxaundecyl)borane

Boric acid (H_3BO_3) 2-butoxyethyl bis[2-[2-(2-methoxyethoxy)ethoxy]ethyl] ester, *10CI*

[74198-17-7]

$H_3C(CH_2)_3OCH_2CH_2OB[(OCH_2CH_2)_3OMe]_2$

M 454.364

Used in brake fluids for automobiles. Moisture-sensitive.

Ger. Pat., 2 831 538, (*1980*); *CA*, **93**, 134858q (*use*)
U.S.P., 4 192 759, (*1980*); *CA*, **94**, 86976y (*synth, use*)

$C_{20}H_{44}B^{\ominus}$ B-01160

Tetrakis(3-methylbutyl)borate(1−), *10CI, 9CI*

Tetraisopentylborate(1−), 8CI. Tetraisoamylborate. Tetraisoamylboride

$[(H_3C)_2CHCH_2CH_2]_4B^{\ominus}$

M 295.378 (ion)

Tetrakis(3-methylbutyl)ammonium salt: [25397-17-5]. *Tetraisopentylammonium tetraisopentylborate. Tetraisoamylammonium tetraisoamylborate.* $C_{40}H_{88}BN$ M 593.952

Reference electrolyte. Large nonhygroscopic cryst. (Me$_2$CO aq.). Sol. Me$_2$CO, MeCN, MeNO$_2$, sl sol. MeOH. Mp 181°.

Coetzee, J. *et al, J. Am. Chem. Soc.*, 1964, **86**, 3403 (*synth, props. use*)
Arrington, D.E. *et al, J. Phys. Chem.*, 1970, **74**, 123 (*use, props*)
Prock, A. *et al, J. Phys. Chem.*, 1970, **74**, 2408 (*props*)
Jansen, M.L. *et al, J. Phys. Chem.*, 1973, **77**, 3089 (*props*)
Lichtin, N.N. *et al, J. Phys. Chem.*, 1980, **84**, 2398; 2946 (*props*)

$C_{20}H_{44}B_2FeN_2Si_2$ B-01161

Bis[(3,4-η)-4,5-diethyl-1,2,2,3-tetramethyl-1-aza-2-sila-5-boracyclopent-3-ene-B^5,N^1]iron

Bis(η^4-4,5-diethyl-1,2,2,3-tetramethyl-1,2,5-azasilaboroline)iron

[81923-59-3]

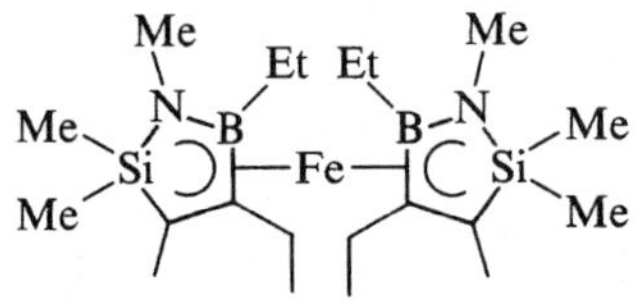

M 446.219

Paramagnetic 16-electron complex. Dark-red cryst. Bp$_{0.001}$ 40° subl.

Köster, R. *et al, Chem. Ber.*, 1982, **115**, 738 (*synth, ms*)

$C_{21}H_9B_3F_{15}N_3$ B-01162

1,3,5-Trimethyl-2,4,6-tris(pentafluorophenyl)borazine, *10CI*

[2723-57-1]

As 2,4,6-Tribromoborazine, B-00015 with

$X = C_6F_5$, $Y = Me$

M 620.728

Solid. Mp 275°. Bp$_9$ 245° subl.

Meller, A. *et al, Monatsh. Chem.*, 1965, **96**, 388 (*synth, ir*)
Wrackmeyer, B. *et al, Chem. Ber.*, 1976, **109**, 3480 (*nmr*)
Nöth, H. *et al, Chem. Ber.*, 1981, **114**, 1150 (*nmr, cmr*)

$C_{21}H_{21}B$ B-01163

Tribenzylborane, *8CI*

Tris(phenylmethyl)borane, *10CI, 9CI*

[1694-84-4]

$(PhCH_2)_3B$

M 284.207

Prepd. from PhCH$_2$MgBr and BF$_3$ in Et$_2$O. Synthetic intermed. Air- and moisture-sensitive solid. Sol. THF, Et$_2$O. Mp 51°. Bp$_2$ 152.4°.

Kacheishvili, G.E. *et al, Zh. Obshch. Khim.*, 1965, **35**, 484 (*synth*)
Ramsey, B.G. *et al, J. Am. Chem. Soc.*, 1972, **94**, 4227 (*uv*)
Zetta, L. *et al, Org. Magn. Reson.*, 1972, **4**, 585 (*nmr*)
Chung, V.V. *et al, Chem. Lett.*, 1976, 209 (*props*)
Manzer, L.E. *et al, Inorg. Chem.*, 1976, **15**, 3114 (*deriv, nmr*)
Jigajinni, V.B. *et al, Tetrahedron Lett.*, 1978, 181 (*use*)

$C_{21}H_{21}B$ B-01164

Tris(4-methylphenyl)borane, 10CI, 9CI

Tri-p-tolylborane, 8CI

[7297-94-1]

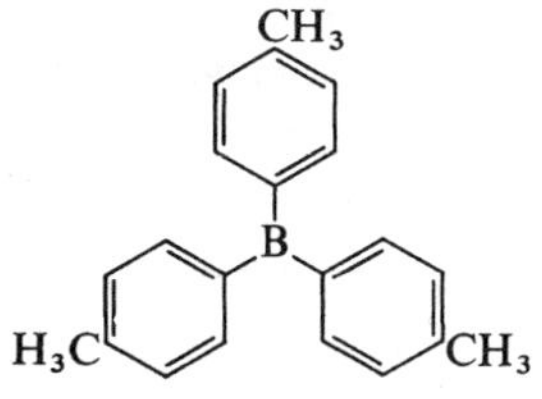

M 284.207

Synthetic intermed. Air-sensitive solid. Sol. THF.

Lapkin, I.I. *et al*, *Zh. Obshch. Khim.*, 1975, **45**, 1511 (*derivs*)
Manzer, L.E. *et al*, *J. Chem. Soc., Chem. Commun.*, 1975, 227 (*deriv*)
Manzer, L.E. *et al*, *Inorg. Chem.*, 1976, **15**, 3114 (*nmr, deriv*)
Ger. Pat., 2 651 775, (*1977*); *CA*, **87**, 135880d (*synth*)
U.S.P., 4 134 923, (*1979*); *CA*, **90**, 121791 (*synth, use*)
Lapkin, I.I. *et al*, *Zh. Obshch. Khim.*, 1981, **51**, 885 (*synth, deriv*)

$C_{21}H_{21}BO_3$ B-01165

Tribenzyloxyborane

Boric acid (H_3BO_3) tris(phenylmethyl) ester, 10CI, 9CI. *Boric acid (H_3BO_3) tribenzyl ester*, 8CI. *Benzyl borate. Tris(phenylmethoxy)borane*

[2467-18-7]

$(PhCH_2O)_3B$

M 332.205

Insect sterilant for house-fly control, bactericide, fungicide, heat stabilizer for halogen-containing polymer blends. Moisture-sensitive liq. d_4^{26} 1.095. Bp_{20} 258-60°, Bp_3 180.4-189.8°. n_D^{25} 1.5546.

▷ED4850000.

Wuyts, M.H. *et al*, *Bull. Soc. Chim. Belg.*, 1939, **48**, 77 (*props*)
Kuskov, V.K. *et al*, *Izv. Akad. Nauk SSSR, Ser. Khim.*, 1956, 733 (*synth, props*)
Borkovec, A.B. *et al*, *J. Econ. Entomol.*, 1969, **62**, 1472 (*use*)
Ogata, Y. *et al*, *J. Chem. Soc. (C)*, 1969, 2413 (*props*)
Ital. Pat., 932 322, (*1972*); *CA*, **86**, 43214r (*use*)
Japan. Pat., 79 138 047, (*1979*); *CA*, **92**, 129960s (*use*)

$C_{21}H_{21}BO_3$ B-01166

Tris(3-methylphenoxy)borane

Boric acid (H_3BO_3) tris(3-methylphenyl) ester, 10CI, 9CI. *Boric acid (H_3BO_3) tri-m-tolyl ester*, 8CI. *m-Tolyl borate. m-Cresyl borate. Tri-m-tolyl borate*

[14750-98-2]

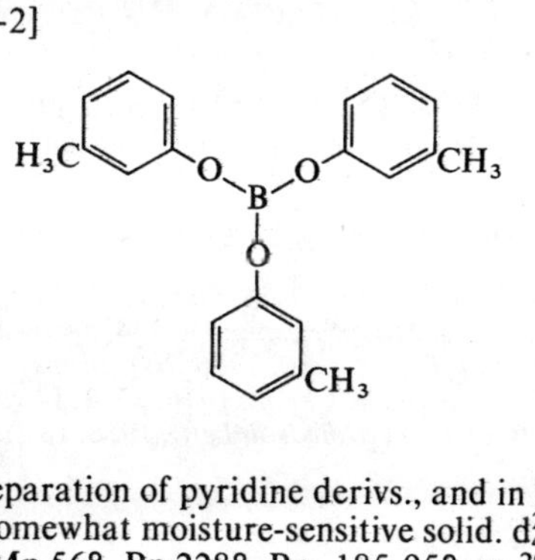

M 332.205

Used for separation of pyridine derivs., and in hydraulic fluids. Somewhat moisture-sensitive solid. d_4^{20} 1.078 (calc.). Mp 56°. Bp 228°, Bp_3 185-95°. n_D^{20} 1.5555 (calc.).

Wuyts, M.H. *et al*, *Bull. Soc. Chim. Belg.*, 1939, **48**, 77 (*props*)
Thomas, L.H. *et al*, *J. Chem. Soc.*, 1963, 612.
Kondratov, V.K. *et al*, *Zh. Prikl. Khim.*, 1971, **44**, 1357; *CA*, **75**, 91736g (*use*)
Fenwick, J.T.F. *et al*, *Inorg. Chem.*, 1975, **14**, 1602 (*props*)
Ger. Pat., 2 531 086, (*1976*); *CA*, **85**, 65537e (*use*)
Sasaki, M. *et al*, *Bull. Chem. Soc. Jpn.*, 1977, **50**, 276.
Japan. Pat., 79 46 299, (*1979*); *CA*, **91**, 92493t

$C_{21}H_{22}BO_2P$ B-01167

(Ethoxycarbonyl)borane triphenylphosphine

(Ethoxycarbonyl)dihydro(triphenylphosphine)boron, 10CI

[78315-29-4]

$Ph_3P—BH_2COOEt$

M 348.187

White cryst. solid. Mp 76-7°.

Wisian-Neilson, P. *et al*, *J. Inorg. Nucl. Chem.*, 1981, **43**, 457 (*synth, nmr, ir*)

$C_{21}H_{23}BNOP$ B-01168

***N*-Ethylcarbamoylborane triphenylphosphine**

[(Ethylamino)carbonyl]dihydro(triphenylphosphine)-boron, 10CI

[78315-27-2]

$Ph_3P—BH_2CONHEt$

M 347.203

White cryst. (Me_2CO aq.). Mp 112-6°.

Wisian-Neilson, P. *et al*, *J. Inorg. Nucl. Chem.*, 1981, **43**, 457 (*synth, nmr, pmr*)

$C_{21}H_{23}BN_2O$ B-01169

Triphenylborane *N,N*-dimethylurea

(N,N-Dimethylurea-N')triphenylboron, 10CI. *N,N-Dimethylurea, compd. with triphenylborane(1:1)*

[69044-16-2]

$Me_2NCONH_2—BPh_3$

M 330.236

Lapkin, I.I. *et al*, *Zh. Obshch. Khim.*, 1978, **48**, 2312.

$C_{21}H_{24}BN$ B-01170

Trimethylborane triphenylamine

(N,N-Diphenylbenzenamine)trimethylboron, 10CI. *N,N-Diphenylbenzenamine, compd. with trimethylborane(1:1)*

[69809-43-4]

$Ph_3N—BMe_3$

M 301.237

Suggested for use in electrophotographic imaging.

Japan. Pat., 78 131 844, (*1978*); *CA*, **91**, P81565s
Japan. Pat., 78 131 845, (*1978*); *CA*, **91**, P81566t

$C_{21}H_{24}BN$ B-01171

Triphenylborane trimethylamine

(N,N-Dimethylmethanamine)triphenylboron, 9CI. *Triphenylborane compd. with trimethylamine(1:1)*, 8CI. *Trimethylamine, compd. with triphenylborane*

[14480-40-1]

Ph_3B—NMe_3

M 301.237

U.S.P., 3 062 708, (*1962*); *CA*, **58**, P2797e
U.S.P., 3 117 850, (*1964*); *CA*, **60**, P16438b
U.S.P., 3 268 401, (*1966*); *CA*, **65**, P14362a
Fr. Pat., 1 458 180, (*1966*); *CA*, **67**, P11973d
Borkovec, A.B. *et al*, *J. Econ. Entomol.*, 1969, **62**, 1472.
Japan. Pat., 78 131 845, (*1978*); *CA*, **91**, P81566t

$C_{21}H_{24}BN$ B-01172

Tris(4-methylphenyl)borane ammonia

Amminetris(4-methylphenyl)boron, 10CI, 9CI. *Tri-p-tolylborane, compd. with NH_3(1:1). Tri-p-tolylborane monoammoniate. Tri-p-tolylborane ammonia*

[13586-42-0]

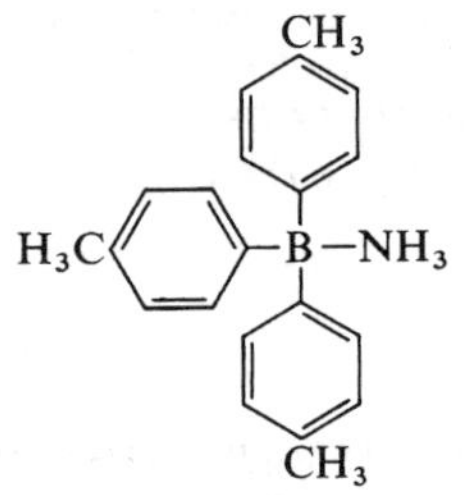

M 301.237

U.S.P., 3 062 708, (*1962*); *CA*, **58**, P2797e
Updegraff, D.M., *J. Infect. Dis.*, 1964, **114**, 304.
U.S.P., 4 134 923, (*1979*); *CA*, **90**, P121791a
Ger. Pat., 3 015 007, (*1980*); *CA*, **94**, P103550z

$C_{21}H_{24}B_3N_3$ B-01173

1,3,5-Trimethyl-2,4,6-triphenylborazine, 10CI, 9CI, 8CI

N,N,N-*Trimethyl*-B,B,B-*triphenylborazole*

[909-21-7]

As 2,4,6-Tribromoborazine, B-00015 with

X = Ph, Y = Me

M 350.871

Insect sterilant. Cryst. (C_6H_6/hexane). Insol. hexane. Mp 268-70°.

Smalley, J.H. *et al*, *J. Am. Chem. Soc.*, 1959, **81**, 582 (*synth*)
Gerrard, W., *Soc. Chem. Ind.*, 1961, **13**, 328; *CA*, **56**, 4939h (*synth*)
Steinberg, H. *et al*, *Organoboron Chemistry, Vol. 2*, Wiley, 1966, 175 (*synth, props*)
Totani, T. *et al*, *Spectrochim. Acta, Part A*, 1969, **25**, 585 (*ir*)
Melcher, L.A. *et al*, *Inorg. Chem.*, 1973, **12**, 601 (*ms*)
Nöth, H. *et al*, *Chem. Ber.*, 1981, **114**, 1150 (*nmr, cmr*)

$C_{21}H_{28}BN_3$ B-01174

[*N*-(Iminophenylmethyl)-*N′*-propylbenzenecarboximidamidato-*N′*,*N*^N^]dipropylboron, 10CI

Dipropylboryl(N-*propylbenzimidoyl*)*benzamidinate*

[67574-91-8]

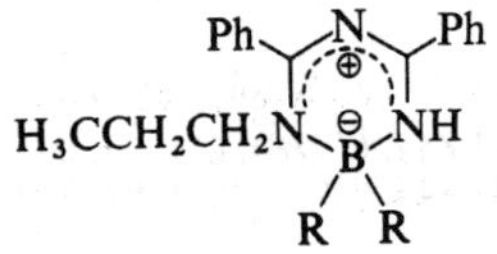

R = $H_3CCH_2CH_2$—

M 333.282

Cryst. Chemically and thermally stable, can be stored in air. Mp 63-6°.

Zolotarev, B.M. *et al*, *Izv. Akad. Nauk SSSR, Ser. Khim.*, 1978, 1312; *Bull. Acad. Sci. USSR, Div. Chem. Sci.*, 1140 (*ms*)
Mikhailov, B.M. *et al*, *Izv. Akad. Nauk SSSR, Ser. Khim.*, 1978, 1385; *Bull. Acad. Sci. USSR, Div. Chem. Sci.*, 1205 (*synth, ms, nmr, pmr*)

$C_{21}H_{29}B$ B-01175

Propylbis(2,4,6-trimethylphenyl)borane, 10CI

Dimesitylpropylborane. Propyldimesitylborane

[74178-57-7]

As Fluorobis(2,4,6-trimethylphenyl)borane, B-01073 with

X = $H_3CCH_2CH_2$—

M 292.270

Prepd. from propylmagnesium bromide and Fluorobis(2,4,6-trimethylphenyl)borane, B-01073. Synthetic intermed. Solid, somewhat sensitive to air. Sol. THF. Mp 69°.

Finocchiaro, P. *et al*, *J. Am. Chem. Soc.*, 1973, **95**, 7029 (*synth*)
Pelter, A. *et al*, *Compr. Org. Chem.*, (*Barton, D.H.R. et al*, Ed.), 1979, **3**, 695 (*rev*)
Brown, N.M.D. *et al*, *J. Organomet. Chem.*, 1980, **185**, 277; 1981, **209**, 1 (*nmr*)
Pelter, A. *et al*, *Tetrahedron Lett.*, 1983, **24**, 625, 627 (*use*)

$C_{21}H_{29}BO$ B-01176

Isopropoxybis(2,4,6-trimethylphenyl)borane

Bis(2,4,6-trimethylphenyl)borinic acid 1-methylethyl ester, 10CI. *Dimesitylborinic acid 1-methylethyl ester. Isopropoxydimesitylborane*

[75260-11-6]

As Hydroxybis(2,4,6-trimethylphenyl)borane, B-01075 with

R = —$OCH(CH_3)_2$

M 308.270

Colourless cryst. solid, stable in air. Sol. most org. solvs. Mp 86-7°. Slowly oxid. in soln.

Brown, N.M.D. *et al*, *J. Organomet. Chem.*, 1980, **193**, 271 (*synth, nmr, cmr*)
Brown, N.M.D. *et al*, *J. Organomet. Chem.*, 1981, **210**, 1 (*nmr*)

$C_{21}H_{30}BN$ B-01177

(Isopropylamino)bis(2,4,6-trimethylphenyl)borane

N-(*1-Methylethyl*)-*1,1-bis(2,4,6-trimethylphenyl)boranamine*, 10CI. *Isopropylaminodimesitylborane*

[75018-26-7]

As (Methylamino)bis(2,4,6-trimethylphenyl)borane, B-01128 with

R = $CH(CH_3)_2$

M 307.285

White cryst. (pentane), stable to air. Sol. most solvs. Mp 83-5°.

Brown, N.M.D. *et al*, *J. Organomet. Chem.*, 1980, **192**, 133 (*synth, nmr, ir*)
Brown, N.M.D. *et al*, *J. Organomet. Chem.*, 1981, **210**, 1 (*cmr*)

$C_{21}H_{30}BN_3O$ B-01178

Dibutyl[*N*-(2-methylphenyl)-*N'*-2-pyridinylureato-*N,N^{N'}*]-boron, 10CI

[66199-95-9]

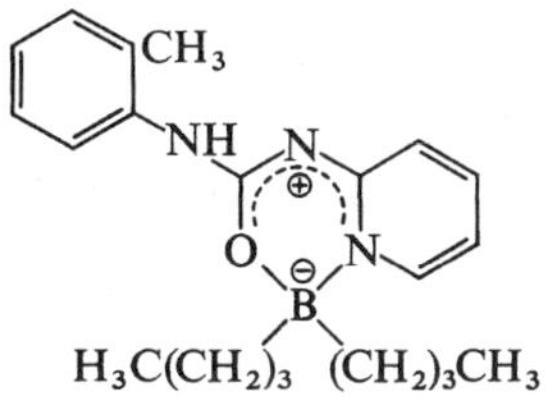

M 351.298

Relatively stable in air. Not dec. by H_2O or alcohols. Mp 80-1°.

Dorokhov, V.A. *et al*, *Izv. Akad. Nauk SSSR, Ser. Khim.*, 1978, 253; *Bull. Acad. Sci. USSR, Div. Chem. Sci.*, 224 (*synth, ir, ms*)

Dorokhov, V.A. *et al*, *Izv. Akad. Nauk SSSR, Ser. Khim.*, 1979, 1340; *Bull. Acad. Sci. USSR, Div. Chem. Sci.*, 1253 (*synth, nmr, uv, ms*)

$C_{21}H_{30}BN_3O$ B-01179

Dibutyl[*N*-(2-methylphenyl)-*N'*-2-pyridinylureato-*N^{N'}*,*O*]-boron, 10CI

[66329-66-6]

As Dibutyl(*N*-phenyl-*N'*-2-pyridinylureato-*N,N^{N'}*)-boron, B-01146 with

R = CH_3

M 351.298

Relatively stable in air, not dec. by H_2O or alcohols. Mp 162-6°.

Dorokhov, V.A. *et al*, *Izv. Akad. Nauk SSSR, Ser. Khim.*, 1978, 253; *Bull. Acad. Sci. USSR, Div. Chem. Sci.*, 224 (*synth, ir, ms*)

Dorokhov, V.A. *et al*, *Izv. Akad. Nauk SSSR, Ser. Khim.*, 1979, 1340; *Bull. Acad. Sci. USSR, Div. Chem. Sci.*, 1253 (*synth, nmr, uv, ms*)

$C_{21}H_{33}B$ B-01180

Tris(bicyclo[2.2.1]hept-2-yl)borane, 10CI, 9CI

Tri-2-norbornylborane, 8CI

[14289-75-9]

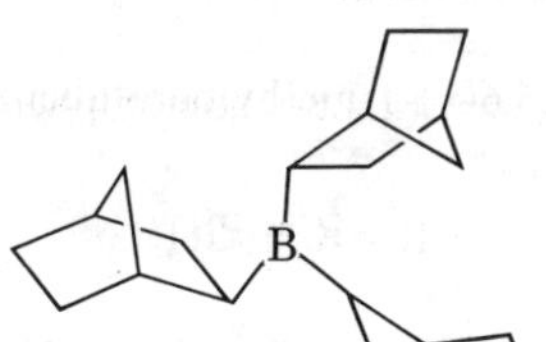

M 296.302

exo,exo,exo-form [22801-27-0]

Obt. by hydroboration of norbornene at 25° with BH_3 in THF. Synthetic intermed., generally prepd. *in situ*, for transfer of a norbornyl group. Air-sensitive. Sol. THF.

Brown, H.C. *et al*, *J. Am. Chem. Soc.*, 1971, **93**, 1506 (*use*)

Hughes, R.J. *et al*, *J. Chem. Soc., Perkin Trans. 1*, 1977, 1172 (*use*)

Hubbard, J. *et al*, *Synthesis*, 1978, 676 (*synth, use*)

Masuda, Y. *et al*, *Bull. Chem. Soc. Jpn.*, 1978, **51**, 901 (*use*)

Ncube, S. *et al*, *Tetrahedron Lett.*, 1979, 1895 (*use*)

Kabalka, G.W. *et al*, *J. Org. Chem.*, 1981, **46**, 4296 (*synth, use*)

$C_{21}H_{33}BN_2$ B-01181

***N,N'*-Diethylboryl-3,3',5,5'-tetramethyl-4,4'-diethyl-2,2'-dipyrrylmethene**

*Diethyl[3-ethyl-5-[(4-ethyl-3,5-dimethyl-2*H*-pyrrol-2-ylidene)methyl]-2,4-dimethyl-1*H*-pyrrolato-*N[1],N[5]*]boron, 10CI*

[69646-58-8]

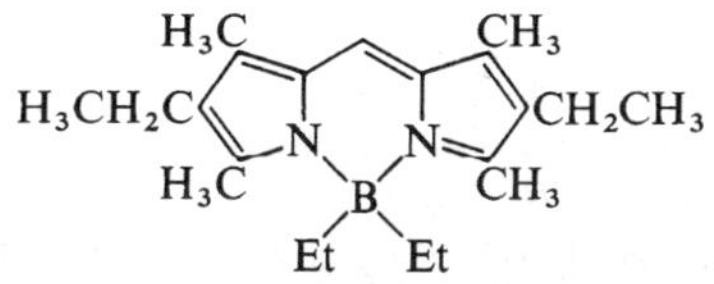

M 324.315

Used as a standard in the detection of solv. adducts of bilinoid pigments. Cryst. (C_6H_6/pet. ether); light-sensitive (may be handled safely in red light). Sol. MeCN, dioxan. Mp 114-7°.

Holzwarth, A.R. *et al*, *Justus Liebigs Ann. Chem.*, 1978, 2002 (*use*)

Braslavsky, S.E. *et al*, *J. Chromatogr.*, 1981, **205**, 85 (*use*)

$C_{21}H_{36}B^{\ominus}$ B-01182

Tricyclopentyl-1-hexynylborate(1−), 10CI, 9CI

1-Hexynyltricyclopentylborate

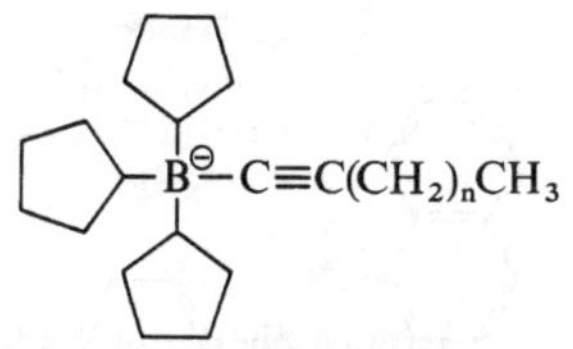

n = 3

M 299.325 (ion)

Li salt: [42126-50-1].

$C_{21}H_{36}BLi$ M 306.266

Prepd. *in situ* from tricyclopentylborane and hexynyllithium. Intermed. in synth. of alkynes and ketones. Air- and acid sensitive. Sol. THF, glyme, diglyme.

Pelter, A. *et al*, *J. Chem. Soc., Chem. Commun.*, 1973, 544 (*use*)

Suzuki, A. *et al*, *J. Am. Chem. Soc.*, 1973, **95**, 3080 (*synth, use*)

Pelter, A. *et al*, *J. Chem. Soc., Perkin Trans. 1*, 1976, 2419, 2428, 2435 (*synth, props, use*)

Hooz, J. *et al*, *Can. J. Chem.*, 1978, **56**, 2786.

Pelter, A. *et al*, *Tetrahedron Lett.*, 1981, **22**, 797.

$C_{21}H_{37}BN_2$ B-01183

Diisopropylbenzenecarboximidamidato-*N,N'*-dibutylboron

[*N,N'*-*Bis(1-methylethyl)benzenecarboximidamidato-N,N']dibutylboron, 10CI*

[69138-05-2]

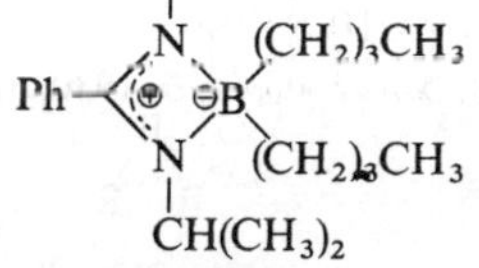

M 328.347

Bp 122-3°.

Mikhailov, B.M. *et al*, *Izv. Akad. Nauk SSSR, Ser. Khim.*, 1978, 2578; *Bull. Acad. Sci. USSR, Div. Chem. Sci.*, 2304 (*synth, nmr*)

Dorokhov, V.A. *et al*, *Dokl. Akad. Nauk SSSR, Ser. Sci. Khim.*, 1979, **245**, 121 (*synth*)

$C_{21}H_{38}B^{\ominus}$ B-01184

Diisopinocamphenylmethylhydroborate

Hydromethylbis(2,6,6-trimethylbicyclo[3.1.1]hept-3-yl)borate(1–), 10CI. *Hydromethyldi-3-pinanylborate(1–)*, 8CI

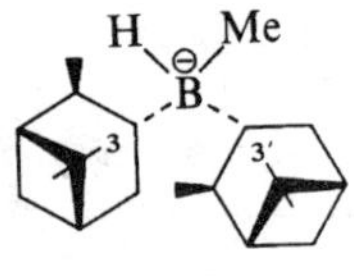

M 301.341 (ion)

Li salt: [34720-85-9].

$C_{21}H_{38}BLi$ M 308.282

Chiral reducing agent. Air- and moisture-sensitive.

(3R,3'R)-form [33393-62-3]

Grundon, M.F. *et al*, *J. Chem. Soc. (C)*, 1971, 2557, 2560 (*synth, use*)
U.S.P., 3 974 183, (*1976*); *CA*, **86**, 72219b
U.S.P., 4 082 810, (*1978*); *CA*, **89**, 109936y (*synth*)

$C_{22}H_{15}BN_2O_2$ B-01185

[1-[(2-Hydroxyphenyl)azo]-2-naphthalenolato(2–)]phenylboron, 10CI

[74302-92-4]

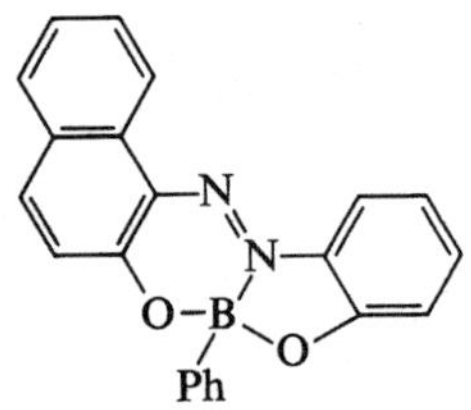

M 350.183

Mp 195°.

Hohaus, E. *et al*, *Z. Naturforsch., B*, 1980, **35**, 316, 319 (*ms, synth*)

$C_{22}H_{16}BN_3O_4S$ B-01186

[4-Hydroxy-3-[(2-hydroxy-1-naphthalenyl)azo]benzenesulfonamidato(2–)]phenylboron, 10CI

[74300-73-5]

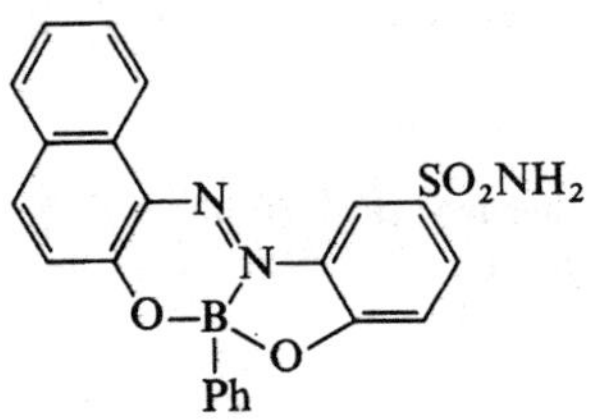

M 429.256

Dark-red cryst. Mp 210-5°.

Hohaus, E. *et al*, *Z. Naturforsch., B*, 1980, **35**, 316, 319 (*synth, ir, nmr*)

$C_{22}H_{19}BN^{\ominus}$ B-01187

Triphenyl-1*H*-pyrrol-1-ylborate(1–), 10CI

As Triethyl-1*H*-pyrrol-1-ylborate(1–), B-00692 with

R = Ph

M 308.209 (ion)

Li salt: [69346-64-1].

$C_{22}H_{19}BLiN$ M 315.150

Used in battery electrolytes.

U.S.P., 4 104 450, (*1978*); *CA*, **90**, 124646t
Klemann, L.P. *et al*, *Proc. Electrochem. Soc.*, 1981, **81**, 179 (*props*)

$C_{22}H_{19}BO_2$ B-01188

Diphenylboryl 1-phenyl-1,3-butanedionate

Diphenyl(1-phenyl-1,3-butanedionato-O,O')boron, 10CI, 9CI

[19426-81-4]

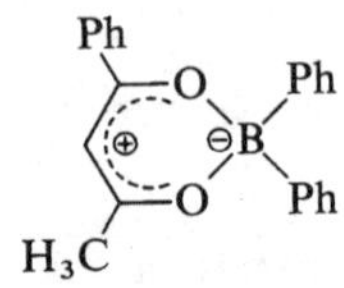

M 326.201

Used as a sensitizer for photoconductors.

Trestianu, H. *et al*, *Tetrahedron*, 1968, **24**, 2499 (*pmr*)
Barabas, A., *J. Coord. Chem.*, 1973, **3**, 91 (*nmr*)
Halm, J.M., *CA*, 1977, **86**, 197878x.
U.S.P., 4 123 268, (*1978*); *CA*, **90**, P95445u

$C_{22}H_{20}B_2Co$ B-01189

Bis[(1,2,3,4,5,6-η)-1-phenylboratabenzene]cobalt, 9CI

Bis(1-phenylborabenzene)cobalt

[36534-31-3]

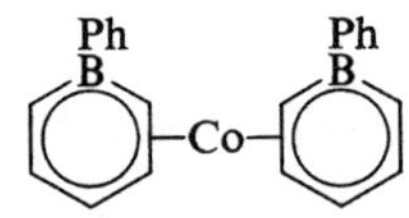

M 364.953

Dark-red plates (hexane) or dark-brown cryst. Mp 70-1°, 191-2°. Paramagnetic.

Herberich, G.E. *et al*, *Chem. Ber.*, 1972, **105**, 3413 (*synth, ms*)
Herberich, G.E. *et al*, *J. Chem. Soc., Dalton Trans.*, 1975, 985 (*esr*)
Herberich, G.E. *et al*, *Chem. Ber.*, 1979, **112**, 607.

$C_{22}H_{20}B_2Fe$ B-01190

Bis[(1,2,3,4,5,6-η)-1-phenylboratabenzene]iron, 9CI

Bis(1-phenylborinato)iron

[54853-79-1]

As Bis[(1,2,3,4,5,6-η)-1-methylboratabenzene]iron, B-00769 with

R = Ph

M 361.867

Red-orange air-stable cryst. (pentane). Mp 160-160.5° (156-157.5°).

Herberich, G. *et al*, *Chem. Ber.*, 1974, **107**, 3780 (*synth, pmr, ir, ms*)
Ashe, A.J. *et al*, *J. Am. Chem. Soc.*, 1975, **97**, 6865 (*synth, nmr, mössbauer, uv, ms*)

$C_{22}H_{20}B_2Os$ **B-01191**

Bis[(1,2,3,4,5,6-η)-1-phenylboratabenzene]osmium, 9CI

Bis(1-phenylborinato)osmium

[60764-69-4]

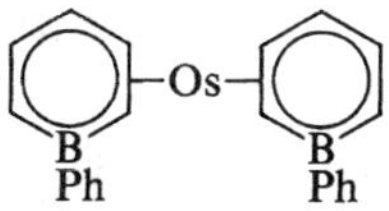

M 496.220

Colourless, air stable cryst. (pentane). Mp 146.5-147°.

Herberich, G.E. *et al*, *Chem. Ber.*, 1976, **109**, 2382 (*synth, pmr, nmr, ms*)

$C_{22}H_{20}B_2Ru$ **B-01192**

Bis[(1,2,3,4,5,6-η)-1-phenylboratabenzene]ruthenium, 9CI

Bis(1-phenylborinato)ruthenium

[60552-36-5]

As Bis[(1,2,3,4,5,6-η)-1-methylboratabenzene]ruthenium, B-00770 with

R = Ph

M 407.090

Air stable, colourless needles (pentane). Mp 136-136.5°. $Bp_{0.01}$ 140-60° subl.

Herberich, G.E. *et al*, *Chem. Ber.*, 1976, **109**, 2382 (*synth, pmr, nmr, ms*)

$C_{22}H_{22}BN_3O_2$ **B-01193**

[5-(Ethylamino)-2-[(2-hydroxy-5-methylphenyl)azo]-4-methylphenolato(2−)]phenylboron, 10CI

[74302-90-2]

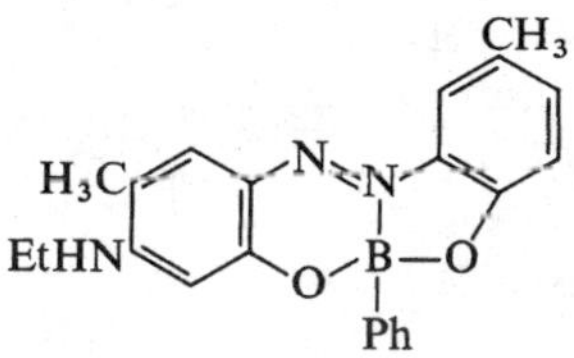

M 371.245

Red-violet cryst. with gold-green lustre. Mp 222°.

Hohaus, E. *et al*, *Z. Naturforsch.*, *B*, 1980, **35**, 316, 319 (*ms, synth, nmr, ir*)

$C_{22}H_{23}BN_2O$ **B-01194**

Hydroxy[2-[[phenyl(phenylamino)methylene]amino]phenyl]-propylboron, 10CI

[70204-53-4]

As Hydroxy[2-[[1-(phenylamino)ethylidene]amino]phenyl]propylboron, B-01032 with

R = Ph

M 342.247

Colourless cryst. Sl. sol. org. solvs. Mp 190-201°.

Dorokhov, V.A. *et al*, *Izv. Akad. Nauk SSSR, Ser. Khim.*, 1979, 174; *Bull. Acad. Sci. USSR, Div. Chem. Sci.*, 163 (*synth, ir*)
Dorokhov, V.A. *et al*, *Izv. Akad. Nauk SSSR, Ser. Khim.*, 1979, 411; *Bull. Acad. Sci. USSR, Div. Chem. Sci.*, 380 (*tautom, ir*)

$C_{22}H_{24}B^{\ominus}$ **B-01195**

Butyltriphenylborate(1−), 10CI

$$H_3C(CH_2)_3BPh_3^{\ominus}$$

M 299.242 (ion)

Li salt: [65859-86-1].
$C_{22}H_{24}BLi$ M 306.183
Potential component of battery electrolytes.

U.S.P., 4 060 674, (*1977*); *CA*, **88**, 107897s (*use*)
Negishi, E. *et al*, *J. Chem. Soc., Perkin Trans. 2*, 1978, 1225 (*synth, cmr*)

$C_{22}H_{24}B_7O_2PRu$ **B-01196**

Dicarbonyl[η^5-nonahydrodicarbanonaborato(4−)](triphenylphosphine)ruthenium, 10CI

closo-*6,6-Dicarbonyl-6-triphenylphosphine-6-ruthenanonahydro-2,3-dicarbanonaborane*

[76601-70-2]

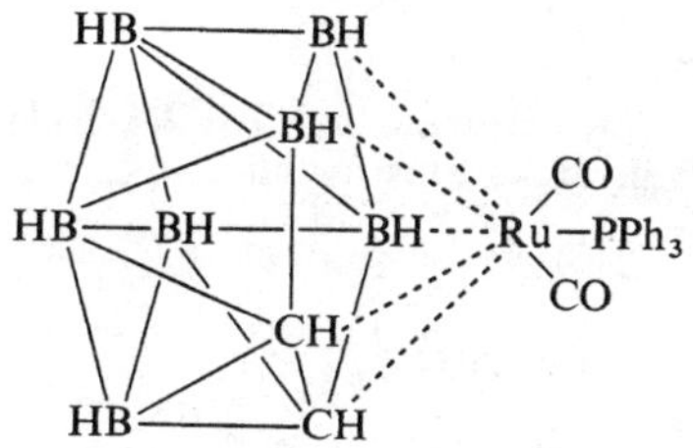

M 528.144

Yellow cryst. (C_6H_6/pentane).

Jung, C.W. *et al*, *J. Am. Chem. Soc.*, 1981, **103**, 811 (*synth, ir, pmr, nmr*)

$C_{22}H_{26}BN$ **B-01197**

Triphenylborane 1-butylamine

1-(1-Butanamine)triphenylboron, 10CI, 9CI. *1-Butanamine compd. with triphenylborane(1:1). 1-Butylamine triphenylborane*

[67616-29-9]

$$Ph_3B—NH_2(CH_2)_3CH_3$$

M 315.264

Wahab, M.A. *et al*, *Egypt J. Chem.*, 1975, **18**, 847 (*ir*)
Wahab, M.A. *et al*, *Egypt J. Chem.*, 1976, **19**, 357 (*synth*)

$C_{22}H_{26}BN$ **B-01198**

Triphenylborane 2-methylpropylamine

(2-Methyl-1-propanamine)triphenylboron, 10CI. *Isobutylamine compd. with triphenylborane(1:1),* 8CI. *Triphenylborane isobutylamine*

[17520-38-6]

$$(H_3C)_2CHCH_2NH_2—BPh_3$$

M 315.264

Solid. Mp 126-7°.

Korshak, V.V. *et al*, *Izv. Akad. Nauk SSSR, Ser. Khim.*, 1963, 1645 (*synth*)
Currell, B.R. *et al*, *J. Organomet. Chem.*, 1967, **8**, 411.
Wahab, M.A. *et al*, *Egypt. J. Chem.*, 1975, **18**, 847; 1976, **19**, 357 (*ir*)

$C_{22}H_{26}BN_3O_2$ B-01199

N^1-[N-(7-Methoxymethyl-3-borabicyclo[3.3.1]non-6-en)-3-yl]-(N-phenylcarbamoyl)pyridon-2-iminate

(2-Imino-N-phenyl-1(2H)-pyridinecarboxamidato-N^{N1},N^2)[[5-(methoxymethyl)-4-cyclohexene-1,3-diyl]-bis[methylene]]boron, 10CI

[76014-65-8]

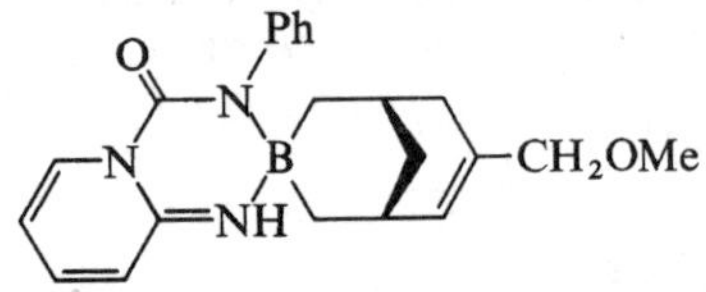

M 375.276

Solid (from yellow soln.). Mp 118-9°.

Buyanovskaya, A.G. *et al*, *Zh. Anal. Khim.*, 1980, **35**, 2106.
Mikhailov, B.M. *et al*, *Izv. Akad. Nauk SSSR, Ser. Khim.*, 1980, 1881; *Bull. Acad. Sci. USSR, Div. Chem. Sci.*, 1355 (*synth, ms, ir, pmr*)

$C_{22}H_{26}BScSi_4$ B-01200

Bis[(1,2,3,4,5-η)-1,3-bis(trimethylsilyl)-2,4-cyclopentadien-1-yl]tetrahydroboratoscandium

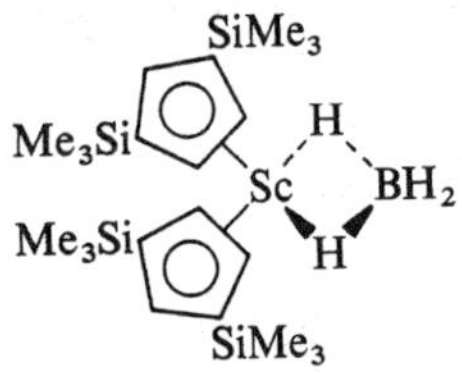

M 458.555

White cryst. Sol. pentane, THF. Mp 82-4°.

Lappert, M.F. *et al*, *J. Chem. Soc., Chem. Commun.*, 1983, 206 (*synth, cryst struct, cmr, nmr, ir*)

$C_{22}H_{30}BN_3$ B-01201

Dibutyl[N-(iminophenylmethyl)benzenecarboximidamidato-N^1,N^N]boron, 10CI

[34629-66-8]

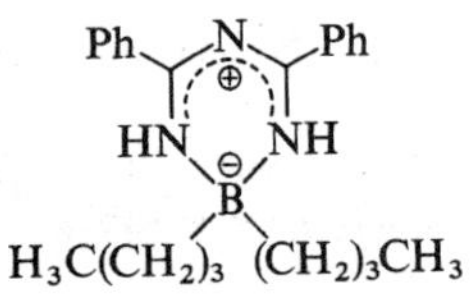

M 347.309

Mikhailov, B.M. *et al*, *Dokl. Akad. Nauk SSSR, Ser. Sci. Khim.*, 1971, **199**, 1328 (*synth*)
Mikhailov, B.M. *et al*, *Zh. Obshch. Khim.*, 1973, **43**, 862.
Mikhailov, B.M. *et al*, *Izv. Akad. Nauk SSSR, Ser. Khim.*, 1978, 1385; *Bull. Acad. Sci. USSR, Div. Chem. Sci.*, 1205.
Zolotarev, B.M. *et al*, *Izv. Akad. Nauk SSSR, Ser. Khim.*, 1978, 1312; *Bull. Acad. Sci. USSR, Div. Chem. Sci.*, 1140 (*ms*)

$C_{22}H_{32}B_4FeMn_2O_6S_2$ B-01202

Hexacarbonylbis[μ-[(1,2,3,4,5-η:1,2,3,4,5-η)-3,4-diethyl-2,5-dihydro-2,5-dimethyl-1,2,5-thiadiborole]](iron)-dimanganese, 10CI

[66921-70-8]

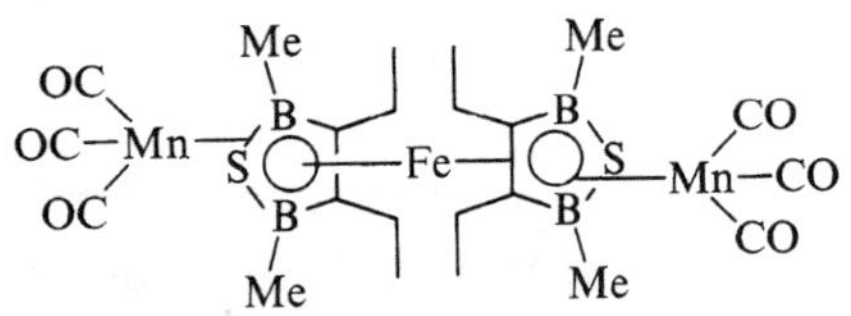

M 665.574

Orange-red cryst. (CS_2). $Bp_{0.1}$ 140-60° subl. Darkens >155°, dec. at 210°.

Siebert, W. *et al*, *Angew. Chem., Int. Ed. Engl.*, 1978, **17**, 527 (*synth, ir, ms, struct*)

$C_{22}H_{33}B$ B-01203

Bis(2-methylcyclohexyl)(2-phenylethenyl)borane, 10CI

[66315-08-0]

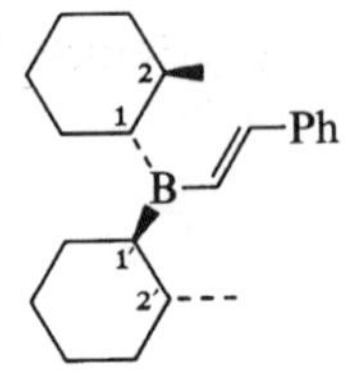

M 308.313

(1RS,1'RS,2RS,2'RS)-(E)-form

(±)-trans,trans,trans-*form*

Formed *in situ* by reaction of phenylethyne with bis(2-methylcyclohexyl)borane. Used in stereoselective synth. of corresp. *E*-alkene. Air-sensitive. Sol. THF.

Yatagai, H. *et al*, *J. Chem. Soc., Chem. Commun.*, 1977, 852 (*synth*)
Pelter, A. *et al*, *Comprehensive Organic Chemistry*, 1979, (*Barton, D.H.R. et al*, Ed.), Pergamon, Oxford, **3**, 695 (*rev*)

$C_{22}H_{34}BClN_2$ B-01204

Dibutyl[2-chloro-N-[2-(1-iminoethyl)cyclohexylidene]benzenaminato-N,N']boron, 10CI

[64578-48-9]

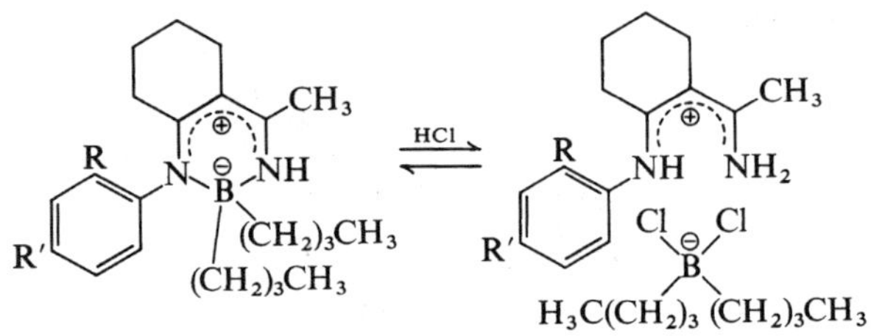

R = Cl, R' = H

M 372.787

Orange liq., relatively stable up to 120°. Bp_1 184°. n_D^{20} 1.5730.

Dorokhov, V.A. *et al*, *Izv. Akad. Nauk SSSR, Ser. Khim.*, 1977, 1843; *Bull. Acad. Sci. USSR, Div. Chem Sci.*, 1704 (*synth, ir, nmr*)

$C_{22}H_{34}BClN_2$ B-01205

Dibutyl[4-chloro-N-[2-(1-iminoethyl)cyclohexylidene]benzenaminato-N,N']boron, 10CI

[64537-96-8]

As Dibutyl[2-chloro-N-[2-(1-iminoethyl)cyclohexylidene]benzenaminato-N,N']boron, B-01204 with

R = H, R' = Cl

M 372.787

Viscous orange liq., relatively stable up to 120°. Bp_3 207-8°. n_D^{20} 1.5768.

Dorokhov, V.A. *et al*, *Izv. Akad. Nauk SSSR, Ser. Khim.*, 1977, 1843; *Bull. Acad. Sci. USSR, Div. Chem. Sci.*, 1704 (*synth, ir, nmr*)

Mikhailov, B.M. *et al*, *Izv. Akad. Nauk SSSR, Ser. Khim.*, 1978, 2574; *Bull. Acad. Sci. USSR, Div. Chem. Sci.*, 2301.

$C_{22}H_{35}BN_2$ B-01206

Dibutyl[*N*-[2-(1-iminoethyl)cyclohexylidene]benzenaminato-*N,N'*]boron, 10CI

*Dibutyl(2-acetimidoyl-*N*-phenyl-1-cyclohexen-1-ylaminato)boron,* 8CI

[24172-91-6]

As Dibutyl[2-chloro-*N*-[2-(1-iminoethyl)-cyclohexylidene]benzenaminato-*N,N'*]boron, B-01204 with

R = R′ = H

M 338.342

Mp 63-6°. Bp_1 186-7°. μ = 2.8D.

B,2HCl: [69109-15-5]. Poorly sol. most org. solvs. Mp 218-21° dec.

Dorokhov, V.A. *et al*, *Dokl. Akad. Nauk SSSR, Ser. Sci. Khim.*, 1969, **187**, 1300 (*synth, ir, nmr*)
Mikhailov, B.M. *et al*, *Izv. Akad. Nauk SSSR, Ser. Khim.*, 1973, 2643.
Dorokhov, V.A. *et al*, *Izv. Akad. Nauk SSSR, Ser. Khim.*, 1977, 1843; *Bull. Acad. Sci. USSR, Div. Chem. Sci.*, 1704 (*synth, ir, nmr*)
Mikhailov, B.M. *et al*, *Izv. Akad. Nauk SSSR, Ser. Khim.*, 1978, 2574; *Bull. Acad. Sci. USSR, Div. Chem. Sci.*, 2301.

$C_{22}H_{37}BN_4$ B-01207

(*N,N''*-Dicyclohexyl-*N'*-2-pyridinylguanidinato-*N'',N^{N'}*)-diethylboron, 10CI

*Diethylboryl 2-[(*N,N'*-dicyclohexylamidino)amino]-pyridinate*

[70426-42-5]

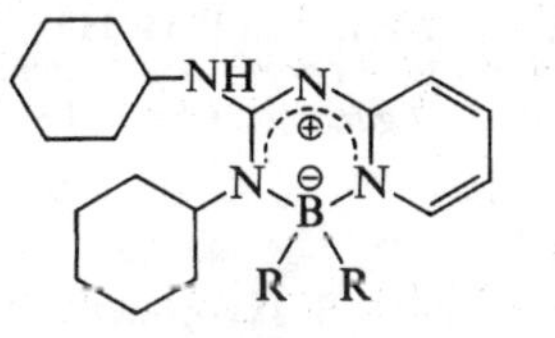

R = Et

M 368.371

Yellow cryst. (EtOH); stable in air and H_2O. Sol. C_6H_6. Mp 103-5°.

Dorokhov, V.A. *et al*, *Dokl. Akad. Nauk SSSR, Ser. Sci. Khim.*, 1979, **245**, 121 (*synth*)
Dorokhov, V.A. *et al*, *Izv. Akad. Nauk SSSR, Ser. Khim.*, 1980, 659; *Bull. Acad. Sci. USSR, Div. Chem. Sci.*, 481 (*synth, ms, ir*)

$C_{22}H_{39}BN_4S$ B-01208

(*N,N'*-Dicyclohexyl-*N''*-2-thiazolylguanidinato-*N,N^{N''}*)bis-(1-methylethyl)boron, 10CI

*Diisopropylboryl 2-[(*N,N'*-dicyclohexylamidino)-amino]thiazolate*

[78798-72-8]

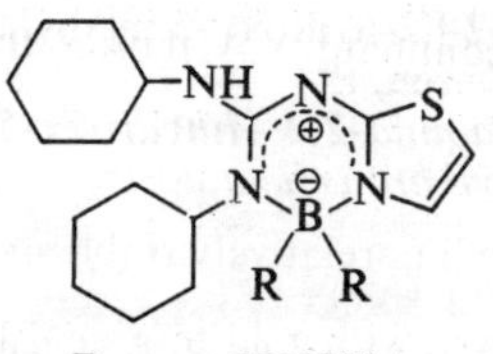

R = $-CH(CH_3)_2$

M 402.447

Yellow cryst. (EtOH); stable in air. Mp 131-4°.

Dorokhov, V.A. *et al*, *Izv. Akad. Nauk SSSR, Ser. Khim.*, 1981, 869; *Bull. Acad. Sci. USSR, Div. Chem. Sci.*, 649 (*synth, nmr, ir, ms*)

$C_{22}H_{41}B$ B-01209

(1-Butyl-1-hexenyl)dicyclohexylborane, 10CI

(5-Decenyl)dicyclohexylborane

As (1-Butyl-1-hexenyl)dichloroborane, B-00691 with

X = Cyclohexyl

M 316.376

***(Z)*-form** [69082-90-2]

Prepd. *in situ* from 5-decyne and dicyclohexylborane. Synth. intermed. Air-sensitive. Sol. THF.

Yatagai, H. *et al*, *J. Chem. Soc., Chem. Commun.*, 1978, 702 (*synth, props*)
Yatagai, H. *et al*, *J. Org. Chem.*, 1979, **44**, 2566 (*pmr*)

$C_{22}H_{42}BO^{\ominus}$ B-01210

2-Furanyltrihexylborate(1−), 10CI

2-Furyltrihexylborate

As 2-Furanyltripropylborate(1−), B-00867 with

R = $-(CH_2)_5CH_3$

M 333.383 (ion)

Li salt: [70415-41-7].
$C_{22}H_{42}BLiO$ M 340.324
Synthetic intermed., prepd. *in situ*.

Akimoto, I. *et al*, *Synthesis*, 1979, 146 (*synth*)

$C_{22}H_{43}B$ B-01211

Di-1-octenyl(1,1,2-trimethylpropyl)borane, 10CI

(2,3-Dimethyl-2-butyl)di-1-octenylborane. Di-1-octenylthexylborane. Thexyldi-1-octenylborane

[62072-21-3]

$(H_3C)_2CHC(CH_3)_2B[(CH{=}CH(CH_2)_5CH_3]_2$

M 318.392

Prepd. *in situ* from 1-octyne and thexylchloroborane followed by reaction with 1-octenyllithium. Synthetic intermed. Air-sensitive oil. Sol. THF.

***(E,E)*-form** [62072-21-3]

***(E,Z)*-form** [75052-87-8]

Zweifel, G. *et al*, *J. Organomet. Chem.*, 1976, **117**, 303 (*ir, nmr, uv*)
Pelter, A. *et al*, *Comprehensive Organic Chemistry*, (Barton, D.H.R. *et al*, Ed.), Pergamon, Oxford), 1979, **3**, 695 (*rev*)
Zweifel, G. *et al*, *J. Am. Chem. Soc.*, 1980, **102**, 5919 (*synth*)

Handle all chemicals with care

$C_{22}H_{43}BO_7$ B-01212

Hydrogen[2,3-dihydroxypropyl hexadecanoato(2–)][1,2,3-propanetriolato(2–)-O^1,O^2]borate(1–), 10CI, 9CI

Glycerol borate palmitate. Diglycerol borate monopalmitate

[52660-31-8]

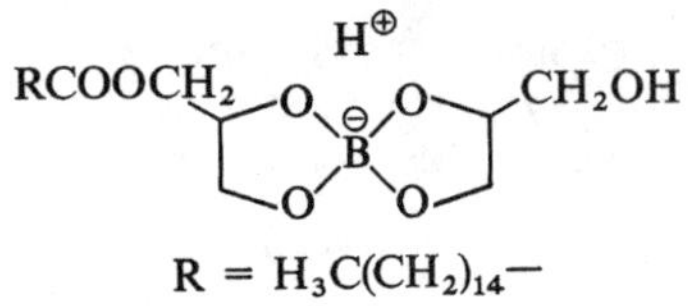

M 430.388

Antistatic agent for plastics; coupling agent for polyolefin moulding compositions contg. inorg. fillers.

Japan. Pat., 73 43 371, (*1973*); *CA*, **81**, 153669x (*use*)
Japan. Pat., 77 89 150, (*1977*); *CA*, **88**, 38553n (*use*)
Japan. Pat., 78 49 051, (*1978*); *CA*, **89**, 90717d (*use*)
Japan. Pat., 78 90 361, (*1978*); *CA*, **89**, 216341x (*use*)

$C_{22}H_{44}BBr$ B-01213

(1-Bromo-1-hexenyl)dioctylborane, 10CI

[67564-13-0]

$[H_3C(CH_2)_7]_2BCBr{=}CH(CH_2)_3CH_3$

M 399.304

Synthetic intermed. Sensitive to air. Sol. THF.

Masuda, Y. *et al*, *Bull. Chem. Soc. Jpn.*, 1980, **53**, 1652.

$C_{22}H_{47}B$ B-01214

Dioctyl(1,1,2-trimethylpropyl)borane, 10CI, 9CI

(2,3-Dimethyl-2-butyl)dioctylborane. Dioctylthexylborane. Thexyldioctylborane

[32327-53-0]

$(H_3C)_2CHC(CH_3)_2B[(CH_2)_7CH_3]_2$

M 322.423

Prepd. *in situ* from thexylborane and 1-octene. Synthetic intermed. for transfer of octyl groups. Air-sensitive oil. Sol. THF.

Bigley, D.B. *et al*, *J. Inorg. Nucl. Chem.*, 1971, **33**, 1157 (*use*)
Brown, H.C. *et al*, *J. Am. Chem. Soc.*, 1975, **97**, 2799 (*synth*)
Hughes, R.J. *et al*, *J. Chem. Soc., Perkin Trans. 1*, 1977, 1172 (*use*)
Kabalka, G.W. *et al*, *J. Organomet. Chem.*, 1977, **125**, 273 (*use*)
Ncube, S. *et al*, *Tetrahedron Lett.*, 1979, 1895 (*use*)

$C_{22}H_{54}B_2N_2$ B-01215

Bis(disiamylborane) ethylenediamine

Tetrakis(1,2-dimethylpropyl)[μ-(1,2-ethanediamine-N:N′)]dihydrodiboron, 10CI

[68307-66-4]

$[(H_3C)_2CHCH(CH_3)]_2BH{-}H_2NCH_2CH_2NH_2{-}BH[CH(CH_3)CH(CH_3)_2]_2$

M 368.302

Useful method for storing disiamylborane, a regioselective hydroborating agent. Can be stored for 15 days at 0° without detectable dec.

Brown, H.C. *et al*, *Inorg. Chem.*, 1979, **18**, 53.

$C_{23}H_{15}BN_2O_3$ B-01216

[2-[(2-Hydroxy-1-naphthalenyl)azo]benzoato(2–)]phenylboron, 10CI

[74300-67-7]

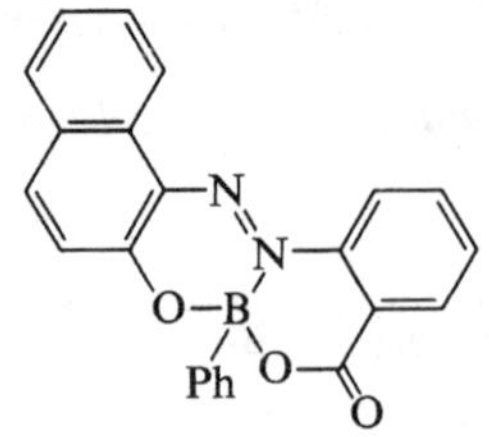

M 378.193

Red-black cryst. Mp 220°.

Hohaus, E. *et al*, *Z. Naturforsch., B*, 1980, **35**, 316, 319 (*ms, synth*)

$C_{23}H_{20}BN$ B-01217

Triphenylborane pyridine

Triphenyl(pyridine)boron, 10CI. Pyridine compd. with triphenylborane. Triphenylborine compd. with pyridine

[971-66-4]

As Tribromoborane pyridine, B-00307 with

X = Ph

M 321.228

Broad-spectrum germicide, fungicide and protozoicide. Solid. Mp 214-5° dec.

Mikhailov, B.M. *et al*, *Dokl. Akad. Nauk SSSR, Ser. Sci. Khim.*, 1956, **109**, 94 (*synth*)
U.S.P., 3 062 708, (*1962*); *CA*, 1963, **58**, 2797e
Roth, H.J. *et al*, *Arch. Pharm.*, (*Weinheim, Ger.*), 1964, **297**, 766
Updegraff, D.M., *J. Infect. Dis.*, 1964, **114**, 304.
Romm, I.P. *et al*, *Khim. Elementoorg. Soedin*, 1976, 15.
Gragg, B.R. *et al*, *J. Organomet. Chem.*, 1977, **132**, 29 (*nmr*)

$C_{23}H_{24}BN_3O_2$ B-01218

[5-(Diethylamino)-2-[(2-hydroxy-4-methylphenyl)azo]phenolato(2–)]phenylboron, 10CI

4-N,N-Diethylamino-2,2′-dihydroxy-4′-methylazobenzene monophenylboron chelate

[74302-89-9]

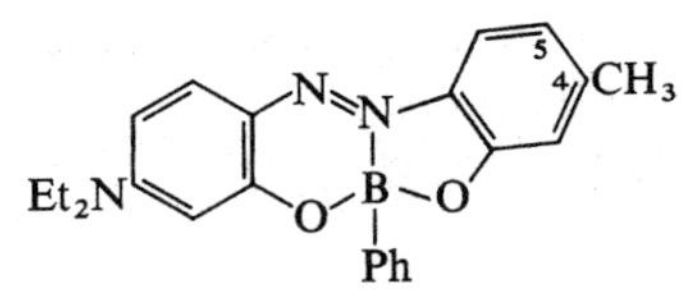

M 385.272

Red-violet cryst. with blue-green sheen. Mp 188°.

Hohaus, E. *et al*, *Z. Naturforsch., B*, 1980, **35**, 316 (*ms*)
Hohaus, E. *et al*, *Z. Naturforsch., B*, 1980, **35**, 319 (*synth, nmr*)

$C_{23}H_{24}BN_3O_2$ B-01219

[5-(Diethylamino)-2-[(2-hydroxy-5-methylphenyl)azo](phenolato(2–)]phenylboron, 10CI

4-N,N-Diethylamino-2,2′-dihydroxy-5′-methylazobenzene monophenylboron chelate

[74302-88-8]

M 385.272

Brown-violet cryst. with gold-to-greenish sheen. Mp 189°.

Hohaus, E. *et al*, *Z. Naturforsch., B*, 1980, **35**, 316 (*ms*)

Hohaus, E. *et al*, *Z. Naturforsch.*, *B*, 1980, **35**, 319 (*synth*, *nmr*)

$C_{23}H_{25}BN_2S$ B-01220

(*N*,*N*-Diethyl-*N'*-phenylthioureato)diphenylborane

(N,N-*Diethyl*-N'-*phenylthioureato*-N',S)*diphenylboron*, *10CI*

[73768-07-7]

As (*N*,*N*-Diethyl-*N'*-phenylthioureato)dimethylborane, B-00856 with

R = Ph

M 372.334

Formed from the reaction of $Et_2NCSNPhSiMe_3$ with Ph_2BCl.

Marringgle, W., *Z. Naturforsch.*, *B*, 1980, **35**, 164 (*synth*)

$C_{23}H_{28}B_9NiP$ B-01221

[Decahydro-7-methyl-π-7-phospha-10-carbaundecaborato(1–)](triphenyl-π-cyclopropenyl)nickel, 8CI

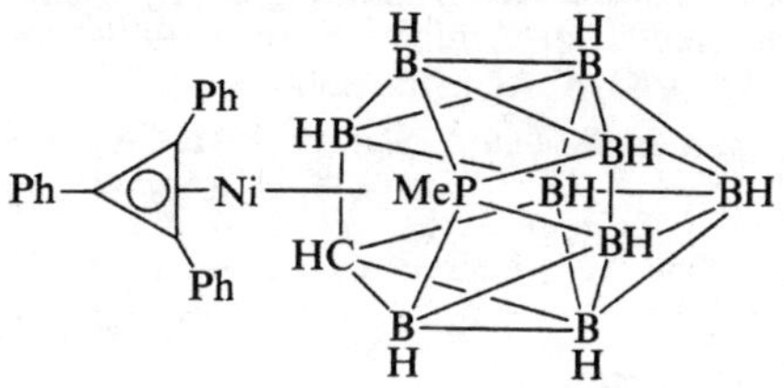

M 491.428

Red cryst. (heptane/C_6H_6). Mp 202-3°.

Welcker, P.S. *et al*, *Inorg. Chem.*, 1970, **9**, 286 (*synth*, *ir*, *ms*, *pmr*)

$C_{23}H_{37}BN_2$ B-01222

Dibutyl[*N*-2-(1-iminoethyl)cyclohexylidene]-4-methylbenzenaminato-*N*,*N'*]boron, 10CI

[64537-93-5]

As Dibutyl[2-chloro-*N*-[2-(1-iminoethyl)cyclohexylidene]benzenaminato-*N*,*N'*]boron, B-01204 with

R = H, R' = CH_3

M 352.369

Viscous orange liq. Bp_1 171°, n_D^{20} = 1.5844.

Dorokhov, V.A. *et al*, *Izv. Akad. Nauk SSSR*, *Ser. Khim.*, 1977, 1843; *Bull. Acad. Sci. USSR*, *Div. Chem. Sci.*, 1704 (*synth*, *nmr*, *ir*)

Mikhailov, B.M. *et al*, *Izv. Akad. Nauk SSSR*, *Ser. Khim.*, 1978, 2574; *Bull. Acad. Sci. USSR*, *Div. Chem. Sci.*, 2301.

$C_{23}H_{40}B^{\ominus}$ B-01223

Tricyclopentyl-1-octynylborate(1–), 10CI, 9CI

1-Octynyltricyclopentylborate

As Tricyclopentyl-1-hexynylborate(1–), B-01182 with

n = 5

M 327.379 (ion)

Li salt: [50543-75-4].

$C_{23}H_{40}BLi$ M 334.320

Prepd. *in situ* from tricyclopentylborane and octynyllithium. Synthetic intermed. Air- and acid-sensitive. Sol. THF, glyme, diglyme.

Pelter, A. *et al*, *J. Chem. Soc.*, *Chem. Commun.*, 1973, 544; 1974, 828 (*use*)

Pelter, A. *et al*, *J. Chem. Soc.*, *Perkin Trans. 1*, 1976, 2419, 2428, 2435 (*synth*, *props*, *use*)

Pelter, A. *et al*, *J. Chem. Soc.*, *Chem. Commun.*, 1977, 913 (*use*)

$C_{23}H_{40}BN_5$ B-01224

(*N*,*N'*-Dicyclohexyl-*N''*-2-pyrimidinylguanidinato-*N*,*N*$^{N''}$)bis(1-methyethyl)boron, 10CI

Diisopropylboryl 2-[(N,N'-*dicyclohexylamidino*)-*amino*]*pyrimidinate*

[74300-85-9]

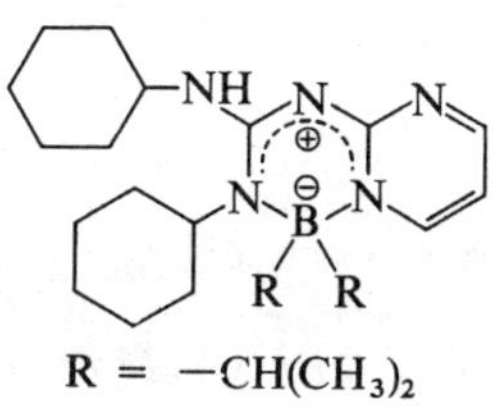

M 397.413

Yellow cryst. (hexane); stable in air. Mp 140-2°.

Dorokhov, V.A. *et al*, *Izv. Akad. Nauk SSSR*, *Ser. Khim.*, 1980, 659; *Bull. Acad. Sci. USSR*, *Div. Chem. Sci.*, 481 (*synth*, *nmr*, *ms*, *ir*, *uv*)

$C_{23}H_{46}B^{\ominus}$ B-01225

1-Octynyltripentylborate(1–), 10CI

Tripentyloctynylborate

$H_3C(CH_2)_5C{\equiv}CB[(CH_2)_4CH_3]_3^{\ominus}$

M 333.426 (ion)

Li salt: [66616-02-2].

$C_{23}H_{46}BLi$ M 340.367

Prepd. *in situ* by addn. of octynyllithium to tripentylborane. Probably oil, air-sensitive. Synth. intermed. Sol. THF.

Pelter, A. *et al*, *J. Chem. Soc.*, *Chem. Commun.*, 1977, 913 (*synth*, *use*)

$C_{24}BF_{20}^{\ominus}$ B-01226

Tetrakis(pentafluorophenyl)borate(1–), 10CI

Tetrakis(*pentafluorophenyl*)*boronate*

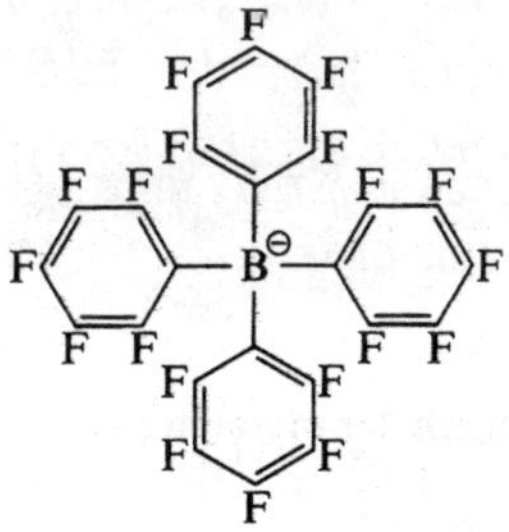

M 679.042 (ion)

Li salt: [2797-28-6].

$C_{24}BF_{20}Li$ M 685.983

Battery electrolyte. Air-stable solid. Sol. H_2O.

Massey, A.G. *et al*, *Proc. Chem. Soc.* (*London*), 1963, 212 (*synth*)

Ger. Pat., 2 901 367, (*1979*); *CA*, **92**, 61799p (*use*)

Horowitz, H.H. *et al*, *Proc. Electrochem. Soc.*, 1981, **81**, 131.

$C_{24}H_8BBr_8O_4^{\ominus}$ B-01227

Bis[3,3′,5,5′-tetrabromo[1,1′-biphenyl]-2,2′-diolato(2−)-*O,O′*]borate(1−), 10CI

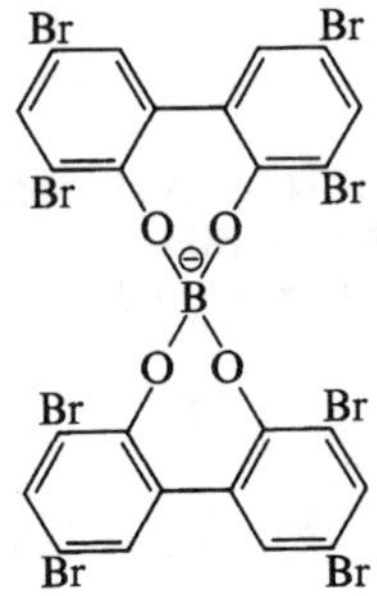

M 1010.367 (ion)

Guanidinium salt: [74210-99-4]. *Hydrogen bis[3,3′,5,5′-tetrabromo[1,1′-biphenyl]-2,2′-diolato(2−)-O,O′]-borate(1−) compd. with guanidine (1:1),* 10CI. *Guanidinium bis(3,3′,5,5′-tetrabromobiphenyl-2,2′-diolato)borate.*

$C_{25}H_{14}BBr_8N_3O_4$ M 1070.445

Fireproofing agent for plastics.

Austrian Pat., 357 336, (*1978*); *CA*, **93**, 151192k (*use*)
Eur. Pat., 5 497, (*1979*); *CA*, **93**, 46174h (*use, synth*)

$C_{24}H_{16}BCl_4^{\ominus}$ B-01228

Tetrakis(4-chlorophenyl)borate(1−), 10CI, 9CI

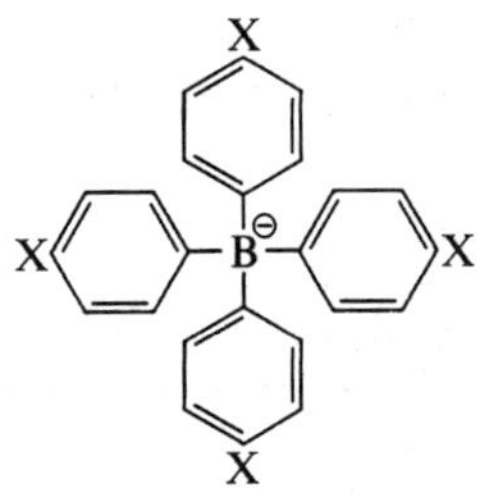

X = Cl

M 457.012 (ion)

K salt: [14680-77-4].

$C_{24}H_{16}BCl_4K$ M 496.111

Used in ion-selective liq. membrane electrodes.

Vandeberg, J.T. *et al*, *Spectrochim. Acta, Part A*, 1971, **27**, 501 (*ir*)
Davies, J.E.W. *et al*, *Lab. Pract.*, 1973, **22**, 20; *CA*, **78**, 66226v (*use*)
Vandeberg, J.T. *et al*, *Org. Magn. Reson.*, 1973, **5**, 57 (*nmr*)
Ammann, D. *et al*, *CA*, 1980, **93**, 21884b (*use*)
Petranek, J. *et al*, *Anal. Chim. Acta*, 1981, **128**, 129.

$C_{24}H_{16}BF_4^{\ominus}$ B-01229

Tetrakis(4-fluorophenyl)borate(1−), 10CI

As Tetrakis(4-chlorophenyl)borate(1−), B-01228 with

X = F

M 391.194 (ion)

Li salt: [68140-33-0].

$C_{24}H_{16}BF_4Li$ M 398.135

Used in battery electrolytes.

U.S.P., 4 139 681, (*1979*); *CA*, **90**, 207210x (*use*)
Ger. Pat., 2 901 367, (*1979*); *CA*, **92**, 61799p (*use*)
Horowitz, H.H. *et al*, *Proc. Electrochem. Soc.*, 1981, **81**, 131.

$C_{24}H_{19}BN_2O_2$ B-01230

[5-Hydroxy-2-(phenylazo)phenoxy]diphenylborane

Diphenyl[4-(phenylazo)-1,3-benzenediolato]boron, 10CI

[74302-93-5]

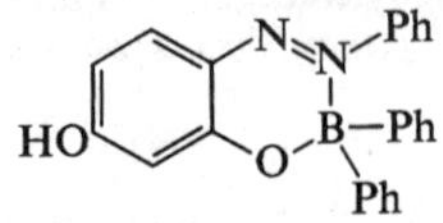

M 378.236

Dark-red cryst. Mp 219°.

Hohaus, E. *et al*, *Z. Naturforsch., B*, 1980, **35**, 319 (*synth*)

$C_{24}H_{20}AgB$ B-01231

Silver(1+) tetraphenylborate(1−), 10CI, 9CI, 8CI

Silver tetraphenylboride

[14637-35-5]

$Ag^{\oplus}Ph_4B^{\ominus}$

M 427.100

Used for gravimetric determination of $Ag^{\oplus}$ and in conductometric determinations of solubility and solubility prod. of Ag salts. Solid.

Mukherji, A.K. *et al*, *Anal. Chem.*, 1959, **31**, 608 (*props*)
Popovych, O., *Anal. Chem.*, 1966, **38**, 117 (*props*)
Costa, G. *et al*, *J. Organomet. Chem.*, 1967, **8**, 339 (*ir*)
Badoz-Lambling, J., *C.R. Hebd. Seances Acad. Sci.*, 1968, **266**, 95 (*props*)

$C_{24}H_{20}BCs$ B-01232

Cesium tetraphenylborate(1−), 10CI, 9CI, 8CI

[3087-82-9]

$Cs^{\oplus}Ph_4B^{\ominus}$

M 452.137

Nmr shift reagent. Used in Cs isotope sepn. Tetrahedral cryst. Sl. sol. H_2O at (2.79×10^{-5}M at 25°).

Arnott, S. *et al*, *Acta Crystallogr.*, 1958, **11**, 449 (*struct*)
Fiedler, H.J. *et al*, *CA*, 1967, **66**, 90840p (*use*)
DeWitte, W.J. *et al*, *Inorg. Nucl. Chem. Lett.*, 1976, **12**, 251 (*nmr*)
Pacey, G. *et al*, *Talanta*, 1980, **27**, 1013 (*use*)
Popovych, O., *Solubility Data Ser.*, 1981, **18**, 52-71.

$C_{24}H_{20}BK$ B-01233

Potassium tetraphenylborate(1−), 10CI, 9CI, 8CI

[3244-41-5]

$K^{\oplus}Ph_4B^{\ominus}$

M 358.330

Catalyst for polymn. of ethylene oxide; contained in anti-fouling paints; electrolyte; gravimetric reagent. Cryst. (Me_2CO aq.). Spar. sol. THF.

Bhattacharyya, D.N., *J. Phys. Chem.*, 1965, **69**, 608 (*props*)
Kirgintsev, A.N. *et al*, *Zh. Prikl. Khim.*, 1970, **43**, 596; *CA*, **72**, 139078m (*synth*)
Vandeberg, J.T. *et al*, *Spectrochim. Acta, Part A*, 1971, **27**, 501 (*ir*)
Vandeberg, J.P. *et al*, *Org. Magn. Reson.*, 1973, **5**, 57 (*nmr*)
Hoffmann, K. *et al*, *J. Organomet. Chem.*, 1974, **67**, 221 (*struct*)
Popovych, O., *Solubility Data Ser.*, 1981, **18**, 17 (*props*)

$C_{24}H_{20}BLi$ B-01234

Lithium tetraphenylborate(1−), 10CI, 9CI, 8CI

[14485-20-2]

$[LiPh_4B]_n$

M 326.173

Polymeric. Used in battery electrolytes. Solid, somewhat sensitive to acid. Sol. THF.

Arnott, S. *et al*, *Acta Crystallogr.*, 1958, **11**, 449 (*struct*)
Bhattachryya, C.L., *J. Phys. Chem.*, 1965, **69**, 608 (*props, synth*)
Revzin, G.E. *et al*, *CA*, 1969, **71**, 3416s (*synth*)
Khol'kin, A.I. *et al*, *CA*, 1976, **85**, 86471u.
Negishi, E. *et al*, *J. Chem. Soc., Perkin Trans. 2*, 1978, 1225 (*cmr, synth*)
Bushby, R.J. *et al*, *J. Chem. Res. (S)*, 1980, 306; *CA*, **94**, 14969a (*synth, purifn*)

$C_{24}H_{20}BNO$ B-01235

[1-[(Methylimino)methyl]-2-naphthalenolato-*N*,*O*]diphenylboron, 10CI

[75095-29-3]

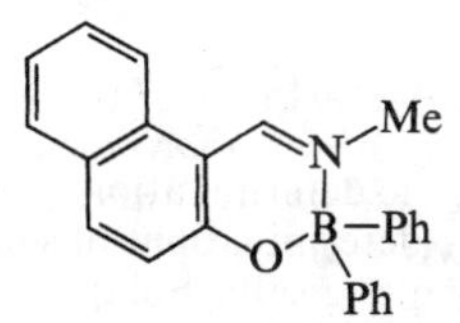

M 349.238

Cryst. (CCl_4). Mp 190°. Dipole moment 4.88D.

Bregadze, V.I. *et al*, *J. Organomet. Chem.*, 1980, **192**, 1 (*ir, uv, struct, synth*)
Kompan, O.E. *et al*, *Zh. Strukt. Khim.*, 1980, **21**, 90.

$C_{24}H_{20}BN_3O$ B-01236

Diphenylboryl-[1-(phenylcarbamoyl)pyridone-2-iminate]

(2-Imino-N-phenyl-1(2H)-pyridinecarboxamidato-N^{N1},N^2)diphenylboron, 10CI

[70440-12-9]

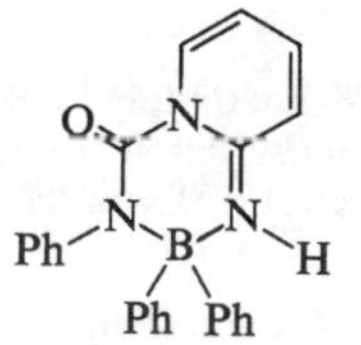

M 377.252

Cryst. Stable in air. Sol. THF, Et_2O, $CHCl_3$. Mp 278-80°.

Dorokhov, V.A. *et al*, *Dokl. Akad. Nauk SSSR, Ser. Sci. Khim.*, 1979, **245**, 121 (*synth*)
Dorokhov, V.A. *et al*, *Izv. Akad. Nauk SSSR, Ser. Khim.*, 1979, 1085; *Bull. Acad. Sci. USSR, Div. Chem. Sci.*, 1014 (*synth, nmr, ir*)

$C_{24}H_{20}BN_3O$ B-01237

Diphenyl(*N*-phenyl-*N*′-2-pyridinylureato-*N*$^{N'}$,*O*)boron, 10CI

O-Diphenylboryl(phenylcarbamoylaminopyridinate)

[70440-09-4]

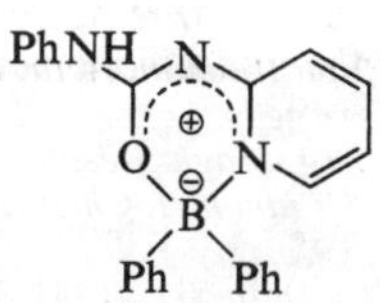

M 377.252

Cryst., rel. stable in air, stable to H_2O and alcohols. Sol. THF. On htg. to 180-90° slowly converts to Diphenyl(*N*-phenyl-*N*′-2-pyridinylureato-*N*,*N*$^{N'}$)boron, B-01238 without melting.

Dorokhov, V.A. *et al*, *Dokl. Akad. Nauk SSSR, Ser. Sci. Khim.*, 1979, **245**, 121 (*synth*)
Dorokhov, V.A. *et al*, *Izv. Akad. Nauk SSSR, Ser. Khim.*, 1979, 1340; *Bull. Acad. Sci. USSR, Div. Chem. Sci.*, 1253 (*synth, nmr, ir, uv*)

$C_{24}H_{20}BN_3O$ B-01238

Diphenyl(*N*-phenyl-*N*′-2-pyridinylureato-*N*,*N*$^{N'}$)boron, 10CI

N-Diphenylboryl(phenylcarbamoylaminopyridinate)

[63720-12-7]

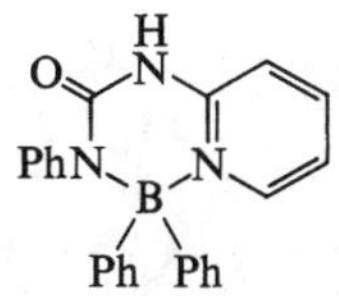

M 377.252

Cryst., rel. stable in air, stable to H_2O and alcohols. Insol. THF. Mp 264-5°.

Handshoe, R. *et al*, *Synth. React. Inorg. Met.-Org. Chem.*, 1977, **7**, 89; *CA*, **87**, 85071t.
Dorokhov, V.A. *et al*, *Dokl. Akad. Nauk SSSR, Ser. Sci. Khim.*, 1979, **245**, 121 (*synth*)
Dorokhov, V.A. *et al*, *Izv. Akad. Nauk SSSR, Ser. Khim.*, 1979, 1340; *Bull. Acad. Sci. USSR, Div. Chem. Sci.*, 1253 (*synth, nmr, ir, uv*)

$C_{24}H_{20}BRb$ B-01239

Rubidium tetraphenylborate(1−), 10CI, 9CI, 8CI

[5971-93-7]

$Rb^{\oplus}Ph_4B^{\ominus}$

M 404.700

Catalyst for polyester resin manuf.; electrolyte. Tetragonal cryst. V. spar. sol. H_2O (2.33×10^{-5} at 25°).

Davies, T. *et al*, *Trans. Faraday Soc.*, 1957, **53**, 19 (*synth, props*)
Arnott, S. *et al*, *Acta Crystallogr.*, 1958, **11**, 449 (*struct*)
Caruso, J.A. *et al*, *J. Am. Chem. Soc.*, 1971, **93**, 1341 (*synth, props*)
Ger. Pat., 2 365 465, (*1975*); *CA*, **83**, 44049r (*use*)
Saleh, N. *et al*, *J. Solution Chem.*, 1979, **8**, 197 (*props*)
Popovych, O., *Solubility Data Ser.*, 1981, **18**, 43 (*props*)

$C_{24}H_{20}BTl$ B-01240

Thallium(1+) tetraphenylborate(1−), 10CI, 9CI, 8CI

[14637-31-1]

$Tl^{\oplus}Ph_4B^{\ominus}$

M 523.615

Flocculent white ppt. V. spar. sol. H_2O (5.29×10^{-5} at 25°). Stable up to 180°.

Wendlandt, W.W., *Anal. Chim. Acta*, 1957, **16**, 216 (*props*)
Costa, G. *et al*, *J. Organomet. Chem.*, 1967, **8**, 339 (*ir*)
Popovych, O., *Solubility Data Ser.*, 1981, **18**, 177.

$C_{24}H_{20}B_2O$ B-01241

Oxybis[diphenylborane], 10CI, 9CI, 8CI

Bis(diphenylboron)oxide. Tetraphenyldiboron oxide

[4426-21-5]

Ph_2BOBPh_2

M 346.041

Cryst. Sol. THF, $CHCl_3$, CH_2Cl_2. Mp 118-9°. ^{11}Bnmr δ 28.5 ppm.

Neu, R. *et al*, *Chem. Ber.*, 1955, **88**, 1761 (*synth*)
Finch, A. *et al*, *Trans. Faraday Soc.*, 1966, **62**, 3314 (*props*)
Nöth, H. *et al*, *Chem. Ber.*, 1966, **99**, 1049 (*nmr*)
Conway, W.R. *et al*, *Synth. React. Inorg. Metal-Org. Chem.*, 1975, **5**, 37 (*synth*)
Gragg, B.R. *et al*, *J. Organomet. Chem.*, 1977, **132**, 29 (*cmr*)
Moehrle, H. *et al*, *Arch. Pharm.* (*Weinheim, Ger.*), 1981, **314**, 580; *CA*, **95**, 115504w (*use*)

$C_{24}H_{21}B_3O_3$ B-01242

Tris(4-ethenylphenyl)boroxin, 10CI

*Tris(*p*-vinylphenyl)boroxin*, 8CI. *Tri(*p*-styryl)boroxine*

[16396-62-6]

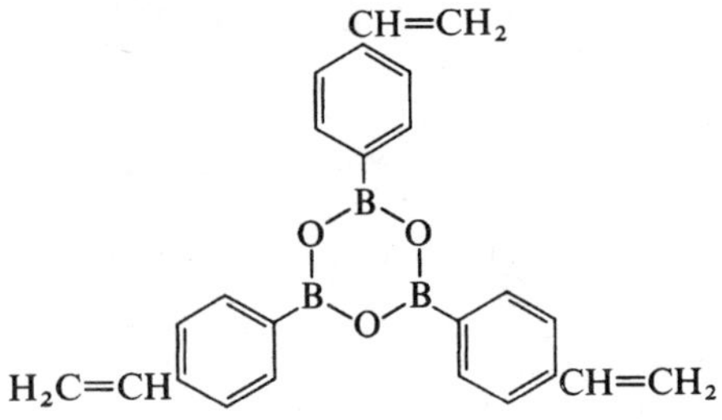

M 389.858

Used in synth. of polymers with chiral cavity. Moisture-sensitive cryst. (CCl_4). Mp 195-6°.

Dale, W.J. *et al*, *J. Org. Chem.*, 1962, **27**, 2598 (*ir, uv, synth*)
Armstrong, D.R. *et al*, *J. Chem. Soc.* (*A*), 1967, 790 (*props*)
Wulf, G. *et al*, *Chem. Ber.*, 1974, **107**, 3364 (*use*)
Rajwar, D.C. *et al*, *Natl. Acad. Sci. Lett.*, 1980, **3**, 33 (*use*)

$C_{24}H_{24}B^{\ominus}$ B-01243

1-Hexynyltriphenylborate(1−), 10CI, 9CI

$H_3C(CH_2)_3C{\equiv}CBPh_3^{\ominus}$

M 323.264 (ion)

Li salt: [42045-41-0].

$C_{24}H_{24}BLi$ M 330.205

Prepd. *in situ* from hexynyllithium and triphenylborane. Synthetic intermed.

Pelter, A. *et al*, *J. Chem. Soc., Chem. Commun.*, 1973, 544 (*use*)
Pelter, A. *et al*, *J. Chem. Soc., Perkin Trans. 1*, 1976, 2419, 2435 (*synth, use*)

$C_{24}H_{24}BN$ B-01244

Ammonium tetraphenylborate(1−), 10CI, 9CI, 8CI

[14637-34-4]

$NH_4^{\oplus}Ph_4B^{\ominus}$

M 337.270

Solid. Almost insol. H_2O, cyclohexane C_6H_6. Mp 110-20° dec.

Davies, T. *et al*, *Trans. Faraday Soc.*, 1957, **53**, 19 (*props, synth*)
Staveley, L.A.K. *et al*, *Pure Appl. Chem.*, 1961, **2**, 271 (*props*)
Costa, G. *et al*, *J. Organomet. Chem.*, 1967, **8**, 339 (*ir*)
Rosenfarb, J. *et al*, *Can. J. Chem.*, 1976, **54**, 3492 (*nmr*)
Oxton, I.A. *et al*, *J. Mol. Struct.*, 1977, **37**, 69 (*ir*)
Westerhaus, W.J. *et al*, *Can. J. Chem.*, 1980, **58**, 1355 (*struct, ir*)

$C_{24}H_{26}BClO$ B-01245

(4-Chlorophenoxy)bis(2,4,6-trimethylphenyl)borane

Bis(2,4,6-trimethylphenyl)borinic acid 4-chlorophenyl ester, 10CI. *Dimesitylborinic acid 4-chlorophenyl ester. Dimesityl(4-chlorophenoxy)borane*

[75260-18-3]

As Hydroxybis(2,4,6-trimethylphenyl)borane, B-01075 with

R = 4-Chlorophenoxy

M 376.732

Colourless cryst. solid, stable in air. Sol. most org. solvs. Mp 100-1°. Slowly oxid. in soln.

Brown, N.M.D. *et al*, *J. Organomet. Chem.*, 1980, **193**, 271 (*synth, nmr, cmr, pmr*)
Brown, N.M.D. *et al*, *J. Organomet. Chem.*, 1981, **210**, 1 (*nmr*)

$C_{24}H_{27}BO$ B-01246

Phenoxybis(2,4,6-trimethylphenyl)borane

Bis(2,4,6-trimethylphenyl)borinic acid phenyl ester, 10CI. *Dimesitylborinic acid phenyl ester. Dimesitylphenoxyborane*

[75260-16-1]

As Hydroxybis(2,4,6-trimethylphenyl)borane, B-01075 with

R = —OPh

M 342.287

Colourless cryst. solid, stable in air. Sol. most org. solvs. Mp 135-6°. Slowly oxid. in soln.

Brown, N.M.D. *et al*, *J. Organomet. Chem.*, 1980, **193**, 271 (*synth, nmr, cmr*)
Brown, N.M.D. *et al*, *J. Organomet. Chem.*, 1981, **210**, 1 (*nmr*)

$C_{24}H_{27}BS$ B-01247

(Phenylthio)bis(2,4,6-trimethylphenyl)borane

Thiobis(2,4,6-trimethylphenyl)borinic acid phenyl ester, 10CI. *Dimesityl(phenylthio)borane*

[77657-18-2]

As Hydroxybis(2,4,6-trimethylphenyl)borane, B-01075 with

R = −SPh

M 358.347

White cryst. (pentane or pentane/Et_2O), sl. affected by atmospheric moisture after 12 hrs. Mp 106-8°.

Davidson, F. *et al*, *J. Organomet. Chem.*, 1981, **204**, 147 (*synth, pmr, cmr, nmr*)
Brown, N.M.D. *et al*, *J. Organomet. Chem.*, 1981, **210**, 1 (*cmr*)

$C_{24}H_{28}BN$ B-01248

(Phenylamino)bis(2,4,6-trimethylphenyl)borane

N*-Phenyl-1,1-bis(2,4,6-trimethylphenyl)boranamine*, 10CI, 9CI. *Dimesitylphenylaminoborane. Anilinodimesitylborane*

[43017-58-9]

As (Methylamino)bis(2,4,6-trimethylphenyl)borane, B-01128 with

R = Ph

M 341.302

White cryst. (hexane/Et_2O). Sol. most org. solvs. Mp 151-2°.

Glogowski, M.E. *et al*, *J. Organomet. Chem.*, 1974, **74**, 175 (*uv*)
Hancock, K.G. *et al*, *J. Organomet. Chem.*, 1975, **90**, 23 (*uv*)
Brown, N.M.D. *et al*, *J. Organomet. Chem.*, 1980, **192**, 133; 1981, **210**, 1 (*synth, ir, nmr*)

$C_{24}H_{28}BS_4^{\ominus}$ B-01249

Tetrakis(2,5-dimethyl-3-thienyl)borate(1−), 10CI

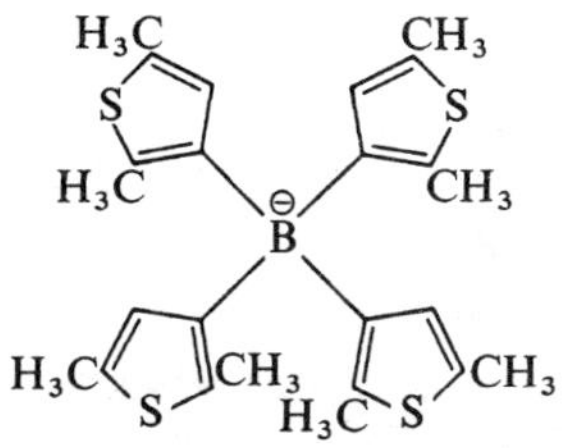

M 455.535 (ion)
Cs salt: [70177-30-9].
$C_{24}H_{28}BCsS_4$ M 588.441
Analytical reagent. No phys. props. reported.

Pacey, G.E. *et al*, *Anal. Chim. Acta*, 1979, **105**, 353.
Pacey, G. *et al*, *Talanta*, 1980, **27**, 1013 (*nmr*)

$C_{24}H_{30}BN$ B-01250

Triphenylborane triethylamine
(N,N-*Diethylethanamine*)*triphenylboron*, 10CI
[5606-04-2]

$Et_3N—BPh_3$

M 343.318
Epoxy resin thermosetting reagent. Solid. Mp 108-12°.

Roth, H.J. *et al*, *Arch. Pharm.* (*Weinheim, Ger.*), 1964, **297**, 766 (*synth*)
Sheikh, S.U., *Bull. Chem. Soc. Jpn.*, 1976, **49**, 2879.
Japan. Pat., 78 123 500, (*1978*); *CA*, **90**, 88329f
Japan. Pat., 79 39 496, (*1979*); *CA*, **91**, 75308u
Japan. Pat., 79 39 497, (*1979*); *CA*, **91**, 75309v
Sheikh, S.U. *et al*, *Proc. Eur. Symp. Therm. Anal., 2nd*, 1981, 446.

$C_{24}H_{33}BN_2$ B-01251

Dibutyl(2-methyl-1,3-diphenyl-1,3-propanediiminato-*N,N′*)-boron, 10CI
[62392-14-7]

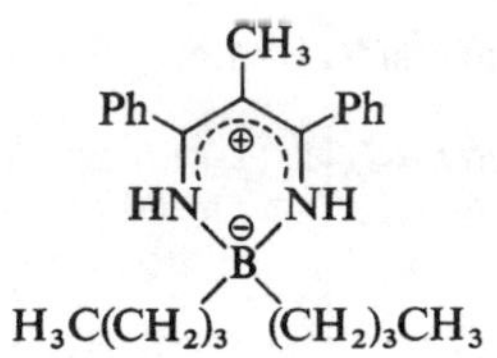

M 360.348
Red cryst. Mp 45-8°. Bp_2 191-2°. High hydrolytic stability.

Mikhailov, B.M. *et al*, *Izv. Akad. Nauk SSSR, Ser. Khim.*, 1976, 2756; *Bull. Acad. Sci. USSR, Div. Chem. Sci.*, 2566 (*synth, uv, ir, nmr*)

$C_{24}H_{41}BN_4$ B-01252

(*N,N″*-Dicyclohexyl-*N′*-2-pyridinylguanidinato-*N″,N^{N′}*)-dipropylboron, 10CI
Dipropylboryl 2-[(N,N′-*dicyclohexylamidino*)*amino*]-*pyridinate*
[70440-06-1]

As (*N,N″*-Dicyclohexyl-*N′*-2-pyridinylguanidinato-*N″,N^{N′}*)diethylboron, B-01207 with

$R = —CH_2CH_2CH_3$

M 396.425
Possesses some antiviral activity. Yellow cryst. (EtOH); stable in air and H_2O. Mp 122-4°.

Dorokhov, V.A. *et al*, *Dokl. Akad. Nauk SSSR, Ser. Sci. Khim.*, 1979, **245**, 121 (*synth*)
Dorokhov, V.A. *et al*, *Izv. Akad. Nauk SSSR, Ser. Khim.*, 1980, 659; *Bull. Acad. Sci. USSR, Div. Chem. Sci.*, 481 (*synth, nmr, uv, ir, ms*)
Lagutkin, N.A. *et al*, *Khim.-Farm. Zh.*, 1982, **16**, 695.

$C_{24}H_{42}B^{\ominus}$ B-01253

Dicyclohexyl-1-hexenyl-1-hexynylborate(1−), 10CI

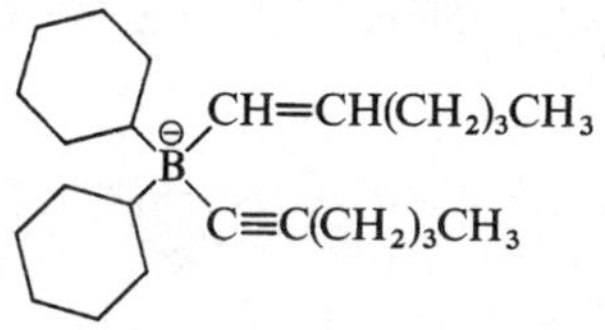

M 341.406 (ion)
Li salt: [67966-24-9].
$C_{24}H_{42}BLi$ M 348.347
Prepd. *in situ* from dicyclohexyl-1-hexenylborane and 1-hexynyllithium. Intermed. in stereoselective synth. of 5,7-dodecadiene.

Negishi, E. *et al*, *J. Chem. Soc., Perkin Trans. 2*, 1978, 1225 (*synth, pmr*)
Zweifel, G. *et al*, *J. Organomet. Chem.*, 1978, **156**, 159 (*synth, use*)

$C_{24}H_{42}B^{\ominus}$ B-01254

Tricyclohexyl-1-hexynylborate(1−), 10CI, 9CI
Hexynyltricyclohexylborate

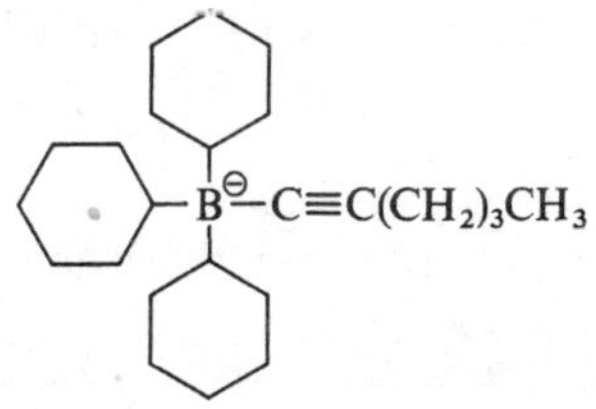

M 341.406 (ion)
Li salt: [42126-51-2].
$C_{24}H_{42}BLi$ M 348.347
Prepd. *in situ* from tricyclohexylborane and hexynyllithium. Intermed. in synth. of alkynes and ketones. Air- and acid-sensitive. Sol. THF, glyme, diglyme.

Pelter, A. *et al*, *J. Chem. Soc., Chem. Commun.*, 1973, 544 (*use*)
Suzuki, A. *et al*, *J. Am. Chem. Soc.*, 1973, **95**, 3080 (*synth, use*)
Pelter, A. *et al*, *J. Chem. Soc., Perkin Trans. 1*, 1976, 2419, 2435 (*synth, props*)

$C_{24}H_{45}BO_7$ B-01255

Hydrogen [2,3-dihydroxypropyl 9-octadecenoato(2−)]-[1,2,3-propanetriolato(2−)-*O*¹,*O*²]borate(1−), 10CI, 9CI
Glyceryl borate (*2:1*) *monooleate. Glycerol monooleoyl-glycerol borate*
[58178-99-7]

As Hydrogen[2,3-dihydroxypropyl hexadecanoato(2−)]-[1,2,3-propanetriolato(2−)-O^1,O^2]borate(1−), B-01212 with

$$R = H_3C(CH_2)_7CH{=}CH(CH_2)_7-$$

M 456.425

Oil-soluble surfactant; used as heat stabilizer for polycarbonates.

Hamanaka, H., *Yukagaku*, 1975, **24**, 542; *CA*, **84**, 76119a (*use*)
Hamanaka, H., *Yukagaku*, 1977, **26**, 268; *CA*, **87**, 203401a (*use*)
Japan. Pat., 78 90 361, (*1978*); *CA*, **89**, 216341x (*use*)

$C_{24}H_{47}BO_7$ B-01256

Hydrogen [2,3-dihydroxypropyl octadecanoate(2−)][1,2,3-propanetriolato(2−)-O^1,O^2]borate(1−), 10CI, 9CI

Glyceryl borate (2:1) monostearate. Glycerol monostearoylglycerol borate

[52660-29-4]

As Hydrogen[2,3-dihydroxypropyl hexadecanoato(2−)]-[1,2,3-propanetriolato(2−)-O^1,O^2]borate(1−), B-01212 with

$$R = H_3C(CH_2)_{16}-$$

M 458.441

Oil-soluble surfactant; used as antistatic agent for plastics, in heat stabilizers for polycarbonates and as corrosion inhibitor for steel in cement.

Japan. Pat., 73 43 371, (*1973*); *CA*, **81**, 153669x (*use*)
Hamanaka, H., *Yukagaku*, 1975, **24**, 542; *CA*, **84** 76119a (*use*)
Ger. Pat., 2 461 359, (*1975*); *CA*, **84**, 93611x (*use*)
Hamanaka, H., *Yukagaku*, 1977, **26**, 268; *CA*, **87**, 203401a (*use*)
Japan. Pat., 78 90 361, (*1978*); *CA*, **89**, 216341x (*use*)

$C_{24}H_{48}B^{\ominus}$ B-01257

Trihexyl-1-hexynylborate(1−), 10CI, 9CI

$$[H_3C(CH_2)_5]_3BC{\equiv}C(CH_2)_3CH_3^{\ominus}$$

M 347.453 (ion)

Li salt: [54816-18-1].
$C_{24}H_{48}BLi$ M 354.394
Prepd. *in situ* from hexynyllithium and trihexylborane. Synthetic intermed.

Pelter, A. *et al*, *J. Chem. Soc., Chem. Commun.*, 1974, 828 (*props*)
Pelter, A. *et al*, *Tetrahedron Lett.*, 1975, 1633 (*props*)
Pelter, A. *et al*, *J. Chem. Soc., Perkin Trans. 1*, 1976, 2419, 2435 (*synth, props*)
Utimoto, K. *et al*, *Tetrahedron Lett.*, 1976, 3969 (*synth*)
Pelter, A. *et al*, *Tetrahedron Lett.*, 1981, **22**, 797 (*props*)

$C_{24}H_{51}B$ B-01258

Trioctylborane, 10CI, 9CI, 8CI

[3248-78-0]

$$[H_3C(CH_2)_7]_3B$$

M 350.477

Prepd. from 1-octene and borane in THF. This method gives a product containg ca. 6% of 2-octyl groups. Synthetic intermed., usually prepd. *in situ*. Used for transfer of 1-octyl group. Air-sensitive, oil. Sol. THF, nonpolar org. solvs. Bp_{12} 189-92°. Heat of formn. ΔH_f −153.04 Kcal mol^{-1}.

Brown, H.C. *et al*, *J. Org. Chem.*, 1957, **22**, 1137 (*synth*)
Bennett, J.E. *et al*, *J. Chem. Soc.*, 1961, 2472 (*props*)
Köster, R. *et al*, *Justus Liebigs Ann. Chem.*, 1965, **682**, 21 (*props*)
Smith, M.B., *J. Organomet. Chem.*, 1974, **76**, 171.
Brown, H.C. *et al*, *J. Org. Chem.*, 1980, **45**, 916 (*props*)
Brown, H.C. *et al*, *Synthesis*, 1980, 153 (*synth*)

$C_{24}H_{51}BO_3$ B-01259

Trioctyloxyborane

Boric acid (H_3BO_3) trioctyl ester, 10CI, 9CI, 8CI. Octyl borate. Trioctyl borate. Trioctyl orthoborate. Tris(octyloxy)borane

[2467-12-1]

$$[H_3C(CH_2)_7O]_3B$$

M 398.475

Heat stabilizer, plasticizer, insect sterilant, catalyst. Moisture-sensitive liq. Sol. non-hydroxylic solvs. d_4^{20}0.855. $Bp_{0.5}$ 182-4°. n_D^{20} 1.4377, 1.4360.

▷ED5630000.

Arbuzov, B.A. *et al*, *Compt. Rend. Acad. Sci. URSS*, 1947, **55**, 411 (*props*)
Gerrard, W. *et al*, *J. Chem. Soc.*, 1955, 3084 (*synth*)
Maijs, L. *et al*, *CA*, 1972, **76**, 51486q (*props*)
Novak, F.I. *et al*, *CA*, 1979, **90**, 167981r (*props*)
U.S.P., 4 211 679, (*1980*); *CA*, **93**, 169189z (*use*)

$C_{24}H_{51}BO_3$ B-01260

Tris(2-ethylhexyloxy)borane

Boric acid (H_3BO_3) tris(2-ethylhexyl) ester, 10CI, 9CI

[2467-13-2]

$$B[OCH_2CH(CH_2CH_3)CH_2CH_2CH_2CH_3]_3$$

M 398.475

Stabilizer for PVC. Moisture-sensitive oil. Sol. non-hydroxylic solvs. d^{24} 0.857. Bp 350-4°, Bp_{20} 227-8°. n_D^{25} 1.4359.

▷ED5800000.

Steinberg, H. *et al*, *Ind. Eng. Chem.*, 1957, **49**, 174 (*synth, props*)
East Ger. Pat., 147 364, (*1981*); *CA*, **95**, 188188f (*use*)

$C_{24}H_{51}BO_{12}$ B-01261

Bis(1,4,7,10-tetraoxaundecyl)(1,4,7,10-tetraoxatetradecyl)-borane

Boric acid (H_3BO_3) 2-[2-(2-butoxyethoxy)ethoxy]ethyl bis[2-[2-(2-methoxyethoxy)ethoxy]ethyl]ester, 10CI

[74198-20-2]

$$H_3C(CH_2)_3O(CH_2CH_2O)_3B[(OCH_2CH_2)_3OMe]_2$$

M 542.470

Used in brake fluids for automobiles. Moisture-sensitive.

Ger. Pat., 2 831 538, (*1980*); *CA*, **93**, 134858q (*use*)
U.S.P., 4 192 759, (*1980*); *CA*, **94**, 86976y (*use*)

$C_{24}H_{51}BO_{12}$ B-01262

(2-Butoxyethoxy)bis(1,4,7,10,13-pentaoxatetradecyl)borane

Boric acid (H_3BO_3) 2-butoxyethyl bis(3,6,9,12-tetraoxatridec-1-yl) ester, 10CI

[74198-23-5]

$$H_3C(CH_2)_3OCH_2CH_2OB[(OCH_2CH_2)_4OMe]_2$$

M 542.470

Used in brake fluids for automobiles. Moisture-sensitive.

Ger. Pat., 2 831 538, (*1980*); *CA*, **93**, 134858q (*use*)
U.S.P., 4 192 759, (*1980*); *CA*, **94**, 86976y (*use*)

$C_{25}H_{21}BN_2O_2$ B-01263

Diphenyl[2-(2-hydroxy-4-methylphenylazo)phenoxy]borane

[2-[(2-Hydroxyphenyl)azo]-4-methylphenolato]diphenylboron, *10CI*

[74300-74-6]

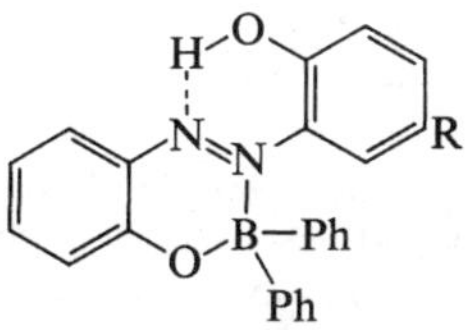

R = CH_3

M 392.263

10CI name refers to alternative tautomer. Dark-red cryst. (C_6H_6). Mp 154°.

Hohaus, E. *et al*, *Z. Naturforsch.*, *B*, 1980, **35**, 316, 319 (*ms*, *synth*)

$C_{25}H_{21}BN_2O_3$ B-01264

Diphenyl[2-(2-hydroxy-4-methoxyphenylazo)phenoxy]borane

[2-[(2-Hydroxyphenyl)azo]-4-methoxyphenolato]diphenylboron, *10CI*

[74300-75-7]

As Diphenyl[2-(2-hydroxy-4-methylphenylazo)phenoxy]-borane, B-01263 with

R = OMe

M 408.263

10CI name refers to alternative tautomer. Black-violet cryst. (C_6H_6). Mp 147°.

Hohaus, E. *et al*, *Z. Naturforsch.*, *B*, 1980, **35**, 316, 319 (*ms*, *synth*)

$C_{25}H_{22}BN_3O$ B-01265

[*N*-(2-Methylphenyl)-*N'*-2-pyridinylureato-*N^N'*,*O*]diphenylboron, 10CI

(O-Diphenylboryl-2-o-tolylcarbamoyl)aminopyridine

[66200-01-9]

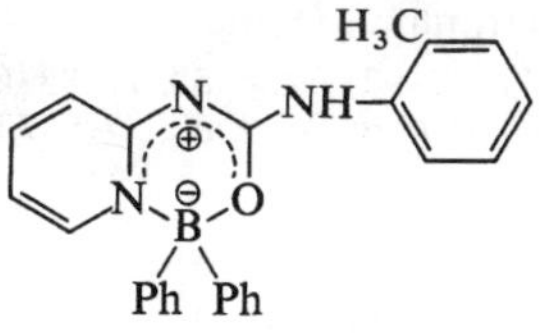

M 391.278

Cryst., relatively air-stable; not dec. by H_2O or alcohols. Mp 244-7°. Isom. before melting.

Dorokhov, V.A. *et al*, *Izv. Akad. Nauk SSSR, Ser. Khim.*, 1978, 253; *Bull. Acad. Sci. USSR, Div. Chem. Sci.*, 224 (*synth*, *ir*)
Dorokhov, V.A. *et al*, *Izv. Akad. Nauk SSSR, Ser. Khim.*, 1979, 1340; *Bull. Acad. Sci. USSR, Div. Chem. Sci.*, 1253 (*synth*, *nmr*)

$C_{25}H_{23}B_2N$ B-01266

Bis(diphenylboryl)methylamine

N-(Diphenylboryl)-N-methyl-1,1-diphenylboranamine, *10CI*, *9CI*

[38762-27-5]

$(Ph_2B)_2NMe$

M 359.083

Solid. Mp 135-9°.

Beck, W. *et al*, *Chem. Ber.*, 1972, **105**, 2883 (*nmr*)
Barlos, K. *et al*, *Chem. Ber.*, 1977, **110**, 2790.
Jonás, K. *et al*, *Chem. Ber.*, 1977, **110**, 2783 (*synth*, *nmr*)

$C_{25}H_{26}B^{\ominus}$ B-01267

1-Heptynyltriphenylborate(1−), 10CI

$H_3C(CH_2)_4C{\equiv}CBPh_3^{\ominus}$

M 337.290 (ion)

Li salt: [78607-90-6].

$C_{25}H_{26}BLi$ M 344.231

Prepd. *in situ* from heptynyllithium and triphenylborane. Synthetic intermed.

Pelter, A. *et al*, *J. Chem. Soc., Chem. Commun.*, 1981, 164 (*props*)

$C_{25}H_{26}BN_3$ B-01268

Guanidinium tetraphenylborate

Hydrogen tetraphenylborate(1−) compd. with guanidine(1:1), *10CI*, *9CI*

[57436-41-6]

$(H_2N)_2C{=}NH_2^{\oplus}BPh_4^{\ominus}$

M 379.311

Forms complexes with crown ethers. Solid. Mp 170-2°.

Madan, K. *et al*, *J. Chem. Soc., Chem. Commun.*, 1975, 427 (*use*)
Kyba, E.P. *et al*, *J. Am. Chem. Soc.*, 1977, **99**, 2564 (*synth*, *use*)
Popovych, O., *Solubility Data Ser.*, 1981, **18**, 91.

$C_{25}H_{29}BO_2$ B-01269

4-Methoxyphenoxybis(2,4,6-trimethylphenyl)borane

Bis(2,4,6-trimethylphenyl)borinic acid 4-methoxyphenyl ester, *10CI*. *Dimesitylborinic acid 4-methoxyphenyl ester. Dimesityl(4-methoxyphenoxy)borane*

[75260-19-4]

As Hydroxybis(2,4,6-trimethylphenyl)borane, B-01075 with

R = 4-Methoxyphenoxy

M 372.313

Colourless cryst. solid, stable in air. Sol. most org. solvs. Mp 115-6°. Slowly oxid. in soln.

Brown, N.M.D. *et al*, *J. Organomet. Chem.*, 1980, **193**, 271 (*synth*, *nmr*, *cmr*)
Brown, N.M.D. *et al*, *J. Organomet. Chem.*, 1981, **210**, 1 (*nmr*)

$C_{25}H_{29}BS$ B-01270

(Benzylthio)bis(2,4,6-trimethylphenyl)borane

Thiobis(2,4,6-trimethylphenyl)borinic acid phenylmethyl ester, *10CI*. *(Benzylthio)dimesitylborane. (Phenylmethylthio)bis(2,4,6-trimethylphenyl)borane*

[77657-17-1]
As Hydroxybis(2,4,6-trimethylphenyl)borane, B-01075 with

R = $-SCH_2Ph$

M 372.374
White cryst. (pentane or pentane/Et_2O), sl. affected by atmospheric moisture after 12 hrs. Mp 76-9°.

Davidson, F. *et al*, *J. Organomet. Chem.*, 1981, **204**, 147 (*synth, pmr, cmr, nmr*)
Brown, N.M.D. *et al*, *J. Organomet. Chem.*, 1981, **210**, 1 (*cmr*)

$C_{25}H_{29}BS$ B-01271

(4-Methylphenylthio)bis(2,4,6-trimethylphenyl)borane
Thiobis(2,4,6-trimethylphenyl)borinic acid 4-methylphenyl ester, 10CI. *Dimesityl(4-methylphenylthio)borane. Dimesityl*(p-*tolylthio*)*borane*
[77657-19-3]
As Hydroxybis(2,4,6-trimethylphenyl)borane, B-01075 with

R = 4-methylphenylthio

M 372.374
White cryst. (pentane or pentane/Et_2O), sl. affected by atmospheric moisture after 12 hrs. Mp 94-6°.

Davidson, F. *et al*, *J. Organomet. Chem.*, 1981, **204**, 147 (*synth, pmr, cmr, nmr*)
Brown, N.M.D. *et al*, *J. Organomet. Chem.*, 1981, **210**, 1 (*cmr*)

$C_{25}H_{30}BN$ B-01272

(Benzylamino)bis(2,4,6-trimethylphenyl)borane
N-*(Phenylmethyl)-1,1-bis(2,4,6-trimethylphenyl)boranamine*, 10CI. *Benzylaminodimesitylborane*
[75018-28-9]
As (Methylamino)bis(2,4,6-trimethylphenyl)borane, B-01128 with

R = $-CH_2Ph$

M 355.329
White cryst. Sol. most org. solvs. Mp 88-90°.

Brown, N.M.D. *et al*, *J. Organomet. Chem.*, 1980, **192**, 133 (*synth, ir, nmr*)
Brown, N.M.D. *et al*, *J. Organomet. Chem.*, 1981, **210**, 1 (*nmr*)

$C_{25}H_{44}BN_5$ B-01273

Dibutyl(*N,N'*-dicyclohexyl-*N''*-2-pyrimidinylguanidinato-*N,N$^{N''}$*)boron, 10CI
Dibutylboryl[2-(N,N'-*dicyclohexylamidino*)*aminopyrimidinate*]
[74300-86-0]

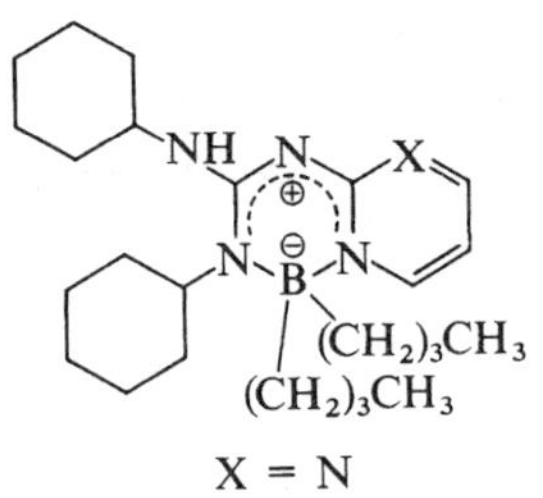

X = N

M 425.466
Possesses some antiviral activity. Yellow cryst. (hexane). Mp 118-21°.

Dorokhov, V.A. *et al*, *Izv. Akad. Nauk SSSR, Ser. Khim.*, 1980, 659; *Bull. Acad. Sci. USSR, Div. Chem. Sci.*, 481 (*synth, ms, ir*)
Lagutkin, N.A. *et al*, *Khim.-Farm. Zh.*, 1982, **16**, 695.

$C_{25}H_{50}B^{\ominus}$ B-01274

1-Heptynyltrihexylborate(1−), 10CI

$[H_3C(CH_2)_5]_3BC{\equiv}C(CH_2)_4CH_3^{\ominus}$

M 361.480 (ion)
Li salt: [74911-29-8].
$C_{25}H_{50}BLi$ M 368.421
Prepd. *in situ* from heptynyllithium and trihexylborane. Synthetic intermed.

Hara, S. *et al*, *Chem. Lett.*, 1980, 221 (*props*)
Pelter, A. *et al*, *J. Chem. Soc., Chem. Commun.*, 1981, 164 (*props*)

$C_{25}H_{70}BIrOP_2Si_6$ B-01275

Carbonyl[tetrahydroborato(1−)-*H,H'*]bis[tris[(trimethylsilyl)methyl]phosphine]iridium, 9CI
Tetrahydroborato(carbonyl)bis[tris(trimethylsilylmethyl)phosphine]iridium
[38437-59-1]

$Ir(BH_4)(CO)[P(CH_2SiMe_3)_3]_2$

M 820.318
Pale-yellow plates (C_6H_6/MeOH). Mp 105-25° dec. ir ν_{CO} 1922 cm^{-1} (nujol).

Hsieh, A.T.T. *et al*, *J. Chem. Soc., Dalton Trans.*, 1972, 1966 (*synth, ir, pmr*)

$C_{26}H_{20}B_2F_4O_6$ B-01276

Tetrafluoro[μ-[[1,1'-(1,3-phenylene)bis[3-(4-methoxyphenyl)-1,3-propanedionato]](2−)]]diboron, 10CI
[67109-24-4]

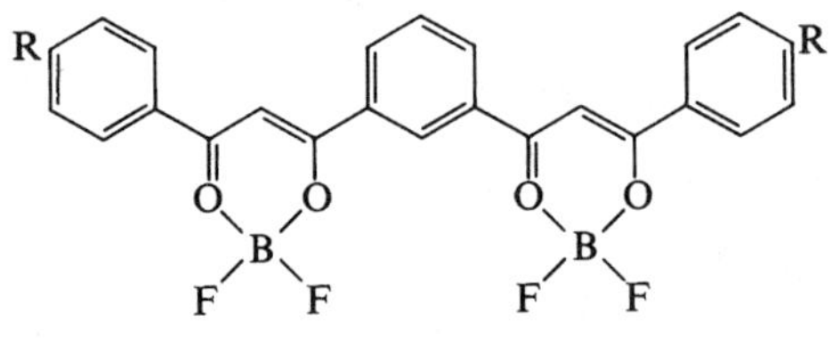

R = OMe

M 526.054
Catalyst for polymerisation of vinylcarbazole. Useful in electrophotographic copying.

Ger. Pat., 2 749 768, (*1978*); *CA*, **89**, 110833u (*use*)
Halm, J.M. *et al*, *Photogr. Sci. Eng.*, 1979, **23**, 252 (*use*)

$C_{26}H_{20}B_2Fe_2O_4$ B-01277

Di-μ-carbonyldicarbonylbis[(1,2,3,4,5,6-η)-1-phenylboratabenzene]diiron, 9CI
Di-μ-carbonyldicarbonylbis(1-phenylborinato)diiron
[54853-77-9]
As Di-μ-carbonylbis[(1,2,3,4,5,6-η)-1-methylboratabenzene]diir B-00988 with

R = Ph

M 529.756
Violet, air-stable leaflets (pentane). Dec. at 150°, subl. at 120° *in vacuo*.

Herberich, G. *et al*, *Chem. Ber.*, 1974, **107**, 3780 (*synth*, *pmr*, *ms*, *ir*)

$C_{26}H_{22}BTl$ B-01278

Di-9*H*-fluoren-9-yl[tetrahydroborato(1–)-*H*,*H'*]thallium

Difluorenylthallium borohydride

[56116-45-1]

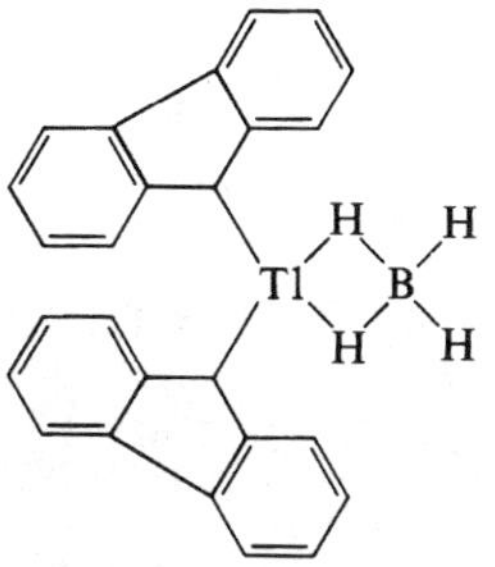

M 549.653

▷Toxic

Kumar, N. *et al*, *CA*, 1974, **82**, 98034e.

$C_{26}H_{26}BTl$ B-01279

Dimethyl(tetraphenylborato-*B*)thallium

Dimethylthallium tetraphenylborate

Me_2TlBPh_4

M 553.684

Solid (Me_2CO). Mp 215° dec.

▷Toxic

Kurosawa, H. *et al*, *Bull. Chem. Soc. Jpn.*, 1967, **40**, 861.

$C_{26}H_{42}B_4Fe_2S_2Zn$ B-01280

Bis-*μ*-[(*η*[5]-2,4-cyclopentadien-1-yl-)bis[*μ*-[(3,4-*η*:3,4-*η*)-3,4-diethyl-2,5-dihydro-2,5-dimethyl-1,2,5-thiadiborole-*B*[2],*B*[5],*S*[1]:*B*[2],*B*[5],*S*[1]]]diiron]zinc, 10CI

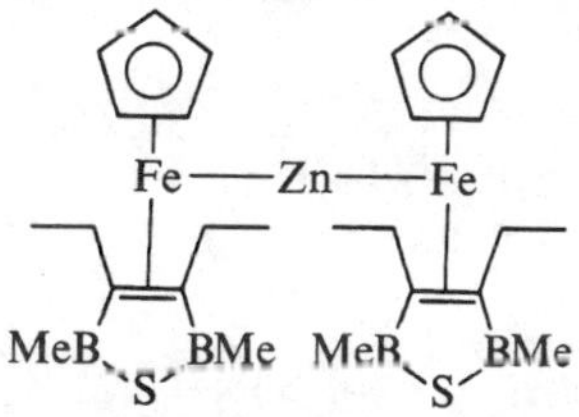

M 639.052

Intermed. for triple decker sandwich compds. Red cryst. (hexane). Mp 182-4°.

Siebert, W. *et al*, *Angew. Chem.*, *Int. Ed. Engl.*, 1980, **19**, 746 (*synth*, *nmr*, *ms*)

$C_{26}H_{42}B_4Fe_3S_2$ B-01281

Bis(*η*[5]-2,4-cyclopentadien-1-yl)bis[*μ*-[(3,4-*η*:3,4-*η*)-3,4-diethyl-2,5-dihydro-2,5-dimethyl-1,2,5-thiadiborole-*B*[2],*B*[5],*S*[1]:*B*[2],*B*[5],*S*[1]]]triiron, 10CI

[74964-92-4]

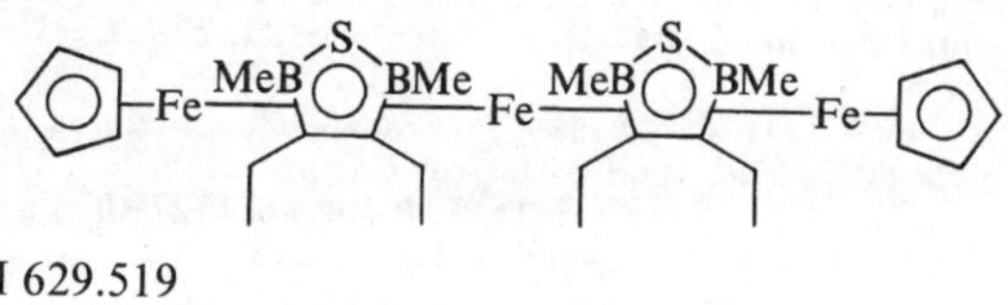

M 629.519

Black needles (toluene). Mp 209-11° dec.

Siebert, W. *et al*, *Angew. Chem.*, *Int. Ed. Engl.*, 1980, **19**, 746 (*synth*, *pmr*, *nmr*, *struct*)

$C_{26}H_{44}B^{\ominus}$ B-01282

Trihexyl(phenylethynyl)borate(1–), 10CI, 9CI

$[H_3C(CH_2)_5]_3BC{\equiv}CPh^{\ominus}$

M 367.444 (ion)

Li salt: [50543-74-3].

$C_{26}H_{44}BLi$ M 374.385

Prepd. *in situ*. Synthetic intermed.

Pelter, A. *et al*, *J. Chem. Soc.*, *Chem. Commun.*, 1973, 544 (*use*)

Pelter, A. *et al*, *J. Chem. Soc.*, *Chem. Commun.*, 1974, 347 (*props*)

Pelter, A. *et al*, *J. Chem. Soc.*, *Perkin Trans. 1*, 1976, 2419, 2435 (*synth*, *props*)

Hara, S. *et al*, *Chem. Lett.*, 1980, 221 (*use*)

Pelter, A. *et al*, *J. Chem. Soc.*, *Chem. Commun.*, 1981, 164 (*use*)

$C_{26}H_{45}BN_4$ B-01283

Dibutyl(*N*,*N''*-dicyclohexyl-*N'*-2-pyridinylguanidinato-*N''*,*N*[N'])boron, 10CI

Dibutylboryl-[2-(N,N'*-dicyclohexylamidino)aminopyridinate*]

[70440-07-2]

As Dibutyl(*N*,*N'*-dicyclohexyl-*N''*-2-pyrimidinylguanidinato-*N*,*N*[N''])boron, B-01273 with

X = CH

M 424.478

Yellow cryst. (EtOH), stable in air and to H_2O. Mp 96-8°.

Mikhailov, B.M. *et al*, *Dokl. Akad. Nauk SSSR*, *Ser. Sci. Khim.*, 1979, **245**, 121.

Dorokhov, V.A. *et al*, *Izv. Akad. Nauk SSSR*, *Ser. Khim.*, 1980, 659; *Bull. Acad. Sci. USSR*, *Div. Chem. Sci.*, 481 (*synth*, *nmr*, *ir*, *uv*, *ms*)

$C_{26}H_{52}B^{\ominus}$ B-01284

Dihexyl-1-octynyl(1,1,2-trimethylpropyl)borate(1–), 10CI, 9CI

Dihexyl-1-octynylthexylborate. Thexyldihexyl-1-octynylborate

As Dibutyl-1-hexynyl(1,1,2-trimethylpropyl)borate(1–), B-01155 with

$R = —(CH_2)_5CH_3$

M 375.507 (ion)

Li salt: [56556-80-0].

$C_{26}H_{52}BLi$ M 382.448

Prepd. *in situ* from octynyllithium and the corresponding trialkylborane. Useful synthetic intermediate.

Pelter, A. *et al*, *Tetrahedron Lett.*, 1975, 1633 (*synth*, *props*)

Pelter, A. *et al*, *J. Chem. Soc.*, *Perkin Trans. 1*, 1976, 2419 (*synth*, *props*)

$C_{26}H_{52}B^{\ominus}$ B-01285

Trihexyl-1-octynylborate(1–), 10CI, 9CI

$[H_3C(CH_2)_5]_3BC{\equiv}C(CH_2)_5CH_3^{\ominus}$

M 375.507 (ion)

Li salt: [50543-73-2].
$C_{26}H_{52}BLi$ M 382.448
Prepd. *in situ*. Synthetic intermed.

Pelter, A. *et al*, *J. Chem. Soc., Chem. Commun.*, 1973, 544 (*props, use*)
Pelter, A. *et al*, *J. Chem. Soc., Perkin Trans. 1*, 1976, 2419, 2428, 2435 (*synth, use, props*)
Pelter, A. *et al*, *J. Chem. Soc., Chem. Commun.*, 1977, 913 (*props*)

$C_{26}H_{54}B_2N_2$ B-01286

Bis(monoisopinocampheylborane) tetramethylethylenediamine
Tetrahydro[*μ*-(N,N,N′,N′-*tetramethyl-1,2-ethanediamine*-N:N′)]*bis*(*2,6,6-trimethylbicyclo*[*3.1.1*]*hept-3-yl*)*diboron*, 10CI. *TMEDA—2*[(*IPC*)*BH*$_2$]
[67826-92-0]

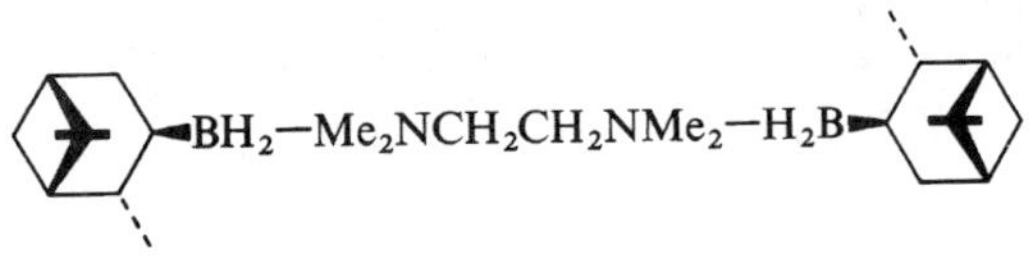

M 416.346
Convenient way of storing the chiral hydroborating agent monoisopinocampheyl borane. Cryst. (pentane or Et_2O). Mp 140-1°.

Brown, H.C. *et al*, *J. Org. Chem.*, 1978, **43**, 4395 (*synth, pmr, nmr*)
Singaram, B. *et al*, *J. Organomet. Chem.*, 1978, **156**, C1 (*synth*)

$C_{26}H_{55}BO_{13}$ B-01287

Bis(1,4,7,10,13-pentaoxatetradecyl)(1,4,7-trioxaundecyl)-borane
Boric acid (H_3BO_3) *2-(2-butoxyethoxy)ethyl bis(3,6,9,12-tetraoxatridec-1-yl)ester*, 10CI
[74198-18-8]

$H_3C(CH_2)_3O(CH_2CH_2O)_2B[(OCH_2CH_2)_4OMe]_2$

M 586.523
Used in brake fluids for automobiles. Moisture-sensitive.

Ger. Pat., 2 831 538, (*1980*); *CA*, **93**, 134858q (*use*)
U.S.P., 4 192 759, (*1980*); *CA*, **94**, 86976y (*synth, use*)

$C_{27}H_{20}BN_3O$ B-01288

Diphenyl[1-(2-pyridinylazo)-2-naphthalenolato]boron, 10CI
Diphenyl[*1-(pyridiyl-2-azo)-2-naphthyloxy*]*borane*
[74302-97-9]

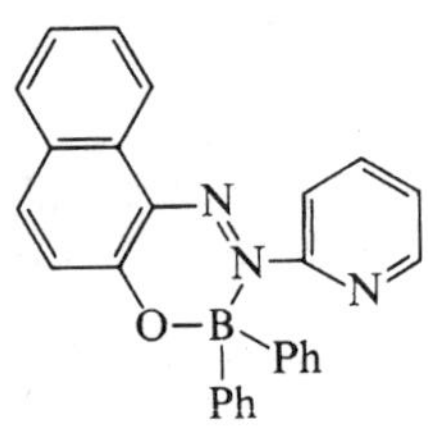

M 413.285
Black-red solid. Mp 190°.

Hohaus, E. *et al*, *Z. Naturforsch., B*, 1980, **35**, 316 (*ms*)
Hohaus, E. *et al*, *Z. Naturforsch., B*, 1980, **35**, 319 (*synth, nmr*)

$C_{27}H_{21}BO_2$ B-01289

(1,3-Diphenyl-1,3-propanedionato-*O,O′*)diphenylboron, 10CI, 9CI
Diphenylboryl 1,3-diphenyl-1,3-propanedionate. 2,2-Dihydro-2,2,4,6-tetraphenyl-1,3,2-dioxaborinium hydroxide inner salt
[19426-82-5]

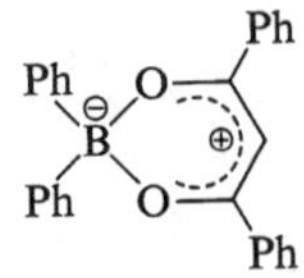

M 388.272
Yellow cryst. (C_6H_6, MeCN or EtOH), stable to atmosphere. Mp 227°.

Neu, R., *Z. Anal. Chem.*, 1960, **176**, 343.
Bally, I. *et al*, *Tetrahedron Lett.*, 1965, 3929 (*uv*)
Dessy, R.E. *et al*, *J. Am. Chem. Soc.*, 1968, **90**, 2005.
Trestianu, A. *et al*, *Tetrahedron*, 1968, **24**, 2499 (*pmr*)
Bally, I. *et al*, *Rev. Roum. Chim.*, 1971, **16**, 739 (*synth*)
Barabas, A., *J. Coord. Chem.*, 1973, **3**, 91.

$C_{27}H_{24}BN_3$ B-01290

Diphenylboryl(*N*-methylbenzimidoyl)benzamidinate
[N-(*Iminophenylmethyl*)-N′-*methylbenzenecarboximidamidato*-N′,N^N]*diphenylboron*, 10CI
[67574-75-8]
As [*N*-(Iminophenylmethyl)-*N′*-propylbenzenecarboximidamidato-*N′*,*N*^{N}]dipropylboron, B-01174 with

R = Ph

M 401.317
Cryst. (hexane). Chemically and thermally stable, can be stored in air. Mp 140-4°.

Zolotarev, B.M. *et al*, *Izv. Akad. Nauk SSSR, Ser. Khim.*, 1978, 1312; *Bull. Acad. Sci. USSR, Div. Chem. Sci.*, 1140 (*ms*)
Mikhailov, B.M. *et al*, *Izv. Akad. Nauk SSSR, Ser. Khim.*, 1978, 1385; *Bull. Acad. Sci. USSR, Div. Chem. Sci.*, 1205 (*synth, ms, pmr, nmr*)

$C_{27}H_{33}B$ B-01291

Tris(2,4,6-trimethylphenyl)borane, 10CI, 9CI
Trimesitylborane, 8CI
[7297-95-2]

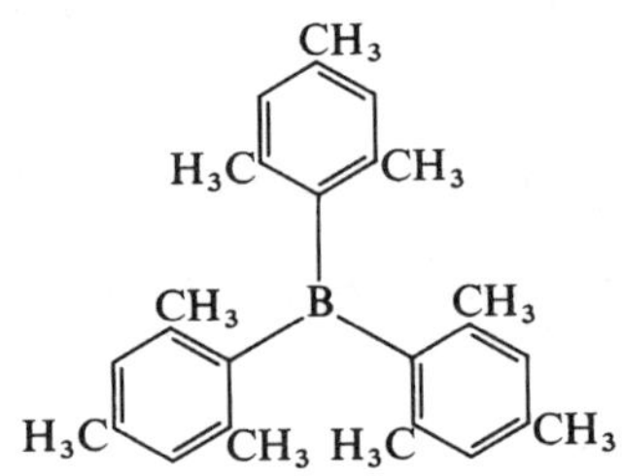

M 368.368
Prepd. from RMgBr and BF_3 in Et_2O. Synthetic intermed. Solid, remarkably resistant to oxidn. Sol. Et_2O, THF. Mp 190.5-191.5°.

Brown, H.C. *et al*, *J. Am. Chem. Soc.*, 1957, **79**, 2302 (*synth*)
Blount, J.F. *et al*, *J. Am. Chem. Soc.*, 1973, **95**, 7019 (*cryst struct*)
Rogozhin, K.L. *et al*, *Izv. Akad. Nauk SSSR, Ser. Khim.*, 1977, 1035; *Bull. Acad. Sci. USSR, Div. Chem. Sci.*, 1977, 949 (*uv*)
Long, M.E., *J. Lumin.*, 1978, **16**, 177; *CA*, **88**, 112790f (*spectra*)
Brown, N.M. *et al*, *J. Organomet. Chem.*, 1981, **209**, 1 (*nmr, synth*)
Ramsey, B.G. *et al*, *J. Org. Chem.*, 1981, **46**, 179 (*deriv, synth*)

$C_{27}H_{37}BN_2$ B-01292

Dibutyl[*N*-(2-iminophenylmethyl)cyclohexylidene]benzenaminato-*N*,*N'*]boron

*Dibutyl(2-benzimidoyl-*N*-phenyl-1-cyclohexen-1-ylaminato)boron*, 8CI

[25513-17-1]

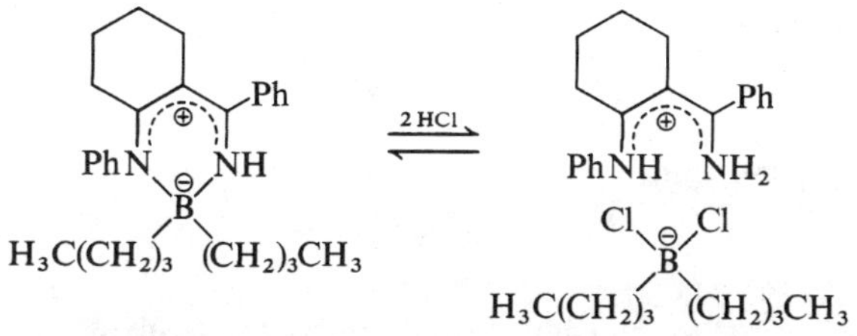

M 400.413

Mp 51-4°. Bp_1 208-10°.

B,2HCl: [69109-16-6].

$C_{27}H_{39}BCl_2N_2$ M 473.335

Poorly sol. most org. solvs. Mp >300° dec.

Dorokhov, V.A. *et al*, *Izv. Akad. Nauk SSSR, Ser. Khim.*, 1977, 1843; *Bull. Acad. Sci. USSR, Div. Chem. Sci.*, 1704 (*synth, ir, nmr*)

$C_{27}H_{57}B$ B-01293

Trinonylborane, 10CI

[4002-27-1]

$$[H_3C(CH_2)_8]_3B$$

M 392.557

Prepd. by hydroboration of 1-nonene in THF at 0°. Synthetic intermed. for transfer of 1-nonyl group. Air-sensitive liq. Sol. THF and other nonpolar solvs. Prep. by hydroboration gives product containing ca. 6% of 2-nonyl groups.

Nikolaev, N.I., *Zh. Fiz. Khim.*, 1977, **51**, 2136.

Pelter, A. *et al*, *Compr. Org. Chem.*, (*Barton, D.H.R. et al*, Ed.), 1979, **3**, 695 (*rev*)

$C_{27}H_{57}BO_3$ B-01294

Trinonyloxyborane

Boric acid (H_3BO_3) trinonyl ester, 10CI, 9CI, 8CI. *Nonyl borate. Trinonyl borate. Trinonyl orthoborate. Tris(nonyloxy)borane*

[2467-14-3]

$$[H_3C(CH_2)_8O]_3B$$

M 440.556

Moisture-sensitive liq. Sol. non-hydroxylic solvs. d_4^{20} 0.858. $Bp_{1.5}$ 228-30°. n_D^{20} 1.4405.

Arbuzov, B.A. *et al*, *Bull. Acad. Sci. USSR, Div. Chem. Sci.*, 1952, 773 (*synth, props*)

Maijs, L. *et al*, *CA*, 1971, **75**, 11843b; 1972, **77**, 157609t (*props*)

Strauss, I. *et al*, *CA*, 1972, **77**, 145170e (*props*)

Maijs, L. *et al*, *CA*, 1973, **79**, 131073e (*pmr*)

$C_{28}H_{16}BF_{12}^{\ominus}$ B-01295

Tetrakis[3-(trifluoromethyl)phenyl]borate(1−), 10CI

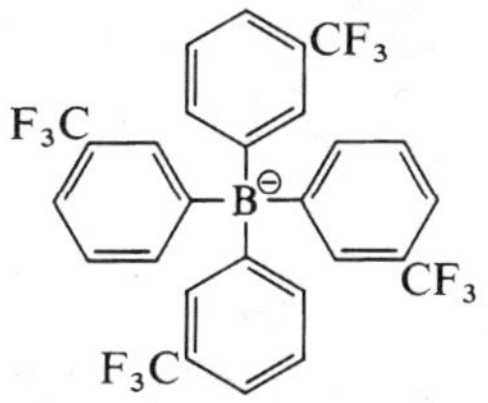

M 591.225 (ion)

Li salt: [70021-22-6].

$C_{28}H_{16}BF_{12}Li$ M 598.166

Used in battery electrolytes.

U.S.P., 4 139 681, (*1979*); *CA*, **90**, 207210x (*use*)

Ger. Pat., 2 901 367, (*1979*); *CA*, **92**, 61799p (*use*)

Horowitz, H.H. *et al*, *Proc. Electrochem. Soc.*, 1981, **81**, 131.

$C_{28}H_{16}BF_{12}^{\ominus}$ B-01296

Tetrakis[4-(trifluoromethyl)phenyl]borate(1−), 10CI

M 591.225 (ion)

Li salt: [70021-21-5].

$C_{28}H_{16}BF_{12}Li$ M 598.166

Used in battery electrolytes.

U.S.P., 4 139 681, (*1979*); *CA*, **90**, 207210x (*use*)

Ger. Pat., 2 901 367, (*1979*); *CA*, **92**, 61799p (*use*)

Horowitz, H.H. *et al*, *Proc. Electrochem. Soc.*, 1981, **81**, 131.

$C_{28}H_{21}BN_2O$ B-01297

Diphenyl[1-(phenylazo)-2-naphthyloxy]borane

Diphenyl[1-(phenylazo)-2-naphthalenolato]boron, 10CI

[74302-95-7]

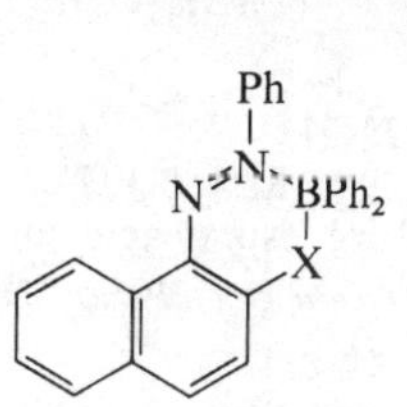

X = O

M 412.297

Dark-red cryst. (EtOH/C_6H_6). Mp 203°.

Hohaus, E. *et al*, *Z. Naturforsch.*, *B*, 1980, **35**, 316, 319 (*ms, synth, ir, nmr*)

$C_{28}H_{22}BN_3$ B-01298

Diphenyl[1-(phenylazo)-2-naphthylamino]borane

Diphenyl[1-(phenylazo)-2-naphalenaminato]boron, 10CI

[74302-98-0]

As Diphenyl[1-(phenylazo)-2-naphthyloxy]borane, B-01297 with

X = NH

M 411.312

Black-violet cryst. Mp 198°.

Hohaus, E. *et al*, *Z. Naturforsch.*, *B*, 1980, **35**, 316, 319 (*ms, synth*)

$C_{28}H_{24}BFe$ **B-01299**

Ferroceniumyltriphenylborate(1−), 8CI

[33340-89-5]

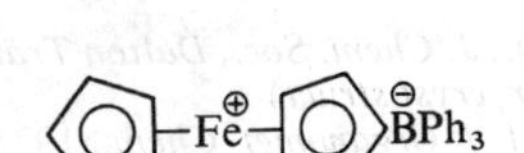

M 427.155

Inner ferrocenium salt. Dark-green solid. Unstable in air, dec. slowly in soln. Mod. sol. $CHCl_3$, CH_2Cl_2, spar. sol. Me_2CO, $MeNO_2$, insol. Et_2O, hydrocarbons. Paramagnetic.

Nesmeyanov, A.N. *et al*, *Dokl. Akad. Nauk SSSR, Ser. Sci. Khim.*, 1971, **198**, 848; **200**, 1365 (*synth*)

$C_{28}H_{24}B_2F_4O_4$ **B-01300**

Tetrafluoro[μ-[[1,1′-(1,4-phenylene)bis[3-(4-ethylphenyl)-1,3-propanedionato]](2−)-O^1,O^3:$O^{1'}$,$O^{3'}$]]diboron, 10CI

[67029-02-1]

As Tetrafluoro[μ-[[1,1′-(1,3-phenylene)bis[3-(4-methoxyphenyl)-1,3-propanedionato]](2−)]]diboron, B-01276 with

$R = CH_2CH_3$

M 522.109

Used in electrophotographic copying apparatus.

Ger. Pat., 2 749 768, (*1978*); *CA*, **89**, 110833u (*synth, use*)
Halm, J.M. *et al*, *Photogr. Sci. Eng.*, 1979, **23**, 252 (*use*)

$C_{28}H_{28}BO_4^{\ominus}$ **B-01301**

Tetrakis(benzyloxy)borate

Tetrakis(benzenemethanolato)borate(1−), 10CI

$(PhCH_2O)_4B^{\ominus}$

M 439.337 (ion)

Na salt: [65460-22-2].
$C_{28}H_{28}BNaO_4$ M 462.327
Moisture-sensitive, esp. in acid conds.

Adams, C. *et al*, *J. Chem. Soc., Perkin Trans. 2*, 1977, 1466, 1472 (*synth, props*)

$C_{28}H_{28}BRh$ **B-01302**

Bis(ethylene)[tetraphenylborato(1−)]rhodium, 8CI

Bis(η²-ethene)(η⁶-tetraphenylborato)rhodium

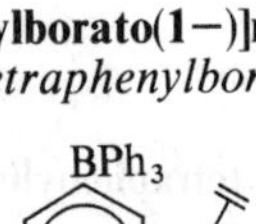

M 478.245

Yellow cryst. (CH_2Cl_2/MeOH).

Schrock, R.R. *et al*, *Inorg. Chem.*, 1970, **9**, 2339 (*synth, ir*)

$C_{28}H_{30}BNO$ **B-01303**

Morpholinium tetraphenylborate

Hydrogen tetraphenylborate(1−) compd. with morpholine(1:1), 10CI, 9CI

[60587-19-1]

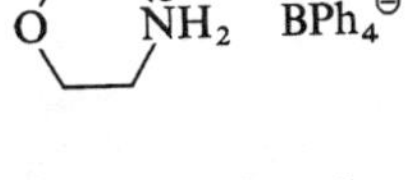

M 407.361

Used as catalyst for crosslinking of epoxy resins.

N-Me: See N-Methylmorpholinium tetraphenylborate, B-01314

Ger. Pat., 2 604 739, (*1976*); *CA*, **85**, 161259 (*use*)
Japan. Pat., 77 56 199, (*1977*); *CA*, **87**, 118785j (*use*)
Japan. Pat., 77 107 066, (*1977*); *CA*, **88**, 106411s (*use*)

$C_{28}H_{32}BN$ **B-01304**

Tetramethylammonium tetraphenylborate

N,N,N-*Trimethylmethanaminium tetraphenylborate(1−)*, 10CI, 9CI

[15525-13-0]

$Me_4N^{\oplus}Ph_4B^{\ominus}$

M 393.378

Catalyst for polymn. of vinyl compds.; electrolyte.

Howick, L.C. *et al*, *Anal. Chim. Acta*, 1958, **19**, 342 (*synth, props*)
Netherlands Pat., 6 600 826, (*1966*); *CA*, **66**, 11401v (*use*)
Barker, B.J. *et al*, *J. Phys. Chem.*, 1974, **78**, 2689 (*props*)
Hoffmann, K. *et al*, *J. Organomet. Chem.*, 1974, **67**, 221 (*struct*)
Czapkiewicz, J. *et al*, *J. Chem. Soc., Faraday Trans. 1*, 1980, **76**, 1663 (*props*)
Popovych, O., *Solubility Data Ser.*, 1981, **18**, 106, 109, 110, 111.

$C_{28}H_{35}BN_2O_2$ **B-01305**

1,5-Cyclooctanediyl[4,4-dimethyl-N-phenyl-N-[(phenylamino)carbonyl]-2-pentenamidato-O,O′]boron, 10CI

[78264-69-4]

As 1,5-Cyclooctanediyl[*N*-phenyl-*N*-[(phenylamino)carbonyl]-2-heptenamidato-*O,O′*]boron, B-01306 with

$R^1 = Ph; R^2 = -C(CH_3)_3$

M 442.407

Formed in reaction of (*tert*-butylethynyl)-9-BBN with PhNCO. Yellow cryst. Mp 178-80°.

Singaram, B. *et al*, *Heterocycles*, 1981, **15**, 231 (*synth, nmr*)

$C_{28}H_{35}BN_2O_2$ **B-01306**

1,5-Cyclooctanediyl[N-phenyl-N-[(phenylamino)carbonyl]-2-heptenamidato-O,O′]boron, 10CI

[78264-70-7]

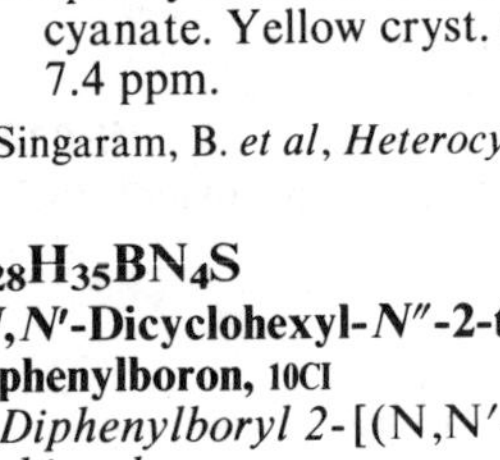

$R^1 = Ph, R^2 = (CH_2)_3CH_3$

M 442.407

Prepd. by reaction of *B*-hexenyl-9-BBN with phenyl isocyanate. Yellow cryst. solid. Mp 128-30°. ¹¹B nmr + 7.4 ppm.

Singaram, B. *et al*, *Heterocycles*, 1981, **15**, 231 (*synth, nmr*)

$C_{28}H_{35}BN_4S$ **B-01307**

(N,N′-Dicyclohexyl-N″-2-thiazolylguanidinato-N,N^N″)diphenylboron, 10CI

Diphenylboryl 2-[(N,N′-dicyclohexylamidino)amino]-thiazolate

[78789-62-5]

As (*N,N′*-Dicyclohexyl-*N″*-2-thiazolylguanidinato-*N,N^N″*)bis(1-methylethyl)boron, B-01208 with

R = Ph

M 470.481
Yellow cryst. (EtOH); stable in air. Mp 159-61°.

Dorokhov, V.A. *et al*, *Izv. Akad. Nauk SSSR, Ser. Khim.*, 1981, 869; *Bull. Acad. Sci. USSR, Div. Chem. Sci.*, 649 (*synth, ir, ms*)

$C_{28}H_{35}BO$ B-01308

4-*tert*-Butylphenoxybis(2,4,6-trimethylphenyl)borane

Bis(2,4,6-trimethylphenyl)borinic acid 4-(1,1-dimethylethyl)phenyl ester, 10CI. *Dimesitylborinic acid 4-(1,1-dimethylethyl)phenyl ester. Dimesityl(4-*tert*-butylphenoxy)borane*

[75260-17-2]

As Hydroxybis(2,4,6-trimethylphenyl)borane, B-01075 with

R = 4-*tert*-butylphenoxy

M 398.394
Colourless cryst. solid, stable in air. Sol. most org. solvs. Mp 175-6°. Slowly oxid. in soln.

Brown, N.M.D. *et al*, *J. Organomet. Chem.*, 1980, **193**, 271 (*synth, nmr, cmr*)
Brown, N.M.D. *et al*, *J. Organomet. Chem.*, 1981, **210**, 1 (*nmr*)

$C_{28}H_{47}BN_2O_2$ B-01309

[*N*-Cyclohexyl-*N*-[(cyclohexylamino)carbonyl]-2-heptenamidato-*O,O'*]-1,5-cyclooctanediylboron, 10CI

[78264-65-0]

As 1,5-Cyclooctanediyl[*N*-phenyl-*N*-[(phenylamino)carbonyl]-2-heptenamidato-*O,O'*]boron, B-01306 with

R^1 = cyclohexyl, $R^2 = -(CH_2)_3CH_3$

M 454.502
Yellow cryst. Mp 138-40°.

Singaram, B. *et al*, *Heterocycles*, 1981, **15**, 231 (*synth, nmr*)

$C_{29}H_{25}BFe$ B-01310

(η^5-2,4-Cyclopentadien-1-yl)[(η^6-phenyl)triphenylborato(1−)]iron, 10CI

[74593-59-2]

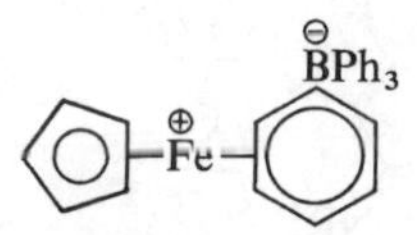

M 440.174
Zwitterionic complex. Mp 275-8° dec.

Slocum, D.W. *et al*, *J. Macromol. Sci., Chem., Part A*, 1981, **16**, 357 (*synth*)

$C_{29}H_{25}BRu$ B-01311

(η^5-2,4-Cyclopentadien-1-yl)[(η^6-phenyl)triphenylborato(1−)]ruthenium, 9CI

[33680-33-0]

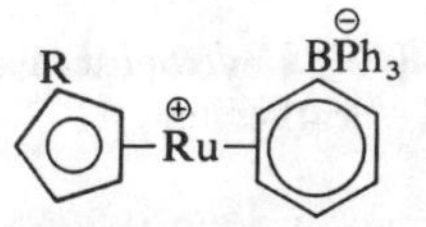

R = H

M 485.397

Zwitterionic complex. Brown cubes (MeOH), prob. air-stable. Sl. sol. most org. solvs. Λ 1.2 Ω^{-1} cm^2 mol^{-1} (Me_2CO).

Kruger, G.J. *et al*, *J. Chem. Soc., Dalton Trans.*, 1974, 1302 (*synth, pmr, ir, cryst struct*)
Haines, R.J. *et al*, *J. Organomet. Chem.*, 1975, **84**, 357 (*synth, ir, pmr*)

$C_{29}H_{26}BN$ B-01312

Pyridinium tetraphenylborate

Hydrogen tetraphenylborate(1−) compd. with pyridine(1:1), 10CI, 9CI

[50328-28-4]

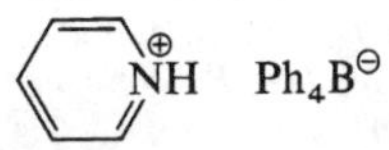

M 399.341
Crosslinking catalyst for epoxy resins. Antifouling agent.

Clements, R. *et al*, *Mol. Struct.*, 1973, **17**, 265 (*ir, raman*)
Japan. Pat., 77 56 199, (*1977*); *CA*, **87**, 118785j (*use*)
Japan. Pat., 77 125 532, (*1977*); *CA*, **88**, 137975h (*use*)
Japan. Pat., 78 34 895, (*1978*); *CA*, **89**, 111372e (*use*)
Popovych, O., *Solubility Data Ser.*, 1981, **18**, 97 (*props*)

$C_{29}H_{29}BO_2$ B-01313

3-(1,3-Diphenyl-1,3-propanedionato)-3-borabicyclo[3.3.1]-nonane

*(1,3-Diphenyl-1,3-propanedionato-*O,O'*)[(5-phenyl-1,3-cyclohexanediyl)bis(methylene)]boron*, 10CI

[76095-69-7]

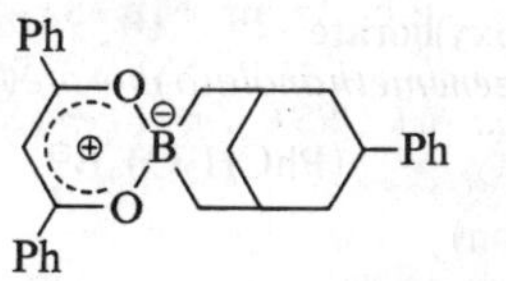

M 420.357
Orange cryst. Mp 180-3°.

Gursky, M.E. *et al*, *J. Organomet. Chem.*, 1980, **199**, 171 (*nmr*)
Gurskii, M.E. *et al*, *Izv. Akad. Nauk SSSR, Ser. Khim.*, 1981, 341; *Bull. Acad. Sci. USSR, Div. Chem. Sci.*, 264 (*nmr*)
Gurskii, M.E. *et al*, *Izv. Akad. Nauk SSSR, Ser. Khim.*, 1981, 394; *Bull. Acad. Sci. USSR, Div. Chem. Sci.*, 312 (*synth, ir, pmr*)

$C_{29}H_{32}BNO$ B-01314

***N*-Methylmorpholinium tetraphenylborate**

Hydrogen tetraphenylborate(1−), compd. with 4-methylmorpholine(1:1), 10CI

[61693-11-6]

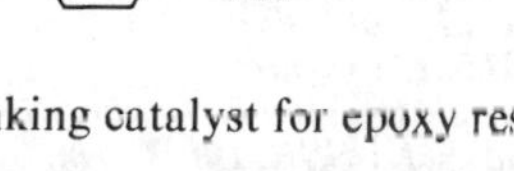

M 421.388
Used as crosslinking catalyst for epoxy resins, and as heat stabiliser.

Japan. Pat., 76 100 200, (*1976*); *CA*, **86**, 90980x (*use*)
Japan. Pat., 78 34 895, (*1978*); *CA*, **89**, 111372e (*use*)

$C_{29}H_{35}BN_2$ **B-01315**
Dibutyl[*N*-(3-imino-1,3-diphenylpropylidene)benzenaminato-*N*,*N'*]boron, 10CI
[64537-99-1]

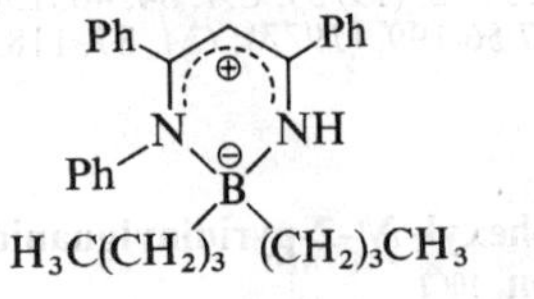

M 422.419
Viscous orange liq., relatively stable to heat. Bp_1 222°. n_D^{20} 1.5969.

Dorokhov, V.A. *et al*, *Izv. Akad. Nauk SSSR, Ser. Khim.*, 1977, 1843; *Bull. Acad. Sci. USSR, Div. Chem. Sci.*, 1704 (*synth, ir, nmr*)
Mikhailov, B.M. *et al*, *Izv. Akad. Nauk SSSR, Ser. Khim.*, 1978, 2574; *Bull. Acad. Sci. USSR, Div. Chem. Sci.*, 2301.

$C_{29}H_{36}BN_5$ **B-01316**
(*N*,*N'*-Dicyclohexyl-*N''*-2-pyrimidinylguanidinato-*N*,*N^N*-'')diphenylboron, 10CI
Diphenylboryl 2-(N,N'*-dicyclohexylamidino)-aminopyrimidinate*
[74313-99-8]
As (*N*,*N'*-Dicyclohexyl-*N''*-2-pyrimidinylguanidinato-*N*,*N^N''*)bis(1-methyethyl)boron, B-01224 with
R = Ph

M 465.447
Cryst. (hexane); stable in air. Mp 183-7°.

Dorokhov, V.A. *et al*, *Izv. Akad. Nauk SSSR, Ser. Khim.*, 1980, 659; *Bull. Acad. Sci. USSR, Div. Chem. Sci.*, 481 (*synth, ms, ir*)

$C_{29}H_{49}BN_4$ **B-01317**
[*N*-[(Cyclohexylamino)(cyclohexylimino)methyl]-*N'*-(1-methylethyl)benzenecarboximidamidato]dipropylboron, 10CI
[70426-41-4]

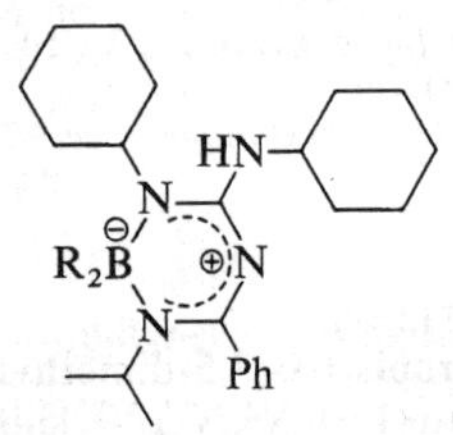

R = $H_3CCH_2CH_2$—

M 464.543
Cryst. (hexane). Mp 114-7°.

Dorokhov, V.A. *et al*, *Dokl. Akad. Nauk SSSR, Ser. Sci. Khim.*, 1979, **245**, 121 (*synth*)
Dorokhov, V.A. *et al*, *Izv. Akad. Nauk SSSR, Ser. Khim.*, 1980, 663; *Bull. Acad. Sci. USSR, Div. Chem. Sci.*, 495 (*synth, nmr, ms, ir*)

$C_{30}H_{21}B$ **B-01318**
Tri-1-naphthylborane, 8CI
Tri-1-naphthalenylborane, 10CI, 9CI. *Tri-1-naphthylborine. Tri-α-naphthylboron*
[6962-78-3]

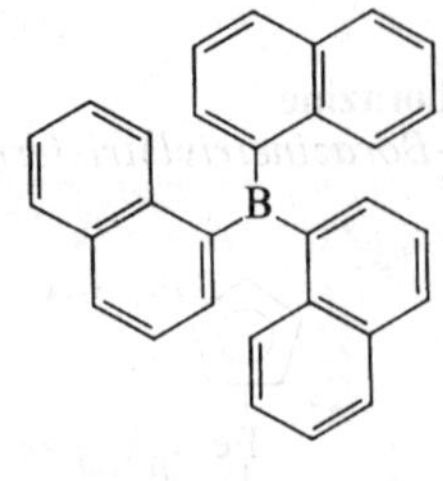

M 392.306
Prepd. from α-naphthylmagnesium bromide and $BF_3 \cdot Et_2O$. Synthetic intermed. Air-sensitive cryst. (Et_2O). Sol. C_6H_6, $CHCl_3$, CCl_4, CS_2. Mp 203°. Blue fluor. Easily oxidised.
C_6H_6 solvate: Solid. Mp 170°.
NH_3 complex:
$C_{30}H_{24}BN$ M 409.336
Sol. C_6H_6. Mp 193-4°.

Krause, E. *et al*, *Ber.*, *B*, 1930, **63**, 934 (*synth, deriv*)
Nakayama, Y. *et al*, *Kogyo Kagaku Zasski*, 1960, **63**, 1477; *CA*, **56**, 11778h (*synth*)
Krakovyak, M.G. *et al*, *Vysokomol. Soedin.*, 1966, **8**, 1890; *CA*, **66**, 18887s (*use*)
DuPont, T.J. *et al*, *J. Am. Chem. Soc.*, 1975, **97**, 6375 (*props*)
Ramsey, B.G. *et al*, *J. Am. Chem. Soc.*, 1977, **99**, 3182 (*props*)
Long, M.E. *et al*, *J. Lumin.*, 1978, **16**, 177; *CA*, **73**, 72398r (*uv*)

$C_{30}H_{27}BFe_3$ **B-01319**
Triferrocenylboron
1,1'',1''''-Borylidynetrisferrocene, 9CI
[12637-02-4]

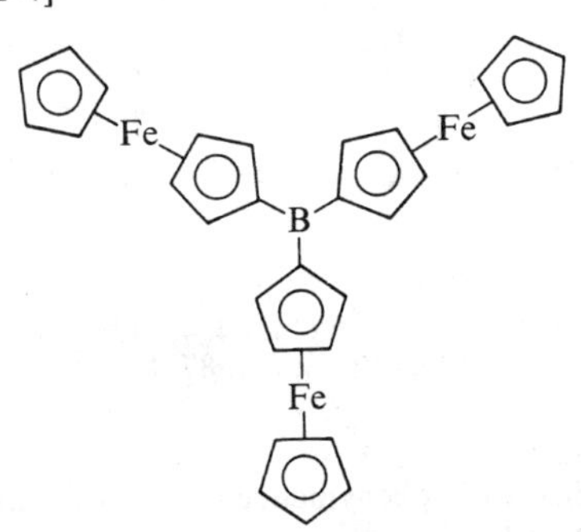

M 565.894
Red cryst.
Ferrocene adduct (*1:3*)*:* Red-orange cryst. (EtOH).
NH_3 adduct:
$C_{30}H_{30}BFe_3N$ M 582.925
Yellow powder (C_6H_6/Et_2O). Mp 180-1° dec.

Post, E.W. *et al*, *Inorg. Chem.*, 1970, **9**, 1670 (*synth*)
Painter, W.J., *Diss. Abs. Int. B*, 1971, **31**, 6482 (*synth*)

$C_{30}H_{27}BRu$ **B-01320**
[(1,2,3,4,5-η)-1-Methyl-2,4-cyclopentadien-1-yl][(η⁶-phenyl-)triphenylborato(1−)]ruthenium, 9CI
[55326-91-5]
As (η^5-2,4-Cyclopentadien-1-yl)[(η^6-phenyl)triphenylborato(1−)]ruthenium, B-01311 with
R = CH_3

M 499.423

Zwitterionic complex. Brown cryst. (MeOH), prob. air-stable.

Haines, R.J. *et al*, *J. Organomet. Chem.*, 1975, **84**, 357 (*synth, ir, pmr*)

$C_{30}H_{30}B_3Fe_3N_3$ B-01321

2,4,6-Triferrocenylborazine

1,1″,1‴′-(2,4,6-Borazinetriyl)trisferrocene, 9CI

[11136-43-9]

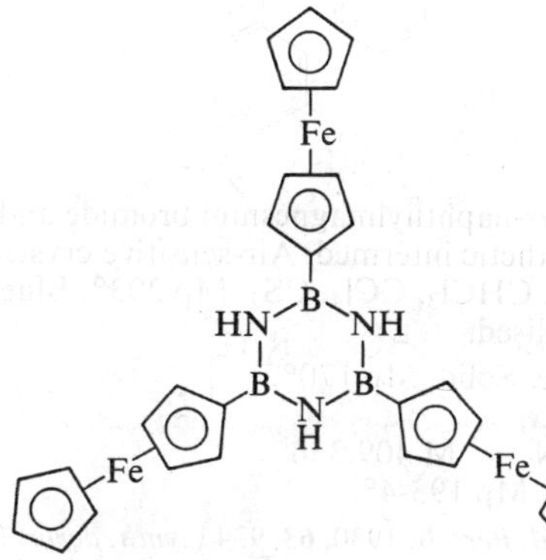

M 632.558

Orange cryst. (toluene). Mp 259°.

Painter, W.J., *Diss. Abs. Int. B*, 1971, **31**, 6520 (*synth*)
Kotz, J.C. *et al*, *J. Organomet. Chem.*, 1971, **32**, 231 (*synth, ms, pmr, uv, ir*)

$C_{30}H_{31}BN_2$ B-01322

2-Ethyl-4-methylimidazolium tetraphenylborate

*Hydrogen tetraphenylborate(1–), compd. with 2-ethyl-4-methyl-1*H*-imidazole(1:1)*, 10CI, 9CI

[53831-70-2]

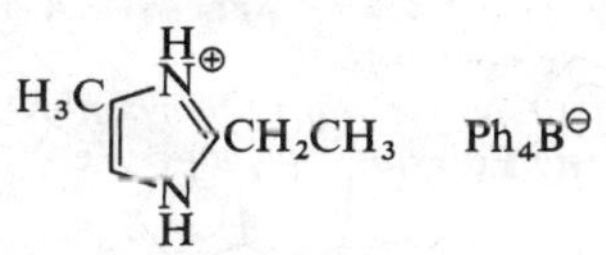

M 430.398

Catalyst for crosslinking of epoxy resins.

Ger. Pat., 2 422 372, (*1975*); *CA*, **84**, 46315u (*use*)
Japan. Pat., 77 56 199, (*1977*); *CA*, **87**, 118785j (*use*)

$C_{30}H_{32}B_2N_2S_2$ B-01323

Tetramethylbis[μ-(*N*-phenylbenzenecarbothioamidato-*N*:*S*)]diboron, 10CI

[63835-75-6]

As Tetramethylbis[*N*-(*N*-phenylethanethioamidato-*N*:*S*)]diboron, B-01147 with

R = Ph, R′ = Me

M 506.336

Liq. Sol. CH_2Cl_2. $Bp_{0.01}$ 144-7°.

Maringgele, W. *et al*, *J. Organomet. Chem.*, 1977, **141**, 241 (*synth, nmr, ms*)
Maringgele, W. *et al*, *Monatsh. Chem.*, 1977, **108**, 751.

$C_{30}H_{36}BN$ B-01324

Triethylammonium tetraphenylborate

Hydrogen tetraphenylborate(1–) compd. with N,N-diethylethanamine(1:1), 10CI, 9CI

[53860-54-1]

$Et_3NH^{\oplus}Ph_4B^{\ominus}$

M 421.431

Catalyst for crosslinking of epoxy resins. Stable for long periods at 25°.

Barker, B.J. *et al*, *J. Phys. Chem.*, 1974, **78**, 2687 (*props*)
Ger. Pat., 2 422 372, (*1975*); *CA*, **84**, 46315u (*use*)
Japan. Pat., 77 56 199, (*1977*); *CA*, **87**, 118785j

$C_{30}H_{37}BN_4$ B-01325

(*N*,*N*″-Dicyclohexyl-*N*′-2-pyridinylguanidinato-*N*″,*N*$^{N'}$)-diphenylboron, 10CI

Diphenylboryl 2-[(N,N′-dicyclohexylamidino)amino]-pyridinate

[70440-08-3]

As (*N*,*N*″-Dicyclohexyl-*N*′-2-pyridinylguanidinato-*N*″,*N*$^{N'}$)diethylboron, B-01207 with

R = Ph

M 464.459

Yellow cryst. (EtOH); stable in air and H_2O. Sol. C_6H_6, pet. ether. Mp 129-30°.

Dorokhov, V.A. *et al*, *Dokl. Akad. Nauk SSSR, Ser. Sci. Khim.*, 1979, **245**, 121 (*synth*)
Dorokhov, V.A. *et al*, *Izv. Akad. Nauk SSSR, Ser. Khim.*, 1980, 659; *Bull. Acad. Sci. USSR, Div. Chem. Sci.*, 481 (*synth, ms, ir*)

$C_{30}H_{38}BO_6P_2Rh$ B-01326

[(η^6-Phenyl)triphenylborato(1–)]bis(trimethyl phosphite-*P*)-rhodium, 9CI

Bis(phosphorous acid)[tetraphenylborato(1–)]rhodium, hexamethyl ester, 8CI. *(Tetraphenylboron)bis(trimethyl phosphite)rhodium*

[33362-35-5]

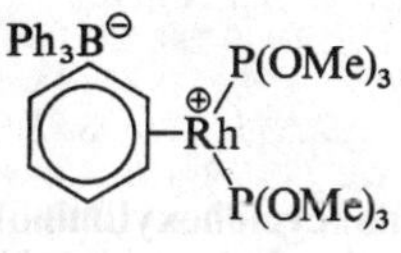

M 670.290

Yellow cryst. (CH_2Cl_2/MeOH). Mp 163-5°.

Haines, L.M. *et al*, *Inorg. Chem.*, 1971, **10**, 1685 (*synth, ir, pmr, cryst struct*)
Nolte, M.J. *et al*, *Acta Crystallogr., Sect. B*, 1974, **30**, 738 (*cryst struct*)

$C_{30}H_{44}B_2Cl_4Ir_2N_{12}$ B-01327

Di-μ-chlorodichlorobis[tris(3,5-dimethyl-1*H*-pyrazolato-*N*[1])hydroborato(1–)-*N*[2],*N*[2′],*N*[2″]]diiridium, 10CI

μ-Dichlorobis[[hydrotris(3,5-dimethylpyrazolyl)-borato]chloroiridium]

[73117-58-5]

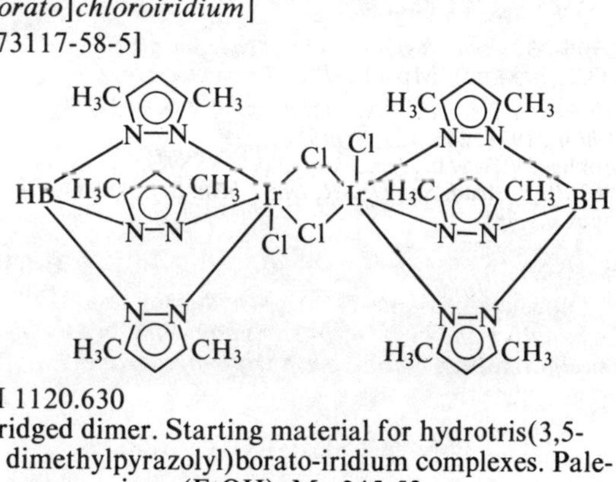

M 1120.630

Bridged dimer. Starting material for hydrotris(3,5-dimethylpyrazolyl)borato-iridium complexes. Pale-orange prisms (EtOH). Mp 245-5°.

Monomer:
$C_{15}H_{22}BCl_2IrN_6$ M 560.315
Unknown.

May, S. *et al*, *Inorg. Chem.*, 1980, **19**, 1582 (*synth, pmr*)

$C_{30}H_{60}B^{\ominus}$ B-01328

1-Hexynyltrioctylborate(1−), 10CI, 9CI

$[H_3C(CH_2)_7]_3BC{\equiv}C(CH_2)_3CH_3^{\ominus}$

M 431.614 (ion)

Li salt: [50543-71-0].
$C_{30}H_{60}BLi$ M 438.555
Prepd. *in situ* from hexynyllithium and trioctylborane. Synthetic intermed.

Pelter, A. *et al*, *J. Chem. Soc., Chem. Commun.*, 1973, 544 (*use*)
Pelter, A. *et al*, *J. Chem. Soc., Perkin Trans. 1*, 1976, 2419, 2428, 2435 (*synth, use*)
Pelter, A. *et al*, *Tetrahedron Lett.*, 1981, **22**, 797 (*props*)

$C_{30}H_{63}B$ B-01329

Tris(decyl)borane, 10CI, 9CI, 8CI
Tridecylborane
[1188-96-1]

$[H_3C(CH_2)_9]_3B$

M 434.638
Prepd. by hydroboration of 1-decene with BH_3·THF. This produces a product containing ca. 6% of 2-decyl groups. Synthetic intermed., generally prepd. *in situ*, for transfer of a 1-decyl group. Air-sensitive oil. Sol. THF, nonpolar solvs. d^{25} 0.807. Bp_5 205-8°. n_D^{25} 1.4543.

Köster, R., *Justus Liebigs Ann. Chem.*, 1958, **618**, 31 (*synth*)
Köster, R., *Justus Liebigs Ann. Chem.*, 1976, 1116 (*use*)
Nikolaev, N.I. *et al*, *CA*, 1979, **90**, 71606w.
Utimoto, K. *et al*, *Tetrahedron*, 1977, **33**, 1945 (*use*)
Vasil'ev, L.S. *et al*, *Izv. Akad. Nauk SSSR, Ser. Khim.*, 1977, 1126; *Bull. Acad. Sci. USSR, Div. Chem. Sci.*, 1031 (*use*)
Kabalka, G.W. *et al*, *J. Org. Chem.*, 1981, **46**, 4296 (*synth*)

$C_{30}H_{63}BO_3$ B-01330

Tri(decyloxy)borane
Boric acid (H_3BO_3) *tris(decyl) ester,* 10CI, 9CI, 8CI. *Tridecyl borate. Tris(decyloxy)borane. Tridecyl orthoborate. Decyl borate*
[20236-81-1]

$[H_3C(CH_2)_9O]_3B$

M 482.636
Insect sterilant, stabilizer for polyamides. Moisture-sensitive liq. Sol. non-hydroxylic solvs. d_4^{20}0.858. $Bp_{0.5}$ 273-6°. n_D^{20} 1.4440.

Arbuzov, B.A. *et al*, *Compt. Rend. Acad. Sci. URSS*, 1947, **55**, 411 (*synth*)
Young, D.M. *et al*, *J. Org. Chem.*, 1961, **26**, 2158 (*synth, props*)
Maijs, L. *et al*, *CA*, 1968, **68**, 117885n (*props*)
Strauss, I. *et al*, *CA*, 1972, **77**, 145170e (*props*)
Maijs, L. *et al*, *CA*, 1972, **77**, 157609t (*props*)
Novak, F.I. *et al*, *CA*, 1979, **90**, 167981r (*props*)

$C_{31}H_{28}BN_3$ B-01331

(*N,N'*-Diphenylguanidine-*N''*)triphenylboron, 10CI
Triphenylborane N,N'*-diphenylguanidine.* N,N'*-Diphenylguanidine, compd. with triphenylborane(1:1)*
[73895-21-3]

PhN=C(NHPh)NH_2—BPh_3

M 453.392

Cryst. (EtOH). Mp 156°.

Lapkin, I.I. *et al*, *Zh. Obshch. Khim.*, 1980, **50**, 368; *J. Gen. Chem. USSR (Engl. Transl.)*, 301
Yuzhakova, G.A. *et al*, *Zh. Obshch. Khim.*, 1981, **51**, 880.

$C_{31}H_{28}BRh$ B-01332

[(2,3,5,6-η)-Bicyclo[2.2.1]hepta-2,5-diene][(η⁶-phenyl)triphenylborato(1−)]rhodium, 10CI
(2,5-Norbornadiene)[tetraphenylborato(1−)]rhodium, 8CI
[31884-67-0]

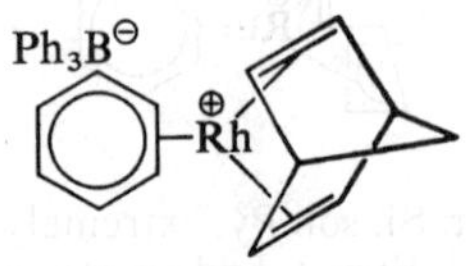

M 514.278
Pale-yellow cryst. (CH_2Cl_2/EtOH).

Schrock, R.R. *et al*, *Inorg. Chem.*, 1970, **9**, 2339 (*synth, ir, pmr*)
Oro, L.A. *et al*, *J. Organomet. Chem.*, 1978, **148**, 81 (*synth, ir*)

$C_{31}H_{28}BTl$ B-01333

Methyl(phenyl)(tetraphenylborato-*B*)thallium
Methyl(phenyl)thallium tetraphenylborate

MePhTlBPh_4

M 615.755
Mp >250°.
▷Toxic

Kurosawa, H. *et al*, *J. Organomet. Chem.*, 1968, **12**, 241.

$C_{31}H_{35}B_2NO_2$ B-01334

5-Diisopropyliminio-2,2,4,4-tetraphenyl-2,4-diborata-1-oxa-3-oxoniacyclopentane
[*μ*-[[*Bis(1-methylethyl)amino*]*carbonyl*-C:O]]-*μ*-*hydroxytetraphenyldiboron,* 10CI, 9CI
[71781-96-9]

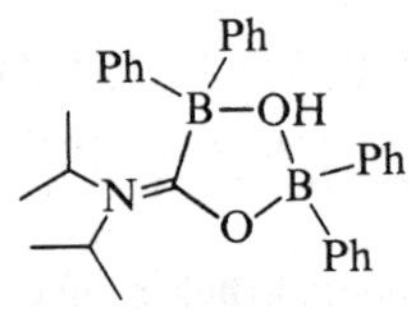

M 475.243
Forms stable cryst. 1:1 complexes with simple amides as well as with amines. Solid (Et_2O/hexane). Mp 179°.

Fletcher, A.S. *et al*, *J. Chem. Soc., Chem. Commun.*, 1979, 347, 573 (*synth, struct, ir, pmr, nmr*)
Fletcher, A.S. *et al*, *Heterocycles*, 1982, **18**, 107.

$C_{31}H_{53}BN_4$ B-01335

[*N*-[(Cyclohexylamino)(cyclohexylimino)methyl]-*N'*-(1-methylethyl)benzenecarboximidamidato]bis(2-methylpropyl)-boron, 10CI
[74323-74-3]
As [*N*-[(Cyclohexylamino)(cyclohexylimino)methyl]-*N'*-(1-methylethyl)benzenecarboximidamidato]dipropylboron, B-01317 with

R = $(H_3C)_2CHCH_2$−

M 492.597

Cryst. (hexane). Mp 71-6°.

Dorokhov, V.A. *et al*, *Izv. Akad. Nauk SSSR, Ser. Khim.*, 1980, 663; *Bull. Acad. Sci. USSR, Div. Chem. Sci.*, 495 (*synth, spectra*)

$C_{32}H_{31}BRu$ B-01336

[(1,2,3,5,6-η)-2,5-Cyclooctadien-1-yl][(η⁶-phenyl)triphenylborato(1−)]ruthenium

[65297-47-4]

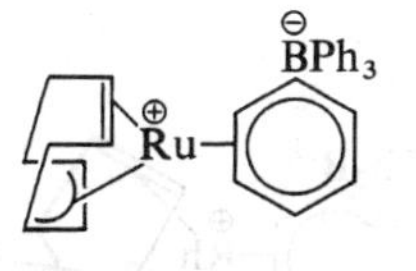

M 527.477

Pale-yellow cryst. Sl. sol. Py, extremely insol. other org. solvs. Mp 200°. Struct. and spectra also determined for the more sol. $PhBF_3$ analogue.

Ashworth, T.V. *et al*, *J. Chem. Soc., Chem. Commun.*, 1977, 937 (*synth, struct*)
Ashworth, T.V. *et al*, *J. Chem. Soc., Dalton Trans.*, 1977, 1809 (*synth, ir*)

$C_{32}H_{31}BS_2$ B-01337

[2-(1,3-Benzodithiol-2-yl)-1-phenyl-1-heptenyl]diphenylborane, 10CI

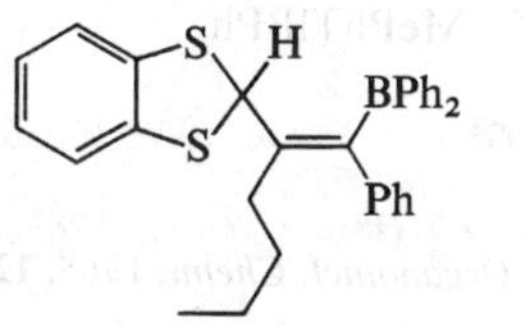

(*E*)-*form*

M 490.527

Formed in reaction of lithium heptynyltriphenylborate with 1,3-benzodithiolium tetrafluoroborate. Synthetic intermed.

(*E*)-*form* [78605-83-1]
Major isom. formed during prepn. (ca. 84%).

(*Z*)-*form* [78605-87-5]
Minor isom. (ca. 16%).

Pelter, A. *et al*, *J. Chem. Soc., Chem. Commun.*, 1981, 164 (*synth, pmr*)

$C_{32}H_{32}BRh$ B-01338

[(1,2,5,6-η)-1,5-Cyclooctadiene][(η⁶-phenyl)triphenylborato(1−)]rhodium, 9CI

(1,5-Cyclooctadiene)[(tetraphenylborato(1−)]rhodium, 8CI

[31974-01-3]

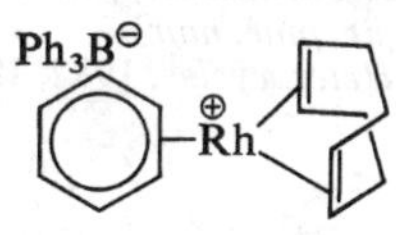

M 530.320

Pale-yellow cryst. Spar. sol. org. solvs.

Schrock, R.R. *et al*, *Inorg. Chem.*, 1970, **9**, 2339 (*synth, ir*)

$C_{32}H_{35}B_9Pd$ B-01339

(1,2,3,4-Tetraphenyl-1,3-cyclobutadiene)[nonahydro-1,2-dimethyl-1,2-dicarbaundecaborato(2−)]palladium, 8CI

[12118-93-3]

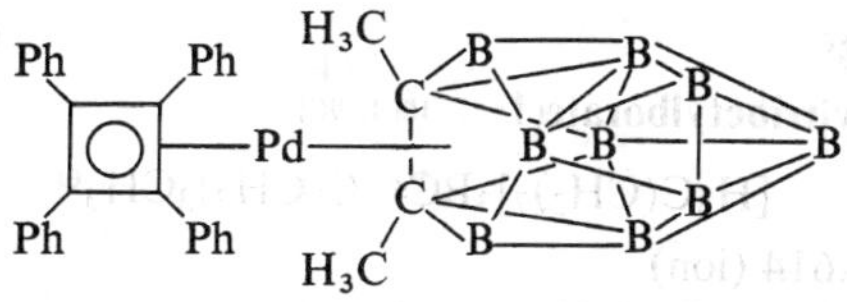

M 623.339

Cryst. (CH_2Cl_2/hexane). Mp >325° dec.

Wegner, P.A. *et al*, *J. Chem. Soc., Chem. Commun.*, 1966, 861 (*synth, ir, pmr, nmr*)

$C_{32}H_{38}B_2Fe_2N_2O_4$ B-01340

Di-μ-carbonyldicarbonylbis[(3,4,5-η)-1-(1,1-dimethylethyl)-2,3-dihydro-3-methyl-2-phenyl-1*H*-1,2-azaborol-3-yl-*B*², *N*¹]diiron, 10CI

*Bis(1-*tert*-butyl-2,3-dihydro-3-methyl-2-phenyl-1*H*-1,2-azaborol-3-yl)tetracarbonyldiiron*

As Di-μ-carbonyldicarbonylbis[(3,4,5-η)-1-(1,1-dimethylethyl)-2,3-dihydro-2-methyl-1*H*-1,2-azaborol-3-yl-*B*²,*N*¹]diiron, B-01149 with

R = Ph, R′ = CH_3

M 647.977

cis-form [72444-61-2]
Red-brown cryst. (CH_2Cl_2).

trans-form [75109-70-5]
Red cryst. (toluene). Dec. >150°. Isomerises to *cis*-form at 90°.

Schulze, J. *et al*, *Angew. Chem., Int. Ed. Engl.*, 1980, **19**, 54 (*synth, ir, pmr, nmr*)
Schulze, J. *et al*, *Chem. Ber.*, 1980, **113**, 2348 (*synth, ir, pmr, nmr, struct*)

$C_{33}H_{28}BTl$ B-01341

Methyl(phenylethynyl)tetraphenylborato-*B*)thallium

PhC≡CTlMeBPh₄

M 639.777

Mp >250°.

Kurosawa, H. *et al*, *J. Organomet. Chem.*, 1968, **12**, 241 (*synth, ir*)

$C_{33}H_{34}BN$ B-01342

Phenyltrimethylammonium tetraphenylborate

N,N,N-Trimethylbenzenaminium tetraphenylborate(1−), 10CI. *Phenyltrimethylammonium tetraphenylboranate*

[71111-17-6]

$PhNMe_3^{\oplus}Ph_4B^{\ominus}$

M 455.448

Solid. Mp 217-8°.

Arbuzov, B.A. *et al*, *Izv. Akad. Nauk SSSR, Ser. Khim.*, 1979, 1631; *Bull. Acad. Sci. USSR, Div. Chem. Sci.*, **28**, 1508 (*synth*)
Caruso, J.A., *J. Solution Chem.*, 1979, **8**, 197 (*nmr*)

$C_{34}H_{25}B$ **B-01343**
Pentaphenylborole, 10CI, 9CI, 8CI
[20195-59-9]

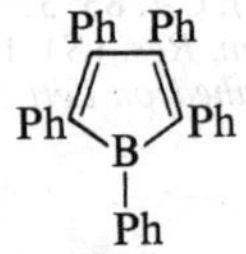

M 444.382
A highly reactive nucleus, readily undergoing oxidn., partial protolysis and the Diels-Alder reaction with dienophiles of low reactivity. Dark-blue cryst. solid (C_6H_6/EtOH); water-sensitive. Mp 120°, 175° dec., 243-5°. In early reports erroneously descr. as pale-yellow.
Cyclopentadienylcobalt complex: [75673-97-1]. *(η^5-2,4-Cyclopentadien-1-yl)[(1,2,3,4,5-η)-1,2,3,4,5-pentaphenyl-1*H*-borole]cobalt.*
$C_{39}H_{30}BCo$ M 568.409
Orange needles (Et_2O). Sol. toluene, CH_2Cl_2, Me_2CO, Et_2O, insol. pentane. Mp 325-9°.
Tricarbonyliron complex: see Tricarbonyl[(1,2,3,4,5-η)-1,2,3,4,5-pentaphenyl-1H-borole]iron, B-01360
Dicarbonylnickel complex: [60840-46-2]. *Dicarbonyl[(1,2,3,4,5-η)-1,2,3,4,5-pentaphenyl-1*H*-borole]nickel.*
$C_{36}H_{25}BNiO_2$ M 559.092
Wine-red light-sensitive cryst. (CH_2Cl_2); air-sensitive in soln. Sol. toluene, CH_2Cl_2. Mp 214-6° dec.
(Cyclooctadiene)platinum complex: [75673-98-2]. *[(1,2,5,6-η)-1,5-Cyclooctadiene][(1,2,3,4,5-η-)-1,2,3,4,5-pentaphenyl-1*H*-borole]platinum.*
$C_{42}H_{37}BPt$ M 747.644
Yellow cryst. (pentane/CH_2Cl_2). Sol. CH_2Cl_2, $CHCl_3$, CS_2, toluene, THF, Me_2CO. Mp 198°.

Braye, W.H. *et al*, *J. Am. Chem. Soc.*, 1961, **83**, 4406 (*synth*)
Eisch, J.J. *et al*, *J. Am. Chem. Soc.*, 1969, **91**, 4575 (*synth, use, nmr*)
Eisch, J.J. *et al*, *J. Am. Chem. Soc.*, 1975, **97**, 4436.
Eisch, J.J. *et al*, *J. Organomet. Chem.*, 1977, **127**, C9.
Herberich, G.E. *et al*, *Angew. Chem., Int. Ed. Engl.*, 1977, **16**, 42 (*props*)
Herberich, G.E. *et al*, *J. Organomet. Chem.*, 1980, **195**, 253 (*synth, derivs, pmr, ms*)

$C_{34}H_{34}BN_3$ **B-01344**
4-(Diethylamino)benzenediazonium tetraphenylborate(1–), 10CI
[77569-52-9]

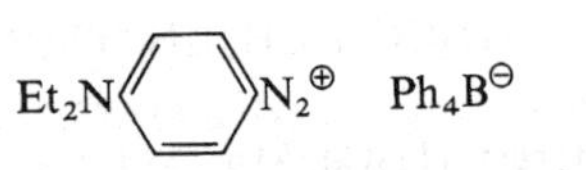

M 495.473
Used in photoimaging.

Rehorek, D. *et al*, *J. Prakt. Chem.*, 1980, **322**, 872 (*props*)
Ger. Pat., 146 350, (*1981*); *CA*, **95**, 71100e (*use*)

$C_{34}H_{34}BN_3$ **B-01345**
(*N,N'*-Diphenylguanidine-*N''*)tris(2-methylphenyl)boron, 10CI
Tri-o-tolylborane N,N'*-diphenylguanidine.* N,N'*-Diphenylguanidine, compd. with tris(2-methylphenyl)borane(1:1)*
[73895-24-6]

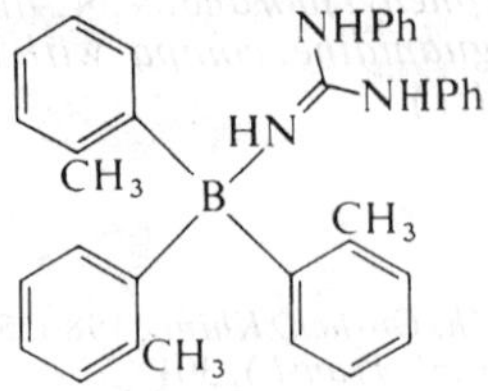

M 495.473
Cryst. (EtOH).

Lapkin, I.I. *et al*, *Zh. Obshch. Khim.*, 1980, **50**, 368; *J. Gen. Chem. USSR* (*Engl. Transl.*), 301
Yuzhakova, G.A. *et al*, *Zh. Obshch. Khim.*, 1981, **51**, 880.

$C_{34}H_{34}BN_3$ **B-01346**
(*N,N'*-Diphenylguanidine-*N''*)tris(3-methylphenyl)boron, 10CI
*Tri-*m*-tolylborane* N,N'*-diphenylguanidine.* N,N'*-Diphenylguanidine, compd. with tris(3-methylphenyl)borane(1:1)*
[73895-23-5]
M 495.473
Cryst. (EtOH).

Lapkin, I.I. *et al*, *Zh. Obshch. Khim.*, 1980, **50**, 368; *J. Gen. Chem. USSR* (*Engl. Transl.*), 301
Yuzhakova, G.A. *et al*, *Zh. Obshch. Khim.*, 1981, **51**, 880.

$C_{34}H_{34}BN_3$ **B-01347**
(*N,N'*-Diphenylguanidine-*N''*)tris(4-methylphenyl)boron, 10CI
*Tri-*p*-tolylborane* N,N'*-diphenylguanidine.* N,N'*-Diphenylguanidine, compd. with tris(4-methylphenyl)borane(1:1)*
[73895-22-4]
M 495.473
Cryst. (EtOH).

Lapkin, I.I. *et al*, *Zh. Obshch. Khim.*, 1980, **50**, 368; *J. Gen. Chem. USSR* (*Engl. Transl.*), 301 (*synth*)
Yuzhakova, G.A. *et al*, *Zh. Obshch. Khim.*, 1981, **51**, 880.

$C_{34}H_{34}BN_3O_3$ **B-01348**
(*N,N'*-Diphenylguanidine-*N''*)tris(2-methoxyphenyl)boron, 10CI
*Tris(*o*-methoxyphenyl)borane* N,N'*-diphenylguanidine.* N,N'*-Diphenylguanidine, compd. with tris(2-methoxyphenyl)borane(1:1)*
[73895-26-8]

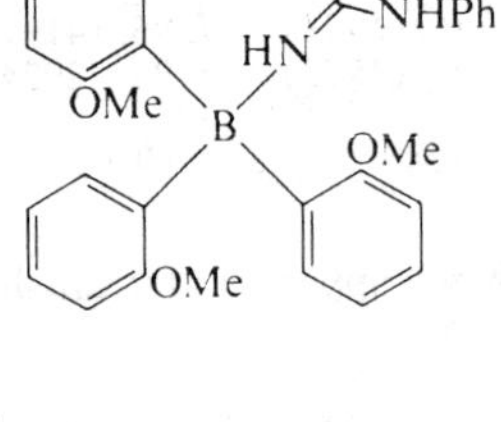

M 543.471
Cryst. (EtOH).

Lapkin, I.I. *et al*, *Zh. Obshch. Khim.*, 1980, **50**, 368; *J. Gen. Chem. USSR* (*Engl. Transl.*), 301

Yuzhakova, G.A. *et al*, *Zh. Obshch. Khim.*, 1981, **51**, 880.

$C_{34}H_{34}BN_3O_3$ B-01349

(*N,N'*-Diphenylguanidine-*N''*)tris(4-methoxyphenyl)boron, 10CI

Tris(p-*methoxyphenyl*)*borane* N,N'-*diphenylguanidine*. N,N'-*Diphenylguanidine, compd. with tris*(*4-methoxyphenyl*)*borane*(*1:1*)

[73895-27-9]

M 543.471

Cryst. (EtOH).

Lapkin, I.I. *et al*, *Zh. Obshch. Khim.*, 1980, **50**, 368; *J. Gen. Chem. USSR* (*Engl. Transl.*), 301

Yuzhakova, G.A. *et al*, *Zh. Obshch. Khim.*, 1981, **51**, 880.

$C_{34}H_{35}B_{10}IrNOP$ B-01350

Benzonitrilecarbonyl(triphenylphosphine)[7-phenyl-1,7-dicarba-*closo*-dodecaborane(12)]iridium, 9CI

[70700-96-8]

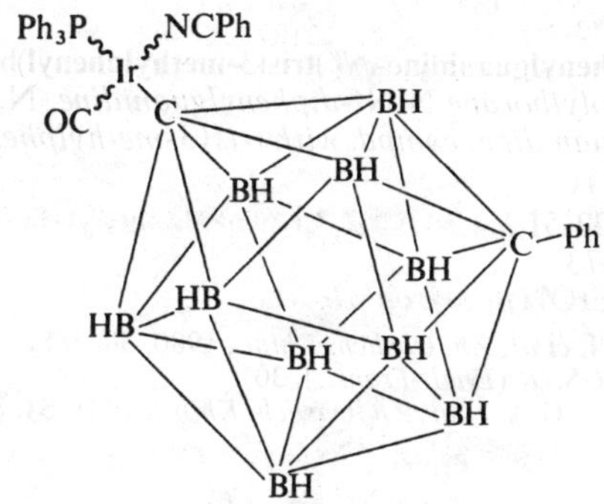

M 804.950

Stereochem. not assigned. Effective catalyst for homogeneous hydrogenation of terminal olefins and acetylenes. Yellow cryst. (C_6H_6/hexane). Mp 191°. ir ν_{CN} 2258w, ν_{CO} 1980vs cm^{-1} (nujol).

Longato, B. *et al*, *Inorg. Chim. Acta*, 1979, **33**, 189 (*synth, ir*)

Morandini, F. *et al*, *J. Organomet. Chem.*, 1982, **239**, 377 (*use*)

$C_{35}H_{39}BN_2$ B-01351

4-Diethylamino-2-methylphenylammonium tetraphenylborate

Hydrogen tetraphenylborate(*1–*), *compd. with* N⁴,N⁴-*diethyl-2-methyl-1,4-benzenediamine*(*1:1*), *10CI*

[78802-91-2]

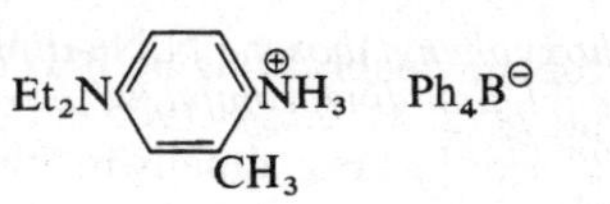

M 498.517

Used in photographic developing.

Japan. Pat., 81 06 235, (*1981*); *CA*, **95**, 141970q (*use*)

European Pat., 29 722, (*1981*), *CA*, **95**, 229247v (*use*)

$C_{36}H_{34}BCuP_2$ B-01352

Bis(triphenylphosphine)copper(1+) tetrahydroborate(1–), 10CI, 9CI, 8CI

[34010-85-0]

$$(Ph_3P)_2Cu^{\oplus}\ BH_4^{\ominus}$$

M 602.968

Used in photoimaging materials. Selective reducing agent for aldehydes and ketones.

Reichle, W.T., *Inorg. Chim. Acta*, 1971, **5**, 325.

U.S.P., 3 859 092, (*1975*); *CA*, **83**, 35746p

Clarke, S.J. *et al*, *J. Chem. Res.* (*S*), 1981, 17 (*use*)

Fleet, G.W.J. *et al*, *Tetrahedron Lett.*, 1980, **21**, 4031 (*use*)

$C_{36}H_{34}BCuP_2$ B-01353

[Tetrahydroborato(1–)-*H,H'*]bis(triphenylphospine)copper, 9CI

Bis(*triphenylphosphine*)*copper*(*I*) *borohydride*

[16903-61-0]

$$(Ph_3P)_2CuBH_4$$

M 602.968

Reducing agent for converting RCOCl → RCHO. Cryst. ($CHCl_3$/EtOH). Sol. H_2O, org. solvs. Mp 164° dec., 177° dec. Forms C_6H_6 solvate, Mp 187°.

Davidson, J.M., *Chem. Ind.* (*London*), 1964, 2021 (*synth*)

Cariati, F. *et al*, *Gazz. Chim. Ital.*, 1965, **95**, 3 (*synth*)

Lippard, S.J. *et al*, *J. Am. Chem. Soc.*, 1967, **89**, 3929 (*cryst struct*)

Fleet, G.W.J. *et al*, *Tetrahedron Lett.*, 1978, 1437 (*use*)

Sorell, T.N. *et al*, *J. Org. Chem.*, 1980, **45**, 3449 (*synth, use*)

Fieser, M. *et al*, *Reagents for Organic Synthesis*, Wiley, 1967-83, **8**, 47.

$C_{36}H_{39}BN_2O_2$ B-01354

1,5-Cyclooctanediyl[*N*-1-naphthalenyl-*N*-[(1-naphthalenylamino)carbonyl]-2-heptenamidato-*O,O'*]boron, 10CI

[78264-64-9]

As 1,5-Cyclooctanediyl[*N*-phenyl-*N*-[(phenylamino)carbonyl]-2-heptenamidato-*O,O'*]boron, B-01306 with

$$R^1 = \text{1-naphthyl},\ R^2 = -(CH_2)_3CH_3$$

M 542.526

Prepd. by reaction of *B*-hexenyl-9-BBN with 1-naphthyl isocyanate. Yellow cryst. Mp 129-31°. ^{11}B nmr + 8.6 ppm.

Singaram, B. *et al*, *Heterocycles*, 1981, **15**, 231 (*synth, nmr*)

$C_{36}H_{48}BN$ B-01355

Tetrapropylammonium tetraphenylborate

N,N,N-*Tripropyl-1-propanaminium tetraphenylborate*(*1–*), *10CI, 9CI*

[15556-39-5]

$$(H_3CCH_2CH_2)_4N^{\oplus}Ph_4B^{\ominus}$$

M 505.592

Electrolyte. Cryst. (H_2O). Mp 248°.

Lind, J.E. *et al*, *J. Phys. Chem.*, 1966, **70**, 3610 (*struct, synth*)

James, C.J. *et al*, *J. Solution Chem.*, 1975, **4**, 91 (*props*)

Gros, M. *et al*, *J. Electroanal. Chem. Interfacial Electrochem.*, 1978, **89**, 29; *CA*, **89**, 12543e.

Popovych, O., *Solubility Data Ser.*, 1981, **18**, 112, 113, 114.

$C_{36}H_{71}BNiP_2$ B-01356

Hydro[tetrahydroborato(1–)-*H,H'*]bis(tricyclohexylphosphine)nickel, 9CI

Hydridobis(tricyclohexylphosphine)nickel tetrahydroborate. Bis(tricyclohexylphosphine)hydridonickel borohydride

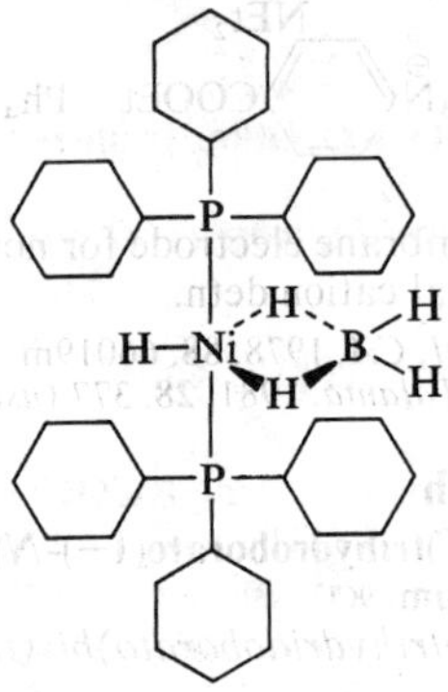

M 635.404

trans-form [24899-12-5]

Bright-yellow cryst., stable at r.t. Fairly stable in air, but soln. dec. in a few minutes. Sol. C_6H_6. Mp 121-5° dec.

Green, M.L.H. *et al*, *J. Chem. Soc., Dalton Trans.*, 1969, 1287 (*synth*, *ir*, *pmr*)
Inorg. Synth., 1977, **17**, 88 (*synth*, *ir*, *nmr*)
Saito, T. *et al*, *J. Chem. Soc., Dalton Trans.*, 1978, 482 (*struct*)

$C_{36}H_{71}BP_2Pd$ B-01357

Hydro[tetrahydroborato(1–)-*H,H'*]bis(tricyclohexylphosphine)palladium, 9CI

Bis(tricyclohexylphosphine)hydridopalladium tetrahydroborate.
Hydridoborohydridobis(tricyclohexylphosphine)palladium

[30916-06-4]

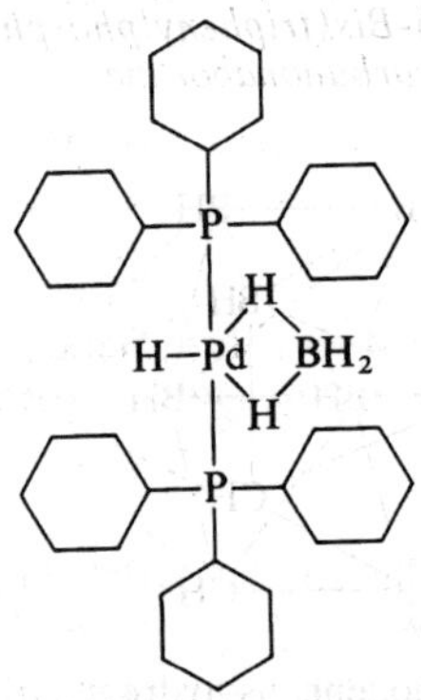

M 683.134

Precursor of catalyst for oligomerization of butadiene. Cryst. (EtOH aq.). Mp 135-8° dec.

Green, M.L.H. *et al*, *J. Chem. Soc. (A)*, 1971, 469 (*synth*, *ir*, *pmr*)
Green, M.L.H. *et al*, *J. Chem. Soc., Dalton Trans.*, 1974, 269 (*use*)
Inorg. Synth., 1977, **17**, 88 (*synth*, *ir*, *pmr*)

$C_{36}H_{75}BO_3$ B-01358

Bis(octadecyloxy)hydroxyborane

Boric acid (H_3BO_3) dioctadecyl ester, 10CI. *Bis(octadecyl) borate*

[72241-51-1]

$HOB[O(CH_2)_{17}CH_3]_2$

M 566.797

Used as antistatic agent for vinyl chloride polymers. Moisture-sensitive.

Japan. Pat., 79 86 546, (*1979*); *CA*, **92**, 23543t (*use*)
Japan. Pat., 81 10 539, (*1981*); *CA*, **95**, 44166z (*use*)

$C_{36}H_{75}BO_3$ B-01359

Tridodecyloxyborane

Boric acid (H_3BO_3) tridodecyl ester, 10CI, 9CI, 8CI. *Dodecyl borate. Tridodecyl borate. Tris(dodecyloxy)borane. Tridodecyl orthoborate*

[2467-15-4]

$[H_3C(CH_2)_{11}O]_3B$

M 566.797

Plasticizer for polycarbonates. Moisture-sensitive liq. d_4^{27} 0.845. $Bp_{3.5}$ 287-8°. n_D^{25} 1.4472.

▷ED5000000.

Steinberg, H. *et al*, *Ind. Eng. Chem.*, 1957, **49**, 174 (*synth*, *props*)
Potuis, S.P. *et al*, *CA*, 1970, **72**, 67624m (*use*)
Maijs, L. *et al*, *CA*, 1973, **79**, 131073e (*pmr*)
U.S.P., 4 211 679, (*1980*); *CA*, **93**, 169189z

$C_{37}H_{25}BFeO_3$ B-01360

Tricarbonyl[(1,2,3,4,5-η)-1,2,3,4,5-pentaphenyl-1*H*-borole]iron, 10CI

Pentaphenylborole iron tricarbonyl

[60840-47-3]

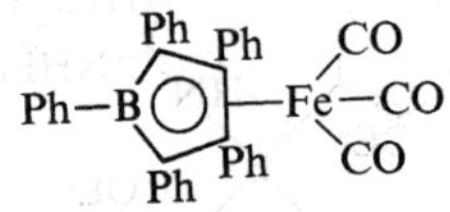

M 584.260

Pale-yellow cryst. (pentane/CH_2Cl_2). Sol. toluene, CH_2Cl_2; stable to H_2O. Mp 218-20°. Sensitive to light and air in soln.

Herberich, G.E. *et al*, *Angew. Chem., Int. Ed. Engl.*, 1977, **16**, 42 (*synth*, *pmr*, *ir*, *ms*)

$C_{37}H_{34}BOP_2Rh$ B-01361

Carbonyl[tetrahydroborato(1–)-*H,H'*]bis(triphenylphosphine)rhodium, 9CI

[36606-35-6]

$(Ph_3P)_2Rh(CO)(BH_4)$

M 670.338

Thought to contain doubly bridging BH_4 moiety. Yellow cryst. ir ν_{BH} 2452, 2416, ν_{CO} 1955 cm^{-1} (nujol). Stable in air as solid but dec. rapidly in soln.

Flynn, B.R. *et al*, *J. Chem. Soc., Chem. Commun.*, 1971, 1615 (*synth*, *ir*, *pmr*)

$C_{37}H_{40}BN_3$ B-01362

Tris(2,5-dimethylphenyl)borane *N,N'*-diphenylguanidine
Tris(2,5-dimethylphenyl)(N,N'*-diphenylguanidine-*N'')*boron*, 10CI. N,N'*-Diphenylguanidine compd. with tris(2,5-dimethylphenylborane(1:1). Tris(2,5-xylyl)-borane compd. with 1,3-diphenylguanidine*
[73895-25-7]

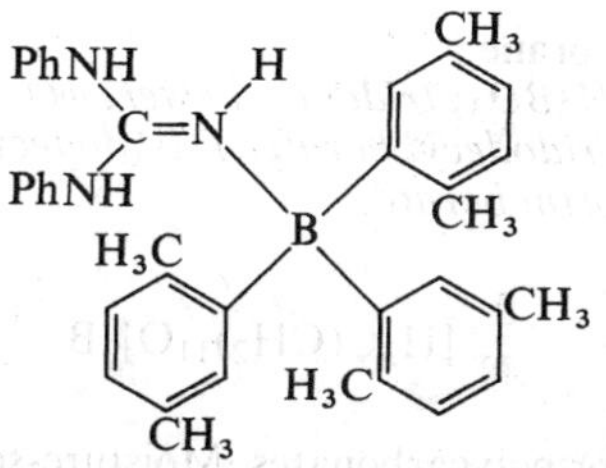

M 537.553
Mp 109° dec.

Lapkin, I.I. *et al*, *Zh. Obshch. Khim.*, 1980, **50**, 368 (*synth, ir, pmr*)
Yuzhakova, G.A. *et al*, *Zh. Obshch. Khim.*, 1981, **51**, 880.

$C_{37}H_{40}BN_3O_3$ B-01363

(*N,N'*-Diphenylguanidine-*N''*)tris(2-ethoxyphenyl)boron, 10CI
Tris(o*-ethoxyphenyl)borane* N,N'*-diphenylguanidine.* N,N'*-Diphenylguanidine, compd. with tris(2-ethoxyphenyl)borane(1:1)*
[73895-28-0]

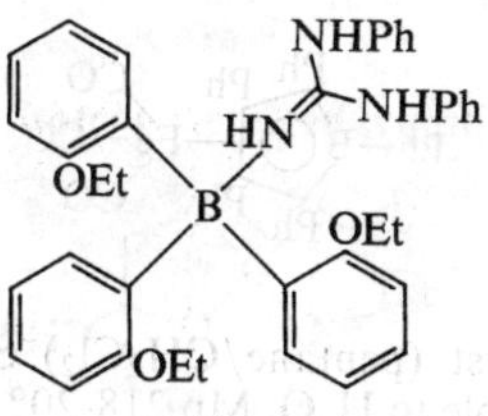

M 585.551
Cryst. (EtOH).

Lapkin, I.I. *et al*, *Zh. Obshch. Khim.*, 1980, **50**, 368; *J. Gen. Chem. USSR* (*Engl. Transl.*), 301
Yuzhakova, G.A. *et al*, *Zh. Obshch. Khim.*, 1981, **51**, 880.

$C_{37}H_{40}BN_3O_3$ B-01364

(*N,N'*-Diphenylguanidine-*N''*)tris(4-ethoxyphenyl)boron, 10CI
Tris(p*-ethoxyphenyl)borane* N,N'*-diphenylguanidine.* N,N'*-Diphenylguanidine, compd. with tris(4-ethoxyphenyl)borane(1:1)*
[73895-29-1]
M 585.551
Cryst. (EtOH).

Lapkin, I.I. *et al*, *Zh. Obshch. Khim.*, 1980, **50**, 368; *J. Gen. Chem. USSR* (*Engl. Transl.*), 301
Yuzhakova, G.A. *et al*, *Zh. Obshch. Khim.*, 1981, **51**, 880.

$C_{37}H_{41}BN_2O_2$ B-01365

4-(Ethoxycarbonyl)-3-diethylaminophenylammonium tetraphenylborate
Hydrogen tetraphenylborate(1−), compd. with 2-(diethylamino)ethyl 4-aminobenzoate(1:1), 10CI
[65281-20-1]

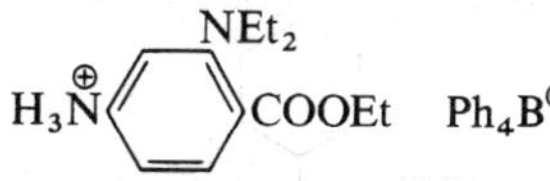

M 556.553
Used as liq. membrane electrode for novocaine and pharmaceutical cation detn.

Hopartean, E. *et al*, *CA*, 1978, **88**, 66019m.
Negoiu, D. *et al*, *Talanta*, 1981, **28**, 377 (*use*)

$C_{38}H_{33}BNOP_2Rh$ B-01366

Carbonyl[(cyano-*C*)trihydroborato(1−)-*N*)]bis(triphenylphosphine)rhodium, 9CI
Carbonyl(cyanotrihydridoborato)bis(triphenylphosphine)rhodium
[36606-39-0]

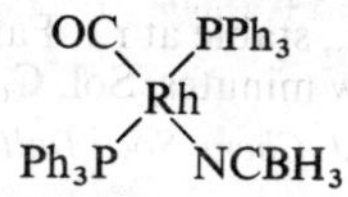

M 695.348
Air-stable yellow solid. Sol. $CHCl_3$, C_6H_6, insol. hexane, alcohols. ir ν_{BH} 2360, ν_{CN} 2192, ν_{CO} 1997 cm^{-1} (nujol).

Vaska, L. *et al*, *J. Chem. Soc., Chem. Commun.*, 1971, 1615 (*ir, synth*)
Inorg. Synth., 1974, **15**, 72 (*synth, ir*)
Brady, R. *et al*, *Inorg. Chem.*, 1976, **15**, 1485 (*uv*)

$C_{38}H_{39}B_7P_2Ru$ B-01367

[(4,5,6,7,8,9-η)-Nonahydro-6,8-dicarbanonaborato(4−)]bis(triphenylphosphine)ruthenium, 10CI
Hyper-closo*-6,6-Bis(triphenylphosphine)-6-ruthenanonahydro-2,3-dicarbanonaborane*
[61081-79-6]

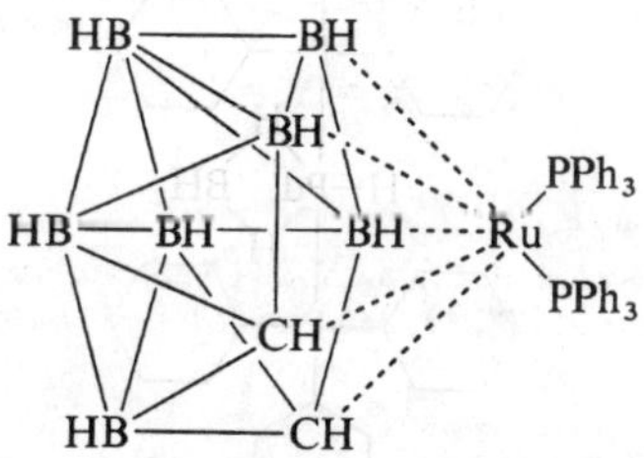

M 734.414
Catalyst for homogeneous hydrogenation of alkenes. Shiny black needles (CH_2Cl_2/EtOH). Mp 209-12°.

Jung, C.W. *et al*, *J. Am. Chem. Soc.*, 1981, **103**, 811 (*synth, pmr, nmr*)

$C_{38}H_{41}AgB_8P_2$ B-01368

Bis(triphenylphosphine)[(7,8,9,-η)undecahydro-5,6-dicarbadecaborato(1−)]silver
[74354-45-3]

$$(Ph_3P)_2Ag(C_2B_8H_{11})$$

M 754.037
Struct. not known, probably monomeric. Cryst. ($CHCl_3$/hexane). Loses PPh_3 on recryst., giving Triphenylphosphine[(7,8,9-η)undecahydro-5,6-dicarbadecaborato(1−)]silver, B-01140.

Colquhoun, H.M. *et al*, *J. Chem. Soc., Chem. Commun.*, 1980, 192 (*synth, nmr*)

$C_{38}H_{42}B_9ClP_2Ru$ B-01369

Chlorohydrobis(triphenylphosphine)[(7,8,9,10,11-η)undecahydro-7,8-dicarbaundecaborato(2−)]ruthenium

[67328-91-0]

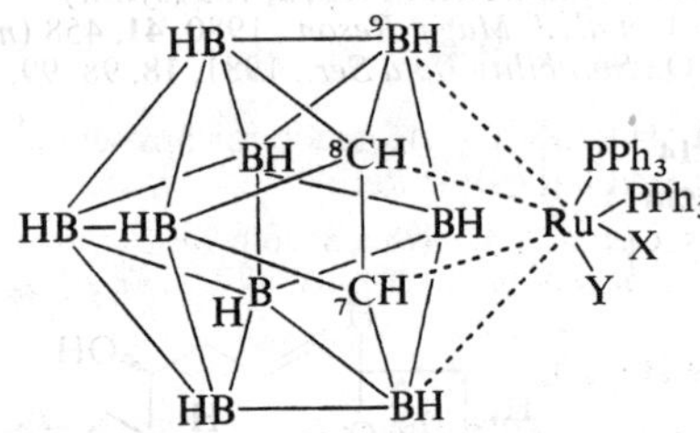

X = Cl, Y = H

M 794.510

Orange-red cryst.

Wong, E.H.S. *et al*, *Inorg. Chem.*, 1978, **17**, 2863 (*synth, ir*)

$C_{38}H_{42}B_9ClP_2Ru$ B-01370

Chlorohydrobis(triphenylphosphine)[(7,8,9,10,11-η)-undecahydro-7,9-dicarbaundecaborato(2−)]ruthenium

[67350-80-5]

M 794.510

Yellow cryst. Spar. sol. org. solvs.

Wong, E.H.S. *et al*, *Inorg. Chem.*, 1978, **17**, 2863 (*synth, ir*)

$C_{38}H_{43}B_9P_2Ru$ B-01371

Dihydrobis(triphenylphosphine)[(7,8,9,10,11-η)undecahydro-7,8-dicarbaundecaborato(2−)]ruthenium

[60172-21-6]

As Chlorohydrobis(triphenylphosphine)[(7,8,9,10,11-η)undecahydro-7,8-dicarbaundecaborato(2−)]ruthenium, B-01369 with

X = Y = H

M 760.065

Colourless cryst.

Wong, E.H.S. *et al*, *J. Chem. Soc., Chem. Commun.*, 1976, 257; *Inorg. Chem.*, 1978, **17**, 2863 (*synth, pmr, nmr, ir*)

$C_{38}H_{43}B_9P_2Ru$ B-01372

Dihydrobis(triphenylphosphine)[(7,8,9,10,11-η)-undecahydro-7,9-dicarbaundecaborato(2−)]ruthenium, 9CI

[2,2-Bis(triphenylphosphine)-2,2-dihydro-2-ruthena]-undecahydro-1,7-dicarbaundecaborane

[60172-17-0]

M 760.065

Stereochemically nonrigid. Light-blue cryst. (dioxan or toluene/EtOH), grey-white when pure. Reversibly forms a dark-blue dehydro complex on heating.

Wong, E.H.S. *et al*, *J. Chem. Soc., Chem. Commun.*, 1976, 257; *Inorg. Chem.*, 1978, **17**, 2863 (*synth, struct, pmr, nmr, ir*)

Marder, T.B. *et al*, *J. Am. Chem. Soc.*, 1981, **103**, 2988 (*pmr, nmr*)

$C_{38}H_{54}B_2N_6O_4$ B-01373

(*N,N'*-Hexamethylenyl-1,6)-di[7-methoxymethyl-3-borabicyclo[3.3.1]non-6-en-3-yl-(*N*-carbamoylpyridon-2-iminate)]

[μ-[[N,N'-1,6-Hexanediylbis[2-imino-1(2H)-pyridinecarboxamidato]](2−)]]bis[[5-(methoxymethyl)-4-cyclohexene-1,3-diyl]bis(methylene)]diboron, 10CI

[76009-48-8]

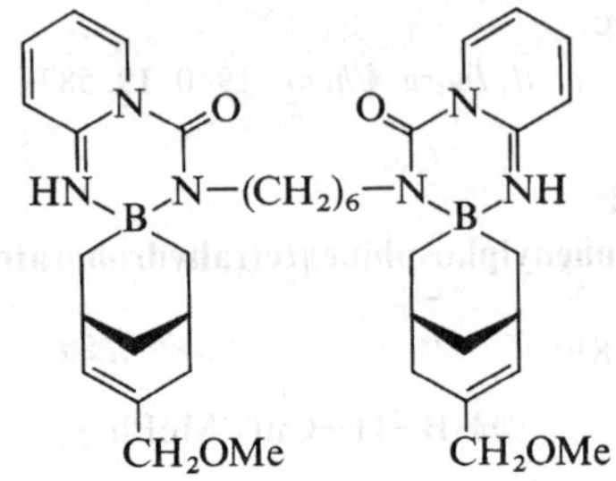

M 680.502

Mp 146-9°.

Buyanovskaya, A.G. *et al*, *Zh. Anal. Khim.*, 1980, **35**, 2106.

Mikhailov, B.M. *et al*, *Izv. Akad. Nauk SSSR, Ser. Khim.*, 1980, 1881; *Bull. Acad. Sci. USSR, Div. Chem. Sci.*, 1355 (*synth, ir, pmr*)

$C_{38}H_{72}BO_8^{\ominus}$ B-01374

Bis[2,3-dihydroxypropyl hexadecanoato(2−)]borate(1−), 10CI

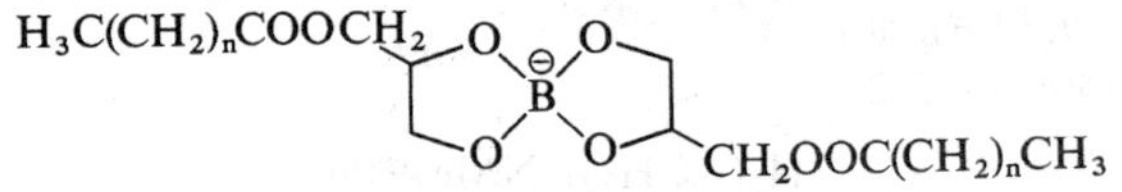

n = 14

M 667.792 (ion)

H deriv.: [65074-58-0]. *Hydrogen bis[2,3-dihydroxypropyl hexadecanoato(2−)]borate(1−)*, 10CI.

$C_{38}H_{73}BO_8$ M 668.800

Coupling agent, surfactant.

Japan Pat., 77 89 150, (*1977*); *CA*, **88**, 38553n (*use*)

Japan. Pat., 78 39 413, (*1978*); *CA*, **90**, 123486d (*use*)

$C_{39}H_{41}B_9OP_2Ru$ B-01375

Carbonylbis(triphenylphosphine)[(7,8,9,10,11-η)undecahydro-7,8-dicarbaundecaborato(2−)]ruthenium, 9CI

[56465-05-5]

As Chlorohydrobis(triphenylphosphine)[(7,8,9,10,11-η)undecahydro-7,8-dicarbaundecaborato(2−)]ruthenium, B-01369 with

XY = CO

M 786.060

Yellow cryst. (EtOH/CH_2Cl_2).

Wong, E.H.J. *et al*, *Inorg. Chem.*, 1978, **17**, 2863 (*synth, ir*)

$C_{39}H_{41}B_9OP_2Ru$ B-01376

Carbonylbis(triphenylphosphine)[(7,8,9,10,11-η)-undecahydro-7,9-dicarbaundecaborato(2−)]ruthenium, 9CI

[60172-19-2]

M 786.060
Yellow cryst. (CH_2Cl_2/EtOH).

Wong, E.H.S. *et al*, *Inorg. Chem.*, 1978, **17**, 2863 (*synth, ir*)

$C_{39}H_{43}AgBP$ **B-01377**

Tris(methyldiphenylphosphine)[tetrahydroborato(1–)-*H*]silver

[72347-03-6]

$$H_3B{-}H{-}Ag(PMePh_2)_3$$

M 661.420
Cryst. (CH_2Cl_2/pentane). Sol. CH_2Cl_2, $CHCl_3$. Mp 90-3° dec.

Bommer, J.C. *et al*, *Inorg. Chem.*, 1980, **19**, 587 (*synth*)

$C_{39}H_{43}BCuP$ **B-01378**

Tris(methyldiphenylphosphine)[tetrahydroborato(1–)-*H*]-copper

[63371-86-8]

$$H_3B{-}H{-}Cu(PMePh_2)_3$$

M 617.098
An example of monodentate $BH_4^{\ominus}$ ligand. Cryst. (CH_2Cl_2/pentane). Sol. CH_2Cl_2, $CHCl_3$. Mp 119-21° dec.

Bommer, J.C. *et al*, *Inorg. Chem.*, 1980, **19**, 587 (*synth*)
Takusagawa, F. *et al*, *J. Am. Chem. Soc.*, 1981, **103**, 5165 (*cryst struct, neutron struct*)

$C_{39}H_{54}BN$ **B-01379**

Dodecyltrimethylammonium tetraphenylborate

N,N,N-*Trimethyl-1-dodecanaminium tetraphenylborate(1–)*, 10CI, 9CI

[40638-32-2]

$$H_3C(CH_2)_{11}NMe_3^{\oplus}Ph_4B^{\ominus}$$

M 547.672
Used to set up an impedance of the interface between two immiscible electrolytic solns.

Seta, P. *et al*, *C.R. Hebd. Seances Acad. Sci.*, 1972, **275**, 1231 (*use, props*)
Schuhmann, D. *et al*, *J. Colloid Interface Sci.*, 1979, **69**, 448 (*use, props*)
Schuhmann, D. *et al*, *Physicochem. Hydrodyn.*, 1980, **1**, 57; *CA*, **92**, 205953j (*use, props*)

$C_{40}H_{36}B_2N_2S_2$ **B-01380**

Tetraphenylbis[*μ*-(*N*-phenylethanethioamidato-*N*:*S*)]diboron, 10CI

[63835-77-8]

As Tetramethylbis[*N*-(*N*-phenylethanethioamidato-*N*:*S*)]diboron, B-01147 with

$$R = CH_3, R' = Ph$$

M 630.478
Solid. Sol. CH_2Cl_2. Mp 100°. $Bp_{0.001}$ 150° subl.

Maringgele, W. *et al*, *J. Organomet. Chem.*, 1977, **141**, 241 (*synth, nmr, ms*)
Maringgele, W. *et al*, *Monatsh. Chem.*, 1977, **108**, 751.

$C_{40}H_{56}BN$ **B-01381**

Tetrabutylammonium tetraphenylborate, 8CI

N,N,N-*Tributyl-1-butanaminium tetraphenylborate(1–)*, 10CI, 9CI

[15522-59-5]

$$(H_3CCH_2CH_2CH_2)_4N^{\oplus}Ph_4B^{\ominus}$$

M 561.699
Electrolyte, used as cond. std. for sp. resistance measurements.

Artamonov, B.P. *et al*, *Zh. Prikl. Khim.*, 1974, **47**, 588 (*synth, purifn*)
Odom, J.D. *et al*, *Org. Magn. Reson.*, 1974, **6**, 360.
Selig, W., *Mikrochim. Acta*, 1980, **2**, 133 (*synth*)
Zozulin, A.J. *et al*, *J. Magn. Reson.*, 1980, **41**, 458 (*nmr*)
Popovych, O., *Solubility Data Ser.*, 1981, **18**, 98, 99, 100, 101.

$C_{40}H_{60}BO_{14}^{\ominus}$ **B-01382**

Aplasmomycin

[61230-25-9]

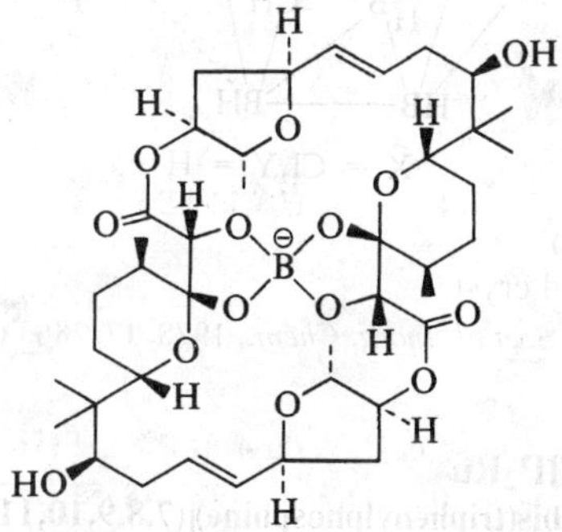

M 775.716 (ion)
Isol. from *Streptomyces griseus*. Shows antibiotic activity against gram-positive organisms. Belongs to small group of boron-containing antibiotics. Closely related structurally to Boromycin.

▷CD8835000.

Na salt:
$C_{40}H_{60}BNaO_{14}$ M 798.705
Needles (MeOH aq.). Mp 283-5° dec. $[\alpha]_D^{22}$ +225° (c, 1.24 in $CHCl_3$).

Okami, Y. *et al*, *J. Antibiot.*, 1976, **29**, 1019 (*isol, ir, uv, ms, pmr, cmr, struct*)
Nakamura, H. *et al*, *J. Antibiot.*, 1977, **30**, 714 (*struct*)
Corey, E.J. *et al*, *J. Am. Chem. Soc.*, 1982, **104**, 6816, 6818 (*synth*)

$C_{41}H_{36}BO_2$ **B-01383**

3-*tert*-Butyl-5-triphenylmethyl-1,2-benzenediolatodiphenylboron

[*3-(1,1-Dimethylethyl)-5-(triphenylmethyl)-1,2-benzenediolato(2–)-O,O'*]*diphenylboron*, 10CI

[73774-36-4]

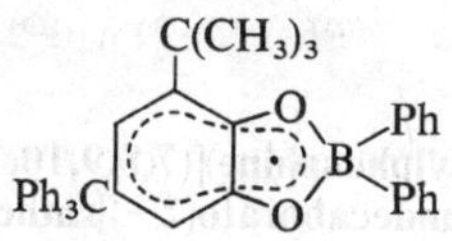

M 571.544
A stable paramagnetic species.

Stegmann, H.B. *et al*, *Tetrahedron Lett.*, 1979, 3689 (*synth, esr*)

$C_{41}H_{39}BP_2Ru$ **B-01384**

(η^5-2,4-Cyclopentadien-1-yl)[tetrahydroborato(1–)-*H*]bis(triphenylphosphine)ruthenium, 10CI

[79408-42-7]

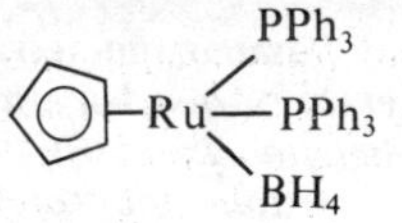

M 705.587
Ash-grey or white cryst. (Et_2O). Mp 95-103° dec., 145-50°.

Blackmore, T. *et al*, *J. Chem. Soc. (A)*, 1971, 2376 (*synth, ir, pmr*)
Wilczewski, T. *et al*, *J. Organomet. Chem.*, 1981, **215**, 87 (*synth, pmr, ms*)

$C_{42}H_{78}B_2N_2$ B-01385

Ethylenediamine bis(diisopinocampheylborane)
[*μ-(1,2-Ethanediamine-*N:N')]*dihydrotetrakis*(*2,6,6-trimethylbicyclo*[*3.1.1*]*hept-3-yl*)*diboron*, 10CI. *1,2-Ethanediamine, compd. with bis*(*2,6,6-trimethylbicyclo*[*3.1.1*]*hept-3-yl*)*borane*(*1:2*)
[68366-09-6]

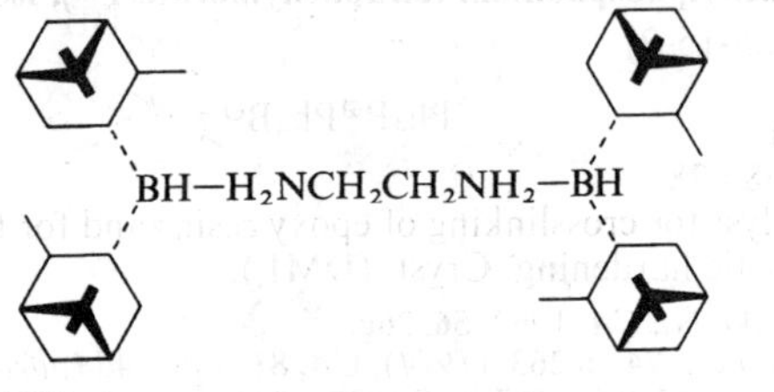

M 632.712
Liq., stable for at least 30 days at 0° in THF.

Brown, H.C. *et al*, *Inorg. Chem.*, 1979, **18**, 53 (*synth, nmr*)

$C_{42}H_{80}BO_8^{\ominus}$ B-01386

Bis[2,3-dihydroxypropyl octadecanoato(2−)]borate(1−), 10CI
As Bis[2,3-dihydroxypropyl hexadecanoato(2−)]borate(1−), B-01374 with

n = 16

M 723.899 (ion)
H deriv.: [52660-30-7]. *Hydrogen bis*[*2,3-dihydroxypropyl octadecanoato*(*2−*)]*borate*(*1−*), 10CI.
$C_{42}H_{81}BO_8$ M 724.907
Surfactant.

Japan. Pat., 78 39 413, (*1978*); *CA*, **90**, 123486d (*use*)
Japan. Pat., 78 90 361, (*1978*); *CA*, **89**, 216341x (*use*)
Japan. Pat., 80 15 480, (*1980*); *CA*, **93**, 238852z (*synth, use*)

$C_{43}H_{38}BP$ B-01387

Methyltriphenylphosphonium tetraphenylborate(1−), 10CI, 9CI, 8CI
[15318-09-9]

$Ph_3PMe^{\oplus}Ph_4B^{\ominus}$

M 596.557
Used as flame retardant for polycarbonates. Cryst. Mp 197°.

Naldini, L., *Gazz. Chim. Ital.*, 1960, **90**, 1231 (*synth, props*)
Wittig, G. *et al*, *Tetrahedron*, 1962, **18**, 1023 (*synth*)
Rundel, W. *et al*, *Justus Liebigs Ann. Chem.*, 1965, **686**, 88 (*synth*)
Schiemenz, G.P., *J. Organomet. Chem.*, 1973, **52**, 349 (*nmr*)
Nesterov, L.V. *et al*, *Zh. Obshch. Khim.*, 1980, **50**, 39; *CA*, **92**, 214490h.
U.S.P., 4 269 762, (*1981*); *CA*, **95**, 44242w (*use*)

$C_{43}H_{62}BN$ B-01388

Butyltris(3-methylbutyl)ammonium tetraphenylborate
N-*Butyl-3-methyl-*N,N-*bis*(*3-methylbutyl*)*-1-butanaminium tetraphenylborate*(*1−*), 10CI, 9CI. *Butyltriisopentylammonium tetraphenylborate*(*1−*), 8CI
[16742-92-0]

$[(H_3C)_2CH_2CH_2CH_2]_3N(CH_2)_3CH_3{}^{\oplus}Ph_4B^{\ominus}$

M 603.780
Has elec. cond. props.

Rosenfarb, J. *et al*, *J. Solution Chem.*, 1976, **5**, 311, 345 (*use*)
Bicknell, R.T.M. *et al*, *J. Chem. Soc., Faraday Trans. 1*, 1980, **76**, 637 (*props*)
Popovych, O., *Solubility Data Ser.*, 1981, **18**, 84 (*props*)

$C_{44}H_{64}BN$ B-01389

Tetrakis(3-methylbutyl)ammonium tetraphenylborate
*3-Methyl-*N,N,N-*tris*(*3-methylbutyl*)*-1-butanaminium tetraphenylborate*(*1−*), 10CI, 9CI. *Tetraisopentylammonium tetraphenylborate*, 8CI
[25397-16-4]

$[(H_3C)_2CHCH_2CH_2]_4N^{\oplus}Ph_4B^{\ominus}$

M 617.806
Electrolyte.

Denat, A. *et al*, *J. Chim. Phys. Physicochim. Biol.*, 1973, **70**, 319; *CA*, **78**, 158546c (*props*)
Carman, P.C., *J. Solution Chem.*, 1978, **7**, 845 (*props*)
Sakamoto, I. *et al*, *Electrochim. Acta*, 1981, **26**, 197 (*props*)

$C_{45}H_{45}B_{10}IrOP_2$ B-01390

Carbonyl(7-phenyl-1,7-dicarbadodecaboran(12)-1-yl)bis(triphenylphosphine)iridium, 9CI
Carbonylbis(*triphenylphosphine*)[*7-phenyl-1,7-dicarba-*closo*-dodecaborane*(*12*)]*iridium*
[57527-11-4]

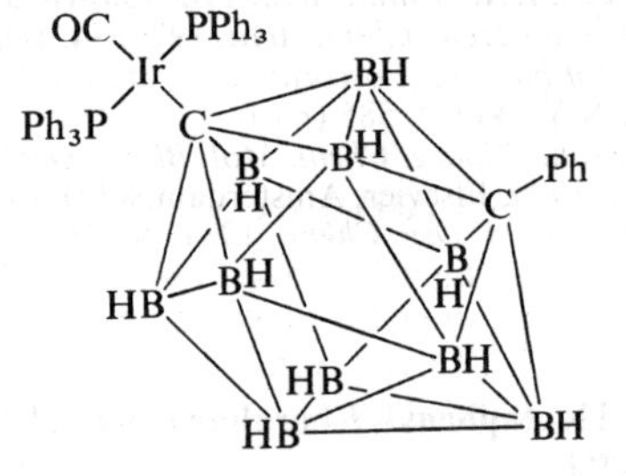

M 964.117
Orange cryst. (C_6H_6/hexane), solid and soln. stable under inert atm. Sol. C_6H_6, CH_2Cl_2. Mp 132-3° dec. ir ν_{CO} 1970 cm^{-1} (nujol). Readily takes up molecular hydrogen but is not an active hydrogenation catalyst.

Longato, B. *et al*, *Inorg. Chem.*, 1976, **15**, 650 (*synth, ir*)

Suggestions for new Entries are welcomed. Please write to the Editor, Dictionary of Organometallic Compounds, Chapman and Hall Ltd, 11 New Fetter Lane, London EC4P 4EE

$C_{45}H_{74}BNO_{15}$ **B-01391**
Boromycin
[34524-20-4]

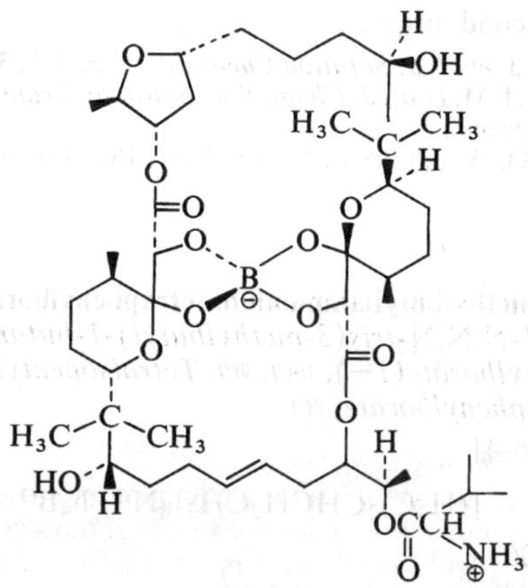

M 879.887
Produced by *Streptomyces antibioticus*. Antibiotic active against gram-positive bacteria. Cryst. (MeOH). Mp 223-8° dec. $[\alpha]_D$ +63.5° (c, 0.55 in $CHCl_3$). First known natural product containing boron. No UV absorption between 210 and 400 nm.
▷ED7325000.

Desvalyl: [70658-70-7]. O^{27}*-De(2-amino-3-methyl-1-oxobutyl)boromycin*, 10CI. *Devalylboromycin. Desvalinoboromycin.*
$C_{40}H_{64}BO_{14}^{\ominus}$ M 779.747 (ion)
Prod. by *S.* spp.

Hutler, R. *et al*, *Helv. Chim. Acta*, 1967, **50**, 1533 (*isol, struct*)
Dunitz, J.D. *et al*, *Helv. Chim. Acta*, 1971, **54**, 1709 (*struct*)
Pache, W., *Antibiotics*, (*Corcoran, J.W. et al*, Ed.), 1975, Springer, N.Y., Vol. 3, 585 (*rev*)
Prelog, V., *Front. Bioorg. Chem. Mol. Biol.*, (*Ovchinnikov, Y.A. et al*, Ed.), 1979, Elsevier, Amsterdam, 87 (*rev*)
Chen, T.S.S. *et al*, *J. Org. Chem.*, 1981, **46**, 2661 (*synth*)
Tyler, P.C. *et al*, *J. Am. Chem. Soc.*, 1981, **103**, 6243 (*synth*)

$C_{48}H_{35}B$ **B-01392**
1,2,3,4,5,6,7-Heptaphenyl-7-borabicyclo[2.2.1]hepta-2,5-diene, 10CI, 9CI
[56859-33-7]

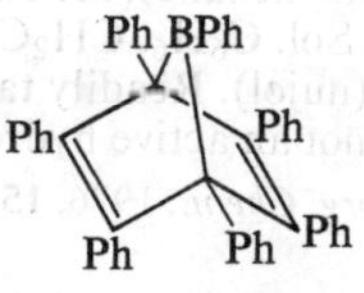

M 622.615
Air-sensitive cryst. (Et_2O). Sol. CCl_4. Mp 210° dec. λ_{max} 318 nm (ϵ 10 000)(Et_2O). Undergoes a thermal[1,3]-suprafacial sigmatropic rearrangement and disrotatory ring-opening to yield a borepin.

Eisch, J.J. *et al*, *J. Am. Chem. Soc.*, 1975, **97**, 4436 (*synth*)
Eisch, J.J. *et al*, *J. Organomet. Chem.*, 1977, **127**, C9 (*synth*)

$C_{48}H_{40}AsB$ **B-01393**
Tetraphenylarsonium(1+) tetraphenylborate(1−), 10CI, 9CI, 8CI
[15627-12-0]

$Ph_4As^{\oplus}$ $Ph_4B^{\ominus}$

M 702.576
Used in catalysts for cross-linking of epoxy resins, and in photoduplication compositions. Inhibits phosphorylation in mitochondria. Used for estimation of solvent activity coefficients. Needle-like cryst. (MeCN). Spar. sol. most solvs. Mp 293-5°.

Wittig, G. *et al*, *Justus Liebigs Ann. Chem.*, 1952, **577**, 26 (*synth, props*)
Popov, A.I. *et al*, *J. Am. Chem. Soc.*, 1953, **81**, 2043 (*synth, props*)
Cox, B.G. *et al*, *Aust. J. Chem.*, 1974, **27**, 477 (*use*)
Abraham, M.H. *et al*, *J. Chem. Soc., Faraday Trans. 1*, 1976, **72**, 955 (*use*)
Phelps, D.C. *et al*, *Biochem. Biophys. Res. Commun.*, 1977, **79**, 1242 (*pharmacol*)
Popovych, O., *Solubility Data Ser.*, 1981, **18**, 182 *et seq.*; *CA*, **94**, 198420u *et seq.*.

$C_{48}H_{40}BP$ **B-01394**
Tetraphenylphosphonium tetraphenylborate(1−), 10CI, 9CI, 8CI
[15525-15-2]

$Ph_4P^{\oplus}Ph_4B^{\ominus}$

M 658.628
Catalyst for crosslinking of epoxy resins and for thermoplastic hardening. Cryst. (DMF).

Hiroyuki, N., *CA*, 1962, **56**, 26g.
Japan. Pat., 74 26 263, (*1974*); *CA*, **81**, 136294d (*purifn, use*)
U.S.P., 3 947 395, (*1976*); *CA*, **85**, 48414v (*use*)
Popovych, O., *Solubility Data Ser.*, 1981, **18**, 228, 229, 231 (*props*)

$C_{49}H_{50}BN_3$ **B-01395**
Tris[(4-dimethylamino)phenyl]methylium tetraphenylborate
N-*[4-[Bis[4-(dimethylamino)phenyl]methylene]-2,5-cyclohexadien-1-ylidene]*-N-*methylmethanaminium tetraphenylborate(1−)*, 10CI, 9CI
[41687-02-9]

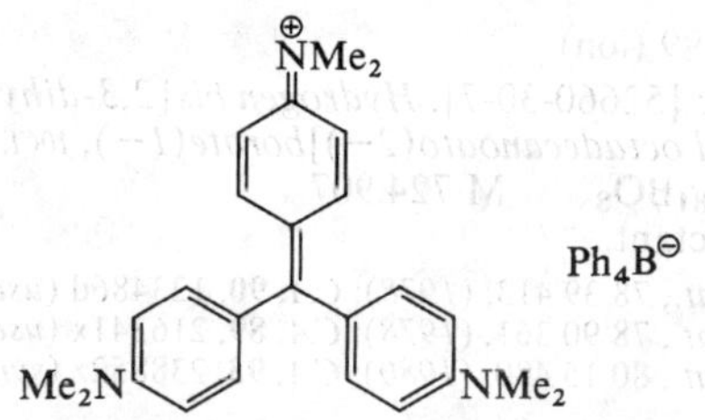

M 691.764
Used as an electrolyte.

Banateanu, Gh., *CA*, 1973, **79**, 93425m (*ir*)
Selig, W., *Mikrochim. Acta*, 1980, **2**, 133 (*synth*)
Vanysek, P., *J. Electroanal. Chem. Interfacial Electrochem.*, 1981, **121**, 149; *CA*, **94**, 216550t.

$C_{50}H_{44}BP_2Rh$ **B-01396**
[1,2-Ethanediylbis(diphenylphosphine)-*P,P'*][(η^6-phenyl)triphenylborato(1−)]rhodium, 10CI
[1,2-Bis(diphenylphosphino)ethane][(η^6-phenyl)triphenylborato(1−)]rhodium
[72332-27-5]

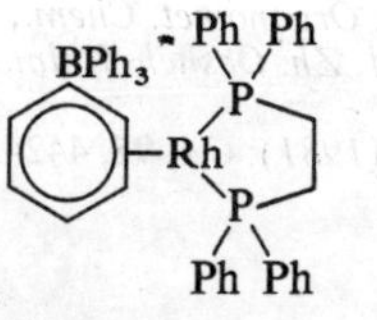

M 820.561
Effective catalyst for methylacetylene oligomerisation and allene polymerisation. Orange, air-stable cryst. Mp 230° dec. Forms CH_2Cl_2 solvate. Adds CO_2 reversibly.

Albano, P. *et al*, *Inorg. Chem.*, 1980, **19**, 1069 (*synth, cryst struct*)
Albano, P. *et al*, *J. Organomet. Chem.*, 1980, **190**, 243 (*use*)

$C_{54}H_{111}BO_3$ B-01397

Trioctadecyloxyborane

Boric acid (H_3BO_3) trioctadecyl ester, 10CI, 9CI, 8CI. *Octadecyl borate. Trioctadecyl borate. Trioctadecyl orthoborate. Tris(octadecyloxy)borane*

[5337-41-7]

$[H_3C(CH_2)_{17}O]_3B$

M 819.279

Plasticizer, stabilizing agent, vulcanizing agent for polymeric materials. Solid, slowly hydrolysed by moisture. Sol. non-hydroxylic solvs. Mp 41-4°. $Bp_{0.3}$ 339-43°.

▷ED5620000.

Steinberg, H. *et al*, *Ind. Eng. Chem.*, 1957, **49**, 174 (*synth*)
Japan. Pat., 76 107 348, (*1976*); *CA*, **86**, 18473s (*use*)
U.S.P., 4 211 679, (*1980*); *CA*, **93**, 169189z (*use*)

$C_{55}H_{52}BO_2P_2Rh$ B-01398

[(2,2-Dimethyl-1,3-dioxolane-4,5-diyl)bis(methylene)]bis[diphenylphosphine]-*P,P′*][(η^6-phenyl)triphenylborato(1−)]-rhodium, 10CI

[4,5-Bis(diphenylphosphinomethyl)-2,2-dimethyldioxolan]rhodium η^6-tetraphenylborate. DIOP(η^6-Tetraphenylborato)rhodium

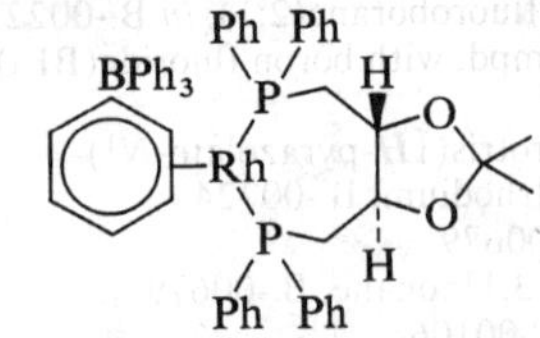

(*R,R*)-*form*

M 920.678

(*R,R*)-*form* [79790-95-7]

Catalyst precursor for asymmetric hydrogenations. Yellow solid, stable in air for several hours. Sol. CH_2Cl_2. Mp 129-30°. Forms solvates with alcohols.

Dang, T.P. *et al*, *J. Organomet. Chem.*, 1975, **91**, 105 (*synth, use*)
Chauvin, Y. *et al*, *J. Organomet. Chem.*, 1978, **146**, 311 (*synth*)
Brown, J.M. *et al*, *J. Organomet. Chem.*, 1981, **216**, 263 (*synth, nmr*)

Name Index

Molecular Formula Index

CAS Registry Number Index